Science

CONCISE ENCYCLOPEDIA OF SOLID STATE PHYSICS

CONCISE ENCYCLOPEDIA OF SOLID STATE PHYSICS

Edited by
Rita G. Lerner
American Institute of Physics, New York, New York

George L. Trigg
American Physical Society, Ridge, New York

Foreword by
Conyers Herring
Bell Telephone Laboratories, Murray Hill, New Jersey

ADDISON-WESLEY PUBLISHING COMPANY, Inc.
Advanced Book Program/World Science Division
Reading, Massachusetts

London · Amsterdam · Don Mills, Ontario · Sydney · Tokyo

CONCISE ENCYCLOPEDIA OF SOLID STATE PHYSICS

Library of Congress Cataloging in Publication Data
Main entry under title:

Concise encyclopedia of solid state physics.

"Articles chosen for this volume have been selected from the Encyclopedia of physics"–Pref.
Includes bibliographies and index.
1. Solid state physics–Dictionaries. I. Lerner, Rita G. II. Trigg, George L. III. Encyclopedia of physics.
QC176.A3C66 1983 530.4'1'0321 '82-18162
ISBN 0-201-14204-X
ISBN 0-201-14205-8 (pbk.)

ABCDEFGHIJ-AL-898765432

Manufactured in the United States of America

CONTENTS

FOREWORD

Solid-state physics is the study of what may be called the material properties of solid bodies.

The term material properties refers to those properties that characterize any large quantity of a particular material, whatever its shape (e. g., any large volume, or a large area of an interface between two substances). Thus, for example, solid-state physics includes such things as the measurement and explanation of the dielectric constant of a crystal, but it does not include the study of the electrostatic field distribution when a specimen of a particular shape is acted on by a particular arrangement of external charges.

The dividing line between "solids" and other substances is vague, and its accepted position has varied a little over the years. To the layman and to early scientists, a solid has simply been something relatively unyielding, or in a more refined concept, something that returns to its original configuration when weak deforming forces are removed. But it has been found that nearly all bodies undergo at least a slight plastic deformation when strained, and we also know that glassy materials, which are very hard and elastic when cold, soften continuously as they are heated. In the nineteenth century these facts, combined with natural philosophers' systematization of the many properties of crystals, led to a greater and greater identification of solids with crystalline materials. In 1912 and 1913 von Laue and the Braggs verified, by x-ray diffraction, the concept that crystals consist of atoms stacked in regular arrays, a concept that had been developed in some depth as a theoretical speculation, out of roots going back to Newton and earlier, during the whole nineteenth century. Following this verification, and especially after the advent of quantum mechanics in 1925, concepts based on crystalline order contributed so much to the understanding of crystals that it became fashionable simply to define solids as crystalline or polycrystalline materials. In recent decades, an increasing interest in amorphous materials and in the role of defects in crystals has led the solid-state physics community to consider as its domain all materials in which atoms spend most of their time vibrating about fixed sites, whether regularly or irregularly arranged. Indeed, since there is an increasing interaction between the literature of solid-state physics and that devoted to the atomic understanding of liquids, the enlarged concept of a field called the physics of condensed matter is increasingly used.

Today, solid-state physics as we have defined it accounts for about a third of all research publication in physics. For most of the first half of the twentieth century the corresponding fraction was fairly constant at a somewhat smaller value, of the order of a fifth, and in the nineteenth century the fraction seems to have been still smaller. The proportion of current research papers that are purely theoretical is less than a third, lower than for any of the other major subfields of physics.

Of the types of materials studied, metals rank first in terms of volume of literature, followed in order by inorganic insulators, inorganic semiconductors, noncrystalline solids, and organic insulators. The kinds of properties of any of these materials that can be studied are so diverse that only a few gross groupings of them can be indicated here. One convenient distinction is that between the properties of surfaces and interfaces, to which about a fifth of current solid-state research papers are devoted, and bulk properties. Another useful distinction separates properties that can be adequately described in terms of positions and motions of entire atoms from those that depend in an essential way on the quantum-mechanical behavior of electrons. Although the distinction between these "atom-unit" and "electronic" properties is often rather fuzzy (e. g., electronic motions determine interatomic forces), it is real enough to have divided the solid-state research community into groups that often feel quite remote from each other. The most prolific areas of atom-unit study are, in rough order, crystal structure (often considered to belong to a separate science, crystallography), plastic deformation (which merges into the fields of metallurgy and materials science), structure and kinetics of surfaces and interfaces, and the production and motion of defects and impurities. In the area of electronic properties, the most prolific areas are, again in rough order, magnetism and magnetic resonance, optical properties, electronic conduction, electronic properties of surfaces, and superconductivity. The areas named account for nearly four fifths of current research publication; the remainder falls into many smaller subfields. Some of the latter, like neutron diffraction and scattering, Mössbauer effect, calorimetry, elasticity, and dielectric and ferroelectric polarization, have important ties to both the atom-unit and the electronic groups.

Fundamental to much of solid-state physics is the understanding of the geometrical nature of crystalline order. This understanding, already developed in the nineteenth century, led at once to a working out of the constraints crystalline symmetry imposes on the tensors describing various material properties, properties whose interrelationships could be further illuminated by thermodynamics.

In the last half-century it has come to be recognized that crystalline order is most appropriately described in terms of correlations between the situation at one point $\mathbf{r}$ of space and that at another point $\mathbf{r}'$: when $\mathbf{r}'$ and $\mathbf{r}$ are near each other such correlations are called *local order*, whereas if the correlations persist for arbitrarily large distances of $\mathbf{r}'$ from $\mathbf{r}$ they are described as *long-range order*. A crystal has a periodic long-range order, e. g., in its mass density or its electron density, whereas an amorphous substance has only local order. In recent decades it has been found that some crystals can have periodic long-range order with two incommensurate periods in the same direction; e. g., the electron spin density in certain antiferromagnetic crystals can have an approximately sinusoidal spatial variation with a wavelength incommensurate with the average interatomic spacing.

Having come to understand ideal crystalline order, we are led next to examine the simplest types of departure from such order, i. e., *localized imperfections*. Special interest naturally attaches to those imperfections that, because of the geometrical nature of the crystal lattice, cannot be eliminated by any purely local shifts of atomic positions, etc. Such "irreducible imperfections" can be classified by dimensionality as point, line, or planar defects. The simplest point defects correspond to having too many or too few atoms of one type or another in a given number of cells of the crystal: a missing atom is called a *vacancy*; an extra atom is called an *interstitial* if it is squeezed in between the normal atomic sites, or an *interstitialcy* if two or more atoms share symmetrically the space that one would normally occupy; an atom of a species not normally present in the crystal is an *impurity*, and may be further classified as interstitial (if it is simply additional to the atoms of the perfect crystal) or substitutional (if it replaces one of the latter). Any of these types of point imperfections can be moved to a neighboring location by displacements of atoms in its vicinity (hence it can diffuse by the action of thermal agitation), but it cannot be created or annihilated except at a boundary.

The most important line imperfections are *dislocations*. Suppose we were to travel around a closed loop in a crystal lattice, avoiding any imperfections, and count how many net lattice spacings we have traversed in each direction of the compass. If the crystal is perfect everywhere, the net (integer) number of lattice translations traversed in each direction must return to zero after we have returned to our starting point. If after such return the net number of lattice translations is not zero, there must be an imperfection in the crystal structure somewhere in the region enclosed by the circuit, and since this imperfection must be encountered in any attempt to shrink the circuit to a point, via any route, the imperfection must be localized in the neighborhood of a line that cannot end inside the crystal, i. e., the line must either extend from boundary to boundary, or form a closed loop. Imperfections of this type are called dislocations. They play an essential role in plastic deformation, since the boundary between a region of a crystal in which a slip by one lattice spacing has occurred and a region in which it has not must be a dislocation. It turns out, too, that dislocations often play an important role in many other phenomena besides plastic deformation.

Planar imperfections include *grain boundaries*, which separate crystalline regions of different orientation, and *stacking faults*, where the registry of adjacent parallel planes, though still such as to make a good fit locally, is in some respect different from that in the perfect crystal. The atomic arrangement at a grain boundary is usually somewhat disordered, but for what is called a *twin boundary* there is no disorder, the boundary being simply a plane of atoms that is shared in common by the crystal lattices of the two grains, which thereby have a special orientational relationship.

The dynamical properties of crystals, whether treated by classical mechanics (often a good approximation for nuclear motions) or quantum mechanics (necessary for the electrons). can become very complicated to describe because many particles interact simultaneously. But considerable understanding can be achieved by use of approximations that bypass these complications: if the vibrations of the atoms are treated as harmonic, they can be resolved into normal modes; if the electrons are assumed to move independently of each other except for the constraint of the exclusion principle, their behavior can be described in terms of wave functions of single electrons. In both cases the symmetry and regularity of the crystal lattice provide a very important framework for the description of the normal modes or the electronic eigenstates. That is, these can be taken in the form of traveling waves, with a modulation having the periodicity of the crystal lattice and with a fundamental wave vector lying within a polyhedron that for historical reasons is called the *first Brillouin zone*; waves with wave vectors outside this polyhedron can always be treated as waves with wave vectors inside it, with a superposed modulation having the periodicity of the lattice. In the vibrational case, a quantum of occupation of such a normal mode is called a *phonon*: a quantum state of this type for an electron is called a *Bloch wave*.

The energy for a Bloch wave or the frequency of a phonon is a continuous though multivalued function of its wave vector inside the first Brillouin zone, and at the boundary of this polyhedron joins continuously to its value at the opposite boundary. Thus in general the spectrum of electron energies or phonon frequencies consists of an infinite or a finite number, respectively, of continuous bands, which may be separated by forbidden gaps. This fact is especially important in the electronic case: the ground state of a crystal, in the independent-electron approximation, is one in which all Bloch-wave states that have energies below some limiting value, the *Fermi energy*, are occupied while all higher ones are empty. If the Fermi energy lies in the forbidden gap between two bands, i. e., if there are just enough electrons to exactly fill an integral number of allowed energy bands, the crystal will be an insulator. If the electron count is sufficient to fill only part of its highest occupied band, or if the Fermi energy is in a range where two or more bands overlap, the crystal will be a metal. In the latter case the Bloch-wave states having energies equal to the Fermi energy will have wave vectors lying on a surface in the first Brillouin zone known as the *Fermi surface*. It turns out that the size, shape, and topology of this surface have an important influence on many properties of metals.

Another important property of Bloch-wave states is their response to uniform electric and magnetic fields. It turns out that wave packets made of these states are accelerated by the electrostatic and Lorentz forces just as free particles would be, except that the ratio of force to acceleration is an *effective mass*, dependent on the crystal potential, in general anisotropic and often negative. The states at the bottom of any band have positive effective masses, those at the top negative; from this it follows that the "hole" left when an electron is removed from near the top of an otherwise full band behaves like a positive charge with a positive mass.

Since the electrostatic interactions of the different electrons in a solid are in fact very strong, we might well question the utility of concepts like those just described, which treat different electrons one at a time. For some problems these concepts do indeed break down, but it can be shown that with relatively minor modifications these concepts have a validity for interacting crystal electrons that is exact in a certain asymptotic limit. That is, the low-lying excited quantum states of a real metal—the states that will be appreciably occupied at low temperatures—can be put into a one-to-one correspondence with those of the independent-electron model described earlier, hence can be described in terms of equivalent one-electron states, or *quasiparticles*, the excitation consisting in occupation of a few such states just outside the Fermi surface and the vacating of a corresponding number just inside. For an insulator such a description—a representation of excited states in terms of electrons near the bottom of the normally empty bands and holes near the top of the normally filled ones—can be shown to be even more accurate.

The quasiparticles just described, and the phonons mentioned previously, are special cases of a more general concept, that of *elementary excitations*. These are approximate excited quantum states of a many-body system that form a continuous family describable in relatively simple physical terms, and whose difference from true eigenstates is small in the sense that their level widths are much smaller than their energies of excitation; some authors prefer to restrict the term to cases where the ratio of these two quanties approaches zero as the excitation energy decreases. The most important type of elementary excitation, after those named, is the *magnon* or *spin wave*, an excitation in which the direction of magnetization in a ferromagnetic material, or that of a sublattice moment in an antiferromagnetic, is spatially nonuniform and propagates as a wave. Another important elementary excitation, though not asymptotically exact in the sense mentioned above because its minimum excitation energy is rather large, is the *plasmon*, a wave of charge accumulation. Other more exotic types of excitations are sometime discussed in the literature. All elementary excitations obey either Fermi or Bose statistics (electrons and holes are Fermi; phonons, magnons, and plasmons are Bose).

Many phenomena that involve small but finite energies of excitation of a solid can be well described in terms of these various kinds of elementary excitations and interactions between them. A systematic though quasi-phenomenological description of these interactions, for electron and hole quasiparticles, is given by Landau's *Fermi-liquid theory*. Examples of effects due to the interactions of these and other kinds of elementary excitations are scattering of the elementary excitations; renormalization of their energies; binding of an electron to a hole in an insulator to form what is called an *exciton*; and hybridzation of different kinds of excitations (phonons with light waves, magnons with phonons, etc.).

The most spectacular of the effects produced by the interaction of elementary excitations is the phenomenon of *superconductivity*, the complete disappearance of electrical resistance that occurs for most metals at extremely low temperatures. Discovered in 1911, this behavior was considered the outstanding mystery of solid-state physics until 1957 when Bardeen, Cooper, and Schrieffer showed that it could be understood as the result of an association of electrons into pairs, mediated usually by their coupling to phonons, and a preferential condensation of these pairs into a single particular pair state. This condensation causes the quantum nature of the pairs to manifest itself as a phase variable whose variation is gradual over macroscopic dimensions and in terms of which macroscopic quantities like current densities can be described.

The imperfections in crystal structure that were discussed earlier have characteristic consequences for the vibrational motion and electronic levels of solids. By destroying the perfect periodicity of the crystal, they can scatter elementary excitations that travel as waves (phonons, electrons, magnons, etc.) and they can bind such excitations into localized states. Thus we can have localized vibrational modes, localized magnon modes, and localized electronic levels; all of these are possible not only at point defects or impurities, but also at dislocations, planar defects, or surfaces. Two fields in which most phenomena are dominated by electronic impurity or defect centers are luminescence and semiconductor physics, the former because of the wealth of possibilities presented by optical transitions at the many possible types of impurities in many possible hosts, the latter because impurities in a nonmetal that bind an electron or hole only weakly can easily be thermally excited to produce free electrons or holes and thus to make the crystal a conductor.

Besides elementary excitations, there are a number of other even more general theoretical tools that are very valuable for an understanding of the properties of solids. One such, thermodynamics, has already been mentioned. Another is the general thermodynamic and quantum-statistical theory of irreversible processes, and especially its branch known as *linear-response theory*. This discipline describes the time-dependent behavior of any measurable property of a system after it has been stimulated out of an equilibrium state by an infinitesimal time-varying perturbation of some parameter. Particularly important is the fact that if either the in-phase or the out-of-phase part of the response is known at all frequencies, the other can be calculated (Kramers–Kronig theorem). This is invaluable, for example, in the analysis of optical absorption and dispersion. Another useful aspect of linear-response theory is the relating of the dissipative response in a process like electrical conduction to the magnitude of certain fluctuations in thermal equilibrium.

In solid-state physics, as indeed in condensed-matter physics generally, the occurrence of sharp phase transitions has fascinated theorists for many decades. These are clearly cooperative phenomena, since the interaction of neighboring elements of a material with each other sometimes leads to correlations between properties measured at points remote from each other, i. e., to long-range order as we defined it earlier. Such order can occur in atomic positions, in the alternation of different chemical constituents, or in the magnetization of atomic spins. When one phase changes to another, the change in physical properties is sometimes completely discontinuous (first-order phase change, with a latent heat) and sometimes continuous, though with discontinuous or singular derivatives (higher-order phase change). In recent years sophisticated methods, including especially an application of renormalization-group concepts borrowed from elementary-particle theory, have been used to elucidate details of the variation of various physical properties as the parameters of a system (temperature, magnetic field, etc.) approach values at which a higher-order phase transition will take place.

All the traditional types of physical measurements—mechanical, acoustic, calorimetric, electrical, magnetic, etc.—are used and indeed are essential in the study of solids. In addition, a number of specifically modern tools have come to be especially useful. Some examples of these are lasers, important both for their high resolution and for their great intensity; synchrotron radiation, which fills the gap between optical and x-ray spectroscopy; radiofrequency and microwave fields, which can, among other things, probe the constituents of solids in exquisite detail by electron spin resonance, nuclear magnetic resonance, double resonances, cyclotron resonance, and dimensional resonances associated with the orbits of the electrons of metallic slabs in a magnetic field; diffraction and inelastic scattering of neutrons, which yield information about crystal structure, magnetic order, and dispersion relations of phonons and magnons; ultralow temperatures; ultrahigh-vacuum technology, a necessary prerequisite to the controlled study of surfaces; ultrahigh magnetic fields, now widely available thanks to superconducting magnets; positron annihilation, which yields information both on electron momenta and on the properties of defects that can trap positrons; electron microscopy, which can make visible dislocations and other defects in crystalline films and can reveal great surface detail; scattering of fast ions from atoms of a solid, and their relatively free passage along channels between atomic planes in a crystal; tunneling of electrons through thin insulating layers between conductors or superconductors, a rich source of information about impurities, elementary excitations, etc.; the Mössbauer effect, which gives information about magnetic fields and electron densities at nuclear sites and about lattice vibrations.

Among the other areas of physics, the ones whose subject matters most often overlap solid-state physics (as measured, e. g., by the frequency of research papers that could legitimately be assigned to either field, or by the frequency with which workers switch from the one field to the other) are optics and atomic, molecular, and electron physics. As for areas outside physics, the strongest overlaps are with materials science (metallurgy, ceramics), chemistry (physical chemistry or chemical physics, polymer science), and electrical engineering (solid-state devices). Though overlap with the earth and planetary sciences is much smaller in total volume, the common area of properties of solids at very high pressures is an important one. In many of the areas of overlap, about the only distinction between solid-state physics on the one hand and chemistry or engineering sciences on the other is that the former tends to emphasize simple and understandable cases, while the latter venture farther afield into complex problems and systems of technological importance.

Less quantifiable but sometimes more spectacular are the influences different fields have on each other through instrumentation on the one hand and major novel concepts on the other. Solid-state physics has been both donor and beneficiary of many such influences. For example, it has received very powerful experimental tools from nuclear physics in the Mössbauer effect and pile neutrons; it has made various contributions by way of repayment, perhaps most notably by making possible the development of solid-state counters, but also by various other contributions to techniques of measuring properties of nuclei, especially their magnetic-dipole and electric-quadrupole moments. Elementary-particle physics has enriched the outlook of solid-state theorists through the notions of Green's functions and renormalization groups; experimental particle physics has benefited, as have many other areas of science and technology, from the invention of superconducting magnets. The concept of superfluidity due to fermion pairing, first developed by solid-state physicists to account for superconductivity, has found application in the theory of heavy nuclei and in the theory of neutron stars. Moreover, from origins in solid-state physics, semiconductor electronics, with its progeny microelectronics and optical electronics, has become a major branch of engineering, and through the collaboration of electrical engineers, materials scientists, and mathematicians has brought us into the computer age, which has revolutionized all fields of science and technology, including solid-state physics itself.

CONYERS HERRING

PREFACE

The aim of this work is to provide, in one volume, a convenient and readily accessible set of resource materials in the area of solid state physics. This is an area of physics which has fairly well defined boundaries, although no field of physics can ever be neatly categorized in such a way that there is no overlap with other fields. The articles chosen for this volume have been selected from the Encyclopedia of Physics which we edited; readers who wish to explore other aspects of condensed matter physics, or to consult articles on topics which provide the foundations of physics in several areas, are referred to the parent volume.

RITA G. LERNER
GEORGE L. TRIGG

MAIN ENTRIES

CONCISE ENCYCLOPEDIA OF SOLID STATE PHYSICS

Acoustoelectric Effect

Esther M. Conwell

The acoustoelectric effect is the appearance of a dc electric field when an acoustic wave propagates in a medium containing mobile charges. It was first named and discussed, theoretically, by R. H. Parmenter in 1953. As pointed out by G. Weinreich, who first detected it experimentally, it is an example of the general phenomenon of "wave–particle drag," of which the operation of a linear accelerator and the motion of driftwood toward a beach are other examples. Although first proposed for metals, it is only significant in semiconductors. It can be quite strong in piezoelectric semiconductors, such as CdS, ZnO and GaAs, but is also seen in nonpiezoelectric semiconductors. In the piezoelectric case, the wave and mobile charges interact through the electric field arising from the strain associated with the wave. This interaction is significant when the electric field is longitudinal, i.e., parallel to the wave propagation direction, and in what follows we assume that the type of wave (i.e., longitudinal or shear) and sample orientation have been chosen to make this the case. In nonpiezoelectric media the interaction is through the shift of carrier energy induced by the strain, the so-called deformation potential. At the frequencies ordinarily used, $\lesssim 1$ GHz, the latter interaction is much smaller than that in piezoelectric materials. In either case, however, because of the interaction the passage of an acoustic wave through the medium causes a periodic spatial variation of the potential energy of the charge carriers. If the mean free path l of the carriers is small compared to the acoustic wavelength λ, as is the case for most of the usual (ultrasonic) frequency range, this results in a bunching of the carriers in the potential-energy troughs. Since the wave is propagating, it drags the bunches along with it. This is the origin of the acoustoelectric field and clearly also causes attenuation of the wave. The effect is stronger, the stronger the bunching. Thus it is very weak in ordinary metals, where space-charge effects prevent appreciable bunching. It is enhanced in those nonpiezoelectric semiconductors where space-charge effects are minimized by either (1) the presence of both positively and negatively charged carriers or (2) the existence of different groups of carriers (many-valley band structure) whose energy is affected differently by the strain so that they bunch in different phases of the wave.

The foregoing discussion suggests that if an external electric field were applied to the sample to give the carriers a drift velocity, v_d, greater than the wave velocity, v_s, the carriers should drag the wave, i.e., the wave should be amplified. This conjecture was verified experimentally on CdS samples at frequencies of 15 and 45 MHz by Hutson, McFee, and White in 1962. They found acoustic gain for shear waves at fields greater than 700 V/cm, at which field v_d equals the shear wave velocity. For not too large acoustic wave amplitudes it was possible to account quite well for the size of the gain and its variation with frequency, etc., with a linear phenomenological theory taking into account the currents and space charge produced by the piezoelectric fields that accompany the acoustic wave. The gain, called *acoustoelectric gain,* is found to be low at low frequencies (less than the conductivity relaxation frequency σ/ϵ), where the carriers can redistribute themselves quickly enough to essentially cancel out the piezoelectric field. It peaks at the frequency for which the acoustic wavelength is of the order of the Debye length, where the bunching is optimum. In a fairly strong piezoelectric like CdS acoustoelectric gains as high as 40 dB/cm have been found, leading to consideration of this effect for practical use as an amplifier (*acoustoelectric amplifier*).

The theory and effects considered so far are linear in the sound amplitude provided it is small, i.e., small enough to bunch only a small fraction of the carriers. At large sound amplitude new effects appear. When two acoustic waves are present, the interaction of the piezoelectric field of one with the bunched carriers of the other may result in the generation of the difference or sum frequency of the two. As a special case of this, for a single large wave the interaction of its piezoelectric field with its own carrier bunches results in the generation of a dc current, called the *acoustoelectric current,* flowing in a direction opposite to the usual or Ohmic current, and may result in the generation of the second harmonic.

When an outside acoustic wave is not introduced, application to a highly conducting sample of CdS or ZnO, for example, of a field high enough to make $v_d > v_s$ causes a large amplification of the thermal equilibrium acoustic waves or flux present in the sample. This gives rise to unusual behavior, the exact nature of which depends on the details of sample inhomogeneity and contacts. One possibility is that immediately after the high field is applied a dc current will flow of the expected magnitude for the Ohmic resistance, i.e., the resistance displayed for $v_d < v_s$, but in a short time the current will drop to a much smaller value and remain there. The smaller value is due to the opposing acoustoelectric current arising from the flux amplification. Another frequently seen possibility is the onset of strong current oscillations with a period equal to the length of the sample divided by v_s. These oscillations are due to the creation close to the cathode of a domain or narrow region of high acoustic flux density, which typically moves down the sample with velocity v_s. When it exits at the anode, the current rises to its Ohmic value, strong flux generation begins again at the cathode, and the process is repeated.

The above discussion has been couched in terms appropriate for $l \ll \lambda$, since this is the case for most of the experimental situations studied. However, both linear and nonlinear processes have been studied with a microscopic theory that does not make this restriction. In this theory acoustic gain, for example, may be thought of as due to an excess of stimulated phonon emission by the carriers over absorption.

See also PIEZOELECTRIC EFFECT; SEMICONDUCTORS.

BIBLIOGRAPHY

N. G. Einspruch, "Ultrasonic Effects in Semiconductors," in *Solid State Physics* (F. Seitz and D. Turnbull, eds.), Vol. 17, p. 217. Academic Press, New York, 1965.

J. H. McFee, "Transmission and Amplification of Acoustic Waves in Piezoelectric Semiconductors," in *Physical Acoustics* (W.

Mason, ed.), Vol. IV, part A, p. 1. Academic Press, New York, 1966.

H. N. Spector "Interaction of Acoustic Waves and Conduction Electrons," in *Solid State Physics* (F. Seitz and D. Turnbull, eds.) Vol. 19, p. 291. Academic Press, New York, 1966.

R. Bray, "A Perspective on Acoustoelectric Instabilities," *IBM J. Res. Devel.* **13,** 487 (1969). See also other articles in this volume, pp. 494–510.

N. I. Meyer and M. H. Jorgensen, "Acoustoelectric Effects in Piezoelectric Semiconductors with Main Emphasis on CdS and ZnO," in *Festkörper Probleme X* (O. Madelung, ed.), p. 21. Vieweg, Braunschweig, Germany, 1970.

E. M. Conwell and A. K. Ganguly, "Mixing of Acoustic Waves in Piezoelectric Semiconductors," *Phys. Rev.* **B4,** 2535 (1971).

Adsorption

J. G. Dash

The surfaces of all solids and liquids adsorb foreign molecules from the atmosphere. The adsorbed molecules change most of the chemical and physical properties of the underlying substrate: Adhesion, catalysis, corrosion, fracture, lubrication, and wear are affected by the topmost molecular layers on a surface. An understanding of these changes involves the study of the films themselves.

The forces of attraction that cause adsorption are the relatively weak and long-range interactions that exist between all atoms, molecules, and macroscopic objects. These "dispersion" forces are due to the attractions between the electric dipole moments induced in each body by the fluctuating fields of their neighbors. Dispersion forces between neutral atoms and molecules are also known as "van der Waals forces," and they are responsible for the condensation of vapors to liquid or solid phases at sufficiently low temperature. Temperature plays a similar role in adsorption, lower temperatures favoring the adsorbed state and decreasing the vapor pressure of the surface film.

The states of surface films depend on their thickness, temperature, and composition, as well as the structure, uniformity, and constitution of the substrate. As a result of this interplay films display a great variety of distinctive regimes. Monolayer films exhibit two-dimensional analogs of the familiar vapor, liquid, and solid phases of bulk matter, as well as others that have no three-dimensional equivalents. In thicker films some layers may behave with separate thermodynamic individuality while the remainder is diffuse. On irregular surfaces films possess a distribution of properties, different regions reacting to the local surface environment. The various regimes are experimentally studied under controlled laboratory conditions. They can be examined by one or more of a wide variety of experimental techniques, including calorimetry, nuclear resonance, electron and neutron diffraction, molecular scattering, and Mössbauer spectroscopy. Development and application of suitable methods for film studies has proceeded rapidly since the 1960s. The measurements have provided the impetus for considerably greater theoretical interest.

Theoretical attention is focused on the special film–substrate combinations and their experimental parameters, but also on more general questions involving the physics of lower-dimensional systems. The physical properties of lower-dimensional systems have interested theorists for many years, since it has been shown that in one- and two-dimensional matter there can be no perfectly ordered states or structures; in other words, two-dimensional crystals, magnets, superconductors, and superfluids cannot exist (except at absolute zero). The reason for this sweeping conclusion is the special richness of low-lying modes of excitation in lower-dimensional phase space, excitations which cause a breakdown of the long-range order at all finite temperatures. However, relatively recent considerations have called these well-accepted ideas into question. It has been shown theoretically that there are at least two forms of long-range order, and that only one of them is destroyed in two dimensions while the second can persist. Applied specifically to the case of crystalline order, the new theory predicts that the long-range "topological order" of directional correlations between atomic positions can persist to finite temperatures while the long-range positional correlations disappear above $T = 0$.

In the following sections several recognized regimes are described, taken in approximate order of increasing surface density.

1. Two-dimensional gases. When a uniform surface is sparsely covered, the adsorbed molecules can act like a two-dimensional gas. Its essential characteristics are adhesion to the substrate, which at sufficiently low temperatures leads to a "freezing out" of higher states of motion normal to the surface, and low surface density, which is equivalent to weak or negligible interactions between the molecules. Beyond these essentials the gas regimes embrace a wide range of properties depending on the atomic structure of the surface, i.e., its "texture." Surfaces composed of relatively small and densely packed atoms are relatively smooth to the adsorbed molecules, which are then free to move about the surface. Examples of very smooth surfaces are the dense (111) planes of close-packed metals and the basal plane of graphite. Low-density helium films on graphite exhibit heat capacities resembling classical two-dimensional gases at temperatures of 4 K and above, thus indicating that the atoms have considerable thermal mobility on the surface and are not strongly affected by interactions with each other. Films on more strongly textured surfaces are less mobile, and the atoms spend longer times in definite adsorption sites before evaporating to the three-dimensional gas phase or hopping to a neighboring site. This restricted mobility can be described in terms of a narrower band structure of the atoms, according to a two-dimensional band model analogous to the conventional theory of band structures of electrons in metals. If the energy bands have narrow widths and are separated by wide gaps, the atoms are relatively immobile; the film is "localized." At the other extreme the band structure is equivalent to a two-dimensional mobile gas, and a continuous range of intermediate mobility is possible. Theoretical band structures of helium on graphite yield wide bands and comparable width gaps, consistent with the thermal measurements. Heavier gases have quite narrow bands, and fragmentary experimental results are consistent with these indications of restricted mobility.

2. *Imperfect gases and phase condensation*. Interactions between the adsorbed atoms are more important at low temperatures and high surface densities. The effects of the interactions begin as small corrections to the equation of state of the low-density "ideal" film (see point 1). These changes are analogous to the deviations occurring in three-dimensional (3D) "imperfect" gases, conventionally described by a series expansion in the density, the so-called virial expansion. The second virial coefficients of several film–substrate combinations have been obtained experimentally by analysis of calorimetric data or vapor-pressure isotherms, and they are in fair to excellent agreement with theoretical coefficients calculated from empirical free-atom pair potentials. Phase condensation in monolayers to two-dimensional (2D) liquid or solid phases occurs in many monolayer systems when cooled to low temperatures. Critical temperatures are somewhat less than half of the critical temperatures of the bulk phases for most of the noble gases and other simple molecular gases. The principal exceptions are ^{4}He on graphite surfaces, for which T_c (2D) is nearer to a third of T_c (3D), and ^{3}He, which appears to have no bound liquid state.

3. *Two-dimensional solids*. Several monolayer systems have two-dimensional solid phases at high density and low temperature. The exemplary systems are ^{3}He and ^{4}He on basal-plane graphite, both of which exhibit classical signatures such as heat capacities varying as T^2, the two-dimensional analog of the well-known Debye T^3 law. The magnitudes of their coefficients indicate that the monolayer solids have elastic constants approximately equal to those of two-dimensional solid helium, but with some significant shifts believed due to substrate structure (see point 4). The He films when heated undergo continuous melting transitions, in contrast to the abrupt phase changes of bulk solid. Other films show a variety of melting behaviors: Ne, CH_4, and Xe on graphite have sharp triple points, while Ar and Kr have continuous transitions. Several possible causes for the differences have been proposed, but there is no satisfactory theory.

4. *Registered phases*. In some films the substrate structure imposes a regularity on the atomic arrangement in the monolayer. The simplest cases are denoted as 1×1 structures, in which adsorbed atoms have the same density and symmetry as substrate atoms. Various arrangements, classified as $n\times m$ structures, are results of the competition between substrate–atom and atom–atom interactions. Low-energy-electron-diffraction studies have disclosed structures as complex as 7×15. The number of distinct structures that could conceivably occur in a macroscopic film is extremely large, but energy considerations reduce the actual number of phases to a very few. The possible phase transitions between registered phases of different symmetries and between registered and nonregistered phases are very interesting for their relevance to theoretical models. Particular interest is attached to the order–disorder transition of the registered phases of ^{3}He and ^{4}He on graphite, which exhibit strong heat capacity peaks shaped like the power-law singularities of the appropriate theoretical model. Other registered phases of special interest include N_2 and H_2 on graphite, which have triangular arrangements with lattice spacings $\sqrt{3}$ greater than the graphite mesh; Xe on Pd and Ni facets, which have subtle superlattice structures; and polar molecular gases such as Br_2 and $FeCl_2$, in which orientational ordering appears to take place.

5. *Multilayer films*. Films of several layers thickness display a variety of habits. In some examples the one or two layers closest to the substrate behave as relatively distinct two-dimensional solids while a topmost third layer is effectively a two-dimensional gas. In other films there may be as many as five distinct layers evidenced by the stepwise character of the vapor pressure during their formation. Individuality of layers occurs only at sufficiently low temperatures, where the thermal energy is less than the difference in substrate attraction between one layer and its neighbors. For many systems there is a transition between layer formation and cluster growth, generally occurring at a few layers thickness. The instability of layer formation *vis á vis* cluster formation is an important phenomenon in all types of films. It occurs in many systems involving strongly cohesive metal films as well as those of weakly bound van der Waals molecules such as helium.

See also CATALYSIS; ISING MODEL; ORDER–DISORDER PHENOMENA; PHASE TRANSITIONS; SURFACES AND INTERFACES; THIN FILMS.

BIBLIOGRAPHY

J. G. Dash, "Two Dimensional Matter," *Scientific American* **228**(5), 30 (May, 1973). (E)

W. A. Steele, *The Interaction of Gases with Solid Surfaces*. Pergamon Press, Oxford, 1974. (A)

J. G. Dash, *Films on Solid Surfaces*. Academic Press, New York, 1975. (A)

K. L. Chopra, *Thin Film Phenomena*. McGraw-Hill, New York, 1969. [especially Chap. IV (nucleation and growth)]. (A)

Allotropy and Polymorphism

F. J. Di Salvo

The atomic structure of some solids changes when the external conditions, such as pressure or temperature, are varied. This phenomenon is called allotropy when it occurs in an element, and polymorphism when it occurs in a compound. Three common examples of allotropy are presented below.

SULFUR

Solid sulfur consists of rather flat S_8 molecular rings that are stacked on top of one another. When heated to 95°C the molecules change orientation, forming a differently stacked structure. Sulfur at room temperature is called rhombic sulfur, and above 95°C, monoclinic sulfur (after the shape of their respective crystallographic unit cells). Monoclinic sulfur melts at 120°C.

Sulfur can also exist in an amorphous form. When liquid sulfur is heated to several hundred degrees centigrade, most of the S_8 molecules break open and join with others to form long sulfur chains. If this liquid is rapidly cooled to room temperature, the chains remain intact and are randomly packed together to form a rubbery solid. At room temperature amorphous sulfur will very slowly change back into rhombic sulfur. Rhombic sulfur is the stable, or equilibrium, form of sulfur at room temperature. By other preparation methods a number of other metastable forms of sulfur can be obtained at room temperature. Consequently, sulfur has a large number of allotropes; however, rhombic and monoclinic sulfur are the only equilibrium forms (in their respective temperature ranges of stability *and* at atmospheric pressure).

IRON

At room temperature iron has a body-centered cubic (bcc) structure; the unit cell is shown in Fig. 1a. (The structure can be visualized by imagining space to be filled with closely packed cubes. At each corner, where eight cubes come together, place an iron atom and then put another in the center of each cube.) When iron is heated to 910°C its structure changes to face-centered cubic (fcc); a unit cell is shown in Fig. 1b. (In this structure an iron atom is placed at each cube corner and one iron atom on each face of the cube, where two cubes touch.) Iron changes back to the bcc structure at 1390°C and melts at 1536°C.

The allotropy of iron is very important for the production of steels. Carbon is moderately soluble in fcc iron, the carbon atoms occupying some of the holes between the iron atoms in this structure (at the cube center in Fig. 1b). However, the solubility of carbon is much lower in bcc iron. If iron containing several weight percent of carbon is cooled from 1100°C (fcc phase) to room temperature, the carbon not soluble in bcc iron forms a compound, Fe_3C. The Fe_3C exists in small platelike regions dispersed in bcc iron. Fe_3C is called cementite, since it makes the iron much stronger. Iron prepared in this manner is called carbon steel.

CARBON

Carbon exists in two structural forms: graphite and diamond. Graphite has a layered structure with weak interlayer bonding. Because of the weak interlayer bonds the layers slide easily over each other. Consequently graphite is used as a lubricant and in pencil lead. Graphite becomes diamond under high pressure (greater than 10 000 atm). With Ni as a catalyst, small diamonds can be manufactured at 1200°C and high pressure. These diamonds are used in making grinding wheels and cutting tools, since diamond is a very hard material.

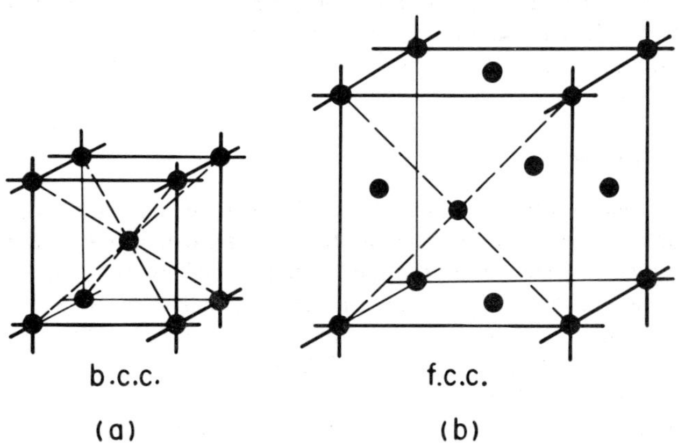

FIG. 1.(a) A unit cell of body-centered cubic (bcc) iron. Iron atoms are spheres that touch along the dotted lines (body center to cell edge). (b) A unit cell of face-centered cubic (fcc) iron. The cubic unit cell is larger than the bcc cell because the iron atoms now touch from the face center to the cell edge.

CONCLUSION

Obviously the properties of some materials are quite affected by a change in their structure, and a knowledge of the allotropic forms of materials has been important to the development of many technologies.

See also CRYSTAL SYMMETRY; METALLURGY.

BIBLIOGRAPHY

B. Meyer, *Elemental Sulfur*. Wiley (Interscience), New York, 1965. (I)

A. L. Ruoff, *Introduction to Materials Science*. Prentice-Hall, Englewood Cliffs, NJ, 1972. (I)

W. J. Moore, *Physical Chemistry*, 3d ed. Prentice-Hall, Englewood Cliffs, NJ, 1963. (A)

Alloys

P. L. Leath

An alloy is a macroscopically homogeneous mixture of metals or (as in the case of carbon steel) a metallic mixture of metals and nonmetals. Most, but not all, alloys are metallic (for one exception, indium antimonide is a semiconductor). Most pairs of metals are miscible (i.e., form *binary alloys*) at some concentrations, although there are many notable exceptions (e.g., indium is insoluble in gallium). Since there are 70 elemental metals, the subject of alloys is immense, and an enormous variety of electronic and other physical properties is possible. The subject has now expanded even further with recent interest in ternary (three-component) alloys, tertiary (four-component) alloys, etc.

Alloys are often classified into ordered (or stoichiometric) and disordered alloys. The ordered alloys have the symmetry of a Bravais lattice with a multiatomic unit cell. Their structure is specified by giving the location of each atom in the unit cell. Some alloys exist essentially only as ordered alloys over the corresponding very narrow ranges of composition necessary for stoichiometry; these alloys are called *intermetallic compounds*. Other alloys (such as β-brass) have ordered phases at stoichiometric concentrations when the temperature is below a phase transition temperature but are disordered otherwise. (That is, they undergo an order–disorder phase transition.)

Disordered alloys, also called *solid solutions*, occur usually over appreciable ranges of composition. Most common are *substitutional* alloys, where the various types of atoms randomly occupy the normal sites of a lattice. But there are also *interstitial* alloys (such as carbon in γ-iron), where the solute atoms are small enough to occupy randomly the interstices between the normal lattice sites of the host metal; and recently there has been interest in *amorphous* alloys, where the atoms are not on sites of a regular lattice but are randomly placed, as in a liquid or glass. The occupations of the sites in a disordered alloy by the atomic types may be purely random but generally there is some degree of short-range order; that is, the species occupation of a particular site may be dependent on the occupation of the neighboring sites (e.g., in the disordered phases of brass the copper atoms are more likely to have zinc than copper nearest neighbors).

Certain principal variables that qualitatively give the alloy structures and phases were pointed out in the classic work of Hume-Rothery and Jones. It is, however, only rarely possible to use these few variables to predict detailed behavior of alloy phases. Clearly the relative sizes of the atoms constitute a vital factor in alloys because the volume-dependent potentials in the cohesive energy are an order of magnitude larger than the interatomic rearrangement potentials. This size factor is especially important in interstitial alloys and certain intermetallic compounds (e.g., interstitial alloys are generally not formed when the ratio of the radii of the solute atoms to those of the host atoms exceeds about 0.6). Electrochemical differences are such that generally we find only intermetallic compounds or very restricted ranges of solubility for elements widely separated in the electrochemical series.

When size and electrochemical factors allow solid solutions, the alloy structure can in some cases be directly related to the electron density or electron-to-atom ratio. According to the Hume-Rothery rules, in nearly free-electron alloys the stable crystal structure at a particular electron-to-atom ratio will be that which minimizes the energies of the electrons in the crystal potential; thus the position of the Fermi surface relative to the Brillouin zone faces is an essential ingredient. These rules seem to work qualitatively well for the d-band transition metal alloys (especially copper, silver, and gold alloys), but they fail for the more nearly free-electron alkali metal alloys because of electrochemical differences. Clearly the d bands play an important role.

Only recently have basic calculational methods been developed to predict the physical behavior of alloys accurately from first principles. The pseudopotential, orthogonalized plane wave (OPW), augmented plane wave (APW), and the Korringa–Kohn–Rostoker (KKR) Green's function methods of calculating electronic energy band structure of metals beginning with the atomic potentials (the atomic potentials in alloys look very much the same as those atomic potentials do in the respective pure metals) are now in many cases being directly applied somewhat successfully in the calculation of such physical properties of alloys as energy band structure, crystal structure, lattice vibration spectra, electrical and thermal resistivity, and magnetic and superconducting properties. In those cases where the potentials of the constituent atoms do not vary greatly the average potential may be used; this is called the *virtual crystal* (or rigid band, or common band) *approximation*. In the cases of strong disorder, when the potentials differ greatly, the *average t-matrix approximation* (ATA) and the *coherent potential approximation* (CPA), which are capable of producing the separate energy bands for each atomic species, are used. Although these calculations have been somewhat successful, such effects as charge transfer between atoms and atomic cluster effects are often important but are not included in the simple approximations just mentioned.

Disordered alloys are dramatically different from ordered alloys and pure metals in their electrical resistance at low temperatures. In ordered metals there is a striking decrease in resistance with decreasing temperature that is absent in disordered alloys. For example, in very pure disordered brass at liquid helium temperatures the electrical resistance is about half its room-temperature value, in contrast to drops by factors of about 10^{-4} in comparable ordered metals. This phenomenon is caused by electronic scattering off of the disorder or off of those regions where the periodicity is destroyed.

Finally, the physical properties of alloys are often greatly affected by heat and mechanical treatment, which may introduce or eliminate such defects as vacancies, dislocations, or grain boundaries. For example, wrought alloys, which have been hot or cold worked and hence are generally very anisotropic and fibrous in contrast to cast alloys, which are generally crystalline, are generally more ductile. The effect of such defects is only understood qualitatively, although progress is rapidly being made.

See also ELECTRON ENERGY STATES IN SOLIDS AND LIQUIDS; METALS.

BIBLIOGRAPHY

C. S. Barrett and T. B. Massalski, *Structure of Metals*, 3d ed. McGraw-Hill, New York, 1966. (I)

R. J. Elliott, J. A. Krumhansl, and P. L. Leath, "The Theory and Properties of Randomly Disordered Crystals and Related Physical Systems," *Rev. Mod. Phys.* **46**, 465–543 (1974). (A)

M. Hansen and K. Anderko, *Constitution of Binary Alloys* 2nd ed. (1958); R. P. Elliott, 1st suppl. (1965); F. A. Shunk 2nd suppl. (1969). McGraw-Hill, New York. (A compendium of data on specific alloys.)

V. Heine and D. Weaire, "Pseudopotential Theory of Cohesion and Structure," in *Solid State Physics* (F. Seitz, D. Turnbull, and H. Ehrenreich, eds.), Vol. 24, pp. 249–463. Academic, New York, 1970. (A)

N. F. Mott and H. Jones, *The Theory and Properties of Metals and Alloys*. Oxford, London and New York, 1936. (E)

P. S. Rudman, J. Stringer, and R. I. Jaffee, eds., *Phase Stability in Metals and Alloys*. McGraw-Hill, New York, 1967. (I)

Anelasticity

A. S. Nowick

The term anelasticity, although once used loosely to refer to nonelastic behavior, was given a more specific meaning by C. Zener in 1946; this meaning has since been widely adopted. Anelasticity is a generalization of Hooke's law of elasticity, which allows for time-dependent effects. Hooke's law may be stated as $\epsilon = J\sigma$, where ϵ is strain, σ is stress, and J is the compliance constant. In anelasticity, the instantaneous response and single-valuedness inherent in Hooke's law are discarded. However, two restrictions are retained: (a) *linearity*, in the sense that doubling the stress doubles the strain at each instant of time; and (b) a *unique equilibrium relationship*, which means that to every value of stress there corresponds a unique value of strain that is attained if sufficient time is allowed. The simplest relation between stress and strain and their time derivatives that obeys these conditions is

$$J_R\sigma + \tau J_U\dot{\sigma} = \epsilon + \tau\dot{\epsilon}, \quad (1)$$

involving three constants τ, J_R, and J_U. Any material that obeys Eq. (1) is called a *standard anelastic solid.* Equation (1) can be solved under conditions of constant stress, say σ_0 (which constitutes a "creep" experiment), to give

$$\frac{\epsilon(t)}{\sigma_0} = J_U + (J_R - J_U)\left[1 - \exp\left(\frac{-t}{\tau}\right)\right]. \quad (2)$$

From this equation the meaning of the constants becomes clear: J_U, called the unrelaxed compliance, corresponds to the instantaneous response, $\epsilon(0)/\sigma_0$, at $t=0$; J_R, the relaxed compliance, is ϵ/σ_0 as $t\to\infty$; τ is the relaxation time (at constant stress). Figure 1 shows this creep behavior, as well as the time-dependent recovery, or "elastic aftereffect," which takes place after the stress is removed. Equation (1) can also be solved under conditions of constant strain to obtain an exponentially decreasing stress, describing a "stress-relaxation experiment." The most important manifestation of anelasticity, however, occurs in the dynamical case, where stress and strain are both periodic, with the strain lagging behind the stress by a phase angle ϕ. Then we can express a complex compliance by $J^* = \epsilon/\sigma \equiv J_1 - iJ_2$ where J_1 is the real part, which is in phase with the applied stress, and J_2 the imaginary part, which lags σ by $\pi/2$. Using Eq. (1), we can express J_1 and J_2 as functions of the angular frequency ω:

$$J_1(\omega) = J_U + (J_R - J_U)/(1+\omega^2\tau^2), \quad (3)$$

$$J_2(\omega) = (J_R - J_U)\omega\tau/(1+\omega^2\tau^2). \quad (4)$$

Equations (3) and (4) are the celebrated Debye equations.

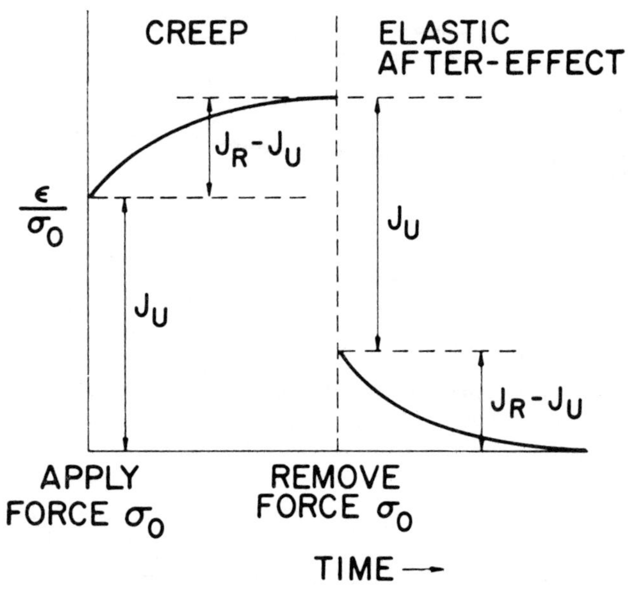

FIG. 1. Behavior of the standard anelastic solid upon application of a static stress σ_0 (creep), and upon the subsequent release of this stress (elastic aftereffect).

The function $J_2(\omega)$ when plotted versus $\log(\omega\tau)$ gives a symmetrical peak centered about $\log(\omega\tau) = 0$ (i.e., $\omega\tau = 1$), which is called a Debye peak. Figure 2 shows the variation of both J_1 and J_2 with the variable $\log(\omega\tau)$. The phase angle ϕ by which ϵ lags behind σ is given by $\tan\phi = J_2/J_1$. This quantity, which also takes the form of a Debye peak, is often called the internal friction, since it is a measure of the energy dissipated per cycle.

Since the Debye peak depends on the variable $\omega\tau$, it can be traced out either by varying the frequency, ω, or by changing τ. Although a continuous variation of the vibration frequency is sometimes possible, the usual experimental methods make it preferable to work at one frequency. It is therefore quite valuable to have a method for tracing out a Debye peak by varying τ while keeping ω constant. This is possible when τ is controlled by a thermally activated process involving an Arrhenius-type relation

$$\tau = \tau_0 \exp(Q/kT) \quad (5)$$

where Q is the activation energy for the process, τ_0 is the preexponential constant, k is Boltzmann's constant, and T is the absolute temperature. In fact, it then turns out that a plot of $\tan\phi$ (or J_2) versus T^{-1} gives a symmetrical peak like that in Fig. 2 except for a change in scale factor. If the peak is then obtained at two or more different frequencies, the activation energy is readily obtained from the shift of the peak location with frequency. Figure 3 shows an example of this type of plot, which is most important in studying anelastic phenomena.

Many phenomena in solids are describable in terms of the equations of the standard anelastic solid. Often, however, this simple model is insufficient to describe the behavior of a material. For example, the internal friction may show a superposition of two or more Debye peaks instead of a single one, or in other cases a single peak may be obtained that is broader than a Debye peak. To treat such cases, it is necessary to introduce a spectrum of relaxation times in place

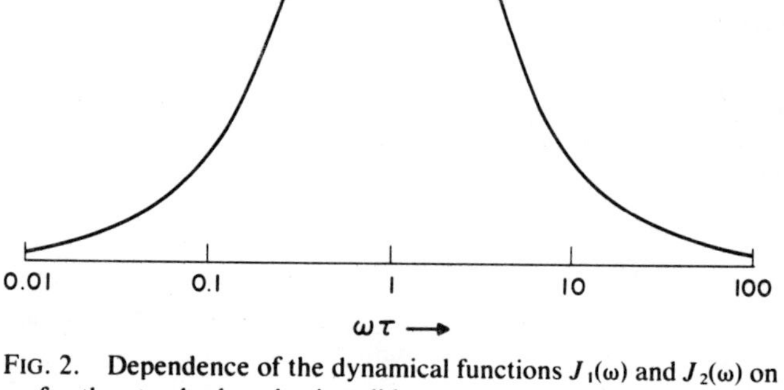

FIG. 2. Dependence of the dynamical functions $J_1(\omega)$ and $J_2(\omega)$ on $\omega\tau$ for the standard anelastic solid.

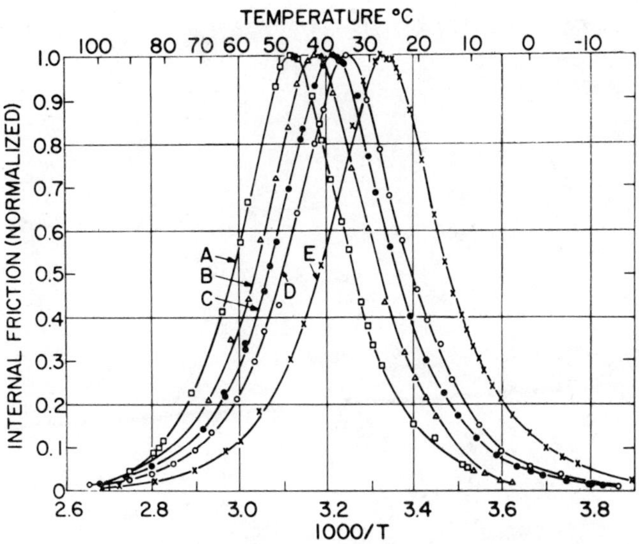

FIG. 3. A series of internal friction peaks for an Fe–C alloy as a function of $1/T$ for five different frequencies: A, 2.1; B, 1.17; C, 0.86; D, 0.63; and E, 0.27 Hz. From C. Wert and C. Zener, Phys. Rev. **76**, 1169 (1949).

of the single relaxation time of the standard anelastic solid. Thus, instead of a single exponential in the creep function of Eq. (2) there may be a summation of terms with different τ values and weighting factors. Correspondingly, Eq. (4) becomes a sum of Debye peaks. Such behavior is called a discrete relaxation spectrum. In more complex cases there may be a continuous variation in τ, a continuous spectrum, described by an appropriate distribution function. In either of these situations, the display of data in the form of a plot of internal friction (tanϕ) versus T^{-1} is widely used and interpreted.

The physical origins of anelasticity are very varied and encompass almost all aspects of solid-state physics. In crystalline materials, anelastic behavior can result from any of the following mechanisms:

1. Point-defect relaxations: redistribution of point defects (whose symmetry is lower than that of the crystal) into sites that become preferential in the presence of a stress field.
2. Dislocation relaxations: motion of dislocation segments, present either from growth or from plastic deformation, in a variety of ways with the aid of jogs, kinks, and impurity atoms on the dislocation lines.
3. Grain-boundary relaxation: viscous sliding of one grain over another in a polycrystalline material.
4. Phonon relaxation: change in the frequency distribution of phonons (lattice vibrations) due to stress.
5. Magnetic relaxations: magnetoelastic coupling via magnetostriction of a ferromagnetic material, giving rise to a number of different relaxation processes.
6. Electronic relaxations: change in the energetics of the electronic configuration produced by stress leading to a redistribution of both free and bound electrons in various materials.

Glasses are also capable of showing a variety of relaxations, many of which are similar in origin to those found in crystalline materials.

In amorphous polymers a major relaxation is associated with the glass transition and attributed to large-scale rearrangements of the main polymer chain. Secondary relaxations, at lower temperatures, are due to side groups that are capable of independent hindered rotations.

See also ELASTICITY; RELAXATION PHENOMENA; RHEOLOGY.

BIBLIOGRAPHY

R. De Batist, *Internal Friction of Structural Defects in Crystalline Solids*. North-Holland, Amsterdam, 1972.

J. D. Ferry, *Viscoelastic Properties of Polymers*. Wiley, New York, 1961.

R. R. Hasiguti and N. Mikoshiba, eds., *Internal Friction and Ultrasonic Attenuation in Solids*. Univ. Tokyo Press, Tokyo, 1977.

D. Lenz and K. Lücke, eds., *Internal Friction and Ultrasonic Attenuation in Crystalline Solids*, Vols. I and II. Springer, Berlin and New York, 1975.

W. P. Mason, ed., *Physical Acoustics*, Vols. 1–11. Academic, Press, New York, 1964–1975.

N. G. McCrum, B. E. Read, and G. Williams, *Anelastic and Dielectric Effects in Polymeric Solids*. Wiley, New York, 1967.

A. S. Nowick and B. S. Berry, *Anelastic Relaxation in Crystalline Solids*. Academic Press, New York, 1972.

C. Zener, *Elasticity and Anelasticity of Metals*. Univ. of Chicago Press, Chicago, 1948.

Auger Effect

Bernd Crasemann

An atom that contains an inner-shell vacancy becomes deexcited through a cascade of transitions that are due to two kinds of competing processes: x-ray emission and radiationless, or Auger, transitions. In either process, the original vacancy is filled by an electron from a higher-energy level. In radiative transitions, the released energy is carried off by a photon; in radiationless transitions, this energy is instead transferred through the Coulomb interaction to another atomic electron, which is ejected. The emitted electron is called a *K-LL* Auger electron, for example, if a *K*-shell vacancy is filled by an *L*-shell electron, and another *L* electron is ejected.

Direct experimental evidence for the existence of radiationless transitions was gained by Pierre Auger through cloud-chamber experiments reported in 1923. X rays traversing the chamber produced photoelectrons; the tracks of these photoelectrons increased if more energetic x rays were used. In addition to the photoelectron tracks, Auger observed numerous short tracks, each of which originated at the same point as a photoelectron track (Fig. 1). The length of the short tracks did not change with x-ray energy, but depended only on the kind of gas that was placed in the cloud chamber. Auger was able to show that the length of many of the short tracks corresponded to the energy that would be released in K-*LL* radiationless transitions. In some cases,

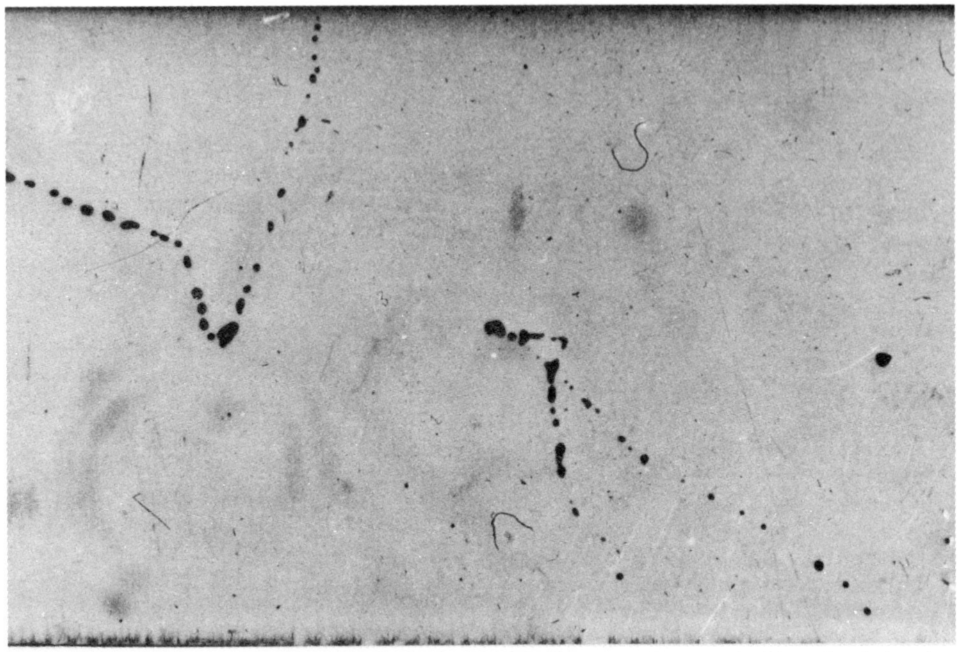

FIG. 1. Photoelectrons and Auger electrons from krypton ionized with 60-keV x rays, photographed in a cloud chamber by Pierre Auger *ca.* 1923. (Courtesy Prof. P. Auger.)

Auger found additional tracks, caused by electrons emitted in a second (*L-MM* or *L-MN*) step of the radiationless deexcitation cascade of an atom.

The quantum-mechanical theory of radiationless transitions was formulated by G. Wentzel in 1927. From perturbation theory, the nonrelativistic matrix element for a direct Auger transition is

$$D=\iint \psi^{*}_{n''l''j''}(1)\psi^{*}_{\infty l_A j_A}(2)\left|\frac{e^2}{|\mathbf{r}_{12}|}\right|\psi_{nlj}(1)\psi_{n'l'j'}(2)\,d\tau_1 d\tau_2,$$

where the quantum numbers n, l, j characterize electrons that are identified schematically in Fig. 2. The state of the continuum (Auger) electron is labeled by $\infty l_A j_A$. In the physically indistinguishable exchange process, described by a matrix element E, the roles of electrons nlj and $n'l'j'$ are interchanged (Fig. 2). The total radiationless transition probability per unit time is

$$w_{fi}=\hbar^{-2}|D-E|^2,$$

if the continuum-electron wave function is normalized so as to correspond to one electron emitted per unit time. The matrix elements D and E can be separated into radial and angular factors. Evaluation of the angular factors depends on a choice of the appropriate angular-momentum coupling scheme. If spin-orbit interaction is neglected, the initial and final two-electron (or two-hole) states can be expressed for different values of the total angular momentum J in the (*LSJM*) representation of Russell–Saunders coupling. For the heavier atoms, inner-shell states are expressed more realistically in *j-j* coupling.

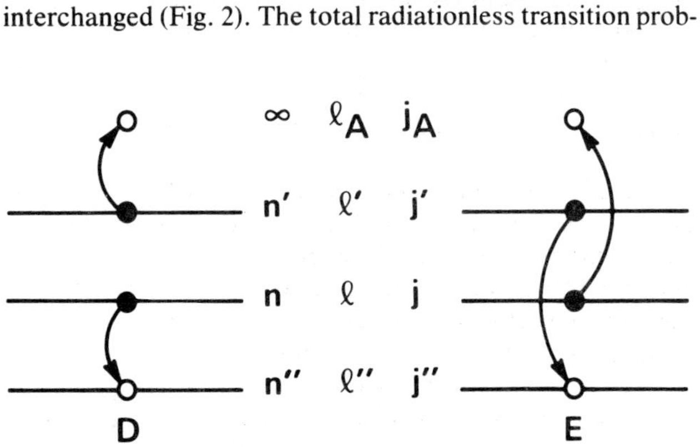

FIG. 2. Energy levels involved in the direct (D) and exchange (E) Auger processes, and notation for the principal, orbital-angular-momentum, and total-angular-momentum quantum numbers that characterize the pertinent electron states.

The selection rules governing radiationless transitions require that the total angular momentum and the parity of the final-state system (ion plus emitted electron) must be the same as of the initial-state ion. A large number of different transitions is generally possible from any given initial state; for example, 2784 matrix elements are required to describe the radiationless decay of an L_3 vacancy in a high-Z atom.

The relative probability that a K-shell hole is filled by an Auger process, rather than by x-ray emission, ranges from 0.999 for the lightest elements ($Z\leq 5$) to 0.02 for uranium. For vacancies in other shells, the Auger transition rate is generally several orders of magnitude greater than the radiative rate, and thus, essentially determines the lifetime of the vacancy. Auger rates can be very fast. Particularly intense are *Coster–Kronig transitions* that shift a hole to a higher subshell within one major shell. Thus, N_1-$N_{4,5}N_{4,5}$ "super-Coster–Kronig" transitions produce a $4s$ level width $\Gamma>60$ eV at $Z\approx 50$, which corresponds to an N_1 vacancy mean life of $<10^{-17}$ s. The limits of validity of per-

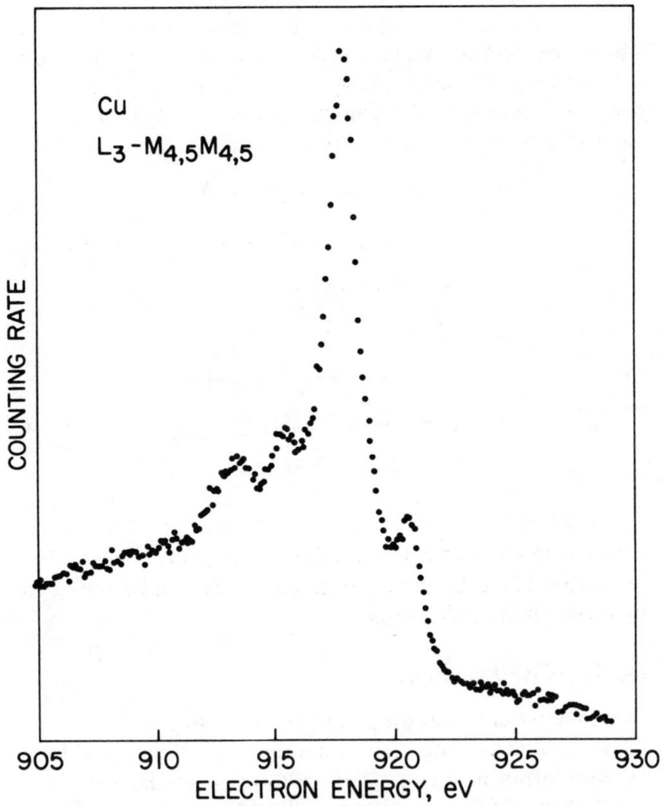

FIG. 3. Energy spectrum of L_3-$M_{4,5}M_{4,5}$ Auger electrons from metallic Cu. The separate peaks result from multiplet splitting due to various couplings of the two final-state holes. (Courtesy Dr. Lo I Yin, NASA Goddard Space Flight Center.)

turbation theory are strained when transition rates of such magnitude are calculated.

The Auger-electron energy is

$$E_{\infty l_A} = E_{n''l''} - E_{nl} - E_{n'l'}.$$

The subscripts pertain to the states indicated in Fig. 2; $E_{n''l''}$ and E_{nl} are (absolute values of) neutral-atom binding energies, while $E_{n'l'}$ is the binding energy of $n'l'$ electron *in the presence of an nl vacancy*. The energy of electrons from solid samples can additionally include a contribution from extraatomic relaxation, typically of the order of 10 eV.

Measurements of electron spectra from radiationless transitions (Fig. 3) and the theory of Auger processes constitute an active field of research, with relevance to fundamental atomic and solid-state theory as well as to surface physics, chemistry, and materials science.

See also ELECTRON ENERGY STATES IN SOLIDS AND LIQUIDS.

BIBLIOGRAPHY

P. Auger, "The Auger Effect," *Surf. Sci.* **48,** 1 (1975). (E) A first-hand account of the discovery of radiationless transitions.

W. Bambynek, B. Crasemann, R. W. Fink, H.-U. Freund, H. Mark, C. D. Swift, R. E. Price, and P. Venugopala Rao, "X-Ray Fluorescence Yields, Auger, and Coster–Kronig Transition Probabilities," *Rev. Mod. Phys.* **44,** 716 (1972). (I) A comprehensive review of the subject.

E. H. S. Burhop and W. N. Asaad, "The Auger Effect," *Adv. At. Mol. Phys.* **8,** 164 (1972). (I) A review of theory and experiment.

B. Crasemann (ed.), *Atomic Inner-Shell Processes*. Academic Press, New York, 1975. (A) This treatise deals with radiative and radiationless transitions, higher-order processes, electron spectrometry, and related topics.

K. S. Sevier, *Low-Energy Electron Spectrometry*. Wiley, New York, 1972. (A) A comprehensive monograph.

Catalysis

Richard P. Messmer

The classical definition of catalysis is simple: It is that phenomenon in chemical kinetics whereby a modification in the rate of a chemical reaction is accomplished by the addition of a chemical substance which is itself not consumed during the reaction. The word *catalysis* is a combination of two Greek words meaning "break down" and was coined in 1835 by the Swedish chemist Berzelius (who was a prolific coiner of technical terms, and has many others, including isomer, monomer, polymer, and protein to his credit).

The simplicity of the above definition is rather deceptive, however, in that it fails to connote the diversity of examples, the complexity of the subject, or the importance, both scientifically and practically, of this basic chemical phenomenon. Although in principle catalysis can refer to either an increase or diminution in the rate of a chemical reaction brought about by a chemical agent, one in practice refers to the agent which accelerates a reaction as a *catalyst* and to an agent which diminishes a reaction rate as an *inhibitor*. Actual examples of true inhibitors or negative catalysts are rather rare. Most substances which appear at first sight to be true inhibitors are found to be consumed in reaction when studied more carefully. Nonetheless, these substances are often referred to conveniently, if not completely accurately, as inhibitors.

Two important factors in the determination of a good catalyst are its *activity* and *specificity*. In general, from a given set of chemical species, the reactants, a number of reactions may occur yielding different product chemical species. Some of these resultant chemical species will be the desired products, others will not. The activity of a catalyst is defined in terms of the specific reaction which yields the desired products. Namely, it is the rate at which this single reaction, which yields the desired products, proceeds. Clearly, even though the activity of a catalyst may be very high the catalyst may not be of much practical use if the desired product is only a small fraction of the total products. Hence, it is of the utmost importance to achieve selectivity in a catalyst. Selectivity is defined as a ratio of reaction rates, namely, the rate of the desired reaction over the sum of rates of all the undesirable reactions.

Catalysis may be arbitrarily classified as biological, in which case the catalysts are called enzymes, and nonbiological. In nonbiological catalysis it is useful to make the distinction between *homogeneous* and *heterogeneous* catalysis. In the former case, the reactants and catalyst are a mixture in a single physical phase, i.e., a gas or a one-phase liquid. For the latter case, the catalyst is in one phase, most often a solid, and the reactants are in the second phase (e.g., a liquid) in contact with the catalyst.

HOMOGENEOUS CATALYSIS

Gas-Phase Reactions

Although many interesting reactions could be discussed, some particularly important examples of gas-phase catalysis which occur in the chemistry of the stratosphere will be used as examples. Two catalytic cycles have been identified which can destroy ozone in the stratosphere. In one case the molecule NO, nitric oxide, acts as a catalyst; in the second, the catalyst is the chlorine atom, Cl. The proposed mechanisms of these reactions are

$$\begin{aligned} &NO + O_3\,(\text{ozone}) \rightarrow NO_2 + O_2,\\ &O_3 + h\nu \rightarrow O_2 + O, \qquad \text{Case I}\\ &NO_2 + O \rightarrow NO + O_2;\end{aligned}$$

$$\begin{aligned} &Cl + O_3 \rightarrow ClO + O_2,\\ &O_3 + h\nu \rightarrow O_2 + O, \qquad \text{Case II}\\ &ClO + O \rightarrow Cl + O_2.\end{aligned}$$

The overall reaction in each case is $2O_3 + h\nu \rightarrow 3O_2$, from which it is clear that the catalyst is not consumed in the reactions. The catalytic agent in each case can be provided by atmospheric pollutants.

Liquid-Phase Reactions

Acid–base catalysis provides for a very large class of synthetically and commercially important reactions. A simple example of an acid-catalyzed reaction is the esterification reaction in which a carboxylic acid (A) combines with an alcohol (B) in the presence of an acid catalyst to form an ester (C) and H_2O. The generally accepted mechanism is:

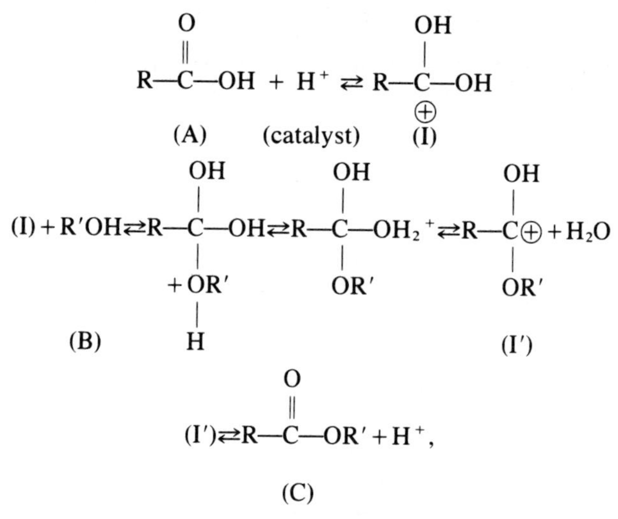

where I and I′ are intermediates known as carbonium ions.

A wide variety of homogeneous catalytic reactions are based on transition-metal species as the catalyst. The commercially successful Wacker process for converting ethylene (C_2H_4) to acetaldehyde (CH_3CHO) has stimulated a great deal of interest in homogeneous metal catalysis. Under certain reaction conditions in the Wacker process the species $PdCl_4^{-2}$, a transition-metal complex in solution, is considered to be the catalytic agent. Another reaction of considerable industrial importance is the Oxo reaction in which CO and H_2 are added to olefins (e.g., ethylene) to form aldehydes. In this case the catalyst is thought to be $HCo(CO)_4$, derived from a cobalt carbonyl molecule.

HETEROGENEOUS CATALYSIS

Although by definition heterogeneous catalysis may take place at the interface between any two phases (e.g., liquid–gas, liquid–liquid), the dominant interest and at present the most important applications arise from solid–gas and solid–liquid catalytic systems in which the solid acts as the catalyst. It is in this area of heterogeneous catalysis, specifically solid–gas systems with a solid as the catalyst, that a great deal of interest among solid-state physicists has recently been kindled.

There are two main features, however, which differentiate the solids of catalytic interest from those of more traditional solid-state interest. First, the solids of catalytic interest are often multiphase, multicomponent, amorphous materials of rather ill-defined structure in contrast to well-defined, single-component crystalline solids. Secondly, the important aspect of the solid in catalysis is its surface properties (i.e., surface structure, defect and surface chemistry), instead of the bulk properties, which have been the chief interest in most of previous solid-state work. Some examples of solid catalysts are silica (SiO_2), alumina (Al_2O_3), silica-aluminas, zeolites (crystalline aluminosilicates), and carbides, oxides, and sulfides of transition metals in polycrystalline or amorphous form (e.g., Cr_2O_3, NiO, V_2O_5). Furthermore, many important solid catalysts are formed by the deposition of a transition metal or a combination of metals on a *support* or *carrier*. Common support materials are the silica-aluminas or alumina. The supported material is finely dispersed on the support. *Dispersion* is defined as the fraction of atoms of the supported material which is exposed at the surface. In some supported catalysts, the dispersion often approaches unity.

An important concept in discussing heterogeneous catalysis is that of the *active site*. In general, not every surface site on a solid catalyst will be equivalent. There will be a distribution of surface sites, only some of which are capable of chemically adsorbing (chemisorbing) a reactant molecule. The chemisorbed species may then undergo a reaction with another reactant molecule at or near the active site, and then the resultant products will desorb from the surface into the gas or liquid phase. Some active sites on compound solids, such as oxides or sulfides, can arise from impurity and defect sites on the surface. In the case of supported metal catalysts some reactions depend markedly on the metal particle size; other reactions seem to be almost independent of this factor. It has become common to refer to the former reactions, which presumably depend strongly on certain active sites, as *demanding* reactions, whereas the latter, less structure-sensitive reactions are called *facile*.

A practical advantage in using heterogeneous catalysis instead of homogeneous catalysis to carry out a given reaction is the ease of separation of products from the catalyst system.

Catalysis by Nonmetals

Although there are innumerable examples of this type of catalysis, only two—chosen for their historical importance—will be mentioned. The first example is that of the catalytic cracking of heavy petroleum stocks by silica-alumina catalysts. In silica-rich compositions the catalyst material is an amorphous solid solution. It is an acid catalyst and is thought to function by creating carbonium ions from both saturated and unsaturated hydrocarbons which in turn undergo bond-breaking and/or rearrangement to form more valuable products. This catalyst was used for many years by the petroleum industry.

The second example is Ziegler-Natta catalysis. In this case the catalyst consists of a titanium trichloride–aluminum alkyl mixture, which catalyzes the polymerization of propylene into a stereoregular polymer. Stereoregular is a term which means that the units in the polymer occur in a specific orientation and not at random. It is this amazing specificity of the reaction which makes it so important. The active sites of the catalyst are believed to be the Ti^{3+} ions in α-$TiCl_3$ crystallites.

Catalysis by Supported Metals

There are two general classes of catalysts which can be considered. In the first, the support is considered to be inert and the catalysis is carried out strictly by the metal on the support. In the second, the support as well as the metal takes part in the catalytic reaction; this is known as *bifunctional* catalysis.

Hydrogenation and hydrogenolysis are important reactions which are carried out by a variety of transition metals, with varying efficiency. Hydrogenation, for example, transforms unsaturated hydrocarbons such as ethylene (C_2H_4) to saturated species such as ethane (C_2H_6). Silica-supported Ni, Pd, Pt, and Rh are active hydrogenation catalysts. Hydrogenolysis involves the rupture of carbon–carbon bonds and the formation of carbon–hydrogen bonds. As an example, the hydrogenolysis of ethane results in the formation of methane (CH_4), for which silica-supported Ni, Ru, and Os are particularly active catalysts.

A good example of bifunctional catalysis is the isomerization of *n*-pentane (D) to isopentane (E),

$$H_3C\text{—}CH_2\text{—}CH_2\text{—}CH_2\text{—}CH_3$$

(D)

$$\xrightarrow{\text{catalyst}} H_3C\text{—}\underset{}{\overset{\displaystyle CH_3\\ |}{CH}}\text{—}CH_2\text{—}CH_3,$$

(E)

by platinum supported on alumina. The mechanism of the reaction is believed to involve the following: *n*-pentane (D) is dehydrogenated to 1-pentene (F) on the platinum, the 1-pentene is then isomerized to isopentene structures (such as G) at the acidic sites of the alumina, and these latter species (such as G) are then hydrogenated at the platinum site to obtain the isopentane product (E). Bifunctional catalysis is very important in the petroleum industry.

$$H_2C{=}CH_2{-}CH_2{-}CH_2{-}CH_3 \qquad H_3C{-}\overset{\overset{\large CH_3}{|}}{C}{=}CH{-}CH_3$$

(F) (G)

Although it has been estimated that catalysis in one form or another is responsible for as much as \$100 billion of the annual United States gross national product, it is nonetheless one of the most poorly understood phenomena of nature on a scientific basis. In order to understand heterogeneous catalytic reactions properly, it is necessary to have detailed information about reaction mechanisms. This kind of information, which would make heterogeneous catalysis more of a science and less of an art, is extremely difficult to obtain for a number of reasons. As examples, the nature, structure, and variety of active sites on a catalyst are poorly understood as are the electronic and structural aspects of *promotors* (chemical agents which stabilize or enhance the activity of the catalyst) and *poisons* (agents which reduce or destroy catalytic activity) at active sites. Thus, the lack of structural information (i.e., the detailed nature and disposition of atoms and molecules) for the catalyst and the catalyst–reactants system has plagued the study of catalysis. It would appear that true progress in the basic understanding of heterogeneous catalysis will come about when new or improved physical and chemical techniques are able to yield the requisite structural information. Alternatively, and perhaps more importantly, progress will be made by making new catalysts with well-defined and -characterized active sites. This latter philosophy is behind a great deal of recent effort in developing heterogeneous catalysts by attaching transition-metal homogeneous catalysts to polymers.

ENZYME CATALYSIS

The word enzyme is derived from two Greek words meaning "in yeasts," as the important work of Buchner in 1897 involved the conversion of D-glucose into ethanol and carbon dioxide by using a cell-free extract of yeast. In living systems there are a myriad of enzymatic reactions with each enzyme specific to a particular chemical reaction. The two characteristic features of enzyme catalysis are the activity and amazing specificity of the enzyme in comparison to laboratory catalysts. The latter feature in particular is critical to a stable biological system.

There are many analogies to laboratory catalysis; for example, the active-site concept is very important and there are the anologs of promotors (coenzymes) and poisons. Enzyme catalysis represents an immense area of scientific activity. Although many enzymes are extremely complex and rather poorly understood, there are certain favorable cases where the detailed structure of the enzyme and substrate are known, as in the case of trypsin, for example. Under the latter circumstances such systems represent perhaps some of the best understood examples of catalysis.

See also ADSORPTION; SURFACES AND INTERFACES.

BIBLIOGRAPHY

M. Boudart, in *Interactions on Metal Surfaces*, p. 275, R. Gomer (ed.). Springer-Verlag, New York, 1975. Introduction to the basic concepts of heterogeneous catalysis. (E)

The Physical Basis for Heterogeneous Catalysis. E. Drauglis and R. I. Jaffee (eds.). Plenum Press, New York, 1975. Contains a number of articles which discuss physical techniques such as low-energy-electron diffraction (LEED) and photoemission for probing clean and adsorbate-covered single-crystal faces of metals. Attempts are made to relate such studies to the understanding of certain aspects of heterogeneous catalysis. (A)

J. C. Bailar, *Catal. Rev.* **10**(1), 17 (1974). Reviews some of the recent effort to attach catalytically active transition-metal complexes to polymers, in order to convert homogeneous catalysts to heterogeneous ones. (E)

R. Ugo, *Catal. Rev.* **11**(2), 225 (1975). Discusses some of the contributions of organometallic chemistry to the understanding of surface reactions. (A)

J. H. Sinfelt, *Adv. Chem. Eng.* **5**, 37 (1964). Review of some important examples of bifunctional catalysis. (I)

R. L. Burwell, Jr. and M. Boudart, in *Investigation of Rates and Mechanisms of Reactions*, Chap. 12, E. S. Lewis, (ed.). Wiley, New York, 1974. Discusses mechanistic studies of heterogeneous catalytic reactions. (I)

R. M. Stroud, *Scientific American* **231**(1), 74 (1974). An excellent discussion of several proteolytic enzymes. (E)

D. W. McKee, *Carbon* **8**, 623 (1970). An interesting example of heterogeneous catalysis, in which the catalysts (metal oxides) and one of the reactants (graphite) are solids and the product is a gas (CO_2). (I)

P. B. Weisz, *Adv. Catal.* **13**, 137 (1962). A review of polyfunctional heterogeneous catalysis; good discussion of mass transport considerations. (I)

Ceramics

Lionel M. Levinson

The conventional understanding of ceramics encompasses the production and use of clays and silicate materials for pottery, bricks, and refractories. Modern ceramics also include a wide variety of inorganic nonmetallic materials with precisely controlled properties which play critical roles in the function of numerous technological devices. The need for materials with closely specified characteristics has transformed the ancient ceramic art into a modern-day branch of materials science and engineering.[1] We will briefly list some of the modern technological ceramics and then examine more closely two particular ceramic systems of interest.

In Table I we present a selection of technical ceramic materials along with some special features and uses of the ceramic. It should be emphasized that the table makes no pretence at completeness and that the use of a particular material in a given application will usually depend on a spectrum of material properties in addition to the particular feature listed. Of the various polycrystalline ceramics in Table I it is clear that the utility of some lies in the ease of fabrication and low cost inherent in ceramic processing. For example, single-crystal alumina could provide superior microcircuit substrates and the use of single-crystal lead zirconate titanate would result in the fabrication of better ultrasonic transducers. There are, however, some ceramic devices in which the ceramic has properties essentially not feasible in the single-crystal analog of the ceramic. In this case the ceramic microstructure produces the new effect. We shall ex-

Table I A List of Some Modern Technical Ceramic Materials

Material	Outstanding Property	Use
Alumina, Al_2O_3	High-temperature strength Low electrical loss	Furnace ware, thermocouples Microcircuit substrates
Beryllia, BeO	High thermal conductivity	Electronic component heat sink
Magnesia, MgO	High-temperature resistivity	Heater element insulation
Ferrite, MFe_2O_4 (M = Mn, Fe, Ni, Zn, Cu, Mg, etc.)	Ferrimagnetic insulator	Computer memories, high-frequency transformers
Garnet, $R_3Fe_5O_{12}$ (R = rare earth)	Low microwave loss, ferrimagnetic	Microwave devices
Zinc oxide, ZnO	Nonlinear conductor	Overvoltage transient suppression
Insulating barium titanate, $BaTiO_3$	High dielectric constant	Capacitors
Conducting barium titanate, $BaTiO_3$	Positive temperature coefficient of resistivity	Relay, sensor
Lead zirconate titanate (PZT), $Pb(Zr, Ti)O_3$	Piezoelectric	Ultrasonic transducers

amine two such examples: (1) the high-frequency ferrite and (2) the ZnO varistor.

MANGANESE ZINC FERRITE

The double oxide spinel $MO{\cdot}Fe_2O_3$, where M can be (among others) Mn, Zn, or ferrous iron, is a magnetic ceramic material which combines the properties of high magnetic permeability with high electrical resistivity. The latter property results in low eddy-current loss so that the material can be used at high (tens of kilohertz) frequencies. A typical application of such ferrites is in the tuned circuit used for frequency synthesis in Touch Tone telephones.[2] The ferrite must have high quality factor Q, low sensitivity to variations in the ambient temperature, and good stability with time. Since Q is the ratio of energy stored ($\propto$ permeability μ) to energy dissipated ($\propto$ conductivity σ), the highest possible value of μ/σ is desired. To maximize μ the $Mn{:}Zn{:}Fe^{2+}{:}Fe^{3+}$ ratio is chosen so that the magnetic anisotropy and magnetostriction is very small. Ideal properties require close control of the purity of the raw materials and processing techniques such as milling and firing atmosphere. A typical processing flow chart is shown in Fig. 1.

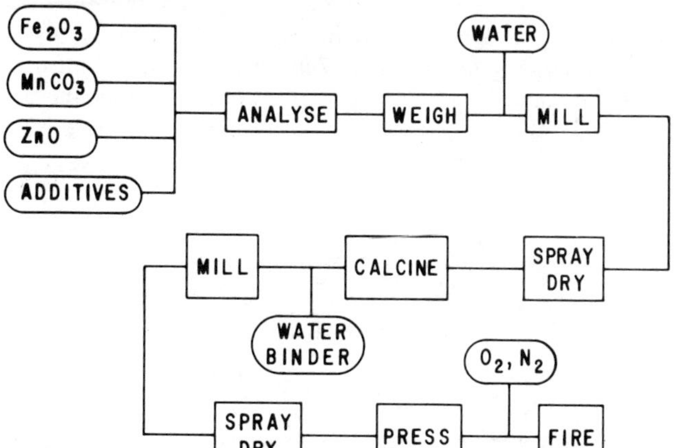

FIG. 1. Typical flow chart for processing of manganese zinc ferrite.

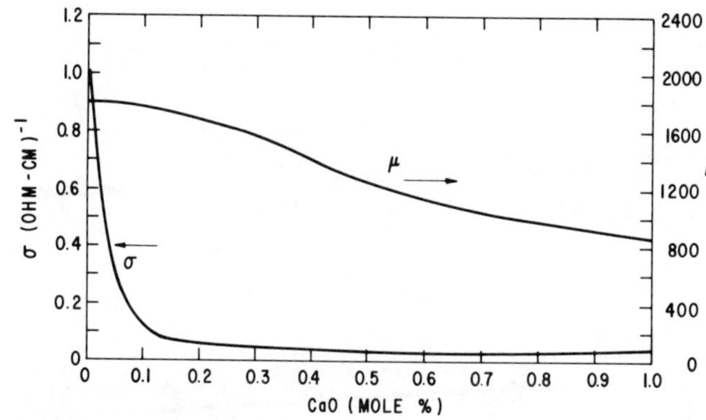

FIG. 2. Conductivity σ and permeability μ as a function of CaO addition to manganese zinc ferrite.

The necessary presence of both Fe^{2+} and Fe^{3+} in manganese zinc ferrite produces an increase in the ferrite conductivity by allowing easy electron transfer from Fe^{2+} to Fe^{3+} sites. This in turn produces an unwanted high-frequency loss by eddy-current effects. This loss, while unavoidable in single-crystal material, can be controlled in fine-grain ceramics by the addition of controlled impurities. CaO and SiO_2 have been found[3] to segregate to the grain boundaries to form a low-conductivity phase, thereby substantially decreasing eddy-current losses. Figure 2 gives the effect of small amounts of CaO on μ and σ. Since μ decreases only

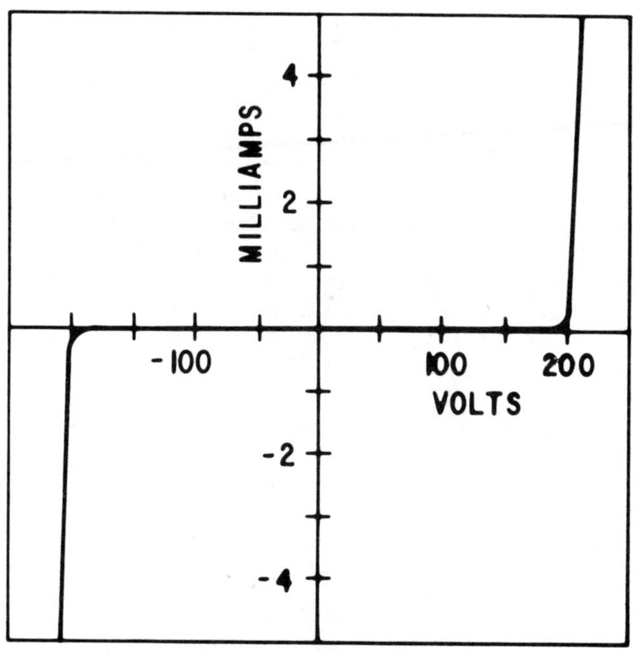

FIG. 3. Current–voltage characteristic of a ZnO varistor.

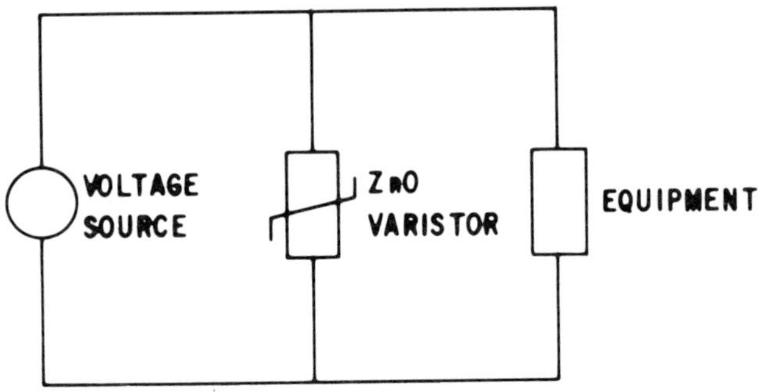

FIG. 4. Use of ZnO varistor to protect electrical equipment.

slightly with CaO content while σ drops precipitously, a few tenths mol% CaO addition substantially improves the performance of the ceramic ferrite.

ZnO VARISTORS

Zinc oxide is an Ohmic semiconductor. However, it has been found[4] that ZnO ceramics containing a few mol% of other oxide additives (for example, Bi_2O_3, Co_2O_3, MnO_2, and Sb_2O_3) exhibit a marked electrical nonlinearity in their current–voltage characteristic. Figure 3 gives a typical current–voltage characteristic of one of these devices. At voltages below a critical value the device acts as an insulator; above that voltage it acts as a conductor. In the vicinity of the turn-on voltage a 5% increase in voltage will typically cause the current to increase by an order of magnitude!

Figure 4 indicates how the ZnO varistor is used to protect electrical and electronic equipment from overvoltage transients due to switching surges or lightning strokes. The varistor is connected in parallel with the load and its turn-on voltage chosen so that it is insulating in normal use. The device conducts when a high-voltage surge in excess of the turn-on voltage appears on the line. This shunts the current and prevents the surge from reaching the protected equipment.

The origin of the varistor effect lies in the ceramic microstructure of this device.[5] The ZnO grains are conducting, but during sintering they become surrounded with a thin ($\lesssim$ 10Å) adsorbed layer of oxide material derived from the oxide additives in the ceramic. This results in the creation of electrical depletion layers at the grain boundaries which inhibit electron flow at low applied voltages. When increasing voltage is applied to the varistor, very high fields appear across the thin insulating depletion layers since the interior of the ZnO grains is conducting and will not sustain a voltage. At a critical applied voltage[6] (about 3 volts per grain boundary) it becomes possible for electrons to tunnel through the depletion layers from one conducting ZnO grain to a neighboring grain, thereby giving rise to the observed high nonlinearity in the electrical characteristics.

See also CONDUCTION; MAGNETIC MATERIALS.

REFERENCES

1. W. D. Kingery, H. K. Bowen, and D. R. Uhlmann, *Introduction to Ceramics*. Wiley, New York, 1976.
2. F. J. Schnettler, "Microstructure and Processing of Ferrites" in *Physics of Electronic Ceramics*, Part B, p. 833. L. L. Hench and D. B. Dove (eds.). Marcel Dekker, New York, 1972.
3. T. Akashi, *Trans. Jpn. Inst. Metals* **2,** 171 (1961).
4. M. Matsuoka, *Jpn. J. Appl. Phys.* **10,** 736 (1971).
5. L. M. Levinson and H. R. Philipp, *J. Appl. Phys.* **46,** 1332 (1975).
6. G. D. Mahan, L. M. Levinson and H. R. Phillip, *J. Appl. Phys.* **50,** 2799 (1979).

Channeling

J. A. Golovchenko

When beams of fast charged particles, such as those obtained from ion accelerators or radioactive decay, are directed into single crystals, it is experimentally observed that various reaction probabilities depend dramatically on the relative angular orientation of the target and incident beam. For example, consider an angularly well-collimated 1-MeV proton beam impinging on a thin silicon crystal. For angles of incidence $\lesssim$10 mrad to the ⟨110⟩ axial direction, the measured Rutherford backscattering yield from collisions with silicon atoms is reduced to ~3% of the value found at larger angles of incidence where the results are similar to those obtained for an amorphous silicon target. For other positive particles, similar reductions in yield of all processes requiring small impact parameters (e.g., nuclear reactions, *K*-shell ionization) are observed near low-index axial or planar directions.

The explanation of the foregoing and other closely related phenomena relies on the concept of channeling. By examining the possible classical trajectories of positive particles in a crystal, we find that there exists a class of motions that have a remarkable stability against approaching target atoms at small impact parameters. Particles executing these motions are called *channeled.* To illustrate the most important aspects of channeling, we need only consider the basic influence that a closely packed atomic string (row) or plane of atoms has in determining these trajectories.

Figure 1 shows the string case. For a particle approaching the string at a small angle, a series of large-impact-parameter, small-angle scattering events leads to an overall repulsion from the vicinity of the string. As the angle of incidence is increased, the distance of closest approach to the string, r_{min}, decreases along with the number of individual atomic deflections that make up the string collision. Finally, at a sufficiently large incident angle, only a few atomic collisions are involved with such large scattering angles that the essential stability of trajectories, determined collectively by

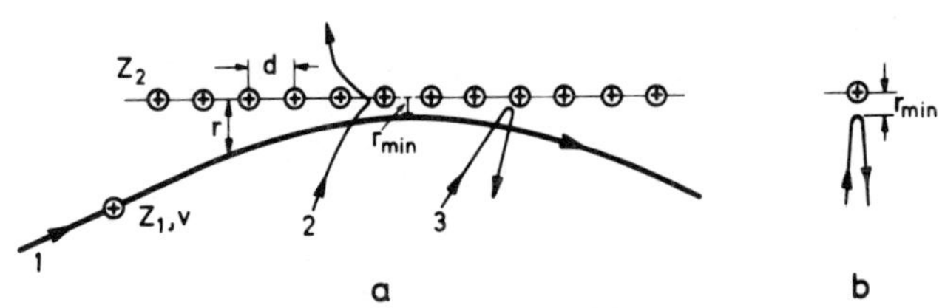

FIG. 1. Single-string collision. (a) Longitudinal view. Trajectory 1 is channeled, 2 and 3 are not. (b) The string is perpendicular to the plane of the paper here.

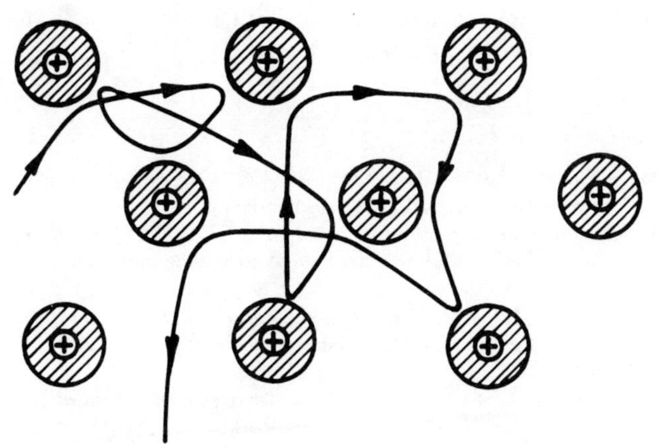

FIG. 2. Motion in a transverse plane showing many-string collisions. Strings are directed into the plane of paper, and only the motion perpendicular to the strings is shown.

many atoms on the string, is lost. Up until this limit is reached, the particle is channeled.

Since the crystal is made up of many parallel strings along the axial direction, the overall penetration of the channeled particle is determined by a succession of many string collisions. This is illustrated in Fig. 2, where a typical channeling trajectory is viewed in a plane transverse to the string direction. Note that although it is possible for a trajectory to be trapped in the "channel" formed by several adjacent strings of atoms, in general an axially channeled particle can wander quite freely from string to string in the motion transverse to the strings.

A similar description applies to the case of planar channeling. Here, for sufficiently small angles of approach, a positive particle is gently repelled from the vicinity of the plane of atoms and again small-impact-parameter events are suppressed. In this case, the extended trajectory will be confined between two parallel atomic planes and an oscillatory motion results.

The preceding physical description may be extended theoretically to provide further insight into the channeling process and qualitative estimates of various experimental quantities. The following discussion is generally confined to the string case, although analogous results may be deduced for the planar case, where channeling effects are somewhat less pronounced.

Channeling can be described within classical mechanics. This is, at first, surprising, since the quantity $2Z_1Z_2e^2/\hbar v$ is small compared to unity in many cases of interest. Z_1 and Z_2 are the atomic numbers of the projectile and target atoms, respectively; v is the particle velocity; and $\hbar$ is Planck's constant divided by 2π. In this situation, a classical description of an *isolated* atomic collision in terms of impact-dependent scattering angles is not possible, and a quantum-mechanical approach is required. However, to determine the applicability of a classical description for channeled motion, an analysis of an entire string collision is required. Such a procedure shows that the spread of suitably constructed wave packets obeying classical mechanics is small enough to render all the classical variables meaningful for heavy channeled particles (e.g., protons), even for $v \rightarrow c$. Indeed, channeling is basically a classical concept.

The application of classical mechanics to the crystal-penetration problem shows that channeling may be associated with an approximately conserved quantity, the transverse energy $E_\perp$. For an axially channeled particle of mass M and energy $E(=\frac{1}{2}Mv^2)$ with an instantaneous angle and distance from the ith string ψ_r and $\mathbf{r}-\mathbf{r}_i$, $E_\perp$ is given by

$$E_\perp = \frac{1}{2} m v^2 \psi_r^2 + \sum_i U(|\mathbf{r}-\mathbf{r}_i|). \qquad (1)$$

The continuum potential $U(r)$ is given by

$$U(r) = \frac{1}{d}\int_{-\infty}^{\infty} V[(r^2+Z^2)^{1/2}]\, dZ. \qquad (2)$$

Here d is the distance between atoms on the string, which is taken to be in the Z direction, and $V(R)$ is the screened Coulomb-potential energy of interaction between the projectile and target atom at separation R. The first term in Eq. (1) is the kinetic energy of the particle motion projected onto a plane perpendicular to the strings, while the second term is an average potential energy obtained by smearing the atomic potential V along the string direction. The stability of channeled motion referred to earlier corresponds to conservation of $E_\perp$ with the velocity parallel to the strings being nearly unchanged during a string collision.

Using simple analytical estimates for V and examining limiting particle trajectories, we find that conservation of $E_\perp$[i.e., govĕrned motion in the transverse Hamilton, Eq. (1)], breaks down for $E_\perp \gtrsim 2Z_1Z_2e^2/d$. Putting this restriction in Eq. (1) leads to a characteristic limiting angle of approach for axial channeling of

$$\psi_r\,(\max) = \psi_1 \equiv \left(\frac{2Z_1Z_2e^2}{Ed}\right)^{1/2}. \qquad (3)$$

This is a central result of the theory. It applies at energies E so high that $\psi_1 < a/d$. Here, a is the Thomas–Fermi screening length $[a = 0.885a_0/(Z_1^{2/3}+Z_2^{2/3})^{1/2}$ where a_0 is the Bohr radius], which measures the range of $V(R)$ for fast collisions. At lower energies, ψ_1 overestimates $\psi_r(\max)$, which now depends in more detail on the form of $V(R)$.

Upon entering the crystal, each particle in a well-collimated beam of energy E will acquire a value of $E_\perp$ [see Eq. (1)] given by $E_\perp = E\psi^2 + U(r_0)$, where ψ is the angle between the beam and the axis directions and r_0 is the point of entry in the transverse plane. The resulting distribution in $E_\perp$ is strongly peaked about $E\psi^2$ since $U(r)$ is quite small over most of the transverse space. Thus as ψ approaches ψ_1, the distribution in $E_\perp$ is weighted toward the highest values of $E_\perp$, and the yield of small-impact-parameter interactions increases. For $\psi = 0$, only particles entering the crystal within an area $\sim \pi\rho^2$, with ρ^2 the host-atom mean-square thermal-vibration amplitude perpendicular to the strings, will be able to interact strongly with target atoms. The channeling dip measured relative to random penetration ($\psi \gg \psi_1$) will be of order $\chi_{\min} = \pi\rho^2$/(transverse unit cell area). Most of the particles, i.e., the fraction $1-\chi_{\min}$ ($\approx 0.95-0.99$ for strong axes) will be well channeled away from the vibrating string atoms.

In transmission experiments through thin crystals, this component of the beam can exit from the back of the crystal within ψ_1 of the axis, the random multiple scattering of such particles being greatly reduced. It is also found that the energy loss due to electronic excitation is reduced by up to about a factor of two for channeled particles. This reduction arises mainly from the low density of target electrons along the path of low-$E_\perp$ trajectories. Coupled with the large reduction in energy lost to recoiling atoms, this also explains the dramatic increase found in the range of channeled particles in thick targets.

To account for effects observed in thick targets, the foregoing theory must be extended to include dechanneling (i.e., those processes that cause $E_\perp$ to change during penetration). For example, multiple scattering on target electrons and variations in the continuum potential resulting from thermal vibrations cause $E_\perp$ to fluctuate, and the statistical accumulation of these fluctuations generally leads to an increase in the nonchanneled component of the beam with increasing depth. For the example discussed earlier, 1-MeV protons→Si ⟨110⟩, the depth of penetration at which half the protons have dechanneled is ~15μm at room temperature.

A few applications of the channeling effect deserve special attention. The first is the lattice location of impurity atoms in cyrstals. If the impurity atoms are in an atomic position of high symmetry, it will normally be found that a characteristic fraction of these atoms lies in a given set of strings or planes and therefore will not interact with channeled particles. By studying this fraction along different crystallographic directions, the impurity site can often be deduced simply. In these studies, the energy transfer in a large Rutherford scattering or reaction products from a nuclear reaction are commonly used as signals from the channeled-particle–impurity interaction. This technique has been extensively used for locating impurities and evaluating lattice disorder introduced by ion implantation.

It is also possible to measure directly the lifetime of short-lived (10^{-16}–10^{-18} s) compound nuclei by using channeling. Here a nucleus sitting in a lattice site is excited by the absorption of a particle from an external (unchanneled) beam. The particle momentum absorbed by the nucleus causes it to recoil from the lattice site. If the nucleus decays by emission of a positively charged particle, a study of the angular distribution of these decay products ejected from the target, near axial or planar directions, indicates the extent to which the nucleus has recoiled into the channel. Knowing the recoil velocity (from kinematics), we can then deduce the nuclear lifetime. This is called the "blocking lifetime" technique because of the blocking action the strings have on preventing charged-particle emission from the target along the axial direction in the zero-lifetime case.

Finally, to indicate the scope of the channeling concept, we note that channeling effects have been observed for particles with energies ranging from the kilo-electron-volt to the giga-electron-volt region. Negative particles are also subject to the conditions of governed motion, but because of the attractive nature of the particle–string interaction, the persistence of these motions can be greatly impaired. Furthermore, for the case of light particles (e.g., electrons and positrons) quantum-mechanical considerations may need to be included (particularly for $E \lesssim 1$ MeV) for an overall description of the crystal penetration. To the extent that these considerations are required, this subject lies outside the scope of the channeling concept.

BIBLIOGRAPHY

D. S. Gemmell, *Rev. Mod. Phys.* **46,** 129 (1973); a review article covering virtually all areas of channeling. (A)

W. Gibson, *Ann. Rev. Nucl. Sci.* **25,** 465 (1975). The application of the "blocking lifetime" technique to compound nuclei is reviewed. (A)

J. Lindhard, *Mat.-Fys. Medd. Danske Vid. Selsk.* **34,** No. 14 (1965). This is the most in-depth theoretical treatment of channeling. (A)

J. W. Mayer, L. Eriksson, and J. A. Davies, *Ion Implantation in Semiconductors*. Academic, New York, 1970. The role of channeling in ion implantation, the measurement of lattice damage, and lattice-location studies are discussed with the stress on applications in semiconductor technology. (A)

Circuits, Integrated

Robert A. Henle

The invention of the transistor in 1948 opened the way to the rapid development of a series of increasingly sophisticated and useful device structures. In less than 30 years, transistors have become the basis of a major industry, which today produces complex semiconductor integrated circuits consisting of up to tens of thousands of identifiable devices.

Silicon integrated circuits use active devices of two basic types: the field-effect transistor, FET,[1] and the bipolar transistor.

The FET operates on a simple and easily understood principle: applying a voltage to a capacitively coupled electrode creates an electric field, which alters the number of charge carriers in a semiconductor and thus modulates its conductivity. The electrode has been realized as a *p-n* junction[2] (junction FET, J-FET), as a Schottky barrier (metal–silicon FET, MESFET), or as a plate separated from the semiconductor by an insulating layer (insulated gate FET, IGFET). This last structure is usually a metal plate insulated from the surface of the semiconductor by silicon dioxide (metal-oxide-semiconductor FET, MOSFET). In recent devices the metal has generally been replaced with polycrystalline silicon; the term MOSFET is generally applied loosely to these silicon gate FET's as well. FET's can be *n* channel or *p* channel. The *n*-channel FET consists of a source and drain, which are *n*-doped regions on a *p*-type substrate. The source and drain are separated by a channel, usually *p* type, which when reverse-biased prevents conduction from source to drain. Applying a positive bias to the gate attracts electrons into the channel. These electrons make the channel appear temporarily *n* doped and allow conduction from source to drain. This device is an *n*-channel enhancement-mode MOSFET.

Devices are also fabricated in *n*-type substrates with a *p*-type source and drain. These require a negative gate bias to attract holes into the channel to initiate conduction, and are known as *p*-channel enhancement-mode MOSFET's. Because electrons have higher mobility than holes, *n*-channel devices perform better than *p*-channel devices.

The bipolar transistor, unlike the FET, does not rely on capacitive coupling. For an *n-p-n* transistor, it consists of a thin *p* region between two *n* regions. These regions correspond to the emitter, base, and collector of the transistor. In operation, the emitter–base junction is forward-biased such that electrons flow from the emitter region into the base. The base collector junction is usually reverse-biased; that is, the *n* region is positive with respect to the base *p* region. The collector junction therefore collects the electrons that do not recombine in the base region. Current and voltage gain between base input and collector output can be obtained by using the base electrode as the input.

A *p-n-p* transistor operates similarly except that the emitter injects holes into the base region. Again, because electrons are more mobile than holes, *n-p-n* transistors with the same dimensions perform at higher frequencies than *p-n-p* transistors.

Until about 1959, all semiconductor circuits were fabricated with discrete devices such as those just discussed. These were used, much like vacuum tubes, in circuits that also contained other, passive components such as diodes (simple *p-n* junctions), resistors, and capacitors. The introduction of transistors made it possible to operate the circuits at much lower voltage and power levels, and offered significant improvements in reliability. Integrated electronics as we know it today had its start in 1958–59,[3] when government-sponsored and industrial research sparked the development of semiconductor integrated circuits. Whereas germanium had dominated the technology in the 1950s, silicon now won out, because it grows a natural oxide that serves as an insulator, as a passivation layer, and, most important, as a diffusion mask.

Integrated circuits are fabricated[4] by a series of processes that form interconnected active and passive circuit components on a single piece of monocrystalline material—usually a thin silicon wafer, perhaps 80 mm in diameter and $\frac{1}{3}$ mm thick. Tens of thousands of devices can be fabricated and interconnected simultaneously. It is this batch fabrication capability that makes integrated circuits so inexpensive. The following briefly describes the principal steps in fabricating the typical integrated device shown in Fig. 1.

The first step is the production of single-crystal material, most commonly by the Czochralski method.[5] Wafers are cut from single crystals, and lapped and polished to a mirror finish. An epitaxial film[6]—that is, a film whose lattice structure is an extension of the substrate crystal—is grown on the surface, to a thickness of several micrometers. This film is generally made to have conductivity opposite to that of the substrate, so that a *p-n* junction occurs at the interface; this is desirable because it provides an isolating junction. On the epitaxial film is grown an oxide layer (SiO_2) several thousand angstroms thick.[7] The oxide layer in turn is coated with a film of photosensitive resist; the resist is polymerized by exposure to ultraviolet light through a mask having the desired pattern; and the wafer is rinsed in a developer, which removes the unexposed film. After the remaining resist is baked, the wafer is etched; where the resist has been removed, the etchant leaves openings in the oxide layer. The

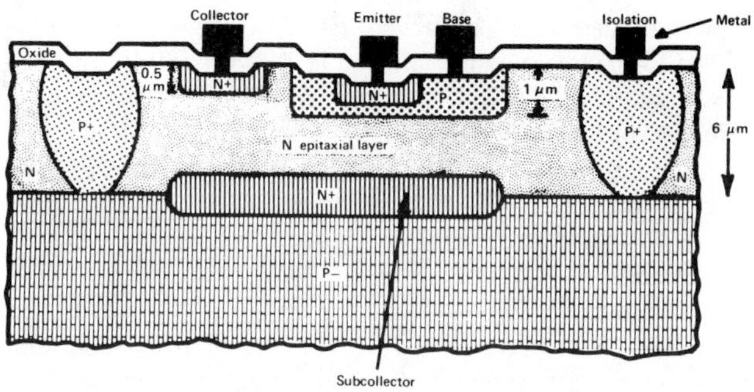

FIG. 1. A typical integrated device.

wafer is now ready for ion implantation or diffusion. With the oxide layer acting as a diffusion mask, materials such as phosphorus or boron are ion implanted[8] or diffused through the openings in the oxide layer. This process—oxidation, the photolithographic process, and diffusion—is repeated to create multiple-junction structures, which can be formed into transistors, resistors, diodes, interconnections, and isolation regions.

The electronic components thus made must be provided with electrical interconnections and contact metallurgy to connect them to other components. Diffused channels in the silicon wafer are used for some interconnections, but metal-to-semiconductor contacts are ultimately required. They are usually made by evaporating or sputtering a thin metal film (less than 1 µm) on the surface of the wafer; silicon oxide is generally used as the first insulating layer. Openings in the oxide, called via holes, permit contact to the silicon surface. The photoresist serves as an etch mask for forming the interconnecting lines. Multiple layers of wiring can be formed by evaporating, sputtering, or otherwise depositing insulating layers over the first layer of metallization and then repeating the process of opening contact areas, metallizing, and going through the photolithographic process.

By extensions of the basic processes that have been de-

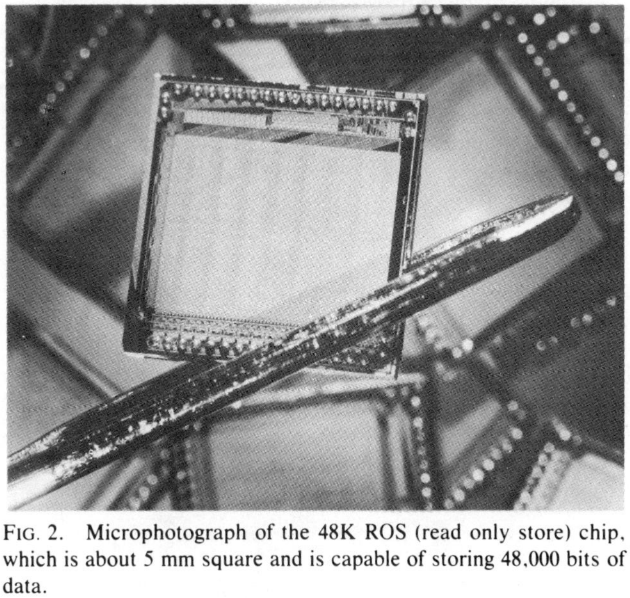

FIG. 2. Microphotograph of the 48K ROS (read only store) chip, which is about 5 mm square and is capable of storing 48,000 bits of data.

scribed, a variety of bipolar and FET transistors and diodes can be formed. Resistors are made by contacting diffused areas that have controlled resistivity; the resistance of a defined area is determined by its length and width. Capacitors are fabricated from biased *p-n* junctions or thin-film structures. Components are isolated by means of reverse-biased *p-n* junctions; another method, oxide isolation, is used in some newer structures.[9]

Once processed, the wafers are cut into chips (Fig. 2), which may measure 5 mm on a side. A chip may contain thousands of components interconnected to form an electronic function such as a microprocessor, or thousands of bits of an electronic memory. The chips are attached to a protective package, which provides the interconnecting leads for external connection. Chips can be attached to packages with wire bonds,[10] beam leads,[11] beam tape,[12] or solder pads.[13]

At present, the minimum line width and line spacings used by industry are about 3 μm. Improved lithographic methods, in particular the use of electron beams to expose the resist, are expected to reduce this dimension to 1 μm or below.[15] By 1980, it is projected that a chip 8 mm square may contain as many as a million components. Even further increases are possible before fundamental barriers are reached.

Integrated circuits have revolutionized computers, television, radio, and electronic products, and have made possible lower costs, higher reliability, savings in size, weight, and power, and better performance. All this explains why dollar shipments of integrated circuits exceeded those of discrete devices in 1976.[14]

See also SEMICONDUCTORS, CRYSTALLINE; TRANSISTORS.

REFERENCES

1. L. J. Selvin, *The Field-Effect Transistor.* McGraw-Hill, New York, 1965.
2. J. G. Linvill and J. F. Gibbons, *Transistors and Active Circuits.* McGraw-Hill, New York, 1961.
3. A. P. Stern, "Preface to the Integrated Electronics Issue," *Proc. IEEE* **52,** 1395 (1964).
4. R. M. Warner (ed.), *Integrated Circuits: Design Principles and Fabrication.* McGraw-Hill, New York, 1965.
5. R. B. Herring, "Silicon Wafer Technology—State of the Art 1976," *Solid State Technol.* **19,** 37–42 (1976).
6. B. E. Watts, "The Preparation and Properties of Elemental Semiconductor Thin Films," *Thin Solid Films* **18,** 1–23 (1973).
7. W. A. Pliskin, "Separation of the Linear and Parabolic Terms in the Steam Oxidation of Silicon," *IBM J. Res. Develop.* **10,** 198–206 (1966).
8. R. Payne, R. Scavuzzo, K. Olson, J. Nacci, and R. Moline, "Fully Ion-Implanted Bipolar Transistors," *IEEE Trans. Electr. Dev.* **ED-21,** 273–278 (1974).
9. A. H. Agajanian, "A Bibliography on Semiconductor Device Isolation Techniques," *Solid State Technol.* **18,** 61–65 (1975).
10. E. F. Koshinz, "Semiconductor: Wire Bond and Face Bonding Considerations and Comparisons," *Proc. SAE Microelectronic Packaging Conference,* November 20-22, 1968, Palo Alto, Calif., pp. 94–100.
11. Special Issue on Beam-Leads, *Western Electric: The Engineer,* December 1967.
12. "Tape-Carrier Packaging Boasts Almost Unlimited Potential," *EDN,* October 20, pp. 30–37 (1974).
13. P. A. Totta, "Flip Chip Solder Terminals," *Proc. Electronics Components Conference,* May 10-12, 1971, Washington, D.C., pp. 275–284.
14. "IC's to Pass Discretes in 1976 Shipments," *Electronics* **49,** 35 (May 27, 1976).
15. I. E. Sutherland, C. A. Mead, and T. E. Everhart, "Basic Limitations in Microcircuit Fabrication Technology," February 1976, Rand Corporation, Santa Monica, California, #WN-9409-ARPA.

Color Centers

Paul W. Levy

Normally transparent crystals and glasses often appear colored because they contain color centers. More specifically, the colored appearance is attributable to absorption bands associated with defects, impurities, or more complex imperfections. Many different types of color centers are formed when the charge or valence state of these imperfections is changed. The defects or imperfections may be incorporated into the lattice in one charge state and not produce observable absorption bands. Subsequently, if the charge state is altered, most often by exposure to ionizing radiation, absorption bands are formed. Alternatively, color centers are often formed by impurities that were introduced into the lattice in their normal valence state. Familiar examples of solids that appear colored because they contain color centers include:

Crystals and glasses colored by exposure to nuclear radiation, to x rays, or to sunlight for a long period of time (solarization).

Colored minerals—e.g., ruby is Al_2O_3 with Cr impurity; smoky quartz usually contains Al and alkalis and the formation of the color usually requires exposure to ionizing radiation. The many colored fluorites are due to the presence of different rare-earth ions and other impurities.

Colored glasses—e.g., "stained" glass windows are made of different glasses with a variety of impurities.

COLOR-CENTER PROPERTIES

Almost all of the properties of color centers may be illustrated by describing F centers in alkali halides. Historically, this was the first center investigated in any detail and has been studied more extensively than any other center. All of the alkali halides, such as NaCl, KCl, and LiF, become colored when exposed to sufficiently large doses of x rays or nuclear radiations. The most commonly occurring bands, induced in a typical alkali halide by x-ray irradiation at liquid-nitrogen temperature, are shown in Fig. 1. In many alkali halides the most prominent band is in the visible. Much of the original research on these bands was done in Germany, and they were labeled F bands after the German word for color, *Farbe*. Likewise, the center associated with the F band was labeled the F center. The nature of the F center was established by studying additively colored crystals that were prepared by exposing pristine samples to alkali-metal vapor. For example, KCl crystals that have been heated in potassium vapor and rapidly cooled to room temperature

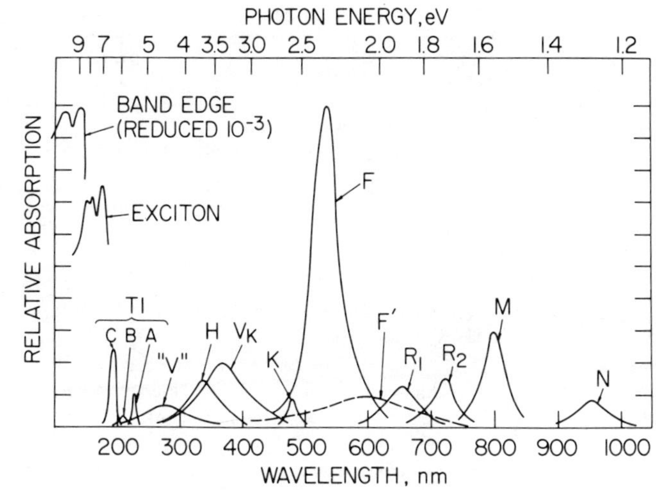

FIG. 1. Principal color-center absorption bands in a "typical" alkali halide. The energies and intensities of bands other than the exciton bands have been chosen to approximate the spectra observed after x-ray irradiation near liquid-nitrogen temperature, roughly 88 K. Exciton bands are observed only near 0 K.

contain a prominent absorption band that is identical to the largest band induced by irradiation. During exposure to the potassium vapor potassium atoms attach themselves to the KCl lattice. To maintain the lattice-ion and defect concentrations appropriate for the crystal and vapor temperature, each potassium ion incorporated into the lattice surface is accompanied by the introduction of a Cl^- ion vacancy into the lattice. To maintain charge neutrality, the electron released from each potassium atom, as it becomes a lattice ion, is transported into the crystal where it is trapped by a Cl^- ion vacancy. The configuration of the trapped electron and negative-ion vacancy is the F center. It, and most other color centers, contain a ground state and at least one excited state. The F-band absorption is due to a transition between the ground state and the next higher state. The lifetime of the excited state is usually quite short and radiative transitions from the upper to the lower state are often observed as luminescence. Also, particularly at higher temperatures, the center can return to the ground state by nonradiative transitions.

Additional features of F centers and other defect-related color centers are illustrated by measurements made on additively colored crystals. If a crystal is cooled rapidly to room temperature after exposure to a hot alkali-vapor environment and then cleaved into thin sections it is found that the F-center concentration is a maximum at the surface and diminishes toward the interior. A uniform distribution of F centers is obtained by heating an additively colored crystal for a comparatively longer time period in a vacuum. This clearly illustrates the introduction of F centers at the surface and the relatively rapid diffusion of centers in the crystal. Additively colored crystals that are slowly cooled, in contrast to the rapidly cooled ones just described, exhibit additional absorption bands that occur on the long-wavelength side of the F band and with one tenth, or less, of its intensity. One of these bands is called the M band and it has been identified with a center formed by two adjacent F centers along ⟨110⟩ directions. Its properties were established by optical absorption measurements on strained and unstrained crystals that utilized light polarized in different directions with respect to the crystal axes. In a similar way other bands have been attributed to R centers, which consist of three adjacent F centers in an equilateral-triangle arrangement with the sides along ⟨110⟩ directions. This arrangement is energetically favored over a linear three-F-center configuration along a single ⟨110⟩ direction. The appearance of multidefect M and R centers in slowly cooled additively colored crystals is attributable to the coagulation of F centers by diffusion processes. These multiple-defect centers form only at temperatures high enough to ensure mobility but low enough to prevent the complete dissociation of previously formed multiple defects. Pure additively colored alkali halides contain very weak absorption bands on the short-wavelength side of the F band. These are called L bands and usually are observed only at liquid-nitrogen or lower temperatures. It has been shown that they are coupled in some manner to the F center. However, they are not well understood.

Additional electronic properties of F centers can be demonstrated by exposing additively colored crystals to light that is absorbed by the F band (termed F light). As the exposure to light increases, the F band diminishes and an additional broad band, lying principally to the long-wavelength side of the F band, appears. This broad band is called the F′ band (F-prime band) and is formed by the capture of a second electron by existing F centers. During exposure to F light a fraction of the F centers absorb photons and are excited to an excited state lying slightly below the conduction band. The lifetime of this state is sufficiently long to allow the electron to be excited thermally into the conduction band. Once in the conduction band the electron is mobile and "wanders about" until it is captured by a negative-ion vacancy to recreate an F center, is captured by an F center to become an F′ center, or is removed from the conduction band in some other way. The F′-center formation process may be reversed by exposing the crystal to light that is absorbed by the F′ center or by raising the crystal temperature. Both processes release electrons from the F′ center into the conduction band, which leads to the reestablishment of F centers. Bleaching experiments such as these have provided much of the existing information on the nature of electron-trap centers and their interaction with mobile charges in the crystal. Bleaching measurements also have been very useful in studies of the electrical antimorphs of the electron centers, namely, the (electronic) hole centers.

The properties of the alkali-halide centers described in this article, and other more commonly occurring centers, are included in Table I.

The electronic nature of F, M, and R centers, or more specifically the fact that they are formed by trapping of electrons on defects, is readily understood for additively colored crystals. As described earlier, the electron released by each metal atom as it is incorporated with the surface of the lattice is trapped on the negative-ion vacancy concomitantly formed to maintain electronic neutrality. These and other considerations suggested that there should be centers that could be described as the electronic and ionic antimorphs of the electron centers, i.e., centers formed by trapping (electronic) holes on positive-ion vacancies. Historically, answers to this and related questions were more readily obtained by studies on crystals colored by exposure to radiation, for reasons that are difficult to describe briefly, than by investigating additively colored crystals prepared by heating them in halogen vapor. In the latter case we would expect the halogen atom to join the lattice and release an electronic hole (capture an electron), which in turn is trapped (by the release of an electron) on a positive-ion vacancy. The resulting configuration (i.e., a hole trapped on an isolated positive-ion vacancy) is often referred to as the V center. However, it has *not* been established that the simple V center described here actually exists. As described later, the hole centers that have been identified are usually more complicated.

To understand the formation of color centers by radiation it is essential to describe the more important processes that occur in solids during exposure to energetic radiation, such as ultraviolet and visible light, x rays, gamma rays, electrons, alpha particles, and neutrons. Most important is the formation of electron–hole ionization pairs, which are often referred to simply as ion pairs. Electron–hole ionization pairs are created whenever the incident radiation is sufficiently energetic to excite lattice electrons, i.e., valence-band electrons, into the conduction band. Once formed, the electron member of the electron–hole ionization pair can "wander about" independently of the concomitantly formed hole. Almost all of the ionization charges enter into recombination. Usually ionization holes are trapped on defects, impurities, or other lattice imperfections more rapidly than ionization electrons are trapped on "opposite-sign" lattice defects. Hole centers usually have large cross sections for electron–hole recombination processes. Thus recombination accounts for a very large fraction of all of the ionization pairs produced during irradiation. The charges that escape recombination during irradiation remain trapped (e.g., to form various electron and hole centers) when the irradiation is terminated.

To retain electrical neutrality the total number of trapped electrons and holes must remain very nearly equal. However, large local charge imbalance may exist in certain circumstances. For example, when samples are subjected to low-energy electron bombardment from one direction, the trapped charges are unequally distributed and discharges can occur between regions of opposite charge. The resulting discharge patterns are known as Lichtenberg figures. However, for radiation-induced color-center formation the total number of electrons trapped in electron-type color centers is approximately equal to the number of holes trapped in hole centers.

The nature of the more simple hole-trapping centers is more complex than the nature of the antimorph electron centers. To achieve local charge neutrality the missing positive lattice-ion charge must be compensated by a trapped hole. To a large extent the electronic nature of hole centers can be understood if it is remembered that the electronic hole is a convenient fiction and that holes are in reality "missing electrons." A hole trapped on a positive-ion vacancy is

Table I Principal alkali-halide color-center absorption bands[a]

Classical[b] notation	Defect associated with given band, properties of band, etc.
α	Negative lattice-ion vacancy
F	One electron on a single negative lattice-ion vacancy
F′	Two electrons on a single monovalent negative-ion vacancy
K	Transition(s) to level(s) above the F-band absorption level
$L_{1,2,3}$	Very weak absorption associated with F centers; origin not known
M	Two adjacent F centers in ⟨110⟩ directions
N	Four adjacent F centers; arrangement not certain
$R_{1,2}$	Three adjacent F centers in "triangle" arrangement
M^+	One electron on two adjacent negative lattice-ion vacancies (ionized M center)
F_A	F center adjacent to substitutional alkali-ion impurity
Z_1	One electron on coupled substitutional divalent impurity and adjacent positive-ion vacancy
β	Positive lattice-ion vacancy
V (or V_1)	The postulated antimorph of the F center, consisting of one hole on a single isolated positive lattice-ion vacancy (or an adjacent negative ion); it is not certain that this center has been observed.
$V_{1,2,\ldots}$	Bands in V region on the short-wavelength side of the F center; often assumed to be produced by hole centers; may be identical with "identified" hole centers.
V_K	One hole trapped on a pair of adjacent negative lattice ions to produce a single ionized two-atom molecule, e.g., Cl_2^-; aptly called the "self-trapped hole."
I	Ionized lattice atom in an interstitial site
H	One hole trapped on a pair of negative lattice atoms centered on a single lattice-atom site, i.e., one hole, one lattice atom, and one interstitial; often described as a "split interstitial molecular ion on a single lattice atom site," e.g., a Cl_2^- molecular ion on a Cl^- site.
H_A	H center adjacent to a substitutional positive-ion impurity
V_A	V center adjacent to a substitutional negative-ion impurity such as Br^- or OH^- on a Cl^- site
Exciton bands	Bands observed only near 0 K and associated with transitions between the "Bohr atom" levels of bound electron–hole pairs, i.e., excitons.
Colloid or "X" bands	Bands arising from the combined absorption and scattering of colloid particles

[a] Unless otherwise stated, the absorption bands given in this table refer to ground-state to first-excited-state transitions.

[b] Originally the letter designations, such as F in F band, referred to specific absorption bands. More recently, it has been common to use the letter designation to refer to the center responsible for a given absorption band. This often causes confusion—e.g., there is a K band but not a K center. This multiple usage, and the use of other systems for designating absorption bands and/or defects that are often similarly ambiguous, require the reader of color-center literature to pay particular attention to nomenclature.

shared by each of the adjacent negative ions. Thus in the portion of the crystal containing a hole on an isolated positive-ion vacancy there would be local charge neutrality. However, as mentioned earlier it is *not* certain that this conceptually simple center actually exists. A discussion of reasons why it does not exist or has not been observed is too long to include in this article. The more commonly occurring and established centers formed by trapped holes are described in what follows.

Among the various kinds of V centers that definitely do occur in alkali halides, the V_K center has been studied most extensively. This center assumes an unexpected and surprising configuration. It can enter into a number of color-center formation processes and is comparable to the F center in controlling radiation-induced color-center formation at all temperatures below which it is stable, very roughly 100 K. The V_K center is aptly described and referred to as a "self-trapped hole." The center is formed when an ionization-created hole attaches itself to a pair of adjacent halogen atoms to create a halogen negative ion. For example, in KCl, the V_K centers are Cl_2^- ions oriented in the six equivalent ⟨110⟩ directions. The V_K centers are unique. The available evidence indicates that they are formed in the initially unperturbed or defect-free regions of the lattice. In contrast, other centers are formed at one or more existing lattice defects. The processes controlling the formation of these centers, i.e., the factors determining exactly where in the crystal a V_K center will be formed or, if it is formed in a mobile state, where it will localize, are not understood in detail.

The properties of self-trapped holes have been very extensively studied—especially in alkali halides—by optical absorption using polarized and unpolarized light, electron spin resonance, luminescence, etc. The characteristics of this center are illustrated by the numerous processes that take place at different temperatures. At temperatures appreciably less than 100 K for most alkali halides the V_K centers are stable and oriented in ⟨110⟩ directions. At higher temperatures, specific to each type of crystal, the defect axis may change from the initial ⟨110⟩ to any one of the ⟨110⟩ directions at an angle of 60° to the original; reorientation to the ⟨110⟩ directions at 90° from the original direction apparently does not occur. At even higher temperatures, where the reorientation occurs very rapidly, the defect becomes quite mobile and usually enters into several different kinds of interactions. It is likely to form any number of other hole centers, such as H or V_A centers. Also, it can interact with electron centers to undergo electron–hole center neutralization or recombination. During recombination, light (i.e., luminescence) may be emitted, or energy may be released as heat or phonon emission and, according to one proposed mechanism, provide the energy for vacancy creation.

A second type of commonly occurring hole center is the H center. It should be carefully contrasted with the V_K center. The V_K center is a negative molecular ion, such as Cl_2^-, formed by hole trapping in the normal or unperturbed lattice. In contrast, the H center can be regarded as a similar Cl_2^- molecular ion formed by hole trapping on an interstitial negative atom and an associated lattice atom. In its stable configuration the H center is a negative ion-pair, e.g., Cl_2^- on a Cl^- lattice-atom site.

An additional commonly occurring hole center is formed when H centers couple with other lattice defects, such as substitutional impurities, to form centers usually labeled H_A centers. Often the A atom is identified, e.g., an H center coupled to a Li atom is labeled H_{Li}.

Next, the *exciton* bands will be described. They are observed only at temperatures near 0 K and occur at wavelengths somewhat longer than those involved in the "band edge" valence-band to conduction-band transitions. (See Fig. 1.) To understand exciton bands, recall first that electron–hole ionization pairs produced at high temperatures dissociate immediately after they are formed and the charges move independently of each other. At low temperatures, however, some ionization pairs are formed in circumstances permitting them to remain coupled. More precisely, the electron remains coupled to a valence-band atom. The observed energies of the exciton absorption bands can be reasonably well calculated from this model. At low temperatures excitons appear to have little tendency to dissociate. However, there is considerable evidence from which to conclude that they are quite mobile. Those observed well after they were formed are stabilized (i.e., localized or trapped) on an appropriate defect. Exciton bands that were observed in a crystal at low temperatures disappear from the absorption spectra as the crystal temperature is raised. This disappearance can be attributed to the dissociation of the excitons with the individual charges entering into the processes usually associated with ionization. Alternatively, they could have become mobile and interacted with lattice defects, surface states, etc. Such interactions produce luminescence or, alternatively, result in the release of an amount of energy slightly less than the band gap energy. In turn, the released energy could enter into an atom displacement process, a chemical reaction at the crystal surface, or some similar process.

CENTERS FORMED BY OPTICALLY ABSORBING IMPURITIES

A large variety of light-absorbing impurities may be incorporated into alkali halides, oxides, and other crystals. The more common examples include:

Alkali impurities in alkali halides, e.g., Na in KCl, KBr, etc.
Alkaline earths in the same hosts, e.g., Ca and Sr in KCl, LiF, etc.
Metals such as Pb, Ca, Tl.
Molecular ions like SO_4^-, CN^-, OH^-.
Nonmetals, including S^{2-}, O^{2-}.

In fact, a very large fraction of the conceivable ions can be incorporated in the alkali halides and simple oxides.

Crystals prepared with specific impurities are described by the *colon notation*. Thus NaCl with lead impurity is specified as NaCl:Pb. This is ambiguous if the Pb was introduced as a compound, e.g., $PbCl_2$. In terms of this example, it has become increasingly common to specify NaCl with lead introduced as a metal by the notation NaCl:Pb, and when the lead is incorporated as $PbCl_2$ by the notation NaCl:$PbCl_2$.

In the development of color-center physics, measurements on alkali halides containing thallium substituted for the alkali metals played a pivotal role. The Tl^+ ion produces three easily observed absorption bands, usually called A, B, and C bands. These three bands occur on the long-wavelength side of the valence-band to conduction-band transition, usually called the edge absorption. There is appreciable evidence for a fourth band, the D band, which lies very close to or on the edge absorption. The properties of the A, B, and C bands have been deduced theoretically by considering how the electronic energy levels of the free Tl^+ ions are modified by embedding the ion in the host alkali-halide crystal. The A and C bands are attributable to transitions that are allowed in the free Tl^+ atom. The B band is associated with a normally forbidden electronic transition in which the forbiddenness is removed by interaction with lattice phonon modes. Inasmuch as this interaction increases with increasing temperature, the B-band absorption increases with increasing temperature. This is a unique property, observed only with these and similar atomic impurity bands. The origin of the D band is not unequivocally established. It has been attributed to transitions between the ground state and levels above the A-, B-, and C-band levels and, alternatively, to transitions from the ground state to the conduction band. A large fraction of the properties of the impurity-related absorption bands are very similar to the defect-related bands. Both types are described well by the configuration coordinate formalism described later.

Many transparent but colored minerals and various kinds of colored glass are examples of materials colored by the presence of specific light-absorbing atoms. Ruby is Al_2O_3 (corundum) containing Cr^{3+} atoms substitutionally replacing Al^{3+} atoms. Emerald is $Be_3Al_2Si_6O_{18}$ (beryl) containing Cr^{3+} in Al sites. Aquamarine is beryl containing Fe^{2+} and Fe^{3+} impurities in Al sites. Citrine is quartz given a yellow color by the presence of Fe^{3+} ions, most likely in Si^{4+} sites. Examples of colored glasses include the common blue cobalt glass that obviously includes Co and pink-tinted spectacle lenses, which are made from didymium glass, which contains a mixture of rare-earth elements. The ubiquitous beer bottle's brown color is due to Fe^{3+} ions, and the light green of wine bottles is ascribable to Fe^{2+} ions.

Almost all colored crystals, in contrast to host crystals that do not contain other than negligible quantities of impurities, exhibit absorption bands in the visible. A large fraction of these bands can be attributed to transitions between atomic energy levels in the impurities, which are often split by crystal-field effects. Ruby and emerald are in this category. The absorption bands in another large group, which includes aquamarine, are due to molecular-orbital and/or charge-transfer transitions. A small number of crystals are colored because they possess valence-band to conduction-band transitions. Cinnabar (HgS) is a good example in this category. Finally there are a number of crystals that appear colored because they contain defect-related color centers of the type described earlier for the alkali halides, i.e., they contain color centers formed by charge trapping on lattice defects. Smoky quartz and some, but not all, colored fluorites are in this group.

To reduce the existing confusion, it is necessary to mention three types of crystals that exhibit color not attributable to any of the processes mentioned so far, i.e., not caused by color centers. First, halite, which is the natural form of NaCl and is usually called rock salt, occasionally exhibits an intense color ranging from very light blue to a deep blue-black purple. Often the color is layered along (110) planes. These blue crystals exhibit a *de facto* absorption band in the red that is generally attributed to a combination of absorption and scattering from colloid Na metal particles. This blue coloring can be induced by exposing NaCl crystals to large doses of energetic radiation, particularly above room temperature, and/or exposing crystals to room light after irradiation. Second, crystals can appear colored if they contain numerous microinclusions of a second mineral, e.g., the color in blue quartz is attributed to microscopic rutile (TiO_2) inclusions. Third, crystals that are composites of regularly arranged microcrystals appear to be colored because they diffract light. The classic example is opal. This mineral consists of nearly equal-size spheres of a hydrated silica located on a uniformly spaced three-dimensional lattice.

OPTICAL PROPERTIES OF COLOR CENTERS AND THE CONFIGURATION COORDINATE DIAGRAM

A large fraction of the optical properties of color centers can be described by utilizing a configuration-coordinate diagram such as Fig. 2. Such diagrams describe the electronic and vibrational, or thermal energy, levels of simple electron or hole centers. To a moderately good approximation such centers can be described by quantum-mechanical calculations based on the Born–Oppenheimer approximation. This procedure separates the solution into an electronic part and a vibrational part. The solution of the vibrational part is obtained by using a "spring-constant" potential which leads to harmonic-oscillator wave-function solutions. The solution of the electronic part, for the lowest two electronic states, is represented by the curved envelopes, properly called "adiabatic potential energy curves." Superimposed on the electronic states are the vibrational or phonon states, shown as closely spaced lines. Usually the lowest electronic level is an *s* state and the next highest level a *p* state. At temperatures approaching 0 K the system can be described by the *s*-state envelope, the *p*-state envelope, and the lowest-lying phonon levels in each state. At low temperatures the absorption transitions will originate from the lowest-lying levels at the minimum in the ground state and terminate on the phonon levels of the excited-state envelope which lie directly above in the configuration-coordinate diagram. Such transitions are indicated schematically in Fig. 2 as the absorption transition. The resulting absorption band will be quite narrow, since all of the absorbing transitions have nearly the same energy. At higher temperatures the transitions originate and terminate from wider distributions of phonon states, and as the temperature increases the absorption bands become wider. Transitions from upper to lower electronic states result in the emission of light, i.e., luminescence. Similarly, the luminescent emission is narrow at low temperatures and increases with increasing temperature. This approach to the optical properties of color centers provides a good description of absorption, luminescence, and

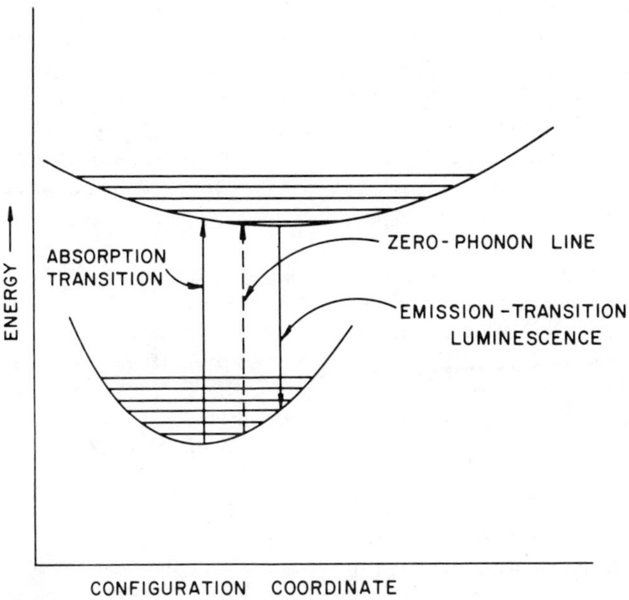

FIG. 2. A typical color-center configuration-coordinate diagram. Many quantitative optical properties of color centers can be deduced from such diagrams. Typical absorption, luminescence, and a "zero-phonon" absorption transition are shown.

many other properties of color centers. The phonon contribution is well approximated by a Boltzmann distribution for all temperatures except those near 0 K. Near 0 K accurate calculations require the inclusion of discrete phonon levels and the zero-point energy. Also, for most calculations the electronic envelope can be regarded as parabolic; in the pure harmonic-oscillator approximation it is precisely parabolic. Calculations made with this model lead to the following expressions for the temperature dependence of the absorption band peak E_0 and the full width of the band at half maximum U_0 (usually both E_0 and U are given in electron volts):

$$E_0(T) = C_1 - C_2 \coth(\hbar\omega/2kT), \tag{1}$$

$$U^2(T) = 8 \ln 2(\hbar\omega)^2 S \coth(\hbar\omega/2kT) \tag{2}$$

where T is the temperature in degrees Kelvin, C_1 and C_2 are constants, $\omega/2\pi$ is the effective vibrational frequency in s^{-1}, $\hbar = h/2\pi$ where h is the Planck constant, k is the Boltzmann constant, and S is the Huang–Rhys factor. The last-mentioned factor is difficult to explain briefly. It is a coupling constant that is a measure of the relative horizontal displacement of the adiabatic potential energy curves along the configuration-coordinate axis and is approximately equal to the mean vibrational quantum number in the transition.

The optical absorption of a crystal containing color centers can be calculated by considering it as an assemblage of a large number of absorbing centers. The result is known as *Smakula's formula*, which relates the measured optical absorption coefficient α (in cm^{-1}) to the concentration of color centers N:

$$Nf = 0.87 \times 10^{17} \frac{n}{(n^2+2)^2} U\alpha_m \tag{3}$$

where f is the oscillator strength, which must be determined independently for each color center and is approximately 0.9 for most alkali-halide centers; n is the index of refraction of the host crystal; U is the full width at half maximum of the absorption band (in eV); and α_m is the absorption coefficient at the peak of the band. This formula is based on the assumption that the absorption band is precisely Gaussian shaped. The numerical coefficient changes by roughly 10% for other common shapes. Precisely Gaussian-shaped bands are observed in many cases and represent cases where the adiabatic potential energy curves for the centers can be approximated by simple harmonic oscillators. In many cases the actual potential cannot be represented by simple parabolic curves and the observed absorption and emission bands are not well described by Gaussian-shaped bands; in fact, the observed bands may be quite asymmetric.

The usefulness of the configuration-coordinate diagram approach and the behavior of many color centers at low temperatures is illustrated by zero-phonon line phenomena. A good example is the R-center absorption in LiF at liquid-helium temperatures, i.e., near 4 K. The absorption spectrum consists of a number of very sharp lines superimposed on a broad underlying continuum. The continuum resembles the spectrum observed at higher temperatures but is slightly narrower and shifted in accord with the peak-energy and bandwidth formulas given earlier. The sharp lines are not observed at higher temperatures. They arise from transitions between individual low-lying phonon states in the upper and lower levels. The most prominent sharp-line transition is between the lowest-lying states, and since these do not include lattice vibration or phonon modes (the vibrational quantum number is zero) it is called the "zero-phonon line." This transition is illustrated in Fig. 2. It is often observed when the other sharp lines are obscured.

RADIATION-INDUCED COLOR-CENTER FORMATION

Most studies on color centers were undertaken to determine the properties of the different types of centers. More specifically, they were designed to identify the atomic and electronic configuration of each center and to determine its properties. A less emphasized but equally interesting aspect of color-center physics is the study of radiation-induced color-center formation. Curves of color-center concentration versus irradiation time or dose depend on a large number of physical parameters. The more important are

1. The nature of the material being irradiated. Some aspects of the color-center formation kinetics are common to most materials. Other aspects depend on the material type. In other words, there are features peculiar to each of the following: alkali halides, oxides, mixed ionic and covalent crystals (such as nitrates and bromates), covalent crystals, semiconductors, glasses, glass-ceramics, and organic glasses.
2. Sample temperature.
3. The type of radiation producing the coloring (e.g., infrared, visible, ultraviolet, x rays, gamma rays, electrons or beta particles, protons, alpha particles); the various elements in different charge states, which include fission fragments, molecules both neutral and

charged, and the nuclear particles, such as mesons.
4. Intensity and total exposure or irradiation time, including differences between pulsed and continuous or steady-state radiations.
5. The presence or absence of impurities or defects in the lattice. In the alkali halides the coloring curves are strongly influenced by dislocations and dislocation-related defects. Plastically deformed alkali halides color more readily than unstrained ones. Impurities are important or control the coloring many materials.

Until recently most color-center formation studies were made by irradiating a crystal with x rays, or other radiation, and then making measurements at a later time. If the crystal temperature was low enough to suppress all temperature-dependent decay occurring during and after irradiation, most of the color-center versus dose or irradiation-time curves are linear or monotonic increasing and are true representations of the color-center formation kinetics. However, at higher temperatures and particularly at room temperature, the center concentrations change both during and after irradiation and the measured coloring curves do not accurately describe the color-center growth kinetics. In many crystals, particularly the alkali halides, the coloring curves contain three distinct stages. The initial part, stage I, increases monotonically at a continuously decreasing rate until a low- or zero-slope plateau is reached. The plateau region is called stage II. The irradiation time included in stage II varies from close to zero to very large values. The following region, stage III, when it is observed, usually is concave upward and continues until the irradiation is terminated. At very high doses some materials exhibit growth beyond stage III which flattens out to a plateau, or occasionally reaches a maximum and then decreases. Sometimes this last region is referred to as stage IV.

First- and second-stage coloring curves obtained by making measurements at some—often controlled—time after irradiation (e.g., at room temperature) are crudely approximated by an expression consisting of the sum of a saturating exponential and a linear component. However, the stage-I and -II color-center versus irradiation-time curves obtained by making measurements during irradiation are very accurately described by the expression

$$a(t) = \sum_{i=1}^{n} A_i(1 - e^{-a_i t}) + \alpha_L t. \tag{4}$$

The number of saturating exponential terms, n, usually does not exceed four. Also, the possibility that this expression is merely a good empirical curve-fitting formula is almost certainly eliminated by the occurrence of stage-I curves containing one exponential component or two unambiguously resolvable exponential components. The recording of color-center absorption curves during irradiation requires rather unique instrumentation. However, growth curves describable by Eq. (4) have been obtained by making measurements using either ^{60}Co gamma rays or 1–3-MeV electron irradiations.

Measurements made after the irradiation is terminated show two interesting features. First, with one or two exceptions, when the irradiation is terminated the measured color-center concentrations undergo abrupt changes. In most crystals and glasses the coloring decreases rapidly. In others, one or more centers increase and in some they increase initially and then decrease. Often the changes occurring after irradiation can be resolved unambiguously into decreasing exponential components or into combinations of increasing saturating exponential and decreasing exponential components.

An understanding of the various coloring curves that are obtained during irradiation and the changes occurring after irradiation can be obtained by considering the coloring of alkali halides (LiF is often an exception). If the measurements are made during irradiation, the first- and second-stage coloring is accurately described by (4).

This can be regarded as an empirical equation. However, it can be readily derived from relatively simple considerations. The primary coloring process is the conversion of precursors into color centers by charge capture. For example, vacancies are precursors in the sense that they can be converted to F centers by electron capture. Usually a crystal will contain a given concentration of precursors, for each type of color center, prior to irradiation. During irradiation the precursor concentration may increase (or in rare cases decrease) by radiation-induced processes. Also, once a color center is formed it may be converted back into a precursor by recombination with opposite-sign charges. Thus, during irradiation F centers can be converted to vacancies by recombination with holes. Also, the changes occurring after irradiation can be attributed to the thermal untrapping of charges from color centers or to recombination with opposite-sign charges thermally released from traps. Both processes give rise to exponential color-center decay, as is observed. When all of these processes are included, the solution of the differential equation describing the coloring can be expressed as Eq. (4).

The introduction of defects during irradiation is attributable to two separate processes. If the radiation, including gamma rays, is sufficiently energetic, defects are created directly by elastic collision processes with lattice atoms. For each atom "knocked out" of its normal position one vacancy and one interstitial are created. To induce these defects by x rays and gamma rays the energies of the bombarding quanta must exceed well-defined values. However, in the alkali halides defect formation is observed during bombardment with x rays that are much too low in energy to produce defects by collision processes. This phenomenon is often called *ionization damage*. It is believed that the energy required for the defect-creation process is obtained from electron–hole recombination. An electron can combine with a V_K center to form a self-trapped exciton, which decays nonradiatively to produce an F-center–H-center pair that is separated by a large enough distance to be stable. Similar electronic processes can occur at vacancies, impurities, dislocations, and related defects.

In most alkali halides the coloring process is strongly influenced by the presence of impurities, defects, and dislocations. Certain impurities increase the stage-I and -II x-ray coloring and depress or do not influence the late-state coloring. Other impurities suppress all stages. Crystals that have been quenched from near the melting point exhibit increased coloring that can be attributed to elevated precursor concentrations. The role of dislocations in the coloring process is clearly demonstrated by making F-center concentration versus irradiation-time measurements on crystals plastically deformed during or prior to irradiation. Both stage-I and -II coloring increase with increasing strain. In fact, the parameters obtained by fitting Eq. (4) to the coloring curves measured on plastically deformed crystals divide into two groups. Some parameters are linear functions of strain and others are proportional to the square root of the strain. This behavior is in accord with current ideas on defects introduced into crystals by strain and/or plastic deformation.

ADDITIONAL INFORMATION ON COLOR CENTERS

The information on color centers given in this article must be regarded as a very much abbreviated overview of the subject. Many aspects of color-center physics have not been mentioned. Examples are electron spin resonance (ESR); electromechanical and magneto-optical properties; luminescence, which includes fluorescence, phosphorescence, and thermoluminescence; ionic, electronic, and photo conductivity; diffusion of defects producing color centers; thermal annealing; dielectric properties; and strain-induced reorientation. The biblography includes books that contain advanced level imformation on almost all aspects of color-center physics.

See also CRYSTAL DEFECTS; ELECTRON ENERGY STATES IN SOLIDS AND LIQUIDS; ELECTRON–HOLE DROPLETS IN SEMICONDUCTORS; EXCITONS; LUMINESCENCE; RADIATION DAMAGE IN SOLIDS; RADIATION INTERACTION WITH MATTER; THERMOLUMINESCENCE.

BIBLIOGRAPHY

J. H. Crawford, Jr., and L. M. Slifkin, eds., *Point Defects in Solids*, Vol. I, *General and Ionic Crystals*. Plenum, New York, 1972.

W. B. Fowler, ed., *Physics of Color Centers*. Academic, New York, 1968.

J. J. Markham, *F Centers in Alkali Halides*. Academic, New York, 1966.

J. H. Schulman and W. D. Compton, *Color Centers in Solids*. MacMillan, New York, 1962.

A. M. Stoneham, *Theory of Defects in Solids*. Clarendon Press, Oxford, London, and New York, 1975.

Conduction

S. Kirkpatrick

Conduction processes in condensed matter consist of the transport of heat, electric charge, mass, or some combination of the three, in response to an imposed temperature gradient, electric field, or density gradient. (Convection, distinguished among conduction processes by the fact that macroscopic motion of all the atoms or molecules occurs, also contributes to transport in fluids, but will not be discussed here.) Conductivities vary more widely among materials than do other physical properties. While the densities of the liquids and solids which form at room temperature vary over roughly one decade, electrical conductivities vary by 20 decades or more from metals to insulators, and may be infinite in superconductors at sufficiently low temperatures. At cryogenic temperatures, thermal conductivities also exhibit a range of many decades, with metals offering the highest thermal conductivities, and insulators the lowest.

The usefulness of specific materials is very often due to their high or low conductivities. Discovery of new materials with extreme conductivities frequently opens novel applications, e.g., superconducting magnets for producing very large magnetic fields. But here we shall be concerned with conduction, especially at low temperatures where the largest variations are observed, as a probe of the microscopic particlelike excitations which carry the heat, charge, or mass currents. These particles are most commonly phonons (long-wave length mechanical vibrations) or electrons, but ions may contribute and Cooper pairs carry the lossless current in superconductors. Sometimes carriers are not available for conduction in the equilibrium material, but must first be created. In photoconduction, this occurs by excitation of a carrier out of a filled electronic level when a quantum of light is absorbed. Carriers may also be injected into a material by quantum-mechanical tunneling through a thin insulating barrier.

A semiclassical description, first employed in the kinetic theory of gases, is convenient for discussing conduction by electrons or ions. A mobility μ is defined as the mean carrier drift velocity induced by a unit applied force. For electrons,

$$\mu \equiv \langle v \rangle / e\mathscr{E} = \tau / m^*,$$

where τ is a relaxation time (10^{-14} s is typical for electrons in metals at room temperature), and m^* the effective mass of an electron moving in the attractive potential of the ions in the material. The conductivity σ is obtained from the associated electric current $j^{(e)}$ by

$$\sigma \equiv j^{(e)}/\mathscr{E} = ne^2\tau/m,$$

with n being the density of carriers and e being the electronic charge. Diffusion processes, in which a particle current $j^{(d)}$ is induced by a density gradient, are characterized by a diffusion coefficient D:

$$j^{(d)} = -D\nabla n.$$

The two types of process are intimately connected. The Einstein relation,

$$\mu = eD/k_BT,$$

holds for any particles which obey classical Maxwell–Boltzmann statistics, for example, electrons and holes in semiconductors, or ions in electrolytes.

The thermal conductivity κ is defined from the heat current j^q which flows in response to a temperature gradient:

$$j^{(q)} \equiv -\kappa\nabla T.$$

A useful expression for κ is

$$\kappa = \tfrac{1}{3}C_V v^2 \tau,$$

where C_V is the specific heat at constant volume, and v is the instantaneous velocity of the particle. For electrons, v is the Fermi velocity; for phonons, the speed of sound is appropriate in this expression. The ratio of the thermal to the electrical conductivity in metals, known as the Lorentz number,

$$\kappa/\sigma = (\pi^2/e)(k_B^2 T/e^2),$$

is of order 2×10^{-8} W Ω/K^2 near room temperature. This expression, although derived from classical arguments, is a quite general result, and remains true to good accuracy even at very low temperatures, where Fermi–Dirac statistics must be used to describe the electrons.

Both n and τ can vary widely with temperature and from one material to another. In insulators, n vanishes as the temperature tends to zero. If E is the excitation energy required to create a free carrier, then n is proportional to $\exp(-E/k_BT)$. "Arrhenius plots" of $\ln(\sigma)$ against $1/T$ yield straight lines for most semiconductors, and provide a means of determining E. Heat is transported principally by the phonons in insulators. These scatter only from defects or sample boundaries at low temperatures, so in this limit $\kappa\propto C_V\propto T^3$ is observed.

In metals, n and v are not sensitive to temperature, but τ can be. Phonon scattering alone gives $\tau\propto T^{-5}$, while scattering from impurities gives a constant contribution to τ. As a result, in a nonsuperconducting metal σ tends to a constant at low temperatures while $\kappa\propto T$.

In noncrystalline solids, such as insulating or semiconducting glasses, localized electronic and vibrational excitations introduce novel transport properties. The dominant electrical conduction mechanism at low temperatures may involve "hopping" of carriers between spatially separated long-lived states, assisted by the absorption of phonons. The characteristic distance and phonon energy involved in hopping change in a complicated way with decreasing temperature. The result, first explained by N. F. Mott, is

$$\ln \sigma\propto T^{-1/4}.$$

Localized vibrations give rise to a phonon scattering time $\propto T^{-1}$ in glasses. As a result, κ is proportional to T^2 at the lowest temperatures.

Two additional mechanisms should be mentioned which do not fit into the kinetic theory framework but may have a dramatic effect on conduction. These are the Mott transition and percolation. In the first, Coulomb repulsion between the carriers overcomes the natural tendency for electrons to tunnel from one atom to the next, leading to an insulating system with no free carriers. In the second, an inhomogeneous or composite material may contain metallic regions which are not connected to each other sufficiently often for the material as a whole to conduct. Conduction first occurs when the fraction of metal present exceeds from finite value, known as a percolation threshold.

See also DIFFUSION; INSULATORS; METAL-INSULATOR TRANSITION; METALS; PHOTOCONDUCTIVITY; TUNNELING; SEMICONDUCTORS, CRYSTALLINE; SUPERCONDUCTIVITY THEORY; TRANSPORT PROPERTIES.

Constants, Fundamental

B. N. Taylor

INTRODUCTION

Four main topics will be touched on in this article: (1) The motivation for "the romance of the next decimal place," or why the fundamental physical constants are important and why their determination to ever greater levels of accuracy can have a profound effect on physics; (2) how a self-consistent set of "best values" of the fundamental constants is obtained, with emphasis on the 1973 least-squares adjustment (the most recent comprehensive study carried out); (3) new developments in the fundamental constants field since the 1973 adjustment and their impact on the recommended set of best values resulting from that adjustment; and (4) future trends—where the field is heading over the next five to ten years.

ORIGINS

Three distinct "sources" of fundamental constants may be identified. The first is physical theory (and its application to the real world), such as Maxwell's theory of electromagnetism, Einstein's theories of relativity, quantum mechanics, and quantum electrodynamics (QED). The speed of light in vacuum, c, the Planck constant, h, the fine structure constant, α, and the Rydberg constant for infinite mass, R_∞, fall within this category.

The second is the fundamental particles such as the electron and proton. The elementary charge, e, the electron rest mass, m_e, and the magnetic moment of the proton (in units of the nuclear magneton), μ_p/μ_N, are all examples of quantities that characterize a basic property of nature's elementary building blocks.

The third is conversion factors. Although not true fundamental constants in the sense of the first two categories, these quantities nevertheless play a critical role in the fundamental-constants field, since knowledge of their values is essential to many fundamental-constant determinations. Examples in this category include the local value of the gravitational acceleration, g, which relates mass to force; the ratio of the old x-ray unit of length or kilo-x-unit to the angstrom, Λ; and the ratio of various so-called as-maintained electrical units to their International System of Units (SI) definitions, for example, A_{NBS}/A, where A_{NBS} is the National Bureau of Standards (NBS) ampere as maintained via standard cells and standard resistors.

It should also be noted that $\alpha=[\mu_0c^2/4\pi](e^2/\hbar c)$ and $R_\infty=[\mu_0c^2/4\pi]^2(m_ee^4/4\pi\hbar^3c)$ are examples of constants that are actually combinations of other quantities, but are considered fundamental constants in their own right since the combination always appears in theoretical equations in the same way; and that e and m_e are examples of constants used as basic measurement units, for example, the mass of the muon is expressed as (approximately) 207 electron masses. (See footnote b to Table I for an explanation of the significance of the brackets and parentheses in the foregoing equations.)

IMPORTANCE

There are at least four reasons why the fundamental physical constants play a critical role in science and technology, and thus must be known to as high an accuracy as possible. First, accurate values of the constants are required for critical comparison of theory and experiment; and it is only such comparisons that enable our understanding of the physical world to advance. A closely related idea is that by comparing the numerical values of fundamental constants obtained from experiments in the different fields of physics, the self-consistency of the basic theories of physics themselves can be tested.

Second, determining the fundamental constants to ever greater levels of accuracy fosters the development of state-of-the-art measurement methods that may have wide application. Determining the next decimal place is never trivial and usually requires an entirely new measurement technology. A recent example is the determination of the Avogadro constant, N_A, which necessitated the development of techniques to measure the lattice spacing of single crystals in absolute units (meters) to an accuracy of a few tenths of a part per million (ppm). The end result has been the extension of our length scale to the picometer range.

Third, the fundamental constants are the obvious key to the development of a reproducible and invariant system of measurement units, a major goal of metrology. If our measurement system could be based on fundamental constants, then there would be no need for artifacts such as the platinum–iridium cylinder kept in a vault at the International Bureau of Weights and Measures (BIPM), Sèvres, France, which defines the kilogram.

Finally, values of the fundamental constants are required for computations and measurements throughout science and technology—for example, the calculation of the excited states of atoms and molecules, which play an important role in air pollution and nuclear fusion.

As an example of the significant role improved measurements of fundamental constants can play in increasing our understanding of physical theory, we consider the determination of $2e/h$ using the ac Josephson effect in superconductors, which was first reported in 1967. By irradiating a Josephson thin-film tunnel junction with X-band microwaves (8–12 GHz) and using the phenomenon of microwave-induced constant-voltage current steps, the following value was obtained:

$$2e/h = 4.835976(12) \times 10^{14}\ \mathrm{Hz/V_{NBS}}\ (2.4\ \mathrm{ppm}), \qquad (1)$$

where the subscript NBS means measured in terms of the unit of voltage maintained at the National Bureau of Standards. (The first number in parentheses is the one-standard-deviation uncertainty in the last digits of the quoted value; the second is its fractional equivalent in ppm.) The difference between this value and the previous best value, that resulting from the 1963 least-squares adjustment of the fundamental constants of Cohen and DuMond, was 35 ± 10 ppm. As we shall see, the cause of the discrepancy was the use by Cohen and DuMond in their 1963 work of a value of α derived from the early 1950s measurement of the deuterium fine-structure splitting, which subsequently turned out to be incorrect.

The Josephson $2e/h$ determination is significant mainly because a reliable *indirect* value of α independent of quantum electrodynamic theory can be derived from it, and this value can in turn be used to compare QED theory and experiment critically and unambiguously. This is in marked contrast to the situation that existed prior to 1967 when no such value was available and tests of QED were mainly checks of internal consistency.

The equation relating α to $2e/h$ may be written

$$\alpha^{-1} = \left[\frac{c}{4R_\infty} \cdot \frac{1}{(\Omega_{\mathrm{NBS}}/\Omega)} \cdot \frac{\mu_p'}{\mu_B} \cdot \frac{(2e/h)_{\mathrm{NBS}}}{\gamma_p'(\mathrm{low})_{\mathrm{NBS}}}\right]^{1/2} \qquad (2)$$

where $\Omega_{\mathrm{NBS}}/\Omega$ is the ratio of the NBS as-maintained ohm to the SI ohm; μ_p'/μ_B is the magnetic moment of the proton in units of the Bohr magneton (throughout, the prime means for protons in a spherical sample of H_2O); and $\gamma_p'(\mathrm{low})_{\mathrm{NBS}}$ is the gyromagnetic ratio of the proton obtained by the so-called low (magnetic) field method measured in terms of the NBS as-maintained unit of current. (It should be noted that the subscript NBS may be replaced by the more general lab.) The value of α derived from Eq. (2) in 1968 using the best data available was

$$\alpha^{-1} = 137.03608(26) \quad (1.9\ \mathrm{ppm}). \qquad (3)$$

Included among the quantities that require an accurate value of α for comparing theory and experiment are the anomalous magnetic moment of the electron and muon, a_e and a_μ; the energy levels in hydrogenlike atoms, especially the $n=2$ Lamb shift ($2^2S_{1/2}$–$2^2P_{1/2}$ splitting); and the ground-state hyperfine splitting in hydrogen, muonium (an electron bound to a positive muon), and positronium (a bound electron–positron pair). Of particular interest in the late 1960s was the hydrogen hyperfine splitting (hfs). This quantity, which is essentially the energy difference between a hydrogen atom in which the electron and proton spins are aligned and one in which they are in opposite directions, can be measured to the extraordinary accuracy of 1 part in 10^{12} using the hydrogen maser. In contrast, the theoretical QED equation for the hydrogen hfs, which involves only well-known constants and α, is limited to an accuracy of a few ppm because of the difficulty in calculating some of the terms in the equation from theory.

The most uncertain term is $\delta_N^{(2)}$, the proton polarizability contribution, which arises from the various excited states or internal structure of the proton. In 1968, theoretical calculations predicted $\delta_N^{(2)} = 0 \pm 5$ ppm. This was in conflict with what was implied by the value of α accepted at that time, which, as noted earlier, was derived from a measurement of the fine-structure splitting in deuterium. When this value of α was used to calculate a theoretical value of the hydrogen hfs, when this result was compared to the hydrogen-maser value, and when their difference was assumed to arise solely from the existence of a polarizability term, it was found that $\delta_N^{(2)} = 43 \pm 9$ ppm. This meant that the probability for $\delta_N^{(2)}$ to be as small as predicted by direct calculation was only 1 in 20 000, an obvious inconsistency. In contrast, when the value of α derived from the Josephson-effect measurement of $2e/h$ [Eq. (3)] was used in place of the deteurium value, it was found that $\delta_N^{(2)} = 2.5 \pm 4.0$ ppm, in keeping with the theoretical calculations. Hence the Josephson-effect value of α removed a discrepancy that during the 1960s was termed one of the major unsolved problems of QED. This example further illustrates the overall unity of science—a low-temperature solid-state physics experiment has provided information about the excited states of the proton—as well as how precise measurements of the fundamental constants can illuminate apparent inconsistencies in our physical description of nature.

RELATIONSHIPS

Although not discussed above, there are actually a number of ways to obtain indirect values of α from $2e/h$ in addition to Eq. (2). For example, in Eq. (2), the proton gyromagnetic ratio may be replaced with

$$\gamma_p'(\text{low})_{\text{lab}} = \frac{F_{\text{lab}}K^2(\mu_p'/\mu_N)}{M_p} \tag{4}$$

where F_{lab} is the Faraday constant measured in terms of a particular laboratory's as-maintained ampere; $K \equiv A_{\text{lab}}/A$; and M_p is the atomic mass of the proton. Furthermore, the Faraday constant is related to the Avogadro constant and elementary charge by $F = N_A e$, and K may be obtained from direct force-balance measurements as well as from the equation

$$K^2 = \gamma_p'(\text{low})_{\text{lab}}/\gamma_p'(\text{high})_{\text{lab}} \tag{5}$$

where $\gamma_p'(\text{high})_{\text{lab}}$ is the proton gyromagnetic ratio as obtained by the so-called high (magnetic) field method.

It should now be clear that complex relationships can exist among groups of fundamental constants and conversion factors, and that a particular constant may be obtained either by direct measurement or indirectly by appropriately combining other directly measured constants. If the direct and indirect values have similar accuracy, both must be considered in order to arrive at a best value for that quantity. However, each of the various routes that can be followed to a particular constant, both direct and indirect, provides a somewhat different numerical value. The best way to handle this situation is by the mathematical technique known as least squares.

LEAST SQUARES

The least-squares method furnishes a coherent procedure for calculating best "compromise" values of the constants from all of the available data. It takes into account all possible routes and yields a single value for each constant by weighting the different routes according to their relative uncertainties. The weights themselves are obtained from the *a priori* uncertainties assigned to the individual direct measurements that constitute the original set of data.

Least-squares adjustments (or studies) of the fundamental physical constants provide a self-consistent set of best values for use by science and technology in any given epoch. More important, however, is that the critical data review that must accompany such studies forces the reviewer to reassign uncertainties to individual experiments on an equal basis, thus making possible the ready identification of discrepancies among different measurements and calculations. This identification, in turn, can stimulate new experimental and theoretical work.

Least-squares studies of the constants were pioneered in the late 1920s by R. T. Birge and continued by others, notably J. W. M. DuMond and E. R. Cohen. The two most recent reviews were those of B. N. Taylor, W. H. Parker, and D. N. Langenberg in 1969, based on their Josephson-effect determination of $2e/h$; and of Cohen and Taylor in 1973, carried out under the auspices of CODATA (Committee on Data for Science and Technology). Their recommended values were officially adopted for international use by CODATA at its eighth general assembly in September 1973 and are those currently in use; values of selected constants are given in Table I.

THE 1973 ADJUSTMENT

To carry out a least-squares adjustment, the data are usually divided into two groups: (1) the more precise data or *auxiliary constants*, which have uncertainties sufficiently small that they can be considered as exactly known—for example, the Rydberg constant, which now has an uncertainty of only 0.01 ppm; and (2) the less precise or *stochastic input data*, which in the 1973 adjustment of Cohen and Taylor had uncertainties larger than about 0.2 ppm. A subset of constants is then chosen in terms of which all of the stochastic input data can be individually expressed, if necessary, with the aid of the auxiliary constants. It is actually the constants constituting this subset that are directly subject to adjustment. In the 1973 study, these so-called adjustable constants were taken to be $\alpha, K, N_A, \bar{R}, \Lambda$, and μ_μ/μ_p. Here, $K \equiv A_{\text{BIPM}}/A$ is the ratio of the BIPM as-maintained ampere to the SI ampere; $\bar{R} \equiv \Omega_{\text{BIPM}}/\Omega$ is the ratio of the BIPM as-maintained ohm to the SI ohm; Λ is the ratio of an x-ray wavelength expressed in angstroms (10^{-10} m) to the same wavelength expressed in kilo-x-units; and μ_μ/μ_p is the ratio of the magnetic moment of the muon to that of the proton. With just these six quantities and certain auxiliary constants, a series of equations, generally known as *observational equations*, were formed for all 27 finally selected items of stochastic data. These 27 observational equations were then solved (with the aid of a computer) for the least-squares adjusted values of the six adjustable constants and their uncertainties. Optimal values in the least-squares sense for other constants not directly subject to adjustment were then calculated from the six adjustable constants. (But this does not apply to the auxiliary constants, since for the purpose of the least-squares adjustment they are assumed to be exactly known.)

The quantities taken as auxiliary constants for the 1973 adjustment were $2e/h$ obtained from the Josephson effect; the speed of light as determined by directly counting the frequency of a stabilized laser; various local values of the gravitational acceleration; the g factor of the electron and muon as obtained from measurements of their anomalous magnetic moments; several magnetic moment ratios—that of the electron to proton, μ_e/μ_p, as well as μ_p/μ_B and μ_p'/μ_B; various atomic masses; and the Rydberg constant is obtained from spectroscopic data. The uncertainties of these

Table I Recommended values of selected physical constants[a]

Quantity[b]	Symbol	Numerical value	Uncertainty (ppm)	Units
Speed of light in vacuum	c	299792458(1.2)	0.004	$m \cdot s^{-1}$
Fine-structure constant, $[\mu_0 c^2/4\pi](e^2/\hbar c)$	α	7.2973506(60)	0.82	10^{-3}
	α^{-1}	137.03604(11)	0.82	
Elementary charge	e	1.6021892(46)	2.9	10^{-19} C
Planck constant	h	6.626176(36)	5.4	10^{-34} J·s
Avogadro constant	N_A	6.022045(31)	5.1	10^{23} mol^{-1}
Electron rest mass	m_e	9.109534(47)	5.1	10^{-31} kg
Proton rest mass	m_p	1.6726485(86)	5.1	10^{-27} kg
Ratio of proton mass to electron mass	m_p/m_e	1836.15152(70)	0.38	
Faraday constant, $N_A e$	F	9.648456(27)	2.8	10^4 $C \cdot mol^{-1}$
Rydberg constant, $[\mu_0 c^2/4\pi]^2(m_e e^4/\pi\hbar^3 c)$	R_∞	1.097373177(83)	0.075	10^7 m^{-1}
Bohr radius, $\alpha/4\pi R_\infty$	a_0	5.2917706(44)	0.82	10^{-11} m
Free electron g factor, or electron magnetic moment in Bohr magnetons	$g_e/2=\mu_e/\mu_B$	1.0011596567(35)	0.0035	
Free muon g factor, or muon magnetic moment in units of $[c](e\hbar/2m_\mu c)$	$g_\mu/2$	1.00116616(31)	0.31	
Bohr magneton, $[c](e\hbar/2m_e c)$	μ_B	9.274078(36)	3.9	10^{-24} $J \cdot T^{-1}$
Proton gyromagnetic ratio	γ_p	2.6751987(75)	2.8	10^8 $s^{-1} \cdot T^{-1}$
Proton magnetic moment in Bohr magnetons	μ_p/μ_B	1.521032209(16)	0.011	10^{-3}
Ratio of electron and proton magnetic moments	μ_e/μ_p	658.2106880(66)	0.010	
Proton magnetic moment in nuclear magnetons	μ_p/μ_N	2.7928456(11)	0.38	
Nuclear magneton, $[c](e\hbar/2m_p c)$	μ_N	5.050824(20)	3.9	10^{-27} $J \cdot T^{-1}$
Ratio of muon and proton magnetic moments	μ_μ/μ_p	3.1833402(72)	2.3	
Ratio of muon mass to electron mass	m_μ/m_e	206.76865(47)	2.3	
Muon rest mass	m_μ	1.883566(11)	5.6	10^{-28} kg
Compton wavelength of the electron, $h/m_e c$	λ_C	2.4263089(40)	1.6	10^{-12} m
Gravitational constant	G	6.6720(4)	615	10^{-11} $m^3 \cdot s^{-2} \cdot kg^{-1}$
Molar volume of ideal gas at STP	V_m	22.41383(70)	31	10^{-3} $m^3 \cdot mol^{-1}$
Molar gas constant, $V_m p_0/T_0$ ($T_0 \equiv 273.15$ K; $p_0 \equiv 101325$ Pa $\equiv 1$ atm)	R	8.31441(26)	31	$J \cdot mol^{-1} \cdot K^{-1}$
		8.20568(26)	31	10^{-5} $m^3 \cdot atm \cdot mol^{-1} \cdot K^{-1}$
Boltzmann constant, R/N_A	k	1.380662(44)	32	10^{-23} $J \cdot K^{-1}$
Stefan–Boltzmann constant, $\pi^2 k^4/60\hbar^3 c^2$	σ	5.67032(71)	125	10^{-8} $W \cdot m^{-2} \cdot K^{-4}$

[a] From the 1973 least-squares adjustment of Cohen and Taylor.
[b] In cases where formulas for constants are given (e.g., R_∞), the relations are written as the product of two factors. The second factor, in parentheses, is the expression to be used when all quantities are expressed in cgs units, with the electron charge in electrostatic units. The first factor, in brackets, is to be included only if all quantities are expressed in SI units.

quantities were mostly in the range of 0.004–0.08 ppm.

The 31 items of stochastic data initially considered by Cohen and Taylor for inclusion in the 1973 adjustment were one value for $\bar{R}$ as determined using a calculable capacitor; three independent force-balance measurements of K; two determinations of F, using silver for one and benzoic and oxalic acids for the other; four determinations of γ_p'(low) and two of γ_p'(high); two determinations of μ_p'/μ_N using, respectively, an omegatron and a mass spectrometer; three measurements of Λ, and two of the product $N_A\Lambda^3$ (one based on Si and one on calcite, $CaCO_3$); two determinations of the electron Compton wavelength, $\lambda_C = h/m_e c$, from electron–positron annihilation; six values of α as obtained from measurements of the electron anomalous moment, the hydrogen hfs, and the fine structure in hydrogen; three determinations of μ_μ/μ_p—two based on muon precession and one from muonium; and one value of the hyperfine splitting in muonium as obtained by appropriately combining a number of separate measurements. The uncertainties assigned to these quantities ranged from 0.2 to 33 ppm.

The following is an example of a typical observational equation, that used for including in the 1973 adjustment the measured values of the proton gyromagnetic ratio as determined by the high-field method:

$$\alpha^2 K^{-2} \bar{R}^{-1} = \left[\frac{4R_\infty}{c(\mu_p'/\mu_B)(2e/h)_{\text{BIPM}}} \right] \cdot \{\gamma_p'(\text{high})_{\text{BIPM}}\}. \quad (6)$$

It should be noted that all of the terms in the square brackets are auxiliary constants and that the braces indicate numerical value only. Equation (7) for the elementary charge is an example of how a quantity not directly subject to adjustment, that is, one that is not an adjustable constant, is obtained from the six adjustable constants:

$$e = [(\mu_0 c/4)(2e/h)_{\text{BIPM}}]^{-1} \cdot \alpha K \bar{R}. \quad (7)$$

Again, all of the quantities in the brackets are auxiliary constants.

Critical analysis of the input data and deciding what uncertainty should be assigned each measurement is the main problem in successfully carrying out a least-squares adjustment. The weight, W_i, of each stochastic datum is related to the datum's *a priori* assigned uncertainty, σ_i, by $W_i = 1/\sigma_i^2$. Thus, if one value of a particular datum has half the uncertainty of another, it carries four times as much weight in the adjustment. The uncertainty problem is made especially difficult because in most experiments sufficient data are taken to reduce the random or statistical uncertainty to negligible amounts, and the final uncertainty assigned the experiment is determined primarily from estimates of the systematic uncertainties. These frequently arise from effects that the experimenter knows little about, and hence their estimation is often subjective and only derivable from educated guesses.

Another problem area is deciding how to handle "discrepant" data, that is, measurements that differ from each other by statistically significant amounts in comparison with their σ_i. Such data cannot be included in an adjustment uncritically because the inconsistencies imply either incorrect uncertainty assignments or the presence of unknown measurement errors.

When confronted with such a situation, the constants reviewer can in general either include the inconsistent data, but only after expanding (increasing) their σ_i so that they are no longer discrepant; or decide, on as sound a theoretical and experimental basis as possible, which of the inconsistent data are least reliable, and discard them. These two approaches clearly imply that different reviewers might treat the same data differently, thereby obtaining a somewhat different set of best values.

In their 1973 adjustment, Cohen and Taylor not only expanded the uncertainties of certain like items of stochastic data in order to obtain better internal consistency among them, but also eventually discarded 4 of the 31 items originally considered for inclusion in the adjustment because they were found to be highly inconsistent with the remainder of the data. In particular, one determination of λ_C, both measurements of F, and one value of α derived from a hydrogen fine-structure measurement were not used. Similarly, to improve the self-consistency of like measurements that were only marginally in agreement, they expanded the *a priori* assigned uncertainties of the four determinations of γ_p'(low) by the (multiplicative) factor 1.43; of the six remaining x-ray-related measurements (Λ, $N_A\Lambda^3$, λ_C) by the factor 1.28; and of the nine remaining QED-related measurements (α, μ_μ/μ_p, muonium hfs) by 1.40. In each case, the value of the statistic χ^2 determined the expansion factor.

As an example of the pitfalls of discarding data, it is instructive to compare the 1969 adjustment of Taylor *et al.* with the 1973 adjustment of Cohen and Taylor. The most critical problem in the 1969 adjustment was the internal inconsistency among the five values of μ_p'/μ_N then available, and the inconsistency between two of these values and the one available measurement of the Faraday constant. After much thought and analysis, Taylor and colleagues decided to discard the two "high" values of μ_p'/μ_N and to retain the remaining three and F.

The most difficult problem in the 1973 adjustment also had to do with μ_p'/μ_N and F. In the intervening four years, however, two new, very accurate (sub-ppm) μ_p'/μ_N determi-

Table II Comparison of recommended values of selected physical constants[a]

Quantity[b]	Value, 1973 adjustment, and ppm uncertainty	Value, 1969 adjustment, and ppm uncertainty	Change 1973 – 1969 (ppm)	Value, 1963 adjustment, and ppm uncertainty	Change 1973 – 1963 (ppm)
α^{-1}	137.03604(11) 0.82	137.03602(21) 1.5	+0.15	137.0388(6) 4.4	−20
e	1.6021892(46) 2.9	1.6021917(70) 4.4	−1.6	1.60210(2) 12	+56
h	6.626176(36) 5.4	6.626196(50) 7.6	−3.0	6.62559(16) 24	+88
m_e	9.109534(47) 5.1	9.109558(54) 6.0	−2.6	9.10908(13) 14	+50
N_A	6.022045(31) 5.0	6.022169(40) 6.6	−21	6.02252(9) 15	−79
μ_p/μ_N	2.7928456(11) 0.38	2.792782(17) 6.2	+23	2.79276(2) 7.2	+34
F	9.648456(27) 2.8	9.648670(54) 5.5	−22	9.64870(5) 5.2	−25

[a] From the 1973 least-squares adjustment of Cohen and Taylor; the 1969 adjustment of Taylor, Parker, and Langenberg; and the 1963 adjustment of Cohen and DuMond.

[b] The units for e are 10^{-19} C; for h, 10^{-34} J·s; for m_e, 10^{-31} kg; for N_A, 10^{23} mol^{-1}; and for F, 10^4 C·mol^{-1}.

nations were completed which showed that the high values of μ_p'/μ_N discarded in 1969 were more nearly correct than the three retained low values, and that it was probably F that was in error and should be the quantity discarded. This shift in outlook accounts for the large changes in the recommended values for certain constants as given in the 1969 and 1973 adjustments. These changes are readily apparent in Table II, where the values of selected constants resulting from the 1973 adjustment are compared with their counterparts from the 1969 adjustment. (For additional comparison purposes, values from the adjustment published by Cohen and DuMond in 1963 are also included. The changes that occurred between the 1963 and 1969 values were primarily due to the change in α resulting from the measurement of $2e/h$ using the Josephson effect.) The table clearly shows that our knowledge of the numerical values of the fundamental constants continually improves as new measurements become available; that the constants are so interrelated that a significant shift in the value of one will usually give rise to large shifts in the values of others; and that no set of recommended constants such as is given in Table I should be taken as final.

To further reinforce this last point, Fig. 1 shows the variation of the accepted value of the electron mass resulting from several different adjustments carried out since 1950. Clearly, significant changes from one adjustment to another are not uncommon.

RECENT DEVELOPMENTS

Several significant fundamental-constants experiments and theoretical calculations have been completed since the 1973 adjustment was carried out. Included in the more precise data or auxiliary-constants category are improved values for the g factors of both the electron and muon as determined from new measurements of their anomalous magnetic

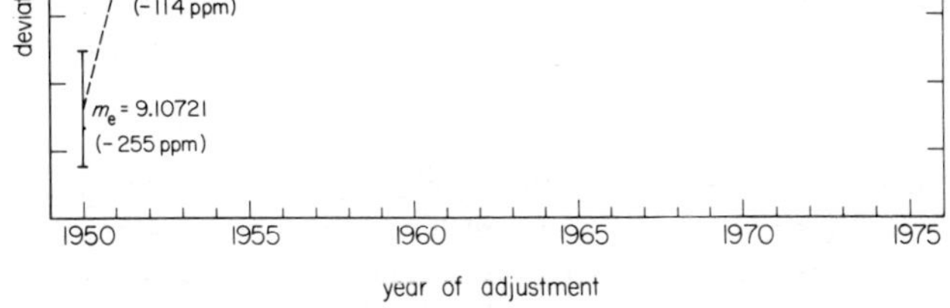

FIG. 1. Variation of the accepted value of the electron mass resulting from least-squares adjustments carried out in the years indicated. The quantity shown is the ppm difference between the various adjusted values of m_e and the 1973 value. (The deviations are the numbers in parentheses.) The actual resulting values of m_e (in units of 10^{-31} kg) are also given. The error bars are the plus and minus one-standard-deviation uncertainties for each value.

moments; the ratio $\bar{R} \equiv \Omega_{\mathrm{BIPM}}/\Omega$, again using a calculable capacitor; the ratio μ_p'/μ_B using a hydrogen maser; and R_∞ as obtained from Doppler-free laser saturation spectroscopy on hydrogen and deuterium. But in each case (as we would hope, at least for the auxiliary constants), the new values are in quite reasonable agreement with the old. Thus, although the uncertainties of these quantities have been reduced by between a half and one order of magnitude, the impact on the 1973 recommended values will not be extensive.

Included in the less precise or stochastic data category are two current-balance determinations of the ampere with uncertainties of 6 and 8 ppm, respectively; two Faraday determinations with uncertainties of 3 and 8.5 ppm using, respectively, silver and 4-aminopyridine; a 0.4-ppm measurement of the proton gyromagnetic ratio by the low-field method, and a 1.0-ppm determination by the high-field method; a 1-ppm measurement of N_A using a combined optical-x-ray interferometer; and a 20-ppm determination of the gas constant based on acoustic interferometric measurements of the velocity of sound in argon. In this same category, but based on experiments related to QED, may be included a 0.2-ppm value of α as obtained from the new measurement of a_e (see above); and a 0.9-ppm α value as obtained from an improved theoretical expression for a particular fine-structure splitting in atomic helium and an earlier experimental measurement of the same splitting.

The two new ampere determinations are highly consistent with the three previously available measurements, which were used in the 1973 adjustment, and are of comparable accuracy. Thus, they have little potential for causing significant changes in the 1973 recommended values. Similarly, although the new gas-constant determination has resulted in a value 159 ppm higher than that given in Table I, because R is uncoupled from most of the other constants, the effect of this large change (if supported by additional experiments), will be confined to thermodynamic quantities such as the Boltzmann and Stefan–Boltzmann constants.

The remaining new determinations may be expected to have a much broader impact because of their closer ties to a larger number of constants. For example, the 0.4-ppm value of γ_p'(low) yields a 0.2-ppm value of α via Eq. (2), in good agreement with the 1973 recommended value but four times more accurate. It also yields a 0.5-ppm value of $K_d \equiv A_{BIPM}/A$ via the new 1-ppm value of γ_p'(high) and Eq. (5), which is 7 ppm less than the 1973 recommended value but five times more accurate.

Similarly, the 1-ppm determination of N_A, which is 9 ppm higher than the 1973 recommended value and five times more accurate, has a great deal of potential for altering many of the 1973 best values, especially those for particle masses, because of its unprecedented accuracy—it literally makes all six of the x-ray-related experiments included in the 1973 adjustment obsolete.

There are two disturbing aspects of the new measurements that should be understood before the next least-squares adjustment is carried out, probably in 1981. First, the two new Faraday determinations are in agreement with, but are even more accurate than, the two Faraday measurements discarded in 1973. This clearly makes that decision suspect and implies that F could change by as much as 20–30 ppm.

Second, a number of the new determinations are highly inconsistent among themselves, even though of apparent high accuracy. If, for example, it is assumed that the new value of γ_p'(low) is reliable because of its low uncertainty and good agreement with the two new QED α values, and the two values of μ_p'/μ_N used in 1973 are reliable because they are in excellent agreement and have sub-ppm accuracy, then it may be concluded that the recently measured values for the following quantities must change by the amounts indicated if they are to be consistent:

N_A must increase by 3.4 ppm or 3.9 standard deviations (σ);
F must decrease 8.2 ppm or 2.9σ;
γ_p'(high) must decrease 3.6 ppm or 3.6σ; and
K must decrease 4.5 ppm or 1.5σ.

At present, there is no experimental evidence to indicate that these various determinations could be in error by such large amounts. If the situation remains unchanged, the next adjustment will be most difficult indeed!

TRENDS

The fundamental-constants field is ever evolving; new experiments are undertaken with each major new advance in measurement science. Especially great progress was made immediately after World War II as a direct result of the technological advances made during the war in the fields of microwaves and electronics. More recently, the laser has made possible significant improvements in the determination of a number of constants, especially R_∞, c, N_A, and γ_p'(low). It is perhaps useful to examine what new experiments are currently on the horizon and what measurement accuracies might be expected.

Speed of Light

The present value is limited by the current SI definition of the meter based on ^{86}Kr. A new definition based on an appropriate molecularly stabilized laser could enable c to be determined to an accuracy of a few parts in 10^{11} by direct frequency counting. However, it is likely that an adopted value of c (consistent with the presently accepted value) will be used to redefine the meter and that it will then never be necessary to determine c again.

Rydberg Constant

Once the meter has been redefined, it should be possible to determine R_∞ to a few parts in 10^{11} by various Doppler-free laser spectroscopy techniques, perhaps using atoms in high Rydberg states.

Fine-Structure Constant

Improvements in the low-temperature single-electron storage technique for determining the electron anomalous magnetic moment, and the exact calculation of the coefficient of the sixth-order term $[(\alpha/\pi)^3]$ in the theoretical QED expression for a_e, should yield a value of α accurate to a few parts in 10^8. A second value of equal accuracy should result from a new parts-in-10^8 measurement of γ_p'(low) and Eq. (2). A somewhat less accurate result may also be obtained from the electron Compton wavelength, λ_C, as determined by measuring electron–positron anihilation radiation and using the relation $\alpha^2 = 2R_\infty\lambda_C$.

Ampere

The present accuracy of so-called absolute-ampere realizations using current-force balances, about 5–10 ppm, is not significantly different from the accuracy achieved when experiments of this type were first carried out at the turn of the century. However, the application of modern technology—lasers, superconductivity, cryogenics, and automation—should lead to experiments with between one and two orders of magnitude improvement in accuracy.

Gravitational Constant

At present, there exists no verified theoretical equation that relates G to any other fundamental constant. Furthermore, because the gravitational force between laboratory-size objects is so weak, direct experimental determinations of G are extremely difficult. It is, therefore, our least well-known constant, with an uncertainty of 615 ppm (see Table I; this result is based on work carried out from 1925 to 1940). However, several new experiments are now under way that have the potential for one to two orders of magnitude improvement in accuracy. The novel accelerating-table, torsion-pendulum approach pioneered by J. W. Beams and colleagues at the University of Virginia, and significantly improved by workers at NBS, is now nearing completion at the latter institution. An accuracy of 10 ppm is believed feasible. Modern technology, including a precision air bearing, digital electronics, and automation, has contributed significantly to the success of the experiment. A preliminary result with an accuracy of 210 ppm was reported in 1975.

Ratio of Proton Mass to Electron Mass

There are at least two approaches under consideration that have the potential for achieving an accuracy of a few parts in 10^8 for m_p/m_e. One is to attempt to apply the low-temperature single-electron storage technique used to determine a_e to the proton and other ions. The second is based on ultraprecise (parts in 10^{11}) Doppler-free laser spectroscopy on two different isotopes, such as hydrogen and deuterium.

Theory

In very few instances is the determination of an improved value of a fundamental constant limited by theory—usually the opposite situation obtains. The most outstanding counterexample is the hydrogen hfs. As noted previously, it is known to an accuracy of 1 in 10^{12}. In contrast, the theoretical expression for the hydrogen hfs is nearly seven orders of magnitude less well known. Thus, the potential exists for obtaining a value of α accurate to parts in 10^9 or 10^{10} (or better), provided the necessary advances in theory can be made.

The research program outlined here should take until the late 1980s to complete.

"NUMEROLOGY" AND TIME VARIATIONS

No article on the fundamental physical constants would be complete without at least a brief discussion of why the constants have the values they do, and whether they are in fact really constant. For example, why does the ratio of the rest mass of the proton to that of the electron happen to be 1836.15...? Why does α^{-1} happen to equal 137.036...? Why not other, perhaps simpler, values? Three rather different points of view have evolved about the constants, especially their dimensionless ratios.

One viewpoint is that the numbers are not at all arbitrary but can be calculated from some as yet unknown basic theory (or theories) in much the same way that the anomalous magnetic moments of the electron and muon can be calculated from QED. For example, in 1969 Wyler derived an expression for α involving only simple integers and π based on the volumes of certain bounded spaces associated with the invariance group of a relativistic quantum wave equation, and giving a value for α in excellent agreement with the current best value. However, the physical basis for the derivation is not clear and it is safe to say that there does not yet exist a physically meaningful "derivation" of an accurate value for any fundamental constant.

The second viewpoint is not aimed at explaining the values of the physical constants but simply notes that if they were terribly different from their currently observed values, we would not be here to measure them—life as we know it on earth, and perhaps the existence of the universe itself, depends on complex physical processes that require the constants to have their present values.

An extreme statement of the third point of view is that the fundamental constants change with time as the universe evolves and therefore the currently observed values are not particularly significant. This idea had its origins in 1937 when Dirac noted that the ratio of the electric to gravitational forces between the electron and proton is of the order 10^{40}, very nearly equal to the age of the universe expressed in an appropriate atomic time unit, for example, the time required for light to travel a distance equal to the classical electron radius. He speculated that either the equality is accidental or it indicates a causal relationship between electromagnetism, gravitation, and cosmology. If such a relationship is in fact true, then since the universe is continually aging, the equality would remain unchanged only if one or more of the fundamental constants entering the two ratios changed, for example, G or e.

These speculations have stimulated a considerable amount of theoretical and experimental work over the last 40 years, the latter mainly involving astronomical observations. As a result, stringent limits have been placed on the possible time dependence of a number of physical constants. For example, it has been shown that h changes less than 4 parts in 10^{13} per year; that G changes less than 1 part in 10^{11} per year; and that α changes less than 1 part in 10^{12} per year. Clearly, for a long time to come, the changes in the values of the fundamental constants caused by man's ineptitude (see Table II and Fig. 1) will dominate any changes that might be caused by Nature's laws!

CONCLUSION

The following quotation by a famous physicist of the early decades of this century seems to be a quite fitting conclusion to this article:

> Why should one wish to make measurements with ever increasing precision?
> Because the whole history of physics proves that a new discovery is quite likely to be found lurking in the next decimal place.

F. K. Richtmyer, 1931

BIBLIOGRAPHY

E. R. Cohen and B. N. Taylor, "The 1973 Least-Squares Adjustment of the Fundamental Constants," *J. Phys. Chem. Ref. Data* **2,** 663 (1973). (I–A)

D. N. Langenberg and B. N. Taylor, eds., *Precision Measurement and Fundamental Constants.* National Bureau of Standards Special Publication 343, U.S. Govt. Printing Office, Washington, DC, 1971. (I–A)

Metrology and Fundamental Constants, lectures presented at the International School of Physics "Enrico Fermi," Summer Course LXVIII (12–24 July 1976, Varenna Italy), North-Holland, Amsterdam, 1980. (E–I)

F. D. Rossini, *Fundamental Measures and Constants for Science and Technology.* CRC Press, Cleveland, OH, 1974. (I)

J. H. Sanders and A. H. Wapstra, eds., *Atomic Masses and Fundamental Constants*, Vols. 4 and 5. Plenum, New York, 1972 and 1976. (I–A)

B. N. Taylor, W. H. Parker, and D. N. Langenberg, *The Fundamental Constants and Quantum Electrodynamics.* Academic Press, New York, 1969. Also published in *Rev. Mod. Phys.* **41,** 375 (1969). (I–A)

Critical Points

J. D. Litster

At most commonly observed phase transitions, such as the melting of a solid or boiling of a liquid, the properties of matter change discontinuously; such phase transitions are called first-order transitions and are not yet fundamentally understood. However, if the pressure on a fluid, such as water, is increased, the boiling temperature will also increase. At the same time the density difference between the liquid and vapor phases becomes smaller, and vanishes continuously as the liquid–gas critical point (at temperature T_c and pressure P_c) is approached. This is a different kind of phase transition, commonly called second order, in that a material property (here, the density difference between the liquid and vapor phases) changes continuously; this property is commonly referred to as an order parameter. Near the critical point anomalous behavior (divergence) is observed in a number of thermodynamic derivatives, such as the compressibility and the specific heat. A variety of superficially unrelated second-order phase transitions have been found to be quite analogous to that of the pure fluid near its critical point. With the order parameter given in parentheses, these include ferromagnets (magnetization), antiferromagnets (sublattice magnetization), certain separating binary mixtures (concentration of components in the two phases), ferroelectrics (spontaneous polarization), and superconductors and superfluids (condensate wave function), to name some of the more extensively studied ones.

The classical models for these phase transitions all predicted the same behavior in the vicinity of the critical point as was given by Landau's 1937 theory [1] of phase transitions. The anomalous quantities diverged according to power laws (e.g., the compressibility diverged as $|T-T_c|^{-1}$) and the specific heat was supposed not to diverge, but rather to have a discontinuity. Some materials, such as ferroelectrics and superfluids, were found to follow the classical theory closely, but most showed power-law divergences with different exponents than predicted and also specific heats that diverged approximately logarithmically. A great deal of experimental work [2] showed that a wide variety of second-order phase transitions had almost identical behavior, contrary to the classical expectations. The theoretical situation was summarized by Fisher [3]. It was known that the Landau approach failed because it failed to consider properly the divergent order-parameter fluctuations near the critical point, but none of a wide variety of theoretical methods [4] was particularly satisfying. The classical theory assumed the free-energy surface in the critical region to be an analytic function of the state variables; relaxing this analyticity requirement at the critical point led to predicted scaling laws [5]—relationships among the various critical exponents and a universal form for the equation of state expressed in reduced variables. These predictions were verified experimentally and an explanation for the universal behavior proposed by Kadanoff [6]. He suggested that critical behavior was determined by the correlation length for fluctuations in the order parameter; when this exceeded the range of the forces responsible for the phase transition, the material properties were insensitive to the detailed nature of these forces and determined only by geometrical considerations, such as the symmetry group of the order parameter and the dimensionality of the space.

A breakthrough was Wilson's use of the renormalization-group approach to express these ideas quantitatively and calculate critical exponents close to those observed experimentally. As a result, we have now a quantitative understanding of the role of critical fluctuations and an appreciation of the importance of the effects of spatial dimensionality and symmetry of the order parameter on critical phenomena.

Current research involves theoretical calculation and experimental checks of various predictions from the renormalization-group approach, including studies of higher-order critical points and systems of reduced dimensionality. There is much interest in dynamical critical behavior, which is not yet so well understood as static properties. Many physicists are also hopeful that the theoretical and experimental approaches that have proved so fruitful in attacking critical phenomena may be helpful in elucidating other problems in condensed-matter physics and perhaps even the as yet intractable one of turbulence.

See also FERROELECTRICITY; FERROMAGNETISM; FLUCTUATION PHENOMENA; PHASE TRANSITIONS; SUPERCONDUCTIVITY THEORY.

REFERENCES

1. L. D. Landau, "On the Theory of Phase Transitions" (English translation), in *The Collected Papers of L. D. Landau* (D. Ter Haar, ed.). Gordon & Breach, New York, 1965. (E)
2. For a review of the situation in 1967, see P. Heller, *Rep. Prog. Phys.* **30,** 731 (1967). (I) A more recent review of magnetic systems is by L. J. de Jongh and A. R. Miedema, *Adv. Phys.* **23,** 1 (1974). (I)
3. M. E. Fisher, *Rep. Prog. Phys.* **30,** 615 (1967). (A)
4. See the extensive discussions of various theoretical approaches in the multivolume series *Phase Transitions and Critical Phenomena* (C. Domb and M. S. Green, eds.). Academic, New York, 1972, 1974. (A)
5. B. Widom, *J. Chem. Phys.* **43,** 3898 (1965); R. B. Griffiths, *Phys. Rev.* **158,** 176 (1967). (I)
6. L. P. Kadanoff *et al., Rev. Mod. Phys.* **39,** 395 (1967). (E)
7. K. G. Wilson, *Phys. Rev.* **B4,** 3174, 3184 (1971). (A) M. E. Fisher, *Rev. Mod. Phys.* **46,** 597 (1974). (I)

Cryogenics

A. C. Anderson

The field of cryogenics involves the production and application of low temperatures and is generally associated with the use of various liquefied gases. It is a broad field. In the industrial sector cryogenics includes the preparation and storage of food and biological materials, the production and delivery of liquid oxygen to medical and metallurgical facilities, the transport and storage of liquefied natural gas, and the liquefaction of hydrogen for the space program. One current area of developmental work is directed toward the application of low-temperature techniques to the generation, transmission, storage, and utilization of electrical energy. The metals employed have small or negligible electrical resistance and hence provide a considerable reduction in power loss.

In experimental physics, cryogenic techniques are employed either because a phenomenon of interest occurs only at low temperature, or because a physical property being studied becomes less complex and more amenable to theoretical interpretation. Phenomena that occur only at low temperatures include superconductivity, the superfluid phases of liquid ^{3}He and liquid ^{4}He, and the effects of the motion of atoms and molecules that persists after thermal motion has been suppressed through a reduction in temperature. Since these phenomena can be understood only in terms of quantum mechanics, cryogenics provides the opportunity for a rigorous test of the modern theory of quantum mechanics applied to condensed matter.

A well-documented example of the use of cryogenics to isolate and simplify physical properties, and hence facilitate theoretical interpretation, has been the study of nonmetallic crystals. Thermal expansion, thermal conductivity, and specific heat can be described in terms of collective atomic vibrations or lattice waves called phonons. At low temperatures those phonons that are of greatest importance have the velocity of a sound wave and a wavelength large compared to the atomic spacing in the solid. The phonon behavior is then rather independent of the detailed atomic arrangement and bonding, and the phonon may be treated as a classical sound wave. This permits the development of a simple theory that still contains quantization, and that may be tested quantitatively against experimental data. Historically this procedure has been of great importance in the development of our knowledge of the physics of solids. Cryogenics continues in this manner to provide the physicist with a highly effective experimental tool.

In measuring low temperatures, the absolute or Kelvin scale of temperature is generally used. On this scale the freezing and boiling points of water are approximately 273 and 373 K, respectively. To appreciate the Kelvin scale and in fact to understand the importance of cryogenics to the physicist, it is necessary to recognize that it is often the ratio of two temperatures, not the difference, that controls a physical or chemical process. Therefore the temperature interval 0.001–0.01 K may contain as much interesting physics as the much "larger" interval from 100 to 1000 K.

The calibration of a low-temperature thermometer is based on a variety of phase transitions, such as the temperature of the triple point of hydrogen or the equilibrium vapor pressure of liquid helium. The temperatures of the phase transitions have been estimated through much exacting experimental work; the accuracy of the determinations is constantly being improved. A thermometer generally utilizes an electrical or magnetic property as the thermometric parameter. Examples are the electrical resistivity of metals and semiconductors or the weak magnetic susceptibility of certain salts and metals. Each type of thermometer has been carefully studied to be certain that impurities, strain, and other variables do not influence the calibration. A variety of thermometers, some precalibrated, are available commercially for use down to 0.01 K.

Providing a research laboratory with appropriate refrigeration facilities has been greatly simplified in recent years. Liquid nitrogen at 77 K and liquid helium at 4 K are available from commercial sources or may be liquefied on site using commercial liquefiers. Gaseous ^{4}He is extracted from natural gas wells in several countries and therefore is sufficiently abundant, at present, that many laboratories vent the evaporated ^{4}He to the atmosphere. Temperatures between 4 K and 300 K are obtained by partial thermal isolation of the experimental samples from the liquid cryogen. Temperatures below 4 K are produced with liquid ^{4}He by rapidly pumping away the evaporating gas. Since helium, unlike other liquids, does not solidify during this process, a liquid bath at a temperature of about 1 K can be produced. Below 2 K, liquid ^{4}He is a superfluid having an extremely large effective thermal conductivity. As a result, the helium bath has the desirable property of being nearly isothermal. On the other hand, a superfluid film creeps up the walls of the cryostat, a process that is often the most important contribution to the liquid loss rate.

The lighter isotope of helium, ^{3}He, has become available as a by-product in the manufacture of nuclear devices. As a liquid, ^{3}He has a higher vapor pressure than ^{4}He and is not a superfluid above 0.003 K. The liquid may be pumped to 0.25 K. The price of ^{3}He dictates that it be used in a closed system. The evaporated gas is either collected for reuse or returned to the cryostat to form a closed-cycle, continuously operating refrigerator. Pumped ^{4}He at 1–2 K is required to absorb the heat as the ^{3}He gas is reliquefied.

A more important closed ^{3}He cycle has recently been developed in which the liquid ^{3}He is dissolved in a small stationary bath of liquid ^{4}He. This process absorbs heat and therefore provides refrigeration. Commercial ^{3}He–^{4}He dilution refrigerators are available that provide temperatures down to 0.01 K continuously. Dilution refrigerators are simple to operate and may be constructed in a variety of configurations to match the experimental problem.

Refrigeration may be provided for temperatures below 0.01 K but not, at present, in a continuous fashion. The ^{3}He–^{4}He dilution process may be used to roughly 0.002 K by not returning the ^{3}He gas to the cryostat. Alternatively, pure ^{3}He can be solidified by the application of a pressure of about 2.9×10^6 Pa (29 atm). Since the freezing characteristics of pure ^{3}He below 0.3 K are such that it cools with the application of pressure, compression and solidification of liquid ^{3}He can provide temperatures to about 0.002 K. Still lower temperatures can be obtained by the use of slightly paramagnetic materials. Application of a large magnetic field causes such a material to warm. The evolved heat may be absorbed by a dilution refrigerator. Removal of the magnetic field then causes the paramagnetic material to cool below the temperature of the refrigerator. The successful application of this technique of adiabatic demagnetization to provide temperatures below 0.001 K is still in its infancy.

It is to be emphasized that the temperatures mentioned in the foregoing refer to the sample or device that is to be cooled. The temperature of the refrigerator or refrigerating material may be much lower, since the transport of heat to the refrigerator becomes an increasingly severe problem with decreasing temperature. It is therefore important to minimize any heat influx to the sample. This is done by providing several stages of stepwise reduction in temperature between 77 and 0.02 K to reduce heat transport by thermal conduction. In addition, and especially at temperatures below 1 K, heat influx from vibration and radiation must also be eliminated. For radiation this includes the electromagnetic spectrum from light waves to radio waves, and many low-temperature laboratories are therefore constructed inside copper-shielded rooms.

See also HEAT CAPACITY; MAGNETIC COOLING; PARAMAGNETISM; THERMAL EXPANSION.

BIBLIOGRAPHY

Bulletin of the International Institute of Refrigeration. (A)

E. M. Codlin, *Cryogenics and Refrigeration* (A Bibliographical Guide). MacDonald, London, 1968; Part 2, IFI/Plenum, New York, 1970.

F. E. Hoare, L. C. Jackson, and N. Kurti, *Experimental Cryophysics.* Butterworth, London, 1961. (A)

O. V. Lounasmaa, *Experimental Principles and Methods Below 1 K.* Academic, New York, 1974. (A)

K. Mendelssohn, *The Quest for Absolute Zero*. McGraw-Hill, New York, 1966. (E)
H. M. Rosenberg, *Low Temperature Solid State Physics*. Oxford, London and New York, 1963. (A)
M. W. Zemansky, *Temperature Very Low and Very High*. Van Nostrand, Princeton, NJ, 1964. (E)

Crystal and Ligand Fields

John P. Fackler, Jr.

In the late 1920s and early 1930s it was recognized that transition-metal-ion spectra and magnetic properties in the condensed state did not conform to predictions made for the gas phase. Dirac and Heisenberg theory could not account for the magnetic moments of metal ions in crystal lattices, nor for the electronic transitions which produce the beautifully varied colors of transition-metal salts such as $NiCl_2 \cdot 6H_2O$ or $Co(ClO_4)_2 \cdot 6H_2O$.

Hans Bethe, a young physicist familiar with Eugene Wigner's application of the principles of group theory to the description of the vibrational behavior of molecules, changed this situation by applying Wigner's procedures to the "term splitting" of ions. He assumed that anions such as F^-, Cl^-, SO_4^{2-}, etc., could be represented by point negative charges which induce a lowering of the symmetry of the electrostatic field surrounding the metal ion. The reduction in the effective symmetry from spherical to cubic (octahedral or tetrahedral) produces a splitting of ion terms. Thus free-ion terms such as ^1D or ^3F no longer had meaning for the ion in the crystal lattice.

Magnetic moments of metal ions in crystal lattices fail to follow the spin-orbit coupling model developed by Van Vleck, based on the earlier work of Hund. In this model, the effective magnetic moment μ_{eff} should depend on the total orbital (L) and spin (S) angular momenta of the ion in question:

$$\mu_{eff} = [L(L+1) + 4S(S+1)]^{1/2}. \tag{1}$$

A nickel(II) ion with its ^3F ground term is calculated by (1) to display a μ_{eff} of 4.5. Instead, typical values are ~3.2. Bethe solved this problem when he recognized that the ^3F ground term in spherical symmetry leads to the three terms $^3A_2$, $^3T_2$, and $^3T_1$ in cubic symmetry. He was able to show that in an octahedral geometry the $^3A_{2g}$ term, which has no orbital angular momentum, must be the ground term. Thus with L effectively reduced (quenched) in the Van Vleck relationship because of the influence of the crystalline field of octahedral symmetry, the μ_{eff} approaches the "spin-only" value of 2.82.

Van Vleck and co-workers developed the experimental evidence needed to establish the Wigner–Bethe formalism now called crystal-field theory. Although this theory gained wide acceptance among physicists concerned with the optical spectra and magnetic properties of ions, chemists, guided particularly by L. Pauling, did not adopt it until the mid-1950s. The idea that a cyanide ion ever could produce an effective crystal field substantially larger than the most electronegative ion, fluoride, and give an observed splitting of the Fe^{3+} ground term in $Fe(CN)_6^{3-}$ which is larger than in FeF_6^{3-} clearly was preposterous. Certainly the valence-bond sharing of electrons between the metal and ligands more effectively described the chemistry of these and related molecular species.

Furthermore, in the mid-1930s R. S. Mullikan introduced molecular-orbital techniques to describe the bonding properties of coordination compounds, a theory that does adequately deal with the question of chemical bonding.

Unfortunately, however, none of these techniques readily lead to quantitative interpretation. In a largely ignored (by

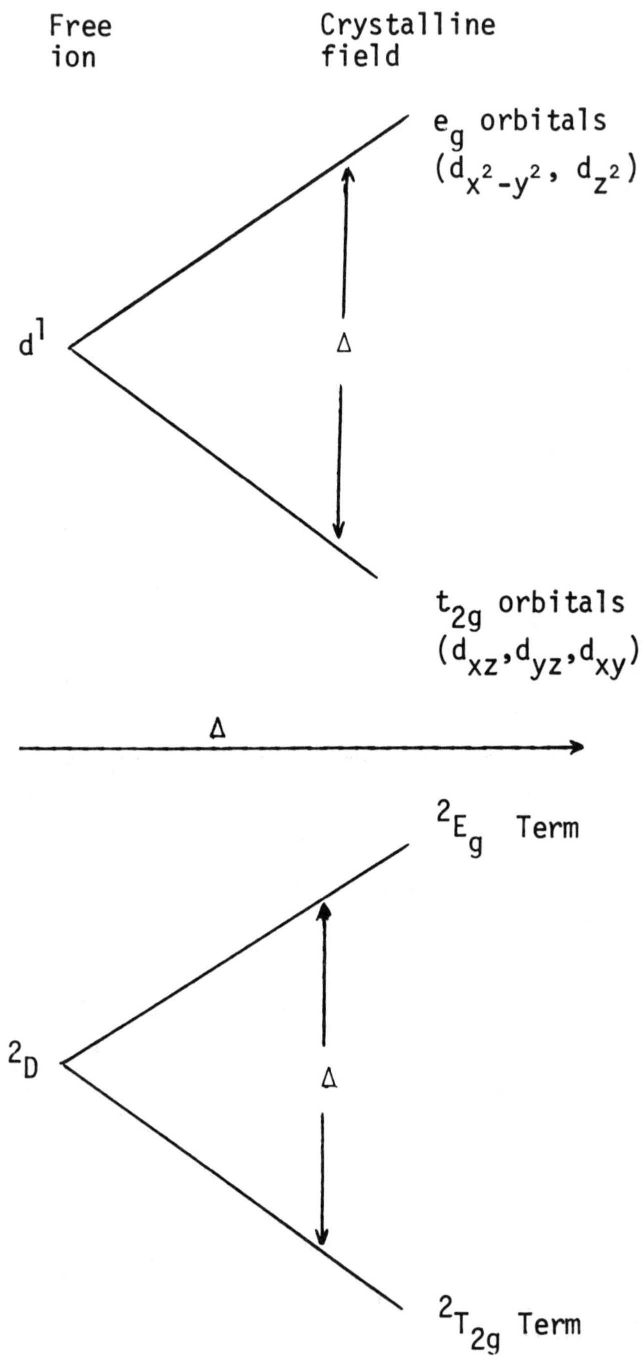

FIG. 1. Orbital splitting diagram for a d^1 configuration in an octahedral O_h field.

chemists) paper in 1935, Van Vleck showed that the various theories lead to the same qualitative conclusion provided that the parameters employed are treated experimentally and are not assumed to be fundamentally calculable constants. By far the easiest of these theories to use is the modified crystal-field theory.

Group theory establishes that the ^2D, ^3F, ^4F, ^5D, and ^6S ground terms found for transition-metal ions become associated with $D \rightarrow E_g + T_{2g}$, $F \rightarrow A_{2g} + T_{1g} + T_{2g}$, and $S \rightarrow A_{1g}$ terms in octahedral symmetry, maintaining the spin multiplicity. The five d orbitals, degenerate in spherical symmetry, split into two new sets of orbitals, the t_{2g} (triplet) and e_g (doublet) orbitals, separated by an energy defined to be Δ_{oct} or $10Dq$ (Fig. 1). A single electron in these orbitals (^2D ground term) produces ^2E_g and $^2T_{2g}$ terms, also separated by $10Dq$. Crystal-field theory suggests that the orbitals directed toward the ligands (the e_g set in octahedral symmetry) are destabilized relative to the orbitals pointing in between the ligands (the t_{2g} set in octahedral symmetry). Hence for a d^1 ion such as Ti^{3+} surrounded by six ligand atoms producing an effective octahedral geometry, the ground term $^2T_{2g}$ and excited state ^2E_g term are $10Dq$ apart. This latter parameter generally has a value of from 10,000 to 30,000 cm^{-1}. The absorbance of visible radiation is associated with transitions such as the $^2T_{2g} \rightarrow {}^2E_g$ transition found for $Ti(H_2O)_6{}^{3+}$.

For a d^2 ion the situation is somewhat more complex than for a d^1 ion in an octahedral field (Fig. 2). Group theory leads to a splitting of the ^3F ground term as indicated. The "strong-field" configuration establishes the ordering of the terms in this extreme. Between this and the free ion is the weak-field situation which merges smoothly into the strong-field configuration with increasing $10Dq$. Hence for a d^2 ion the $^3T_{1g}$ ground term prevails under all values of $10Dq$. Electronic transitions of the $d \rightarrow d$ variety occur between this term and the spin-triplet excited-state terms $^3T_{2g}$, $^3T_{1g}(P)$, and $^3A_{2g}$.

Unfortunately, attempts to calculate $10Dq$ by crystal-field approximations fail. Furthermore, the strength of the absorbances can be estimated only by introducing new relationships and parameters which take into account the sharing of electrons between the metal ion and its ligands. The ligand-field theory, commonly used today to describe the interactions between the metal ion and its ligands, results from these modifications. B. N. Figgis has stated, "The term ligand field theory has been employed to cover all aspects of the manner in which an ion or atom is influenced by its nearest neighbors."

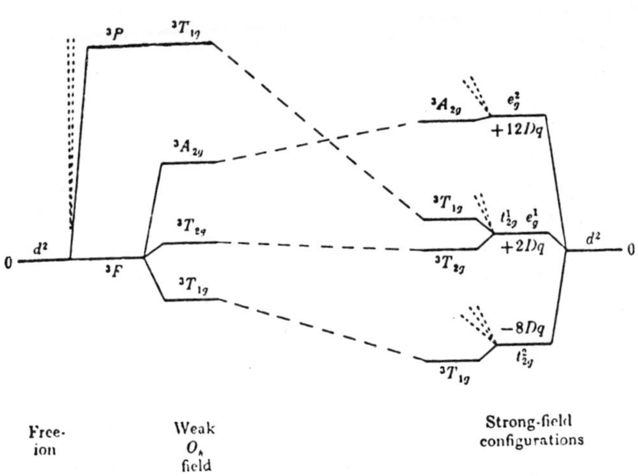

FIG. 2. Partial correlation diagram (spin-triplet terms only) for d^2 configuration in an octahedral O_h field.

Ligand-field theory leads to an understanding of the relative size of $10Dq$ for various ligands and metal ions. Furthermore, it leads to an adjustment in the separation between free-ion terms as a result of the nephelauxetic (or cloud-expanding) ability of certain types of ligands for a given metal ion. Thus with relatively few parameters readily associated with chemical behavior, electronic properties of transition-metal compounds can be understood. Furthermore, symmetry plays a dominant role in our description of these properties.

See also MAGNETIC MOMENTS.

BIBLIOGRAPHY

C. J. Ballhausen, *Introduction to Ligand Field Theory*. McGraw-Hill, New York, 1962. (I)

F. A. Cotton, *Chemical Applications of Group Theory*. Wiley, New York, 1974. (I)

J. P. Fackler, *Symmetry in Coordination Chemistry*. Academic Press, New York, 1971. (E)

J. P. Fackler (ed.), *Symmetry in Chemical Theory*. Dowden, Hutchinson and Ross, Allentown, Pa., 1972. (E)

B. N. Figgis, *Introduction to Ligand Fields*. Wiley, New York, 1966. (I)

M. Gerloch and R. C. Slade, *Ligand-Field Parameters*. Cambridge University Press, London, 1973. (A)

Crystal Binding

J. C. Phillips

Solids are often classified according to the nature of their binding. Four broad types of binding are covalent, ionic, metallic, and molecular.

The simplest of these is ionic binding, which arises when there is substantial charge transfer from one or more cations to one or more anions. As an example, consider NaCl (table salt). In the crystal the electronic configuration is Na^+Cl^-. The electrostatic interaction energy of point charges centered on the ions is called the Madelung energy; in NaCl this energy is about −9 eV per atom pair. The cohesive energy of NaCl is −8 eV per atom pair; about + 1 eV per atom pair arises from closed-shell repulsion (exclusion principle). Many naturally occurring minerals have ionic binding with complex crystal structures, e. g., mica $[KAl_3Si_3O_{10}(OH)_2]$. Ions are rather incompressible, so that ionic radii empirically determined from interatomic distances in simple ionic crystals can be used to rationalize complex ionic structures.

Metallic binding ideally occurs when the valence electrons of metallic (cationic) atoms such as Na or Al are weakly bound to the ion cores. In the solid these electrons then form a nearly free electron gas, with a cohesive energy of about 1 eV per valence electron. In this case there is no classical electrostatic interaction (as in ionic crystals) that can account for the observed cohesion. Instead, quantum-mechanical exchange forces (between electrons of parallel spin) or dynamical correlation (avoidance) between electrons of antiparallel spin takes place; these interaction energies are large in metals just because the electrons are spread out away from the ionic centers.

Three kinds of metals should be distinguished: "normal" metals, with *s* and *p* valence electrons only; transition metals (with *d* electrons); and rare-earth metals (unfilled *f* shell). Only the *s* and *p* valence electrons can become free and contribute fully to cohesion, but the *d* electrons contribute substantially and the *f* contribution in the rare earths is small. Because there are in some elements more *d* electrons than *s* and *p* electrons in normal metals, the elements with the largest cohesive energies are those in the middle of the transition series. The 3*d*, 4*d*, and 5*d* electrons contribute with increasing effectiveness to cohesion. Tungsten, in the middle of the 5*d* series, has the greatest cohesion of any element (9 eV per atom).

Covalent bonding describes sharing of valence electrons in pairs, as in classical chemical theories. Atoms with $N=4$ to $N=7$ s–p valence electrons often form covalent bonds, and in the elemental crystals the coordination configurations are such that each atom tends to have $8-N$ nearest neighbors. The classical covalent crystals are those with $N=4$ (tetrahedral bonding with sp^3 hybrids, in chemical terms), including diamond and Si. With the exception of elements from the first period, covalent binding energies are about $(8-N)$ eV per atom.

Molecular binding arises because of van der Waals interactions between, e. g., benzene molecules in solid benzene. It is weak and is typically of the order of 0.1 eV per molecule or less.

In most crystals the binding is best described as a mixture of ionic, covalent, and metallic interactions. We can distinguish between ionic and metallic binding on the one hand and covalent binding on the other because covalent crystals (such as grey Sn, with the diamond structure) are about 20% less dense than their metallic morphotropes (white Sn, in this example). When there is more than one kind of atom in the crystal there will always be some ionic contribution to the binding. For example, there are many tetrahedrally coordinated compounds with eight valence electrons per atom pair, such as III–V and II–VI compounds; here the bonding is predominantly covalent (tetrahedral coordination) but it contains an ionic component, which chemists associate with the heat of formation of the compound from the elements. An example of mixed metallic–ionic bonding is provided by transition-metal carbides (e. g., WC), which contain solids with the highest known melting points.

See also METALS; RARE EARTHS; TRANSITION ELEMENTS.

BIBLIOGRAPHY

A. Navrotsky and J. C. Phillips, "Ionicity and Phase Transitions," *Phys. Rev.* **11B**, 1583 (1975). (A)

J. C. Phillips, "Covalent-Ionic and Covalent-Metallic Transitions," *Phys. Rev. Lett.* **27**, 1196 (1971) (A)

J. C. Phillips, "Chemical Bonds in Solids," in *Treatise on Solid State Chemistry* (N. B. Hannay, ed.), Plenum, New York, 1974. (I)

E. P. Wigner and F. Seitz, "Qualitative Analysis of the Cohesion in Metals," *Solid State Phys.* **1**, 97 (1955). (E)

Crystal Defects

George D. Watkins

A crystalline solid with no defects would consist of a perfectly periodic three-dimensional array of unit cells, each containing the identical arrangement of atoms. Such an idealized crystal, however, cannot exist. In a real crystal, there are always structural imperfections.

If a perfect crystal could exist, it would be very strong mechanically. Aside from this desirable feature, however, it might be rather dull. Indeed, many of the important properties of crystalline solids—electrical, optical, mechanical—are determined not so much by the properties of the perfect crystal, but rather by its imperfections: The semiconductor industry is based on the ability to dissolve trace amounts of chemical impurities in silicon or germanium to control the conductivity level and type. The optical absorption and luminescence of solids is controlled primarily by impurities, and other simple lattice imperfections. The ductility and strength of structural materials are determined by imperfections and, as a result, can be modified and controlled by the addition of impurities, heat treatment, cold work, etc.

In many respects, therefore, it is often instructive to consider the host crystal as simply the matrix (or solvent) for the defects, which, in turn, play the dominant role in determining the physical properties of the material. The basic structural defects can be classified as follows:

1. Point defects. These include *lattice vacancies* (lattice sites which are missing an atom), *interstitials* (extra atoms not on regular lattice sites), and *foreign atoms* (chemical impurities).
2. Line defects, such as *dislocations*.
3. Surface defects, such as *grain boundaries* and *stacking faults*.

LATTICE VACANCIES AND INTERSTITIALS

In any real crystal, there must always be a finite number of lattice vacancies and interstitials (see Fig. 1). This can be seen simply as follows: The external surface of a crystal provides an inexhaustible source of vacancies (the empty space beyond) and interstitial atoms (the last atomic layer.) The equilibrium vacancy and interstitial concentrations in the interior of the crystal therefore represent the finite solubility of each of these point defects in the lattice matrix (solvent) as they dissolve in from the surface. For a monatomic metal, for instance, simple thermodynamic arguments lead to the expression for the equilibrium solubility concentration of each species,

$$c_i \sim \exp(-E_i/kT), \qquad (1)$$

where E_i is the energy required to transfer each defect into the crystal (its formation energy). At any finite temperature T, there is therefore always a finite stable concentration,

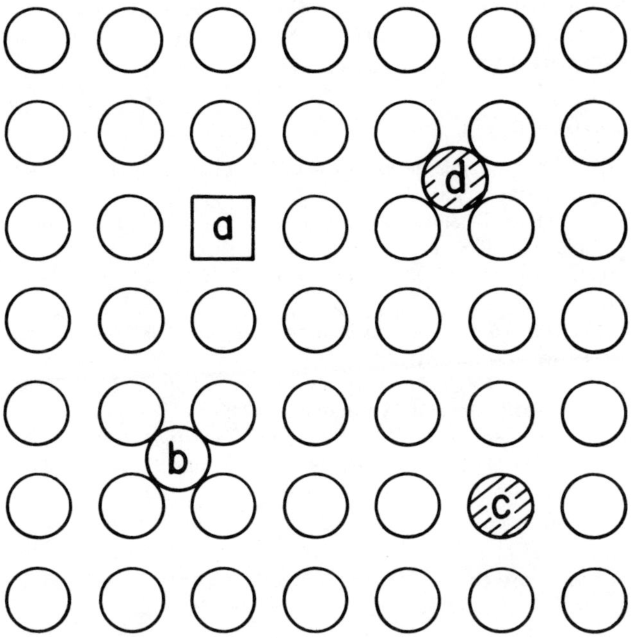

FIG. 1. Simple point defects: (a) vacancy; (b) interstitial; (c) substitutional impurity; (d) interstitial impurity.

which increases exponentially as the temperature is raised. For most materials, E_I, the energy to form an interstitial, is greater than E_V, that for the vacancy, and the vacancy tends to be the dominant defect. Typically, near the melting point, a crystal may have a vacancy concentration of 0.01–0.1%.

In insulators and compound materials (more than one atom species per unit cell), the thermodynamics can be somewhat more complicated in that defects can carry charge, and chemical reactions with crystal components in the gas phase must also be considered (e.g., an oxide in an oxygen atmosphere). However, a proper treatment still leads to equilibrium concentrations of the form of Eq. (1), but where E_i may require a more general definition.

The presence of vacancies and interstitial atoms in a crystal provides a mechanism by which mass transport (diffusion) can occur in the lattice. The vacancy provides a missing site into which a neighboring atom can jump. When the atom jumps, the vacancy has moved, now occupying the original site of the neighbor. Similarly, the interstitial may move by jumping to a nearby empty interstitial site, or by pushing a neighbor into an adjacent interstitial site and taking its place. Both defects therefore stir up the lattice atoms as they randomly jump from one lattice position to another. The jump rate for a defect will have a temperature dependence of the form

$$\nu_i \sim \exp(-U_i/kT), \qquad (2)$$

where U_i represents an energy barrier that the atoms must overcome in order to accomplish the jump. The contribution of a defect to atomic diffusion in the crystal is therefore proportional to its jump rate (2) times the number of the defects (1) leading to a temperature dependence for the diffusion coefficient:

$$D \sim \exp[-(E_i + U_i)/kT]. \qquad (3)$$

In nonmetals, vacancies and interstitials may also be charged. In this case, defect motion produces electrical (*ionic*) conductivity which is also proportional to (3). This is often the dominant source of electrical conductivity in ionic-compound insulators such as alkali halides and ceramic oxides.

At temperatures well below the melting temperature, the *equilibrium* concentration of vacancies and interstitials may be small, because of the exponential temperature dependence, Eq. (1). However, vacancies and interstitials are often present greatly in excess of this amount as a result of (a) quenching from a high-temperature equilibrium, (b) mechanical distortion of the crystal, and (c) radiation damage, where the incoming particle displaces atoms. In these cases, return to equilibrium (annealing) may be extremely slow, being governed by the migration rate (2) which is also exponentially dependent on temperature.

The presence of vacancies and interstitials in semiconductors can alter the *electronic* properties of the material in that they serve to trap and scatter free electrons and holes. Electronic transitions of the trapped charges may also give rise to optical absorption and luminescence bands in semiconductors and insulators.

FOREIGN ATOMS

Chemical impurities may dissolve in a crystal as extra atoms (interstitial) or by replacing normal lattice atoms (substitutional). Even the purest crystals inevitably contain trace amounts of impurities. In addition, impurities may be introduced intentionally by addition during crystal growth, by diffusion into the crystal from the surface at elevated temperatures, or by ion bombardment, where a high-energy ion of the impurity is driven in through the surface of the material. Transmutation doping has also been used, where a nuclear reaction converts a host atom into an impurity. Like vacancies and interstitials, impurities may trap and donate electrons or holes and alter the electronic and optical properties of the crystal. In addition, they may introduce some of their characteristic atomic optical absorption or luminescent spectra to the crystal. Their elastic interactions with dislocations are important in determining the strength of materials.

DISLOCATIONS

A perfect crystal should be very strong, the critical shear stress necessary to slide one perfect atom plane past another being comparable to the shear modulus G of the crystal. Contrary to this, however, it is found experimentally that the purer and more "perfect" the crystal is made, the softer it usually tends to become. The critical shear stress in a well-annealed single crystal may be only 10^{-5}–$10^{-4}G$. The reason for this is the presence of *dislocations*.

Two basic types of dislocations are illustrated in Fig. 2. In Fig. 2a, the *edge* dislocation can be viewed as having been formed by slicing the crystal with a knife partway in from the top and inserting an extra half plane of atoms in the cut. In Fig. 2b, the *screw* dislocation could be produced by slicing part way in from the right and shearing the crystal parallel

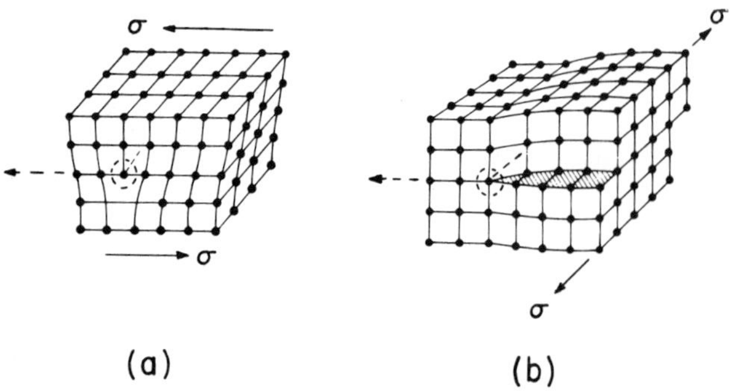

FIG. 2. (a) An *edge* dislocation; (b) a *screw* dislocation (after B. Henderson, *Defects in Crystalline Solids*. Edward Arnold, London, 1972). Applied shear stress σ causes the upper half of the crystal to slip one atom spacing as the dislocation line moves through the crystal to the left.

to the slice by one atomic spacing. For both, the imperfection is the line which terminates the cut, the surrounding material being a strained but otherwise perfect crystal. As shown in the figure, applied shear stress σ allows the upper part of the crystal to slip one atomic spacing as the dislocation moves to the left. A reduction in the critical shear stress arises because now only a single line of atoms is required to move at a time, rather than the whole plane of atoms together.

Dislocations are not thermodynamically stable, the energy to form one in a bulk crystal being too great. However, they are almost always present in crystals as a result of accidents in growth. Typically, densities for an as-grown crystal may be $\sim 10^{12}/m^2$. The rapid and chaotic manner in which atoms fall into place during growth makes the formation of lattice misfits of this type likely. Once formed on a small scale, they can propagate as the crystal grows. Also vacancies and interstitials, present in high concentrations during growth from the melt, can nucleate dislocation loops as they condense out upon cooling.

Dislocation motion can be impeded by the presence of point defects, precipitates, or other dislocations in or near its slip plane. Interaction with point defects and precipitates is an important mechanism for the strengthening of constructional metals such as steel, where controlled additions of impurities and subsequent heat treatments to disperse the impurities or allow precipitation are used to tie up the dislocations to the desired degree. *Work hardening* is another mechanism for strengthening materials. Here dislocation movement under stress generates more dislocations, ultimately leading to a massive tangle of "pinned" dislocations.

SURFACE DEFECTS

The interface between two regions of different crystal orientation is called a *grain boundary*. These can originate in growth as two separately growing regions merge. Grain boundaries can also form by regrouping of dislocations during anneal. *Stacking faults* are the result of an error in layer growth. A stacking fault for a crystal normally growing in the layer order *ABCABCABC* might be *ABCBCABC*.

See also COLOR CENTERS; CRYSTAL GROWTH; DIFFUSION; ELECTRON ENERGY STATES IN SOLIDS AND LIQUIDS; LUMINESCENCE.

BIBLIOGRAPHY

A. H. Cottrell, *Dislocations and Plastic Flow in Crystals*. Oxford Univ. Press, Oxford, 1953. (I–A)

J. H. Crawford, Jr. and L. M. Slifkin, eds., *Point Defects in Solids*, Vols. 1–3. Plenum Press, New York, 1972. (I)

B. Henderson, *Defects in Crystalline Solids*. Edward Arnold, London, 1972. (I)

D. Hull, *Introduction to Dislocations*, 2nd ed. Pergamon Press, Oxford, 1975. (E–I)

F. A. Kröger, *The Chemistry of Imperfect Crystals*, 2nd ed., Vols. 1–3. North-Holland, Amsterdam, 1973. (A)

F. Seitz, in *Imperfections in Nearly Perfect Crystals*, Chap. 1 (W. Shockley, J. H. Hollomon, R. Maurer, and F. Seitz, eds.). Wiley, New York, 1952. (E, I)

Crystal Growth

R. L. Parker

Among the many techniques for growing crystals, the following may be mentioned as representative examples: growth from the melt (solidification), growth from the vapor phase, growth from solution (aqueous, molten salts, other solvents), growth by precipitation from solid solution, growth by chemical (vapor) transport reaction, epitaxial growth from vapor or solution, electrocrystallization, and growth from the solid phase (strain annealing). It is useful, in classifying the various growth processes, to consider the thermodynamic phase diagram for the system; it specifies the regions of equilibrium between the desired solid and the mother medium from which the crystal grows, which may be liquid, vapor, solution, or also solid. Thus, in melt growth of a single-component system at constant pressure, the temperature of the liquid is gradually lowered so that a portion of it is at the freezing point; further cooling may then cause nucleation of the new solid phase, which, if properly controlled, may form a seed upon which further precipitation of the liquid can take place in an orderly way, yielding a single crystal.

Probably the most characteristic and fundamental problem in crystal growth is understanding, predicting, and controlling the crystal morphology or form. By morphology is meant not only the external forms of crystals, important and fascinating as these can be (as, for example, in snow crystals), but more generally the complete history of the solid–liquid, solid–vapor, or solid–solid interface morphology during the entire growth process. It is this morphology that determines the microstructure, the distribution of impurities, and the dislocation and imperfection substructure of the crystal, which in turn control the useful properties of the crystalline material. By no means is the prediction of morphology an easy matter. The interface morphology is controlled simultaneously and in varying proportions by bulk transport effects (flow of heat and/or solute) and by interface kinetic effects. Heat flow is clearly important in, for example, the freezing of a pure liquid, since the latent heat must be removed. Solute flow is clearly important in, for example, the precipitation from a solid solution. Interface kinetics, which refers to the structure, on a molecular level, of the surface and to the relative ease or difficulty that a molecule may have in moving from the mother medium to a relatively final position in the crystal surface, may be controlling for particular materials and for particular molecular and crystal structures. Thus, both macroscopic and microscopic theory are involved in understanding crystal growth mechanisms.

Unless a crystal seed is present, the new phase must start by a nucleation process. Nucleation is the formation of very small particles of new phase in the mother medium and may take place when the appropriate thermodynamic variable (temperature, concentration, pressure, etc.) has departed sufficiently from the equilibrium value. It is generally assumed to take place by a random fluctuation process (heterophase fluctuation) analogous to the better-known density fluctuations in a gas; dimers, trimers, and higher species are assumed to form, with the larger sizes having short lifetimes and dilute concentrations. There is found, by an appropriate balancing of the free energy of formation of a particle against its surface energy, to be a critical embryo size for nucleation; particles greater than this size have a tendency to grow, and those below it tend to dissolve. The dependence of the number of these critical nuclei on temperature or on solute concentration is extremely sensitive, so that only a few degrees change in undercooling can cause many orders of magnitude change in nucleation rate. While the preceding discussion is for "three-dimensional" nucleation, "two-dimensional" nucleation of a monolayer disk may take place from the random fluctuation of adsorbed particles on an otherwise plane crystal surface, and provides one kind of interface kinetic mechanism.

Model calculations of nucleation, interface structure, and interface kinetics have been made for many years. Perhaps the best-known model for crystals is the BCF or Burton–Cabrera–Frank model of growth at the emerging step of a screw dislocation at the crystal surface. Recently, the great capacity and speed of modern electronic computers have made it possible to calculate atom-by-atom condensation onto crystal surfaces for a model crystal surface of more than 1600 lattice sites in size, using the Monte Carlo technique. In this technique, atoms are deposited at random sites on an initially flat crystal, but their evaporation depends on the number of their first neighbors. Not only interface structure, but growth rates can be calculated. It is possible to confirm and make more precise predictions of surface roughness that were made previously by statistical mechanics (Ising model).

It is often observed that crystal forms are not stable shapes. An example is the quite common dendritic or tree-shaped form, which exhibits periodic branching and subbranching, as in snow crystals. This results from the unstabilizing point effect of diffusion (of heat or solute) acting on a small protuberance. This effect can be counteracted by the additional surface energy thereby produced, and it can for particular materials be overcome by the size and/or anisotropy of the interface kinetic effect (sluggish kinetics). Morphological stability theory has been applied to this phenomenon, using the transport equations and appropriate boundary conditions, and has been highly successful in such areas as the breakdown of planar to cellular growth of alloy crystals in the presence of undercooling caused by solute rejection, and in such other areas as ice cylinder growth from supercooled water. It seems likely that the ultimate calculation of crystal form will use this macroscopic theory, with boundary conditions determined by computer modeling on the atomic scale.

When the mother medium is a fluid (liquid or gas), fluid flow and convection can greatly influence crystal growth rates and morphology. Although often overlooked in the past, these effects are due to the breakup of relatively long-range diffusion fields of heat or solute. These fields are translated away and are replaced by relatively short-range diffusion fields. The result is that the fluid of bulk concentration or temperature is now quite close to the interface, and is only separated from it by a thin hydrodynamic boundary layer; this will tend to increase growth rates. Thus, modern crystal growth technology (for example, large Czochralski machines pulling rotating dislocation-free silicon crystals from the convecting melt) must concern itself with hydrodynamics.

Crystalline perfection or freedom from defects is not solely due to the growth process, for subsequent processing, annealing, handling, and the like often alter the concentration and types of defects. However, the growth process is certainly the key aspect. Among the defects that may be introduced are point defects (vacancies, interstitials, foreign atoms), line defects (dislocations), planar defects (stacking faults, twins, low-angle boundaries), and three-dimensional defects (striations, cellular growth, voids, inclusions). An example of controlling defects is the use of substantial temperature gradients in directional solidification to remove constitutional supercooling and consequently to prevent cellular growth by morphological instability. Another is to control dislocation density by use of a bottlenecked seed at which dislocations may emerge on the crystal surfaces. Of the many crystal characterization techniques, x-ray topography is rapidly growing as a major tool for structural defects such as dislocations. Electrically or optically active defects are often detected by measuring appropriate physical properties such as resistivity or absorption.

Finally, while the physicist may be interested in crystals as good specimens in which to measure physical properties or because the mechanisms of growth are complex and fascinating, it cannot be ignored that substantial recent advances in the technology of crystal growth are due to immense demand for production of industrial and consumer products. These demands exist in the electronics, optical, and communications industries for such materials as Si for transistors, GaP for light-emitting diodes, magnetic garnet for bubble domains, and ruby for lasers. In a typical year's issues of the *Journal of Crystal Growth,* more than 100 different compositions of new or improved crystals may be found.

See also CRYSTAL DEFECTS; CRYSTAL SYMMETRY.

BIBLIOGRAPHY

D. Elwell and H. J. Scheel, *Crystal Growth from High Temperature*

Solutions, Academic Press, New York, 1975.
Journal of Crystal Growth, North-Holland, Amsterdam, 1967–present.
R. A. Laudise, *The Growth of Single Crystals*, Prentice-Hall, New Jersey, 1970.
R. L. Parker, "Crystal Growth Mechanisms," in *Solid State Physics* (H. Ehrenreich, F. Seitz, and D. Turnbull, eds.), **25,** Academic Press, New York, pp. 151–299, 1970.

Crystal Symmetry

Martin J. Buerger

INTRODUCTION

Homogeneity

Crystals are anisotropic homogeneous solids. The homogeneity is on the scale of a cluster of a small number of atoms. It implies that the solid consists of clusters related to each other by direct or opposite isometries, and that the environments of homologous points of all clusters are the same.

Coincidence Operations

Every cluster is related to any other cluster by one of four simple operations: translation, rotation about an axis, reflection across a plane, or inversion through a point; or by acceptable combinations of these four. (The condition for two different operations to form an acceptable combination is that each of the component operations transforms the other into itself.)

In every crystal the clusters are related to each other at least by the operations of translations. Accordingly, any other operations, called *symmetry operations*, must be consistent with translations. It is readily demonstrated that this requires every rotation operation to be through an angle of $2\pi/n$, where $n = 1,2,3,4$, or 6.

Each of the operations relating the crystal clusters generates a group. As an aid to describing the operations it is convenient to designate the locus that remains unmoved under the operations of the cyclical group as the *symmetry element* of that group. The symmetry element of a group of rotations is the axis A of the rotation; of a group of reflections is the mirror plane m of the reflection; of an inversion is the point i through which the operation inverts; etc.

Any of the permissible operations that relate a pair of atomic clusters in a particular homogeneous solid must also relate all clusters in the solid. If the solid contains operations beyond the generally required translations, the solid is said to be *symmetrical* with respect to each of the symmetry elements corresponding to the groups of operations relating its atomic clusters.

Crystallographic Groups

Two kinds of symmetries are especially important in crystallography: *point groups* and *space groups*. Point groups are so named because in each group one point remains unmoved under the operations of the group; they contain no translations. Space groups are so named because in each group all three-dimensional space remains invariant under the operations of the group; they contain pure translations and may also contain operations with translation components.

Since all crystals are ideally homogeneous, every crystal must conform to one of the space-group symmetries. Nevertheless it is useful to consider the point groups because many of the properties of crystals are not dependent on its translations. For example, the appearance of a crystal conforms to its point-group symmetry because the eye, even when aided by the microscope, cannot see the tiny translations. Furthermore, the point groups, which are easy to construct, lead to a simple way of enumerating all space groups.

THE CRYSTALLOGRAPHIC POINT GROUPS

Symbolism for Symmetry Elements

The symbols universally used by crystallographers to designate symmetry elements are the Hermann–Mauguin, or *international*, symbols. These are shown in Table I for both point groups and space groups. These symbols replace the 1890 point-groups symbols of Schoenflies, which cannot be suitably extended to space groups. The equivalences are shown in Table II.

Chirality

Any operation or group of operations that relates objects of the same chirality (i.e., objects related by direct isometry, which can be regarded arbitrarily as all right-handed or all left-handed) is said to be of the *first sort*. Any operation or

Table I Hermann–Mauguin symbols for the symmetry elements of crystallographic groups

Point-group symbols	Isogonal space-group symbols
1	1
2	2 2_1
3	3 3_1 3_2
4	4 4_1 4_2 4_3
6	6 6_1 6_2 6_3 6_4 6_5
$\bar{1}$ $(=\tilde{2})$ = i, an inversion center	$\bar{1} = i$
$\bar{2}$ $(=\tilde{1})$ = m, a mirror	Various glide planes[†]
$\bar{3}$ $(=\tilde{6})$ = $3+i$	$\bar{3} = 3+i$
$\bar{4}$ $(=\tilde{4})$ Not decomposable	$\bar{4}$
$\bar{6}$ $(=\tilde{3})$ = $\dfrac{3}{m}$, i.e., $3 \perp m$	$\bar{6} = \dfrac{3}{m}$

[†]Glide planes

Symbol	Translation component
m	0
a	$\frac{1}{2}\underset{\sim}{a}$
b	$\frac{1}{2}\underset{\sim}{b}$
c	$\frac{1}{2}\underset{\sim}{c}$
n	$\frac{1}{2}(\underset{\sim}{a}+\underset{\sim}{b})$ or $\frac{1}{2}(\underset{\sim}{b}+\underset{\sim}{c})$ or $\frac{1}{2}(\underset{\sim}{a}+\underset{\sim}{c})$
d	$\frac{1}{4}(\underset{\sim}{a}+\underset{\sim}{b})$ or $\frac{1}{4}(\underset{\sim}{b}+\underset{\sim}{c})$ or $\frac{1}{4}(\underset{\sim}{a}+\underset{\sim}{c})$ or $\frac{1}{4}(\underset{\sim}{a}+\underset{\sim}{b}+\underset{\sim}{c})$

Table II The distribution of point-group symmetries over the crystal systems and their possible lattice types

Point-group symbols		Crystal system		
Schoenflies	Hermann–Mauguin†	Name	Geometry	Bravais-lattice type
C_1	1	Triclinic	$a \neq b \neq c$	P
C_i	$\bar{1}$		$\alpha \neq \beta \neq \gamma$	
C_2	2	Monoclinic§	$a \neq b \neq c$	P, A (or B)
C_s	m		$90° = \alpha = \beta \neq \gamma$	
C_{2h}	$\frac{2}{m}$			
C_3	3	(Trigonal subdivision)		P, R
D_3	32			
C_{3v}	$3m$			
C_{3i}	$\bar{3}$			
D_{3d}	$\bar{3}\frac{2}{m}$ $(\bar{3}m)$			
		Hexagonal	$a = b \neq c$	
			$90° = \alpha = \beta \neq \gamma = 120°$	
C_6	6	(Hexagonal subdivision)		P
D_6	622			
C_{6h}	$\frac{6}{m}$			
C_{6v}	$6mm$			
C_{3h}	$\bar{6}$			
D_{3h}	$\bar{6}m2$ or $\bar{6}2m$			
D_{6h}	$\frac{6}{m}\frac{2}{m}\frac{2}{m}$ $(6/mmm)$			
D_2	222	Orthorhombic	$a \neq b \neq c$	P, I, A (or B or C), F
C_{2v}	$mm2$		$90° = \alpha = \beta = \gamma$	
$D_{2h} \equiv V$	$\frac{2}{m}\frac{2}{m}\frac{2}{m}$ $(2/mmm)$			
C_4	4	Tetragonal	$a = b \neq c$	P, I
D_4	422		$90° = \alpha = \beta = \gamma$	
C_{4h}	$\frac{4}{m}$			
C_{4v}	$4mm$			
S_4	$\bar{4}$			
D_{2d}	$\bar{4}m2$ or $\bar{4}2m$			
D_{4h}	$\frac{4}{m}\frac{2}{m}\frac{2}{m}$ $(4/mmm)$			
T	23	Cubic (isometric)	$a = b = c$	P, I, F
O	432		$90° = \alpha = \beta = \gamma$	
T_h	$\frac{2}{m}\bar{3}$			
T_d	$\bar{4}3m$			
O_h	$\frac{4}{m}\bar{3}\frac{2}{m}$ $(m\bar{3}m)$			

† Hermann–Mauguin symbols in parentheses are abbreviated forms in common use.
§ The geometry and Bravais lattice types for monoclinic crystals are for the first setting, in which the unique axis is c. For the second setting of monoclinic crystals the unique axis is b; the geometry is $a \neq b \neq c$, $90° = \alpha = \gamma \neq \beta$; its lattice types are P, A (or C).

group that relates objects of opposite chirality is said to be of the *second sort*. Pure rotations relate objects of the same chirality; they are called *proper rotations*. Rotoinversions and rotoreflections are called *improper rotations*; they relate a sequence of objects that are alternately right-handed and left-handed. The objects of homogeneous patterns are either all of the same chirality or they are half right-handed and half left-handed.

Monaxial Groups with Proper Rotations

In Hermann–Mauguin notation a proper rotation axis is labeled by the numerical value of its n. Accordingly there are five such crystallographic point groups, namely, 1,2,3,4, and 6, shown in Fig. 1.

Groups with Proper Rotations about Intersecting Axes

It can readily be shown by Euler's construction, or by multiplying together matrices that represent rotations, that a rotation through an angle α about an axis A, followed by a rotation through β about B, is equivalent to a rotation through $-\gamma$ about an axis C. In the product notation for combining the elements of a group, this can be expressed as

$$A_\alpha B_\beta = C_{-\gamma}. \tag{1}$$

This implies that

$$A_\alpha B_\beta C_\gamma = 1 \tag{2}$$

where the right-hand side of (2) is the identity operation, indicating no net motion, expressed in Hermann–Mauguin notation. Since

$$\alpha = \frac{2\pi}{n_1}, \quad \beta = \frac{2\pi}{n_2}, \quad \gamma = \frac{2\pi}{n_3} \tag{3}$$

in crystal symmetry, it is obvious that if there are two rotations about intersecting axes, a rotation about a third intersecting axis is implied. (Exceptionally, the third rotation is one of the other two with an opposite rotation.)

Euler's construction allows the angular relations between rotation axes to be obtained from a spherical triangle in which axes A,B, and C occur in a clockwise sense at arc distances $A \wedge B = c$, $B \wedge C = a$, $C \wedge A = b$, and with $A = \alpha/2$, $B = \beta/2$, $C = \gamma/2$. For this spherical triangle, the sum of its angles is

$$\frac{\alpha}{2} + \frac{\beta}{2} + \frac{\gamma}{2} \geq \pi. \tag{4}$$

With the aid of (3) this can be expressed as

$$\frac{1}{n_1} + \frac{1}{n_2} + \frac{1}{n_3} \geq 1. \tag{5}$$

The only distinct solutions of (5) are

$$2,2,n_3 \quad 2,3,3 \quad 2,3,4 \quad 2,3,5.$$

The crystallographic possibilities for these solutions are the following.

4 "dihedral" sets: 2,2,2 2,2,3 2,2,4 2,2,6

1 "tetrahedral" set: 2,3,3

1 "octahedral" set: 2,3,4

The angles a,b, and c between the rotation axes can be found from relations like

$$\cos a = \frac{\cos(\alpha/2) + \cos(\beta/2)\cos(\gamma/2)}{\sin(\beta/2)\sin(\gamma/2)}. \tag{6}$$

The arrangements of these six axial combinations are shown in Fig. 2.

The Axial Frames of Crystallography

The five monaxial point groups and the six sets of polyaxial point groups just demonstrated are the only crystallographic point groups with proper rotations only. These 11

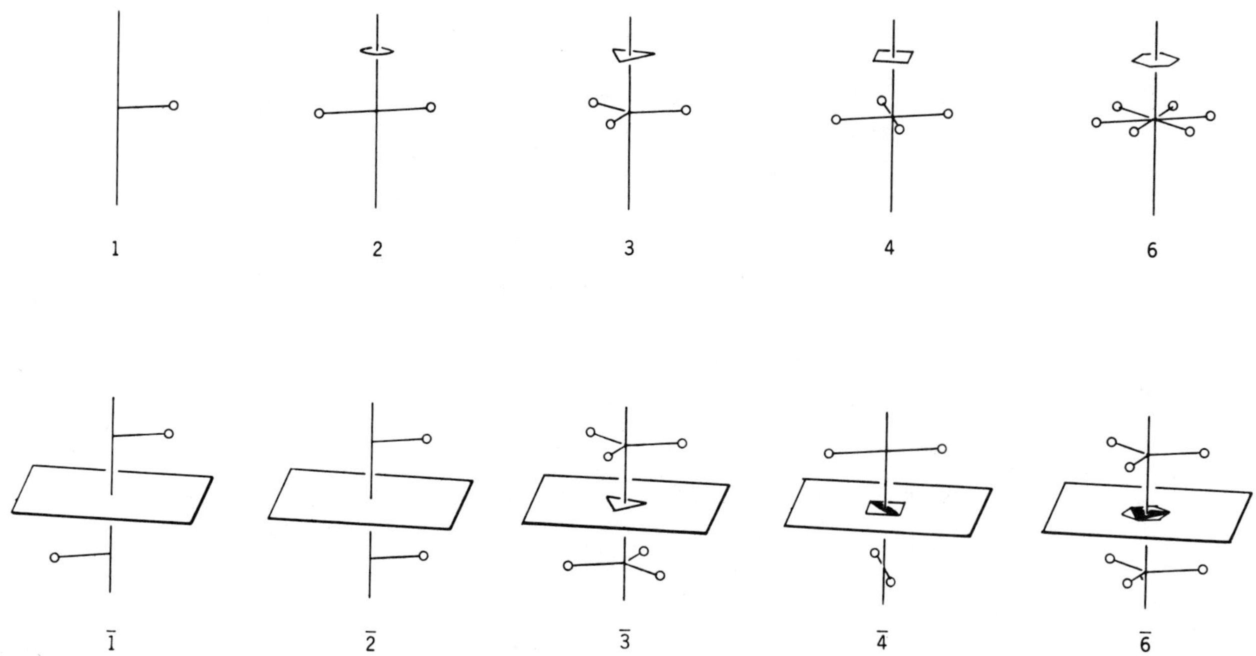

FIG. 1. Top: The symmetry axes of the five crystallographic point groups with proper rotations. Bottom: The symmetry axes of the five crystallographic point groups with improper rotations.

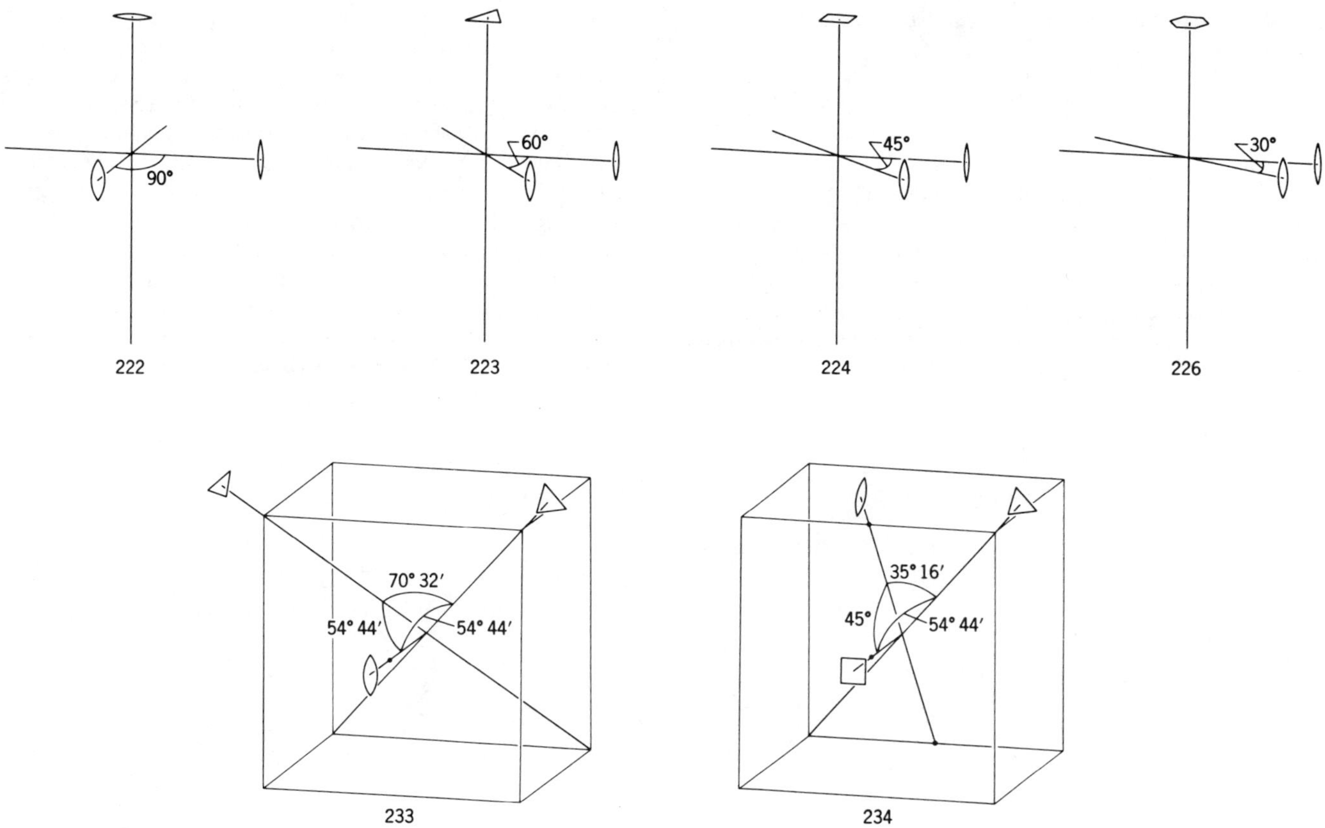

FIG. 2. The six combinations of crystallographic proper rotation axes. [From M. J. Buerger, *Elementary Crystallography* (Wiley, New York, 1963).]

arrangements of axes provide axial frames on whose geometries all other angular relationships in other crystallographic groups are based.

Groups with Improper Rotations

Corresponding to the five monaxial groups with proper rotations are five with improper rotations, symbolized by the rotoinversion axes $\bar{1},\bar{2},\bar{3},\bar{4}$, and $\bar{6}$. As listed in Table I and shown in Fig. 1, all but $\bar{4}$ can also be represented by other symmetry elements.

A proper and an improper rotation axis of the same n can occur along the same line. When these n and $\bar{n}$ axes are written as a fraction $n/\bar{n}$ and reduced, there arise three new groups: $2/\bar{2}\rightarrow 2/m$; $[3/\bar{3}\rightarrow 3/(3+i)\equiv\bar{3}]$; $4/\bar{4}\rightarrow 4/m$; $6/\bar{6}\rightarrow 6/(3/m)\rightarrow 6/m$.

Intersecting axes can be combined as were the three proper rotations, provided either that proper and improper axes coincide, as just noted, or that there are two improper axes and one proper axis. These intersecting combinations generate the other point groups having improper rotations listed in Table II.

The 32 Point Groups

The arrangements of symmetry elements in the 32 point groups are illustrated in Fig. 3. The "point" of the point group is assumed to be at the center of the sphere whose outline is the dotted circle; if there are intersecting symmetry elements, the point of intersection is at the center of the circle. A reflection plane is represented by the line where it intersects the sphere's surface. An n-fold axis is shown as a polygon with n vertices at the point where the axis intersects the sphere's surface. A tiny circle in the center of a diagram indicates that the point group has a center of symmetry.

TRANSLATION GROUPS

Lattices

In three-dimensional space a translation can be represented by its components on three noncoplanar vectors $\mathbf{t}_1,\mathbf{t}_2$, and $\mathbf{t}_3$. For crystallographic applications, let $\mathbf{t}_1$ be the shortest translation in some direction. A set of points equivalent by translation $\mathbf{t}_1$ constitute a line parallel to $\mathbf{t}_1$. Let $\mathbf{t}_2$ be chosen so that it translates the points of this line to the points of a parallel line that is next adjacent to it in some direction; then let $\mathbf{t}_3$ be chosen so that it translates a point of the plane parallel to $\mathbf{t}_1$ and $\mathbf{t}_2$ to a point in the next parallel plane. When $\mathbf{t}_1$, $\mathbf{t}_2$, and $\mathbf{t}_3$ are chosen in this way they are called *conjugate translations*. There are an infinite number of sets of such translations in a homogeneous solid. Any translation of the solid can then be described by

$$\mathbf{t}_{uvw}=u\mathbf{t}_1+v\mathbf{t}_2+w\mathbf{t}_3, \qquad (u,v,w,\text{ integers}). \qquad (7)$$

The translations $\mathbf{t}_{uvw}$ form a group if u,v, and w run over all integers. If the translations of this group operate on an arbitrary point, the resulting collection of points is called a *lattice*. (This is the only correct meaning of "lattice" in crystallography.)

Cells

The parallelepiped, three of whose adjacent edges are conjugate translations, is known as a *primitive cell*. The adjective "primitive" implies that to each cell of this kind there corresponds one point of the lattice. The infinite number of primitive cells of a given crystal all have the same volume. A particular kind of primitive cell, known as the *reduced cell*, is characterized by having the three shortest translations.

Except as noted below, the cell edges of the reduced cell offer a natural coordinate system to which the geometrical features of a crystal can be referred. When a crystal has point-group symmetry other than 1 or $\bar{1}$, however, there are certain cases for which it is mathematically simpler to use as the coordinate system the edges of a more symmetrical cell that contains more than one set of translationally equivalent atoms. These cells are called *multiple cells*, specifically *double, triple*, and *quadruple* cells, according as the cell contains 2,3, or 4 sets of translationally equivalent atoms. Such cells have, in addition to a lattice point at the cell origin, one or more additional lattice points at various special positions in the cell. The characteristics of these cells and their designations are given in Table III.

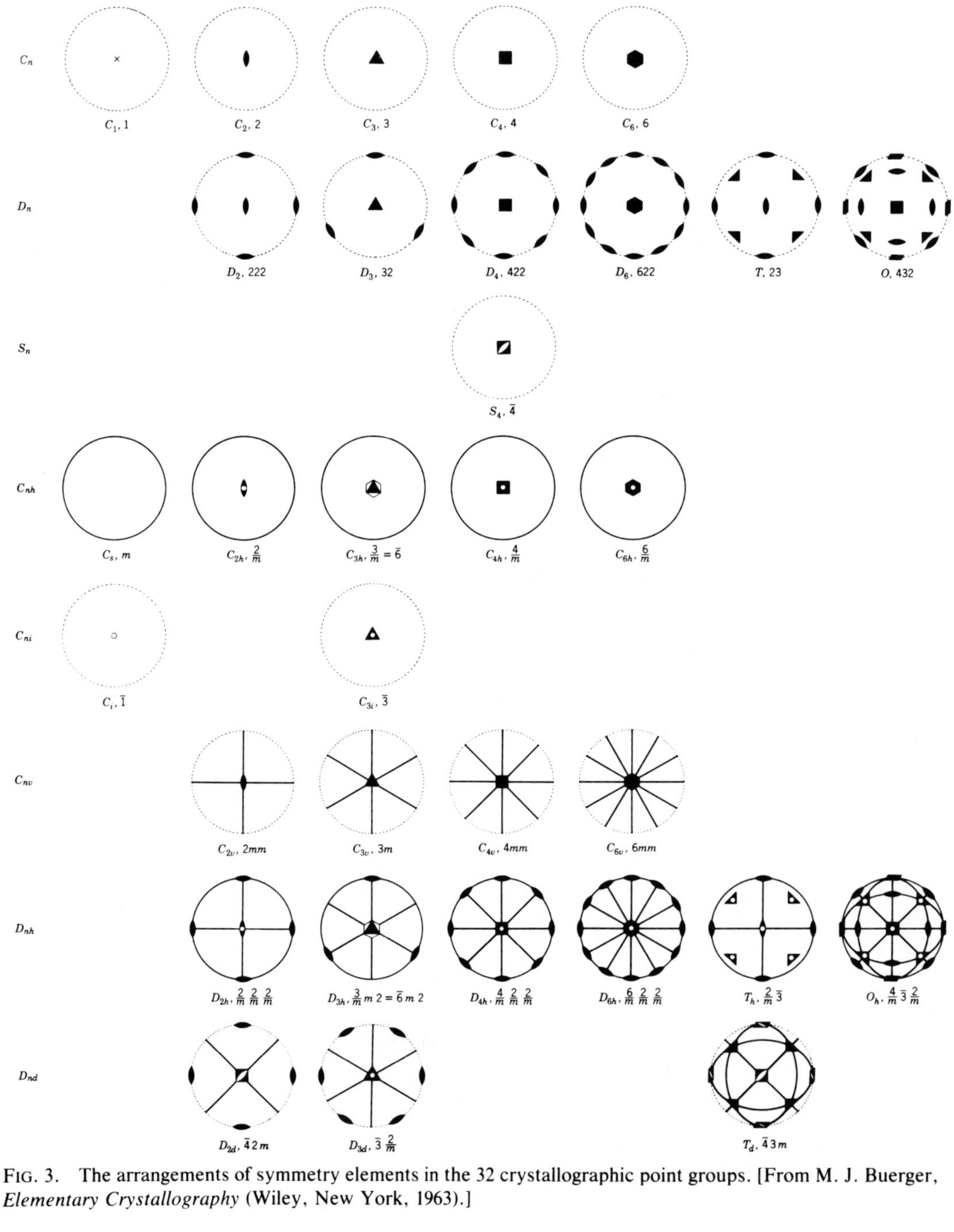

FIG. 3. The arrangements of symmetry elements in the 32 crystallographic point groups. [From M. J. Buerger, *Elementary Crystallography* (Wiley, New York, 1963).]

Table III Characteristics of crystallographic cells

Symbol	Significance of symbol	Coordinates of lattice points in fractions of cell edges *a, b, c*	Cell multiplicity
P	"Primitive"	000	1
I	"Innenzentriert": body-centered	000; $\frac{1}{2}\frac{1}{2}\frac{1}{2}$	2
A	"*A*-centered": point at center of *bc* face	000; $0\frac{1}{2}\frac{1}{2}$	
B	"*B*-centered": point at center of *ca* face	000; $\frac{1}{2}0\frac{1}{2}$	
C	"*C*-centered": point at center of *ab* face	000; $\frac{1}{2}\frac{1}{2}0$	
R	"Rhombohedral" referred to hexagonal axes	000; $\frac{2}{3}\frac{1}{3}\frac{1}{3}$; $\frac{1}{3}\frac{2}{3}\frac{2}{3}$	3
F	"Face-centered" on *A*, *B*, and *C* faces	000; $0\frac{1}{2}\frac{1}{2}$; $\frac{1}{2}0\frac{1}{2}$; $\frac{1}{2}\frac{1}{2}0$	4

When an appropriate cell has been selected, it is customary to label its three adjacent edges a,b, and c, in such a way as to define a right-handed system of coordinates. The angles between cell axes are designated $\alpha = b \wedge c$, $\beta = c \wedge a$, $\gamma = a \wedge b$.

Crystal Systems

The symmetry of a crystal requires a specialization of its lattice. It turns out that sets of several different point-group symmetries place the same constraints on the lattice, so that, when multiple cells are taken into account, only six kinds of symmetrically different cell edges are required for the coordinate systems of the 32 crystal classes. Crystals that can be referred to the same kind of coordinate system are said to belong to the same *crystal system*. The distribution of the point groups over the crystal systems is given in Table II.

Rational Lines and Directions

An aspect of crystal geometry is said to be rational if it can be expressed in integral numbers of lattice translations. A direction is rational if it has the form

$$\mathbf{r} = u\mathbf{a} + v\mathbf{b} + w\mathbf{c} \qquad (u,v,w,\text{ integers}). \tag{8}$$

The three integers u,v,w, are known as the *indices of the direction*. Since sets of three integers are used for various indices, to indicate that they refer to a direction they are traditionally written between square brackets. (There is a tendency, especially among metallurgists, to use angular brackets for this purpose.)

Rational Planes

A plane is rational if it contains three lattice points (in which case it contains an infinite number). The translations of the lattice require such a plane to be one of an infinite stack of equally spaced parallel planes whose spacing is designated d. The equation of the nth plane from the origin is

$$hx + ky + lz = n \tag{9}$$

where h,k, and l are integers and x,y, and z are fractions based on a,b, and c, respectively, as units. A rearrangement of (9) for $n=1$ provides the intercept form for the plane of the stack nearest the origin:

$$\frac{x}{1/h} + \frac{y}{1/k} + \frac{z}{1/l} = 1. \tag{10}$$

This plane has intercepts on the a,b, and c axes of $1/h, 1/k$, and $1/l$, respectively, so the stack cuts these units into h,k, and l parts, respectively. These three integers, placed between parentheses, are the *indices of the stack of planes*. (The designation of rational planes by these indices was proposed by Whewell in 1825 and by Grassman in 1829; the integers became known as "Miller indices" when they were popularized by Miller's 1839 book, *A Treatise on Crystallography*.)

SPACE GROUPS

General Features

Every crystal has a group of translations that are consistent with the symmetry of its point group, specifically one of the Bravais-lattice types of Table II. Its symmetry may include rotational operations with translation components parallel to the axis of rotation, and its reflectional operations may have translation components parallel to the reflection plane, from Table I. The angles of the rotational component of the screws, and the angles between various axes and planes, are not affected by the translations, however; so that the resulting angular geometry of every combination of symmetry elements is *isogonal* with the angular geometry of the corresponding translation-free symmetry elements in some possible crystallographic point group. From the point of view of group theory, each of the 230 space groups is *isomorphous* with one of the 32 point groups. This relation makes it possible to derive every space group by considering, in turn, the combinations of permissible symmetry operations that are isogonal with the symmetry operations of the appropriate point groups.

Groups with Only Operations of the First Sort

For each rotation operation that generates a symmetry element in point-group symmetry, there are several isogonal operations in space-group symmetry with different translation components. The rotation is accompanied by a translation component τ parallel to the axis A such that n translations τ are equal to an integral number p of lattice translations t along the axis:

$$n\tau = pt \qquad (n,p,\text{ integers}). \tag{11}$$

Thus the space-group operation is characterized by two parameters: n, which is the same as in the isogonal point group; and p, which may be zero. Since these are, in general, helicoidal motions, the symmetry elements are known as *screw axes*. They are designated by the general symbol n_p; the various possibilities are shown in the upper right-hand part of Table I. When $p/n<\frac{1}{2}$, the screw is right-handed; when $p/n>\frac{1}{2}$, it is left-handed; when $p/n=\frac{1}{2}$, it is neutral. The number of threads in the screw is p for $p/n\geq\frac{1}{2}$, but $n-p$ for $p/n\leq\frac{1}{2}$.

A translation $\mathbf{t}'$ normal to the axis A of the screw produces a translation-equivalent operation A' as well as a nonequivalent but equal screw operation about parallel axes B and C located along the perpendicular bisector of AA' and at a distance of t' cotα/2 from AA'. These screw operations accumulate along axes B and C, producing screw axes parallel with A. The resulting sets of symmetry elements, repeated by the lattice translations, constitute infinite sets of parallel axes, which have been called "flocks of axes." An interesting way of appreciating the values of the n's for the A,B, and C axes is as follows: If the sphere and its spherical triangle on which the reasoning of (1) through (5) were based is allowed to expand to infinite radius, the three axes may be allowed to become parallel so that the spherical triangle becomes a plane triangle. Then inequalities (4) and (5) become equalities. The solutions for the n's of the three parallel axes and their separations become

$$2,2,[\infty] \quad \text{with} \quad a:b:c=[\infty:\infty]:1,$$

$$2,3,6 \quad \text{with} \quad a:b:c=2:\sqrt{3}:1,$$

$$2,4,4 \quad \text{with} \quad a:b:c=\sqrt{2}:1:1,$$

$$3,3,3 \quad \text{with} \quad a:b:c=1:1:1.$$

The possible space groups isogonal with a particular point group differ from each other by various characteristic translations. Such a set of space groups can be formed by combining the translations of the several Bravais-lattice types permitted to the point group with the different symmetry elements isogonal with each symmetry element of the point group. As a simple example, to derive the space groups isogonal with point group 3, it is seen from Table II that the Bravais-lattice types consistent with this point group are P

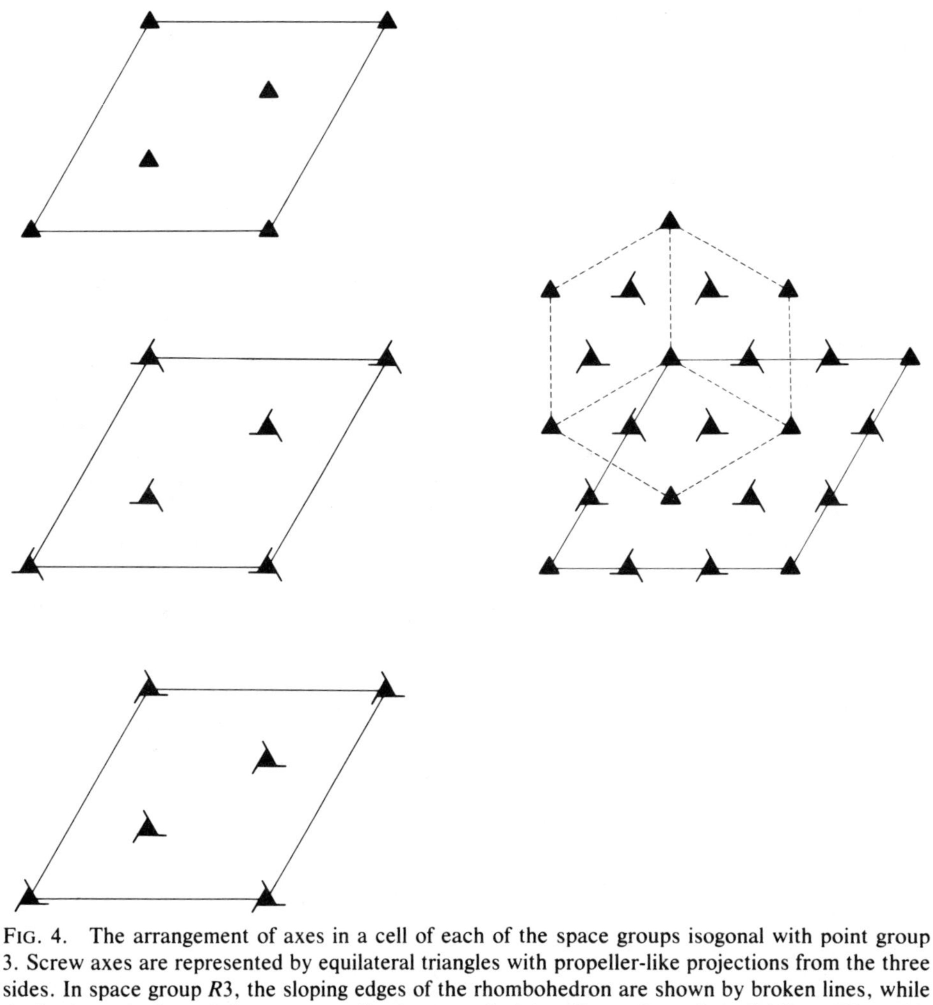

FIG. 4. The arrangement of axes in a cell of each of the space groups isogonal with point group 3. Screw axes are represented by equilateral triangles with propeller-like projections from the three sides. In space group $R3$, the sloping edges of the rhombohedron are shown by broken lines, while the horizontal edges of the corresponding hexagonal triple cell are shown by solid lines. [From M. J. Buerger, *Elementary Crystallography* (Wiley, New York, 1963).]

and R. Table I shows that the space-group symmetry elements isogonal with point group 3 are 3, 3_1, and 3_2. The combination of the two lattice types with the three symmetry elements yields $P3$, $P3_1$, $P3_2$, $R3$, ($R3_1$ and $R3_2$), seen in Fig. 4. These combinations of symbols are, in fact, the Hermann–Mauguin symbols for the space groups. The last two symbols represent the same space group as $R3$ because, if the operations of any one of the symmetry elements 3, 3_1, or 3_2 is combined with the translations of R, the other two symmetry elements are generated in the B and C positions.

Groups with Operations of the Second Sort

The space-group elements of the second sort are listed in the lower right-hand part of Table I. Only the symmetry elements isogonal with point-group symmetry element m call for special comment. In space groups a reflection may have a translation component in the reflecting plane, since the translation and reflection each transform the other into itself. The resulting *glide reflection* operation generates a *glide plane* as a symmetry element. A sequence of two glide operations is the same as a translation, so the translation component can only be parallel to a lattice translation. The Hermann–Mauguin symbol for a glide plane reveals its translation component; the six possibilities are shown at the lower right in Table I.

Symbolism and Construction of Space Groups

Since a space group differs from the point group with which it is isogonal only in its characteristic translations, its symbol must specify these translations. The symbol always consists of two major parts. The first part is the symbol of the particular Bravais-lattice type that specifies the lattice transformations of the space group. This is one of the several symbols listed in the last column of Table II that are aligned horizontally with the first symbol in column 1 for the isogonal point group. Next there follows a part that has the same general form as the symbol for the correct isogonal point group given in column 2, but with the appropriate isogonal-substituent symbols given in Table I. A space group is completely specified by its symbol and indeed can be constructed from it.

To deduce the geometrical arrangements of the symmetry elements requires a knowledge of the results of combining the various symmetry operations of space groups with the translations of the lattice and with each other. This process is easily carried out with the aid of the *algebra of operations*. After some of the simpler space groups have been derived, the more complicated ones can be deduced from them by a stepwise augmentation.

Equivalent Positions

If a point is placed in a cell, in general the symmetry elements of the space group require the cell to have other symmetrically related points. This symmetrical collection is known as a *set of equivalent positions*, or an *equipoint*. If the location of the original point is unspecialized, the set is said to occupy the *general position* of the space group. Each one of the N operations in the cell generates one of the N points of the general position.

If a point in the general position is shifted until it comes to occupy any one of several specific kinds of symmetry elements (noted later), then a set of s points (characteristic of the kind of symmetry element) coalesce on the symmetry element to form one point. Such a set of equivalent positions is called a *special position*. The number of its points per cell is reduced from the N of the general position to $M = N/s$, an integer known as the *multiplicity* of the equipoint. The coalescence factor s is 2 for inversion centers and mirrors; it is n for a pure n-fold rotation axis and p for a screw axis of p threads. There is no reduction in multiplicity for points on a glide plane or on a screw axis with one thread.

Every space group has a general position. A few space groups have no special positions because they do not have appropriate symmetry elements; for example, $P2_12_12_1$ has only screw axes. Most space groups have several special positions and these may have various multiplicities, depending on the kinds of symmetry elements in the group.

A point placed in the general position has three variable coordinates, x,y, and z. The coordinates of the remaining points of the set can be obtained from this initial one by operating on its coordinates with vectors that can be derived for the individual symmetry elements of the group. Points in special positions have coordinates that are specializations of those of the general position. In general, points on mirrors have two free parameters; those on pure rotation axes and on screws have one free parameter; while points on symmetry centers and on $\bar{n}$ points have no free parameters.

Crystal Structures

The pattern of matter in a crystal consists of sets of integral numbers of atoms in each cell distributed over the available equipoints of its space group. In this way the integral subscripts of a Daltonian compound are accommodated by the integral number of available locations in the cell.

Ideally, the translations of the lattice provide that all regions of the crystal have the same composition. If the conditions of growth, such as temperature, pressure, and composition of the nutrient material, vary with time, however, the composition of the crystal may vary from cell to cell due to the vicarious replacement of one atomic species by another, or of several atomic species by several others, in such a way that the replacing set has certain properties similar to the replaced set.

Finally, we note that a crystal structure is called a "lattice" only by those not acquainted with symmetry theory.

See also CRYSTALLOGRAPHY, X-RAY.

BIBLIOGRAPHY

M. J. Buerger, *Introduction to Crystal Geometry*. McGraw-Hill, New York, 1971. (E)

M. J. Buerger, *Elementary Crystallography*. MIT Press, Cambridge, Mass., 1977. (I)

J. J. Burckhardt, *Die Bewegungsgruppen von endlichen Ordnung*. Birkhäuser, Basel, 1947. (I)

E. S. Federow, *Symmetry of Crystals* (transl. by David and Katherine Harker). ACA Monograph 7, 1971. (A)

N. F. M. Henry and K. Lonsdale, *International Tables for X-Ray Crystallography*, Vol. I. Kynoch, Birmingham, 1952. (I)
H. Hilton, *Mathematical Crystallography*. Oxford (Clarendon Press), London and New York, 1903. (I)
I. Janssen, *Crystallographic Groups*. North-Holland, Amsterdam, 1973. (A)
P. Niggli, *Geometrische Kristallographie des Discontinuums*. Borntraeger, Leipzig, 1919. (I)
F. C. Philips, *An Introduction to Crystallography*, 4th ed. Wiley, New York, 1971. (E)
A. V. Shubnikov and V. A. Koptsik, *Symmetry in Science and Art* (David Harker, ed.). Plenum, New York, 1974. (I)
A. Speiser, *Die Theorie der Gruppen von endlichen Ordnung*. Birkhäuser, Basel, 1956. (A)

Crystallography, X-Ray

Martin J. Buerger

INTRODUCTION

Background

Until about the time of World War II, experimental crystallography was carried out largely by the observation of crystal faces and forms with the aid of the reflecting goniometer and by the study of crystal optics with the aid of the polarizing microscope. Chiefly with these tools the point-group symmetries, crystal systems, and axial ratios of many crystals were established.

Since the discovery of the diffraction of x rays by crystals, this kind of information and much more (especially the absolute cell geometry, the space group, and the locations of atoms in the cell) are now routinely determined by diffraction using various kinds of radiation, especially x radiation but also beams of neutrons or electrons. In this article attention is specifically given to the use of x rays, partly because the well-established techniques of x-ray diffraction are practiced in all laboratories doing crystallographic research, and partly because x-ray diffraction lacks some of the complications of neutron and electron diffraction.

It is now obvious that the symmetry theory of the nineteenth century makes it possible to use the diffraction tool of the twentieth century. Without both of these, our understanding of the nature of crystalline matter would be in a very primitive state.

Because Laue's and Bragg's descriptions of the diffraction of radiation by three-dimensional gratings are treated in most elementary physics texts, they are not discussed here.

Friedel's Law

In 1913 Georges Friedel pointed out that it is impossible to determine from the symmetry displayed by an x-ray photograph whether a crystal has a center of symmetry. This prevents making an x-ray determination of the point-group symmetry of a crystal and, in turn, of the space-group symmetry, with the aid of simple qualitative diffraction experiments. Such experiments involve an insertion of a center of symmetry into the diffraction symmetry of the crystal's point group, so that only the 11 centrosymmetrical point-group symmetries (called the *Friedel symmetries*) of crystals can be determined. It is now known that this is because of the inability of an x-ray diffraction beam to leave a record of its phase; only its intensity is recorded.

The consequences of Friedel's law are avoided if the usually imperceptible effects of anomalous dispersion can be determined. The true symmetry of the space groups can also be determined by the Patterson synthesis, described later.

Space-Group Extinctions

In Paul Niggli's 1919 book *Geometrische Kristallographie des Discontinuums* he showed that when a crystal cell contains any translations that are submultiples of the cell edges, the spacings of certain classes of reflections appear to be halved. This is the same as saying that certain classes of x-ray reflections are missing. These *extinctions*, as they are called, occur in the general reflections hkl due to the translations of nonprimitive cells; they occur in classes $hk0$, $h0l$, $0kl$, and hhl due to the translation components of glide planes, and in the classes $h00$, $0k0$, $00l$, and $hh0$ due to the translation components of screw axes.

Diffraction Symbols

The qualitative characteristics of the symmetry revealed by diffraction can be condensed into a short formula called the *diffraction symbol*. This has three parts: (i) the symbol for the Friedel point group in Hermann–Mauguin notation; (ii) the symbol for the Bravais-lattice type; and (iii) a sequence corresponding, item by item, to the Friedel symbol, in which there are set down the Hermann–Mauguin symbols for glide planes and screw axes (detected by classes of missing reflections), while every other item of the Friedel symbol is entered as a short dash. This list is a complete record of all diffraction characteristics sought and all diffraction characteristics found in the qualitative record.

As an example, suppose that the observed diffraction symmetry provides the Friedel symmetry as $2/m\ 2/m\ 2/m$; that the reflections of the class hkl show no systematic extinctions; that the reflections of the class $hk0$ are extinguished when $h+l$ is odd; and that there are no other systematic extinctions inconsistent with these. The diffraction symbol is

$$\underbrace{2/m\ 2/m\ 2/m}_{\text{Friedel group}} \qquad \underset{\text{lattice type}}{P} \qquad \underbrace{-\ -\ n}_{\text{glide planes and screw axes}}$$

This diffraction symbol is consistent with two space groups, $Pm2_1n$ and $Pmmn$.

There are only 122 diffraction symbols for the 230 space groups. Of these, 61 each uniquely characterize only one space group. Each of the remaining 61 are consistent with 2 or 3 space groups. Except when the effects of anomalous dispersion can be observed, there are no qualitative methods by which these sets of 2 or 3 possibilities can be distinguished.

The Reciprocal Lattice

To every lattice there corresponds a related lattice, called the *reciprocal lattice*, that has properties useful in experimental x-ray diffraction by crystals. Its relation to the crystal lattice can be demonstrated as follows: If the edges of the primitive cell are the translations $\mathbf{t}_1$, $\mathbf{t}_2$, and $\mathbf{t}_3$, the volume of the cell is

$$V=\mathbf{t}_1\times\mathbf{t}_2\cdot\mathbf{t}_3. \qquad (1)$$

All primitive cells have the same volume. In particular, the volume of the cell between two adjacent planes of the stack (hkl), one of which is the first plane from the origin to have integral intercepts (Fig. 1) on $\mathbf{t}_1$, $\mathbf{t}_2$, and $\mathbf{t}_3$ (specifically A on $\mathbf{t}_1$, B on $\mathbf{t}_2$, and C on $\mathbf{t}_3$), is the area of this plane times the spacing d to the next plane of the stack. If $\mathbf{n}$ is taken as the unit vector in the direction of d, this volume is

$$V=(A\mathbf{t}_1-C\mathbf{t}_3)\times(B\mathbf{t}_2-C\mathbf{t}_3)\cdot d\mathbf{n} \qquad (2)$$

$$=(BC\mathbf{t}_2\times\mathbf{t}_3+CA\mathbf{t}_3\times\mathbf{t}_1+AB\mathbf{t}_1\times\mathbf{t}_2)\cdot d\mathbf{n}. \qquad (3)$$

Let integers

$$BC\equiv h,\ CA\equiv k,\ AB\equiv l. \qquad (4)$$

Then (3) can be written

$$V=(h\mathbf{t}_2\times\mathbf{t}_3+k\mathbf{t}_3\times\mathbf{t}_1+l\mathbf{t}_1\times\mathbf{t}_2)\cdot d\mathbf{n}. \qquad (5)$$

This has the same form as (1). It can be simplified by defining three new vectors:

$$\mathbf{t}_1^*=\frac{\mathbf{t}_2\times\mathbf{t}_3}{V},\qquad \mathbf{t}_2^*=\frac{\mathbf{t}_3\times\mathbf{t}_1}{V},\qquad \mathbf{t}_3^*=\frac{\mathbf{t}_1\times\mathbf{t}_2}{V}. \qquad (6)$$

When these are substituted in (5) there results

$$1=(h\mathbf{t}_1{}^*+k\mathbf{t}_2{}^*+l\mathbf{t}_3{}^*)\cdot d\mathbf{n}. \qquad (7)$$

This is a relabeled equivalent of (2) in which the cross product is a vector normal to the plane and therefore parallel to $d\mathbf{n}$, so that the two vectors on the right of (7) are parallel. The sum in (7) is the set of components of a vector $\mathbf{t}_{hkl}^*$. If absolute values of the two terms of the product in (7) are set down, it is seen that

$$1=t_{hkl}^*d_{hkl}. \qquad (8)$$

The asterisk (pronounced "star") in (6)–(8) is used to indicate a feature in the reciprocal lattice.

If the sequence (1) through (8) is carried out for the reciprocal lattice rather than the crystal lattice, all items in (1) are starred, and (8) becomes

$$1=t_{hkl}d_{hkl}^*, \qquad (9)$$

so that the star can be affixed to either t or d. This reciprocity constitutes the basic relation between a lattice and its reciprocal. Other relations include (6) and its equivalent with the starred and unstarred items interchanged, as well as angular relations of the following type and their interchanged equivalents:

$$\alpha^*=\frac{\cos\beta\,\cos\gamma-\cos\alpha}{\sin\beta\,\sin\gamma}. \qquad (10)$$

A symmetrical angular relation is

$$\frac{\sin\alpha^*}{\sin\alpha}=\frac{\sin\beta^*}{\sin\beta}=\frac{\sin\gamma^*}{\sin\gamma}. \qquad (11)$$

Each of the Bravais-lattice types P,A,B,C, and R have reciprocals of their own type, but the reciprocal of the Bravais-lattice type I is F and of F is I.

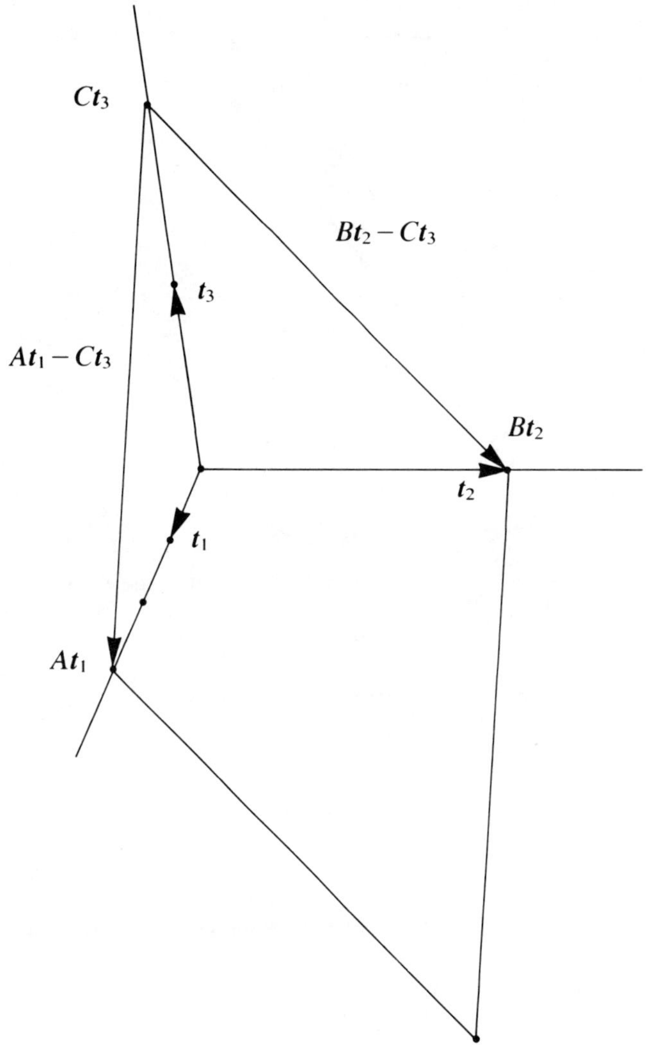

FIG. 1. The three axes $\mathbf{t}_1$,$\mathbf{t}_2$, and $\mathbf{t}_3$ of a primitive cell, and the first plane from the origin to make rational intercepts $A\mathbf{t}_1$, $B\mathbf{t}_2$, and $C\mathbf{t}_3$ on these axes.

Application of the Reciprocal Lattice

Bragg's law can be manipulated as follows:

$$\sin\theta=\frac{n\lambda}{2d(hkl)}=\frac{1/nd(hkl)}{2/\lambda}=\frac{1/d_{nh\ nk\ nl}}{2/\lambda}. \qquad (12)$$

In the right-hand expression for $\sin\theta$, Bragg's order of reflection, n, is combined with $d(hkl)$. The resulting $nh\ nk\ nl$ are *not* the "Miller" indices of a plane, since the three numbers are not prime to one another. They are the *indices of a reflection*; reflection indices are distinguished by being written without parentheses.

A geometrical interpretation of the right-hand part of (12), due to Ewald, is illustrated in Fig. 2. The primary x-ray beam is along the direction SO and the origin of the reciprocal lattice is O. Whenever a crystal is so oriented that any reciprocal-lattice point P^*, at the end of a vector of length $1/d$, comes to lie on the surface of a sphere (known as the *Ewald sphere*), Bragg's law is satisfied and a reflected beam develops in the direction SP^* that makes an angle 2θ with the x-ray beam.

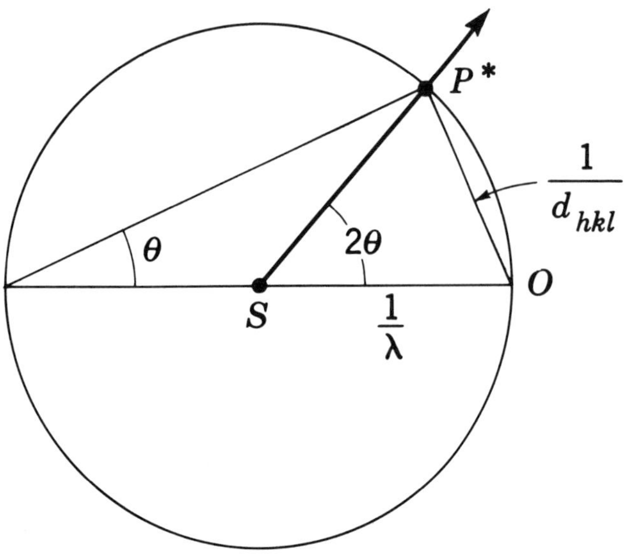

FIG. 2. Ewald's interpretation of Bragg's law with the aid of the reciprocal lattice. [From M. J. Buerger, *Contemporary Crystallography* (McGraw-Hill, New York, 1970).]

EXPERIMENTAL METHODS OF X-RAY CRYSTALLOGRAPHY

Less Suitable Methods

The Laue, powder, and rotating-crystal methods are treated in almost every account of x-ray diffraction by crystals. Unfortunately these methods are inherently unsuitable for crystal-structure analysis and are avoided whenever possible. Except for a few spots having the smallest values of θ, a Laue photograph cannot be indexed because all other spots contain the superposed records of *hkl* and one or many reflections with multiple indices *nh nk nl*. Therefore the space group of the crystal cannot be determined, nor can data be derived for *F*(*hkl*) from a Laue photograph. Neither the powder method nor the rotating-crystal method gives any information about the symmetry of the crystal, nor can their diffraction records be indexed with certainty.

The Weissenberg Method

If a crystal is rotated about a rational axis, the diffraction directions lie along generators of a nest of Laue cones coaxial with the axis. In the Weissenberg instrument, each cone intersects a coaxial cylindrical film in a circle that, when the film is unrolled, becomes a straight line known as a *layer line*. With the cone containing the incoming x-ray beam counted as the zero cone, the *n*th cone contains reflections with indices *hkn* if the rotation axis is the crystallographic *c* axis. In the Weissenberg method the desired cone is allowed to reach the film through a circular slot in a cylindrical metal screen called the *layer-line screen*. The cylindrical camera can be translated parallel to its axis, and this translation is coupled to the oscillation of the crystal about this axis (Fig. 3). In this process the diffraction spots are spread out on the two-dimensional surface of the film so that their two variable indices *h* and *k* are determinable. The coordinates *x* and *z* of a spot are readily transformed into indices *h* and *k*.

The photographs taken with a Weissenberg camera are almost universally recorded by *equi-inclination*, a technique in which the primary beam enters along a generator of the cone whose reflections *hkn* are being photographed. This technique has the advantage that the shapes of rows of the reciprocal lattice as recorded on the Weissenberg film are the same regardless of the value of *n*.

With the introduction of the Weissenberg method in 1924, it became possible for the first time to determine with one instrument not only the Friedel symmetry of the crystal, but also the indices and intensities of all reflections *hkl*.

The Precession Method

In the precession instrument, a crystal is held in one gimbal and a flat film in another gimbal, as shown in Fig. 4. The unmoved points of these gimbals are aligned along the incoming primary x-ray beam, so that the beam encounters the crystal first. The vertical and horizontal axes of the two gimbals are coupled and a rational axis of the crystal is set parallel with the normal to the film surface. These axes are then inclined at an angle μ to the x-ray beam and caused to precess about it. The nest of Laue cones coaxial with the rational direction of the crystal then also precesses about the x-ray beam. If the rational axis is the crystallographic *c* axis, the reflections of the *n*th cone are *hkn*. The reflections of any chosen cone are allowed to reach the film through an annular opening in a flat layer-line screen that is attached to, and precesses with, the crystal.

Precession photographs are undistorted pictures of the reciprocal lattice, level by level. Accordingly they not only reveal the Friedel symmetry of each level, but the indices of all reflections are obvious by mere inspection. Indeed, indexing for space-group determination is unnecessary since, for example, a glide plane parallel to a level can be detected by simple comparison of the appearances of the zero level and an *n*th level; if a glide plane is present, alternate lines of spots on the zero level are missing in a direction along the translation component of the glide plane.

Diffractometer Methods

Several instruments are currently used to record the intensities of x-ray reflections with the aid of photon counters.

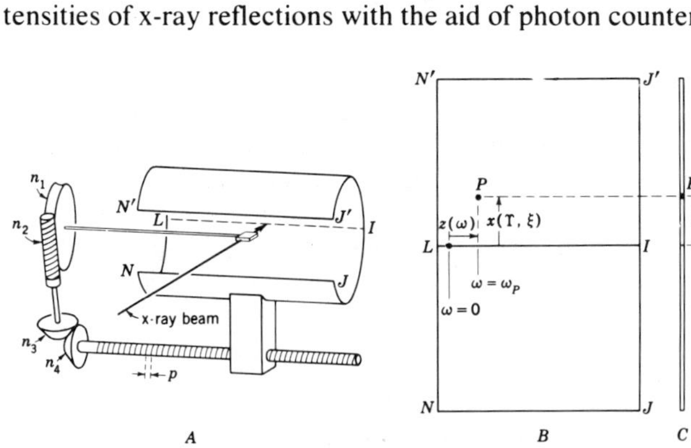

FIG. 3. Diagram of a Weissenberg instrument showing (A) the coupling of the camera translation to the crystal rotation; (B) the resulting position of a reflection on the unrolled film; (C) the position of the corresponding spot along a layer line of a rotating-crystal photograph. [From M. J. Buerger, *X-Ray Crystallography* (Wiley, New York, 1942).]

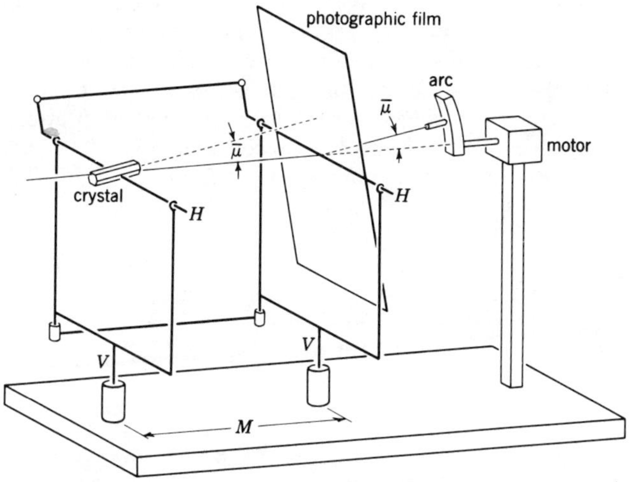

FIG. 4. Diagram of a precession instrument. The motor causes the normal to the photographic film to precess around the direction of the x-ray beam; the coupling of the crystal to the film causes a chosen rational direction of the crystal to precess around the x-ray beam also. [From M. J. Buerger, *Contemporary Crystallography* (McGraw-Hill, New York, 1970).]

Two versions have come into common use in the United States, one with Weissenberg geometry, shown in Fig. 5, and the four-circle diffractometer, shown in Fig. 6. The photons are usually counted by means of proportional counters or scintillation counters.

Diffractometers furnish more precise measurements of the diffraction intensities than the photographic methods and so are preferred for this part of the crystal-structure analysis. Formerly, diffractometers were set and operated manually. It is now usual to operate them automatically, and to arrange that their data records be presented in a form suitable for

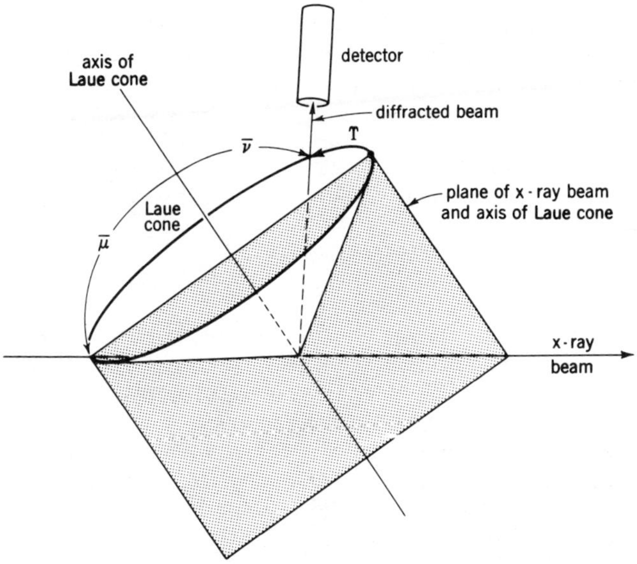

FIG. 5. Diagram of the single-crystal diffractometer with Weissenberg geometry. [From M. J. Buerger, *Contemporary Crystallography* (McGraw-Hill, New York, 1970).]

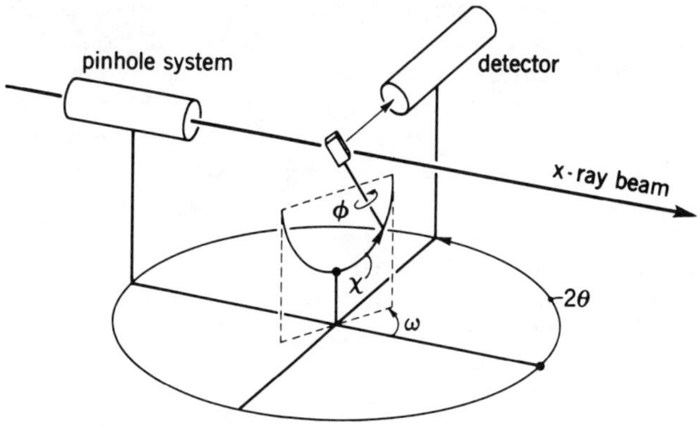

FIG. 6. Diagram of the four-circle diffractometer. [From M. J. Buerger, *Contemporary Crystallography* (McGraw-Hill, New York, 1970).]

the computer that corrects the raw data for various experimental factors.

FOURIER TRANSFORMATIONS

A crystal structure and its diffraction amplitudes are Fourier mates, so that each can be found if the other is known. The diffraction amplitude $F(hkl)$ is given in terms of the electron density $\rho(xyz)$ by

$$F(hkl) = V\int_0^1\int_0^1\int_0^1 \rho(xyz)\exp[i2\pi(hx+ky+lz)]\,dx\,dy\,dz. \quad (13)$$

Here x,y, and z are coordinates for the electron density expressed in terms of fractions of the cell edges a,b, and c, respectively. It is usually convenient to regard the electron density as composed of a set of j atoms whose scattering powers are f_j, expressed in units of the scattering power of a free electron. The amplitude is then given by

$$F(hkl) = \sum_j f_j(hkl)\exp[i2\pi(hx_j+ky_j+lz_j)]T_j. \quad (14)$$

Here x_j,y_j, and z_j are coordinates of the atoms in fractions of the cell edge, and T_j is an exponential temperature factor.

To recover the electron density $\rho(XYZ)$ from the $F(hkl)$'s, the relation is

$$\rho(XYZ) = V^{-1}\sum_h\sum_k\sum_l F(hkl)\exp[-i2\pi(hX+kY+lZ)]. \quad (15)$$

Any crystal can be solved for its electron density by making use of (15) if the diffraction amplitudes $F(hkl)$ are fully known. As seen in (13) and (14), however, these are complex quantities, and all that can be determined by experimental methods are the magnitudes $|F(hkl)|^2$. The phases are not recorded by any technology known at this time, so (15) cannot, in fact, be solved directly. The ways used to circumvent this difficulty are noted later.

THE RAW DATA AND THEIR TREATMENT

The Integrated Reflection

It is usual to measure the radiation reflected by a crystal by the *integrated reflection*, which is given by

$$\frac{E(hkl)\omega}{I_0} = \left(\frac{e^4}{m^2c^4}\right)\lambda^3N^2\,\Delta V\,Lp|F(hkl)|^2. \quad (16)$$

Here $E(hkl)$ is the energy reflected in the orders hkl by a tiny fragment of crystal composed of N cells and volume ΔV (so small that absorption and extinction can be ignored) as the crystal is rotated with angular velocity ω in a beam of radiation whose intensity is I_0 and whose wavelength is λ;L and p are the Lorentz and polarization factors, while e,m, and c are universal constants. The left-hand side of (16) has dimensions of power×radians, and represents the power of the reflection hkl integrated over its angular range. Since the terms before L are fixed for the entire experiment, (16) can be written as

$$E(hkl)\omega/I_0 = KLp|F(hkl)|^2 \qquad (17)$$

where K is a constant for the experimental setup.

Measurement of Intensities

Integrated reflections can be measured on photographs made with a precession or Weissenberg instrument. This is easy if the instrument is fitted with an integrating film cassette. This attachment shifts the film by a uniform amount at the end of each cycle so that the spot caused by each reflection overlaps its duplicates over a small area. The composite spot has a plateau whose photographic density, measured by a photometer, can be converted to an x-ray exposure that is proportional to the integrated intensity.

Better values of the integrated reflections can be obtained with a single-crystal diffractometer. The reflections can be automatically integrated by counting while the crystal is uniformly rotated through the region of the Bragg reflection; alternatively, the crystal may be held fixed at equally spaced angular intervals for equal times while counting, and the results summed. In either case the background is measured and subtracted.

Correction of Raw Data

To transform the integrated reflections into $|F(hkl)|^2$'s, they must be corrected for the Lorentz factor, polarization, absorption, and extinction. The actual corrections are made by digital computers. The Lorentz, polarization, and absorption factors may be somewhat complicated, but offer no serious problems, although the absorption may be quite severe. The effects of extinction are most difficult to allow for.

DEDUCTION OF THE CORRECT ARRANGEMENT OF ATOMS

The Heavy-Atom Method

In spite of the phase problem, the structures of certain crystals are solvable by Fourier synthesis of the electron density. Such crystals are characterized as consisting of a small number of heavy atoms among many lighter ones. In not unfavorable cases it is possible to find the locations of the heavy atoms since they are present in small numbers and so usually occupy equivalent positions with symmetry-restricted coordinates. Furthermore, it is nearly always possible to locate the heavy atoms with the aid of the Patterson function.

The phases and amplitudes of many of the $F(hkl)$'s of (14) are dominated by the heavy atoms. These phases can then be used as preliminary phases of the coefficients in (15), from which a fair approximation of the electron density can be obtained. This reveals not only the heavy atoms, but some of the light ones as well. From this improved set of atoms an improved set of phases can be computed by (14), which can then be used in another Fourier summation (15). This iteration eventually comes to an end because no phases computed by (14) need to be changed.

The heavy-atom method has been responsible for the solution of a great many structures. Different authorities have suggested conditions under which the method works well. One is that $\Sigma f_H^2 \approx \Sigma f_L^2$; another is $\Sigma Z_H/\Sigma Z_L \geqslant 0.3$.

The Patterson Function

In 1934 A. L. Patterson showed that the centrosymmetrical function

$$P(uvw) = V^{-2} \sum_h \sum_k \sum_l |F(hkl)|^2 \cos[i2\pi(hu+kv+lw)] \qquad (18)$$

has high values at coordinates uvw whenever the vector from the origin to uvw is the same as a vector between two atoms p and q in the crystal cell; i.e., when

$$u = \pm(x_q - x_p), \quad v = \pm(y_q - y_p), \quad w = \pm(z_q - z_p). \qquad (19)$$

The magnitude of $P(uvw)$ is proportional to $f_p f_q$, so that $P(uvw)$ is large when p and q are both heavy atoms. Thus the vector distance between heavy atoms is easy to determine, and if the atoms are related by symmetry they can be easily located in the cell.

While the Patterson function provides the interatomic distances in a crystal structure, it does not, in itself, solve the structure in a direct way. The function suffers from the serious defect that if there are n atoms in the crystal cell, there are n^2 high values in the cell of $P(uvw)$. The clutter due to this overlap increases as n^2, so the usefulness of the function decreases as the number of atoms in the cell increases.

In 1950 it was shown that, in principle, a Patterson function can be solved for the electron density, and later *image-seeking functions* were presented for this purpose. Many crystal structures have been solved by a variety of this function known as the *minimum function*. Such solutions take place in Patterson space, which is a variety of crystal space. These have not been as popular as the direct methods for solving crystal structures in Fourier space, which are now noted.

Direct Methods in Fourier Space

In 1947 David Harker and J. S. Kasper presented some inequalities, based upon Schwarz's inequality, that related the phases of certain reflections. Two years later they solved the crystal structure of $B_{10}H_4$ with the aid of such inequalities. This was merely the first of many contributions by many x-ray crystallographers that culminated in many kinds of direct methods of solving the phase problem. Many structures, especially those of organic crystals, have been solved recently by what is known as the *symbolic addition method*, introduced by J. Karle and I. Karle in 1963. This method, which is especially suited to crystals composed of atoms having about the same scattering powers, is usually successful for crystals having cells containing up to 100 atoms.

REFINEMENT

An approximately solved structure becomes more exactly defined by successive Fourier synthesis, but the final synthesis suffers from errors due to the arbitrary termination of the infinite Fourier series. It also suffers from lack of knowledge of the exact temperature factor. Corrections can be applied by several devices, the most important of which is the *difference Fourier synthesis*. This uses coefficients ($F_{obs} - F_{cal}$) for each reflection *hkl*. The result of this synthesis is the difference between the observed and calculated electron densities, which shows the magnitudes and directions in which the model must be changed to bring it into conformity with the actual structure.

The progress of refinement is followed by computing at each stage the discrepancy index, a residual

$$R = \frac{||F_{obs}| - |F_{cal}||}{|F_{obs}|}. \qquad (20)$$

The value of R is usually about 5% for an acceptably refined structure, and may be as small as 2% for a very well-refined structure. It is 82.8% for a completely wrong model if centrosymmetrical, and 58.6% if noncentrosymmetrical. A preliminary model that has $R \approx 45\%$ is regarded as promising.

The use of least squares in fitting an approximately correct model to the observed amplitudes was first suggested by E. W. Hughes in 1941 and applied to refining the structure of melamine. Least-squares refinement became popular with the rise of electronic digital computers, and currently is the standard way of finishing the details of the structure. Because the structures of most inorganic and the simpler organic structures are overdetermined by the large numbers of their diffraction-amplitude data, and because modern computers have large capacities, it is usual not only to refine the three coordinates of each atom, but also to determine the three principal axes of the ellipsoids of thermal motion of each atom and their orientations with respect to the cell axes.

See also CRYSTAL SYMMETRY.

BIBLIOGRAPHY

A. V. Azaroff, *Elements of X-Ray Crystallography*. McGraw-Hill, New York, 1968. (I)

M. J. Buerger, *X-ray Crystallography*. Wiley, New York, 1942. (E)

M. J. Buerger, *Vector Space*. Wiley, New York, 1959. (I)

M. J. Buerger, *Crystal-Structure Analysis*. Wiley, New York, 1960. (I)

M. J. Buerger, *The Precession Method*. Wiley, New York, 1964. (E)

M. J. Buerger, *Contemporary Crystallography*. McGraw-Hill, New York, 1970. (E)

A. I. Kitaigorodskii, *The Theory of Crystal Structure Analysis* (trans. by David and Katherine Harker). Consultants Bureau, New York, 1961. (A)

H. Lipson and W. Cochran, *The Determination of Crystal Structures*. Cornell Univ. Press, Ithaca, NY, 1966. (I)

M. M. Woolfson, *Direct Methods in Crystallography*. Oxford (Clarendon), London and New York, 1961. (I)

M. M. Woolfson, *An Introduction to X-Ray Crystallography*. Cambridge, London and New York, 1970. (I)

Cyclotron Resonance

C. C. Grimes

Cyclotron resonance denotes the resonant absorption of power from an alternating electromagnetic field by charged particles orbiting about a steady magnetic field. The resonance occurs when the frequency of the alternating field is equal to the orbital frequency of the particles. This phenomenon has been extensively employed to study the properties of electrons and holes in solids, and electrons and ions in plasmas. It forms the basic principle utilized in cyclotron particle accelerators.

When a charged particle moves in a magnetic field, it is acted on by the Lorentz force, which continuously rotates the component of the particle's velocity that is perpendicular to the magnetic field. Consequently, the particle moves along a helical orbit with the axis of the helix parallel to the field. The particle's orbital motion is then periodic, having an angular frequency $\omega_c = qB/m^*c$, which is called the cyclotron frequency. Here q/m^* is the charge-to-mass ratio for the particle, B is the magnetic flux density, and c is the velocity of light. If an electromagnetic field having angular frequency equal to ω_c acts on an orbiting particle in a direction that accelerates it, then the particle continually gains energy from the electromagnetic field. The transfer of energy from an electromagnetic field to charged particles is employed in cyclotrons and leads to the resonant absorption of power observed in cyclotron resonance. For a particle to experience a well-defined resonant interaction, it must spend enough time in an orbit to complete a substantial fraction of a revolution about the magnetic field. This condition is expressed by $\omega_c\tau \geq 1$ where τ is the characteristic time between scattering events. The scattering time is limited in solids by the density of impurities, crystal defects, and lattice vibrations, and in plasmas by the density of ions, neutral atoms, and molecules. Examples of typical scattering times, magnetic fields, and frequencies are given below.

Historically, the first observation of cyclotron resonance was in the ionosphere where it was observed that the absorption of radio waves had a peak at a frequency of 1500 kHz. This absorption peak was identified as due to cyclotron resonance of electrons moving in the earth's magnetic field, which has an average flux density of 0.5 G in the ionosphere. Resonance at such a low frequency is possible because the density of scatterers is very low and the electron scattering time is relatively long ($\tau \approx 10^{-5}$ s). Similarly, in gaseous plasmas it is relatively easy to achieve suitable conditions to study cyclotron resonance of electrons or even ions. Since the charge-to-mass ratio of ions is much smaller than for electrons, ion cyclotron resonance occurs at proportionately lower frequencies for the same magnetic field. Cyclotron resonance of electrons and ions in gaseous plasmas constitutes a useful diagnostic tool and provides a means by which energy can be coupled into a plasma to heat it.

Cyclotron resonance has been widely applied to the study of the energy band structure of solids. Its utility lies in the fact that the shapes of the constant-energy surfaces in momentum space (Fermi surfaces) for holes and electrons can be deduced from the variations of the effective masses m^* with magnetic field direction relative to the crystal axes. Typical scattering times in crystals of high purity and perfection cooled to a few degrees Kelvin are $\tau \approx 10^{-11}$ s. In semiconductors m^* is typically $0.1m$ where m is the electron mass. To satisfy the condition $\omega_c \tau \geq 1$, microwave frequencies of 10^{10}–10^{11} Hz are commonly employed in conjunction with magnetic fields of 1–10 kG. In ionic crystals, electrons interact strongly with optical phonons to form polarons, which have frequency-dependent effective masses. Polarons have been studied by cyclotron resonance at frequencies from the microwave region to the infrared region in conjunction with magnetic fields ranging from a few kilogauss to well above 100 kG.

In semimetals and metals the skin effect restricts high-frequency electromagnetic wave penetration to a very thin layer at the surface of a specimen. To achieve cyclotron resonance, the static magnetic field is applied parallel to the specimen surface, so that orbiting electrons and holes near the surface will repeatedly pass through the skin layer and be accelerated by the electromagnetic field for a brief interval during each orbital period. Cyclotron resonance studies of metals and semimetals have yielded a wealth of information on Fermi surface shapes, the density of states at the Fermi surface, the contribution of electron–phonon interactions to the density of states, and the Fermi velocities, as well as some information on the anisotropy of electron scattering times.

See also FERMI SURFACE.

BIBLIOGRAPHY

C. Kittel, *Introduction to Solid State Physics*, 4th ed., pp. 342, 380. Wiley, New York, 1971. (I)

B. Lax and J. G. Mavroides, in *Solid State Physics* (F. Seitz and D. Turnbull, eds.), Vol. 11, p. 261.) Academic Press, New York, 1960. (I,A)

Deformation of Crystalline Materials

R. Bullough

When a crystalline body is subjected to applied forces it will, in general, undergo a change of shape. If the forces are small this shape deformation can be entirely elastic, in which case the atomic structure also suffers a homogeneous deformation (referred to as a homogeneous lattice deformation) that is identical to the shape deformation; removal of the forces after such a deformation results in the body reverting to its original shape. Such a deformation may be termed a simple elastic deformation. On the other hand, if the applied forces are sufficiently large then the formation and movement of extended lattice defects, such as dislocations and twin lamellae, will occur, which ensures that a lattice-invariant or plastic deformation occurs and contributes to the total shape deformation of the body. Removal of the forces now results in the body only partly recovering its original shape. Such a deformation may be termed an elastic–plastic deformation. A complete survey of the physical processes that contribute to the onset and development of the plastic deformation and its detailed contribution to the total shape deformation in, generally, a polycrystalline body is not feasible in this article; it will therefore suffice merely to highlight some of the more important factors involved and refer the reader to the general references for appropriate amplification.

The onset of observable bulk plastic deformation is defined by the *yield strength* of the material, which is simply the applied stress required to produce some specified plastic deformation, usually a few tenths of a percent. This essentially macroscopic concept of yielding implies the *simultaneous sustained* movement of large numbers of dislocations. A single dislocation will glide under an applied (resolved) shear stress, referred to as the Peierls stress for a pure crystal or as the dislocation flow stress, that is much lower than the bulk yield stress and, in fact, if all the preexisting dislocations were simply driven out of the crystal by the applied stress, the total plastic strain would be quite small for annealed materials. It follows that some mechanism for the continuous generation of dislocations is essential and that the macroscopic yield stress is that stress required to generate the dislocations rather than simply to move individual dislocations. Several dislocation sources have been suggested, of which the Frank–Read source is probably the most important, since it clearly demonstrates how certain arrangements of dislocations can interact under the applied stress to generate, purely geometrically, a large number of dislocations and thereby help to define, in dislocation terms, the observed slip lines often associated with plastic deformation.

However, it is certain that the Frank–Read source is not the only athermal source of dislocations, and it is probably not even the most prevalent. Dislocations can be rather easily emitted from various internal stress concentrations, such as surface imperfections and grain boundary irregularities; the precise atomic mechanisms involved have not, as yet, been clearly identified. Since the free energy per atom spacing along a dislocation is about 7 eV, it is quite impossible for thermal fluctuations to generate dislocations spontaneously. Special configurations of mobile dislocations have been identified with observable macroscopic features of the plastic deformation. For example, when the mobile dislocations, which may be causing slip on two or more intersecting crystallographic planes, are confined to a broad band, such a band is referred to as a *deformation band*. If the dislocation motion is confined to a single set of parallel crystallographic planes (single slip) and moreover occurs in a band approximately normal to the slip planes we have a *kink band*. In polycrystalline materials the strain associated with the region of the body where plastic flow begins will often lead to effective softening adjacent to the region, with the result that the plastic zone propagates as a band through the entire crystal. Such deformation bands are often easily visible and are referred to as *Lüders bands*. The plastic deformation of a polycrystalline aggregate involves each grain plastically deforming subject to compatibility with its adjoining grains. Such successful accommodation requires, in general, at least five operative slip systems in each grain to ensure the requisite arbitrary shape changes. Large plastic deformation will often severely distort the grains out of their original shape and tend to rotate the operating slip planes toward the direction of deformation, producing preferred orientations of the grains. In addition to dislocation motion causing the plastic deformation, a further contribution to the lattice-invariant deformation, which is usually conceptually subsumed in the total plastic deformation, can arise by the formation of mechanical twin lamellae. The essential relation between the deformation arising from the dislocation movement and that arising by twinning is perhaps clarified by regarding the twin boundary as a special "surface" dislocation whereby it separates *volumes* of crystal that differ only by a lattice *rotation*, in contrast to the slip dislocation, which is a line that separate regions of different lattice *translation*.

Although dislocations cannot be formed by thermal fluctuations, temperature will assist the movement of dislocations and at elevated temperatures the deformation will continue to increase even under constant applied stress; this phenomenon is called *creep* and the steady-state nature of creep arises from a dynamic balance between the deforming dislocation network and its simultaneous recovery by various dislocation climb processes.

As deformation proceeds, the dislocations multiply and interact with each other. This interaction combined with the overall increase in dislocation density ensures that, in general, the *flow stress* (stress required to maintain plastic deformation) will increase—the material will work-harden. In tensile deformation, however, the load-bearing capacity of the body can reach a maximum, since the effect of the reduction in the cross-sectional area is eventually greater than the increase of flow stress due to work-hardening. At greater loads the specimen will "neck" and finally fracture; the *ductility* of the material is defined by the strain or elongation to fracture.

A complete microscopic theory of plastic deformation of crystalline solids does not yet exist. There is a reasonable understanding of the mechanics of single dislocations and very simple dislocation configurations, and a mathematically sophisticated continuum theory of plasticity has been developed; there is, however, no formal bridge between the two. To achieve such a connection we require a mathematical framework for the deformation behavior of a complex dislocation network. Some progress using the continuous distribution theory of dislocations with its associated non-Riemannian representation of the imperfect crystal has been made, but the complete bridge is certainly not yet built to permit a full microscopic understanding of deformation.

See also ANELASTICITY; CRYSTAL DEFECTS; RHEOLOGY.

BIBLIOGRAPHY

A. S. Argon, ed., *Physics of Strength and Plasticity*. MIT Press, Cambridge, Mass., 1969. (I,A)

B. A. Bilby, *Prog. Solid Mech.* **1,** 329 (1960). (A)

R. Bullough, "Microplasticity," in *Surface Effects in Crystal Plasticity*, pp. 5–14. NATO Adv. Study Inst., Sept. 1975, Hohegeiss, Germany (I)

R. Bullough and J. A. Simmons, "On the Deformation of an Imperfect Solid," in *Physics of Strength and Plasticity*, p. 47. MIT Press, Cambridge, Mass., 1969. (A)

A. H. Cottrell, *Dislocations and Plastic Flow in Crystals*. Oxford (Clarendon Press), London and New York, 1953. (E,I)

R. de Wit, *Solid State Physics* **10,** 249 (1960). (E)

F. C. Frank and W. T. Read, *Phys. Rev.* **79,** 772 (1950). (I)

J. Friedel, *Dislocations*. Addison-Wesley, Reading, Mass., 1964. (I)

R Hill, *The Mathematical Theory of Plasticity*. Oxford (Clarendon Press), London and New York, 1950. (A)

J. P. Hirth and J. Lothe, *Theory of Dislocations*. McGraw-Hill, New York, 1968. (I)

U. F. Kocks, A. S. Argon, and M. F. Ashby, *Thermodynamics and Kinetics of Slip*. Pergamon, New York, 1975. (I,A)

K. Kondo, RAAG Memoirs of the Unifying Study of Basic Problems in Engineering and Physical Sciences by means of Geometry, **1,** 458 (1955). (A)

E Kröner, *Arch. Rat. Mech. Anal.* **4,** 273 (1960). (A)

F. A. McClintock and A. S. Argon, eds., *Mechanical Behavior of Materials*. Addison-Wesley, Reading, Mass., 1965. (E)

W. T. Read, *Dislocations in Crystals*. McGraw-Hill, New York, 1953. (E)

de Haas–van Alphen Effect

M. S. Dresselhaus and G. Dresselhaus

The de Haas–van Alphen effect refers to a very small magnetic-field-dependent oscillatory term in the diamagnetic susceptibility χ. This oscillatory term can be observed under suitable conditions at high magnetic fields B. The de Haas–van Alphen phenomenon is important because the period of this oscillatory effect is proportional to $1/B$ and can be very accurately related to extremal cross sections of Fermi surfaces. The effect can also be used to determine other electronic parameters, such as the cyclotron effective mass and the effective g factor for conduction electrons and holes.

This effect, first discovered in 1930 by W. J. de Haas and P. M. van Alphen, later developed into the dominant technique for the detailed study of the Fermi-surface topology for metals and semimetals. Many of the sensitive experimental techniques contributing to the widespread application of this effect were developed in the pioneering work of Shoenberg. The theoretical breakthrough was Onsager's observation that the extremal Fermi-surface cross-sectional area S_0 is simply related to the de Haas–van Alphen period. A quantitative theory based on this idea was later presented by Lifshitz and Kosevich.

Application of the Bohr–Sommerfeld quantization condition to the **k**-space orbit of electrons perpendicular to the magnetic field leads to the relation

$$S_0 = 2\pi e B_n (n+\gamma)/c\hbar \tag{1}$$

from which the de Haas–van Alphen period is defined as

$$P = \left| \frac{1}{B_n} - \frac{1}{B_{n-1}} \right| = \frac{2\pi e}{c\hbar S_0}, \tag{2}$$

where B_n is a resonant magnetic field corresponding to quantum number n, e is the charge on the electron, $\hbar$ is the Dirac–Planck constant, and c is the speed of light. The significance of this resonant magnetic field can be understood in terms of the energy levels of an electron in a magnetic field. In the effective-mass approximation, the magnetic energy levels are given by

$$E_n(k_B) = \frac{\hbar^2 k_B{}^2}{2m_B{}^*} + \frac{\hbar e B}{m_c{}^* c}, \tag{3}$$

in which $\mathbf{k}_B$ is the wave vector along the magnetic field $\mathbf{B}$, $m_B{}^*$ is the effective mass component along $\mathbf{B}$, and $m_c{}^*$ is the cyclotron effective mass corresponding to motion perpendicular to $\mathbf{B}$. Each magnetic subband n has an energy minimum at $k_B = 0$ where the density of states becomes infinite. According to Eq. (3) the energy of each magnetic subband extremum increases linearly with B. Because of the singularity in the density of states, a series of resonant responses is observed, each resonance corresponding to the passage of a magnetic energy subband extremum through the Fermi level,

$$E_F = \frac{\hbar e B_n}{m_c{}^* c}(n + \tfrac{1}{2}), \tag{4}$$

so that B_n specifies the magnetic field where subband n becomes unoccupied (occupied) with increasing (decreasing) B. From this point of view, for every value of B_n, the density of states at the Fermi level becomes infinite, thereby introducing an oscillatory magnetic-field-dependent term in the density of states. Thus any observable depending on the density of states also exhibits an oscillatory magnetic field dependence with a period $1/B$. We refer to such oscillations in the electrical conductivity as the Shubnikov–de Haas effect, in honor of its discovery in 1930 by Shubnikov and de Haas. The corresponding oscillatory phenomena have since been found in a large number of observables, including temperature of an adiabatically isolated specimen (magnetothermal oscillations), the Hall constant, the sound attenuation coefficient, the velocity of sound, the thermoelectric power, and the optical reflectivity.

For the de Haas–van Alphen effect (and the phenomena related to this effect) to be observable, it is necessary for the electronic motion to be dominated by the magnetic field. This is necessary in order to exploit the singularity in the magnetic-field-dependent density of states at each magnetic subband extremum. In contrast, there are no singularities in the zero-field density of states; in this case only a single nonresonant threshold is present at the energy-band extremum. The magnetic field dominates the electronic motion when B is large enough for a carrier to complete an electron orbit before scattering. This condition is expressed mathematically as

$$\omega_c \tau \gg 1, \tag{5}$$

where the cyclotron frequency ω_c is related to B according to

$$\omega_c = eB/m_c{}^* c \tag{6}$$

and τ is the mean time between collisions. Thus to observe the de Haas–van Alphen effect it is necessary to use high magnetic fields to maximize ω_c, and to carry out the measurements at low temperatures to minimize the collision probability and thereby maximize τ. Since the condition $\omega_c\tau \gg 1$ can be satisfied for many materials, the de Haas–van Alphen and related effects can be used to determine Fermi-surface parameters for a large number of materials.

The detailed interpretation of de Haas–van Alphen data to yield electronic parameters is made on the basis of the Lifshitz–Kosevich theory, which gives the oscillatory component of the electronic thermodynamic potential for volume V as

$$(\Omega)_{\rm osc} = 2VkT\left(\frac{eB}{hc}\right)^{3/2}\left(\frac{\partial^2 S}{\partial k_B{}^2}\right)^{-1/2}_{k_m} \sum_{j=1}^{\infty} \frac{\exp(-2\pi^2 jkT_D/\beta^* B)}{j^{3/2}\sinh(2\pi^2 jkT/\beta^* B)}$$

$$\times \cos\left[\frac{jhcS}{eB} - 2\pi j\gamma \pm \frac{\pi}{4}\right]\cos\left[\frac{j\pi g m_s{}^*}{2m_0}\right], \tag{7}$$

in which the cross-sectional area S has an extremum at $k_B = k_m$, β^* is the effective double Bohr magneton $\beta^* = e\hbar/m_c{}^* c$, $m_c{}^*$ is the cyclotron effective mass $m_c{}^* = \hbar^2(\partial S/2\pi\partial E)_{E_F}$ at the Fermi energy E_F, T_D is the Dingle temperature related to the width $\hbar/\tau$ of the Landau levels and given by $kT_D = \hbar/\pi\tau$, g is the effective g factor, $m_s{}^*$ is the spin effective mass, and m_0 is the free-electron mass. The upper or lower signs in the phase correspond to a maximum or minimum, respectively, in the extremal cross-sectional area of the Fermi surface. The parameter γ enters the theory from the quantization condition given in Eq. (1). The magnetic susceptibility is obtained from Eq. (7) through differentiation: $\chi = -(\partial^2\Omega_{\rm osc}/\partial B^2)_T$. The oscillations in χ arise from the magnetic-field-dependent cosine function in Eq. (7), and the periodicity in $1/B$ follows from the argument of this function. Since the de Haas–van Alphen period is related to the extremal Fermi-surface cross-sectional area perpendicular to $\mathbf{B}$, the shape of a given Fermi surface can be determined by variation of the magnetic field direction relative to the crystalline axes. Furthermore, the hyperbolic sine function gives rise to an exponential temperature dependence of χ, the exponential terms depending on the parameters $m_c{}^*$ and T_D. Careful temperature-dependent measurements of the amplitude of the de Haas–van Alphen oscillations thus can be analyzed to yield explicit values for $m_c{}^*$ and T_D. However, the effective-mass determinations using the de Haas–van Alphen effect are not nearly as precise as are the very accurate measurements of Fermi-surface cross-sectional areas.

Introduction of electron spin into the electronic problem results in a spin splitting of the magnetic subbands. In this case, whenever a magnetic subband extremum associated with either spin orientation (parallel or antiparallel to the applied field) crosses the Fermi level, a resonant response will be achieved. Measurement of $\chi(B)$ can therefore be used to obtain the effective g factor for conduction electrons and holes at the Fermi energy. With improved experimental de Haas–van Alphen techniques, it has also been possible to analyze de Haas–van Alphen line shapes to provide information on relaxation times and relaxation processes in metals and semimetals.

Recent experimental application of the de Haas–van Alphen effect to the free-electron-like alkali metals has opened up the exciting possibility of direct experimental measurements of "many-body effects" in an electron gas. The experiments have been interpreted to yield Landau Fermi-liquid parameters for the spin correlation or interaction function. This determination is possible because in Fermi-liquid theory the effective g factor as measured by a de Haas–van Alphen experiment differs from the corresponding quantity as determined by a microwave electron-spin resonance measurement.

The de Haas–van Alphen and related effects today represent the most accurate and widely applicable means for the determination of Fermi surface properties of metals and semimetals. Though mature, the field remains active.

See also CYCLOTRON RESONANCES; FERMI SURFACE.

BIBLIOGRAPHY

H. Alles, R. J. Higgins, and D. H. Lowndes, *Phys. Rev. Lett.* **30,** 705 (1973) (provides flavor of current work in this field).

W. J. de Haas and P. M. van Alphen, Leiden Commun. **208a,** 212a (1930) (of historical interest).

C. Kittel, *Introduction to Solid State Physics*, 3rd ed., p. 290. Wiley, New York, 1966. (E)

B. Knecht, D. L. Randles, and D. Shoenberg, *Proc. of the 13th Int. Conf. on Low Temperature Physics*. LT13 **4,** 37 (1972); J. M. Perz and D. Shoenberg, *Proc. of the 14th Int. Conf. on Low Temperature Physics*, LT14 **3,** 141 (1975) (provides flavor of current work in this field).

I. M. Lifshitz and A. M. Kosevich, Zh. Eksperim, Teor. Fiz. **29,** 730 (1955) [English translation: Soviet Phys.-JETP **2,** 636 (1956)] (a classic paper in the field). (A)

L. Onsager, *Philos. Mag.* **43** 1006 (1952) (of historical interest).

D. Shoenberg, *Phil. Trans. Roy. Soc. London* **A245,** 1 (1952) (a classic paper in the field). (I)

W. Shubnikov and W. J. de Haas, Leiden Commun. **207,** 210 (1930) (of historical interest).

J. M. Ziman, *Principles of the Theory of Solids*, p. 274. Cambridge University Press, Cambridge, 1964. (E)

Diamagnetism

Edward A. Boudreaux

The phenomenon of diamagnetism is a consequence of induction, whereby a substance experiences polarization when placed in a magnetic field. In this respect it is similar to the analogous phenomenon of paramagnetism, except that the induced magnetic polarization in diamagnetism is negative and is some hundred to a thousand times smaller than in paramagnetism. Furthermore, diamagnetic substances contain no permanent magnetic dipoles, whereas normal paramagnetic materials do. Also, there is neither a normal temperature dependence nor a magnetic field dependence associated with diamagnetism.

In discussing inductive magnetic effects it is customary to invoke the definition $K = I/H$, where K is the magnetic susceptibility per unit volume of substance, I is the intensity of magnetization of the substance, and H is the strength of the magnetic field. For diamagnetism I is negative and smaller than H, and thus the susceptibility is negative and small (i.e., $\sim 10^{-6}$ emu). The measurement of diamagnetic susceptibilities is attained through the observation of an apparent mass change which the substances exhibit when placed in a sufficiently strong magnetic field. In a diamagnetic material this is observed as a net mass loss resulting from the magnetic force tending to repel the substance from the field.

According to the atomic theory of matter, diamagnetism is a consequence of the orbital motion of electrons, and hence is a universal property of all matter. In fact the majority of substances are observed to be diamagnetic, while, relatively speaking, the net nondiamagnetic behavior of matter is rather rare in nature. Common substances such as wood, plastics, water, sand (SiO_2), nitrogen, CO_2, NaCl, $CaCO_3$, Al_2O_3, etc., and the millions of organic compounds and most biological substances are all diamagnetic. In graphite and certain metals an anomalous large diamagnetism is observed, which is also found to be somewhat temperature and magnetic field dependent. This is called "Landau diamagnetism," after the famous physicist, and results from the large delocalization of the orbital motion of electrons in such systems.

Much of the fundamental theory of diamagnetism and its applications were established by noted scientists such as Langevin, Oxley, Stoner, Pascal, Van Vleck, and others, within the first third of the 20th century. Langevin, for example, is credited with formulating the theory of atomic diamagnetism, while Pascal derived a systematic empirical means for calculating diamagnetic susceptibilities of organic compounds. Van Vleck applied quantum mechanics in extending the theory of diamagnetism from atoms to molecules. Thus he was able to show that in diamagnetic molecules a "temperature-independent paramagnetism" (TIP) must also be inherently present.

The sum total of these effects are outlined in the following flow diagram:

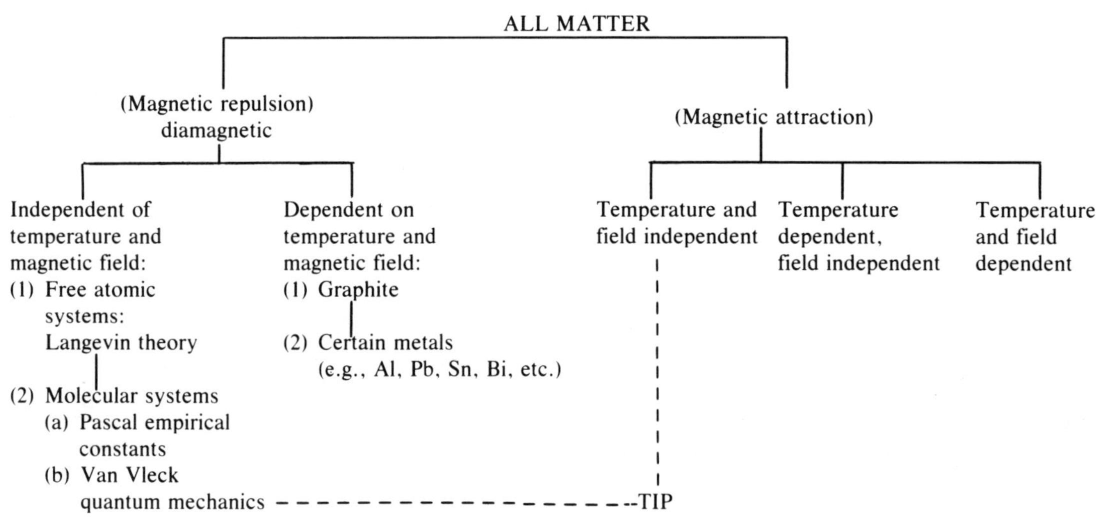

Although the application of static diamagnetic susceptibilities is not currently utilized to any significant extent, diamagnetism still remains an important diagnostic tool to the researcher.

See also PARAMAGNETISM.

BIBLIOGRAPHY

L. N. Mulay and E. A. Boudreaux, *Theory and Applications of Molecular Diamagnetism*. Wiley-Interscience, New York, 1976.

L. N. Mulay, "Techniques of Magnetic Susceptibility," in *Physical Methods of Chemistry*. (A. Weissberger and B. W. Rossiter, eds.). Wiley-Interscience, New York, 1972, Part I, Vol. IV.

L. N. Mulay, *Magnetic Susceptibility* (reprint monograph). Wiley, New York, 1966.

Ya. G. Dorfman, *Diamagnetism and the Chemical Bond*. Edward Arnold, London, 1965.

P. W. Selwood, *Magnetochemistry*, 2nd ed. Interscience, New York, 1956.

J. H. Van Vleck, *The Theory of Electric and Magnetic Susceptibilities*. Oxford, London, 1932.

E. C. Stoner, *Magnetism and Atomic Structure*. Methuen, London, 1926.

Dielectric Properties

Nicholas Bottka

The term *dielectric* applies to the material properties governing the interaction between matter and an electromagnetic field. Induced or permanent electric polarization or magnetization of matter as a function of a static or an alternating electric, magnetic, or electromagnetic field constitutes the *dielectric properties* of the material.

The macroscopic Maxwell equations describe the response of a system to an external electromagnetic field, or probe, described in space and time by the complex field vectors $\mathbf{E}(\mathbf{r}, t)$ and $\mathbf{H}(\mathbf{r}, t)$. The electric displacement $\mathbf{D}(\mathbf{r}, t)$ and the magnetic flux density $\mathbf{B}(\mathbf{r}, t)$ are material-related parameters and, in general, are not proportional to $\mathbf{E}(t)$ and $\mathbf{H}(t)$ at arbitrary times, i.e., the response of the system depends on the past history of the source $\mathbf{E}(t)$. This causality is expressed by the Fourier transform of the Maxwell relation (in cgs electrostatic units)

$$\mathbf{D}(\mathbf{k}, \omega) = \mathbf{E}(\mathbf{k}, \omega) + 4\pi i \mathbf{J}(\mathbf{k}, \omega)/\omega \tag{1}$$

where $\mathbf{k}$, the wave vector (or propagation direction) of the electromagnetic probe, will have complex components if a wave of angular frequency ω is absorbed by the medium. $\mathbf{J}$ is the current density; it describes all the material properties.

The electromagnetic field can be a time-varying *longitudinal* or *transverse* field; its interaction with a system must conserve both energy and momentum. The electromagnetic wave in optical reflectivity is an example of a transverse probe, coupling directly to the transverse current-density fluctuation of the electrons. Slow neutrons and fast electrons are examples of longitudinal probes, in that they couple directly to elementary excitations in the medium, e.g., density fluctuations in a solid such as phonons and plasmons.

Ideally the external probe is weakly coupled to the material so that the system response can be represented in terms of the properties of the excitations in the absence of the probe. One then speaks of a *linear* response of the system (as opposed to nonlinear interactions), and Eq. (1) is redefined as

$$\begin{aligned}\mathbf{D}(\mathbf{k}, \omega) &= \mathbf{E}(\mathbf{k}, \omega) + 4\pi i\, \boldsymbol{\sigma}(\mathbf{k}, \omega)\cdot\mathbf{E}(\mathbf{k}, \omega)/\omega \\ &= \boldsymbol{\epsilon}(\mathbf{k}, \omega)\cdot\mathbf{E}(\mathbf{k}, \omega),\end{aligned} \tag{2}$$

where $\boldsymbol{\epsilon}$ and $\boldsymbol{\sigma}$ are the complex *dielectric function* and *conductivity* of the material, respectively. Analogously one can define a complex magnetic *permeability* $\boldsymbol{\mu}$ relating $\mathbf{B}$ to $\mathbf{H}$. In general $\boldsymbol{\epsilon}$ is a tensor and represents the response of the system to a perturbing field.

For a transverse probe $\mathbf{E} = \mathbf{E}_0 \exp(i\mathbf{k}\cdot\mathbf{r} - i\omega t)$ in an isotropic nonmagnetic medium, the Maxwell relations for the linear case in Eq. (2) yield the dispersion relation

$$\mathbf{k}\cdot\mathbf{k} = \omega^2 \boldsymbol{\epsilon}(\mathbf{k}, \omega)/c^2. \tag{3}$$

When $\boldsymbol{\epsilon}$ is complex, Eq. (3) can be satisfied only by complex $\mathbf{k}$; all solutions correspond to damped waves in a passive medium. The converse is not necessarily true; i.e., even if $\boldsymbol{\epsilon}$ is real there is a solution with complex $\mathbf{k}$ corresponding to an evanescent wave (this wave is damped, but there is no dissipation of energy since $\boldsymbol{\epsilon}$ is real). If $\boldsymbol{\epsilon}$ and $\mathbf{k}$ are real, Eq. (3) reduces to the *Maxwell relation* $n^2 = \epsilon$, relating the refractive index n to the dielectric constant ϵ of the nonabsorbing medium. For optical frequencies, $\boldsymbol{\epsilon}$ is usually assumed to be independent of $\mathbf{k}$. The $\mathbf{k}$ dependence of $\boldsymbol{\epsilon}$ gives rise to such phenomena as the *anomalous skin effect* in metals and the interaction of light with excitons.

Historically, one of the greatest scientific achievements of our times has been the understanding of the correlation between the experimentally observed *macroscopic* quantities $\boldsymbol{\epsilon}$, $\boldsymbol{\mu}$, and $\boldsymbol{\sigma}$ and the underlying *microscopic* nature of matter. A class of interactions known as *dipole interactions* contributed substantially to this understanding. For an isotropic atomic system consisting of N elementary dipoles, the macroscopic polarization $\mathbf{P}$ (electric-dipole moment per unit volume) is related to the microscopic average dipole moment $\langle \mathbf{p} \rangle = \langle q\mathbf{r} \rangle$ (in general, represented as a sum of bound electronic, atomic, and permanent dipole moments) by

$$\mathbf{P} = N\langle \mathbf{p} \rangle = N\alpha\mathbf{E}' = (\boldsymbol{\epsilon} - 1)\mathbf{E}/4\pi = \boldsymbol{\chi}\mathbf{E}/4\pi, \tag{4}$$

where q is the charge and $\mathbf{r}$ is the displacement from equilibrium of the elementary particle; α is the *polarizability* and $\boldsymbol{\chi}$ is the *susceptibility* of the system. $\mathbf{E}'$ is the *local electric-field* strength; in addition to the external field it includes the contribution from induced and permanent dipoles. An expression analogous to Eq. (4) exists for magnetic materials relating the macroscopic magnetization $\mathbf{M}$ to the magnetic susceptibility $\boldsymbol{\chi}_M$ and the magnetic field.

For a system of noninteracting oscillating dipoles, the classical model for polarizability yields

$$\boldsymbol{\epsilon} - 1 = 4\pi q^2 \sum_s (\omega_s^2 - \omega^2 + i\gamma_s\omega)^{-1} N_s/M_s, \tag{5}$$

where N_s is the number of particles per unit volume of mass M_s in the system, ω_s is the natural resonant frequency, and γ_s is the damping constant of the dipole oscillator. The sum over s indicates the possibility of various dipoles. Figure 1 shows the behavior of the real and imaginary parts of $\boldsymbol{\epsilon} - 1$ as a function of frequency in the vicinity of ω_0 for a system of N identical noninteracting oscillators. Strong absorption is indicated by the peak in the imaginary part of $\boldsymbol{\epsilon}$. The frequency dependence of the real part of the dielectric function describes the classical dispersion characteristics of the dielectric medium. Far from ω_0 the real part of $\boldsymbol{\epsilon}$ rises with increased frequency; this behavior is called *normal dispersion* in contrast to the *anomalous dispersion* in the half-width region of the spectral line where the real part of $\boldsymbol{\epsilon}$ falls with increased frequency.

In the case of electronic polarizability where $\omega_0 \simeq 10^{15}\ \mathrm{sec}^{-1}$, the strong absorption takes place in the visible region of the spectrum. In the case of ionic polarizability [q and M in Eq. (5) refer to an ion] where $\omega_0 \simeq 10^{13}\ \mathrm{sec}^{-1}$, resonant absorption takes place in the infrared region.

Although the dispersion formula of classical physics correctly predicts the general appearance of a spectral line, neither the intensity nor the resonance frequency can be derived from such classical considerations. Moreover, the classical picture leads to a number of unexplained physical catastro-

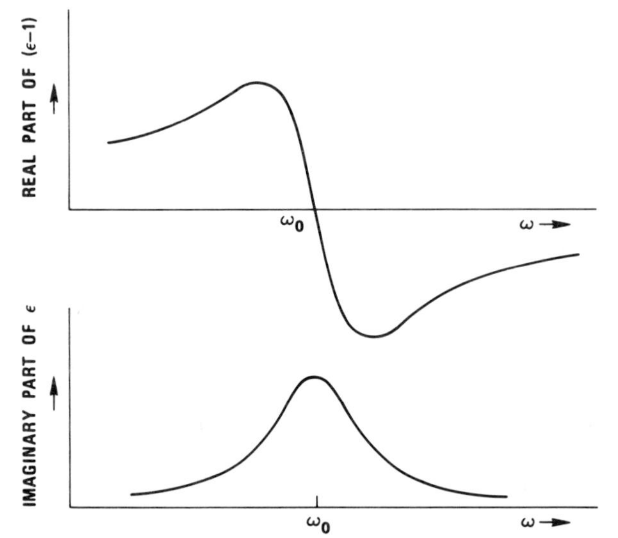

FIG. 1. The real and the imaginary parts of the dielectric function ϵ as a function of frequency near the resonance frequency ω_0.

phes. The more realistic quantum-mechanical calculation leads to the dispersion formula

$$\epsilon - 1 = (8\pi\hbar^{-1}) \sum_{f \neq j} \omega_{jf} |\langle \mu_{jf} \rangle|^2 (\omega_{jf}^2 - \omega^2 + i\gamma\omega)^{-1}(N_j - N_f), \quad (6)$$

where the resonance frequency now corresponds to a transition between two stationary quantum-mechanical energy states E_j and E_f, and $\langle \mu_{jf} \rangle$ is the matrix element of the transition coupling the electromagnetic field to the initial and final quantum-mechanical states j and f; N_j and N_f represent the number of atoms in the lower- and the higher-energy state, respectively.

In addition to the bound electronic and atomic polarizability described above, there are cases where free conduction electrons also contribute to absorption, such as in metals and semiconductors. This occurs at low frequencies, anywhere from microwave to infrared. In such a regime the dielectric function takes the form

$$\epsilon_{\text{free}} = 1 - 4\pi q^2 (N/M)/(\omega^2 - i\gamma\omega) \quad (7)$$

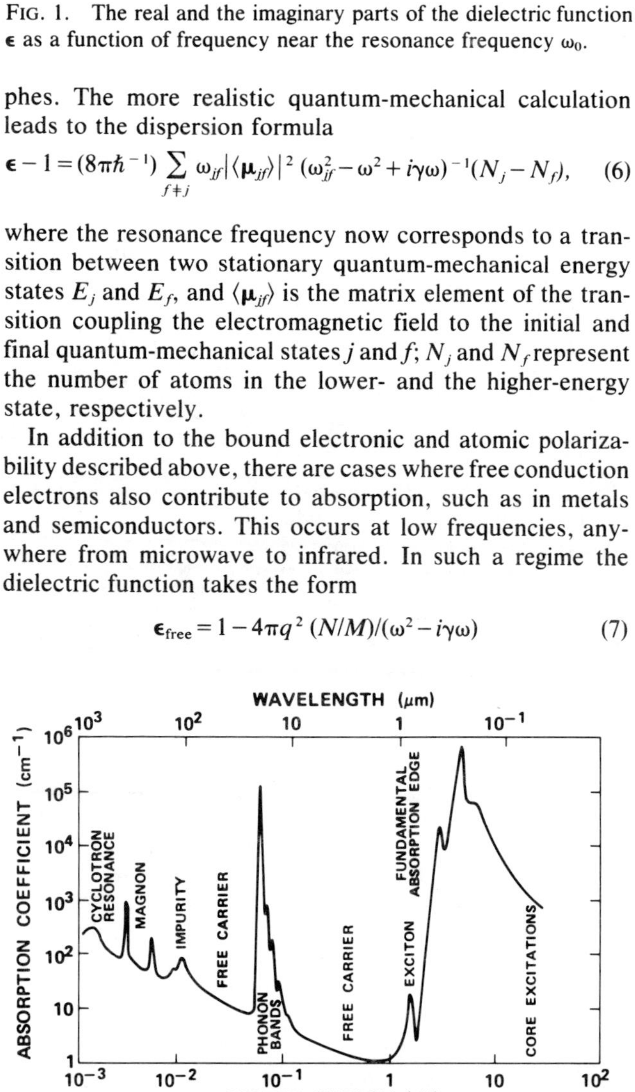

FIG. 2. The absorption spectrum of a hypothetical solid showing the frequency regime where the various excitations give rise to structure in the dielectric function.

In conductors and semiconductors, the damping constant γ_s is inversely proportional to the mobility of the free carrier type (which for intrinsic semiconductors can be both electrons and holes). At these frequencies the optical absorption coefficient is proportional to the imaginary part of ϵ, and one speaks of a "λ^2" dependence of the *free-carrier absorption*. For regions of low absorption, the real part of ϵ in Eq. (7) goes through zero at $\omega_p = (4\pi N q^2/M)^{1/2}$. This is the frequency where the free-electron gas undergoes collective density fluctuations known as *plasma oscillations*. The quanta of energy associated with these oscillations is called the *plasmon*. For most insulators and semiconductors, and many metals, the free-electron model is approximately valid in the frequency range $\omega_v < \omega < \omega_c$, where ω_v is the maximum interband transition of the valence electrons and ω_c is the threshold for interband transitions of the core electrons.

In summary, the many excitations discussed above can be put into perspective by considering the absorption spectrum of a hypothetical solid shown in Fig. 2. The strength and the frequency of the excitation give an insight into the nature of matter and its dielectric properties.

See also EXCITONS; PHONONS; PLASMONS.

BIBLIOGRAPHY

C. J. F. Böttcher, *Theory of Electric Polarization*. Elsevier, Amsterdam, 1952. (I).

A. R. von Hippel, *Dielectrics and Waves*. Wiley and Sons, New York, 1954. (I).

J. N. Hodgson, *Optical Absorption and Dispersion in Solids*. Chapman and Hall, London, 1970. (A).

D. Pines, *Elementary Excitations in Solids*. Benjamin, Advanced Book Program, Reading, Mass. 1963. (A).

Diffusion

H. B. Huntington

Diffusion is basically the process by which a concentrated quantity is spread out over a wider extent. As applied to matter in three dimensions it is the homogenization that occurs from random (or nearly random) motion in the microcosmos, apart from convective flow.

The basic quantity in diffusion, D, is called the *diffusivity* or the *diffusion constant* and is defined empirically by Fick's first law

$$\mathbf{F}_i = -D \nabla c_i \quad (1)$$

where c_i and $\mathbf{F}_i$ are, respectively, the concentration and flux of the ith constituent. (The more basic scientific approach is to recognize that the flux is proportional to the gradient of the chemical potential rather than of the concentration.) Fick's second law is a heat-flow-type equation that results from combining the first law with the continuity equation. It is applied to determine D from measurements on specimens of simple geometry.

Diffusion occurs in all phases of matter [1]. In gases the kinetic theory analysis shows that D varies as the square of the mean free path. In crystalline solids atoms move by jumps of fixed length related to the lattice constant. As a first approximation the random walk analysis gives for D

$$D = \frac{1}{6}\lambda^2\nu \qquad (2)$$

where λ is the (fixed) length of the step or jump and ν is the frequency of jumping. Liquids and amorphous solids present an intermediate situation that is less easily modeled but does appear to depend strongly on the "free-volume" character of the structure. In condensed phases the diffusional jump is almost always associated with the presence of some sort of a structural defect, such as a vacancy or an interstitial. The presence of these defects keeps the jumping process from being completely random and introduces a correlation factor, f, usually less than 1, into the expression for D in Eq. (2).

The study of diffusion in solids [2–7] has both technological and scientific importance. In the field of materials science the term has become synonymous with atom movements. As such it plays a role in all heat treatments: homogenization, aging, grain coarsening, annealing, precipitation, sintering, segregation, tempering, and the like. From the scientific side the study of diffusion provides much basic information on the energies of formation and motion of lattice defects, since diffusion data frequently show an Arrhenius-type dependence on temperature, indicating an activated process for the atom motions.

The techniques for measuring the diffusivity are numerous and depend on a wide variety of material properties. The use of radioisotopes followed by sequential sectioning has on the whole proved to be the most direct, precise, and reliable method. There are, however, several interesting techniques, such as internal friction, nuclear magnetic resonance, and Mössbauer studies, that give atom movement data by determining the jump time of the atoms rather than the overall distances that they travel.

Basic information on the mechanisms for atom motion can be obtained by measurements other than temperature dependence. The influence of pressure on diffusion shows the volume of the activated complex. The effect of impurity enhancement on diffusion can rule out certain mechanisms. Comparative studies of the motion of different radioactive isotopes of the same element can be used to help determine the extent to which a diffusing atom moves alone or as a partner in a cooperative process. Comparison between diffusion measurements and those of ionic conductivity in the appropriate materials can be used to determine, through the application of the Nernst–Einstein equation, the correlation coefficient f involved.

See also CRYSTAL DEFECTS.

REFERENCES

1. W. Jost, *Diffusion in Solids, Liquids and Gases*. Academic, New York, 1960. (I)
2. A. S. Novick and J. J. Burton, eds., *Diffusion in Solids: Recent Developments*. Academic, New York, 1975. (I)
3. H. I. Aaronson, ed., *Diffusion* (ASM Seminar for 1972). American Society for Metals, Metals Park, Ohio, 1973. (I)
4. L. A. Girifalco, *Atomic Migration in Crystals*. Blaisdell, New York, 1964. (E)
5. P. G. Shewman, *Diffusion in Solids* McGraw-Hill, New York, 1963. (I)
6. David Lazarus, "Diffusion in Metals," in *Solid State Physics* (F. Seitz, D. Turnbull, and H. Ehrenreich, eds.), Vol. 10 (1960); N. L. Peterson, "Diffusion in Metals," Vol. 22 (1968). Academic, New York. (I)
7. J. R. Manning, *Diffusion Kinetics for Atoms in Crystals*. Van Nostrand, Princeton, NJ, 1968. (A)

Elasticity

W. E. Bron

Interest in the elastic and inelastic properties of matter dates back to Galileo and has spawned in the interim an overwhelming literature on the mathematical, engineering, and physical aspects of the problem. Much of this body of work ignores the atomic nature of real solids by noting that displacements with wavelengths λ long compared to the interionic spacing a are influenced only weakly by the individual ions and can, accordingly, be treated in the continuum limit. Recently, however, a major interest in the elastic properties of solids has arisen as a result of their applicability to the determination of the interionic forces present in lattice vibrations. It is with the modern goal in mind that this survey is written.

In the long-wavelength limit, $\lambda \gg a$, the equation of motion of a displacement wave can be simply expressed in terms of Hooke's relation between the macroscopic parameters stress and strain. Since most solids are anisotropic, Hooke's law is a tensor relationship of the form

$$\sigma_{ik} = c_{ikjl}\epsilon_{jl} \tag{1a}$$

in which all indices take on the values x, y, z or alternatively 1, 2, and 3, and summation over repeated indices is implied; σ_{ik} are elements of the stress tensor, ϵ_{jl} elements of the strain tensor, and c_{ikjl} are variously called the elastic stiffness constants, the elastic moduli, or the elastic constants. (Their dimension is newtons per square meter.) In order to shorten the cumbersome indexes, a convention exists to transform them to matrix notation as follows:

tensor notation (c)	xx yy zz 11 22 33	yz zy 23 32	xz zx 13 31	xy yx 12 21
matrix notation (C)	1 2 3	4	5	6

such that Hooke's law becomes

$$\sigma_m = C_{mn}\epsilon_n . \tag{1b}$$

This transformation follows from the symmetries

$$c_{ikjl} = c_{kilj} = c_{iklj} = c_{jlik} ,$$

the first three of which are always present; the last follows from the existence of an elastic potential. In the most general case there remain 21 independent elements of the C tensor.

In the matrix notation Eq. (1) explicitly becomes

$$\sigma_1{}^{\mathrm{P}} = C_{11}\epsilon_1 + C_{12}\epsilon_2 + C_{13}\epsilon_3 + C_{14}\epsilon_4 + C_{15}\epsilon_5 + C_{16}\epsilon_6 ,$$

$$\sigma_2{}^{\mathrm{P}} = C_{21}\epsilon_1 + C_{22}\epsilon_2 + C_{23}\epsilon_3 + C_{24}\epsilon_4 + C_{25}\epsilon_5 + C_{26}\epsilon_6 ,$$

$$\sigma_3{}^{\mathrm{P}} = C_{31}\epsilon_1 + C_{32}\epsilon_2 + C_{33}\epsilon_3 + C_{34}\epsilon_4 + C_{35}\epsilon_5 + C_{36}\epsilon_6 ,$$

$$\sigma_4{}^{\mathrm{S}} = C_{41}\epsilon_1 + C_{42}\epsilon_2 + C_{43}\epsilon_3 + C_{44}\epsilon_4 + C_{45}\epsilon_5 + C_{46}\epsilon_6 ,$$

$$\sigma_5{}^{\mathrm{S}} = C_{51}\epsilon_1 + C_{52}\epsilon_2 + C_{53}\epsilon_3 + C_{54}\epsilon_4 + C_{55}\epsilon_5 + C_{56}\epsilon_6 ,$$

$$\sigma_6{}^{\mathrm{S}} = C_{61}\epsilon_1 + C_{62}\epsilon_2 + C_{63}\epsilon_3 + C_{64}\epsilon_4 + C_{65}\epsilon_5 + C_{66}\epsilon_6 . \tag{2}$$

In Eq. (2) the superscripts P and S refer to pressure (stress) and shear stress, respectively; ϵ_1, ϵ_2, and ϵ_3 refer to simple dilational displacements, whereas ϵ_4, ϵ_5, and ϵ_6 refer to shear displacements.

It follows that

$$C_{mn} = \frac{\partial \sigma_m}{\partial \epsilon_n} . \tag{3}$$

For crystals with cubic lattice symmetry

$$C_{11} = C_{22} = C_{33}, \quad C_{44} = C_{55} = C_{66}, \quad C_{12} = C_{23} = C_{31},$$

and

$$C_{14} = C_{15} = C_{16} = C_{24} = C_{25} = C_{26} = C_{34} =$$
$$C_{35} = C_{36} = C_{45} = C_{46} = C_{56} = 0. \tag{4}$$

Young's modulus Y is defined as the longitudinal stress $\sigma_1{}^{\mathrm{P}}$ per unit strain ϵ_1 when the body is not constrained in the transverse directions, i.e., when ϵ_2 and ϵ_3 are free to change. Accordingly, for a cubic crystal

$$Y_{\mathrm{cub}} = \frac{(C_{11} - C_{12})(C_{11} + 2C_{12})}{C_{11} + C_{12}} .$$

The bulk modulus B, defined in terms of the pressure P and the volume V as

$$B = -V \frac{dV}{dP} ,$$

becomes for a cubic crystal

$$B_{\mathrm{cub}} = (C_{11} + 2C_{12}),$$

and the compressibility $K = 1/B$. The shear modulus (or modulus of rigidity) S becomes

$$S_{\mathrm{cub}} = C_{44} .$$

The corresponding elastic moduli for an isotropic body are

$$Y_{\mathrm{iso}} = \frac{C_{44}(3C_{12} + 2C_{44})}{C_{12} + C_{44}} ,$$

$$B_{\mathrm{iso}} = C_{12} + C_{44} ,$$

$$S_{\mathrm{iso}} = C_{44} .$$

If the forces between lattice ions are central, the elastic constants of isotropic bodies obey the further relations

$$C_{12} = C_{66}, \quad C_{23} = C_{44}, \quad C_{31} = C_{55},$$
$$C_{14} = C_{56}, \quad C_{25} = C_{64}, \quad C_{36} = C_{45}. \tag{5}$$

For cubic crystals this becomes the well-known Cauchy relation; i.e., $C_{12} = C_{44}$.

Interest in the elastic constants in regard to lattice dynamics stems from the fact that the propagation of elastic waves in solids can be written in terms of these quantities. We define the density of the medium as ρ, and u, v, and w as the lattice displacements caused by the elastic wave in a small incremental volume Δx, Δy, and Δz in, respectively, the x, y, and z directions of the volume element. The elastic wave is generated by a stress gradient across the incremental volume. The equation of motion of the medium in, say, the x direction is then

$$\rho \frac{\partial^2 u}{\partial t^2} = \frac{\partial \sigma_1}{\partial x} + \frac{\partial \sigma_6}{\partial y} + \frac{\partial \sigma_5}{\partial z} , \tag{6}$$

and similarly for v and w. From Eq. (3) and the conditions (4) it follows for cubic crystals that

$$\rho \frac{\partial^2 u}{\partial t^2} = C_{11} \frac{\partial \epsilon_1}{\partial x} + C_{12} \left(\frac{\partial \epsilon_2}{\partial x} + \frac{\partial \epsilon_3}{\partial x} \right) + C_{44} \left(\frac{\partial \epsilon_6}{\partial y} + \frac{\partial \epsilon_5}{\partial z} \right). \quad (7)$$

The elements of the strain tensor are, of course, themselves related to the displacements. Specifically

$$\epsilon_1 = \frac{\partial u}{\partial x}, \quad \epsilon_2 = \frac{\partial v}{\partial y}, \quad \epsilon_3 = \frac{\partial w}{\partial z},$$

$$\epsilon_4 = \frac{\partial v}{\partial z} + \frac{\partial w}{\partial y}, \quad \epsilon_5 = \frac{\partial u}{\partial z} + \frac{\partial w}{\partial x}, \quad \epsilon_6 = \frac{\partial u}{\partial y} + \frac{\partial v}{\partial x}.$$

Therefore Eq. (7) can be rewritten as

$$\rho \frac{\partial^2 u}{\partial t^2} = C_{11} \frac{\partial^2 u}{\partial x^2} + C_{44} \left(\frac{\partial^2 u}{\partial y^2} + \frac{\partial^2 u}{\partial x^2} \right) + (C_{12} + C_{44}) \left(\frac{\partial^2 v}{\partial x \partial y} + \frac{\partial^2 w}{\partial x \partial z} \right). \quad (8$$

Similar equations of motion exist for $\partial^2 v/\partial t^2$ and $\partial^2 w/\partial t^2$.

Solutions to these equations can be written in the form

$$u = u_0 \exp[i(kx - \omega t)], \quad (9)$$

in which k is the wave vector of the wave and ω its angular frequency. The magnitude of the k vector is related to the wavelength by $k = 2\pi/\lambda$, and ω is related to the linear frequency ν by $\omega = 2\pi\nu$. The velocity of propagation is given by $v = \nu\lambda = \omega/k$. Solutions exist for both longitudinal (l) and transverse (t) waves. That is, respectively, waves for which the displacement is in the same direction as the propagation direction and (two) waves in which the displacement is perpendicular to the propagating direction. The velocities depend then on the propagation direction. For waves propagating along the [100], [110], and [111] symmetry directions of cubic crystals the relations determined by substituting Eq. (9) into Eq. (8) and displayed in Table I can be used to determine the propagation (sound) velocity of the displacement waves.

The relations of Table I are, in fact, the ones that permit accurate measurements to be made of the elastic constants. Modern methods of determining the elastic constants proceed by measuring the sound velocity propagating in the body under observation, in various directions and polarizations, until enough relations of the type of Table I are obtained to specify the number of independent elastic constants. In these experiments piezoelectric transducers (such as quartz crystals) are excited to produce ultrasonic pulses whose time of flight across a known length of material yields the desired velocity measurement.

In modern (classical) models for the lattice vibrations of real crystals the ions of the lattice are replaced by nondeformable masses ("cores") and the outer electrons are replaced by deformable "shells." The cores and shells are interconnected by springs that simulate the interionic and polarization forces. In lowest order the springs are considered to be harmonic; i.e., the force exerted by the spring is directly proportional to its change in length. The solutions to the equations of motion of such systems of particles and forces also result in displacement waves whose wave vector and frequency vary over a wider range than those of the waves discussed earlier. The form of these waves is, however, related to that of those of Eq. (9). In fact, in the harmonic limit and in the limit of long wavelength (i.e., small k vectors) the solutions obtained from the lattice of discrete masses and springs reduce to the sound waves in elastic media. This means that the elastic constants, as defined earlier, can be used in part to define the properties of the springs in the dynamical model. This procedure is extensively and successfully performed at this time.

Experimentally the elastic constants are found to depend on temperature. In general they tend to decrease in magnitude with increasing temperature. This behavior results in part from the expansion of the crystal lattice with temperature. Clearly, as the lattice expands the interionic forces change. In addition, however, increasing temperature results in an increase in the average amplitude of vibration of the lattice ions. The combined effects have been shown to be the major source of the observed temperature dependence of the elastic constants, at least in simple solids.

So far the discussion has been limited to small lattice displacement and lattice vibrations in the "harmonic" limit for which the elastic energy stored in the lattice is described by terms up to second order in the strain components. [Within the elastic limit the lattice energy is the integral over the product of the stress and the strain, and hence, according to Eq. (1a), is proportional to the square of the strain components.] Effects resulting from even higher-order products of the strain components are described in elasticity theory by higher-order elastic constants. Third-order elastic constants can be measured through sound-velocity measurements on statically stressed solids, and the relationships between these and even higher-order elastic constants can be derived.

See also ANELASTICITY; DEFORMATION OF CRYSTALLINE MATERIALS; LATTICE DYNAMICS; RHEOLOGY.

Table I Sound velocities

[100] direction	[110] direction	[111] direction
$\rho v_l{}^2 = C_{11}$	$\rho v_l{}^2 = \frac{1}{2}(C_{11} + C_{12} + 2C_{44})$	$\rho v_l{}^2 = \frac{1}{3}(C_{11} + 2C_{12} + 4C_{44})$
$\rho v_t{}^2 = \rho v_{t'}^2 = C_{44}$	$\rho v_t{}^2 = C_{44}$	$\rho v_t{}^2 = \rho v_{t'}^2$
	$\rho v_{t'}^2 = \frac{1}{2}(C_{11} - C_{12})$	$= \frac{1}{3}(C_{11} - C_{12} + C_{44})$

BIBLIOGRAPHY

Theory of Elasticity

L. D. Landau and E. M. Lifshitz, *Theory of Elasticity*. Addison-Wesley, Reading, Mass., 1959. (A)

H. Leipholz, *Theory of Elasticity*. Nordhoff, Leyden, 1974.

A. E. H. Love, *The Mathematical Theory of Elasticity*, 1st ed. Dover, New York, 1892. (A)

S. P. Timoshenko and J. N. Gordier, *Theory of Elasticity*, 1st ed. McGraw-Hill, New York, 1934. (I)

Application to Solid-State Physics

N. W. Ascroft and N. D. Mermin, *Solid State Physics*. Holt, Rinehart & Winston, New York, 1976. (I)

C. Kittel, *Introduction to Solid State Physics*, 4th ed., Chapter 4. Wiley, New York, 1971. (E)

C. Zener, *Elasticity and Anelasticity of Metals*. Univ. of Chicago Press, Chicago, 1948. (E)

Measurement and Values of Elastic Constants

O. L. Anderson, *Physical Acoustics* (W. P. Mason, ed.), Vol. IIIB, pp. 43–95. Academic, New York, 1965. (E)

R. Truell, C. Elbaum, and B. B. Chick, *Ultrasonic Measurements in Solid State Physics*. Academic, New York, 1969. (E)

Higher-Order Elastic Constants

R. A. Cowley, *Adv. Phys.* **12,** 421 (1963). (I)

R. C. Hollinger and G. R. Barsch, *J. Phys. Chem. Solids* **37,** 845 (1976). (I)

J. M. Ziman, *Electrons and Phonons*, Chapter 3. Oxford (Clarendon Press), London and New York, 1960.

Application to Lattice Dynamics

W. Cochran and R. A. Cowley, *Handb. Phys.* **25/2a,** 59 (1967). (A)

J. A. Reissland, *The Physics of Phonons*. Wiley, New York, 1973.(I)

Electroluminescence

M. N. Kabler

Electroluminescence is the efficient generation of light in a nonmetallic solid by an applied electric field. Electroluminescence is "cool" light in the sense that the brightness is far above that characteristic of the temperature alone; thus incandescent light is excluded.

The primary electronic states of nonmetallic solids comprise two bands of allowed states separated by a forbidden gap where only states due to impurity atoms or lattice imperfections can exist. At normal temperatures the higher or conduction band (CB) is empty except for a few mobile electrons, while the lower or valence band (VB) is filled with electrons except for a few vacant states called holes, which are also mobile. Because of interactions with thermal vibrations, any electrons in the CB immediately fall to its low-energy edge and any holes in the VB rise to its high-energy edge. When excess electrons and holes are produced and brought into proximity by the action of an applied electric field, the electrons spontaneously fall into or recombine with the holes. A recombination event releases energy comparable to the band gap, which is dissipated as heat or radiated as electroluminescence.

p-n JUNCTION ELECTROLUMINESCENCE

The most effective structure for producing electroluminescence is the *p-n* junction. Such a junction is illustrated schematically in Fig. 1, which indicates how electron energy varies as a function of distance perpendicular to the junction both with and without an applied electric field. The device is commonly called a light-emitting diode, or LED. Semiconductors selectively doped with impurity atoms can, at normal temperatures, exhibit high conductivities arising either from extra electrons or extra holes. In the former case the extra electrons come from donor states D near the CB edge, and the material is termed n type; in the latter case holes are created when VB electrons are trapped in acceptor states A, and the material is p type. When n-type and p-type regions are made contiguous with each other, a *p-n* junction is formed. Being nearly void of mobile electrons and holes, the interface region acts as an insulator and is the location of a strong electric field. Under a constant forward bias of a few volts, applied through suitable conducting contacts, electrons and holes are swept in opposite directions across the junction, where they can recombine.

The energy of an emitted photon is within the range of transparency of the material, that is, no higher than the band gap. Although many materials have band gaps in the ultraviolet, efficient LEDs operating beyond the yellow-green range are not yet available. This is largely due to difficulties in producing *p-n* junctions in the wider-band-gap materials.

LUMINESCENT PROCESSES

In general, electrons and holes can recombine near atomic impurities as well as in regions of the unperturbed lattice. The color of the luminescence and the relative probability of producing light instead of heat are characteristic of the particular impurities involved. In many semiconductors, luminescent transitions from the bottom of the CB to the top of the VB cannot take place without the intervention of a lattice vibration or imperfection in order to satisfy the law of conservation of momentum. Such luminescent transitions are termed indirect, and the material is said to have an indirect edge. An important example is GaP, with a band gap in the green spectral region near 2.3 eV (540 nm). In other materials with so-called direct edges this transition can occur without restriction; GaAs is an example, its band gap being in the infrared near 1.5 eV. Because of the momentum-conservation requirement, an indirect luminescent transition is relatively unlikely in pure materials and thus electroluminescence is inefficient. However, a crystal with an indirect edge can be doped with impurity atoms which sequentially trap both electrons and holes, thereby providing a radiative recombination path up to several orders of magnitude more efficient. Such a path is included in Fig. 1, where I represents the localized impurity state.

LIGHT EMITTING DIODES

Commercial activity has concentrated on developing efficient electroluminescent devices in the visible spectral range. Most materials with band gaps in this range, partic-

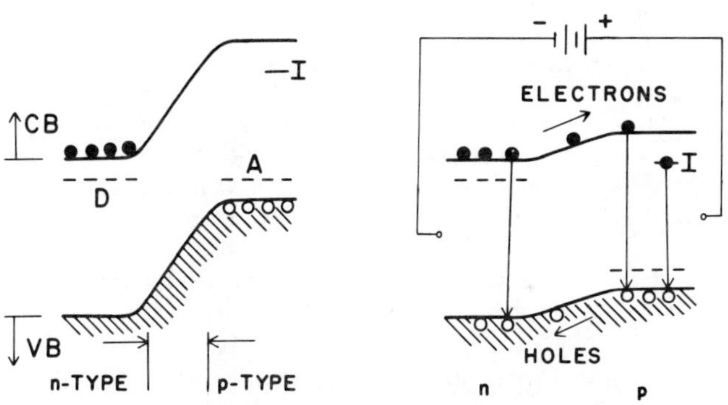

FIG. 1. Schematic representation of electron energies in a *p-n* junction with no bias voltage (left), and with a forward bias in the 1 V range (right). The vertical arrows represent electroluminescent processes, one of which occurs through an impurity level *I*.

ularly toward the violet, cannot be made to incorporate *p-n* junctions, and only compounds and alloys of elements from groups IIIA and VA of the periodic table have thus far yielded LEDs of broad utility.

For the green spectral region, N doping of GaP has proved quite effective in mitigating the inefficiency due to the indirect edge while at the same time retaining an emission wavelength near the band-gap energy. N is isoelectronic with P, for which it substitutes in the lattice. Thus the N normally carries no net charge, and its attractive potential is short range; it traps a conduction electron in a strongly localized orbit and is, for this reason, very effective in providing an interaction through which momentum is conserved during recombination with an approaching hole.

Other dopants, for example, Zn and O incorporated in the lattice as nearest-neighbor pairs, can produce efficient red luminescence in GaP. However, considerable success in the yellow-to-red range has been achieved by N doping of GaAs-GaP alloys. The alloy composition can be chosen to give a particular band-gap energy and emission color. Impurity doping to increase the efficiency of emission is not required for the red and near infrared, since the corresponding alloys have direct edges. The same is true for alloys of AlAs and GaAs.

In many of these *p-n* junction devices the efficiency with which electrical energy is converted to light inside the material can be 10% or greater at room temperature. However, because of geometrical constraints, internal reabsorption, and internal reflection at the semiconductor–air interface, only a small fraction of the photons created actually can emerge into a useful beam. External efficiencies of commercial LEDs generally lie in the 0.1–5% range. The efficiencies of luminescent processes themselves invariably fall at higher temperatures, and thus light output is limited by internal heating due to nonradiative processes. The prominent semiconductors Si and Ge find little utility as LEDs, because their band edges lie in the infrared and are indirect.

LASER DIODES

Electroluminescent lasers can be made by fabricating special *p-n* junction configurations. Since light amplification requires population inversion, the states near the bottom of the conduction band must be more than half filled with electrons and states near the top of the valence band must be more than half empty, i.e., more than half filled with holes. This requires large hole and electron currents into the recombination region and constriction of this region to a small volume in order to maintain high concentrations. An example of an LED which can be driven into laser oscillations is GaAs heavily doped to increase electron and hole concentrations. However, relatively high threshold current densities are required. The most efficient laser diodes are based on the heterojunction, which is a *p-n* junction similar to that of Fig. 1 except that there are two different materials with different band gaps on opposite sides of the junction. This effectively restricts recombination to that side of the junction having the smaller band gap, as well as providing a refractive-index gradient which aids in confining the light.

Heterojunctions are formed by depositing thin layers of the different semiconductors on each other, usually from the liquid phase. When contiguous layers differ in lattice parameter, for example, InAs on GaAs, there occurs near the interface an array of defects or dislocations which act as nonradiative recombination centers and thereby drastically lower the luminescent efficiency. Thus heterojunctions are usually constructed from alloys with nearly identical lattice parameters. A prime example is the AlAs-GaAs system, in which the lattice parameter changes by only 0.14% throughout the entire range of composition. A continuous-wave laser comprising two heterojunctions from AlAs-GaAs alloys is depicted alongside an ordinary sewing needle in Fig. 2. This laser is emitting at a wavelength of 750 nm, and the thickness of the emitting region is only about 500 nm.

HIGH-FIELD ELECTROLUMINESCENCE

It is generally possible to inject electrons or holes from one material into another if a sufficiently high electric field

FIG. 2. The minute, star-like luminescence from an AlAs-GaAs heterojunction laser glowing beneath the eye of an ordinary sewing needle. This laser is similar to lasers designed for optical communications. (Photograph courtesy of RCA Laboratories.)

is applied across the interface. Two characteristic examples of electroluminescence originating in this way are the reverse-biased *p-n* junction and certain luminescent insulators containing metallic particles.

If a voltage of polarity opposite that shown in Fig. 1 is applied to a *p-n* junction, the top of the VB on the *p* side can be raised considerably above the bottom of the CB on the *n* side. Electrons from the *p* side can tunnel horizontally through the forbidden gap to the *n* side and into CB states having high kinetic energies. If this kinetic energy is higher than the band-gap energy, the electron can collide with a normal lattice atom and create a new electron–hole pair which can, in turn, recombine radiatively. In practice a considerable amount of energy is lost as heat, and the reverse-biased *p-n* junction is therefore not an efficient electroluminescent source.

The second class of high-field process is exemplified by luminescent ZnS in powder form and doped with Cu_2S. Inside the small ZnS crystallites, the Cu_2S precipitates on lattice imperfections to form submicroscopic, needlelike, electrically conducting particles. The external electric field is strongly intensified near the sharp ends of the needles, where electrons or holes are emitted into the ZnS on alternate half cycles of the applied field. The holes are trapped at impurity atoms, and electrons subsequently arrive to produce recombination luminescence. Cu is the principal luminescent impurity, but many others can play various roles in the process, and most visible colors can be produced. Large electroluminescent panels of this type have been constructed which sandwich the luminescent powder between a metallic base and a transparent conducting electrode such as tin oxide, but their efficiencies and lifetimes are too low to be competitive for general space illumination.

APPLICATIONS

There is now available a wide range of LEDs, which fill an increasingly large fraction of the commercial market for small, low-power sources operating as indicator lights and in symbolic displays. Pocket computers are a conspicuous example. The low voltages required to drive LEDs are compatible with transistor circuitry, a considerable advantage. It is probable that panels of the ZnS type will also gain acceptance for display applications. As the reliability and lifetime of laser diodes improves they will find numerous uses where bright, narrow light beams are required. Optical communications is a particularly promising area, since laser diodes are ideal sources for coupling into fiber-optic waveguides. Their size and operating wavelengths match the diameters and maximum-transmission wavelengths of optical fibers (see Fig. 2), and their output can be modulated at frequencies in the 10^8-Hz range. The less stringent source requirements of many fiber-optic systems can be met by current LEDs. Electroluminescent devices will probably find a role in the interior illumination market, but not before further advances are made in the field of materials science.

See also CRYSTAL DEFECTS; ELECTRON ENERGY STATES IN SOLIDS AND LIQUIDS; LUMINESCENCE; SEMICONDUCTORS, CRYSTALLINE.

BIBLIOGRAPHY

H. Kressel, I. Ladany, M. Ettenberg, and H. Lockwood, "Light Sources", *Physics Today*, p. 38, May 1976.

H. J. Queisser and U. Heim, "Optical Emission from Semiconductors", in *Annual Review of Materials Science*, (ed. by R. A. Huggins, R. H. Bube, and R. W. Roberts), Vol. 4, p. 125, 1974.

C. H. Gooch, *Injection Electroluminescent Devices*, Wiley, N.Y., 1973.

R. J. Elliott and A. F. Gibson, *An Introduction to Solid State Physics and Its Applications*, Barnes and Noble, N.Y., 1974.

Electron and Ion Impact Phenomena

Edward W. Thomas

An energetic electron or ion impacting on a surface transfers energy to the lattice of the solid and to the free electrons in the solid by a successive series of collisions. As a result, certain particles are ejected from the surface; familiar examples are electron ejection (secondary-electron emission) and ejection of surface atoms (known as sputtering). The projectile itself may suffer sufficient angular deviation in a collisional scattering event so that it becomes directed out of the surface and may be observed as a reflected primary particle. The deposition of energy into the solid will cause various displacements of the lattice structure, known collectively as radiation damage, and it may excite electrons which subsequently decay with the emission of light (luminescence).

The rates at which these phenomena occur are described by a quantity known as a "coefficient," which is the number of relevant events occurring for every primary projectile incident; thus the secondary-electron emission coefficient is the number of secondary electrons ejected for each primary electron incident. Coefficients are strongly dependent on the condition of the surface and the crystal structure of the substrate. For a reliable measurement that can be related to a theoretical concept it is necessary that the target surface be atomically clean, the substrate must be of known crystallographic structure, and the orientation of the surface to the incident projectile direction must be defined.

ELECTRON IMPACT MECHANISMS

Most often studied is the flux of emergent electrons including both scattered primary electrons and ejected secondary electrons. Figure 1 shows the energy distribution of emergent electrons; region I is due mainly to reflected pri-

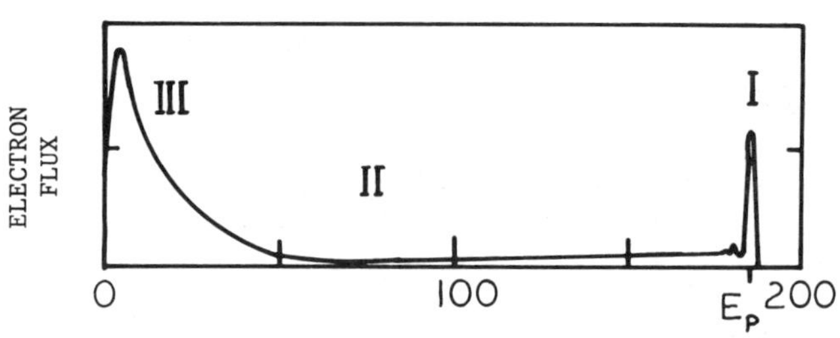

FIG. 1. A typical energy distribution for electrons emerging from a solid as a result of electron bombardment. E_P denotes the energy of the primary electrons which produce the secondary emission.

maries, region III to secondary electrons, and region II includes components from both sources.

Secondary Electron Ejection

The incident primary electron produces excitation and ionization within the solid. Those excited electrons which diffuse to the surface, overcome the potential barrier, and emerge to be detected are called secondary electrons. Figure 1 indicates that true secondary electrons cannot be distinguished from reflected primary electrons; it is conventional to regard the secondary electrons to be those with energies below 50 eV (excepting Auger electrons discussed below). Secondary electrons arise primarily from excitation of outer-shell electrons of the solid and they have an average energy of a few electron volts. Typical peak secondary-emission coefficients of metals and semiconductors are 1 to 1.5 (secondary electrons out per primary electron in); for insulators and intermetallic compounds yields may be as high as 10 or 20.

Primary electrons of sufficient energy may eject inner-shell electrons from the atoms of the solid, so creating a vacancy which must be subsequently filled by an electron falling from a higher level. The energy liberated as the vacancy is filled may be transfered to some other outer-shell electron which thereby becomes ionized, and escapes from the solid, a process known as Auger-electron emission. Auger electrons are energetic with a relatively small energy spread; they are observed in the secondary-electron energy spectrum as discrete high-energy peaks, superimposed on the general continuous background of other secondaries and reflected primaries. The Auger-electron energy is characteristic of the atom from whence it was ejected so that analysis of the Auger spectrum, using an electron energy analyzer, can indicate which atomic species are present in the target.

Electron Reflection

The high-energy peak of electrons emerging from the solid (I in Fig. 1) is due to primary electrons scattered or reflected from close to the surface and much of the lower energy tail (II in Fig. 2) is due to primary electrons scattered from deeper in the solid. Small subsidiary peaks slightly below the incident energy are due to primary electrons which have lost discrete amounts of energy by exciting plasma oscillations in the electron gas. Reflection coefficients increase with the nuclear charge of the substrate and do not vary strongly with the energy of the projectile.

Electron-Induced Desorption

Because an incident electron has a small mass it cannot transfer appreciable energy to the heavy substrate nuclei by collisions and therefore cannot directly eject atoms. An incident electron may, however, ionize the electrons which bond a surface atom to the substrate; removal of the electron destroys the bond and the atom is ejected. This phenomenon is observed in the removal of weakly bonded adsorbed molecules and is consequently known as electron-induced desorption.

Luminescence Phenomena

Impact of electrons on a surface gives rise to a variety of ligh-emission phenomena emanating from the surface, or, if the material is sufficiently transparent, from some distance within.

Broad-band luminescence characteristic of the solid itself is caused by excitation of electrons to a higher level followed by their decay across a forbidden band gap with resulting emission of light. The phenomenon, called cathodoluminescence, is best known for insulators and semiconductors where large gaps exist between bands, but is observed also in metals.

Plasmon radiation is an optical emission from the decay of collective electron oscillations excited in metals; it emanates from the surface. Transition radiation occurs as a result of annihilation of the dipole formed by an electron and its image charge in the metal; this source also is located at the surface. Bremsstrahlung radiation occurs when electrons are decelerated or deflected in the Coulomb field of the atoms and this may emanate from deep in the solid. These three luminescence phenomena are seen only with electrons having energy in excess of tens of keV.

ION IMPACT MECHANISMS

Impact of an ion or atom on a surface exhibits certain complexities associated with the structure of the projectile. Figure 2 represents a potential energy diagram of an ion at a few angstroms from a metal surface. It is possible for an electron (1) to tunnel through the barrier and enter the ground state of the projectile; the excess energy is given to electron (2) causing its ejection by the Auger mechanisms. Thus as an ion approaches a surface it changes its form to a neutral atom and electrons are ejected before the projectile penetrates the surface. Clearly an incident atom cannot undergo this change. After the projectile enters the solid it may lose some of its electrons by collisional ionization but we have no direct monitor of its structure.

Secondary-Electron Emission

Collisional transfer of kinetic energy from the projectile to electrons in the target will result in a secondary-electron emission spectrum similar to that observed with incident primary electrons: There is a large peak of low-energy emergent electrons and subsidiary peaks at high energies due to Auger transitions. One should note that heavy particles transfer energy inefficiently to light electrons and the projectile energy must exceed some thousands of electron volts before the energy transferred to the electrons is sufficient to overcome the surface potential barrier. Thus low-energy incident ions and atoms cannot eject electrons by "kinetic" mechanisms involving transfer of kinetic energy from the projectile.

Incident ions may also eject electrons by "potential" mechanisms related to their structure; the Auger effect described above and illustrated in Fig. 2 is such a mechanism and will obviously occur at any impact energy however small. Impact of neutral atoms cannot produce secondary electrons by this mechanism.

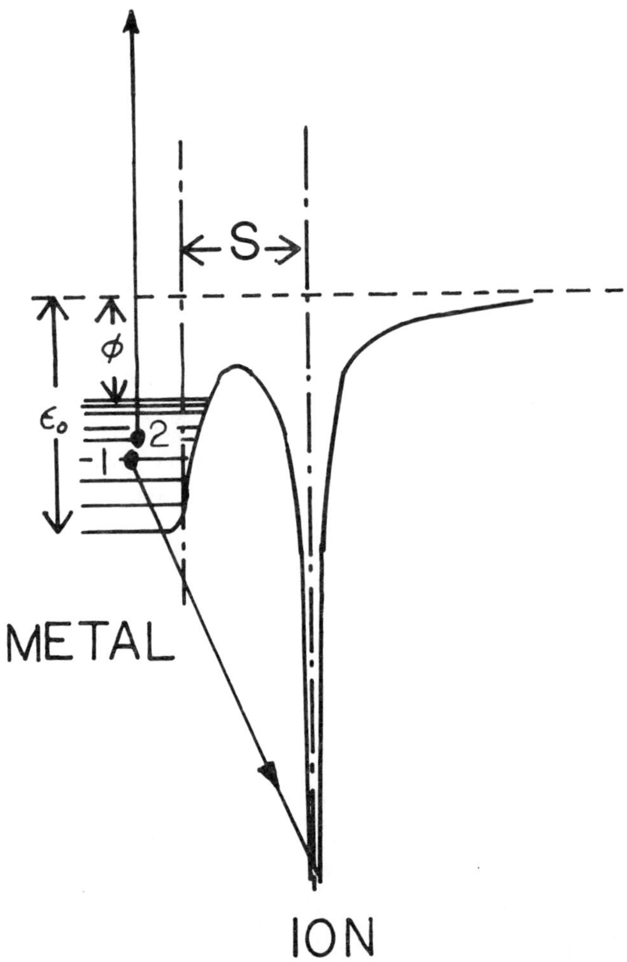

FIG. 2. A potential energy diagram showing an ion at a distance S from a metal surface; the conduction-band electrons lie at energies between the work function ϕ and the bottom of the conduction band ϵ_0. Auger neutralization of the ion occurs by electron 1 falling to the bound state in the ion; electron 2 carries away the excess energy.

Reflection of Ions

Projectile ion collisions with electrons cause energy loss but, because of the substantial mass difference between the particles, no appreciable deviation of the heavy projectile. Only collisions of the projectile ion with lattice atoms can produce the deviation necessary to return the projectile toward the surface. For low-energy impact the projectile recoils from the surface with an energy loss related only to the angle of scattering and the mass of the atom from which it recoiled. Measurement of energy loss may be used to determine the mass of the atoms from which scattering occurs and thereby to analyze the composition of the surface. For high-energy impact there is a continous distribution of recoil energies related to the penetration of projectiles into the solid.

The reflected projectile flux is found to include both ions and atoms. In part the distribution between charge states is related to interaction of the recoiling species with the target surface; Fig. 2 can be used to represent an ion which emerges from a surface and which picks up an electron by the Auger mechanism to become a neutral atom. The distribution of charge states is also partially established while the projectile is in the solid. A small fraction of the reflected projectiles may be excited and will emit normal atomic spectral lines when the projectiles have receded from the surface.

Sputtering

Ion impact on a surface will collisionally transfer energy to the lattice with the result that some lattice atoms may acquire sufficient energy to be ejected or sputtered from the surface. Typically the sputtered particles have an average energy of a few electron volts and are ejected with a coefficient which ranges from 10^{-3} (atoms ejected per ion incident) for a light projectile such as H^+ to 10 or 20 for heavy projectiles such as Pb^+. The ejected particles are found to include both ions and atoms. The distribution between these charge states is related in part to interaction of the emerging particle with the surface. Figure 2 can be used to describe how a sputtered ion can be neutralized as it emerges from the surface. A substantial fraction (up to 10%) of the sputtered particles may emerge as molecules representing the simultaneous ejection of two or more neighboring atoms in the lattice. A fraction of the emerging sputtered particles is found to be excited and will emit atomic line radiation at some distance from the surface.

Luminescence Phenomena

The emission of light from the surface or the bulk occurs through essentially the same mechanisms as for electrons discussed earlier. Intrinsic luminescence is termed ionoluminescence (the counterpart of cathodoluminescence) and is complicated by formation of additional energy levels through radiation damage caused by the ion impact.

Atomic line emission from sputtered or reflected atoms or molecules is not to be termed luminescence as it arises from a region in front of the surface rather than from the bulk material.

CRYSTALLOGRAPHIC EFFECTS

When a simple cubic crystal is "viewed" parallel to the sides of the unit cube it appears highly transparent with the strings of atoms lined up behind one another and large open "channels" exposed between the strings. The majority of projectiles directed along such a channel will lose energy only by collisions with electrons and therefore will penetrate considerable distances. Coefficients for reflection and ejection should be very small. Clearly the magnitude of electron and ion impact coefficients will be closely related to the orientation of the incoming projectile with respect to these crystallographic channels.

APPLICATIONS

Electron and ion impact phenomena find much application in the analysis and modification of surfaces. The utility of the techniques is related to their performance in the vacuum environment where contaminants may be avoided. Under electron bombardment each element in the near surface region produces Auger electrons of characteristic energy, which may be separately detected by use of an electron energy spectrometer; the analysis provides the atomic composition of the surface and is known as Auger electron spectroscopy (AES). The secondary-electron emission mechanism can be used to provide amplification of weak electron currents using materials where the secondary-emission coefficient exceeds unity; this finds application in photomultipliers. Sputtering of a surface by ion impact provides a method for removal of thin layers of material, one atomic layer at a time. Sputtering is used to remove contaminants from a surface (sputter cleaning), to machine a surface (ion milling), or to remove a layer and expose some underlying surface (sputter etching). Mass analysis of the ions sputtered from a surface can provide a qualitative indication of that surface's composition, a technique known as secondary-ion mass spectrometry (SIMS). Energy analysis of slow ions scattered from the surface gives an indication of surface composition (ion scattering spectrometry or ISS).

See also AUGER EFFECT; CHANNELING; LUMINESCENCE; RADIATION DAMAGE IN SOLIDS; SECONDARY ELECTRON EMISSION.

BIBLIOGRAPHY

E. W. McDaniel, *Collsion Phenomena in Ionized Gases*, Chap. 13. Wiley, New York, 1964. (I)

P. D. Townsend, J. C. Kelly, and N. E. W. Hartley, *Ion Implantation, Sputtering and Their Applications*. Academic Press, New York, 1976. (A)

I. A. Abroyan, M. A. Eremeev, and N. N. Petrov. *Sov. Phys. Usp.* **10,** 332 (1967) [*Usp. Fiz. Nauk* **92,** 105 (1967)]. (A)

Electron Beam Technology

Walter Schebesta

Electron beam technology includes all processes where electron beams (EB) are directed at matter with the purpose of altering the structure, composition, or state, or removing some of the material (Table I).

The advantages of EB technology result from the following characteristics:

(i) The EB obtains its energy from the electric field between cathode and anode.

Table I Summary of EB processes and their technical parameters

Process	Acceleration Voltage (kV)	Beam Power (kW)	Power Density (W m^{-2})
Melting	15–35	10–2000	10^8–10^9
Evaporation	10–35	5–500	10^9–10^{10}
Welding	30–170	0.1–100	10^9–10^{11}
Fusion treatment	20–150	1–10	10^8–10^{10}
Drilling	80–170	1–60	10^{10}–10^{13}
Lithography	10–25	10^{-7}–10^{-5}	10^6–10^8
Polymerization	20–100	1–50	10^4–10^6

(ii) The EB can be focused by electromagnetic lenses; power densities up to 10^{13} W m^{-2} can be achieved.

(iii) The natural environment of the process is high vacuum of about 10^{-2} Pa.

ELECTRON BEAM GENERATING AND FOCUSING

A thermionic cathode, mostly a tungsten filament, emits electrons which are accelerated toward the anode; the voltage between cathode and anode is 10–200 kV (Fig. 1). The Wehnelt electrode, situated between cathode and anode, is some 100 V more negative than the cathode. It has two purposes: it controls the beam current and it forms the electric field in a way that the electrons are collimated to a beam, which traverses through an aperture in the anode.

If a focused EB is necessary, behind the anode there are one or more electromagnetic lenses, which consist of current carrying coils around the beam. The magnetic field in the coils forces the trajectories of the electrons to converge.

For special processes the EB must describe certain patterns on the target (e.g., lithography). For this purpose the EB will be deflected by magnetic fields also excited by coils.

PHYSICAL PHENOMENA

The penetration depth of accelerated electrons in material is proportional to the square of the acceleration voltage and inversely proportional to the density of the material; 50-kV electrons (e.g., electrons which are accelerated by a voltage of 50 kV) have a range of 8 μm in steel, while 150-kV elec-

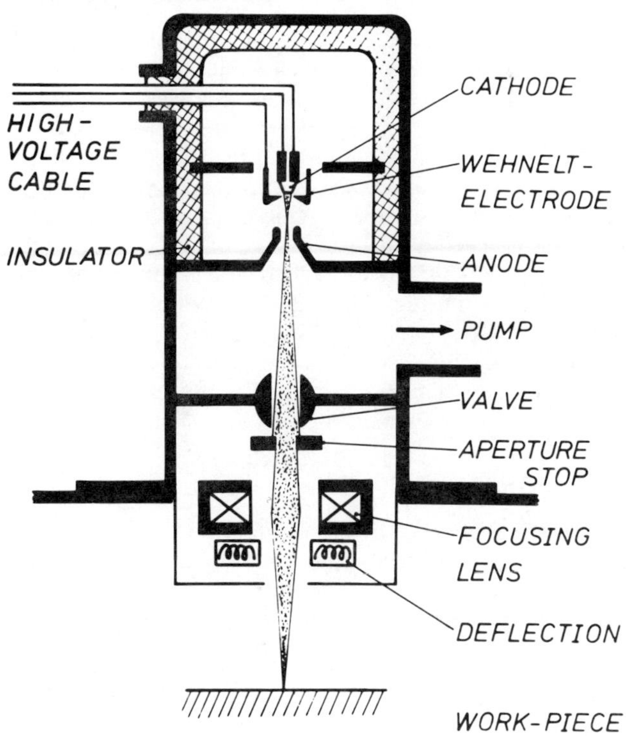

FIG. 1. Schematic diagram of an electron beam gun with a focusing lens and deflection coils for material machining.

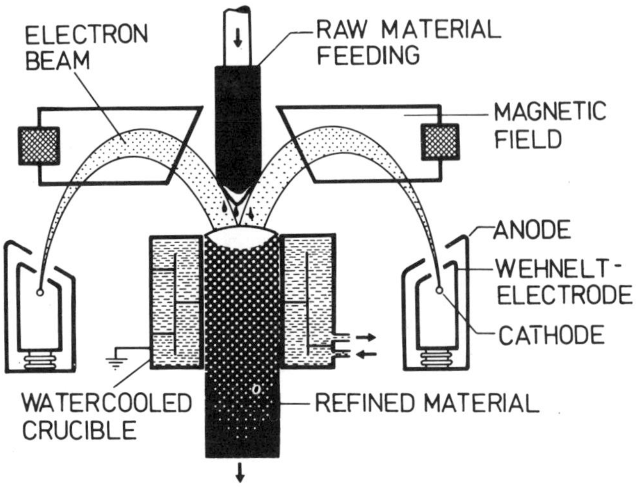

FIG. 2. Schematic diagram of an EB melting furnace for simultaneous bombardment of melt stock and molten pool.

trons have a range of 70 μm, until they are stopped completely. Generally the beam electrons lose their energy by a large number of subsequent collisions with the electrons, atoms, and molecules of the target material. These interactions lead to emission of radiation, to backscattered and secondary electrons, to ionization, and to a temperature rise in the zone of impact. This causes different reactions such as melting, evaporation, etc., dependent on the power density.

EB MELTING

The need for refractory (e.g., with high-melting-point) and reactive (e.g., with high affinity to the metalloids H, N, and O) metals for aerospace and nuclear industry required the development of EB-melting furnaces. The lack of an upper temperature and the absence of impurities due to vacuum (10^{-2} Pa) makes it possible to keep the metal in a molten state, while higher volatile impurities evaporate. The maximum power output of a single gun is about 200 kW; for a higher power level the installations are powered by a cluster of several guns. The EB source is preferably situated outside the line of metal flow in order to avoid vapor-induced breakdown (Fig. 2).

There also exist EB melting furnaces with casting equipment inside the vacuum chamber. The use of 10-MW furnances for steel production is planned.

EB EVAPORATION

Focused EB are used as a heat source for the evaporation of metals in order to produce metal coatings. As the energy is concentrated on the surface, interactions between evaporant and crucible are reduced.

Typical applications are coatings in electronics and of optical lenses, and Al-coatings of plastics and iron sheets.

EB WELDING

A focused EB with a power density of 10^9–10^{11} W m^{-2} impinging on a surface initially forms a thin cavity channel by removing material; the solidifying melting bath around the channel joins the workpieces together. The outstanding advantage is the great depth-to-width ratio (up to 50:1; see Fig. 3). The deep and thin welding seams with almost no distortion of the microstructure in the vicinity of the seam permit a variety of new constructional principles.

For mass production there are machines already in use where the welding is performed at higher pressure than 10^{-2} Pa.

EB FUSION TREATMENT

The surface of a semifinished product is melted in a point or line pattern using a focused EB. The subsequent quenching of the liquid (up to 10^5 °C/s) results in a hardened microstructure with remarkable properties in certain materials. Sometimes metal foils are melted in, like Ni in Al surfaces.

EB DRILLING

The high power density (up to 10^{13} W m^{-2}) of focused EB pulses (100 kV, several mA, 10^{-5}–10^{-2} s) causes the explosive removal of the material. Owing to the short pulses the energy cannot be lost by heat conduction.

With this technique channels for cooling liquid in turbine

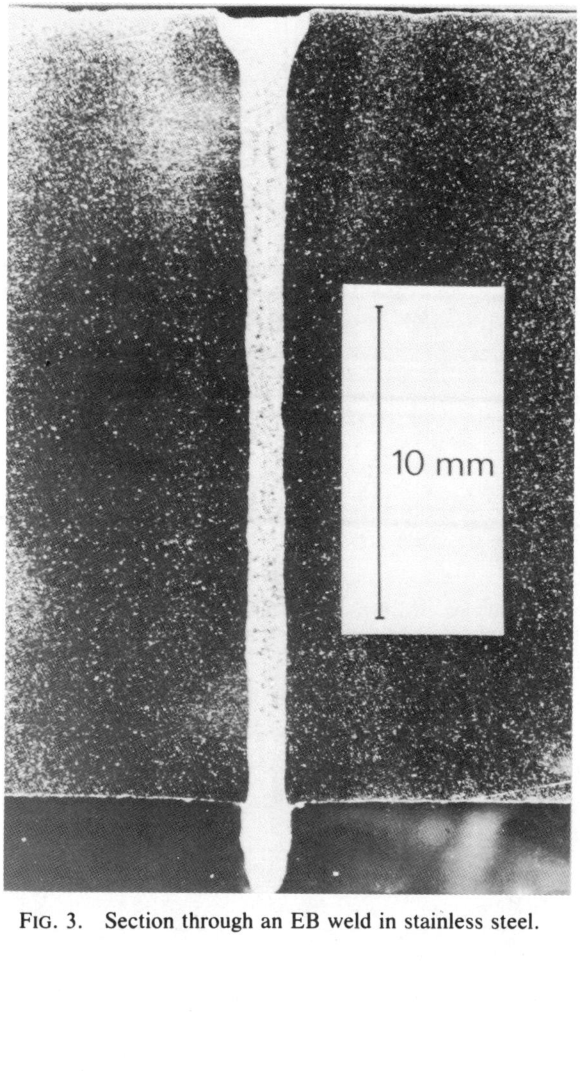

FIG. 3. Section through an EB weld in stainless steel.

vanes (0.3 mm Ø; several mm deep) or cavities (about 50 μm Ø and 10 μm deep) in printing rollers are made.

EB LITHOGRAPHY

In thin-film technology focused EB with some 100-nA beam current are used to draw patterns into photoresists with line widths <1 μm. Photoresists are organic polymers in which EB exposure causes temporary broken bonds which may either increase or decrease the solubtility of the irradiated product in a suitable developer.

EB POLYMERIZATION

The irradiation of organic compounds with EB (10^5–10^7 $W \cdot m^{-2}$) is applied in order to cross-link the molecules.

BIBLIOGRAPHY

J. Pierce, *Theory and Design of Electron Beams*. Van Nostrand, New York, 1949.

O. Winkler and R. Bakish, *Vacuum Metallurgy*. Elsevier, Amsterdam-London-New York, 1971.

A. H. Meleka, *Electron Beam Welding*. McGraw-Hill, London, 1971.

R. F. W. Piece and J. G. Skinner (ed.), "Proceedings of the 13th Symposium on Electron, Ion, and Photon Beam Technology," Colorado Springs, 1975, *J. Vac. Sci. Technol.* **12**(6) (1975).

Electron Diffraction

L. Marton†

Diffraction, in light optics, has been defined as "Any departure of the actual light path from that prescribed by geometrical optics" [1]. Contrary to light, where the discovery of diffraction preceded the wave theory, electron diffraction was discovered as a consequence of a deliberate attempt to prove the wave nature of the electron.

In 1924 Louis de Broglie advanced the, for the times, revolutionary view that electrons (or any corpuscles) are accompanied by waves [2]. The origins of that view can be traced back to Einstein's 1905 theory about the wave–particle duality of light. Starting with this theory of Einstein, de Broglie came to the conclusion that electrons should exhibit the same kind of dual nature as light, where the relation between a particle's mass m_0, its velocity v, and its wavelength λ could be expressed as

$$\lambda = \frac{h}{m_0 v}$$

h being the constant of Planck. Today we usually write the de Broglie relation as

$$\lambda = \frac{h}{p}$$

where p is the momentum of the particle; this formulation implies the necessary relativistic correction.

About a year later a young graduate student at the University of Göttingen, Walter Elsasser, surmised that Clinton J. Davisson and Charles H. Kunsman's work on the scattering of slow electrons by metal surfaces [3] may have shown evidence of the diffraction of the electrons by the crystal lattice, thus supporting de Broglie's view of the wave nature of the electron [4].

In 1927 came definite evidence for the particle–wave duality through the discovery of electron diffraction by crystal lattices. This discovery was made almost simultaneously by C. J. Davisson and Lester H. Germer, using slow electrons diffracted by a nickel single crystal [5], and by George P. Thomson and A. Reid, using fast electrons passing through a thin celluloid film [6]. Their work was followed by rapid confirmation and extensive further development. Quite early it was shown that the angular dependence of the diffracted beams satisfied Bragg's law (originally established for x-ray diffraction):

$$n\lambda = 2d \sin\theta$$

where d is the interatomic distance, θ the diffraction angle and n an integer. The de Broglie wavelength λ, as calculated from the relativistically correct equation

$$\lambda = \frac{h}{[2m_0 eV(1 + eV/2m_0c^2)]^{1/2}},$$

was confirmed with good accuracy. In this equation e is the charge of the electron, c the light velocity, and V the potential difference applied to accelerate the electron [7]. It was also shown that, while the Davisson–Germer experiment corresponded to the Laue diagram of x-ray diffraction, the Thomson–Reid experiment was a counterpart of the Debye–Scherrer diagram of x rays.

Because of easier experimental requirements (lesser vacuum, less stringent control of the accelerating potentials) the high-voltage, transmission method (Thomson) developed much faster than the slow-electron reflection technique. The original diffraction cameras, consisting of an electron source and an aperture, collimating the beam before it passed through the specimen, were soon replaced by the arrangement first suggested by A. A. Lebedev [8]. In this arrangement an axially symmetrical magnetic field, the axis being parallel to the electron beam, is used to focus the beam on the recording medium (usually the photographic emulsion). This procedure has the advantage of replacing a well-collimated beam with a divergent beam and still attaining a higher intensity of the recorded pattern. At the same time this device allows the use of a larger area of the specimen. The accelerating potentials used by different authors were generally in the 10- to 100-kV range, resulting in electron wavelengths corresponding to (12.2–3.7) $\times$ 10^{-12} m (0.0122–0.0037 nm).

The development of the electron microscope introduced a further change in instrumentation. In 1936 H. Boersch showed the relationship between diffraction and image formation in the electron microscope [9]. Following the introduction of the three-stage microscope [10], J. B. LePoole showed how it can be used advantageously for "selected area diffraction" [11]. Prior to that date some electron microscopes used special adapters for diffraction experiments.

In the experiments described so far, diffraction was produced by periodic structures, such as crystal lattices. The

† Deceased.

reason for this is that the discovery of the equivalent to x-ray diffraction antedated by about 10 years the discovery of diffraction on opaque edges. In light optics we are familiar with the two kinds of edge diffractions: the Fraunhofer diffraction, produced by a plane wave front; and the Fresnel diffraction, produced by all other wave fronts. Because of the extreme shortness of the electron wavelength, the interference fringes, caused by the diffraction effect, are so closely spaced that their observation required a high degree of resolution of the electron optical system used for recording them. By 1940 the electron microscope was sufficiently developed for the observation of Fresnel fringes, and they were described almost simultaneously by H. Boersch [12] and J. Hillier [13]. By that time, all workers in electron microscopy agreed that the resolving power, as in light microscopy, is diffraction limited and obeys the Abbé relation

$$\rho = \frac{0.61\lambda}{n \sin\alpha}$$

where ρ is the least resolved distance, n the index of refraction of the medium surrounding the object, and α the semiangle subtended by the object at the lens.

The year 1940 also marks the beginning of another important link between electron diffraction and electron microscopy. Until that time the dominant interpretation of the origin of contrast in electron microscope images was based on mass-thickness scattering by the specimen [14]. B. von Borries and E. Ruska found, however, variations of intensities in chromium oxide smoke images, which could not be interpreted by scattering alone [15]. This finding was followed by the observation of Bragg reflections in images of crystalline specimens [16,17] and led finally to the theory of diffraction contrast in images of crystalline specimens [18,19].

For an understanding of intensity relations in electron diffraction and in electron microscope images of crystalline objects, detailed calculations based on the theory of diffraction phenomena are necessary. There are two theories of electron diffraction: the kinematic (or geometric) theory and the dynamic theory. Both theories were developed originally for x-ray diffraction and adapted for use with electrons. The kinematic theory, in the ideal case, describes the interaction of the electron with a single atom and derives the intensities of the resulting beams. The dynamic theory takes into account the multiple scattering resulting from the presence of many atoms. This theory requires a wave-mechanical treatment, since the scattering produces a background that very considerably modifies the intensity distributions. X rays, which produce much less scattering than electrons, can usually be treated by means of the kinematic theory. In the case of electrons, very thin layers of crystals, roughly below 100 atom layers thick, give diffraction phenomena sufficiently well described by the kinematic theory. Above a thickness of 100 atom layers there is so much scattering that we have to have recourse to the dynamic theory [18–20].

For the interpretation of diffraction contrast in electron microscopy it is necessary to know how many electrons pass through the aperture of the optical system. If, by any chance, the object is so inclined that Bragg reflection throws the electron beam outside the objective aperture, no electrons may reach the image plane. The same is true if a lattice defect in the crystal produces a similar Bragg reflection. The thicknesses employed in electron microscopy require the application of the dynamic theory in calculating image intensities [21].

In the usual applications of the dynamic theory, involving both elastic and inelastic scattering of the incident electrons, the energy losses suffered in the inelastic process are relatively low. They are mostly due to plasmon excitation of the solid specimen and they are in the 10- to 50-eV range. In 1928, however, S. Kikuchi, while observing the diffraction of electrons passing through thin mica sheets, discovered a pattern differing widely from the single-crystal Laue spots or the polycrystalline Debye–Scherrer patterns [22]. He found a system of lines and bands, requiring a modification of the dynamic theory. These Kikuchi patterns consist of a continuous angular distribution of inelastically scattered electrons modulated by lines or band edges at the position of the various Bragg reflections. The energy analysis of the electrons diffracted into a Kikuchi line has shown that, for silicon, the average energy loss was about 450 eV, in contrast to the ten-times-lower loss for the ordinary patterns taken in transmission [23]. Energy losses are considerably higher for ordinary patterns taken in reflection. However, neither the ordinary transmission nor the reflection experiment has shown the surprisingly high loss, accompanied by a very high proportion of inelastically scattered electrons, observed by Kikuchi. In transmission experiments Kikuchi lines appear at specimen thicknesses in the range of 10^{-7}–10^{-6} m (0.1–1 μm).

For three decades high-energy electron diffraction (HEED) remained the only practical approach to the problems accessible by diffraction methods, although L. H. Germer pointed out quite early [24] the advantages of low-energy electron diffraction (LEED). By 1960, however, two of the experimental obtacles to LEED were removed: high-vacuum technology was radically improved to the extent that the surfaces investigated remained stable during the experiment; and the low intensities of the diffraction patterns, which hampered earlier work, could be enhanced by postacceleration of the diffracted electrons. The usual energy range of LEED is from 20 to 500 eV, corresponding to a wavelength range of (274–55) × 10^{-12} m (0.274–0.055 nm). At these low energies the electron–solid interaction is quite strong. Furthermore, LEED is sensitive to the vibrations of atoms, in a manner similar to x-ray diffraction. Because inelastic collision damping, caused by surface plasmon excitation, is quite pronounced, the dynamic theory of diffraction is more useful than the kinematic one. In addition, the so-called perturbation theory is applied, where a structure is assumed. By the use of experience gathered on simple systems, reflectivities can be calculated and compared with the experiment and by successive approximations a structure is arrived at where experiment and theory show a good agreement [25].

A further complication may arise in interpreting the experiments because of refraction effects of the electron wave in entering matter. The refractive index may not modify significantly the results of HEED observations in transmission, although the Debye–Scherrer diagrams of fine crystallites may show a small refraction effect [26]. The interpretation of Kikuchi lines, however, requires knowledge of the internal potential of the solid. In LEED observations refraction effects can be quite marked [20].

The discovery of electron diffraction was followed, within a remarkably short time, by application of the new effect to many other problems. Within the framework of this short survey it is necessary to limit ourselves to an enumeration of some of these applications, without extensive discussion. As with x-ray crystallography, structure determinations have been in the forefront. Because of the limited penetration of electrons, surface structures are much more accessible to electron diffraction investigation than to x rays. Thus surface physics benefits both from HEED reflection observations and from observations by means of LEED. For example, it was found that whereas bulk metals show the same structure on their surface as in their interior, semiconductors such as silicon or germanium have a different arrangement of the atoms on their surface. HEED in transmission is very widely used for the study of thin films of monocrystalline, polycrystalline, or amorphous materials. In the gaseous phase HEED is very useful for the study of molecular structures [27]. Within 10 years after the discovery of electron diffraction, results were published on more than 150 substances and by 1953 Pinsker's book [20] tabulated the structures of about 500 molecular species.

LEED is confined to surface studies. A very wide variety of problems that concern surface structures and their changes due to physical or chemical effects can be solved to a good approximation. Some examples are recrystallization due to temperature changes, adsorption of gases, and oxidation or corrosion. Studies in catalysis afford another good example.

Selected-area diffraction [18], mentioned earlier, allows the investigation of very limited areas of a specimen (of the order of 0.1 μm^2). Two more somewhat unusual applications of electron diffraction are electron holography and electron interferometry. Electron holography has been attempted, but until 1975 no success had been recorded. Electron diffraction on three parallel single crystals, placed at about 0.035 m distance from each other, was used to create the electron optical analog to the Mach–Zehnder interferometer [28], by means of which for the first time electron interferometer fringes could be observed.

See also CRYSTAL DEFECTS; CRYSTALLOGRAPHY, X-RAY; DIFFRACTION; ELECTRON MICROSCOPY; LOW-ENERGY ELECTRON DIFFRACTION (LEED); REFLECTION HIGH-ENERGY ELECTRON DIFFRACTION (RHEED); SURFACES AND INTERFACES.

REFERENCES

1. E. U. Condon and H. Odishaw, eds., *Handbook of Physics*, p. 6–78. McGraw-Hill, New York, 1958.
2. L. de Broglie, Thèse, Paris. Masson, Paris, 1924.
3. C. J. Davisson and C. H. Kunsman, *Science* **64**, 522 (1921).
4. W. Elsasser, *Naturwiss*. **13**, 711 (1925).
5. C. J. Davisson and L. H. Germer, *Nature* **119**, 558 (1927); *Phys. Rev*. **30**, 705 (1927).
6. G. P. Thomson and A. Reid, *Nature* **119**, 890 (1927).
7. Extensive tabulation of λ is given by L. Marton, C. Marton, and W. G. Hall in "Electron Physics Tables," Natl. Bur. Standards Circ. No. 571 (1956).
8. A. A. Lebedev, *Nature* **128**, 491 (1931).
9. H. Boersch, *Ann. Phys*. **26**, 631 (1936); **27**, 75 (1936).
10. L. Marton, *Science* **100**, 318 (1944); *J. Appl. Phys*. **16**, 131 (1945).
11. J. B. LePoole, *Philips Tech. Rev*. **9**, 33 (1947).
12. H. Boersch, *Naturwiss*. **28**, 709, 711 (1940); *Phys. Z*. **44**, 202 (1943).
13. J. Hillier, *Phys. Rev*. **58**, 842 (1940).
14. L. Marton and L. I. Schiff, *J. Appl. Phys*. **12**, 759 (1941).
15. B. v. Borries and E. Ruska, *Naturwiss*. **28**, 366 (1940).
16. J. Hillier and R. F. Baker, *Phys. Rev*. **61**, 722 (1942).
17. R. D. Heidenreich, *Phys. Rev*. **62**, 291 (1942).
18. R. D. Heidenreich, *Fundamentals of Transmission Electron Microscopy*. Wiley (Interscience), New York, 1964.
19. P. B. Hirsch, A. Howie, R. B. Nicholson, D. W. Pashley, and M. J. Whelan, *Electron Microscopy of Thin Crystals*. Butterworth, Washington, 1965.
20. Z. G. Pinsker, *Electron Diffraction* (translated from the Russian by A. A. Spink and E. Feigl). Butterworth, London, 1953.
21. M. J. Whelan, "Electron Diffraction Theory and Its Application to the Interpretation of Electron Microscope Images of Crystalline Materials," *Adv. Electronics Electron Phys*. **39**, 1 (1975).
22. S. Kikuchi, *Proc. Imp. Acad. Japan*, **4**, 271, 275, 354, 471 (1928); *Jpn. J. Phys*. **5**, 83 (1928).
23. W. Hartl and H. Raether, *Z. Phys*. **161**, 238 (1961).
24. L. H. Germer, *Z. Phys*. **54**, 408 (1929); *Bell Syst. Tech. J*. **8**, 591 (1929).
25. J. B. Pendry, "Low Energy Electron Diffraction," in *Techniques of Physics* (G. K. T. Conn and K. R. Coleman, eds.), Vol. 2. Academic Press, New York, 1974.
26. H. Raether, "Elektroneninterferenzen," in *Handbuch der Physik* (S. Flugge, ed.), Vol. 32, pp. 443–551. Springer-Verlag, Berlin and New York, 1957.
27. L. S. Bartell, "Electron Diffraction by Gases" in *Physical Methods of Chemistry* (A. Weissburger and B. W. Rossiter, eds.), 4th ed., p. 125. Wiley (Interscience), New York, 1972; K. Kuchitsu, "Gas Electron Diffraction" in *Molecular Structure and Properties* (G. Allen, ed.), p. 203. MTP, Oxford, 1972; J. Karle, "Electron Diffraction" in *Determination of Organic Structures by Physical Methods* (F. G. Nachod and J. J. Zuckerman, eds.), p. 1. Academic Press, New York, 1973.
28. L. Marton, *Phys. Rev*. **85**, 1057 (1952); L. Marton, J. A. Simpson, and J. A. Suddeth, *Phys. Rev*. **90**, 490 (1953); *Rev. Sci. Instr*. **25**, 1099 (1954).

Electron Energy States in Solids and Liquids

Joseph Callaway

INTRODUCTION

In an atomic or molecular gas at ordinary temperatures and densities, electrons are bound to specific atoms or molecules and occupy states characteristic of isolated systems. Such states have wave functions which decay exponentially at large distances from the home-base atom or molecule, and the energy levels of the system are reasonably sharp. Interactions between electrons on different atoms are, on the average, quite weak.

In contrast, in condensed matter, interactions cause broadening of the energy levels of electrons in outer atomic shells. It is more correct to describe the electron states as belonging to the system as a whole, rather than to any specific atom or molecule. Such states are called band states, and one says that energy bands are formed. At the normal interatomic distances in solids, only the outer shells have broadened into bands. The energies of tightly bound core states are still quite sharp.

These ideas are, in a rough way, independent of the particular atoms or molecules involved, and are valid whether or not the condensed phase is solid or liquid, or is ordered or disordered. Closer investigation indicates that in molecular crystals, such as solid hydrogen, or in crystals of rare gases, solid argon for example, the band of valence electron states is quite narrow, and a molecular or atomic description is quite appropriate. On the other hand, in metals, the broadening is quite advanced and states originating in different atomic levels may be strongly mixed.

While these general concepts do not distinguish between ordered and disordered systems, it has proved much easier to develop a detailed understanding of electron states in crystals than in glassy solids or liquids, and in pure materials rather than alloys. This is because the existence of long-range structural order makes possible enormous simplifications in the mathematical description of electron states. Ordered systems are also often easier to characterize from an experimental point of view, and many types of experiments are possible only for ordered systems because of increased electron mean free path. We shall therefore devote most of this article to electron states in perfect crystalline solids, and reserve only a few paragraphs toward the end for disordered solids and liquids.

ELECTRON STATES IN ORDERED SOLIDS

The essential characteristic which simplifies the description of states in perfect crystals is periodicity. We will ignore here the existence of surfaces. Then there exists an infinite set of direct-lattice vectors $\mathbf{R}_i$,

$$\mathbf{R}_i = n_{i1}\mathbf{a}_1 + n_{i2}\mathbf{a}_2 + n_{i3}\mathbf{a}_3, \qquad (1)$$

in which the $\mathbf{a}_i$ are the three primitive lattice vectors and the n_{ij} are integers, such that the crystal is unaltered if all atoms are displaced by any $\mathbf{R}_i$. We shall consider the electron states from a single-particle point of view in which each electron has its own wave function. This wave function is the solution of a Schrödinger equation

$$H\psi = E\psi, \qquad (2a)$$

in which the Hamiltonian is periodic,

$$H(\mathbf{r}+\mathbf{R}_i) = H(\mathbf{r}). \qquad (3)$$

It is a fundamental property of solutions of the Schrödinger equation for a periodic Hamilton that (Bloch's theorem)

$$\psi(\mathbf{r}+\mathbf{R}_i) = \exp i\mathbf{k}\cdot\mathbf{R}_i\, \psi(\mathbf{r}): \qquad (4)$$

If the coordinate of an electron is changed by a direct-lattice vector, the only change in the wave function is multiplication by a phase factor. The quantity $\mathbf{k}$ is called the wave vector and can be regarded as a quantum number specifying the electron state. It is customary to incorporate $\mathbf{k}$ into the notation for the wave function so that it is written $\psi_\mathbf{k}(\mathbf{r})$ or $\psi(\mathbf{k},\mathbf{r})$. It follows from (4) that the wave function can be written in the form

$$\psi(\mathbf{k},\mathbf{r}) = \exp(i\mathbf{k}\cdot\mathbf{r})\, u(\mathbf{k},\mathbf{r}), \qquad (5)$$

in which $u(\mathbf{k},\mathbf{r})$ has the same periodicity as the Hamiltonian,

$$u(\mathbf{k},\mathbf{r}+\mathbf{R}_i) = u(\mathbf{k},\mathbf{r}). \qquad (6)$$

Thus, $u(\mathbf{k},\mathbf{r})$ repeats itself in each unit cell of the crystal, and the wave function ψ has the form of a plane wave modulated in each cell by the periodic function u.

It is useful to define a set of reciprocal-lattice vectors $\mathbf{K}_s$ through

$$\mathbf{K}_s = n_{sl}\mathbf{b}_l + n_{s2}\mathbf{b}_2 + n_{s3}\mathbf{b}_3, \qquad (7)$$

where the n_{si} are integers and the $\mathbf{b}$'s are related to the $\mathbf{a}$'s by

$$\mathbf{b}_i\cdot\mathbf{a}_j = 2\pi\,\delta_{ij}. \qquad (8)$$

It is easy to find $\mathbf{b}$'s which obey (8); for example,

$$\mathbf{b}_1 = \frac{2\pi(\mathbf{a}_2\times\mathbf{a}_3)}{\mathbf{a}_1\cdot(\mathbf{a}_2\times\mathbf{a}_3)}, \qquad (9)$$

and the other b's are obtained by cyclic permutation. All direct- and reciprocal-lattice vectors satisfy

$$\mathbf{K}_s\cdot\mathbf{R}_i = 2\pi\times\text{integer}. \qquad (10)$$

The reciprocal lattice may or may not look like the direct lattice; for example, if the direct lattice is simple cubic, so is the reciprocal lattice, but if the direct lattice is body-centered cubic, the reciprocal lattice is face-centered cubic, and vice versa. Equation (10) has the following implication: If $\mathbf{k}$ and $\mathbf{k}'$ are two wave vectors which differ by a reciprocal-lattice vector, $\mathbf{k}'=\mathbf{k}+\mathbf{K}_s$, then

$$\exp i\mathbf{k}\cdot\mathbf{R}_i = \exp i\mathbf{k}'\cdot\mathbf{R}_i \qquad (11)$$

for all $\mathbf{R}_i$. Hence $\psi(\mathbf{k},\mathbf{r})$ and $\psi(\mathbf{k}',\mathbf{r})$ behave identically under all lattice translations according to (4) and may be considered to be the same function. For this reason, wave vectors can be restricted to be in a region of momentum space called the Brillouin zone (abbreviated BZ) such that no two interior points differ by a reciprocal-lattice vector. Two examples of BZ's are shown in Fig. 1.

We adopt the convention that wave vectors $\mathbf{k}$ are restricted to a BZ. When this is done, there will be many solutions of the Schrödinger equation which have the same wave vector but different energies. We designate these different, orthogonal, solutions by an index, usually in the order of increasing energy; thus (2a) is written in a more explicit way as

$$H\psi_n(\mathbf{k},\mathbf{r}) = E_n(\mathbf{k})\,\psi_n(\mathbf{k},\mathbf{r}). \qquad (2b)$$

The energies of states with a given value of the index n form a continuous manifold when regarded as functions of $\mathbf{k}$ inside a BZ. This manifold is called an energy band, and n in (2b) is referred to as the band index.

As an example, calculated energy bands for the semiconductor silicon are shown in Fig. 2.

Band states can be regarded as occupied at $T = 0$ K in order of increasing energy according to the rules of Fermi–Dirac statistics, until the required number of electrons per atom has been accommodated. If the energy of the highest occupied state falls inside an energy band, the substance is a metal. The energy of this state is the Fermi energy

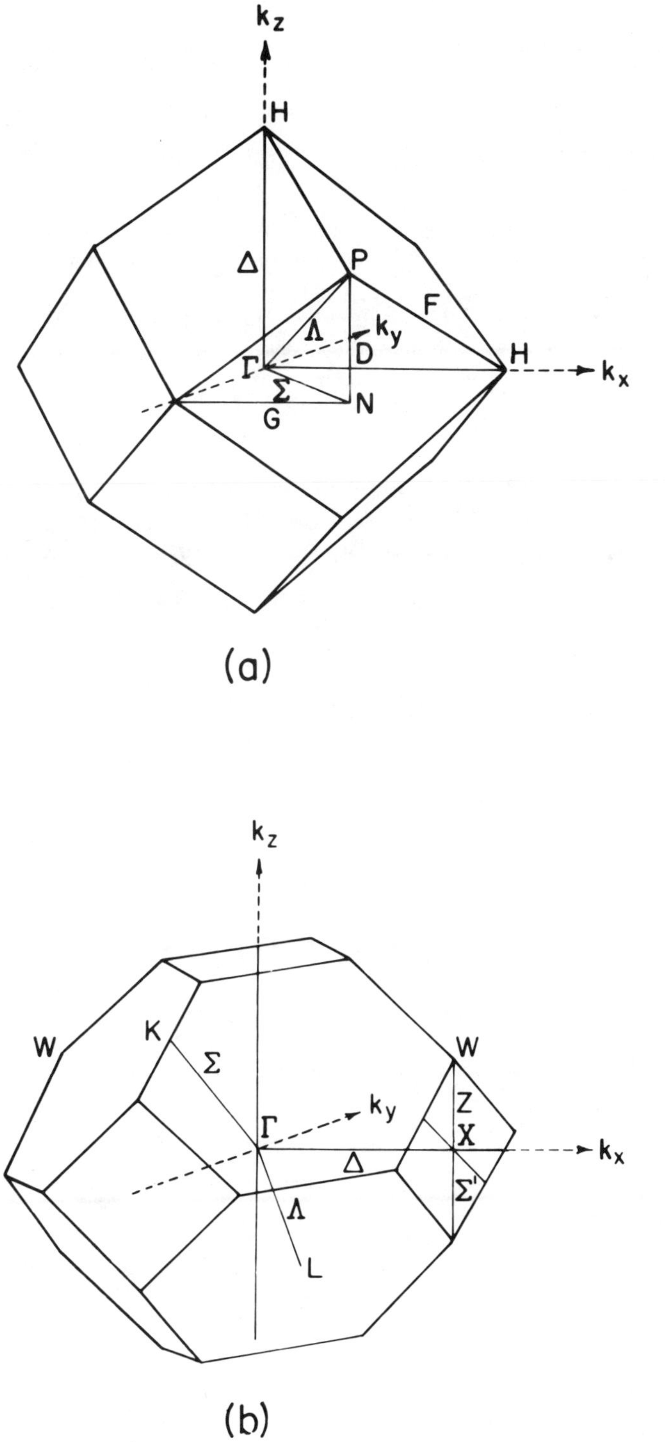

FIG. 1. Brillouin zones for the body-centered-cubic lattice (a), and the face-centered-cubic lattice (b). Points and lines of symmetry are indicated according to the accepted notation.

E_F. In the (excellent) approximation in which the wave vector $\mathbf{k}$ is treated as continuous, the solutions $\mathbf{k}_F$ of

$$E_n(\mathbf{k}_F) = E_F \tag{12}$$

(for all relevant bands) define the Fermi surface. The Fermi surface can simply be considered to be the surface(s) bounding the occupied region of $\mathbf{k}$ space. There are several experimental methods which can be used to determine the Fermi surface and so check theoretical calculations. Some of the most important ones include the de Haas–van Alphen effect, the radio-frequency size effect, and the magnetic field dependence of the electrical resistivity and of ultrasonic attenuation. Limitations of space preclude much discussion of these techniques here, but additional information can be found in the references.

If there are no allowed states immediately adjacent in energy to the highest occupied state, the material is a semiconductor or an insulator. The minimum energy separation between the highest occupied state and lowest empty state is the "band gap," which determines the temperature dependence of the electrical conductivity of a pure semiconductor. If the minimum separation between full and empty states is determined for states of the same $\mathbf{k}$, one has the "direct" band gap, which is of great importance in the study of optical properties.

It is found in many applications that electrons in solids move in weak applied external electric ($\mathbf{E}$) and magnetic ($\mathbf{B}$) fields according to the simple equation

$$m^*\left(\frac{d\mathbf{v}}{dt} + \frac{1}{\tau}\right) = -e(\mathbf{E} + \mathbf{v} \times \mathbf{B}), \tag{13}$$

in which e is the magnitude of the electronic charge, τ is a relaxation time which roughly describes the effect of collisions, and m^* is known as the effective mass. The quantity m^* is, in general cases, a tensor; that is, acceleration and force are not necessarily in the same direction. It is possible to determine the effective-mass tensor from the derivatives of the energy band function. Suppose that only a single band

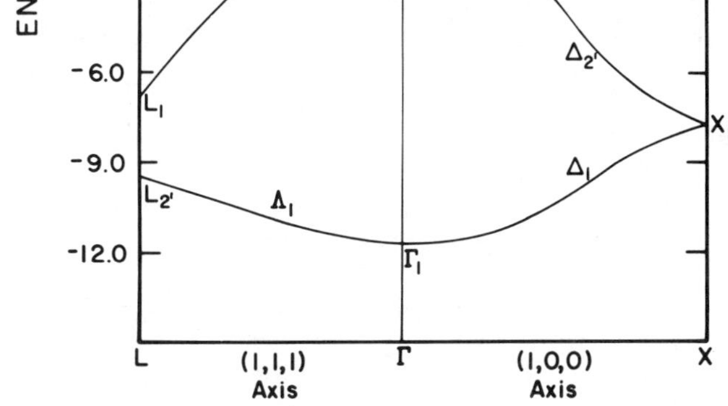

FIG. 2 Energy bands calculated for the semiconductor silicon for two directions of the wave vector $\mathbf{k}$. The points Γ, X, L, etc., on the horizontal axis refer to the Brillouin zone of Fig. 1(b). The cross hatched region is the band gap. Bands are labeled according to symmetry properties of the wave functions (see references for details).

need be considered. Then one has

$$\left(\frac{1}{m^*}\right)_{\alpha\beta} = \frac{1}{\hbar^2}\frac{\partial^2 E(\mathbf{k})}{\partial k_\alpha \partial k_\beta}, \qquad (14)$$

in which α, β are Cartesian indices. A variety of experimental measurements are available to determine m^*, and values smaller than the free electron mass by factors of the order of 100 are found in some materials (e.g., InSb).

It is possible for regions of $\mathbf{k}$ to be important, where $E(\mathbf{k})$ has negative ("downward") curvature. In this case m^* is negative, but it is customary to consider the mass to be positive and to delete the minus sign in (13) so that one describes the motion of a positive charge. Such effective positive charges are known as holes. The concept of a hole is particularly important in semiconductors, where states near the top of a full valence band are described in this way.

Another important concept is the "density of states." This measures the fractional number of states which have energies in a certain interval. Specifically let $G(E)\, dE$ be the number of states of a given spin direction which have energies between E and $E+dE$. $G(E)$ is given by the expression

$$G(E) = \frac{\Omega}{(2\pi)^3} \sum_n \int d^3k\, \delta(E - E_n(\mathbf{k})), \qquad (15)$$

in which the sum includes all bands and the integral runs over the Brillouin zone. The delta function serves as a counter, registering a unit contribution from those bands (n) and wave vectors ($\mathbf{k}$) such that $E_n(\mathbf{k}) = E$. The quantity Ω in the multiplicative constant is the volume of the unit cell. Since the volume of the Brillouin zone is $(2\pi)^3/\Omega$, the constant normalizes the integral so that the contribution from each band to the integral of $G(E)$ over all energies is unity. A very simple case in which (15) may be explicitly evaluated is that of a parabolic band characterized by an effective mass m^* as is found for a nearly free electron system. Suppose $E(\mathbf{k}) = \hbar^2 k^2/2m^*$ (a scalar effective mass) and consider only a single band. The result is

$$G(E) = \frac{\Omega}{4\pi^2}\left(\frac{2m^*}{\hbar^2}\right)^{3/2} E^{1/2}. \qquad (16)$$

This expression will not be valid in a region of energies where an energy band intersects a Brillouin-zone face, for then the expression for $E(\mathbf{k})$ must become more complicated, as some derivatives of the energy must vanish. In fact, it can be shown that the density of states must possess sharp structure at such energies, called "van Hove singularities." Further information can be found in the references.

Electrons in a metal produce a contribution to the specific heat that depends linearly on temperature. This contrasts with the contribution from lattice vibrations, which is proportional to T^3 at low temperatures. At sufficiently low temperatures the linear term dominates the cubic one, and one can determine experimentally the coefficient γ in the relation

$$C_V^{(e)} = \gamma T, \qquad (17)$$

in which $C_V^{(e)}$ is the electronic contribution to the specific heat per atom. It can be shown that γ is given by

$$\gamma = \frac{1}{3}\pi^2 K^2 G(E_F), \qquad (18)$$

in which K is Boltzmann's constant and $G(E_F)$ is the density of states evaluated at the Fermi energy. Hence measurement of γ yields a value for $G(E_F)$ that can be compared with calculations.

CALCULATION AND MEASUREMENT OF ELECTRONIC ENERGIES

There are several rather standard techniques available for the numerical calculation of electronic energies and wave functions from Eq. (2). The most commonly applied methods include (1) the "Green's-function" method, (2) the "augmented-plane-wave" method, (3) the "orthogonalized-plane-wave" method, and (4) the method of "linear combination of atomic orbitals". Space permits only a brief description of some common general features, but details of specific approaches are described in the references. All of the methods involve the expansion of the unknown $\psi_n(\mathbf{k},\mathbf{r})$ as a combination of some previously specified functions $\Phi_s(\mathbf{k},\mathbf{r})$, which are constructed to obey Bloch's theorem explicitly,

$$\Phi_s(\mathbf{k},\mathbf{r}+\mathbf{R}_l) = \exp\,(i\mathbf{k}\cdot\mathbf{R}_l)\,\Phi_s(\mathbf{k},\mathbf{r}).$$

The simplest set of potentially useful functions are plane waves of the form

$$\exp[i(\mathbf{k}+\mathbf{K}_s)\cdot\mathbf{r}]$$

in which $\mathbf{K}_s$ is a reciprocal-lattice vector. In general, one writes

$$\psi_n(\mathbf{k}\cdot\mathbf{r}) = \sum_s c_{ns}(\mathbf{k})\Phi_s(\mathbf{k},\mathbf{r}). \qquad (19)$$

The sum includes, in principle, an infinite number of terms, but of course only a finite subset can be employed in actual calculations. The coefficients $c_{ns}(\mathbf{k})$ have to be determined so that (2) is satisfied. Equation (19) is substituted into (2); the result is multiplied by $\Phi_t^*(\mathbf{k},\mathbf{r})$, where Φ_t is a general member of the basis set, and integrated over the entire crystal. A set of linear homogeneous equations is obtained:

$$\sum_s H_{ts}(\mathbf{k})c_{ns}(\mathbf{k}) = E_n(\mathbf{k}) \sum_s S_{ts}(\mathbf{k})c_{ns}(\mathbf{k}), \qquad (20)$$

in which the H_{ts} are matrix elements of the Hamiltonian on the basis Φ,

$$H_{ts}(\mathbf{k}) = \int \Phi_t^*(\mathbf{k},\mathbf{r}) H \Phi_s(\mathbf{k},\mathbf{r})\, d^3r, \qquad (21)$$

and the S_{ts} are overlap matrix elements,

$$S_{ts}(\mathbf{k}) = \int \phi_t^*(\mathbf{k},\mathbf{r})\Phi_s(\mathbf{k},\mathbf{r})\, d^3r. \qquad (22)$$

The overlap matrix S_{ts} is usually not simply a unit matrix since the functions Φ are not generally orthonormal. However, a variety of standard mathematical techniques exist to enable numerical solution of (20) once the elements of matrices H and S are constructed.

The most important problem encountered in these calculations is the determination of the function $V(\mathbf{r})$, which represents the potential energy of an electron in the crystal. This must include contributions from all the nuclei of the system, which may be assumed to be in their equilibrium positions; the average field of the other electrons; and some representation of the exchange interaction from the Hartree–Fock equations. The latter two terms have to be computed self-consistently. That is, one assumes some initial distribution of electrons; computes the corresponding potential function; then solves the Schrödinger equation to find energy bands and wave functions. The new distribution of electrons must now be used and the process repeated until the electron distributions resulting from two successive stages of the procedure agree within some assigned limit of error. This iterative process is quite time consuming. However, reasonably good results are usually obtained in a single-pass, non-self-consistent calculation if the potential energy is computed by solving Poisson's equation for a charge distribution produced by the superposition of neutral-atom charge densities centered on crystal lattice sites.

Many ingenious techniques have been developed to permit the experimental determination of the properties of electron states. The easiest to understand are optical. The absorption of light can produce a transition between an occupied state in band n and an unoccupied state in band l. Because the wave vector of a light wave in the infrared, visible, or near-ultraviolet regions of the spectrum is quite small on a scale appropriate to crystalline lattices, it is an excellent approximation to suppose that the initial and final states have the same wave vector. Such transitions can be represented by a vertical line on an energy band diagram, and are said to be direct. Indirect transitions also occur in which one or more phonons are emitted or absorbed. The interpretation of the spectra of band-to-band transitions is not so simple as in the case of free atoms because both initial and final states belong to continuous bands. Sharp structures that are easily identifiable are associated with the onset of transitions into a band or with the occurrence of van Hove singularities in the density of states. Ultraviolet photoemission spectroscopy, in which the excited electron escapes from the solid, also provides information concerning the distribution of initial states.

Optical investigations can determine the energy differences between states substantially removed from the Fermi energy. States near the conduction-band minimum or valence-band maximum in semiconductors, and states close to the Fermi surface in metals, can be studied experimentally by applying a strong external magnetic field. There are many interesting effects of which cyclotron resonance is the easiest to understand. Consider Eq. (13) in the absence of an electric field and suppose (1) that the relaxation time τ is long enough so that $1/\tau$ can be neglected, and (2) that the effective mass m^* is a scalar rather than a tensor. Then it is easy to verify that an electron will spiral around the direction of the magnetic field with a frequency

$$\omega_c = eB/m^*. \tag{23}$$

The quantity ω_c is called the cyclotron frequency. Electromagnetic waves of frequency ω_c will be absorbed and in this way m^* can be determined. The discussion is easily extended to allow for a tensor effective mass, and for relaxation. Another effect of great importance is the de Haas–van Alphen effect. The magnetic susceptibility of a metal in a strong magnetic field exhibits oscillations as a function of the field whose period is proportional to $1/B$. The observed constant of proportionality can be used to determine the areas of cross sections of the Fermi surface in a direction perpendicular to the magnetic field.

ELECTRON STATES IN DISORDERED MATERIALS

Our ability to describe electron states in extended systems rests on the applicability of Bloch's theorem. If the Hamiltonian is not periodic because of disorder on a microscopic scale, the wave vector **k** is no longer a good quantum number. We can imagine that the disorder (caused either by random atomic positions in an alloy, or by thermal motions as in a liquid) causes electrons to scatter rapidly between states. Even in pure materials, some such scattering is always present, and gives rise to the electrical resistivity of a metal. In a liquid metal, the scattering is so frequent that the mean free path may be only a few atomic separations. If a state of definite **k** were somehow to be prepared initially, it would soon be destroyed.

A straightforward description of the properties of a disordered system in terms of energy bands is therefore not to be expected. The concept of the density of states, however, remains a useful one. The density of states of a disordered system will not exhibit the sharp structure characteristic of ordered systems in which the periodic lattice structure gives rise to van Hove singularities, but need not be a featureless object. An energy gap between filled and empty regions of a density of states may still persist. It has been shown that in certain amorphous solids, even if the gap is at least partly filled as a result of disorder, states in the vicinity of the gap are localized and therefore will not permit electrical conduction, while states more removed in energy are extended and will conduct. The energy separating conducting and nonconducting states is called the mobility edge.

Much current research is devoted to the development of a quantitative theory of disordered systems. To a large extent this work has emphasized the development of approximations in which the actual disordered system is conceptually replaced by an ordered effective medium. The properties of this medium are determined from the scattering properties of the individual atoms by the condition that, on the average, no further scattering should occur (the coherent potential approximation). An effective, non-Hermitian Hamiltonian results, which can be used for the computation of a density of states.

See also CRYSTAL SYMMETRY; CYCLOTRON RESONANCE; DE HAAS–VAN ALPHEN EFFECT; FERMI SURFACE; LIQUID METALS; SEMICONDUCTORS, CRYSTALLINE.

BIBLIOGRAPHY

A. Elementary

J. S. Blakemore *Solid State Physics*, 2nd ed. Saunders, Philadelphia, 1974.

C. Kittel, *Elementary Solid State Physics*. Wiley, New York, 1966.

B. Intermediate

P. W. Anderson, *Concepts in Solids*. Benjamin, New York, 1963.

N. W. Ashcroft and N. D. Mermin, *Solid State Physics*. Holt, Rinehart and Winston, New York, 1976.

R. J. Elliott and A. F. Gibson *An Introduction to Solid State Physics and Its Applications*. Barnes and Noble, New York, 1974.

C. Kittel, *Introduction to Solid State Physics*, 5th ed. Wiley, New York, 1976.

J. M. Ziman, *Principles of the Theory of Solids*, 2nd ed. Cambridge University Press, London, England, 1972.

C. Advanced

J. Callaway, *Quantum Theory of the Solid State*. Academic Press, New York, 1974.

R. J. Elliott, J. A. Krumhansl, and P. L. Leath, "The Theory and Properties of Randomly Disordered Crystals and Related Physical Systems," *Rev. Mod. Phys.* **46,** 465 (1974).

G. C. Fletcher, *The Electron Band Theory of Solids*. North Holland, Amsterdam, 1971.

W. A. Harrison, *Solid State Theory*. McGraw-Hill, New York, 1970.

W. A. Harrison, *Electronic Structure and the Properties of Solids*. W. H. Freeman and Co, San Francisco, 1980.

W. Jones and N. H. March, *Theoretical Solid State Physics*. Wiley-Interscience, London, 1973.

C. Kittel, *Quantum Theory of Solids*. Wiley, New York, 1963.

J. C. Slater, *Quantum Theory of Molecules and Solids*, Vol. 2 and 3. McGraw-Hill, New York, 1963.

Electron–Hole Droplets in Semiconductors

Jagdeep Shah

A pure semiconductor crystal is devoid of free electrons and holes (vacant electron states) at low temperatures. Excitation of such a semiconductor by photons of appropriate energy creates electrons and holes which interact among themselves via Coulomb interactions. At low densities and temperatures, an electron becomes bound to a hole to form a hydrogen-atom-like complex known as an exciton. In many semiconductors it is possible to create a large density of excitons by intense photoexcitation. In 1968 Keldysh suggested that observations of anomalous photoconductivity in Ge at high excitation intensities might be explained if, at some critical density, the gas of excitons condensed into a metallic liquid phase consisting of electron–hole droplets (EHD). This gas–liquid transition is analogous to condensation of atomic Na vapor into a metallic Na liquid, i.e., an electron is not bound to a particular hole but is free to move independently.

The correctness of this idea was demonstrated by three key experiments. In the first, a rapid increase in the intensity of a luminescence feature of Ge with slight decrease in temperature was interpreted as a first-order phase transition and the line shape of this feature was shown to be that expected for EHD. In the second experiment, small differences in dielectric constants between the metallic droplets and the surrounding exciton gas in an optically excited crystal of Ge scattered a probe laser beam in much the same way as fog scatters light. The angular distribution of the scattered laser light conclusively demonstrated the existence of macroscopic droplets with an average radius of the order of 5 μm. In the third experiment, discrete giant charge pulses were observed from an optically excited reverse-biased *p-n* junction in Ge. These result from dissociation of EHD in the strong electric field at the junction and provided further evidence for the finite size of EHD.

A detailed theory describing this electron–hole liquid has been developed by considering the ground-state energy $E_G(r_s)$ of an electron–hole pair in a degenerate electron–hole plasma as a function of r_s, the interparticle separation in units of the exciton radius. This energy is the sum of the kinetic, exchange, and correlation energies. The kinetic and exchange energies can be calculated analytically, but various approximations have to be used to calculate the correlation energy. Typically $E_G(r_s)$ has the shape shown in Fig. 1, with a minimum at r_0. If this minimum energy lies below the exciton binding energy, i.e., if $\phi_0 = |E_G(r_0)| - |E_x|$ is positive, then a condensate can form because it is stable against emission of an exciton at absolute zero degrees ($T=0$); ϕ_0 is then the binding energy of the condensate. Many factors, such as the anisotropy and the multivalley nature of the band structure, affect ϕ_0; but there is no way to predict the existence of EHD in a semiconductor without a detailed calculation. As a result of the Coulomb interactions between the carriers the band gap within the EHD is renormalized from E_g to $E_g(r_0)$ as is shown schematically in Fig. 2. The chemical potential $\mu(r_0)$ and $E_g(r_0)$ can be simply calculated from $E_G(r_s)$.

The formation of a metallic state in optically excited semiconductors provides a unique opportunity to explore the properties of quantum metals, and a wide variety of experimental techniques have been used to study this novel state of matter. One technique which stands out above all others is the study of photoluminescence spectra of EHD and excitons under a variety of conditions. It is clear from Fig. 2 that at $T=0$ the energies of the lowest- and the highest-energy emitted photons correspond to $E_g(r_0)$ and $\mu(r_0)$, respectively. Since $E_F{}^e + E_F{}^h = \mu(r_0) - E_g(r_0)$, one can obtain the liquid density n_0 from these measurements. Also, ϕ_0 is obtained by measuring the difference in energy between $\mu(r_0)$ and the low-energy edge of exciton emission. One can also obtain these quantities at $T \neq 0$ by analyzing the line shapes; such measurements show that n_0 and ϕ_0 vary with T in a manner expected for a metal or a degenerate plasma, confirming once again the metallic character of EHD.

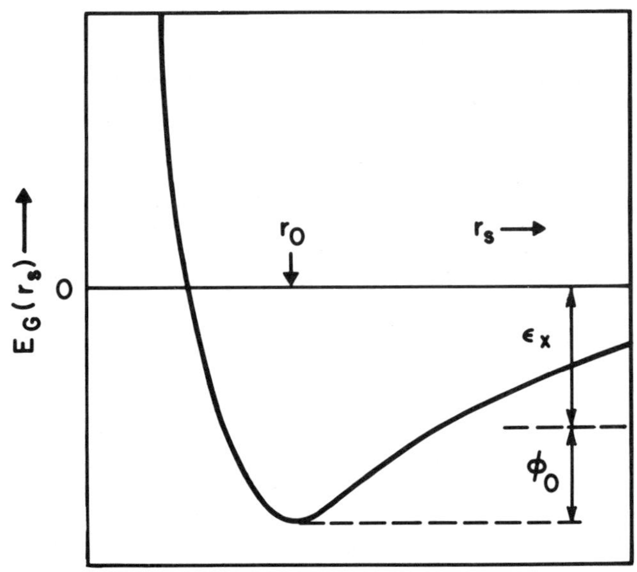

FIG. 1. Schematic representation of ground-state energy per pair versus normalized interparticle separation.

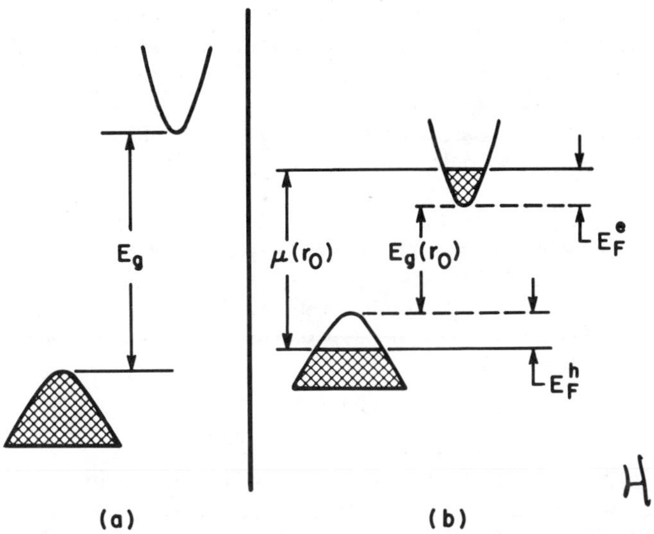

FIG. 2. Schematic energy versus wave-vector representation of the band structure of an indirect-gap semiconductor (a) outside and (b) inside EHD. Shaded region is occupied by electrons. $E_F{}^e$ and $E_F{}^h$ are the electron and hole Fermi energies.

When the source of excitation is removed, the EHD luminescence decays with a characteristic decay time. The *e*-*h* pairs within a droplet decay by radiative recombination, by nonradiative Auger recombination, and also by thermionic emission into the surrounding gas if the temperature is not too low. The finite lifetime of this metallic liquid phase distinguishes it from other liquids. However, assuming quasi-equilibrium between the gas of excitons and the liquid, one can discuss the gas–liquid coexistence curve in the temperature–density phase diagram of EHD, as for any other gas which undergoes a first-order transition to a liquid. A schematic phase diagram for EHD is shown in Fig. 3. The exciton–EHD system has a critical temperature T_c above which liquid cannot form. The region under the coexistence curve is the unstable region in that a plasma created at any point (n,T) in this region will spontaneously break up into droplets of density n_0 and gas of density n_{ex}. The coexistence curves for Ge and Si, including the values of T_c, have been determined from luminescence measurements. Phase diagrams have also been calculated theoretically.

At low temperatures and densities, the photoexcited *e-h* pairs form excitons, corresponding to the lower left region in Fig. 3. For $T<T_c$, as the density of the gas phase is increased beyond the density of gas on the coexistence curve, the gas becomes supersaturated and condensation occurs. Experimentally, there is a well-defined excitation intensity (I_{th}) beyond which the EHD becomes observable, i.e., the EHD luminescence increases extremely rapidly for $I>I_{th}$. It has been shown that the threshold for monotonically increasing intensity is considerably larger than that for monotonically decreasing intensity. This hysteresis is a result of the surface energy σ of EHD and provides a technique of determining σ.

The theory of nucleation of drops from excitons has been developed by including the effect of finite lifetime of drops in the well-known classical theory of homogeneous nucleation. Density fluctuations of excitons create clusters of excitons or embryos. Embryos with radius R less than a critical radius R_c vanish quickly while embryos with $R>R_c$ continue to grow to a macroscopic size limited by the EHD lifetime. The time to form a macroscopic EHD is determined by the time to form a critical embryo and the time it takes for the critical embryo to collect excitons to grow to its final size. At high supersaturation $(I \gg I_{th})$, the second factor dominates, while at small supersaturation, the first factor dominates. Nucleation theory predicts a very strong dependence of nucleation time on supersaturation close to threshold. Time-resolved luminescence spectroscopy on nanosecond time scales provides a direct confirmation of these predictions. This technique also demonstrates that EHD can form not only by nucleation from a supersaturated gas but also through expansion and cooling of a hot dense plasma, i.e., expansion and cooling of (n',T') first creates a liquid of density n_l (Fig. 3) and then leads to phase separation into EHD and excitons.

The droplets can move through the crystal under exciton or phonon pressure or under external forces such as strain gradients, electric field, or photon pressure. Acceleration and deceleration of EHD has been directly observed by Doppler velocimetry. Motion produced by applied electric fields has been used to measure the net charge on EHD in Ge. A remarkable phenomenon has been observed in Ge in the presence of a stress gradient; The EHD accelerate to the point of maximum stress and coalesce into a single large drop whose radius and lifetime are 300 μm and 500 msec, respectively, at high pump intensities. These large drops have also been photographed using infrared image-scanning cameras.

Early work on EHD was done exclusively on Ge and Si. During the last year, however, the existence of the metallic liquid phase has been established in other semiconductors such as CdS and GaP; and there will be others in the future. From a theoretical point of view, it has been shown that the polar character of these semiconductors plays an essential role in the binding of the liquid; their polar character also opens possibilities of new experimental investigations. For the direct-gap semiconductor CdS, the population inversion accompanying the formation of the liquid phase is responsible for the observation of large optical gain and laser action

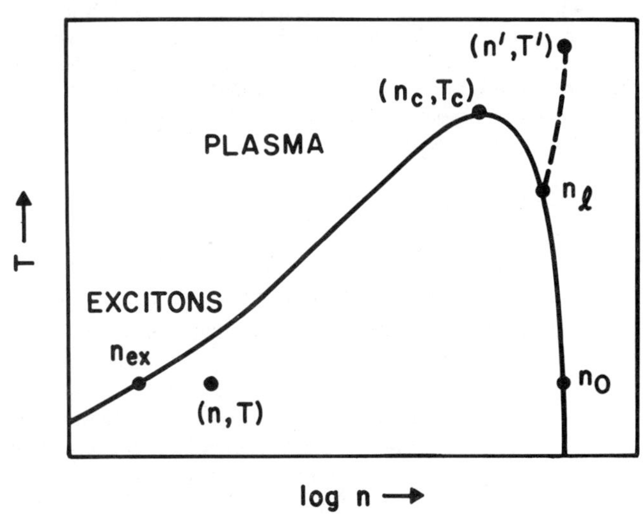

FIG. 3. Schematic phase diagram of EHD.

Table I Binding energy (ϕ_0), liquid density (n_0), critical temperature (T_c), and lifetime (τ) of EHD in various semiconductors

	ϕ_0 (meV)	n_0 (cm^{-3})	T_c (K)	τ (μs)
Ge	1.8	2.4×10^{17}	6.5	40
Si	8.2	3.3×10^{18}	27	0.15
CdS	13	2×10^{18}	≈55	$\sim 10^{-3}$
GaP	14	6×10^{18}	>35	0.035

at low temperatures. Properties of EHD in these semiconductors are shown in Table I. Other subjects which have attracted recent attention are the influence of the exciton-plasma Mott transition on the EHD phase diagram and interactions of phonons with EHD.

A study of electron–hole droplets in semiconductors has also provided increased understanding of phenomena normally associated with other branches of physics of condensed matter. It allows us to investigate, for example, effective ultrahigh densities, quantum and many-body effects in metals, gas–liquid phase transitions, and thermodynamics of multicomponent systems. Nearly 10 years after its conception, electron–hole droplets in semiconductors continues to be an active field of scientific research.

See also ELECTRON ENERGY STATES IN SOLIDS AND LIQUIDS; EXCITONS; LUMINESCENCE; SEMICONDUCTORS, CRYSTALLINE.

BIBLIOGRAPHY

V. S. Bagaev, *Springer Tracts Mod. Phys.* **73,** 72 (1975). (I)
C. D. Jeffries, *Science* **189,** 955 (1975). (E)
Ya. E. Pokrovskii, *Phys. Status Solidi* (a) **11,** 385 (1972). (I)
G. A. Thomas, *Scientific American* **234,** 28 (June 1976). (E)
A two part review of EHD (*Theory* by T. M. Rice and *Experimental Aspects* by J. C. Hensel, T. G. Phillips and G. A. Thomas) can be found in *Solid State Physics,* Vol. 32 (H. Ehrenreich, F. Seitz and D. Turnbull, ed.). Academic Press, New York, 1977. (A)
The preceding reference is an excellent, exhaustive review covering the literature up to mid-1976. A flavor of more recent work may be obtained from: B. Étienne, L. M. Sander, C. Benoit à la Guillaume, M. Voos, and J. Y. Prieur, *Phys. Rev. Lett.* **37,** 1299 (1976); R. F. Leheny and J. Shah, *ibid.* **38,** 511 (1977); J. Shah, R. F. Leheny, W. R. Harding, and D. R. Wight, *ibid.* **38,** 1164 (1977); J. Doehler, J. C. V. Mattos, and J. M. Worlock, *ibid.* **38,** 726 (1977); J. Shah, M. Combescot, and A. H. Dayem, *ibid.* **38,** 1497 (1977); J. Shah, A. H. Dayem, and M. Combescot, *Solid State Commun.* **24,** 71 (1977); G. A. Thomas and T. M. Rice, *Solid State Commun.* **23,** 359 (1977).

Electron Microscopy

Albert V. Crewe

The electron microscope can be defined as a device which uses electrons to obtain a magnified image of a specimen. The original aim of these devices was to overcome the resolution limitation of optical microscopes which is caused by the fact that the wavelength of light is quite large. It was shown by Rayleigh that the maximum resolving power of any microscope is approximately λ, where λ is the wavelength of the radiation. In the case of the visible light, the value of λ is about 5,000 Å or 5×10^{-5} cm. It was later shown by de Broglie that electrons can be considered as waves and the wavelength can be written as $12.5/V^{1/2}$ Å, where V is the accelerating voltage. For example, at 100 kV, the wavelength is about 0.04 Å. Such a short wavelength should enable one to resolve any object of biological or metallurgical significance.

Unfortunately, the wavelength limit has not yet been achieved in electron microscopy because all available lenses have large coefficients of spherical aberration. This coefficient, denoted is C_s, is comparable to the focal length of the lens (usually in the millimeter range). The optimum resolution is given by

$$S = 0.43 C_s^{1/4} \lambda^{3/4} \text{ Å}$$

and is usually a few Å.

There are now three main types of electron microscope.

1. The *conventional electron microscope* was invented by Ruska in 1935 and is the direct analog of an optical microscope. It uses the same type of magnetic lens used in all electron microscopes, namely, an axially symmetric magnetic field produced between parallel pole pieces which have an axial hole to allow passage of the electrons. The entire instrument must be in a vacuum. Although many improvements have been made upon the original design, the basic principles remain the same. As in the case of optical microscopes, one uses thin specimens (in this case, less than 1 μm) and the machine can be used in bright-field or dark-field; and, in addition, it is possible to obtain diffraction patterns. One of the most productive uses of this microscope is in obtaining images of regular structures (such as small crystallites of biological material) with the microscope slightly out of focus. A Fourier transform of the resulting image can provide considerable detail about the unit cell.

The current status of this instrument is that a resolution of 2 Å can readily be obtained.

2. In the *scanning electron microscope* a finely focused beam of electrons is scanned in a rectangular raster over a specimen. When the electrons interact with the atoms of the specimen, they can be reflected (backscattered electrons) or they can cause the ejection of much slower electrons (secondary electrons). Either group can be used to form a picture. The usual method is to use an electron detector to form an electric current proportional to the number of such electrons and use this current to modulate the intensity of an oscilloscope or television which is scanned synchronously with the beam raster. The pictures which are produced by such microscopes are of the surface topography of the specimen and are remarkable in that they give the illusion of three-dimensionality. Much of the development of this microscope took place at Cambridge University (England) in the laboratories of Professor Oatley.

The current status is that a resolution of 30 Å can be achieved with a microscope which uses a field-emission source (see next section).

3. The *scanning transmission electron microscope* can be thought of as a hybrid of the other two. A finely focused beam of electrons is scanned over a thin specimen and the transmitted electrons are used to modulate a synchronously scanned oscilloscope. In order to attain a high enough probe

current, it is necessary to use a field-emission source of electrons. Field emission occurs when a negative voltage is applied to a very sharp point of metal in a very high vacuum. Such sources have a very high brightness.

In current microscopes, the transmitted electrons are divided into two groups, those which have been scattered by the nuclei of the atoms of the specimen (elastically scattered) and those which lose some energy by scattering from the orbital electrons of the atoms (inelastically scattered). Since the number of these electrons depends in different ways on the atomic number of the atom, it is possible to make a rough measurement of that atomic number and also to estimate the mass of a particular region of the specimen. In addition these signals allow one to obtain very high contrast. Most of the development of this type of microscope has taken place at the University of Chicago in the laboratories of Professor Crewe.

The present status is that 2.5-Å resolution has been achieved and isolated heavy atoms (such as uranium or lead) are readily visible. It has also been possible to take many successive pictures of the same atoms to study their motion.

See also ELECTRON DIFFRACTION; FIELD EMISSION.

BIBLIOGRAPHY

A. V. Crewe, "A High Resolution Scanning Electron Microscope" *Scientific American* **224**(4), 1971. (E/I)

P. Grivet, *Electron Optics*. Pergamon Press, Oxford, 1965. (A)

M. E. Haine and V. E. Cosslett, *The Electron Microscope. The present state of the Art*. E. & F. N. Spon Limited, London, 1961. (E/I)

Electron Spectroscopy *see* **Photoelectron Spectroscopy.**

Electron Spin Resonance

A. Rassat

A free electron with spin $\hbar\mathbf{s}$ and magnetic moment $\boldsymbol{\mu}_s = -g_e\beta_e\mathbf{s} = -\gamma_S\hbar\mathbf{s}$ ($\hbar$ is Planck's constant/2π, β_e is the Bohr magneton $=\hbar|e|/2m$, γ_S is the electron magnetogyric ratio $=g_e|e|/2m$, g_e is Landé factor = 2.0023..., e is electron charge, m is electron mass) has in a magnetic field $\mathbf{B}_0$ a "Zeeman energy" $W_Z = -\boldsymbol{\mu}_s\cdot\mathbf{B}_0 = g_e\beta_e\mathbf{B}_0\cdot\mathbf{s} = \gamma_S\hbar\mathbf{B}_0\cdot\mathbf{s}$ with two stationary states, α ($m_s = +\frac{1}{2}$) and β ($m_s = -\frac{1}{2}$), separated by $\Delta W = g_e\beta_e B_0$. $\mathbf{B}_0$ also induces precession of the spin at angular velocity $\boldsymbol{\omega}_0 = \gamma_e\mathbf{B}_0$. In a collection of independent free electrons, spins precess in phase, unless there is inhomogeneity δB in the magnetic field, which after a time τ induces a dephasing $\delta\phi = \tau\delta\omega = \tau\gamma_e\delta B$. If a small oscillatory magnetic field $\mathbf{B}_1$ [conveniently provided by an electromagnetic radiation of frequency $\nu_0(=\omega_0/2\pi)$ and amplitude $B_1 \ll B_0$] is applied perpendicular to the static magnetic field B_0 so that $h\nu_0 = g_e\beta_e B_0$ (resonance condition), transitions are induced between the α and β levels: (electron spin) resonance occurs. Induced absorption and emission having equal probability, the populations of each level tend to become equal; there is no net absorption or emission of radiation, and resonance cannot be detected. However, in most systems, *relaxation* occurs: some mechanism ("spin–lattice relaxation") permits thermal equilibrium to be transfered to the spins with a characteristic time constant T_1, the "spin-lattice relaxation time." If this process is fast enough (but not too fast, because of inherent broadening $\Delta\omega = 1/T_1$), resonance should be detected. "Spin–spin relaxation," with characteristic time constant T_2, must also be considered. It does not modify the level population, but dephases the spins precessing around the resultant magnetic field. It thus broadens the lines by changing locally the magnetic fields. (T_2 may be visualized as the time for an average dephasing $\Delta\phi$ between spins to become of the order of 1 rad.) Now, if relaxation is not too slow and B_1 is not too strong ($\gamma_e^2 \times B_1^2 T_1 T_2 \ll 1$), resonance may be detected.

The basic components of an electron-spin-resonance (ESR) *instrument* are an homogeneous magnetic field, a source of monochromatic electromagnetic radiation, and a device to detect its absorption. For $B_0 \simeq 3300$ G (0.33 T), the frequency is $\nu_0 \simeq 10^{10}$ Hz (wavelength 3 cm, in the microwave range). A klystron is a convenient source. For technical reasons, its frequency is fixed, and resonance condition is obtained by varying the applied static magnetic field B_0. Absorption I of radiation is recorded as a function of B_0 and is maximum at resonance. Figure 1 gives a typical signal. Linewidth ΔB is related to T_2 by $1/T_2 = \gamma_e\Delta B$.

Generally, samples are paramagnetic solids or liquids. Let us study first a crystalline sample containing *perfectly oriented* paramagnetic molecules (or ions), having zero orbital angular momentum, and spin $\frac{1}{2}$ ("one unpaired electron"), diluted in a diamagnetic matrix, so that the electronic spins do not interact appreciably. Resonance occurs when $h\nu_0 = g_e\beta_e B$. $\mathbf{B}$, the local field felt by the electron spin, is now the sum of $\mathbf{B}_0$ and of the magnetic fields due to orbital and spin angular momenta (spin–orbit and spin–spin interactions) in the molecule. Interactions with nuclear spins are separated from other magnetic interactions and give rise to the "hyperfine structure." For instance, if the unpaired electron interacts with one nuclear spin $\hbar\mathbf{I}_N$, the local field depends on the $2I_N+1$ possible stationary orientations of the I_N spin relative to $\mathbf{B}_0$: $2I_N+1$ resonance lines are observed. Separating the hyperfine terms, the resonance con ditions at one orientation (θ,Φ) of $\mathbf{B}_0$ relative to the molecule may be written (if $A_N \ll g_e\beta B_0$)

$$h\nu_0 = g(\theta,\Phi)\beta_e B_0 + A_N(\theta,\Phi)m_{I_N}$$

$$(m_{I_N} = -I_N,\ -I_N+1,\ \ldots,\ +I_N).$$

$g(\theta,\Phi)$ and $A_N(\theta,\Phi)$ are the Landé factor and hyperfine coupling constant of nucleus N with the unpaired electron for direction (θ,Φ) of $\mathbf{B}_0$. If there are different interacting nuclei N, resonance conditions are $h\nu_0 = g\beta_e B_0 + \Sigma_N A_N m_{I_N}$. The number of observed lines is often less than $\Sigma_N(2I_N+1)$ because of some coincidences. (Figure 2 shows a spectrum obtained at two orientations for one electron interacting with one spin $I=1$.) If resonance is studied at all possible orientations, results may be given in the form of an "effective-spin Hamiltonian," $\mathcal{H}'_{\text{spin}} = \beta_e\mathbf{B}_0\cdot\bar{\bar{\mathbf{g}}}\cdot\mathbf{s}' + \Sigma_N \mathbf{I}_n\cdot\bar{\bar{\mathbf{A}}}_n\cdot\mathbf{s}'$ (nuclear terms may be added). $\bar{\bar{\mathbf{g}}}$ and $\bar{\bar{\mathbf{A}}}_N$ are second rank tensors

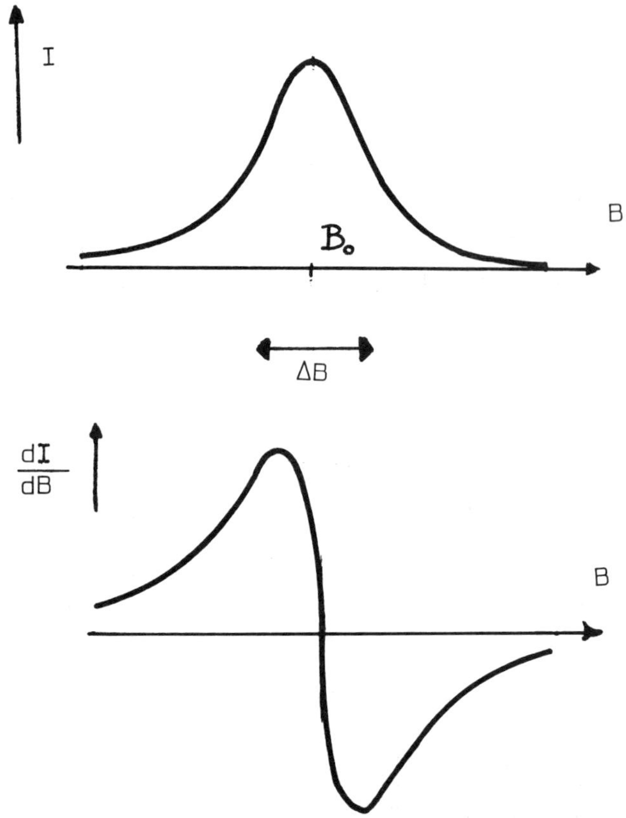

FIG. 1. Resonance curve for a collection of electrons resonating at $g = h\nu_0/\beta_e B_0$. (a) Intensity I of absorption recorded as a function of the applied magnetic field, ν_0 being fixed. (b) Derivative curve dI/dB, the usual presentation of results. Schematically, a single ESR absorption line δ is expected to be Gaussian, $I \simeq \exp[-(B-B_0)^2/(\Delta B)^2]$, in solids, where linewidth ΔB comes from a statistical distribution of local magnetic fields (inhomogeneous broadening) or Lorentzian, $I \simeq 1/[\Delta B^2 + (B-B_0)^2]$, where linewidth ΔB is the damping term of a usual (mechanical) resonance curve (homogeneous broadening). $\Delta B = 1/\gamma_e T_2$.

($\mathbf{B}_0 \cdot \bar{\bar{\mathbf{g}}} \cdot \mathbf{s}' = \sum_u \sum_v B_u g_{uv} s_v'$, $u,v = x,y,z$) which by a suitable choice of Cartesian axes ("principal axes") may be written in diagonal form (as a set of three "principal values"). Except for symmetry reasons, both tensors will not be diagonal in the same axes. Diagonalization of the matrix representation of this Hamiltonian in the basis of the effective electron spin $|s' = \frac{1}{2}, m_{s'}\rangle$ and nuclear-spin $\Pi_N |I_N, m_{I_N}\rangle$ states gives energy levels reproducing exactly the experimental resonance conditions, if $\bar{\bar{\mathbf{g}}}$ and $\bar{\bar{\mathbf{A}}}_N$ have correctly been determined. Independently from their experimental determination, the $\bar{\bar{\mathbf{g}}}$ and $\bar{\bar{\mathbf{A}}}_N$ tensors could in principle be computed from a magnetic Hamiltonian, in the basis of the (electronic and nuclear) kinetic and electrostatic Hamiltonian spinless eigenfunctions. The $\bar{\bar{\mathbf{g}}}$ tensor comes from Zeeman and spin–orbit interactions. The $\bar{\bar{\mathbf{A}}}_N$ tensor comes from two interactions: dipolar interaction with nucleus N,

$$H_{\text{dip}_N} = \sum_i^{\text{all electrons}} (\gamma_S \gamma_N \hbar^2 \mu_0 / 4\pi) \times [3(\mathbf{I}_N \cdot \mathbf{r}_{iN})(\mathbf{s}_i \cdot \mathbf{r}_{iN}) I_{iN}^{-2} - \mathbf{I}_N \cdot \mathbf{s}_i] r_{iN}^{-3}$$

(γ_N is nuclear magnetogyric ratio of nucleus N, μ_0 is free space permeability, $\mathbf{r}_{iN}$ connects electron i and nucleus N); and (Fermi) contact interaction with nucleus N

$$H_{\text{contact}_N} = \sum_i^{\text{all electrons}} (2\mu_0/3)\gamma_S \gamma_N \hbar^2 \delta(\mathbf{r}_{iN}) \mathbf{s}_i \cdot \mathbf{I}_N$$

($\delta(\mathbf{r}_{iN})$ is the three-dimensional Dirac "function"). H_{dip} gives rise to a traceless tensor $\bar{\bar{\mathbf{A}}}'$ ("purely anisotropic") and H_{contact} to an isotropic term $a_N \cdot \bar{\bar{\mathbf{1}}}$ ($\bar{\bar{\mathbf{A}}}_N = a_N \cdot \bar{\bar{\mathbf{1}}} + \bar{\bar{\mathbf{A}}}_N'$).

Let us study a different situation for the same paramagnetic species, still in a rigid diamagnetic matrix but so that the orientations of B_0 relative to the molecule have angular distribution. The observed "immobilized spectrum" $I = F(B_0)$ is the superposition of spectra $I(\theta,\Phi) = f(\theta,\Phi,B_0)$ corresponding to the various orientations: $F(B_0) = \int\int f(\theta,\Phi,B_0)\, \rho(\theta,\Phi) \sin\theta\, d\theta\, d\Phi/4\pi$. For an isotropic distribution ("powder spectrum"), $\rho(\theta,\Phi) = 1$. For an anisotropic distribution, $\rho(\theta,\Phi)$ may be estimated from the experimental spectrum. Anisotropies of $\bar{\bar{\mathbf{g}}}$ and hyperfine $\mathbf{A}$ terms both affect the "powder spectrum," as shown in Fig. 3.

Let us consider now the same paramagnetic species in *dilute fluid solution*. Magnetic interactions are anisotropic and the energy levels vary as the molecules rotate. For a spherical molecule of radius a, this random tumbling can be characterized by one rotational correlation time τ_c, related

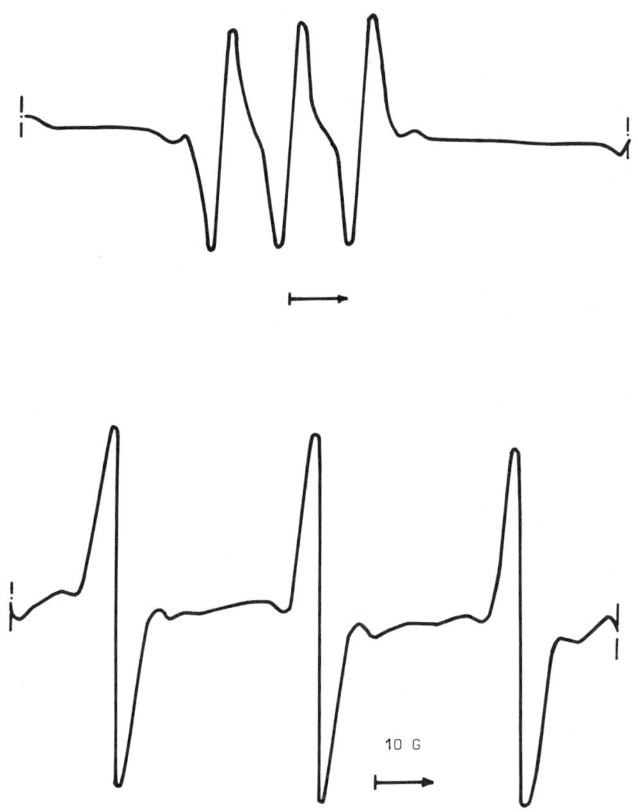

FIG. 2. Observed spectrum at $\nu_0 \sim 10^4$ MHz for a single crystal containing a diluted and oriented nitroxide free radical, at two perpendicular orientations of the magnetic field B_0. In this paramagnetic molecule, the unpaired electron interacts with one nitrogen nucleus (spin $I = 1$). The three lines correspond to $m_I = -1, 0, +1$. $g(\theta,\Phi)$ is measured of the central line $g(\theta,\Phi) = h\nu_0/\beta_e B_0(\theta,\Phi)$. The hyperfine coupling constant $A(\theta,\Phi)$ is measured as the average hyperfine splitting A' and A'' between the lines. If $g\beta_e B_0 \gg A_N$, $A' = A'' = A_N$.

FIG. 3. Spectrum of a $M/100$ frozen solution of a nitroxide radical. Principal axes of $\bar{\bar{g}}$ and $\bar{\bar{A}}$ tensors coincide. The spectrum can be computer simulated with the principal values $A_{xx}=7.1$ G, $A_{yy}=5.6$ G, $A_{zz}=32$ G, and $g_{xx}=2.0089$, $g_{yy}=2.0061$, $g_{zz}=2.002$.

to temperature t and viscosity η of the medium by $\tau_c=4\pi a^3\eta/3kt$ (k is Boltzmann's constant). This random motion has a component at the resonance frequency and induces transitions, thus modifying the spin–lattice relaxation time T_1. Spin–spin relaxation time T_2 is also influenced by random rotation if anisotropic interactions $\delta\omega$ are present: let us simulate this rotation by sudden random jumps between two orientations of resonance field B_0 and $B_0+\delta B$ (at fixed ν_0). τ_c being the average lifetime at each orientation (thus simulating the correlation time). For spins precessing initially in phase, each jump introduces a dephasing $\delta\phi=\pm\tau_c\delta\omega=\pm\tau_c\gamma_e\,\delta B$. If $\tau_c\gg T_2$, the phase difference will be 1 rad after T_2, $1/T_2\simeq\gamma_e\,\delta B$. (In the limiting case $\tau_c\rightarrow\infty$, the "immobilized spectrum" consists of two lines whose separation is of course δB.) If $\tau_c\ll T_2$, dephasing after n random jumps is given by the accumulation of n random dephasing $\delta\phi$: $\Delta\phi^2=n\delta\phi^2=n(\tau_c\delta\omega)^2$. $\Delta\phi\approx1$ rad is reached after T_2/τ_c jumps: $1/T_2\approx(\delta\omega)^2\tau_c=\gamma_e^2(\delta B)^2\tau_c$. In this case, the spectrum is averaged out to a single line, of width $1/T_2$, at the average position between B_0 and $B_0+\delta B$. Spin–spin relaxation may be induced by other mechanisms (such as internal motion in a free radical, or chemical exchange of an electron between molecular systems). The same relationships exist between T_2 and the difference of resonance frequencies and the characteristic lifetime of the processes (see Fig. 4). When different mechanisms operate at the same time, the observed linewidth is the sum of all contributions.

Let us consider now a case of *spin 1* species ("two unpaired electrons") diluted in a diamagnetic environment. There are singlet and triplet states, the lowest having ${}^1E$ and (average) ${}^3E$ energy. The "exchange integral" J is by definition $2J={}^1E-{}^3E$. Electron spin–spin dipolar interaction is nonzero in the triplet state: To first order, it comes from the spin–spin magnetic energy

$$H_{\text{dip}}=\sum_{i}^{\text{all electrons}}\sum_{j<i}[\gamma_s{}^2\hbar^2\mu_0/4\pi][\mathbf{s}_i\cdot\mathbf{s}_j-3(\mathbf{r}_{ij}\cdot\mathbf{s}_i)(\mathbf{r}_{ij}\cdot\mathbf{s}_j)r_{ij}^{-2}]r_{ij}^{-3};$$

$\mathbf{r}_{ij}$ connects electrons i and j. The influence of singlet and triplet levels and of the dipolar interaction on ESR spectra can be simulated by an effective-spin Hamiltonian $\mathcal{H}'$ similar to the one used for spin-$\frac{1}{2}$ systems, but dealing with two effective-spins $\frac{1}{2}\mathbf{s}_1'$ and $\mathbf{s}_2'$, and having "fine-structure" terms: an exchange term $\mathcal{H}'_{\text{spin(exch)}}=2J\mathbf{s}_1'\cdot\mathbf{s}_2'$ and a dipolar term $\mathcal{H}'_{\text{spin(dip)}}=\mathbf{s}_1'\cdot\bar{\bar{\mathbf{D}}}\cdot\mathbf{s}_2'$. $\bar{\bar{\mathbf{D}}}$ is a traceless tensor whose principal values are written $(2D, -D+3E, -D-3E)$ and $D=g_e^2\beta_e^2/r^3$, r being an average distance between the two unpaired electrons. As for spin-$\frac{1}{2}$ systems, various situations occur depending on the state (solid or liquid) of the system and of the relative order of magnitude of the Zeeman, exchange, fine, or hyperfine terms: In rigid matrix (or in viscous solutions, if $\tau_c\gg\hbar/D$), the dipolar interaction gives "fine structure" and also "half-field transitions" at $g\simeq4$. In solutions, the dipolar interaction is averaged out to zero if $\tau_c\ll\hbar/D$, but contributes (very effectively) to the line broadening. Finally, the case of two interacting spins may also be found for spin $\frac{1}{2}$ not perfectly isolated from their neighbors, i.e., for concentrated solutions of radicals in a diamagnetic matrix or liquid. Pairwise interactions give rise to exchange J and dipolar D energies. The total-spin expectation value of a pair of electrons oscillates between 0 and 1 with a period $\tau=h/2J$. By analogy, we may expect this oscillation ("exchange") to induce a spin–spin relaxation with $1/T_2=\pi D^2/\hbar J$. This is indeed the case: as the concentration increases, lines broaden (dipolar broadening) and then narrow (exchange narrowing).

Applications of ESR use information related to the spectral moments: Total *intensity* is proportional to the number of paramagnetic species. Concentration is measured by comparison with a reference, because absolute determination is difficult. Under favorable conditions 10^{11} spins may be detected. The *position* of spectral lines permits identification of the paramagnetic species, of its surroundings, and of its orientation. *Linewidth,* related to T_2, depends on a characteristic time τ and some characteristic difference $\delta\omega$. For instance, τ may be the lifetime of some jump process between positions having resonance at ω_0 and $\omega_0+\delta\omega$, or the rotational correlation time in solution. In this case, the linewidth is a measure of the "effective volume" of the species, of the solvent viscosity and temperature, and of anisotropic interactions, among which dipolar interactions ($\simeq r^{-3}$) pro-

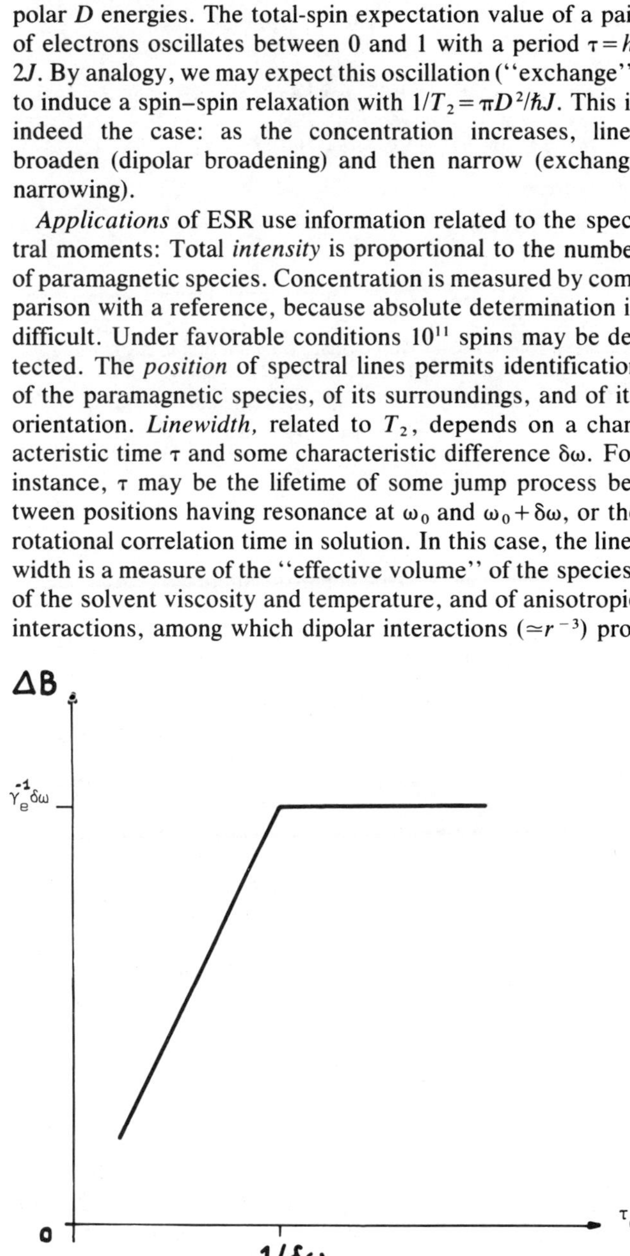

FIG. 4. Schematic ESR linewidth ΔB in solution of a paramagnetic species having an anisotropic interaction $\delta\omega$, as a function of its correlation time τ_c: when $\tau_c\ll1/\delta\omega$, $\Delta B=\gamma_e{}^{-1}\delta\omega^2\tau_c$; when $\tau_c\gg1//\delta\omega$, $\Delta B=\gamma_e{}^{-1}\delta\omega$.

vide geometrical information. T_1 measurements (saturation, pulses) give information complementary to those obtained from T_2. This information is used, for instance, in chemical (structural and dynamical) analysis of free radicals, of excited triplet molecules, of transition-metal ions, and in biophysics, especially in the "spin-label method," where a free radical reveals information about the biomolecule to which it is bound.

See also PARAMAGNETISM; RELAXATION PHENOMENA.

BIBLIOGRAPHY

Elementary

Elementary treatment of ESR may be found in quantum-mechanics textbooks, for instance, in D. Park, *Introduction to the Quantum Theory,* p. 151. McGraw-Hill, New York, 1964. (E)

Introduction

A. Carrington, *Quarterly Reviews* **17,** 67 (1963). (E)

General Texts

N. M. Atherton, *Electron Spin Resonance: Theory and Applications.* Wiley, New York, 1973. (E,I)

A. Carrington and A. D. McLachlan, *Introduction to Magnetic Resonance.* Harper and Row, New York, 1966. (E,I)

G. E. Pake, *Paramagnetic Resonance.* Benjamin, New York, 1962. (I,A)

C. P. Slichter, *Principles of Magnetic Resonance.* Harper and Row, New York, 1963. (I,A)

Specialized

A. Abragam and B. Bleaney, *Electron Paramagnetic Resonance of Transition Ions.* Clarendon Press, Oxford, 1970. (A)

J. R. Bolton, and D. C. Borg, H. M. Schwartz, *Biological Applications of Electron Spin Resonance.* Wiley-Interscience, New York, 1971. (E)

Spin Hamiltonian and Magnetic Interactions

R. McWeeny, *Spins in Chemistry.* Academic Press, New York, 1970. (I)

H. F. Hameka, *Advanced Quantum Chemistry.* Addison-Wesley, Reading, Mass., 1965. (A)

Relaxation

A. Hudson and G. R. Luckhurst, *Chem. Rev.* **69,** 191 (1969). (E)

A. Manenkov and R. Orbach, *Spin-Lattice Relaxation in Ionic Solids.* Harper and Row, New York, 1966. (A)

L. T. Muus and P. W. Atkins (eds.), *Electron Spin Relaxation in Liquids.* Plenum Press, New York, 1972. (A)

Ellipsometry

H. Lüth

Ellipsometry is an optical reflectance technique in which the change of the state of polarization of light upon reflection on a surface is measured, rather than the change of intensity. Even though the method dates back to Drude (1889) [1], ellipsometry has recently attracted considerable attention because of a wide variety of applications in modern solid-state optics, especially in the physics of surfaces and interfaces [2].

The change of the state of polarization upon reflection can be expressed in terms of the ratio of the two complex reflection coefficients $r_\parallel$ and $r_\perp$ for light polarized parallel and perpendicular to the plane of incidence, r being the ratio of reflected and incident electric field strength. The complex quantity

$$\rho = r_\parallel / r_\perp = \tan\psi \exp(i\Delta)$$

defines the two ellipsometric angles Δ and ψ that are measured in ellipsometry.

In contrast to a standard reflectivity (intensity) measurement, the two angles Δ and ψ completely determine the two optical constants n (refractive index) and κ (absorption coefficient) of an isotropic reflecting medium. For the mathematical dependence of n and κ on Δ and ψ calculated from Fresnel's formulas, see, e.g., Ref. [3].

The application of ellipsometry in surface physics is favored because in suitable cases an adsorbed species can be detected in a coverage as low as 1/100 of a monolayer. If used at a fixed wavelength λ, measured changes $\delta\Delta$ and $\delta\psi$ due to adsorption of gases can give information about adsorption kinetics (e.g., sticking coefficients) or absolute coverages [4]. The analysis, mostly done by means of computers, is rather involved because the two measured changes $\delta\Delta$ and $\delta\psi$ are not sufficient to determine all parameters (optical constants, thickness) of one or more adsorbed phases. For the study of adsorption processes, therefore, ellipsometry is most effectively used in combination with other methods, such as Auger electron spectroscopy [4].

Surface electronic structure can be successfully investigated by ellipsometric spectroscopy, in which $\delta\Delta$ and $\delta\psi$ caused by surface treatments (e.g., gas adsorption) are taken at various wavelengths over a given spectral range [5]. Transitions between surface states as they can be found by this type of optical spectroscopy are of considerable importance, e.g., for semiconductor device technology and for corrosion and catalysis studies. Another promising development can be expected from an extension of ellipsometric spectroscopy into the infrared in order to study vibration frequencies of adsorbed gases [6].

The conventional experimental setup for ellipsometry consists of two rotatable polarizers and a fixed quarter-wave plate as compensator to compensate the change of polarization of a parallel light beam due to reflection (null ellipsometry) [7]. The parameters Δ and ψ are calculated from component settings at extinction. For spectroscopy or high-speed measurements, automatic ellipsometers are much more convenient: The extinction settings can be done and read out by means of electronically controlled Faraday rotators in a self-compensating ellipsometer [8]. Polarization modulators [9] or rotating analyzers [10] are used in other types of automatic ellipsometers in which the parameter information is contained in the phase and relative amplitude of the ac component of the transmitted light intensity.

See also SURFACES AND INTERFACES.

REFERENCES

1. P. Drude, *Ann. Phys. Chem.* **36,** 532 (1889). (I)
2. "Proc. Symp. Recent Developments in Ellipsometry (Nebraska,

1968)," *Surface Sci.* **16,** (1969); "Proc. 3d Int. Conf. on Ellipsometry (Nebraska, 1975)," *Surface Sci.* **34,** (1976). (I)
3. G. A. Bootsma and F. Meyer, *Surface Sci.* **14,** 52 (1969). (A)
4. R. Dorn, H. Lüth, and G. J. Russell, *Phys. Rev.* **B10,** 5049 (1974). (I)
5. H. Lüth, *Appl. Phys.* **8,** 1 (1975). (I)
6. M. J. Dignan, B. Rao, M. Moskovits, and R. W. Stobie, *Can. J. Chem.* **49,** 1115 (1971). (A)
7. F. L. McCrackin, E. Passaglia, R. R. Stromberg, and H. L. Steinberg, *J. Res. Natl. Bur. Stand.* **A67,** 363 (1963). (A)
8. H. J. Mathieu, D. E. McClure, and R. H. Muller, *Rev. Sci. Instrum.* **45,** 798 (1974). (A)
9. S. N. Jasperson and S. E. Schnatterly, *Rev. Sci. Instrum.* **40,** 761 (1969). (A)
10. D. E. Aspnes and A. A. Studna, *Appl. Opt.* **14,** 220 (1975). (A)

Excitons

Carson Jeffries

The term exciton is broadly used to describe elementary localized excited states in solids, with the characteristic feature that the excitation can propagate through the crystal lattice. Two limiting cases are recognized depending on the type of solid.

Frenkel excitons. In molecular crystals such as anthracene, in rare-gas crystals, and in some alkali halides, a particular molecule, atom, or ion may be initially in an electronically excited state. But since the coupling to the adjacent lattice atoms is strong this excitation can be rapidly transferred from site to site without motion of the atoms themselves. Indeed these Frenkel excitons are best described as propagating waves of electric polarization. The excitons are experimentally manifested as optical absorption bands or characteristic luminescence radiation.

Mott–Wannier excitons. In semiconducting crystals like Si, Ge, CdS, and Cu_2O, optical excitation promotes an electron from a lower-energy valence band to an upper conduction band, leaving a vacant state, or hole, in the valence band. The electrons and holes are independently mobile, and give rise to the electrical conductivity of the semiconductor. However, an electron and a hole have an attractive electrical interaction: They may combine into a hydrogenlike particle—a neutral mobile excited state. The radius of this Mott–Wannier exciton is much larger than the crystal lattice spacing and the binding energy is correspondingly small. In the simplest cases the exciton displays a set of energy levels approximately like $E \propto n^{-2}$, i.e., like atomic hydrogen. As an example the exciton in Ge has a binding energy of 4 meV and a radius of 115 Å, to be compared to 13.6 eV and 0.5 Å for the H atom. The exciton in Ge may be visualized as a rather large, loosely bound particle (an electron and a hole in orbit about their center of mass) drifting through the crystal lattice; the electron and hole ultimately recombine, giving up characteristic luminescence radiation in the infrared. Being neutral, excitons do not contribute directly to electrical conductivity. They can be trapped on impurity atoms and other lattice defects; the radiation from commercial light-emitting diodes is usually due to the decay of trapped excitons.

Exciton condensation. At sufficient densities and low temperatures excitons form macroscopic droplets of a conducting electron–hole liquid in crystals like Ge and Si, much like the condensation of water vapor into fog droplets. In the liquid phase the electrons and holes are no longer bound into excitons but are free to move independently: The medium is a plasma of constant density, displaying both quantum and classical properties. The droplets have surface tension; they move freely through the crystal and can be accelerated to the velocity of sound by crystal strain gradients.

See also ELECTRON ENERGY STATES IN SOLIDS AND LIQUIDS; ELECTRON–HOLE DROPLETS IN SEMICONDUCTORS; QUASIPARTICLES.

BIBLIOGRAPHY

A. S. Davydov, *Theory of Molecular Excitons*. McGraw-Hill, New York, 1962. (A)

J. J. Hopfield, "Excitons and their Electromagnetic Interactions" in *Quantum Optics*, pp. 340–395, R. J. Glauber (ed.). Academic Press, New York, 1969. (I)

C. D. Jeffries, "Electron-Hole Condensation in Semiconductors," *Science* **189**, 955–964 (1975). (E)

R. S. Knox, *Theory of Excitons*. Academic Press, New York, 1963. (A)

Faraday Effect

Laura M. Roth

In 1845, Michael Faraday discovered that when plane-polarized light passed through lead glass in the direction of a magnetic field, the plane of polarization was rotated. The Faraday effect has since been observed in many media and over a wide range of frequencies. A plane-polarized light wave can be resolved into left and right circularly polarized waves, and the rotation arises from a difference in index of refraction between the two. The effect is usually proportional to the magnetic field, and the rotation per unit field per unit path length is given by the Verdet constant V. For positive V, the rotation is in the sense of the current flowing in the coil producing the field. Thus on reflection the rotation increases, in contrast to the optical activity of quartz and many organic molecules which is due to an intrinsic left or right handedness of the structure, for which the rotation reverses upon reflection. Related effects are the Voigt effect, the Cotton–Moulton effect, the magneto-optic Kerr effect, and the Hall effect.

In atoms and in impurity ions in solids such as lead glass, the Faraday effect is related to the strong dispersion which occurs at frequencies near an optical absorption line. In a magnetic field the lines undergo Zeeman splittings, with a differential dispersion for the two circularly polarized components. For paramagnetic ions the rotation is enhanced by population effects. Much recent interest experimentally has centered on magneto-optic phenomena in semiconductors, where Faraday rotation occurs which is related to electronic transitions across the energy gap. At frequencies above the absorption edge for interband transitions oscillatory effects are seen in large magnetic fields. The Verdet constant of Ge is -0.16 at 1.5 μm, and of Corning 8363 lead glass is 0.17 at 5000 Å, in min/Oe cm.

In solids, free charge carriers also contribute to the Faraday effect. In a plasma, such as the ionosphere, it is the electrons which contribute at radio frequencies. The rotation due to free carriers depends on their mass and density and is proportional to the square of the wavelength. In solids, the electron and hole effective masses can be measured by infrared Faraday rotation when the damping is too large to observe microwave cyclotron resonance. The density of electrons in intragalactic space is extremely small but the path length is enormous so that Faraday rotation of polarized radio waves, for example from pulsars, has been observed and has been used to estimate galactic magnetic fields. Ionospheric Faraday rotation has been measured by satellite.

In magnetic materials the microwave Faraday rotation is due to the tensor property of the magnetic permeability. For ferro- and ferrimagnets the rotation is extremely large and proportional to the magnetization. Metallic ferromagnets do not transmit radiation, but ferrites and antiferromagnets are good magneto-optic materials at microwave frequencies. In the optical region yttrium iron garnet (YIG) is transparent, has a large rotation (100°/cm at 7.9 μm), and displays striking magnetic domain effects.

In applications of the Faraday effect the figure of merit is the ratio of rotation to absorption. Faraday isolators are used in microwave and laser systems to prevent the highly amplified signal from reflecting back to the source. A number of other devices have been suggested in optical laser systems and in integrated optics.

See also HALL EFFECT; KERR EFFECT, MAGNETO–OPTICAL; MAGNETIC MATERIALS.

BIBLIOGRAPHY

C. L. Andrews, *Optics of the Electromagnetic Spectrum*, Chap. 19. Prentice-Hall, Englewood Cliffs, New Jersey, 1960. (E)

L. J. Aplet and J. W. Carson, "A Faraday Effect Optical Isolator," *Appl. Optics* **3**, 544 (1964). (E)

E. Scott Barr, "Men and Milestones in Optics V: Michael Faraday," *Appl. Optics* **6**, 631 (1967).

R. M. Bozorth, "Basic Research and Applications in Magnetism," *IEEE Trans. Magnet.* **MAG5**, 692 (1969). (E)

C. L. Hogan, "The Ferromagnetic Faraday Effect at Microwave Frequencies and Its Applications," *Rev. Mod. Phys.* **25**, 253 (1953). (A)

F. A. Jenkins and H. E. White, *Fundamentals of Optics*, 3rd ed., Chap. 29. McGraw-Hill, New York, 1957. (E)

J. G. Mavroides, "Magneto-Optical Properties," in *Optical Properties of Solids*, F. Abeles (ed.). North Holland, Amsterdam, 1972. (I)

E. D. Palik and B. W. Henvis, "A Bibliography of Magneto-Optics of Solids," *Appl. Optics* **6**, 603 (1967).

G. L. Vershaur, "The Galactic Magnetic Field," in *Galactic and Extragalactic Radio Astronomy*, G. L. Vershaur and K. I. Kellermann (eds.). Springer-Verlag, New York, 1974. (I)

S. Wang, M. Shah, and J. Crow, "Studies of the Use of Gyrotropic and Anisotropic Materials for Mode Conversion in Thin Film Optical Wave Guide Application," *J. Appl. Phys.* **43**, 1861 (1972). (A)

K. A. Wickersheim, "Optical and Infrared Properties of Magnetic Materials," and K. J. Button and T. S. Hartwick, "Microwave Devices," in *Magnetism*, Vol. I, G. T. Rado and H. Suhl (eds.). Academic Press, New York, 1963. (I)

H. J. Zeiger and G. W. Pratt, *Magnetic Interactions in Solids*, Secs. 4.10.3 and 6.10. Clarendon, Oxford, 1973. (A)

Fatigue

Johannes Weertman

Everyone is familiar with the fact that although a paper clip cannot be broken in two by simply bending it in one direction with the fingers, it is possible to break the clip without much trouble by repeatedly bending it back and forth. This is a simple example of the phenomenon of fatigue—the breaking of material by subjecting it to a cyclic stress whose amplitude is significantly smaller than the magnitude of the stress required to cause failure under static conditions.

In an engineering test of the fatigue properties of a metal the number N of stress cycles that a sample can withstand before failure occurs may be measured as a function of the amplitude S of the cyclic stress. If the experimental data are plotted on a graph of cyclic stress amplitude S versus the log of the number N of cycles to failure to obtain what is called an S–N curve, it is most often found that S decreases monotonically with increasing N from a value equal to the static failure stress (at $N=\frac{1}{4}$) to an apparent limiting value (for N larger than about 10^3–10^5) equal to about one half of the static failure stress. This lower value of S is called the fatigue limit or the endurance limit. A cyclic stress whose amplitude is smaller than the fatigue limit will never produce a fatigue failure, regardless of the number of stress cycles. However, in a number of metals and alloys there is no indication that the S–N curve ever levels off and for these materials a fatigue limit does not exist. Even in a material for which a fatigue limit appears to exist there is no complete assurance that if N is increased to a value larger than any of those used in the tests, fatigue failure may not occur under a cyclic stress whose amplitude is smaller than the apparent fatigue limit. There is virtually no frequency effect in fatigue if the temperature is not high relative to the melting point of the material. Whether the cyclic stress is applied at 1 cycle per hour or at 100 cycles per second, failure occurs at the end of the same number of total cycles. Only if the frequency is so very high that heat produced by plastic working cannot be dissipated easily is a frequency effect observed.

In material that contains no preexisting cracks, fatigue failure occurs in two stages. In the first stage small cracks are nucleated. If the material already contains cracks, the first stage of fatigue failure is of course bypassed. Small cracks may be nucleated in the interior of the material by the coalescence of a number of crystal dislocations on intersecting slip planes. It is easier for small cracks to be nucleated at the surface of a metal part or fatigue specimen. The escape at a free surface of the dislocations that create slip bands causes the surface to be roughened. With an electron microscope it can be seen that small notches are produced in the roughened surface that are the equivalent of small surface cracks.

The subsequent growth of a small crack to a large size constitutes the second stage of the fatigue failure process. During each stress cycle a fatigue crack will increase its length by an incremental amount. The rate of growth is given by the Paris equation:

$$da/dN = C(\Delta K)^n$$

where $2a$ is equal to the length of the crack, N now is the number of stress cycles (not the number required to produce failure), C is a constant whose value depends on the material tested, and ΔK is the cyclic stress intensity factor. The term ΔK is a measure of the strength of the stress singularity at the tip of a crack. Depending on the crack geometry and the specimen geometry, the stress intensity factor either is equal to, or is approximately equal to, $\Delta K \cong S(\pi a)^{1/2}$. The exponent n in the Paris equation usually is equal to 4 but it can have values within the range of $n=2$ to $n=6$. In some materials, particularly in aluminum, regularly spaced striations can be seen on the fatigue crack surface. The spacing between the striations is such that it is exactly equal to the experimentally measured crack growth rate, proving that one striation is created in each stress cycle.

Complete failure of a part or test specimen occurs when a crack reaches a length sufficiently large that it will propagate catastrophically under a static stress of the same magnitude as S. The fatigue life of a part is thus determined by how long it takes a fatigue crack to be nucleated if no cracks are present initially, and by how long it takes a small crack to grow to a large size. Generally the growth stage occupies the major fraction of the fatigue life of a specimen or metal part.

When the value of the stress intensity factor ΔK approaches the value required for catastrophic crack propagation under a static load, the fatigue crack propagation rate becomes greater than that predicted by the Paris equation. The Paris equation also breaks down when ΔK is so small that the predicted crack growth rate is smaller than about 3–10 atom distances per stress cycle. In these circumstances fatigue cracks apparently do not propagate at all and thus $da/dN \equiv 0$.

In tests in which a specimen is cycled under conditions of a constant plastic strain amplitude (the plastic strain is equal to the total strain less the elastic strain component) rather than under a constant stress amplitude the number N of cycles required to reach failure is given by the Coffin relationship (which is also called the Coffin–Manson law):

$$N^{1/2}\epsilon_p = C$$

where ϵ_p is the plastic strain amplitude and C is a material constant. The constant C is approximately equal to $\epsilon_s/2$, where ϵ_s is the plastic strain at failure under a static load, i.e., when $N = \frac{1}{4}$.

Placing the surfaces of a part under residual compressive stress, for example by shot peening, increases the fatigue life because cracks will not grow if the total stress is not tensile in part of a stress cycle and because cracks are less likely to be nucleated at a free surface if the stress there is compressive. The fatigue life also is affected by the environment around a specimen. For example, the fatigue crack growth rate of a crack in aluminum is an order of magnitude slower if the specimen is in a high vacuum than if it is in air of 50% humidity.

A phenomenon called static fatigue is sometimes observed in glass and in some inorganic crystalline material. No cyclic loading is involved in this phenomenon. A crack will grow under a static load very slowly until its length reaches the size required for catastrophic failure. Some sort of continuously occurring corrosive reaction at the crack tip evidently is required to produce static fatigue.

See also ANELASTICITY; CRYSTAL DEFECTS; ELASTICITY; MECHANICAL PROPERTIES OF MATTER.

BIBLIOGRAPHY

J. T. Barnby, *Fatigue*. Mill and Boon Ltd., London, 1972. (E)

N. E. Frost, K. J. Marsh, and L. P. Pook, *Metal Fatigue*. Oxford (Clarendon Press), London and New York, 1974. (A)

A. J. Kennedy, *Processes of Creep and Fatigue in Metals*. Wiley, New York, 1963. (A)

Special Technical Publications on fracture and fatigue, published at irregular intervals by the American Society for Testing and Materials, Philadelphia. (A)

Fermi Surface

Bernard R. Cooper

The concept of the Fermi surface in metals is intimately related to the existence of lattice periodicity, the most fundamental property of crystalline solids. The existence of the Fermi surface is a consequence of the presence of lattice periodicity and the fact that electrons obey Fermi–Dirac statistics (i.e., obey the Pauli exclusion principle). Interest in the Fermi surface stems from the fact that it is the most important single entity characterizing the electron states in a metal, and also from the fact that macroscopic thermal, electric, magnetic, and optical properties are determined by, or have important contributions from, the behavior of the electrons at and near the Fermi surface. In this article we define the Fermi surface, particularly relating its existence to the lattice periodicity and to the occupation of electron energy bands in solids. We then briefly describe several of the most important experimental techniques for measuring the properties of the Fermi surface.

For a perfect three-dimensional lattice there are three primitive translations, $\mathbf{a}_1$, $\mathbf{a}_2$, $\mathbf{a}_3$, which define the periodicity of the lattice. That is, the crystalline atomic arrangement looks the same in every respect when viewed from any point $\mathbf{r}$ as when viewed from the point

$$\mathbf{r}' = \mathbf{r} + \mathbf{t}_n, \tag{1}$$

where

$$\mathbf{t}_n = n_1\mathbf{a}_1 + n_2\mathbf{a}_2 + n_3\mathbf{a}_3. \tag{2}$$

Here n_1, n_2, and n_3 are arbitrary integers; and for $\mathbf{a}_1$, $\mathbf{a}_2$, $\mathbf{a}_3$ to be primitive, any two points from which the crystalline atomic arrangement looks the same must be given by (1).

The requirement that the potential acting on an electron moving through a solid is identical at two points $\mathbf{r}$ and $\mathbf{r}'$ related by (1) leads to the requirement that the electronic wave functions obey Bloch's theorem and are labeled by a crystal-momentum quantum number $\mathbf{k}$. So the most general electron wave function in a three-dimensional periodic lattice has the form

$$\psi_k(\mathbf{r}) = u_k(\mathbf{r})e^{i\mathbf{k}\cdot\mathbf{r}}, \tag{3}$$

where u_k is periodic so that

$$u_k(\mathbf{r}) = u_k(\mathbf{r} + \mathbf{t}_n). \tag{4}$$

The crystal momentum $\mathbf{k}$ takes on a discrete set of values determined by periodic boundary conditions. We consider a system in the shape of a parallelepiped of edges N_1a_1, N_2a_2, and N_3a_3; and take the wave function to be identical for locations separated by a translation $\mathbf{t}_N = N_1\mathbf{a}_1 + N_2\mathbf{a}_2 + N_3\mathbf{a}_3$. (That is, we imagine all of space to be filled by identical contiguous parallelepipeds, and require the electronic wave functions to behave identically in each parallelepiped.) Then the discrete values of $\mathbf{k}$ are given by

$$\mathbf{k} = 2\pi\left(\frac{k_1\mathbf{b}_1}{N_1} + \frac{k_2\mathbf{b}_2}{N_2} + \frac{k_3\mathbf{b}_3}{N_3}\right), \tag{5}$$

where k_1, k_2, and k_3 are integers; and $\mathbf{b}_1$, $\mathbf{b}_2$, and $\mathbf{b}_3$ are the primitive reciprocal-lattice vectors given by

$$\mathbf{b}_1 = \frac{\mathbf{a}_2 \times \mathbf{a}_3}{\mathbf{a}_1\cdot(\mathbf{a}_2 \times \mathbf{a}_3)}, \tag{6}$$

with $\mathbf{b}_2$ and $\mathbf{b}_3$ obtained by cyclic permutation of the indices on the $\mathbf{a}$'s.

Reciprocal space is defined as the space in which the vectors $\mathbf{b}_1$, $\mathbf{b}_2$, and $\mathbf{b}_3$ are primitive vectors in the same way that $\mathbf{a}_1$, $\mathbf{a}_2$, and $\mathbf{a}_3$ are in ordinary space. If we consider a vector

$$\mathbf{K}_m = 2\pi(m_1\mathbf{b}_1 + m_2\mathbf{b}_2 + m_3\mathbf{b}_3), \quad \text{with } m_1, m_2, m_3 \text{ integers}, \tag{7}$$

then because the electron wave functions are of the form given by Eqs. (3) and (4) as required by the Bloch theorem for a periodic lattice, it is not possible to distinguish between the wave function with crystal momentum $\mathbf{k}$ and that with $\mathbf{k}+\mathbf{K}_m$. The set $\mathbf{K}_m$ (or strictly speaking $\mathbf{K}_m/2\pi$) are called reciprocal-lattice vectors and define the reciprocal lattice in the same way that the $\mathbf{t}_n$ of Eq. (2) define the real-space crystal lattice. The wave function shows periodicity with $\mathbf{k}$ in reciprocal space (also referred to as $\mathbf{k}$ space) analogous to the periodicity of $u_k(\mathbf{r})$ in real space. Thus all distinct electronic states can be considered by treating only that part of $\mathbf{k}$ space closer to the origin than to any nonzero $\mathbf{K}_m$. This part of $\mathbf{k}$ space is called the first Brillouin zone, and is constructed by setting up planes that perpendicularly bisect the lines connecting the origin to all $\mathbf{K}_m$, and then taking the volume about the origin enclosed by these intersecting planes. Figure 1 shows the first Brillouin zone for the face-centered-cubic (fcc) structure with conventional labeling of points and lines of high symmetry.

There are $N_1N_2N_3$ discrete values of $\mathbf{k}$ contained within the first Brillouin zone, i.e., one discrete value per real-space lattice point. For $\mathbf{k}=0$, there will be a number of discrete electronic energy eigenvalues. For instance, for a noninteracting free electron gas these values are given by $E_m = \hbar^2K_m^2/2m$ for K_m taking on all values given by Eq. (7), including 0. As $\mathbf{k}$ varies across the Brillouin zone, the values of E emanating from each E_m form a semicontinuous band with

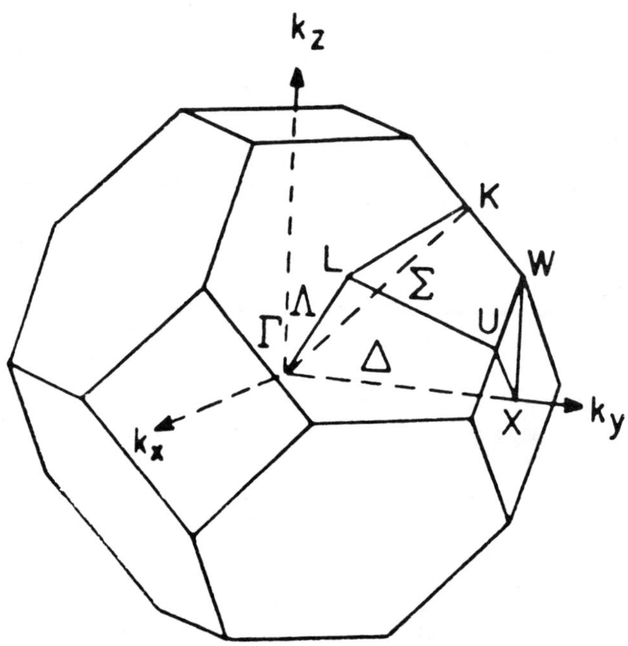

FIG. 1. The first Brillouin zone for the face-centered-cubic structure with points and lines of high symmetry labeled in conventional notation. [after B. Segall, *Phys. Rev.* **125**, 109 (1962)].

band index m. The fact that $\mathbf{k}$ is a good quantum number and that the band index m serves to lump all other quantum numbers (allowing for degeneracies for $\mathbf{k}$ at high-symmetry points and lines), together with the fact that electrons obey Fermi statistics, means that each band can hold two electrons (one with spin up and one with spin down) for a general discrete $\mathbf{k}$ in the Brillouin zone. Thus the first Brillouin zone can hold two electrons per real-space crystal lattice site per band. At absolute zero the band energy states will be filled with electrons up to an energy level, the Fermi energy (E_F), such that all the electrons are accounted for. For a metal there are one or more partially filled bands. Thus there is a surface in reciprocal space separating filled from empty states. This surface is the *Fermi surface*.

The simplest example of a Fermi surface is that for a noninteracting electron gas. The Fermi surface in that case is a sphere, and the radius of that sphere is the Fermi radius. For example, if one had an fcc lattice with one free electron per lattice site, the Fermi sphere would have a volume equal to one-half that of the first Brillouin zone shown in Fig. 1. Copper presents a simple and much studied case for a real metal. The 10 $3d$ electrons per copper atom form filled bands in the solid and do not give a Fermi surface. There is one valence electron per atom, corresponding to the atomic $4s$ state. The band formed from the valence electrons is partially filled and has behavior close to that for free electrons (i.e., the dispersion relationship over much of the Brillouin zone is approximately $E{\sim}k^2$ with an electron effective mass differing somewhat from the free electron mass); and as shown in Fig. 2, the Fermi surface is basically spherical. However, interaction effects between the s and d electrons give rise to "necks" contacting the hexagonal <111> faces of the Brillouin zone.

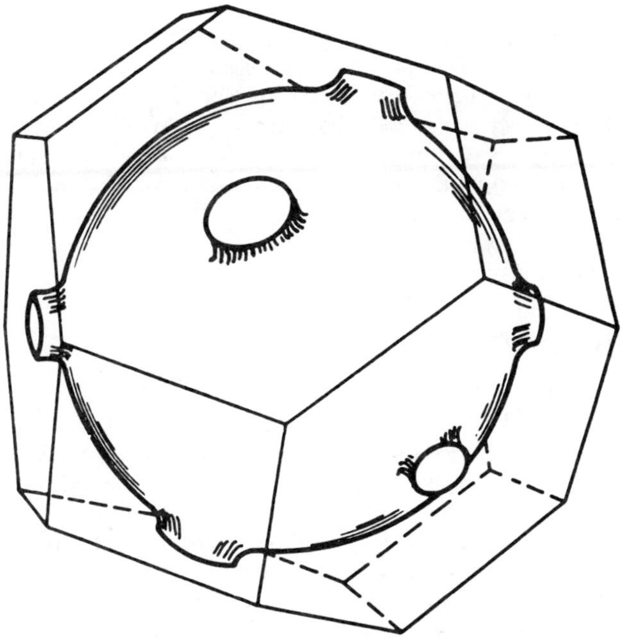

FIG. 2. A sketch of the Fermi surface of copper. The polyhedron represents the Brillouin zone [after B. Segall, *Phys. Rev.* **125**, 109 (1962)].

The Fermi surface has a symmetry appropriate to the particular crystal. Depending on the location in the Brillouin zone of the lowest-energy states and the number of electrons available for filling states, the Fermi surface can be quite complicated and bear no resemblance to a sphere. There may be disconnected surfaces in different parts of the Brillouin zone. Especially for metals with several valence electrons there may be several Fermi surfaces corresponding to partial filling of different bands. One can view the surface either from the direction of the filled states or from that of the unfilled states, the holes. If the filled states occupied the outer part of the Brillouin zone in Fig. 1, then the Fermi surface for electrons would be concave. For example, we could interchange the filled and unfilled states for the copper Fermi surface of Fig. 2. In such a case it is simpler to deal with the convex surface enclosing the unfilled volume, and one then treats the situation in terms of a *hole surface*.

At finite temperature, following the laws of Fermi–Dirac statistics, the electron states with energy lower than E_F have only a finite probability for being filled, and those above E_F have only a finite probability for being unfilled. However, these probabilities differ significantly from unity only within kT of E_F. Since E_F is typically several electron volts and 1 eV is the energy equivalent of $T = 11,606$ K, this results in a slight blurring of the Fermi surface at finite temperatures.

Because of its relationship to the filling of the energy band states, the Fermi surface characterizes the electronic structure of a metal. Also the behavior of electrons at and near the Fermi surface determines, or gives important contributions to, many macroscopic properties of metals such as the electrical transport properties, magnetic behavior, and specific heat. For example, under the influence of a dc electric field, the only electrons that can respond are those at the Fermi surface, since they can be displaced into adjacent empty states with different lattice momenta. Thus the behavior of the electrons at the Fermi surface determines the electrical transport properties of metals. Because of this fundamental relationship to the electronic structure and to the macroscopic properties of metals, over a period of about 25 years there has been a great deal of experimental and theoretical work aimed at characterizing the Fermi surfaces of metals.

The experiments most directly characterizing the Fermi surface are those measuring the sizes and shapes of the orbits of electrons moving on the Fermi surface in a magnetic field. Changing the direction of the magnetic field causes the observed behavior to be governed by different parts of the Fermi surface thereby allowing one to map out the Fermi surface. Such experiments are possible only at very low temperature in very pure materials since the electron must be able to complete its orbit without scattering for the necessary observations to be possible.

Among the most valuable techniques for measuring important features of the Fermi surface are the following. (1) *The de Haas–van Alphen effect.* This involves measuring the periodic fluctuation of magnetic susceptibility in a varying magnetic field. The period of oscillation gives a direct measure of extremal cross-sectional areas of the Fermi surface normal to the magnetic field. By making measurements at all orientations of the crystal relative to the magnetic field, one can reconstruct the Fermi surface almost exactly. The de Haas–van Alphen effect has probably yielded the most precise Fermi-surface measurements. (2) *Magnetoacoustic attenuation*. This involves ultrasonic attenuation in a magnetic field applied perpendicular to the direction of sound propagation. Variation of the attenuation with magnetic field gives extremal diameters of the Fermi surface for sections normal to the applied field. (3) *High-field magnetoresistance*. This involves observing whether the electrical resistance transverse to a high magnetic field saturates or increases indefinitely on increasing H for varying directions of H. Nonsaturation for a given direction implies a multiply connected Fermi surface, i.e., the geometry in directions such as the necks in Fig. 2 where the Fermi surface intersects the Brillouin-zone boundary. Thus one can map out "neck" directions or directions of other multiply connected Fermi surfaces, as opposed to directions of closed Fermi surface such as the spherical portion in Fig. 2. (4) *Azbel'-Kaner cyclotron resonance*. This involves absorption of microwave energy as a function of varying magnetic field applied parallel to the surface of the metal. The period of oscillation in $1/H$ depends on the integral of the inverse of the component of velocity of the electron normal to $\mathbf{H}$ around the circumference of the orbit. If the shape of the Fermi surface is known by other means, this allows one to measure the electron velocity, i.e., the derivative of electron energy with lattice momentum, at the Fermi energy.

Finally, we should point out that once the Fermi surface is known in detail, as is now the case for most metallic elements, this knowledge can be combined with measurements of macroscopic thermal, electrical, magnetic, and optical properties to obtain information about the interactions of conduction electrons with each other, with collective excitations of various types, and with electromagnetic fields. Thus, once the Fermi surface is known, one can use that knowledge to consider effects beyond those treated by the one-electron theory (i.e., theory based on the model of electrons moving independently in an effective potential) for the electronic structure of metals.

See also CYCLOTRON RESONANCE; DE HAAS–VAN ALPHEN EFFECT; ELECTRON ENERGY STATES IN SOLIDS AND LIQUIDS; MAGNETOACOUSTIC EFFECT; MAGNETORESISTANCE.

BIBLIOGRAPHY

R. J. Elliott and A. F. Gibson, *An Introduction to Solid State Physics*. Harper & Row, New York, 1974. (E)

Charles Kittel, *Introduction to Solid State Physics*, 5th ed. Wiley, New York, 1976. (E)

A. P. Cracknell and K. C. Wong, *The Fermi Surface*. Oxford Univ. Press, Oxford, 1973. (I)

J. M. Ziman, *Electrons in Metals*. Taylor & Francis, London, 1970. (I)

Walter A. Harrison, *Solid State Theory*. McGraw-Hill, New York, 1970. (A)

W. A. Harrison and M. B. Webb (eds.), *The Fermi Surface*. Wiley, New York, 1960. (A)

A. B. Pippard, *The Dynamics of Conduction Electrons*, Gordon and Breach, New York, 1965. (A)

Ferrimagnetism

L. Néel

HISTORY

In order to explain ferromagnetism, Pierre Weiss in 1906 conjectured that there exist forces tending to align the magnetic moments of the atoms parallel to one another, and that the effect of these forces is equivalent to a fictitious magnetic field, the molecular field $\mathbf{H}_m$, parallel and proportional to the macroscopic magnetization $\mathbf{J}$:

$$\mathbf{H}_m = N\mathbf{J}. \tag{1}$$

This field is essentially *uniform* and, near absolute zero, aligns *all* the atomic magnets parallel to one another. Furthermore, according to this same theory, at temperatures above the Curie point Θ the material becomes paramagnetic, with a susceptibility χ that varies with temperature according to the relation known as the Curie–Weiss law,

$$1/\chi = (T - \Theta)/C. \tag{2}$$

Weiss's theory makes possible an understanding, in general, of the properties of metallic ferromagnets such as Fe, Co, or Ni; but it fails to explain the properties of magnetite, Fe_3O_4, which has been known the longest of all ferromagnets. For one thing, its saturation magnetic moment per molecule is about $4\mu_B$, whereas one would expect instead to observe a value near $14\mu_B$, equal to the sum of the moments of the ions present ($10\mu_B$ for the two Fe^{3+} ions and $4\mu_B$ for the Fe^{2+} ion). Moreover, in the paramagnetic regime, the inverse $1/\chi$ of the susceptibility does not obey a law linear in T, such as Eq. (2), but follows a rather more complicated relation, of hyperbolic character. The ferromagnetic spinel ferrites, with general formula Fe_2O_3MO, with $M =$ Mn, Co, Ni, etc., behave like magnetite.

In spite of many ingenious efforts, these properties remained unexplained for a long time. In 1947, Verwey and Heilman[1] recognized that the ferromagnetic spinels possess a crystalline structure called *inverse*. Specifically, in a normal spinel, such as Al_2OMgO, the Mg^{2+} ion occupies the tetrahedral A site, surrounded by four O^{2-} ions, and the two Al^{3+} ions occupy the octahedral B sites, surrounded by six O^{2-} ions; whereas in an inverse spinel, the divalent ion occupies one of the B sites while one of the trivalent ions occupies the A site, and the other, the remaining B site.

FUNDAMENTAL HYPOTHESIS

Néel then conjectured[2] that the strongest magnetic forces were exerted between the ions located at A and the ions located at B and that they are *negative*, thus tending to align the magnetic moments of the ions located at A in the *opposite sense* to those of the ions located at B. The result is that in the ferromagnetic spinels at very low temperatures the magnetic moments of the two ferric ions cancel one another and the only magnetic moment observable, for magnetization to saturation, is that of the ion M. This is just what experiment confirms.

Thus, while classical ferromagnetism such as that of iron arises from positive interactions, here the ferromagnetism is due to *negative* interactions, tending to orient the magnetic moments *antiparallel*. It is this characteristic that is intended to be emphasized by calling such substances *ferrimagnetic*, to call to mind the case of the ferrites. We remark that this antiparallel arrangement of magnetic moments was ultimately confirmed[3] by neutron diffraction.

When a translation renders the two sublattices identical, in regard to both the surroundings of the sites and the nature of the ions that occupy them, the ferrimagnetism degenerates into *antiferromagnetism*.

THEORY OF FERRIMAGNETISM

In order to obtain a more exact notion of the variation of the magnetic properties of ferrimagnets, we can adapt to this case the approximation of the molecular field. We divide the crystalline lattice into *two sublattices A and B*, the first formed of the ions located on the A sites, the second of those located on the B sites. *In the absence of interactions*, the magnetizations $\mathbf{J}_A$ and $\mathbf{J}_B$ of the two sublattices in a magnetic field $\mathbf{H}$ are given by the laws of pure paramagnetism:

$$\mathbf{J}_A = J_{sA} B_A(\mathbf{H}/T), \quad \mathbf{J}_B = J_{sB} B_B(\mathbf{H}/T), \tag{3}$$

which, for small values of H, reduce to Curie laws,

$$\mathbf{J}_A = C_A \mathbf{H}/T, \quad \mathbf{J}_B = C_B \mathbf{H}/T; \tag{4}$$

here J_{sA}, J_{sB} and C_A, C_B denote, respectively, the saturation magnetizations at absolute zero and the Curie constants corresponding to the two sublattices. The two functions B_A and B_B, called Brillouin functions, represent the law of magnetization of a perfectly paramagnetic body, without interactions; they depend on the quantum number j and are different when j is not the same for the two sublattices. Explicitly,

$$B = \frac{2j+1}{2j} \coth\left(a\frac{2j+1}{2j}\right) - \frac{1}{2j} \coth\left(\frac{a}{2j}\right),$$

with $a = gj\mu_B H/kt$ (g is the Landé factor, k is the Boltzmann constant).

We then assume that the effect of differing interactions is the same as that of two different molecular fields $\mathbf{H}_A$ and $\mathbf{H}_B$, acting, respectively, on sublattice A and on sublattice B. Thus the molecular field here has only a local character.

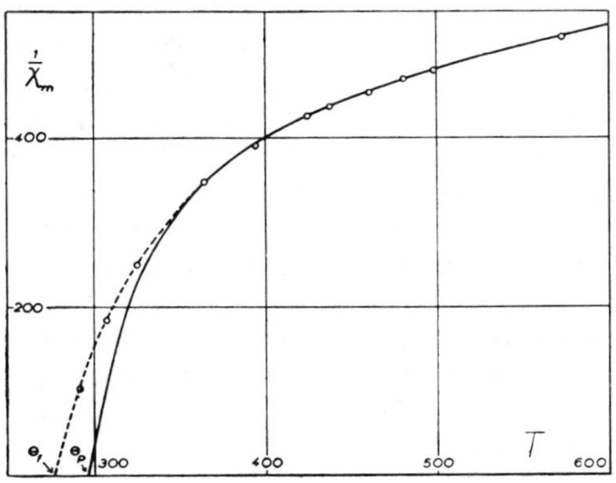

FIG. 1. Thermal variation of the reciprocal of the susceptibility of magnetite (experimental points and fitted curve).

Next we recognize that $\mathbf{H}_A$ and $\mathbf{H}_B$ depend linearly on $\mathbf{J}_A$ and $\mathbf{J}_B$:

$$\begin{aligned} \mathbf{H}_A &= n\alpha\mathbf{J}_A - n\mathbf{J}_B, \\ \mathbf{H}_B &= -n\mathbf{J}_A + n\beta\mathbf{J}_B. \end{aligned} \tag{5}$$

The coefficient $-n$, characterizing the interactions between the magnetic ions belonging to two different sublattices, is essentially *negative*. The coefficients $n\alpha$ and $n\beta$ characterize the interactions within each of the sublattices; α and β are assumed small relative to unity.

In the high-temperature regime, the substance is paramagnetic: The magnetizations $\mathbf{J}_A$ and $\mathbf{J}_B$ are parallel to the applied field $\mathbf{H}$ and in the same sense. On replacing $\mathbf{H}$ in Eqs. (4) by $\mathbf{H}+\mathbf{H}_A$ and $\mathbf{H}+\mathbf{H}_B$, respectively, we obtain the total magnetization

$$\mathbf{J} = \mathbf{J}_A + \mathbf{J}_B. \tag{6}$$

When we eliminate H_A, H_B, J_A, and J_B among Eqs. (4), (5), and (6), we deduce the value of the susceptibility,

$$\frac{1}{\chi} = \frac{T}{C_A + C_B} + \frac{1}{\chi_0} - \frac{\sigma}{T-\theta}, \tag{7}$$

where χ_0, σ, and θ are functions of n, α, and β. The curve representing the temperature variation of the inverse of the susceptibility, $1/\chi$, is a hyperbola. The susceptibility becomes infinite when the temperature decreases toward a value θ_f a little larger than θ. Below θ_f, a spontaneous magnetization appears. The law (7) represents with good precision the thermal variation of the susceptibility of the spinel ferrites of iron, cobalt, and nickel, except in the immediate neighborhood of the Curie point θ_f where the molecular field approximation becomes poor. Below the temperature θ_f, the substance is magnetized in the absence of an external field, solely through the action of the molecular fields. The magnetizations $\mathbf{J}_A$ and $\mathbf{J}_B$, directed opposite to each other, are obtained by solving the system

$$J_A = J_{sA} B_A(\mathbf{H}_A/T), \quad \mathbf{J}_B = J_{sB} B_B(\mathbf{H}_B/T). \tag{8}$$

The resulting spontaneous magnetization is given by the relation (6).

By choosing suitable values of n, α, and β, it is possible to account for the thermal variation of the susceptibility in the paramagnetic region, and, *at the same time*, for the thermal variation of the spontaneous magnetization at temperatures below θ_f (cf. Figs. 1 and 2).

THE COMPENSATION TEMPERATURE

The temperature variation of the resultant spontaneous magnetization of a ferrimagnet can show a behavior very different from that of an ordinary ferromagnet. This is because it is the *difference* of two spontaneous magnetizations, and because of amplifications of the differences between B_A and B_B. Thus one anticipates a variety of forms of behavior as represented in Fig. 3, of which most have been experimentally observed. In particular, according to forms N and V, the spontaneous magnetization J_s disappears a first time at a temperature θ_c, before reaching the Curie temperature θ_f. Actually, it is a case of a change in sign of J_s, as can be easily verified by observing the reversal of orientation or a

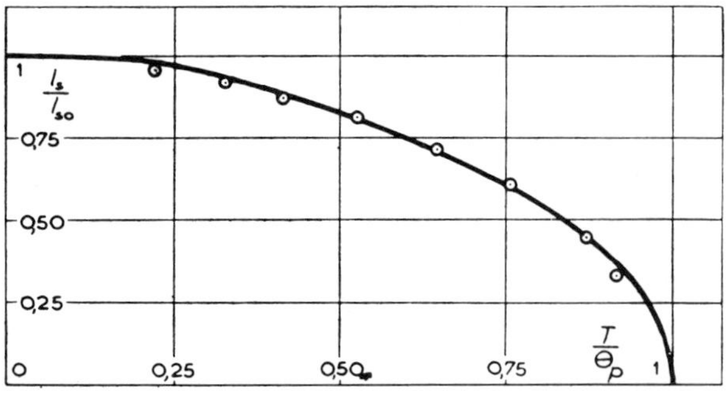

FIG. 2. Thermal variation of the spontaneous magnetization of magnetite (experimental points and curve calculated with the values of n, α, and β deduced from susceptibility).

bar magnet, made of such a substance, in the Earth's magnetic field when the temperature passes through θ_c. The temperature θ_c is that at which the partial spontaneous magnetizations J_A and J_B become equal; it is for this reason that it has acquired the name of compensation temperature. The compound for which its existence was first established[4] is the mixed ferrite $Li_{0.5}Fe_{1.25}Cr_{1.25}O_4$.

Besides the spinel ferrites, other two-sublattice ferrites are well known. We mention only pyrrhotine, Fe_7S_8; the antimonide Mn_2Sb; and maghemite, γ-Fe_2O_3.

GARNET FERRITES

There exist ferrites with structures much more complex than those that we have discussed so far. The most important are the *garnet ferrites* discovered by Bertaut, Forrat, and Pauthenet,[5] with the general formula $Fe_5T_3O_{12}$, where T is a rare earth. These are cubic compounds with *three sublattices a, d, c*. The T^{3+} ions are on c; two Fe^{3+} ions are on a, and the other three are on d, bound to the first by strong negative interactions that give rise to a ferrimagnetic assemblage with a Curie point in the neighborhood of 570 K, very insensitive to the nature of the ions T. When the ions T are not themselves magnetic (as is the case for yttrium and lutecium), the magnetic behavior of the two iron sublattices

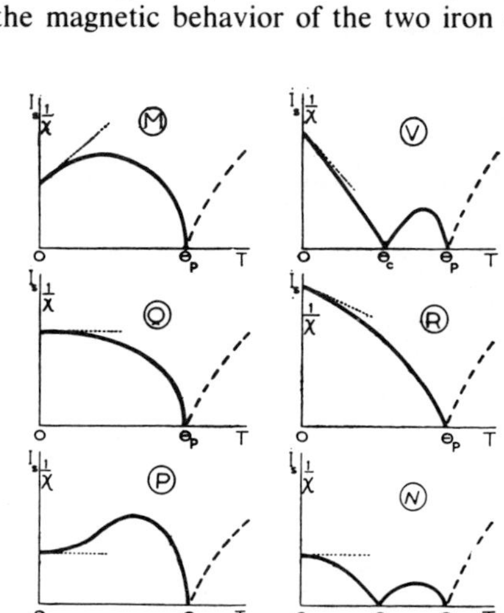

FIG. 3. Different possible types of thermal variation of the spontaneous magnetization of a ferrimagnet.

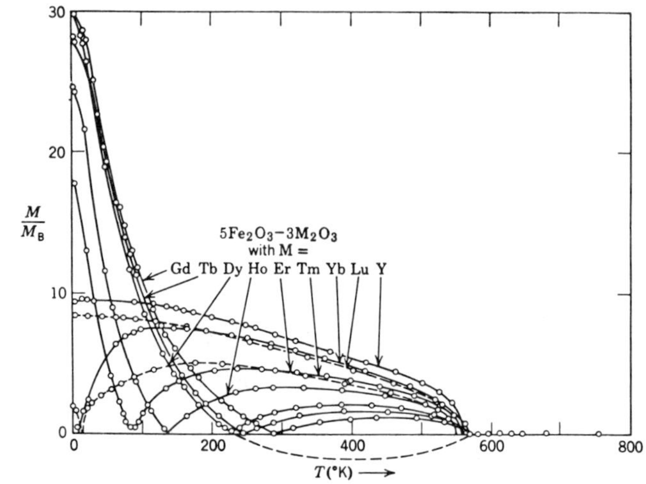

FIG. 4. Thermal variation of the spontaneous magnetization of a series of garnet ferrites.

is exhibited in the pure state. One can, for example, by means of a magnetic field of 100–200 T, decouple the two sublattices and transform the substance into a true ferromagnet. If the ions T are magnetic, they are weakly coupled to the Fe^{3+} ions and are magnetized in the directions opposite to that of the resultant magnetic moment of the latter. This results in the appearance of a compensation temperature that decreases regularly across the series of rare earths, from 290 K for gadolinium to 84 K for erbium. Figure 4 gives the temperature variation of the saturation magnetization of a series of garnet ferrites.

The garnets, which can be obtained as very beautiful crystals, interest theorists greatly because of the wide variety of ions that can be substituted for the Fe and T ions. They also have many applications, for example, yttrium iron garnet (YIG), especially at high frequencies, because they are excellent electrical insulators.

CONCLUDING REMARKS

In the ferrimagnets that we have considered, all the magnetic moments are parallel or antiparallel to one another at very low temperatures; they are called collinear. But there also exist much more complicated configurations, of which the simplest is the triangular configuration of Yafet and Kittel. It shows up in spinel ferrites when the coefficient β is negative and sufficiently large: The B sublattice splits into two secondary sublattices, whose spontaneous magnetizations make an angle less than π.

One special property most often characterizes the ferrimagnets. In ordinary ferromagnets, the interactions arise from the *exchange* of electrons between adjacent magnetic atoms. In contrast, the strong negative interactions that give rise to ferrimagnetism are *superexchange* forces requiring the presence between the two interacting magnetic ions of an anion such as O^{2-}: The electrons are exchanged along the chain $M-O-M$.

See also FERROMAGNETISM; MAGNETIC MATERIALS; MAGNETIC MOMENTS; PARAMAGNETISM.

BIBLIOGRAPHY

For a more thorough discussion and a more complete bibliography, consult Sōshin Chikazumi and Stanley H. Charap, *Physics of Magnetism*. Wiley, New York, 1964.

A. Herpin, *Théorie du Magnétisme*. Presses Universitaires de France, Paris, 1968.

Eckart Kneller, *Ferromagnetismus*. Springer-Verlag, Berlin, 1962.

REFERENCES

1. E. J. W. Verwey and E. L. Heilman, *J. Chem. Phys.* **15**, 154 (1974).
2. L. Néel, *Ann. Phys. (Paris)* **3**, 137 (1948).
3. C. G. Shull, E. O. Wollan, and W. A. Strausser, *Phys. Rev.* **81**, 483 (1951).
4. E. W. Gorter, *Philips Res. Rep.* **9**, 295 (1954); **9**, 321 (1954); **9**, 403 (1954).
5. F. Bertaut and F. Forrat, *C. R. Acad. Sci.* **242**, 382 (1956); R. Pauthenet, *Ann. Phys. (Paris)* **3**, 424 (1958).

Ferroelectricity

J. D. Axe

Ferroelectrics are materials that display an electric polarization in the absence of an externally applied electric field, together with the property that the direction of the polarization may be reversed by an electric field. The study of the properties of ferroelectric materials and the attempt to understand the nature of the ferroelectric state constitute the field of ferroelectricity.

Any macroscopic collection of matter is, to a very high degree of accuracy, electrically neutral, there being very nearly equal numbers of positive and negative charge. However, the positive and negative charges are not necessarily distributed in a homogeneous or symmetric manner. If the "centers of gravity" of the positive and negative charge distributions within a volume of material do not coincide, the material is said to have an electric *dipole moment,* whose magnitude is given by the magnitude of the charge multiplied by the vector distance between the centers of gravity of the positive and negative charges. The dipole moment per unit volume is defined as the *dielectric polarization*. Any material develops a dielectric polarization when placed in an electric field, but a substance which has such a natural charge separation even in the absence of a field is called a *polar* material. Whether or not a material is polar is determined by symmetry considerations. For example, amorphous materials such as liquids and glasses cannot be polar because there exists no unique natural direction for the polarization to lie along. Similarly, only certain crystal symmetries are compatible with a polar charge distribution. Crystal structures are divided into 32 crystal classes which are specified by the number of rotational axes and reflection planes that leave the crystal unchanged. Only crystals with no more than one axis of rotational symmetry without additional planes of reflection symmetry perpendicular to the rotation axis are polar. The polarization must lie along the rotation axis. Thus there are, for example, no polar cubic crystals. Ten of the thirty-two crystal classes are polar.

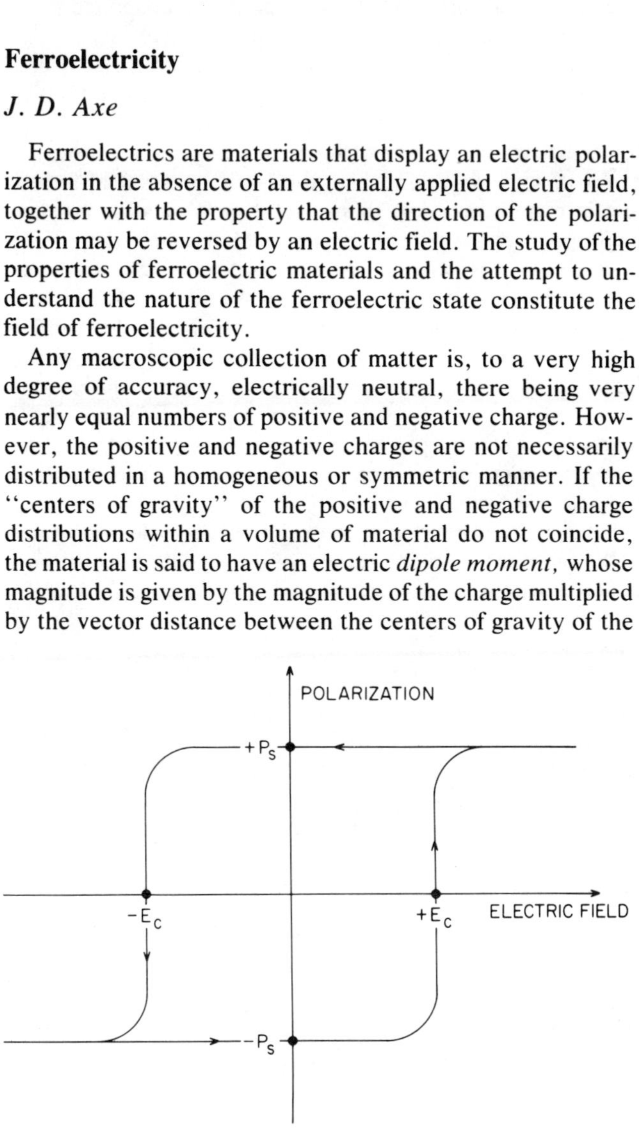

FIG. 1. A typical ferroelectric hysteresis loop showing the behavior of the polarization induced by an applied field. The arrows indicate the direction in which the electric field is scanned.

It is important to note that under ordinary circumstances polar crystals display no net dipole moment. This is because the dipole moment is neutralized by electrical charge that has built up on the surfaces of the crystal by internal conduction or by picking up charges from the ambient atmosphere. (In this respect, ferromagnetism differs from ferroelectricity. There are no readily available magnetic charges to neutralize the magnetic moment of a ferromagnet.) Often a polar crystal can be made to show an observable net electric dipole moment by rapidly changing its temperature. The magnitude of the intrinsic internal dipole moment in general changes with temperature, momentarily disturbing the balance with the compensating surface charges. This phenomenon is known as pyroelectricity, and all polar crystals are, in principle, pyroelectrics.

The term ferroelectric is usually reserved for a small group of polar crystals for which the electric moment can be reversed with an externally applied electric field. The variation of the electric polarization in an applied electric field for a ferroelectric material is shown schematically in Fig. 1. The arrows indicate that as the electric field is decreased from its maximum positive value to zero, the polarization assumes the nonzero value P_s, which is called the *spontaneous polarization,* and is a true measure of the intrinsic electric moment of the sample. As the field is reversed, it passes a critical value $-E_c$ at which the sample switches into a new configuration such that when the field is again removed, the spontaneous polarization is found to be oppositely directed. It is this relatively rare capability of being switched between two equally stable states of opposite polarization which causes certain polar crystals to be classified as ferroelectric.

In order to reverse polarization, a crystal must pass through an unstable nonpolar configuration. That this is possible with the aid of a modest applied electric field suggests that in a ferroelectric the nonpolar state is only slightly less stable than the polar state and might be stabilized by making other changes in the external environment, the temperature or pressure, for example. In fact, this is the case for virtually all known ferroelectrics. Figure 2 shows the spontaneous polarization typical of a well-known ferroelectric material, barium titanate, obtained by making measurements such as

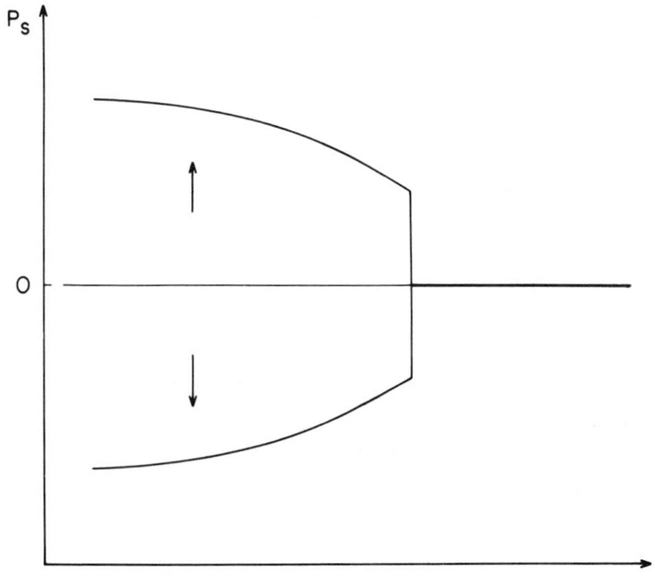

FIG. 2. Phase diagram for a ferroelectric material. Below a certain critical temperature, there are stable polar states differing only in the direction of P_s (indicated by arrows).

those of Fig. 1 at several temperatures. At low temperatures, there is a two-phase region, indicating, as discussed above, that the $+P_s$ and $-P_s$ states are equally stable. (In fact, in a crystal never exposed to a biasing external electric field both phases can coexist in the same sample in the form of oppositely polarized *ferroelectric domains*.) However, above a particular temperature, known as the Curie temperature, T_C, whose value is characteristic of the particular ferroelectric substance, the polar phases are less stable than a high-temperature nonpolar phase ($P_s = 0$). It is possible for a system to have a stable ferroelectric phase at higher temperatures than the nonpolar phase, but the opposite case is much more usual. In some materials, the spontaneous polarization goes smoothly to zero at T_C, rather than vanishing abruptly as it does in barium titanate.

There are various other properties of ferroelectric materials that are directly affected by the loss of symmetry associated with the transition from the nonpolar to polar phase. For example, the nonpolar phase of barium titanate is cubic, i.e., the basic building block of the crystal (the unit cell) is a cube. Since cubic polar crystals are impossible, the unit cell must distort on passing into the polar phase. Careful x-ray measurements show that the unit cell slightly elongates (by about 1%) in the direction of polarization. Thus the ferroelectric phase of $BaTiO_3$ develops not only a spontaneous dielectric polarization but a spontaneous noncubic strain as well. There are other manifestations of the loss of a cubic symmetry as well, such as spontaneous birefringence (the velocity of light is different parallel and perpendicular to the polar axis). These properties show their own characteristic temperature dependences, which are all, however, qualitatively similar to that for the spontaneous polarization shown in Fig. 2.

The most unusual behavior of ferroelectrics occurs in those quantities which measure the response of certain properties to an applied electric field. Figure 3 shows the behavior of the dielectric constant of barium titanate near its Curie temperature. The dielectric constant is essentially a measure of the rate at which the intrinsic polarization increases with applied electric field. The data of Fig. 3 are unusual not only in that they show a pronounced maximum at the Curie temperature, but also in that over the entire range of temperatures the dielectric constant is very large, 1000 or greater. (Ordinary insulating dielectric materials have dielectric constants that are nearly temperature independent and varying in magnitude from about 2 to 10.) Such an anomalously large temperature-dependent dielectric response is typical of what are called "proper" ferroelectrics. As the classification implies, there exist also "improper" ferroelectrics, which have relatively normal dielectric behavior.

There are other response coefficients, for example, piezoelectric constants (the strain response to an applied field) and the electrooptic coefficients (birefringence response to an applied field), which can have unusually large temperature-dependent values near the Curie temperature not unlike that shown in Fig. 3. Most of the technical applications of ferroelectrics utilize one or more of these anomalous properties in some way. The high dielectric constant of ferroelectric materials makes them useful in fabricating high-capacitance ceramic condensers. Materials with large piezoelectric response are useful for sensitive acoustic transducers (devices that transform pressure variations into voltage variations or vice versa). The favorable electrooptic properties are put to use in high-speed optical shutters and light modulators. In principle, it is possible to use the bistable nature of ferroelectric domains as the basis for computer memories, but none has yet been produced with performance characteristics which compare favorably with magnetic memory elements.

On a microscopic level, the origin of ferroelectricity lies in the spontaneous displacement of the equilibrium positions of atoms within the unit cell of a crystal. Once again, $BaTiO_3$

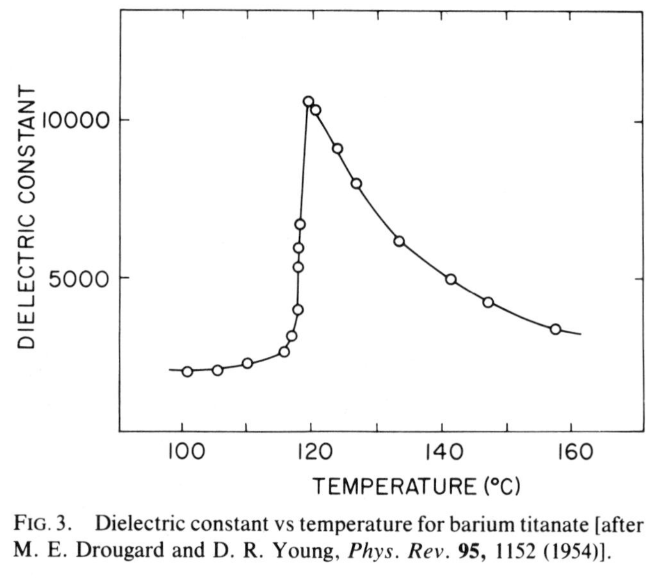

FIG. 3. Dielectric constant vs temperature for barium titanate [after M. E. Drougard and D. R. Young, *Phys. Rev.* **95,** 1152 (1954)].

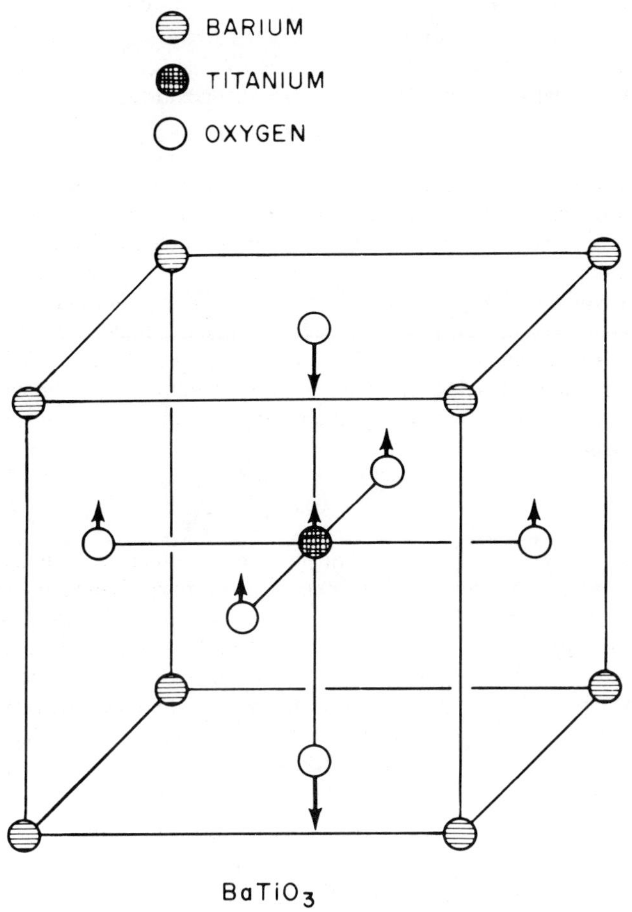

FIG. 4. Schematic representation of barium titanate showing the atomic displacements responsible for ferroelectricity.

provides a simple example. Figure 4 shows the unit cell in the high-temperature, nonpolar phase. The arrows indicate schematically the atomic displacements which occur on passing into the ferroelectric phase. These displacements destroy the cubic symmetry and produce a net dipole moment along the displacement axis.

Recently, a step toward the understanding of the nature of the ferroelectric instability has come with the realization that the displacement pattern which appears spontaneously in a ferroelectric phase often appears also as a mode of oscillation in the parent nonpolar phase. In such cases, the frequency of this particular mode of vibration becomes anomalously low or even tends to zero near the Curie temperature. These so-called *soft modes* can be detected by optical or neutron spectroscopy. Thus the dynamics of certain ferroelectric phase transformations has become closely linked to the study of small oscillations of a crystal lattice, a well-known subject with many established results which can be used to focus on problems of ferroelectricity. The idea of a soft-mode instability has also proven helpful in illuminating the distinction between "proper" and "improper" ferroelectrics, and in understanding many other structural phase transformations as well as certain magnetic ones.

See also. CRYSTAL SYMMETRY; DIELECTRIC PROPERTIES; NEUTRON SPECTROSCOPY; PIEZOELECTRICITY; PYROELECTRICITY.

BIBLIOGRAPHY

W. Cochran, *Adv. Phys.* **9,** 387 (1960). The connection between ferroelectricity and lattice dynamics is pointed out.

Ferroelectric Materials and Ferroelectricity (T. F. Connolly, ed.). Plenum, New York, 1970. Comprehensive indexing makes this the most complete guide to ferroelectric literature to date. (A)

F. Jona and G. Shirane, *Ferroelectric Crystals*. Pergamon, New York, 1962. A very complete review of the state of experimental observations on ferroelectric materials, now somewhat in need of updating. (I)

C. Kittel, *Introduction to Solid State Physics*, 3rd ed., Wiley, New York, 1966, Chap. 13. A good elementary account of the basic phenomena. (E)

J. F. Scott, *Rev. Mod. Phys.* **46,** 83 (1974). Review of modern developments in "soft mode" ferroelectrics.

Ferromagnetism

A. S. Arrott

Coulomb's law governs the interaction between electrical charges at rest. It is a requirement of the principle of relativity that Coulomb's law be modified when the charges are in motion relative to an observer. Though these modifications are usually very small, depending on the ratio of the products of velocities to the square of the velocity of light, their effects have been observed since antiquity. If the velocity of light were infinity, there would be no magnetism. If it were but ten times larger than it is, electrical technology with its generators, transformers, and motors would be most difficult to achieve. Yet the genius, Faraday, was able to create a language of fields and fluxes that describes these workings of magnetic devices without reference to relativity or the velocity of light. Faraday discovered the law of induction that governs the behavior of the transformer. Not only is the transformer a practical device but also it is useful for studying and teaching the properties of magnetic materials.

Consider a magnetic material in the form of a toroid with two coils of wire wound about its surface as shown in Fig. 1a. The toroid has a mean radius R and rectangular cross-section ab. A current i is passed through the primary coil and a voltage V is measured across the ends of the secondary coil. The voltage is proportional to the rate of change of current according to Faraday's law of induction

$$V = -L_{12}\frac{di}{dt}, \tag{1}$$

where the mutual inductance L_{12} depends on the material of the toroid as well as upon its dimensions. In SI units V is in volts, di/dt is in amperes/second, and L_{12} has the units volt-seconds/ampere which is called the henry. For some materials L_{12} can be much larger than for others. One of these is iron. Materials with properties similar to iron are called ferromagnetic. The least ambiguous definition of ferromagnetism is the existence of a spontaneous magnetic flux

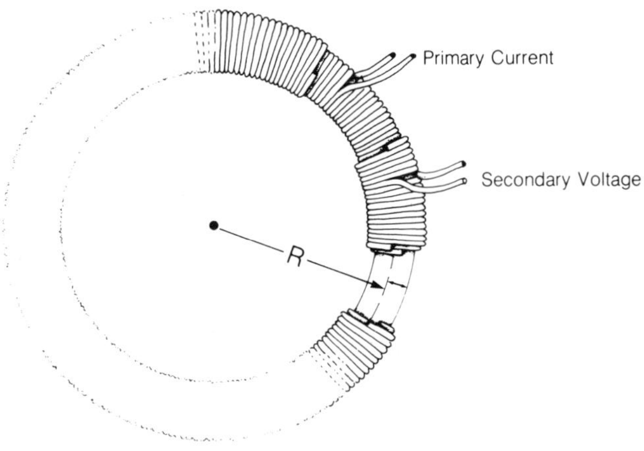

FIG. 1(a). A toroidal transformer. The cross section could be circular, or as assumed in the text a rectangle of area $a \cdot b$.

density within a material. It is possible to deduce this from the behavior of a transformer, but it is more satisfactory to ascertain this by probing the material with neutrons. From the scattering of neutrons one may map out the magnetic flux density on the scale of atomic dimensions. With a transformer one obtains averages of the magnetic flux density over the volume of the material. If the current were changed at a constant rate from large negative to large positive values, the voltage might vary with current as shown in Fig. 1b. The voltage would be proportional to the mutual inductance. The mutual inductance at large values of the current would approach the value it would have if no iron were present in the core of the toroid. In the absence of iron one could say that the primary current i_p creates a magnetic field strength H given approximately by

$$H \cong N_p i_p / 2\pi R \tag{2}$$

where N_p is the number of turns in the primary winding. The

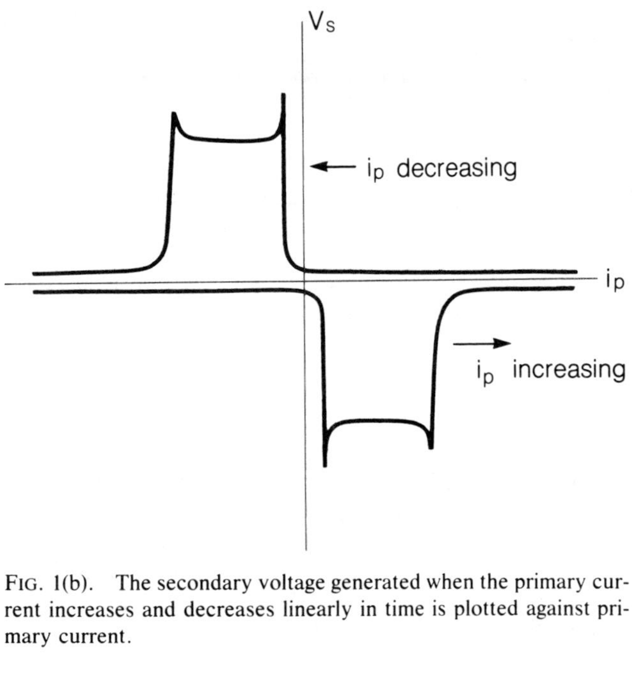

FIG. 1(b). The secondary voltage generated when the primary current increases and decreases linearly in time is plotted against primary current.

magnetic field strength would be almost parallel to the circulating axis of the toroid. The units of H are amperes/meter. The magnetic field strength is said to polarize the vacuum and create a magnetic flux density of magnitude

$$B = \mu_0 H, \tag{3}$$

where μ_0 is the permeability of the vacuum, a concept of convenience in establishing the SI units of B and H. The flux density is measured in webers/(meter)^2, which are also called teslas. The permeability of vacuum is defined to be $4\pi \times 10^{-7}$ tesla meters/ampere [or newtons/(ampere)^2 or henrys/meter]. The changing current produces a changing magnetic field strength which produces a changing magnetic flux density. The changing magnetic flux $\Phi = abB$ produces the voltage in the secondary winding according to Faraday's law in the form

$$V = -N_s \frac{d\Phi}{dt}, \tag{4}$$

where N_s is the number of turns in the secondary. Actually there are electrons accelerating in the primary giving a push to the electrons in the secondary according to Coulomb's law with relativistic corrections, but Faraday's view is very helpful. The unit of flux is the volt-second or the weber. Magnetic flux is defined as an integral of the magnetic flux density B over an area, here, the cross section of the secondary coil, $a \times b$. The mutual inductance of the toroidal transformer with no iron core is then approximately

$$L_{12} \cong N_p N_s \mu_0 (ab/2\pi R); \tag{5}$$

when the iron is present,

$$L_{12} \cong N_p N_s \mu_R \mu_0 (ab/2\pi R), \tag{6}$$

where μ_R is the relative permeability, which can be as large as 10^6 for iron. As μ_R depends on how the magnetic flux density B varies throughout the toroid and how B changes in time in response to the current in the primary, it will depend on that current and also on the past history of currents applied. At high currents (large H) B uniformly curls about the toroid and has contributions from the current in the primary and from the material of the toroid. This is expressed by the relation

$$B = \mu_0 (H + M_s), \tag{7}$$

where M_s is called the saturation magnetization, H is the field which would be present if the iron were not there, and M_s approaches an almost constant value at large values of H. If a large current i_l is reversed, the direction of B, M_s, and H will be reversed. The change in B, ΔB, will be

$$\Delta B = \mu_0(\Delta H + 2M_s) = \mu_0(N_p/2\pi R)\cdot 2i_l + 2\mu_0 M_s.$$

One can determine ΔB experimentally, calculate ΔH from the current, and then obtain the value of M_s. ΔB can be determined from the time integral of the secondary voltage, that is,

$$-\int V dt = N_s \int \frac{d\phi}{dt} dt = N_s \Delta\phi = N_s ab \Delta B. \tag{8}$$

The voltage in the secondary coil results mainly from the

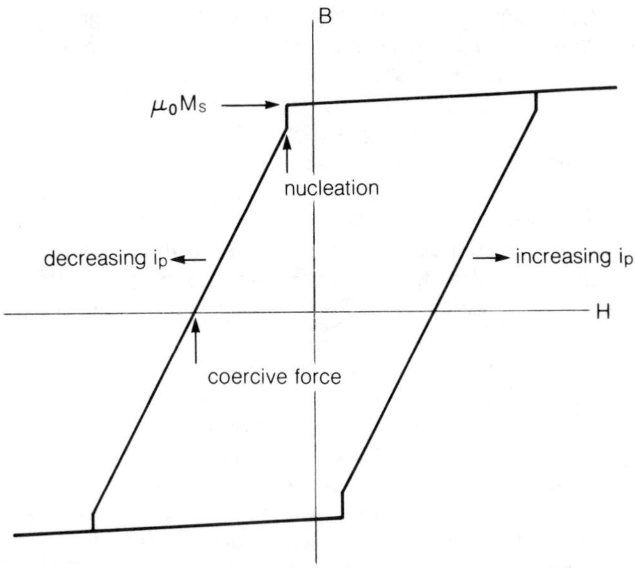

FIG. 1(c). The magnetic flux density, calculated from the integration of Fig. 1b, plotted against the magnetic field, calculated from the primary current.

reorientation of the magnetization pattern in the iron as the primary current changes, but also has a direct contribution from the changing current itself. The time integral can be carried out by using an operational amplifier with capacitive feedback as an analog computer. The integration of Fig. 1b produces Fig. 1c, which is called the major or saturation hysteresis loop. The axes are labeled using average values of $B(=\phi/ab)$ and $H(=N_p i/2\pi R)$. The value of this H for which B is zero is called the coercivity H_c. The value of B for which $H=0$ is called the remanence B_r. For the particular loop shown in Fig. 1, $B_r = B_s$.

The study of ferromagnetism divides neatly into attempts to answer two questions: How does the spontaneous magnetization arise in certain materials such as iron, cobalt, and nickel and several of the rare-earth elements as well as in oxides and other compounds of these and a few other metals? And how does the reversal process take place? The latter question is crucial in understanding the technical use of ferromagnetic materials. Though neither question has been fully answered in any single material, the study of magnetism has contributed significantly to understanding the behavior of electrons in solids and to the development of complex technologies based on the behavior of magnetic materials.

The carrier of magnetism is the electron which can be described as possessing an intrinsic angular momentum (spin) as well as a charge. The intrinsic magnetic moment of an electron (the Bohr magneton, $\mu_B = 9.27 \times 10^{-24}$ joule/tesla) is then a relativistic consequence of a spinning charge. Electrons in atoms combine in such a manner that the spin of the atom is generally less than the sum of the spins of the individual electrons. When atoms combine in solids, the spin of the atom is generally less than when the atoms are separated. The quantum-mechanical symmetry effect known as the Pauli exclusion principle makes it impossible for two electrons to have the same spatial coordinates if they have parallel spins (spins are either parallel or antiparallel for two electrons). Thus electrons with parallel spins avoid one another. This decreases the Coulomb repulsion between them below what it would be on the average for two electrons with antiparallel spins. Thus the net electrostatic binding of electrons to the atom, including the Coulomb attraction to the nucleus, would be greater for the parallel spin electrons. Yet to avoid one another the parallel-spin electrons need more kinetic energy. It is the balance between these effects which determines the net spin of an atom and the spontaneous magnetization of a material.

Calculations of the electronic structure of atoms are sufficiently accurate to account for Hund's rules, which relate the magnetic moments to the number of electrons on the atom.

The moment of an atom is more likely to be preserved in a solid if the electrons responsible for the moment are not appreciably involved in the binding of the solid. This is the case for elements of the first transition series, the lanthanides, and the actinides. The elements of the first transition series keep their atomic moments better in compounds such as oxides and sulfides than they do in metals.

A free manganese atom has five electrons in orbitals of d symmetry, all with their spins parallel. In some solids manganese exhibits a magnetic moment per manganese atom of $5\mu_B$. If this happened in the metal, it would make a most attractive magnetic material with a saturation magnetic flux density equal to 4.5 tesla. It is technologically unfortunate that this does not happen. The highest spontaneous magnetic flux density at room temperature is found in an alloy of iron and cobalt for which $B_s = 2.4$ tesla. The spontaneous magnetic flux density for the ferromagnetic elements iron, cobalt, nickel, and gadolinium are 2.20, 1.82, 0.64, and 2.45 tesla, respectively, in the limit of low temperatures. At room temperatures iron, cobalt, and nickel have B_s values 2.15, 1.80, and 0.61 tesla, respectively.

In some materials the direction of alignment of the magnetic moment varies from atom to atom, e.g., staggered up–down–up–down. If the net magnetic flux density is zero in any microscopic region, such a material is called antiferromagnetic. If a staggered pattern results in a net magnetic flux density, the material is called ferrimagnetic. The arrangement of magnetic moments in ferrimagnetic and antiferromagnetic materials is studied in detail by neutron diffraction.

At sufficiently high temperatures all of these types of magnetically ordered materials become magnetically disordered and exhibit paramagnetism. The disordering temperature is called the Curie temperature. For gadolinium this Curie temperature is just below room temperature. For iron, cobalt, and nickel the Curie temperatures are ~1040 K (~770°C), ~1390 K (~1120°C), and ~630 K (~358°C), respectively.

Perhaps the behavior of Gd metal is best understood. The seven electrons in the $4f$ shell of Gd are aligned parallel in accordance with Hund's rule. The f electrons on one atom are aligned with respect to the f electrons on neighboring atoms through a polarization of the itinerant conduction electrons which are also responsible for the binding of the metal. The degree of this polarization ($\sim\frac{1}{2}\mu_B$ per atom) and the Curie temperature are reasonably well accounted for by first-principles theory aided by modern computing techniques. Present understanding of the spontaneous magnetization and Curie temperature of iron is more qualitative than quantitative.

A general framework for discussing ferromagnetism is provided by the Heisenberg model, which assumes a net spin on each atom and an interaction between spins on neighboring atoms. The energy is written as

$$E_{\mathrm{ex}} = -\sum_{i=1}^{N}\sum_{j=1}^{n} J_{ij}\mathbf{S}_i\cdot\mathbf{S}_j, \tag{9}$$

where J_{ij} is called the exchange parameter and E_{ex} is called the exchange energy. The sums are over all pairs of n near neighbors on a lattice of N sites. The detailed calculation of the statistical-mechanical properties of the Heisenberg model has been a major activity in theoretical physics for half a century. One attempts to calculate the spectrum of magnetic excitations known as spin waves, the temperature dependence of the spontaneous magnetization, and the nature of magnetic fluctuations near the Curie temperature. Comparisons of the predictions of the model with experimental results are often rather satisfactory for a model with but one parameter $J_{ij}=J$ for the nearest neighbors.

The concept of exchange energy plays an essential role in models of the magnetization reversal process in technical magnetization. In these models the magnetization is taken as a continuum vector field which changes direction by at most a small fraction of a radian from atom to atom and does not change in magnitude. The changing of direction is a magnetic strain which increases the (negative) exchange energy by an amount

$$\Delta E_{\mathrm{ex}} = A\int \{[\mathrm{curl}\,\mathbf{M}_s]^2+[\mathrm{div}\,\mathbf{M}_s]^2\}\,dV/M_s^{\,2} \tag{10}$$

where $A=JS^2/a$ (a is an atomic distance) is an exchange constant. The exchange energy, which is responsible for ferromagnetism, resists rapid changes in the magnetization direction. In the toroidal transformer core the magnetization direction tends to curl about the axis of the toroid and to be everywhere parallel to the surfaces. The reason for this is that in addition to the exchange interaction between the spins on neighboring atoms there is a much weaker magnetostatic interaction among all magnetic moments. The magnetostatic interaction is again a relativistic effect between spinning atoms. This energy is a minimum if div $\mathbf{M}_s=0$ everywhere in the material and $\hat{n}\cdot\mathbf{M}=0$ on all surfaces, where $\hat{n}$ is the normal to the surface. If these conditions are not met, each magnetic moment will experience a demagnetizing field H_D which has as its sources div $\mathbf{M}_s$ and $\hat{n}\cdot\mathbf{M}_s$. The curling pattern satisfies div $\mathbf{M}_s=0$ and $\hat{n}\cdot\mathbf{M}_s=0$ while at the same time making only a very small increase in the exchange energy ($|\mathrm{curl}\,\mathbf{M}_s| = M_s/R$, where R is the radius of the toroid). To reverse the sense of the curling pattern in response to a reversal of current in the primary of the toroid both the exchange energy and the demagnetizing field energy must increase in the process. In addition the process will pass through configurations which are unstable, irreversible changes will occur, and hysteresis will result. A complete theoretical description of this process does not exist. Observations on an iron transformer show that the crystal structure plays an important role in the reversal process. The magnetic moment is not entirely due to the spin of the electrons, but also includes a contribution from the orbital motion of the electrons. This is the origin of the magnetocrystalline anisotropy, an energy which depends on the orientation of the magnetization with respect to the crystal axes. If the toroidal core itself is wound of a thin metallic tape in which all the crystal grains are oriented along the tape, the crystalline anistropy as well as the exchange energy and the demagnetizing field energy prefers the magnetization to curl about the toroidal axis. All these energies would be the same if the magnetization pattern were completely reversed. But the energy of interaction of the magnetization with the primary current does change if the primary current is reversed with respect to the magnetization pattern. This is the driving force for reversals. The effect of magnetic anisotropy on the reversal process is to inhibit all processes that take place by a gradual rotation of the magnetization everywhere. What happens is that somehow one or more domain walls nucleate and pass through the cross section of the toroid reversing the magnetization as they go. The walls separate regions aligned in opposite directions. The magnetization in a wall rotates from one direction to the other in a distance of a few hundred atoms for iron at room temperature. The wall has exchange energy because of the rapid rotation, anisotropy energy because the magnetization no longer is confined to preferred directions, and generally some demagnetizing field energy because either div $\mathbf{M}\neq 0$ in the wall or $\hat{n}\cdot\mathbf{M}\neq 0$ where the wall intersects the surfaces. A change in the direction of magnetization can also give rise to mechanical strains through the elongation (or shrinkage) of the atomic lattice in the direction of magnetization. This is called magnetostriction, and it also influences the reversal processes. Furthermore, domain walls interact with crystal imperfections, grain boundaries, and surfaces. Presumably imperfections are also necessary for the nucleation of the domain walls. If the field from the current in the primary which is sufficient to nucleate the walls is greater than the field necessary to move the walls, the material has a "square" hysteresis loop. If the reverse is the case, the field necessary to move the walls influences the form of the hysteresis loop.

For the loop shown in Fig. 1c, the nucleation takes place as in a square-loop material, but the walls reach a stable configuration in that nucleation field. Further changes follow the change in the magnetic field, until the walls disappear, leaving the magnetization reversed. In a part of the loop where walls are present, if the direction of field change is reversed, minor hysteresis loops will be followed. Figure 2 shows the behavior of a typical commercial transformer material including a few minor loops. By cycling the magnetic field with decreasing amplitude it is possible to demagnetize the core reaching a state with no net B for zero applied field. The virgin magnetization curve $OABCDEF$ is then traced out by increasing this magnetic field monotonically. The initial permeability μ_{RI} and the maximum permeability μ_{RM} are indicated in Fig. 2. The loops $DD'D$ or $EE'E$ would correspond to efficient operation of a power transformer, while the loop $AA'A$ might be employed in a communications transformer. In the first case the area of the

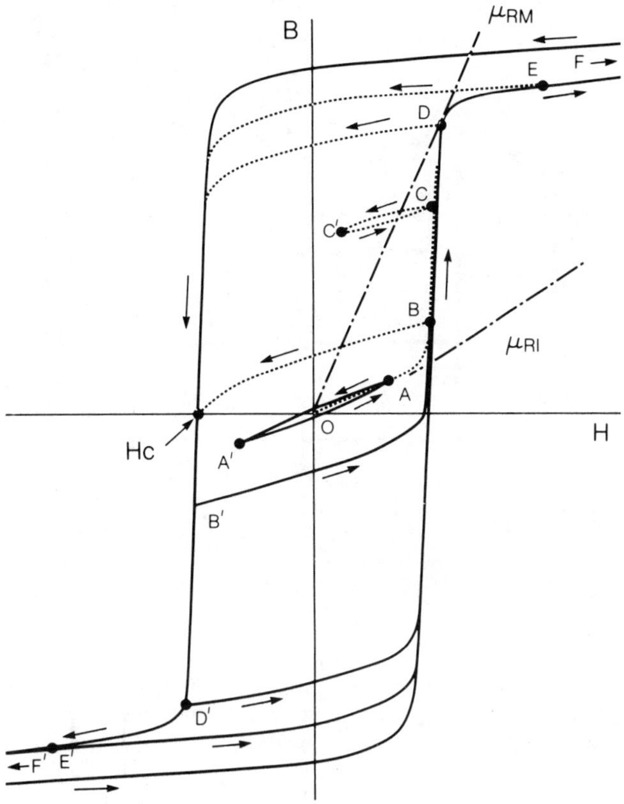

FIG. 2. Hysteresis loops for a commercial transformer material. Portions of the major loop are closed beyond the limits of the diagram (*F* and *F*′).

hysteresis loop contributes to undesirable energy losses, in the second case the area of the loop corresponds to a decrease in the quality of the signal transmission.

The most used magnetic material is polycrystalline motor-grade iron (99.6% Fe) with μ_{RM} about 5×10^3 and a hysteresis loss, ω_H, about 150 joules per cubic meter of core material for one cycle of a loop with a maximum induction of $B_M = 1.2$ tesla. High-grade power transformers use grain-oriented silicon–iron (3.2 wt% Si) with μ_{RM} values over 7×10^4. It is possible to operate with flux densities near 1.9 tesla for fields of 800 A/m ($\mu_0 H = 10^{-3}$ tesla). Core losses for a loop with $B_M = 1.7$ tesla at 60 Hz can be as little as 200 J/m^3/cycle. Only one-quarter of these losses comes from the area of the hysteresis loop as measured slowly.

An alloy with 79Ni-5Mo-16Fe when suitably heat treated has μ_{RM} over 5×10^5 and a hysteresis loss of less than 0.05 J/m^3 at $B_M = 0.5$ tesla. As the superior magnetic materials are metallic electrical conductors, it is necessary to use thin laminations with surface insulation to minimize the effects of eddy currents. At radio frequencies a poorly conducting material such as nickel–zinc ferrite with $\mu_R = 2 \times 10^3$ is suitable. The best materials in each higher frequency range have lower permeabilities, e.g., $\mu_R = 10$ for $3BaO \cdot 2CoO \cdot 12Fe_2O_3$ at 10^9 Hz. The poor conductors show resonant responses that make them useful as components for microwave devices.

The preparation of suitable materials for magnetic circuits has been an important metallurgical activity for almost a century. Materials with square hysteresis loops are well suited for memory elements in computers. A special class of magnetic materials support domain walls in the form of cylinders intersecting both surfaces of thin plates. These are termed magnetic bubbles and form the basis for a new technology of computing devices.

Preparing heterogeneous materials that impede the propagation of domain walls results in permanent magnetic materials. For such a material, one could cut a slice out of the toroidal core, and it would remain magnetized parallel to its sides despite the strong demagnetizing fields arising from $\hat{n} \cdot \mathbf{M}$ at the ends of the piece. In barium ferrite the magnetocrystalline anisotropy and the impedance to domain wall motion are sufficient that even a thin slice will remain magnetized. The material is made of partially sintered small grains. Such magnets are in household use for pinning notes to steel cabinets as well as in many fractional horsepower motors. More powerful permanent magnets are made from heterogeneous alloys of Fe with Al, Ni, Co, and Ti. The strongest permanent magnets to date are compounds of Co and Sm in the ratios between 5 to 1 and 8 to 1.

The stored energy density in a permanent magnet is a maximum if the remanent magnetization $M_r = B_r/\mu_0$ can be maintained against a demagnetizing field as large as $-\frac{1}{2} M_r$. Then the stored energy is $\frac{1}{2}(BH)_{max} = \frac{1}{8} B_r^2/\mu_0$. In barium ferrite and samarium–cobalt materials the remanent magnetization can be so maintained. Though the latter have B_r near 1.0 tesla compared to 0.38 for barium ferrites and hence more than six times the stored energy density, the cost difference is much in favor of barium ferrite, unless space is at a premium. The Fe-Al-Ni-Co-Ti–type alloys have even higher values of B_r, but magnetization reversals start before the ideal value of $\frac{1}{8} B_r^2/\mu_0$ is reached. Their stored energy values are intermediate between the cheap ferrites and the expensive samarium–cobalt alloys.

See also FERRIMAGNETISM; HYSTERESIS; MAGNETIC DOMAINS AND BUBBLES; MAGNETIC MATERIALS; MAGNETS (PERMANENT) AND MAGNETOSTATICS; PARAMAGNETISM.

BIBLIOGRAPHY

D. Hadfield, *Permanent Magnets and Magnetism*. Wiley, New York, 1962.

R. S. Tebble and D. J. Craik, *Magnetic Materials*. Wiley-Interscience, New York, 1969.

Field Emission

L. W. Swanson

Electrons may be extracted from cold conductors or semiconductors by application of a strong electric field. The phenomenon, called *field emission,* which was first reported by R. W. Wood in 1897, occurs at fields of the order of 10^9–10^{10} V/m.

Fowler and Nordheim, in 1928, provided the first generally accepted explanation of field emission in terms of the newly developed quantum mechanics which they applied to the Sommerfeld model for the electronic energy levels in a metal. Application of a high field to the metal (see Fig. 1) produces a triangular potential-energy barrier through which electrons, arriving at the metal surface, may quantum mechanically tunnel. By solving the Schrödinger equation, Fowler and Nordheim obtained the barrier penetration probability $D(E_x)$ as a function of electron kinetic energy in a direction perpendicular to the barrier E_x.

By multiplying $D(E_x)$ by the number of electrons arriving at the surface with energy E_x and integrating over all values of E_x, Fowler and Nordheim derived the "Fowler–Nordheim" equation which relates field emission current I to field F and work function ϕ:

$$I = 6.2 \times 10^6 \{E_F{}^{1/2}/C\,\phi^{1/2}\} F^2 \exp\left\{-\frac{6.8 \times 10^9 \phi^{3/2}}{F}\right\}, \quad (1)$$

where I is in A/m^2; C, ϕ are in volts; F is in V/m; and $C = \phi + E_F$. Equation (1) represents the variation of I with field very well except at high current densities ($>10^{10}$ A/m^2), where space-charge effects become noticeable.

In 1937 E. W. Müller invented the field-emission microscope (FEM), which consists of a sharp conducting point directed at a fluorescent screen deposited on conductivized glass that is maintained about 1–10 kV positive with respect to the point. The resulting projection microscope (see Fig. 2) has a magnification of upwards of 10^5 and a resolution of 20Å, which is enough to image certain single adsorbed atoms.

From the slope of a plot of ln I/F^2 vs $1/F$, an average work function for the emitter surface may be obtained since, according to Eq. (1), we have $\partial \ln(I/F^2)/\partial(1/F) \propto \phi^{3/2}$. When annealed, field emitters develop hemispherical end forms with large facets of single-crystallographic planes; the work functions of these planes may also be obtained by intercepting the electrons emitted from then in a Faraday cup.

In the hands of Gomer and co-workers, at the University of Chicago, the FEM has been a particularly useful tool for adsorption studies, and it can be used to make measurements of heats of desorption, work function changes, and diffusion energies for adsorbates. For example, in an elegant use of

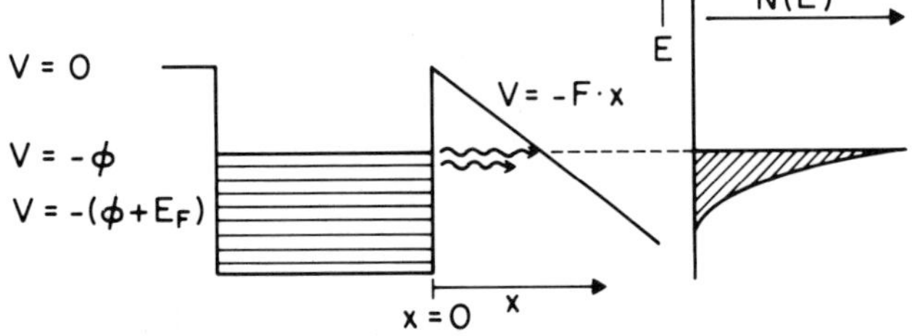

FIG. 1. Sommerfeld model for a metal at $T = 0$ K in the presence of a high field. Electrons are able to tunnel through the potential barrier, especially at the level $V = -\phi$ where the barrier is thinner. The energy distribution curve $N(E)$ is shown. Notice the sharp cutoff at $E = -\phi$ due to the absence of electrons above the Fermi level at $T = 0$ K.

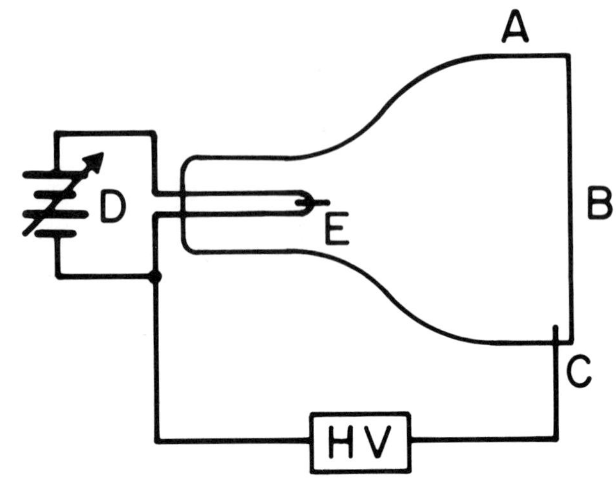

FIG. 2. Simple field-emission microscope. A is an evacuated ($p < 10^{-9}$ Torr) glass envelope, B is a conductivized fluorescent screen on the inside of the envelope, C is an external electrical connection to the screen, and D is a low-voltage heating supply for flash cleaning of the field emitter E.

the FEM the relative binding energies of potassium to several planes of tungsten were measured by relating single-plane work functions to surface coverages at different temperatures.

Useful information about emitter band structure, surface states, broadening of adsorbate energy levels, and even vibrational energy levels may be obtained from studies of the total energy distribution (TED) of field-emitted electrons (Fig. 1). An anomalous hump in the energy distribution for the (100) plane of molybdenum and tungsten has been discovered that has been attributed to the existence of a surface state on this plane which disappears when adsorption takes place. A similar structure in the TED from regions containing a single atom of Ca, Sr, or Ba has also been observed. In this case the structure was due to a resonance-enhanced tunneling when the broadened energy levels of the adsorbate atom overlapped the energy levels of the metal just below the Fermi level of the metal. By comparing the TED of hydrogen and deuterium adsorbed on a tungsten field emitter at 78 K, followed by subsequent warming to 300 K, the detection of vibrational energy levels of molecularly adsorbed species has been accomplished.

The small optical source size and very high current densities of field-emission cathodes make them attractive electron sources for microprobe applications because, for focused beam sizes below about 5,000 Å, field-emission sources provide higher currents than thermionic cathodes. Thus Crewe and co-workers at the University of Chicago were able to develop a field-emission scanning electron microscope (SEM) with a resolution of 5 Å and a beam current one hundred times larger than the corresponding thermionic one; this allows the SEM to scan at a correspondingly higher rate.

Other fine-focus applications of field emission which have recently been developed include high-resolution scanning Auger microscopy, high-density information recording, and electron-beam microfabrication of integrated circuits.

A field-emission-initiated vacuum arc process is being used in flash x-ray devices to produce very large currents in submicrosecond pulses for x-ray generation.

See also ADSORPTION; ELECTRON MICROSCOPY; FIELD-ION MICROSCOPY; TUNNELING; WORK FUNCTION.

BIBLIOGRAPHY

W. P. Dyke, "Advances in Field Emission," *Scientific American* **210** (1), 108 January (1964). (E)

R. Gomer, *Field Emission and Field Ionization*. Harvard University Press, Cambridge, Mass. 1961. (I) An introduction to theory and techniques of field emission.

R. H. Good and E. W. Müller, "Field Emission," *Handbuch der Physik*, Vol. 21, p. 1976, Springer Verlag, Berlin, 1956. (I)

J. W. Gadzuk and E. W. Plummer, "Field Emission Energy Distribution," *Rev. Mod. Phys.* **45**, 487 (1973). (A)

L. W. Swanson and A. E. Bell, "Recent Advances in Field Electron Microscopy of Metals," *Adv. Electr. Electr. Phys.* **32**, 193 (1973). (A)

Field-Ion Microscopy

Erwin W. Müller

The field-ion microscope (FIM) is a powerful tool of surface physics uniquely offering a direct view of a metal surface in atomic details, including means of manipulating and identifying individual atoms. Introduced by Müller in 1951, the FIM is a basically simple device in which the specimen is the cap of a finely pointed needle tip, chemically etched from the material to be studied to a radius between 10 and 200 nm. It is an improvement of the 15-years-older field-emission microscope (FEM), in which the nearly hemispherical tip cap is imaged on a phosphor screen some 10 cm away by radial projection through electrons tunneling out of the surface when a negative potential is applied to the tip to produce a field strength of some 10^9 to 5×10^9 V/m. Because of the strong dependence of field-emission current density upon the work function, the FEM remains a useful device for the study of adsorption, but the large lateral velocity component of the degenerate electron gas in the metal and the large de Broglie wavelength of the field-emitted electrons limit the resolution of the FEM to about 2 nm.

Imaging the surface with ions emitted from a tip at positive potential can achieve true atomic resolution, as the lateral velocity component of the ions may be reduced with the emitter temperature, and the short de Broglie wavelength of the heavy ions also poses no diffraction limit. Imaging with ions became a possibility with the discovery of field desorption in 1941. However, the realization of the field-desorption microscope, imaging by the radial transfer of a single monolayer of adsorbate from the tip to the distant screen, had to wait for 30 more years to the advent of a powerful image intensification device, the microchannel plate.

Fortunately, in 1951 a continuous supply of gas to be field ionized at the tip surface proved feasible, simply by backfilling the evacuated FEM tube with about 1 Pa of hydrogen or another imaging gas. Two other advances had to be made to develop the FIM, by the mid 1950s, to a useful tool for surface research: the mounting of the specimen at the end of a "cold finger" which could be filled with liquid nitrogen, hydrogen, or helium for cryogenic cooling, and the use of field evaporation for preparing the specimen surface. By temporary application of a higher voltage than needed for ionizing the imaging gas, the field at the more protruding parts of the crudely etched tip would become high enough to "field evaporate" the areas of excessive local protuberance, and to obtain finally a geometrically highly perfect, almost hemispherical surface shape that alone would guarantee the proper imaging by the radial projection principle (Fig. 1). Moreover, after the perfectly clean surface had been inspected or photographed, the operator could remove more single atoms or many atomic layers in a perfectly controlled way to bring the interior of the specimen to the surface for a detailed view of bulk features.

The magnification M of the radial projection microscope, easily exceeding a million times, is given by

$$M = R/\beta r,$$

where R is the tip-to-screen distance, r is the tip radius, and β is an image compression factor (≈ 1.5) due to the fact that the emitter is a tip cap rather than a sphere in free space. More important for a microscope is its resolution, expressed as the objectside diameter δ of an atomic image spot made up by some 1000 ions/s arriving at the screen, calculated by Müller and Tsong to be

$$\delta = \delta_0 + \left[\frac{4\hbar\beta r^{1/2}}{(2\kappa meF)^{1/2}} + \frac{16\beta^2 rkT}{\kappa eF} \right]^{1/2},$$

where δ_0 is the diameter of the imaging gas atom of mass m and charge e, and κ (≈ 4 to 8) is a geometrical field factor relating the applied voltage V to the field F,

$$F = V/\kappa r.$$

The first term in the bracket describing the diffraction limit is almost negligible, while the second term is the effect of

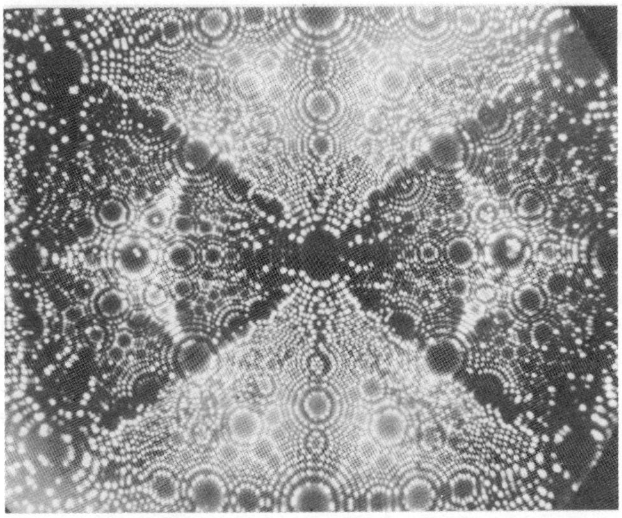

FIG. 1. Field-ion micrograph of a tantalum tip. The random extra bright spots particularly seen to the left and right of the central (011) plane are oxygen interstitials.

the temperature-dependent lateral velocity component of the ions. The best resolution can be expected for helium as the gas with the highest ionization field and allowing the lowest tip temperature, followed by neon, argon, and hydrogen. Typical ionization fields are 4.5×10^{10} V/m for He, 3.6×10^{10} V/m for Ne, and about 2×10^{10} V/m for Ar and H_2. They can be calculated by considering the probability with which a ground-state electron may tunnel out through the potential barrier formed by the Coulombic attraction to the atom core, the externally applied field, and its modification by the image force near the metal surface. Refinements of calculating the transition probability including surface states and noble-gas adsorbate between the ionizing atom and the surface are presently advancing.

A condition for a stable image is that the specimen must not undergo field evaporation at or below the ionization field of the imaging gas. In the absence of a field, the energy to remove an n-fold charged ion of total ionization energy I_n from a metal surface of work function ϕ is

$$Q_0 = \Lambda + I_n - n\phi,$$

where Λ is the sublimation energy of the neutral atom. In an electric field the ion experiences the classical Schottky potential saddle of height $-(n^3e^3F)^{1/2}$, so that the total energy to remove the ion is

$$Q = Q_0 - (n^3e^3F)^{1/2}.$$

At a finite temperature the barrier is overcome by thermal excitation at an evaporation rate

$$\kappa_E = (kT/h)\exp(-Q/kT).$$

Surprisingly, this simple image-force theory describes quite well the experimental evaporation fields of various metals. Refinements have been introduced considering the polarizability of the surface atom at a kink site, field penetration into the surface, and tunneling of highly charged ions through the top of the potential saddle. Not understood are the experimentally established three- to five-fold charges of ions from the highly refractory metals, while the theory predicts doubly charged ions.

Only the refractory metals have evaporation fields at or above the 4.5×10^{10} V/m required for optimum ionizing of helium, but many more metals such as Fe, Ni, Co, Pd, Cu, and Au may be imaged with neon and, at considerable loss of resolution, also with hydrogen and argon. Because of the very low efficiency by which heavy ions excite a phosphor screen, the very faint neon imaging used to be a tour de force until in the past few years the ion-to-electron image converting and thousand-fold intensifying microchannel plate became a standard component of the FIM. While some good images of aluminum and bismuth have been obtained with argon, the low ionization field of this gas, as well as of hydrogen, makes it difficult to retain atomic cleanliness of the specimen when impurity gases may be adsorbed causing field-induced rearrangement and reactions at the surface. By contrast, helium and neon imaging requires only modest background vacuum conditions, as with the high ionization fields the common residual gases are ionized in free space and their ions rejected before they can approach the clean field-evaporated surface.

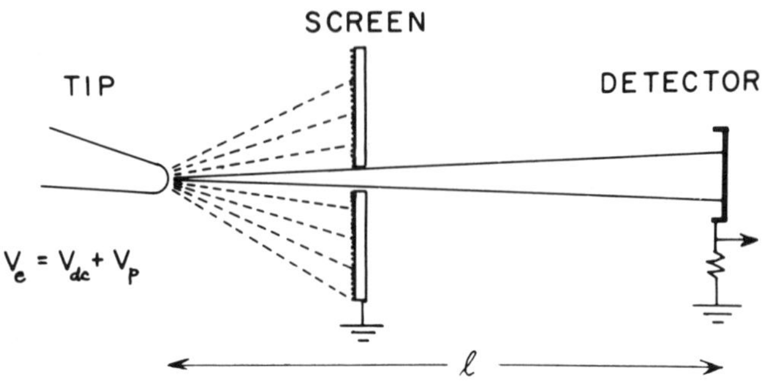

FIG. 2. Scheme of the straight time-of-flight atom probe. At the instant of field evaporation the tip potential is the sum of the dc and the pulse voltages. (From *CRC Critical Rev. in Solid State Sci.* **6**, 85 (1976) by permission of CRC Press.)

Field-ion microscopy assumed a new dimension with the introduction in 1967 of the atom-probe FIM (Fig. 2). A probe hole in the screen and external manipulation of the tip direction permit the operator to select a single atom or a small surface area of interest for mass spectrometrical identification. By superimposition of a nanosecond evaporation pulse of amplitude V_p on the imaging voltage V_{dc}, particles of charge ne and mass m are removed from the surface and accelerated near the tip to a velocity v according to the energy equation

$$\tfrac{1}{2}mv^2 = ne(V_{dc} + V_p).$$

Those selected by aiming at the probe hole pass into a drift space of up to 2.5 m length, where they arrive at an electron-multiplier detector of single-ion sensitivity. The velocity of the particle and hence its identifying m/n ratio are obtained by precisely measuring the time of flight (1 to 20 μs) to the detector. The simple straight time-of-flight atom probe still has a somewhat limited mass resolution because of an uncertainty of the ion energy due to desorption during the finite risetime of the pulse. Unambiguous mass identification ($\Delta M/M < 1/1000$) is achieved with an energy-focusing modification of the drift tube. This advanced instrument (Fig. 3) represents the most sensitive microanalytical tool for surface analysis, requiring only one atom for identification.

Field-ion microscopy has offered new insights into surface phenomena in atomic dimensions and at high fields. Experimenting with one or a few single metal atoms vapor-deposited on perfect net planes gives data on surface migration and binding energies, as well as on atomic interaction in clustering. Adsorbed electronegative common gases remain invisible but may betray their presence by displacing substrate atoms. Atom-probe FIM work revealed that the imaging gases He and Ne are field adsorbed in apex positions and held on each imaged surface atom by dipole forces up to a temperature of 200 K. Upon field evaporation, He may form stable compounds such as WHe^{3+} and $PtHe^{2+}$. In certain alloys, one constituent such as Co in ordered Pt–50% Co may remain invisible, while in dilute alloys the solute may appear as an extra bright spot such as W in Pt–5% W, or look like a vacancy such as Au in Pt–1% Au. Field-ion microscopy of *in situ* produced radiation damage of metals allows the counting of vacancies or the direct viewing of interstitials and their motion to the surface in isochronous

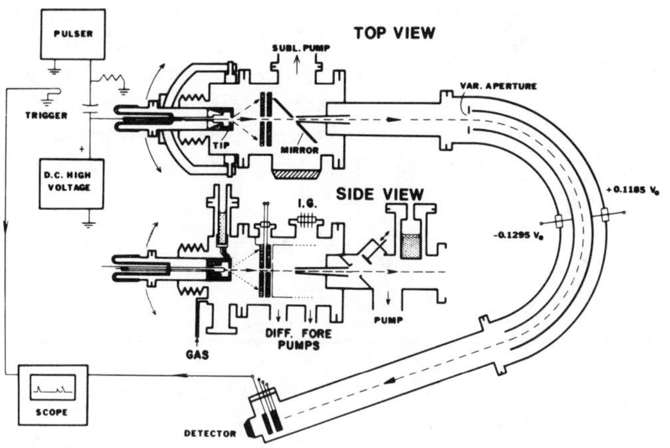

FIG. 3. Advanced time-of-flight atom probe with an electrostatic toroidal sector for compensating energy spread of ions caused by evaporation during the fractional nanosecond of the imperfect pulse front. Drift-path length is 2.5 m. (From Rev. Sci. Instr., **45**, 1053 (1974), by permission of Am. Institute of Physics.)

annealing experiments. Nuclei of precipitates may be analyzed in the atom probe, which also permits depth profiling by controlled layer-by-layer field evaporation. Similarly, segregation of impurity atoms at cores of dislocations and at grain boundaries may be located and identified, promising results useful in embrittlement, corrosion, and fatigue studies.

See also ADSORPTION; CRYSTAL DEFECTS; ELECTRON MICROSCOPY; SURFACES AND INTERFACES.

BIBLIGRAPHY

R. Gomer, *Field Emission and Field Ionization*. Harvard University Press, Cambridge, Mass., 1961. (I)

E. W. Müller, "The Imaging Process in Field Ion Microscopy, from the FEM to the Atom-Probe," *CRC Crit. Rev. Solid State Sci.* **6**(2), 85–109 (1976). (I)

K. M. Bowkett and D. A. Smith, *Field Ion Microscopy*. North Holland, Amsterdam, 1970. (I)

E. W. Müller and T. T. Tsong, *Field Ion Microscopy, Principles and Applications*. Elsevier, New York, 1969. (I)

E. W. Müller and T. T. Tsong, "Field Ion Microscopy, Field Ionization and Field Evaporation," in *Progress in Surface Science*, Vol. 1, part 4 (S. Davison, ed.). Pergamon Press, New York, 1973. (A)

Fluctuation Phenomena

Thomas J. Greytak

By fluctuation we usually mean the deviation of some quantity from its mean or most probable value. Almost all the quantities that a physicist might be interested in studying exhibit fluctuations, some on quite a microscopic scale (e.g., the motion of an atom about its equilibrium lattice position in a crystalline solid). This discussion, however, will be limited to macroscopic quantities that can be treated as thermodynamic variables. Fluctuations in these quantities manifest themselves in several ways: They may limit the precision of measurements of the mean value of the quantity, as is the case with thermal noise in electronic circuits or photon shot noise in light beams; they are the cause of some of the familiar features of our surroundings, such as the blue sky; or they may cause spectacular and unexpected effects, such as the critical opalescence of an apparently simple and homogeneous fluid.

FLUCTUATIONS IN A HYDROSTATIC SYSTEM

The simplest example of thermodynamic fluctuations occurs in a hydrostatic system such as a gas, a liquid, or a simple solid under hydrostatic pressure. In such a system of a fixed number of particles, the internal energy U is related to the principal state variables of temperature, entropy, pressure, and volume by the first and second laws of thermodynamics:

$$dU = T\,dS - P\,dV. \tag{1}$$

Only two of the state variables can be chosen independently; the others are then fixed through the equation of state and an energy equation such as an expression for the heat capacity. Let these variables pertain to a subsystem specified by a certain number N of contiguous particles. N is assumed to be large enough that the definition of local thermodynamic variables makes sense, yet the subsystem must be small enough that its energy is only a very small part of the energy of the total system. In thermal equilibrium the *average* values of the intensive variables T and P are equal to the temperature and pressure of the total system. The average values of S and V will be much smaller than those of the total system and can be related to the average T and P.

At any instant of time there is a finite probability that the subsystem will not be in its most likely state. The probability that a pair of independent variables X and Y will deviate from their mean values by amounts ΔX and ΔY is given by a bivariate Gaussian density that depends only on the three parameters $\langle \Delta X^2\rangle$, $\langle \Delta Y^2\rangle$, and $\langle \Delta X\,\Delta Y\rangle$. These quantities are listed in Table I. Notice that the cross correlation between ΔT and ΔV and between ΔS and ΔP vanishes. If either of these pairs of variables were chosen to be the independent ones, their joint probability density would factor into the product of two single Gaussian densities, one for each variable. In this case the variables are statistically independent as well as independent in the thermodynamic sense: a knowledge of the fluctuation in one gives no additional information about the fluctuation in the other. It should be pointed out, though, that ΔT and ΔV or ΔS and ΔP are statistically independent random variables only when taken at the same instant of time. Inspection of the table will show that the ratio of the root mean square fluctuation in any variable to its mean value, $\langle \Delta X^2\rangle^{1/2}/\langle X\rangle$, is proportional to $N^{-1/2}$. This is why conventional thermodynamics need deal only with the mean values of the variables.

Through expansion of any other thermodynamic variable in terms of two of the principal ones, the fluctuations in that quantity can be found. For example, the fluctuations in the number density, $n \equiv N/V$, are given by

$$\langle \Delta N^2\rangle = \frac{n^3 kT\kappa_T}{N} \tag{2}$$

Table I Fluctuations in a Hydrostatic System[a]

$\langle \Delta T^2 \rangle = \dfrac{kT^2}{C_V}$	$\langle \Delta V^2 \rangle = kTV\kappa_T$
$\langle \Delta S^2 \rangle = kC_P$	$\langle \Delta P^2 \rangle = \dfrac{kT}{V\kappa_S}$
$\langle \Delta T\, \Delta V \rangle = 0$	$\langle \Delta S\, \Delta P \rangle = 0$
$\langle \Delta T\, \Delta S \rangle = kT$	$\langle \Delta V\, \Delta P \rangle = -kT$
$\langle \Delta T\, \Delta P \rangle = \dfrac{\alpha k T^2}{\kappa_T C_V}$	$\langle \Delta V\, \Delta S \rangle = kTV\alpha$

[a] C_V is the heat capacity at constant volume; C_P the heat capacity at constant pressure. The isothermal compressibility is

$$\kappa_T \equiv \frac{-1}{V}\frac{\partial V}{\partial P}\bigg|_T ;$$

the adiabatic compressibility

$$\kappa_S \equiv \frac{-1}{V}\frac{\partial V}{\partial P}\bigg|_S ;$$

the expansion coefficient

$$\alpha \equiv \frac{1}{V}\frac{\partial V}{\partial T}\bigg|_P .$$

and the fluctuations in the internal energy can be shown to be

$$\langle \Delta U^2 \rangle = kT^2 C_V + kTV\kappa_T \left(P - \frac{\alpha T}{\kappa_T} \right)^2 . \qquad (3)$$

In a similar manner all of the entries in Table I can be found from the ones pertaining to a single pair of variables.

FLUCTUATIONS IN OTHER SYSTEMS

The results for the hydrostatic system can be carried over directly to several other systems that are thermodynamically isomorphic to it. To identify the new variables with those of the hydrostatic case we must examine the increment of work done on the system for a given change in the mechanical variables. For the hydrostatic case [Eq. (1)] the work is $(-P)\,dV$; for a magnetic system with magnetization M and an external or applied magnetic field H it is $H\,dM$; for a dielectric body with polarization $\mathcal{P}$ and electric field E it is $E\,d\mathcal{P}$; for an interface of surface tension $\mathcal{S}$ and area A it is $\mathcal{S}\,dA$; for an elastic rod with tension $\mathcal{T}$ and length L it is $\mathcal{T}\,dL$. For any of these systems we need only make the appropriate substitution of variables in Table I in order to find the fluctuations. For example, for the magnetic system $-P \to H$ and $V \to M$, so we find that the fluctuations in T and M are statistically independent and that

$$\langle \Delta T^2 \rangle = \frac{kT^2}{C_M}, \quad \langle \Delta M^2 \rangle = kT \frac{\partial M}{\partial H}\bigg|_T \qquad (4)$$

where C_M is the specific heat at constant magnetization and $(\partial M/\partial H)|_T$ is the isothermal susceptibility.

In a more general system with m independent thermodynamic variables the fluctuations in these variables form a set of m jointly Gaussian random variables. The associated probability density function contains as parameters the m mean square fluctuations $\langle \Delta X^2 \rangle$ and the cross correlations of all possible pairs $\langle \Delta X\, \Delta Y \rangle$. The most important example is probably a binary mixture where $m=3$. If the concentration of one of the components is $c_a \equiv N_a/N$ and the difference in chemical potential of the two components is $\mu' \equiv \mu_a - \mu_b$, then we can show that

$$\langle \Delta c_a\, \Delta T \rangle = \langle \Delta c_a\, \Delta P \rangle = 0, \qquad (5)$$
$$\langle \Delta c_a^2 \rangle = \frac{kT}{N}\frac{\partial c_a}{\partial \mu'}\bigg|_{TP} .$$

It is interesting to note that for this system we cannot choose a set of three conventional variables that are all mutually statistically independent.

TIME EVOLUTION OF THE FLUCTUATIONS

So far only static fluctuations—those that are evaluated at one instant of time—have been considered. The fluctuations are actually random processes that evolve in time, $\Delta n(t)$, for example. (Density fluctuations will be used for illustration since they are the ones most often studied experimentally, but the discussion that follows could be applied to any thermodynamic variable.) A great deal of information about the temporal behavior of the fluctuations is contained in the time correlation function, $\langle n(t)n(t+\tau) \rangle$. It is usual to consider each spatial Fourier component of the variable separately. If $n_{\mathbf{q}}(t)$ is the complex amplitude of the $\mathbf{q}^{\text{th}}$ Fourier component of the density, the relevant correlation function is $\langle n_{\mathbf{q}}(t) n_{\mathbf{q}}^*(t+\tau) \rangle$. In order to calculate such a quantity we use the Onsager "regression of fluctuations" hypothesis, a version of the fluctuation-dissipation theorem, which states that the time dependence of the equilibrium correlation function is the same as the time evolution of an externally induced density disturbance of wave vector $\mathbf{q}$ which is released at $t=0$. This reduces to a well-posed initial-value problem that can be solved *exactly* once the macroscopic equations of motion are known. The details of this type of calculation for simple fluids have been discussed by Mountain. The general result of such calculations is that the equilibrium correlation function is made up of a sum of contributions, one from each hydrodynamic normal mode of the system, each reflecting the time dependence associated with the macroscopic decay of that mode. In the case of the simple fluid there is a sound-wave contribution to $\langle n_{\mathbf{q}}(t) n_{\mathbf{q}}^*(t+\tau) \rangle$ which oscillates at a frequency $\omega = uq$ (u is the velocity of sound) while it decays toward zero because of processes that damp macroscopic sound waves. There is also a thermal diffusion contribution, which has a simple exponential decay at a rate proportional to the thermal conductivity. For a binary mixture there would be a third contribution to the correlation function, another exponential decay at a rate proportional to the concentration conductivity.

This simple picture for the time dependence of the correlation function is based on the assumption that well-defined hydrodynamic normal modes can exist in the system. In a gas when the mean free path l of the molecules becomes comparable to or greater than q^{-1}, this assumption is no longer valid. The coherent atomic motion characteristic of such modes is dissipated by the distribution of individual molecular velocities. In this case a kinetic theory, such as the Boltzmann equation, must be used to calculate the correlation function. Such calculations are discussed by Yip.

MEASUREMENT OF THE FLUCTUATIONS BY SCATTERING SPECTROSCOPY

A major advance in the study of fluctuations occurred with the realization that all scattering experiments on condensed-matter systems, whether they used light, x rays, neutrons, or electrons, measured essentially the same quantity, the dynamic structure factor $S(\mathbf{q},\omega)$. $S(\mathbf{q},\omega)$ is the space and time Fourier transform of the correlation function for the variable that couples to the scattering probe. For the density fluctuations in a classical system it can be written as

$$S(\mathbf{q},\omega) = \int_{-\infty}^{\infty} d\tau \int_V d^3R\, e^{i(\mathbf{q}\cdot\mathbf{R}-\omega\tau)}\langle n(\mathbf{r},t)n(\mathbf{r}+\mathbf{R},t+\tau)\rangle,$$

$$= \int_{-\infty}^{\infty} d\tau\, e^{-i\omega\tau}\langle n_{\mathbf{q}}(t)n_{\mathbf{q}}{}^{*}(t+\tau)\rangle. \qquad (6)$$

Experimentally the scattering geometry singles out one scattering angle, which, through the Bragg condition, determines the wave vector of the fluctuations being observed. A spectrometer then carries out the frequency analysis of the scattered beam. Figure 1 shows the features of $S(q,\omega)$ for a classical gas. The correlation function for such a system was discussed in the previous section. A spectrum of this type can be obtained by the Brillouin scattering of laser light from a monatomic gas. Notice that in the hydrodynamic region there are distinct contributions from the two normal modes of the system, sound waves and thermal diffusion, but in the kinetic region such a distinction no longer makes sense.

Light scattering has been used to study the dynamics of density fluctuations in gases, liquids, superfluids, and solids. It has also been applied to concentration fluctuations in solutions and orientation fluctuations in liquid crystals. Neutron scattering has studied fluctuations in the magnetization of ferromagnets and antiferromagnets, and fluctuations in the local atomic order associated with structural phase transitions. Inelastic x ray and electron scattering are also being applied to studies of similar fluctuation phenomena, but often the fluctuations studied are on too microscopic a scale to be classified as thermodynamic fluctuations.

DIVERGENCE OF THE FLUCTUATIONS

It was pointed out earlier that usually we need only deal with the mean of a thermodynamic variable when discussing macroscopic phenomena, since the amplitude of the fluctuations relative to the mean is always inversely proportional to the square root of the number of particles involved. However, in some circumstances, usually at a second-order phase transition, the constant of proportionality (some thermodynamic derivative) diverges. For example, the isothermal compressibility [Eq. (2)] diverges at the gas–liquid critical point, the isothermal magnetic susceptibility [Eq. (4)] diverges at the Curie temperature of a ferromagnet, and the osmotic compressibility $(\partial c_d/\partial\mu')|_{T,P}$ [Eq. (5)] diverges at the consolute point of a binary mixture. In such cases the fluctuations themselves dominate the dynamics of the system and the equations of motion dealing only with the mean values are no longer valid. Exactly how the medium does behave in these circumstances is not well understood. It is the subject of a currently active field of research known as dynamic critical phenomena. The clearest example of this fluctuation-dominated behavior is the critical opalescence of a simple fluid at its critical point. The density fluctuations scatter so much light that the fluid becomes translucent or opaque. It is interesting that this particular phenomenon, which first attracted physicists to the study of thermal fluctuations at the turn of the century, is still the subject of frontier research in physics.

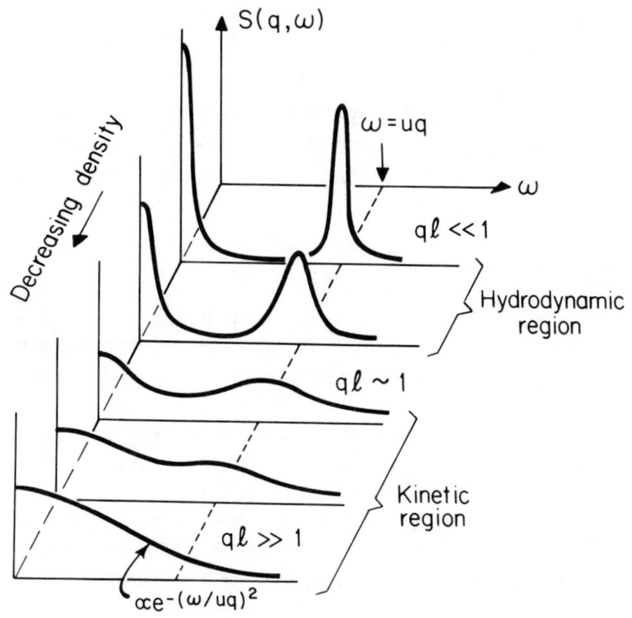

FIG. 1. Dynamic structure factor for density fluctuations in a classical monatomic gas; ℓ is the atomic mean free path and u is the velocity of sound.

See also CRITICAL POINTS.

BIBLIOGRAPHY

Theory of Thermodynamic Fluctuations

L. D. Landau and E. M. Lifshitz, *Statistical Physics*, Chapter 12. Addison-Wesley, Reading, Mass., 1958.

Mathematics of Random Variables and Random Processes

A. Papoulis, *Probability, Random Variables, and Stochastic Processes*. McGraw-Hill, New York, 1965.

Time Dependence of Fluctuations

R. D. Mountain, *Rev. Mod. Phys.* **28**, 205 (1966).

S. Yip, *J. Acoust. Soc.* **49**, 941 (1971).

Scattering Spectroscopy

W. Marshall and S. W. Lovesey, *Theory of Thermal Neutron Scattering*. Oxford, London and New York, 1971.

P. A. Fleury and J. P. Boon, "Laser Light Scattering in Fluid Systems," *Adv. Chem. Phys.* **24**, 1–93 (1973).

Critical Phenomena

H. E. Stanley, *Introduction to Phase Transitions and Critical Phenomena*. Oxford, London and New York, 1971.

Friction

Herbert S. Cheng

Friction is the tangential force required to overcome the resistance in sliding one contacting surface against another under a normal load. It is mainly a surface phenomenon, and depends primarily on the conditions at the surfaces.

COEFFICIENT OF FRICTION

In most cases, friction is approximately proportional to the normal load and therefore is measured by the ratio of its magnitude to the normal load, known as the coefficient of friction f.

The value of f starts from a range of 0.001–0.2 for lubricated surfaces to 0.1–2.0 for dry surfaces, and approaches infinity or very high values for ultraclean surfaces.

For ordinary surfaces, f is not sensitive to the surface roughness. However, for ultrasmooth surfaces f is high because of large cohesive forces, and for very rough surfaces f is again high because of asperity interlocking.

The maximum frictional force required to produce incipient sliding is known as the static friction or static coefficient of friction f_s. After sliding commences, the friction force always acts in the direction opposing the sliding velocity, and is usually slightly less than the static friction. Its coefficient for continuous sliding is known as f_k, the kinetic coefficient of friction. In general, f_k is independent of the sliding speed. For hard materials f_k decreases slightly with the sliding speed, and for some systems this characteristic can lead to an instability during sliding and can produce a stick–slip-type oscillation.

BASIC CONCEPT OF DRY FRICTION

When two dry surfaces with ordinary roughness are pressed against each other, contact usually does not extend over the entire area, but only covers the tip of tall asperities. The total real area of contact is only a small fraction of the entire apparent area. These tips adhere to the opposing surface and form junctions which must be sheared if sliding is to occur. The total force in shearing these junctions constitutes the main part of the sliding friction. Thus, the friction is mainly the product of the shear strength of the material at the junctions and the total area of all junctions at the onset of sliding.

When ultraclean surfaces are sheared, the strong adhesion at the interface causes the junctions to grow indefinitely, and therefore results in seizure or very high coefficient of friction.

Ordinary dry metallic surfaces usually are covered with oxides or other contaminants which reduce both interfacial shear strength and adhesion. Under these conditions, the junction growth is confined. The final real area of contact at the onset depends strongly on the properties of the oxide and the substrate, and is approximately proportional to the normal load. Thus, the coefficient of friction for dry surfaces is primarily a function of material property and surface condition, and representative values can be found in Refs. 1 and 2. Table I lists some typical values for some common metallic surfaces.

EFFECT OF BOUNDARY LUBRICATION

The interfacial shear strength at the junction can be greatly reduced by introducing a surface film through physical ad-

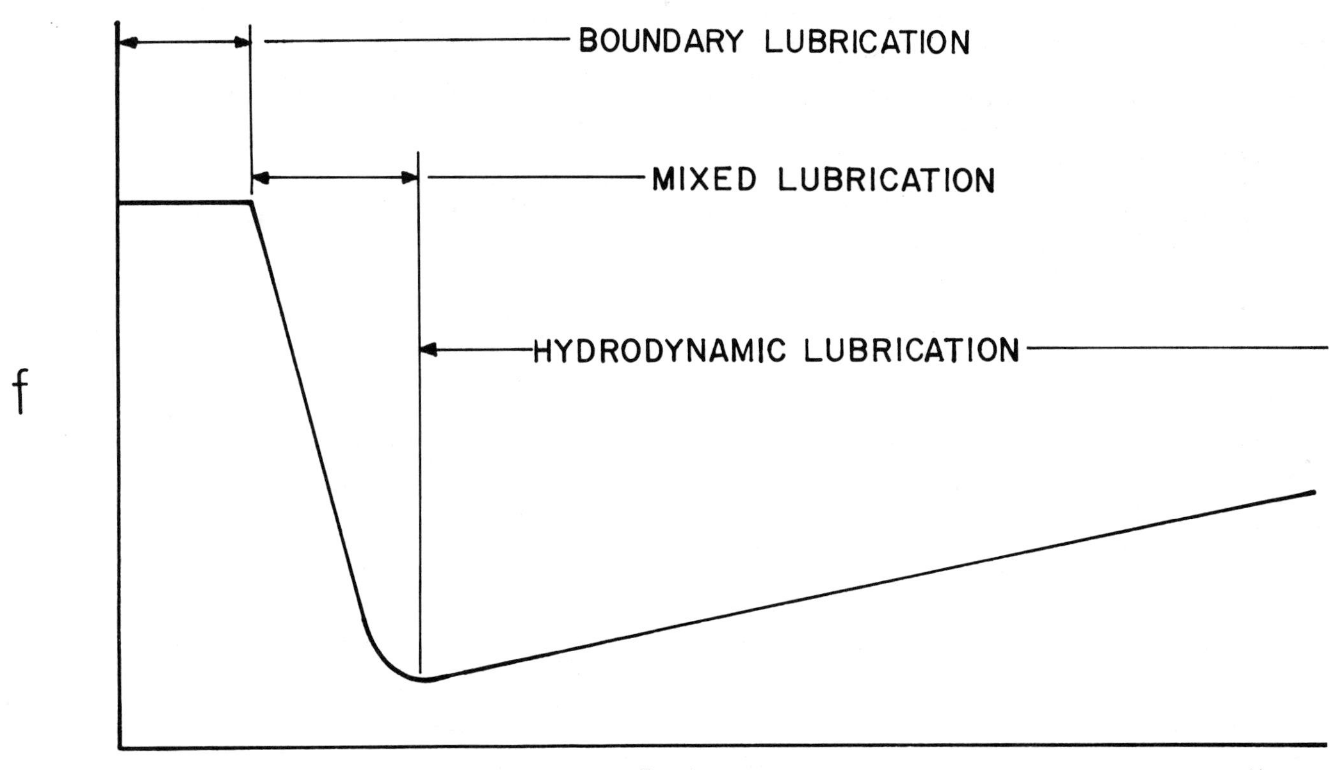

FIG. 1.

Table I Typical Metallic Friction[a]

Metal	On itself	On steel
Aluminum	1.3	0.5
Copper	1.3	0.8
Brass	—	0.5
Bronze	—	0.3
Mild steel	0.6	0.6
Cast iron	0.4	0.4

[a] The values of static frictional coefficients between various metallic surfaces are given by Tabor [1].

sorption or chemisorption of a compound in a lubricant or through chemical reaction with a lubricant or lubricant additives. With the reduced shear strength, the coefficient of friction can be lowered by as much as one order of magnitude from 1.0 to 0.1. Metallic surfaces with an effective boundary lubrication have a value of f_s in the range 0.1–0.2.

EFFECT OF FLUID-FILM LUBRICATION

When the surfaces are completely separated by a self-acting or an externally pressurized lubricant film, the friction is usually governed by the law of shearing a Newtonian liquid instead by shearing the junctions. The local shear stress for these cases is equal to the lubricant viscosity times the local shear rate of the lubricant, which is approximately equal to the ratio of the sliding speed to the film thickness. The total frictional force can be obtained by integrating the shear stress over the entire area. The above method in determining friction is valid as long as (1) the surfaces are completely separated and (2) the flow is Newtonian. The surface separation depends on a lubrication parameter known as the Hersey number, $H = \mu N/P$, where μ is the viscosity, N is the angular speed, and P is the average pressure. A typical variation of the coefficient of friction with H is shown in Fig. 1, and is commonly known as the Streibeck curve. When H is large, f is low because surfaces are completely separated. As H decreases, friction rises rapidly due to partial surface contact. For very small H the friction levels off to a constant value which is governed by the characteristics of the surface film.

In sliding bearings lubricated with mineral oil, water, gas, or other liquids, the flow is normally Newtonian unless the lubricant contains polymeric additives. For these conditions, evaluation of friction based on Newtonian flow is usually valid.

In concentrated Hertzian contacts such as ball-race or roller-race contacts, gears, and cams, friction does not necessarily obey the Newtonian law, and cannot be determined without reliance on empiricism.

See also MECHANICAL PROPERTIES OF MATTER; RHEOLOGY; TRIBOLOGY.

REFERENCES

1. M. J. Neale, *Tribology Handbook*. Wiley, New York, 1973.
2. C. T. Lynch, *Handbook of Material Science*, Volume 1. CRC Press Cleveland, 1974.
3. F. P. Bowden and D. Tabor, *Friction and Lubrication of Solids*. Clarendon Press, Oxford, Part I, 1950, Part II, 1964.
4. E. Rabinowicz, *Friction and Wear of Materials*. Wiley, New York, 1965.

Galvanomagnetic Effects

C. M. Hurd

These occur when a conductor carrying an electric current is placed in a magnetic field. They are the result of the applied electric and magnetic fields' influences on the paths of the itinerant electrons in the substance. They are qualified as adiabatic or isothermal depending on whether heat flows and temperature gradients are allowed to exist, as specified in Table I.

The effect of an electric field **E** and a magnetic flux density **B** upon a moving particle of charge q (such as an electron, where $q = -1.6\times 10^{-19}$ C) is a force **F** given by the force law (in S.I. units):

$$\mathbf{F} = q(\mathbf{E} + \mathbf{v}\times\mathbf{B}).$$

Here **v** is the velocity of the particle in the observer's reference frame, and $\mathbf{v}\times\mathbf{B}$ stands for the vector product of these quantities. The total force **F** thus consists of a velocity-independent part $q\mathbf{E}$, called the electrostatic force, and a velocity-dependent part $q\mathbf{v}\times\mathbf{B}$, called the Lorentz force.

The galvanomagnetic effects arise from a combination of two circumstances: the deflection of the itinterant electrons in a conductor by the Lorentz force, and the constraint on the electron flow produced by the boundaries of the sample. Consider a conducting sample subjected to a steady electric field **E**. Under the influence of the electrostatic force, each itinerant electron has superimposed upon its random thermal motion an acceleration along the direction exactly opposite to **E** (since an electron carries a negative charge). During its mean free path between randomizing collisions, which occur especially with the ions in the substance, each electron therefore contributes to a general current drift along **E**. The Lorentz force produced by a uniform flux density **B**, assumed for generality to be noncollinear with **E**, deflects each moving electron from its unperturbed path. A few will at first be deflected without hindrance, but since electrons are mutually repulsive, and since they are constrained to remain within the bounds of the sample, they eventually pile up against an inside face. This creates an electric field which opposes the Lorentz force. Ultimately, a new dynamical equilibrium is reached among the electrons in which the current drift is maintained along **E** but is now nonuniform because of the established concentration gradients. Since the electrons possess thermal energy as well as charge, a temperature gradient will be set up concomitantly with the electric field. A third effect is the decrease in the effective electrical conductivity that is implied as electrons are deflected from their otherwise forward motion under the influence of **E**. These three features are typical manifestations of a galvanomagnetic effect.

The effects are classified according to the three basic relative orientations of applied electric and magnetic fields:

(1) *Transverse* effects in a *transverse* magnetic field. **B** is perpendicular to the primary current density **J** flowing in the sample, and the galvanomagnetic effect is measured in the direction mutually perpendicular to **B** and **J**.

(2) *Longitudinal* effects in a *transverse* magnetic field. **B** is again perpendicular to **J**, but the galvanomagnetic effect is measured along the direction of **J**.

(3) *Longitudinal* effects in a *longitudinal* electric field. **B** is parallel to **J**, and the galvanomagnetic effect is measured along **J**.

The standard definitions of the galvanomagnetic effects are in terms of specified conditions maintained in an assumedly isotropic conductor. These are given in Table I, where the densities of the magnetic flux, electron current, and heat current are represented by **B**, **J**, and **Q**, respectively, referred to a set of Cartesian axes; T is the temperature at any point in the sample.

Two extreme conditions are distinguished for electrons contributing to any galvanomagnetic effect depending on the

Table 1 Galvanomagnetic Effects in Isotropic Media

Class	Name	Primary fluxes ($\mathbf{B}=\mathbf{B}_z$)	Quantity measured	Defining conditions
1. Transverse effects in a transverse field	Hall (Isothermal)	$\mathbf{J}=\mathbf{J}_x$	E_y/J_x	$J_y=\delta T/\delta x=\delta T/\delta y=0$
	Hall (Adiabatic)	$\mathbf{J}=\mathbf{J}_x$	E_y/J_x	$J_y=\delta T/\delta x=Q_y=0$
	Ettingshausen	$\mathbf{J}=\mathbf{J}_x$	$(\delta T/\delta y)/J_x$	$J_y=\delta T/\delta x=Q_y=0$
2. Longitudinal effects in a transverse field	Electrical transverse magnetoresistivity (Isothermal)	$\mathbf{J}=\mathbf{J}_x$	E_x/J_x	$J_y=\delta T/\delta x=\delta T/\delta y=0$
	Electrical transverse magnetoresistivity (Adiabatic)	$\mathbf{J}=\mathbf{J}_x$	E_x/J_x	$J_y=\delta T/\delta x=Q_y=0$
	Nernst (Isothermal)	$\mathbf{J}=\mathbf{J}_x$	$(\delta T/\delta x)/J_x$	$J_y=Q_x=\delta T/\delta y=0$
	Nernst (Adiabatic)	$\mathbf{J}=\mathbf{J}_x$	$(\delta T/\delta x)/J_x$	$J_y=Q_x=Q_y=0$
3. Longitudinal effects in a longitudinal field	Electrical longitudinal magnetoresistivity	$\mathbf{J}=\mathbf{J}_z$	E_z/J_z	$\delta T/\delta z=0$
	Unnamed	$\mathbf{J}=\mathbf{J}_z$	$(\delta T/\delta z)/J_z$	$Q_z=0$

relative importance of the perturbation produced by the Lorentz force. If this force is only a minor influence on the electron's motion, the particle moves in the *low-field condition*. Alternatively, it is in the *high-field condition* when the Lorentz force is completely dominant so that the scattering effects are negligible by comparison. The combination ambient temperature and applied field strength determines where the prevailing conditions lie.

The transverse effects are generally inversely proportional to the density of itinerant electrons. In a metal, which has typically 10^{29} electrons/m^3, the effects are usually too small to be of more than academic interest, although the magnetoresistance effect can be a source of inefficiency in electromagnetic machinery. In semiconductors, however, where the charge carrier density is never more than 10^{-3} of that found in metals, the effects can be large enough to have practical applications. The Hall and magnetoresistance effects in the semiconducting compounds of elements from groups III and V have found application in magnetometers, susceptibility meters, Hall-effect amplifiers, multiplying elements, displacement transducers, and contactless switches.

Academically, the galvanomagnetic effects have been helpful in furthering our knowledge of conduction processes in both semiconductors and metals. Measurements of the low-field Hall and magnetoresistance effects are made routinely to determine the sign, density, and mobility of the charge carriers in semiconductors. Studies of the dependence of these effects on the magnetic field strength lead to information about the nature of the electron scattering processes in both semiconductors and metals, while studies in the high-field condition, particularly in metals, are an important tool for investigating the substance's electronic structure.

See also CONDUCTION; HALL EFFECT; MAGNETORESISTANCE.

BIBLIOGRAPHY

Galvanomagnetic effects in semiconductors

A. C. Beer, *Galvanomagnetic Effects in Semiconductors*. Academic Press, New York, 1963. (A)

B. Kemp, *Hall Effect Instrumentation*. Sams and Co., Indianapolis, 1963. (E)

E. H. Putley, *The Hall Effect and Related Phenomena*. Butterworth, London, 1964. (I)

E. H. Putley, "The Hall Effect and its Applications," *Contemp. Phys.* **16**, 101–126 (1975). (E)

E. D. Sisson, "The Hall Effect: Devices and Applications," *Anal. Chem.* **43**, 67A–71A (1971). (E)

S. Stricker, "The Hall Effect and its Applications," in *Advances in Electronics and Electron Physics*, Vol. 25, pp. 97–143, L. Marton (ed.) (1968). (I)

Galvanomagnetic effects in metals

H. Fritzche, G. Pearson, and W. Scanlon, "Galvanomagnetic and Thermomagnetic Effects," in *Methods of Experimental Physics*, Vol. 6B, pp. 145, 160, and 166. K. Lark-Horovitz and V. A. Johnson (eds.) (1959). (I)

C. M. Hurd, *The Hall Effect in Metals and Alloys*. Plenum Press, New York, 1972. (A)

J.-P. Jan, "Galvanomagnetic and Thermomagnetic Effects in Metals," *Solid State Phys.* **5**, 1–96 (1957). (A)

W. A. Reed, "Experimental Methods of Measuring High-field Magnetoresistance in Metals," in *Methods of Experimental Physics*, Vol. 11, pp. 1–31, R. V. Coleman (ed.) (1974). (I)

Galvanomagnetic effects in anisotropic materials

P. I. Baranskii, I. S. Buda, I. V. Dakhovskii, and A. G. Samoilovich, "Galvanothermomagnetic Effects in Anisotropic Media," *Phys. Stat. Solidii (B)* **67**, 291–299 (1975). (A)

C. M. Hurd, "Galvanomagnetic Effects in Anisotropic Metals," *Adv. Phys.* **23**, 315–433 (1974). (A)

Glass, Nonmetallic

Brage Golding

Nonmetallic glass (synonymously, amorphous or noncrystalline solids) is a class of materials (i) whose microscopic atomic arrangement exhibits no periodicity or long-range order, in contrast to crystals; and (ii) whose shear viscosity is sufficiently large that macroscopic shapes are maintained for relatively long times, in contrast to fluids. A substance in its amorphous phase is metastable, since its atoms or molecules exist in a higher configurational energy state than in its crystalline phase, but are unable to reach that state because of substantial energy barriers separating the two phases. The amorphous state can be achieved by (1) rapid cooling of a viscous fluid, the traditional preparation method for inorganic glasses such as SiO_2 (vitreous silica), As_2S_3, and sulfur, and for organics such as polystyrene and polymethyl methacrylate; or (2) through circumvention of the liquid state by preparation from the vapor phase, as in vacuum deposition, sputtering, gaseous decomposition, etc., as used in the preparation of amorphous silicon and germanium.

If, upon cooling, a liquid successfully avoids crystallization, i. e., becomes supercooled, it passes through the glass transition temperature T_g, defined as the temperature at which the substance's shear viscosity exceeds approximately 10^{14} P. On cooling through T_g, the specific volume and entropy of a glass decrease abruptly but continuously from values characteristic of the liquid toward values approaching, but not reaching, those of a crystalline state. The rate of cooling through the glass transition region determines the total change in the substance's extensive thermodynamic properties by influencing the overall degree of structural rearrangement.

The structures of many inorganic glasses can be visualized as made up of small molecular groups (e. g., SiO_4 tetrahedra in silica glass), joined together to satisfy all, or nearly all, bonding requirements (each oxygen joined to two adjacent tetrahedra). This continuous random network differs from a crystalline network in that the bonding angles linking the groups are not fixed but have distributions of values about the crystalline angle. The glassy network of vitreous silica can be modified and the softening temperature lowered for glasses of commercial and technological importance by incorporating other oxides (e. g., Na_2O and B_2O_3) into the structure to form the soda silicates and borosilicates. Such glasses are transparent to visible light because optical excitation of electronic transitions begins in the ultraviolet part of the spectrum, whereas excitation of vibrational modes occurs in the infrared.

A glass is an elastically isotropic solid, characterized by two independent elastic constants, and supports a long-wavelength longitudinal and a doubly degenerate transverse acoustic mode. Oxide glasses are generally brittle, exhibit little plastic deformation, and fracture by crack propagation, whereas many organic polymeric glasses are capable of withstanding strains of several hundred percent before breaking. Heat is carried by sound waves but because of the disordered structure the thermal vibrations undergo strong scattering, resulting in low thermal conductivities.

A glass may be viewed as a solid collection of intrinsic structural defects. When subjected to external stimuli such as time-dependent electric or stress fields, the induced rearrangement of defects creates frequency- and temperature-dependent structure in the dielectric and elastic responses, i.e., large dielectric and elastic relaxational loss and dispersion characteristic of a particular glass. The specific heat of glasses reflects the extra atomic degrees of freedom by an enhancement of the T^3 Debye specific heat at low temperatures and the appearance of a quasilinear temperature-dependent defect contribution below 1 K.

See also CRYSTAL DEFECTS; ELASTICITY; GLASSY METALS; SEMICONDUCTORS, AMORPHOUS.

BIBLIOGRAPHY

R. J. Charles, "The Nature of Glass," *Sci. Am.* **217** [3], 127 (1967). (E)

R. H. Doremus, *Glass Science*, Wiley, New York, 1973. (I)

R. W. Douglas and Bryan Ellis, eds., *Amorphous Materials*, Wiley (Interscience), New York, 1972. (A)

V. D. Fréchette, ed., *Non-Crystalline Solids*, Wiley, New York, 1960. (A)

J. A. Prins, ed., *Physics of Non-Crystalline Solids*, North-Holland, Amsterdam, 1965. (A)

L. D. Pye, H. J. Stevens, and W. C. LaCourse, eds., *Introduction to Glass Science*, Plenum, New York, 1972. (I)

Glassy Metals

Pol Duwez

Solid materials are either crystalline or amorphous. The crystalline state is characterized by a periodic repetition of a certain number of atoms constituting a unit cell. The translation of the unit cell in three directions in space defines a crystalline lattice. In the amorphous, or glassy, state there is no long-range periodicity and the atoms are arranged more or less at random, as in the liquid state. The transition from the liquid to the crystalline state is very sharp and, under equilibrium conditions, takes place at a well-defined temperature with an appreciable heat release. By contrast, the transition from the liquid to the amorphous state is very broad and can be observed by a progressive increase in the viscosity of the melt to a value that actually is so high that the material becomes rigid enough to be considered a solid. During this transition the atoms remain in approximately the same configuration, and a glass can be considered a supercooled liquid.

Fused silica and mixtures of silica with other oxides of Al, Na, Mg, Ca, etc., constitute the oldest and most important class of glassy inorganic solids. In addition to the silicates, glassy solids can also be obtained in some mixtures of group-Va, -VIa, and -VIIa elements (chalcogenide glasses). When cooled from the liquid state these materials will not crystallize. Contrary to silicates and chalcogenides, it is very difficult to prevent crystallization of liquid metals and alloys. About 20 years ago, however, new methods were developed to cool liquid alloys at very high rates—of the order of 1 million degrees per second—and under these extreme conditions crystallization of some metallic alloys can be avoided. This new class of amorphous alloys is now referred to as "glassy metals."

The basic principle involved in achieving very high rates of cooling is the extraction of heat from the liquid by conduction, onto a solid substrate of a high-conductiyity metal, such as pure copper. The liquid layer in contact with the substrate is subjected to a very sharp drop in temperature (provided that a good contact is achieved), and solidification proceeds from this layer through the remaining liquid by the motion of a liquid–solid interface that propagates at a rate governed by the heat flow (or diffusion) equation. Since the local rate of cooling decreases as the interface progresses through the liquid, the thickness of the liquid layer, and consequently that of the resulting solid foil, is always limited to rather small values, in practice of the order of 50 μm or less. The liquid alloy must also be spread into a thin layer on the solid substrate as quickly as possible after its initial contact with the cooling surface. This can be achieved by propelling a globule of liquid alloy at rates of 100 m/s and higher against a copper substrate. In another technique the liquid globule is rapidly squeezed between a fast-moving piston and a fixed anvil. Both piston and anvil are lined with copper to ensure fast heat removal.

The basic principle of the piston and anvil technique—cooling from two sides while controlling the thickness of the quenched foil—has been incorporated in the design of what could be called a liquid-metal rolling mill. In this device, the liquid alloy is fed into a gap between two rollers generally made of or lined with copper. By proper adjustment of the mass flow of the liquid alloys, it temperature, the peripheral velocity of the rolls, and the size of the gap, it is possible to obtain a continuous ribbon cooled rapidly enough to be amorphous. For some time the most important drawbacks to large-scale applications of amorphous alloys were the limited size of the quenched foils and the fact that they could not be obtained in a continuous manner. The rolling-mill technique has removed these limitations and glassy metals are now produced in the form of continuous ribbons as wide as 50 mm and of essentially any length.

It is generally recognized that pure metals cannot be quenched from the liquid state into a glassy solid, at least not with the presently achieved rates of cooling which are in the range of 10^6–10^8 °C/s. So all glassy metals are actually alloys containing two or more elements. There is no theory capable of predicting the composition of metallic alloys that are likely to be amorphous after rapid cooling from the liquid state. However, from the large number of experimental results obtained during the last 15 years guide lines can be formulated, and the search for new glassy alloys does not depend entirely on chance. A large number of glassy alloys contain both strongly metallic elements, such as Cu, Ag, and Au, or transition metals Fe, Co, Ni, Pd, or Pt, and non-metallic elements (metalloids) of valence 3 (B), 4 (C, Si, Ge), and 5 (mostly P). Exceptions to this statement have been found in a few binary alloys such as Cu–Zr, Nb–Ni, Ta–Ni, and La–Au. The most favorable compositions in a given alloy system are those located near a eutectic point that corresponds to a relatively low melting temperature when compared with those of the alloying elements. In the Pd–Si system, for example, a eutectic is located at about 16 at.% Si and melts at 800°C, which is much lower than the melting points of Pd (1552°C) and Si (1440°C). Liquid-quenched glassy alloys cover the range between the eutectic and about 20 at.% Si.

Ternary glassy alloys contain either two metals and one metalloid (Pd–Ni–P), or one metal and two metalloids (Fe–P–B). In both cases the sum of the metallic elements amounts to about 70–85 at.% of the total, leaving about 20% for the metalloids. A possible explanation for the metal-to-metalloid ratio is based on a model of the structure of a glassy solid known as the Bernal model. In this model the atoms are assumed to be rigid spheres in close contact but arranged in space in a random fashion. A mathematical analysis of this model shows that in spite of the close packing there are "voids" and the volume fraction of these voids is approximately 19%. If the model is made of spheres of different sizes, it is logical to assume that the smaller spheres would fit inside the voids of the random packing. Since the sizes of metalloid atoms are generally smaller than those of metals, it is logical to assume that they will fill the voids in the Bernal packing, and if so the concentration of the metalloids in a glassy alloy would be close to 19%, which is in satisfactory agreement with most of the results reported so far (from about 15 to 27%). By combining the most favorable range of concentrations of metals and metalloids with the existence of a low melting point in the phase diagram, we can predict whether or not a new alloy is likely to be obtained in the glassy state by quenching from the liquid state.

Experimental methods of studying the structure of crystalline or amorphous solids, as well as that of liquids, are based on the scattering of radiation or particles, giving rise to interference effects. The most common methods involve x-rays, electrons, or neutrons. Both x-ray and electron-diffraction patterns are very similar to those obtained for liquid alloys. A typical x-ray diffraction pattern consists of a series of broad diffraction maxima of decreasing intensity and increasing width with increasing Bragg angle. By using the theory developed for the structure of liquids it is possible to derive, from the diffraction pattern, quantitative information on the distribution of atoms surrounding a given atom. For a more complete description of the structure, it is necessary to assume that the atoms, taken as rigid spheres, are packed according to a given model, to calculate the distribution function with the help of a computer, and to compare it with the measured one. The glassy nature of an alloy can also be determined by transmission electron microscopy. The absence of crystallinity results in a complete lack of contrast, as shown on the right side of Fig. 1. In this case,

FIG. 1. Electron micrograph of a liquid-quenched $Pd_{80}Si_{20}$ alloy. The quenched foil of decreasing thickness from left to right shows a progressive transition from microcrystalline to glassy.

the thickness of the liquid-quenched foil decreases from left to right and as a result the rate of cooling increases from left to right, as evidenced by a progressive decrease in the size of the crystallites.

The physical properties of glassy alloys are quite different from those of crystalline alloys of the same composition. The electrical resistivity of glassy alloys is relatively high—

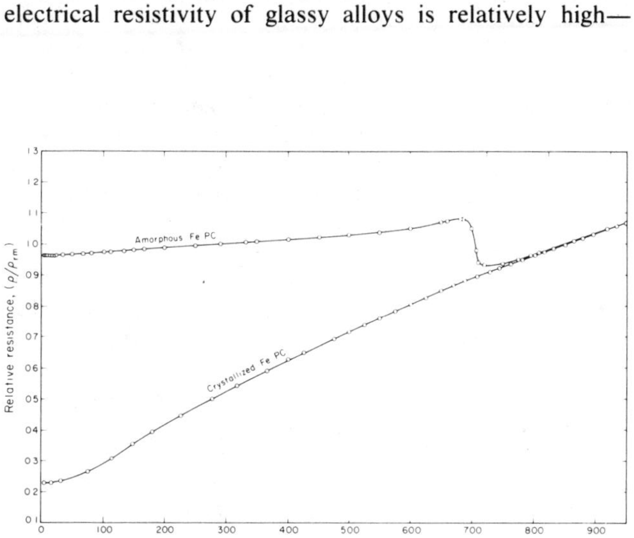

FIG. 2. Variation of electrical resistivity with temperature of an $Fe_{75}P_{15}C_{10}$ glassy alloy.

greater than 100 $\mu\Omega\cdot$cm at room temperature.The temperature coefficient is very small and can be zero or negative—in the range of $\pm 10^{-4}$ K^{-1}. The combination of these two properties leads to a very high residual resistivity at 0 K. The high resistivity of amorphous alloys is related to the scattering of the conduction electrons by a random atomic arrangement. Because of the random structure, however, the phonon contribution to the scattering of electrons is small, hence the small temperature coefficient of resistivity. A typical resistivity–temperature curve is shown in Fig. 2 for an amorphous $Fe_{75}P_{15}C_{10}$ alloy. Notice the small temperature dependence of the resistivity of the amorphous alloy. The rapid drop in resistivity around 675 K is due to the crystallization of the alloy. The residual resistivity at 0 K of the amorphous alloy is about seven times that of the crystalline alloy.

For some amorphous alloys the electrical-resistivity–temperature curve has a minimum below which the resistivity increases with decreasing temperature. This behavior is similar to the Kondo effect observed in crystalline alloys containing a small concentration of an element having a magnetic moment. Contrary to the crystalline case, however, the minimum can occur at a relatively high temperature. For example, in amorphous $Pd_{80}Si_{20}$ containing 7% Cr, the minimum temperature is about 500 K, compared to less than 20 K for Au alloys containing small amounts of Fe. Because of the rapid rise in resistivity below 150 C, the PdSiCr amorphous alloy would constitute a very sensitive low-temperature resistance thermometer.

Since 1954, superconducting amorphous metals and alloys have been obtained in the form of thin films by vapor deposition on a target maintained at liquid-helium temperature. These films are very unstable and crystallize at temperatures as low as 50 K. More recently, superconducting metallic glasses have been obtained by rapid quenching from the liquid states. In this case the alloys can be as thick as 50 μ and, more important, they are very stable at room temperature. Their amorphous-to-crystalline transition temperature is at least 400°C. So far, about 15 liquid-quenched superconducting alloys have been reported. Their transition temperatures range from 1.5 to 8.71 K. These are relatively low compared with some crystalline alloys, which have transition temperatures as high as 23 K. No theory has shown that the absence of a lattice in a metallic glass should lead to a low transition temperature, so that it is possible that new alloys will be found with higher transition temperatures.

The intrinsic properties of amorphous superconductors have been reported for only a limited number of alloys but the results show the effect that the absence of lattice periodicity has on these properties. It is clearly established that the absence of periodicity in amorphous solids is responsible for the drastic effect a magnetic field has on both the transition temperature and the current-carrying capacity of amorphous superconductors. These two inherent characteristics of amorphous superconductors could seriously limit their usefulness in practical applications. High current densities are indeed required for superconducting power transmission lines. The requirements are even more drastic in the case of high-field magnets. This application may become very important when fusion reactors become a reality. To contain the high-temperature plasma, magnetic fields as high as 100 000 G might be required, and this can be achieved only with a superconducting winding. Presently known superconducting crystalline alloys could solve the problem were it not that their properties are drastically degraded by a radiation field such as that expected in fusion applications. For example, the transition temperature of a crystalline superconducting alloy of the Nb_3Sn type will decrease from about 18 K to less than 4 K when subjected to radiation doses expected in fusion reactors. The degradation of properties of a crystalline alloy by radiation is due to the progressive disordering of the crystalline lattice. Since the atomic amorphous alloys are already arranged in a completely disordered configuration (similar to that in a liquid), their physical properties should not be greatly affected when exposed to a high-radiation environment. Some experimental results on this subject have been published and they all tend to prove that the properties of amorphous alloys are not subject to radiation damage.

Ferromagnetism is not limited to crystalline solids, and many glassy alloys are ferromagnetic. Early studies were concerned with alloys of Ni and P obtained by electrodeposition. These are weak ferromagnets with a Curie temperature of 212 K for a P concentration of 17 at.%. Vapor-deposited films of AuCo alloys are also glassy and ferromagnetic, but the largest number of glassy ferromagnets are obtained by liquid-state quenching. Liquid-quenched $Fe_{75}P_{15}C_{10}$ is strongly ferromagnetic, with a Curie temperature of 320°C. The $Fe_{75}P_{15}C_{10}$ alloy is the prototype of a large series of ferromagnetic alloys in which Fe is replaced by several transition metals of the 3d series (Ni, Co, Mn, Cr) and P and C are replaced by other metalloids such as P, B, or Si. Some of these complex amorphous alloys have remarkable magnetic properties. For example, the amorphous alloy (Fe_4Co_{96}) $(P_{16}B_6Al_3)$ has a coercive force of 0.13 Oe and a remanence of 4500 G. This extremely soft ferromagnet, with very low hysteresis losses, is comparable to the high-quality Permalloy and would be much cheaper to produce. In another series of amorphous alloys $(Fe, Co, Ni)_{75}P_{16}B_6Al_3$, the magnetostriction can vary from a positive to a negative value if the Fe content is changed from 7 to 4 at.%.

In addition to their interesting electrical and magnetic properties, some amorphous alloys have very high tensile strengths (about 250 kg/mm^2) and might be used as reinforcing ribbons in composite materials. Other alloys containing Fe, Ni, and Cr plus the glass-former metalloids have excellent corrosion properties comparable to those of the best stainless steels. The justification for using glassy alloys for their strength and corrosion properties will probably be their potentially low cost of fabrication.

Glassy metals are an interesting new class of solid materials from both fundamental and practical viewpoints. The structure and the physical properties of glassy metals are not yet well understood, mostly because most of the theories of metals are based on the existence of a crystalline lattice. The absence of long-range order in glassy metals may require new approaches in the theoretical explanation of their properties.

See also ALLOYS; CONDUCTION; CRYSTAL GROWTH; KONDO EFFECT; LIQUID METALS; MAGNETIC MATERIALS; SUPERCONDUCTING MATERIALS.

BIBLIOGRAPHY

Pol Duwez, *Ann. Rev. Mater.* **6**, 83–117 (1976).
T. Masumoto and R. Maddin, *Mater. Sci. Eng.* **19**, 1–33 (1975).
Carlos A. Pampillo, *J. Mater. Sci.* **10**, 1194–1227 (1975).

Hall Effect

P. R. Emtage

In 1879, E. H. Hall found that when a metal strip which bore a current was placed in a magnetic field, as in Fig. 1a, a voltage was produced across the strip. This transverse voltage is the simplest and most widely useful of the galvanomagnetic effects, since the number and nature of the current carriers can often be found from it.

The Hall voltage comes from a transverse electric field that cancels the sideways deflection of the current carriers by the magnetic field. Suppose the current is carried by electrons of charge $-e$ and drift velocity v_x in the direction of the current. In a magnetic field B_z they are acted on by a transverse force f_y—the Lorentz force—which causes the cyclotron motion and is at right angles to both field and velocity,

$$f_y = ev_xB_z.$$

The electron trajectories are therefore bent toward one side of the conductor; a surface charge forms on the sides and causes an electric field E_y that opposes further deflection—see Fig. 1b. Equilibrium is reached when the total force is zero,

$$ev_xB_z - eE_y = 0. \tag{1}$$

The transverse field E_y should be expressed in terms of measurable quantities, such as electric current, rather than electron velocity. If there are n electrons per unit volume, the current density is

$$j_x = -nev_x,$$

the negative sign being inserted because the charge on the electron is negative. Accordingly from Eq. (1) we find

$$E_y = Rj_xB_z, \tag{2}$$

wherein R is defined as the Hall coefficient, and from the above argument is given by

$$R = -1/ne. \tag{3a}$$

Note that, had we supposed the current to be carried by p positive carriers of charge $+e$, we should have found

$$R = 1/pe. \tag{3b}$$

The sign of the Hall coefficient therefore says whether the current is borne by positive or by negative charges, and its magnitude yields the density of current carriers.

The above results are so simple that they should be regarded with the gravest suspicion. Equations (3) hold only in very high magnetic fields, when each electron makes several cyclotron revolutions between collisions. In low fields no unique electron drift velocity v_x exists, and therefore Eq. (1) cannot be satisfied for all groups of electrons simultaneously. The total transverse current must, however, be zero, and therefore a well-defined low-field Hall coefficient does exist, and in simple cases has the form

$$R = -r/ne,$$

r being a factor of order unity. It is common to define a

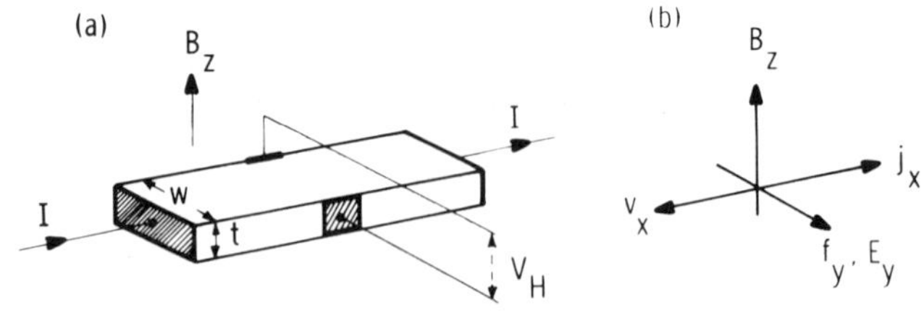

FIG. 1. (a) Configuration in which Hall effect is found. (b) Fields, etc., in the system.

"Hall density of carriers" through

$$n_H \equiv -1/Re. \tag{4}$$

Where other evidence suggests that the electron structure is in fact simple, this definition is a useful guide to the electron density.

MEASUREMENTS

The Hall coefficient R is defined from Eq. (2), and measurements of it are carried out on bar-shaped samples such as that in Fig. 1a. Let the bar have width w and thickness t; if the total current is I, the current density is $j_x = I/wt$. The Hall voltage V_H is measured between electrodes on the sides, and is related to the Hall field through $E_y = V_H/w$. Therefore,

$$R = E_y/j_xB_z = tV_H/IB_z.$$

In practice it is always necessary to find the mean of V_H from four measurements with field and current both forward and reversed, so as to eliminate errors resulting from misalignment of the electrodes, thermoelectric effects, and other galvanomagnetic effects, as well as inhomogeneities or anisotropy in the sample.

Values of the Hall coefficient at room temperature for a variety of elemental conductors are shown in Table I; the last column shows the apparent number of free electrons per atom, n_H being defined by Eq. (4). The results for the polyvalent metals are distressingly irregular, but there is some order amidst the chaos: all the metals have high carrier densities, and the monovalent metals have nearly one free electron per atom; the semimetals show much lower carrier dens-

Table I Hall Coefficients R (at room temperature) and Apparent Number of Free Electrons per Atom, n_H/n_a, for Some Elemental Conductors

Element	Group[a]	$R \times 10^{10}$ (m^3/C)	n_H/n_a
Na	Ia (M)	−2.1	1.17
Cu	Ib (M)	−0.536	1.38
Be	IIa (M)	+2.44	−0.21
Al	IIIa (M)	−3.0	0.31
Cr	IVb (TM)	+3.55	−0.21
Pt	VIII (TM)	−0.244	3.87
As	Va (SM)	−70	1.9×10^{-2}
Sb	Va (SM)	+250	-7.6×10^{-3}
Ge	IVa (SC)	-8×10^{8}	1.7×10^{-9}

[a] In this column M = metal, TM = transition metal, SM = semimetal, SC = semiconductor.

ities, as expected; while Ge, a semiconductor with no free electrons at all at low temperatures, is nearly destitute of current carriers.

The outstanding oddity of Table I is that the Hall coefficients are in many cases positive. It was found in Eqs. (3) that a positive Hall coefficient should correspond to positively charged current carriers; the results therefore conflict with the electron theory of conduction, and for a resolution of this enigma we must turn to the band theory of electrons.

BAND THEORY

There is not space here for more than the briefest description of the band theory. In a semiconductor the valence electrons occupy a complete set or band of states that spans a finite range of energies; no electron in this set can be moved to any other valence state without violating the Pauli exclusion principle, so the valence electrons are "rigid" and can carry no current. There must, however, be states of higher energy, the conduction bands, within which an electron can move—otherwise, the material would be impervious to high-energy electrons.

In a semiconductor the conduction band is separated from the valence band and is normally empty, as in Fig. 2a. In many semimetals the conduction and valence bands overlap somewhat, as in Fig. 2b; to minimize energy, electrons in the upper part of the valence band "fall" into the lower part of the conduction band. In most polyvalent metals the overlap of bands is large, and no clear distinction between conduction and valence bands can be drawn.

The peculiarities of the band picture can be seen if we imagine an electron in state i removed from a filled valence band, as in Fig. 2a; conduction can now occur since other valence electrons can move into the vacant state. Denote by Σ a sum over all valence states, and by Σ' a sum over all states except i. The charge on each valence electron is $-e$; let the energy of each electron relative to the band maximum be $-\epsilon_v$. The energy ϵ_h and charge q_h of the "hole" in the band are

$$\epsilon_h = \Sigma'(-\epsilon_v) - \Sigma(-\epsilon_v) = \epsilon_i,$$

$$q_h = \Sigma'(-e) - \Sigma(-e) = +e.$$

The hole in the valence band therefore acts as a "particle" of positive energy and positive charge, and the Hall coefficient due to its motion is positive. It is from the motion of electrons in incompletely filled bands that so many materials derive positive Hall coefficients.

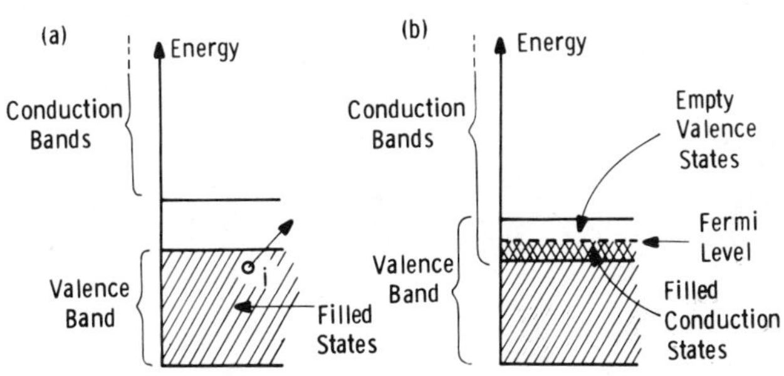

FIG. 2. Electron bands in (a) a semiconductor or insulator; (b) a semimetal.

See also CONDUCTION; ELECTRON ENERGY STATES IN SOLIDS AND LIQUIDS; GALVANOMAGNETIC EFFECTS.

BIBLIOGRAPHY

A. A. Abrikosov, *Introduction to the Theory of Normal Metals*. Academic Press, New York, 1972. (A)

C. Kittel, *Introduction to Solid State Physics*. Wiley, New York, 1966. (E)

E. H. Putley, *The Hall Effect and Related Phenomena*. Butterworths, London, 1960. (I)

Heat Capacity

Norman Pearlman

An intrinsically positive, extensive property of matter, defined as the limiting ratio of heat input ΔQ to temperature increment ΔT, is the heat capacity

$$c = \lim_{\Delta T \to 0} \Delta Q / \Delta T.$$

Constraints, such as constant volume or pressure, are usually indicated by subscripts. Tabulated values are ordinarily heat capacity per mole. The quantity normally measured is c_p, which is related to c_v by

$$c_p - c_v = VTB\beta^2$$

(V is volume, β the thermal expansion coefficient, B the bulk modulus). This is related to energy differences,

$$\Delta E = \int_{T_1}^{T_2} c_v(T)\, dT,$$

and, according to the constraint, to entropy differences,

$$\Delta s_{p,v} = \int_{T_1}^{T_2} \frac{c_{p,v}(T)}{T}\, dT,$$

thereby yielding information on the excitations concerned. For instance, the temperature-independent molar heat capacity

$$C_v = (\nu/2)R$$

(ν is an integer; R the gas constant per mole, $R = N_A k_B$, with N_A Avogadro's number and k_B Boltzmann's constant) observed for many gases and solids at room temperature corresponds to the classical equipartition energy $k_B T/2$ per degree of freedom, of which there are ν per molecule. In solids, $\nu = 6$ accounts for lattice energy in terms of normal-mode harmonic oscillators. Their energy, and hence also their heat capacity, vanishes exponentially as T approaches zero, as required by the third law of thermodynamics. For the quadratic oscillator frequency distribution assumed in Debye's theory,

$$(C_v/R)_{\text{lattice}} = (12\pi^4/5)(T/\theta_D)^3, \quad T \ll \theta_D$$

(θ_D, the Debye temperature, is $\hbar\omega_m/k_B$, with ω_m the maximum oscillator frequency). It is possible to relate θ_D to lattice elastic properties. The electrons in normal metals form a degenerate Fermi gas, and so their heat capacity contribution can be observed only at low temperatures, where $(C_v)_{\text{lattice}}$ has become negligible. The electronic contribution,

$(C_v)_{electronic}$, is found to be linear in T, as calculated in free-electron theory,

$$(C_v/R)_{electronic} = \gamma T/R = (\pi^2/2)(T/T_F).$$

The Fermi degeneracy temperature T_F is related to the Fermi level E_F by $T_F = E_F/k_B$, and thereby to the density of electron states per electron at the Fermi level,

$$D(E_F) = \frac{3}{2} E_F$$

and hence also to the electron effective mass, which is proportional to $D(E_F)$. In superconductors, below the transition temperature T_c the ground state is separated from excited states by an energy gap E_g which, according to BCS theory, is zero at T_c and increases to about $3.5k_BT_c$ at 0 K. In consequence, there is a jump in $(C_v)_{electronic}$ at T_c:

$$(C_v)^{supercond}_{electronic}(T_c)/\gamma T_c = 2.43$$

according to BCS theory. Also, the heat capacity vanishes exponentially rather than linearly as T approaches zero:

$$(C_v)^{supercond}_{electronic}(T)/\gamma T_c = ae^{-bT_c/T}$$

where both a and b vary with T. While in a superconductor the ground state and the energy gap are collective properties of all the electrons due to their interaction with the lattice, a number of interactions (magnetic, crystal field, etc.) involving isolated degrees of freedom also produce a gap, ΔE, between the ground state and the first excited state. In terms of the temperature parameter $T_S = \Delta E/k_B$, the heat capacity of such a mode (called a "Schottky anomaly") is

$$c/k_B = (T_S/T)^2 e^{T_S/T}/(1 + e^{T_S/T})^2.$$

Unlike the contributions described earlier, this is not monotonic. It has a peak at $T = 0.416T_S$, decreasing, as $(T_S/T) \times e^{-T_S/T}$ for $T < T_S$, and as $(T_S/2T)^2$ for $T > T_S$. Since the excited state is empty at $T = 0$ K while its occupancy equals that of the ground state for $T \gg T_S$, the total area under c/T is the entropy difference corresponding to this rearrangement, or $k_B \ln 2$.

See also LATTICE DYNAMICS.

BIBLIOGRAPHY

C. Kittel, *Introduction to Solid State Physics*, 5th ed. Wiley, New York, 1976.

M. W. Zemansky, *Heat and Thermodynamics*. McGraw-Hill, New York, 1968.

Helium, Solid

H. R. Glyde

I. INTRODUCTION

Since helium was first solidified at the famous Kamerlingh Onnes low-temperature-physics laboratories in Leiden by W. H. Keesom[1] on 25 June 1926, this fascinating solid has continued to reveal new and unexpected properties right up to the present. The initial experiments by Sir Francis Simon at Oxford University and by Keesom and their collaborators focused on measuring the melting curve, the specific heat, and the thermal conductivity of solid helium as a test of our early understanding of solids. This pioneering work up to 1957 is elegantly and beautifully reviewed by Domb and Dugdale,[2] a review that stands today along with those of Guyer,[3] Koehler,[4] and Glyde[5] as an excellent introduction to solid helium.

While the helium atom is simple, the solid is not. The atom has a rigid, spherical electron cloud of two electrons tightly bound to a central nucleus of two protons and two neutrons in the heavier isotope, ^{4}He. The lighter isotope, ^{3}He, has one neutron less, leaving a net nuclear spin of $\frac{1}{2}$ (in spin units), which leads to a host of intriguing magnetic properties in solid ^{3}He.

When and why does a collection of these simple atoms solidify? At room temperature we are familiar with helium as a gas, a gas so light that helium-filled balloons rise rapidly in air. If we cool helium, it liquefies at 4.2 K [K, degrees Kelvin, have the same size as degrees centigrade (°C), but the zero is shifted to absolute zero temperature, T = 0 K = −273.16°C]. Even down to lowest temperatures reached today,[6] less than 10^{-3} K = 1 mK, helium remains liquid. To solidify helium we must apply pressure to squeeze the light, highly kinetic atoms together—a pressure of 25 atm or more for ^{4}He, 35 atm or more for ^{3}He. A "phase" diagram displaying the regions of temperature and pressure where helium is solid is shown in Fig. 1. Helium forms the body-centered-cubic (bcc), the hexagonal-close-packed (hcp), and, at high pressure, the face-centered-cubic (fcc) structures. A full discussion of the phases can be found in the books by Keller[7] and Wilks.[8] Recent calculations[9] suggest that above 15,000 atm only the fcc phase forms, which is the structure formed by heavier rare-gas crystals, Ne, Ar, Kr and Xe.[10]

Since a pressure is required to solidify helium, the interatomic binding must be weak. This is expected, for the helium atoms remain neutral and uncharged. However, the electrons on one atom can induce an electric dipole on an adjacent atom, and this leads to a weak dipole–dipole attraction, the van der Waals force. Also when two atoms get close together their rigid electron clouds collide, almost like two billiard balls. This weak attraction and sharp repulsion at close approach is represented by the approximate potential energy curves in Fig. 2. Barker[11] presents a full and au-

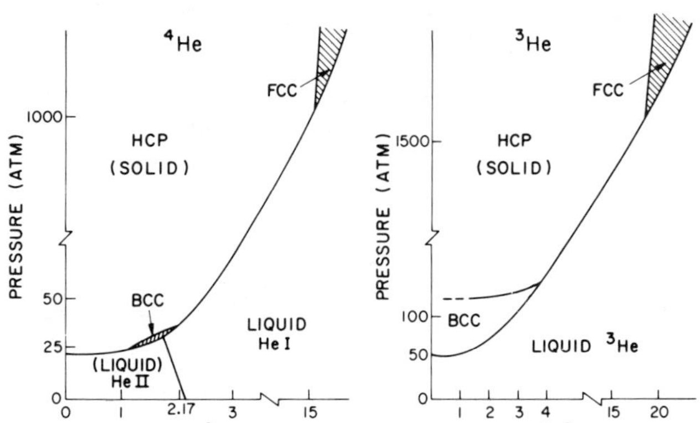

FIG. 1. The phase diagram of helium.

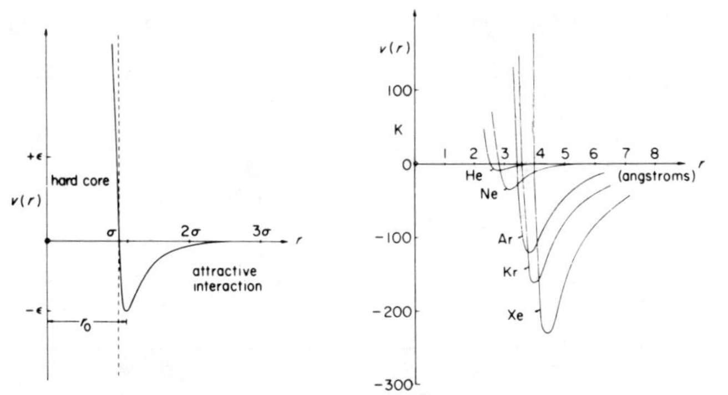

FIG. 2. Approximate potentials $v(r)$ between pairs of rare-gas atoms (from Guyer[3]). Here σ is the hard-core (electron cloud) diameter, $v(\sigma)=0$; r_0 is the minimum of $v(r)$; and ϵ is the well depth. For the Lennard-Jones (12-6) potential σ and ϵ are listed in Table I.

thoritative discussion of modern He-He atom potentials. Computer experiments show that a fluid of hard spheres (billiard balls with no attraction) solidifies under pressure. In helium it is the stiffness of the He-He atom repulsion that is chiefly responsible for solid formation. This means we have a weakly bound solid of highly kinetic atoms. By contrast, most solids form at zero pressure under a strong attractive potential.

II. THE BASIC PHYSICS OF SOLID HELIUM

To illustrate the basic physics of solid helium we focus on the specific heat, C_V. This apparently dull property transmits a wealth of information on the number, energy, and character of the basic excitations in the solid that must be excited as we add heat to raise its temperature. C_V has been measured in the hcp and bcc phases of both ^{3}He and ^{4}He and all results, except the most recent by Greywall[13] on bcc ^{3}He, are reviewed in Refs 4, 5, and 12.

In 1912, Peter Debye[14] proposed his famous approximate model for the specific heat of solids at low temperature due to the heating of the collective vibrations of the atoms, called phonons. This is

$$C_V{}^P = 3R\left(\frac{4\pi^4}{5}\right)\left(\frac{T}{\Theta_D}\right)^3, \quad (1)$$

where $R=1.987$ cal mol^{-1} K^{-1} is the gas constant. In the model Θ_D is a free parameter, the characteristic "Debye temperature" describing the interatomic forces in the solid. Equation (1) holds if $T\lesssim 0.1\Theta_D$ although Θ_D may vary with T if $T\gtrsim 0.02\Theta_D$. ($3R$ is the specific heat of a classical *harmonic* solid at room temperature.) The phonon specific heats of all phases studied obey Eq. (1) although for 15 years until Greywall's measurements bcc ^{3}He was thought an exception.

Fitting Eq. (1) to the observed C_V gives the Θ_D shown in Fig. 3 from which we may calculate the properties displayed in Table I. In Debye's model, the zero-point vibrational energy which remains at $T=0$ K is

$$E_{ZP} = \tfrac{9}{8}\Theta_D. \quad (2)$$

This energy, a purely quantum-mechanical effect, acts like

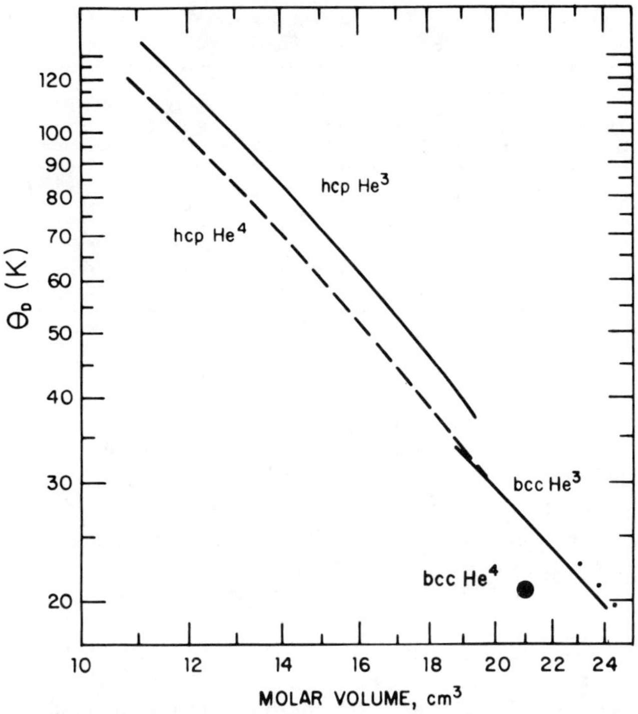

FIG. 3. The Debye temperatures of solid helium as a function of molar volume (after Sample and Swanson, see Ref. 7).

a kinetic energy and clearly dominates the thermal energy. The table also shows that the zero-point energy is roughly equal to the potential energy in helium.

The mean square vibrational amplitude, $\overline{u^2}$, of the atoms vibrating around their lattice points is, in the Debye model at $T=0$ K,

$$\overline{u^2} = 109.2(1/m\Theta_D)\ \text{Å}^2, \quad (3)$$

where m is the atomic mass. From Table I we see that $\overline{u^2}$ is large in solid helium so that the helium atoms are not well localized at their lattice points. Since the atoms cannot get closer together than the core diameter σ, we expect the mean interatom spacing R in the solid to be $R\gtrsim\sigma+(\overline{u^2})^{1/2}$, i.e., the core diameter of σ plus the rms vibrational amplitude. This expected spacing is shown in Table I, and, clearly, the large zero-point vibrational amplitude expands the helium lattice. The potential seen by a helium atom in this expanded solid due to its near neighbors on each side is shown in Fig. 4.

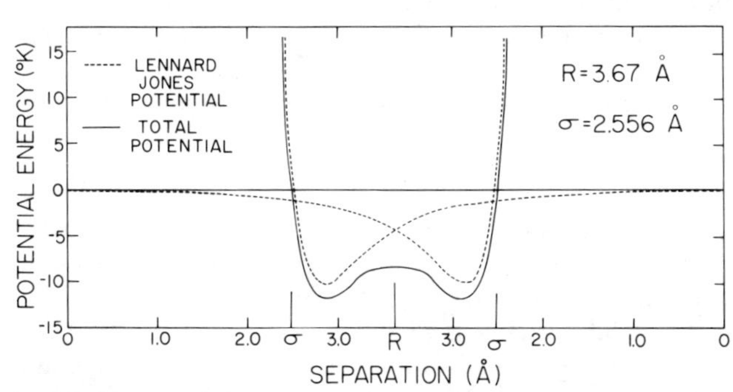

FIG. 4. The potential seen by a helium atom in a linear solid with near-neighbor interactions. Note that the second derivative of the potential at R is negative.

Table I The Observed Debye Temperatures of bcc ^3He ($V=23$ cm^3/mol, $p\approx55$ atm) and hcp ^4He ($V=19$ cm^3/mol, $p\approx60$ atm) at $T=0$ K and for the Heavier Rare-Gas Crystals for $T\to\infty$[a]

Rare-gas crystal	Debye temperature Θ_D (K)	Melting temperature T_m (K)	Zero-point energy (cal/mol)	Thermal energy at T_m (cal/mol)	Potential energy (cal/mol)	δ	rms amplitude $(\overline{u^2})^{1/2}$ (Å)	Potential parameters σ (Å)	Potential parameters ϵ (K)	Expected spacing $R=\sigma+(\overline{u^2})^{1/2}$ (Å)	Observed spacing R_{obs} (Å)	$(\overline{u^2})^{1/2}/R_{obs}$	de Boer parameter Λ
^{3}He(bcc)	22	1.2	49	0.02	−44	1	1.29	2.556	10.2	3.85	3.67	0.35	0.347
^{4}He(hcp)	34	2.6	76	0.14	−76	1	0.90	2.556	10.2	3.45	3.60	0.24	0.302
Ne	66	24.56	149	194.7	−580	0.25	0.29	2.701	36.2	2.99	3.156	0.091	0.0658
Ar	84	83.81	208	524.6	−2030	0.10	0.18	3.403	121	3.58	3.755	0.048	0.0212
Kr	64	161.37	160	969.3	−2810	0.057	0.14	3.640	166	3.78	3.992	0.036	0.0113
Xe	55	202	123	1209	−3950	0.031	0.12	3.94	232	4.06	4.336	0.028	0.0070

[a] The T_m for the heavier rare-gas crystals is at saturated vapor pressure (the triple point). The δ is the ratio of zero-point energy $[(9/8)R\Theta_D]$ to the absolute value of the potential energy. The rms atomic vibrational amplitude $(\overline{u^2})^{1/2}$ at $T=0$ K is obtained from the Debye model [Eq. (3)] and the de Boer parameter from Eq. (4) using the Lennard-Jones potential parameters

Note that in solid helium the atom sits at a point of unstable equilibrium and simple harmonic vibration theory would predict an unstable solid.

At low temperature in solid ^3He there is also a contribution to C_V due to excitation of the nuclear spins[15]:

$$C_V{}^N=3R(J/T)^2,$$

where in bcc ^3He, with the exchange constant $J\approx-10^{-3}$ K, the $C_V{}^N$ is comparable to $C_V{}^P$ at $T\approx0.1$ K. At higher temperature, thermal vacancies contribute to C_V, particularly in bcc ^3He. This contribution cannot be described by standard expressions,[16] which supports predictions that vacancies tunnel rapidly and are not localized in solid helium. Recent direct measurements[17] find a fraction 5×10^{-3} of sites vacant in bcc ^3He (1.8×10^{-3} in hcp ^4He) at melting, 20–50 times that in classical solids. This nonlocalized character is not special to vacancies but holds for all point defects[3,12,15,18] in solid helium.

III. HELIUM AS A QUANTUM SOLID

The large zero-point vibrational amplitude and energy in solid helium results from the uncertainty principle of quantum mechanics. This states that we cannot simultaneously localize an atom at a point (a lattice point here) and make its momentum vanish. There is a trade-off between reducing the momentum (energy) and localizing the particle. To keep the zero-point energy manageable each nucleus has a large vibrational distribution about its lattice point. In quantum mechanics where matter, like light, is described by waves we describe this vibrational distribution by a wave "function" centered at the lattice point. The wavelength λ of this matter wave, the famous de Broglie wavelength, is a measure of the "quantumness" of the particle. This quantumness was articulated by de Boer[19] in his famous "de Boer parameter"

$$\Lambda=\frac{1}{2\pi}\frac{\lambda}{\sigma}=\frac{\hbar}{\sigma(2m\epsilon)^{1/2}}, \tag{4}$$

which is the ratio[20] of the de Broglie wavelength, $\lambda=h/p$ [with momentum $p=(2m\epsilon)^{1/2}$], to the minimum separation between the atoms in the crystal, σ. If λ is comparable to σ, the solid becomes a quantum solid (see Table I). Also, Λ is intimately related to the mean square vibrational amplitude, $\Lambda\approx(\overline{u^2}/R^2)^{1/2}$ (see Table I). Since vibration becomes harmonic when the amplitude is small, large Λ means very anharmonic vibrations of atoms in solid helium. Hence very anharmonic vibrations remaining at $T=0$ K can also be used to identify a quantum solid.

The central issue in the theory of solid helium is finding a suitable wave function to describe the vibrational distribution of the atomic nuclei. This is a nuclear wave function. The pioneering work on this problem by the Nobel laureate Max Born and his student D. J. Hooton is succinctly reviewed by Domb and Dugdale.[2] Since physicists are simple folk, we seek a simple function, a Gaussian, centered on each lattice point. The width of the Gaussian is left as a free parameter. The variational principle of quantum mechanics states that the most accurate wave function is that which minimizes the energy. Hence we minimize the energy to find the free parameter. This energy minimization leads to an equation which must be solved iteratively until self-consistent. This method, although proposed by Born, was first articulated by Nosanow and Shaw[21] in 1962 and by Nosanow[22] in 1966 and forms the basis of the modern theory of solid helium. It has been extended by Koehler[23] and Horner[24] to correlated Gaussian functions necessary to describe the collective, phonon vibration of the nuclei. These latter developments constitute the self-consistent-phonon (SCP) theory formulated independently by Choquard[25] for highly anharmonic crystals and by Bocarra and Sarma[26] for ferroelectric transitions.

Although a Gaussian describes the collective phonon excitations well, it does not describe the short-range correlations in the atomic motion induced by the hard core of the He-He potential. Particularly in the less dense bcc and hcp phases of helium it is vital to modify the Gaussian to account for these short-range correlations—a problem familiar in nuclear-matter studies. There are a number of approaches each with its own merits and drawbacks. These developments of the self-consistent-phonon theory including short-range correlations have been exhaustively reviewed by Guyer,[3] Brandow,[28] Koehler,[4] Horner,[27] Glyde,[5] and Varma and Werthamer.[29] This theory may be used to calculate all the properties of helium: the phonon frequencies, the ground-state energy, pressure and compressibility, the thermal properties, and nuclear exchange.

The most direct test of the SCP theory is by neutron scattering, which measures the phonon frequencies directly. Measurements on bcc, hcp, and fcc ^4He, which show a number of interesting effects, are generally well described by the theory (see Fig. 5) although a more consistent formulation

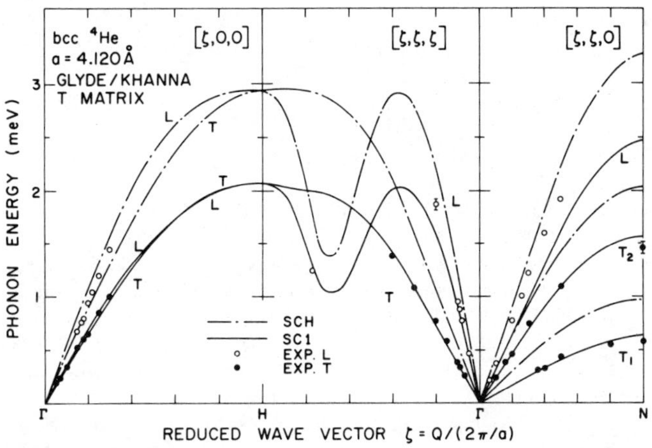

FIG. 5. The phonon-frequency dispersion curves observed in bcc ^{4}He by Minkiewicz *et al.*[34] and calculated by Glyde and Goldman[35] in the self-consistent harmonic and the complete first-order self-consistent approximations.

of short-range correlations valid at all volumes remains an outstanding problem today. This work, light scattering, and sound velocity measurements are reviewed by Koehler,[4] Horner,[27] and Glyde[5] with more recent experiments by Thomlinson *et al.*[30] and Stassis *et al.*[31] The ground-state energy, from which elastic properties are obtained by differentiation, is also well described (see Fig. 6). Thermal conductivity and second-sound propagation in helium is discussed by Ackerman and Guyer[32] and by Beck.[33]

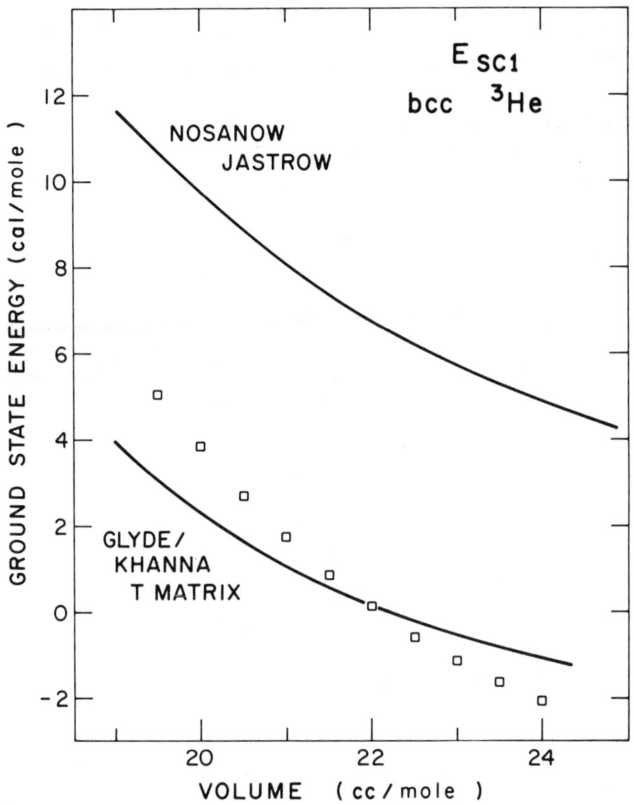

FIG. 6. The ground-state energy of bcc ^{3}He: experiment, Pandorf and Edwards[36]; calculations, Glyde and Goldman.[35]

IV. *THE MAGNETIC PROPERTIES OF SOLID* ^{3}*He*

The nuclear vibrational wave function of a ^{3}He nucleus in solid ^{3}He is so large that it overlaps with the wave functions of neighboring ^{3}He nuclei. This overlap leads to an exchange interaction between the ^{3}He nuclear spins, **I**, which depends on their relative orientation. For a simple Heisenberg exchange interaction between two spins, this energy is

$$H = -2J\mathbf{I}_1 \cdot \mathbf{I}_2, \tag{5}$$

where J is the exchange constant which is proportional to the overlap. Thouless[37] argues that in the lowest-energy state given by the smoothest wave functions J must be negative so that solid ^{3}He becomes an antiferromagnet at low temperature. This sign is confirmed by experiment.[3,12] For example, $J \simeq -0.8$ mK in bcc ^{3}He under low pressure ($V = 24$ cm^3/mol). The overlap leads to a physical exchange of atoms (nuclei and electrons together) by tunneling at a rate $\nu = J/h$ ($\simeq 2 \times 10^7$ Hz at $V = 24$ cm^3/mol), hence the name "exchange interaction."

The wave functions developed in the quantum theory of solid ^{3}He can be used to calculate J explicitly. The calculations to date, which use a wide variety of methods, are reviewed by McMahan[38] and Trickey *et al.*[12] While J is generally found negative, the numerical values, which are extremely sensitive to the many-body nature of wave functions, differ markedly in different approaches. Essentially, numerical accuracy is beyond the ability of our present theory. Also J decreases rapidly with volume as the applied pressure reduces the vibrational amplitudes of the nuclei. This volume dependence is also difficult to predict.

The exchange energy leads to a variety of interesting nuclear magnetic properties. Particularly, the rapid atomic diffusion and tunneling leads to nuclear dynamics usually observed only in liquids. The exchange interaction is also observed through its contribution to all the thermodynamic properties of ^{3}He, particularly the pressure. The studies of these effects have been thoroughly reviewed by Richards,[39] Guyer *et al.*,[15] Trickey *et al.*,[12] and Varma and Werthamer.[29] These show that the contributions of exchange to different properties cannot be reconciled using a simple Heisenberg form (5) of exchange energy. Rather, simultaneous exchange of many neighbors must be included and J is not a simple number but an operator. The more general form of H introduced by McMahan and Wilkins[40] is now used to describe experiments but this is by no means firmly established. As an example of the breakdown, the Heisenberg form predicts a transition to the ordered antiferromagnetic state in bcc ^{3}He at a Néel temperature

$$T_N = -2.748\ J$$

$$\simeq 2.2 \text{ mK (at } V = 24 \text{ cm}^3\text{/mol).}$$

The ordering has been observed[41] at $T_N = 1.1$ mK in zero external magnetic field, a factor of 2 below the prediction. Also magnetization measurements[42] show that the antiferromagnetic state is not alternate spin up, spin down ordering. Perhaps the bcc lattice is distorted by the spin ordering itself. This is an active area where much is still not understood.

See also CRYSTAL BINDING; CRYSTAL DEFFECTS; TUNNELING.

REFERENCES

1. W. H. Keeson, *Comm. Kamerlingh Onnes Lab., Leiden* **184b** (1926); *Helium*, p. 180. Elsevier, Amsterdam, 1942.
2. C. Domb and J. S. Dugdale, *Progress in Low Temperature Physics*, Vol. 2, p. 338, (J. Gorter, ed.). North Holland, Amsterdam, 1957.
3. R. A. Guyer, *Solid State Phys.* **23,** 413 (1969).
4. T. R. Koehler, in *Dynamical Properties of Solids*, Vol. II, Chap. 1, G. K. Horton and A. A. Maradudim (eds.). North Holland, Amsterdam, 1975.
5. H. R. Glyde, in *Rare Gas Solids*, Vol. I, Chap. 8, M. L. Klein and J. A. Venables (eds.). Academic Press, New York, 1976.
6. See, for example, R. B. Kummer, R. M. Mueller, and E. D. Adams, *J. Low Temp. Phys.* **27,** 319 (1977).
7. W. E. Keller, *Helium-3 and Helium-4*. Plenum, New York, 1969.
8. J. Wilks, *The Properties of Liquid and Solid Helium*. Oxford University Press, London, 1967.
9. B. L. Holian, W. D. Gwinn, A. C. Luntz, and B. J. Alder, *J. Chem. Phys.* **59,** 5444 (1973); see also J. P. Franck, *Phys. Rev. Lett. 40,* 1272 (1978).
10. See *Rare Gas Crystals*, Vols. I and II, M. L. Klein and J. A. Venables (eds.). Academic Press, New York, 1976.
11. J. A. Barker, in *Rare Gas Solids*, Vol. I, Chap. 4, M. L. Klein and J. A. Venables (eds.). Academic Press, New York, 1976.
12. S. B. Trickey, W. P. Kirk, and E. D. Adams, *Rev. Mod. Phys.* **44,** 668 (1972).
13. D. S. Greywall, *Phys. Rev. Lett.* **39,** 105 (1976); *Phys. Rev.* **B15,** 2604 (1977).
14. P. Debye, *Ann. Phys.* **39,** 789 (1912).
15. R. A. Guyer, R. C. Richardson, and L. I. Zane, *Rev. Mod. Phys.* **43,** 532 (1971).
16. J. H. Hetherington, in *Proceedings of the International Quantum Crystals Conference*, Colorado State University, p. C-48, J. Sites (ed.) 1977; and *Phys. Rev.* **176,** 231 (1968).
17. B. A. Frass, S. M. Heald, and R. O. Simmons, in *Proceedings of the International Quantum Crystals Conference*, Colorado State University, p. C-73, J. Sites (ed.) 1977.
18. A. Widom and M. G. Richards, *Phys. Rev.* **A6,** 1196 (1972).
19. J. De Boer, *Progress in Low Temperature Physics*, Vol. 2, Chap. 1, J. C. Gorter (ed.). North Holland, Amsterdam, 1957.
20. Here we use the definition of the Boer parameter introduced by Newton Bernardes [*Phys. Rev.* **120,** 807 (1960)] which is $(2\pi\sqrt{2})^{-1}$ times de Boer's definition.
21. L. H. Nosanow and G. L. Shaw, *Phys. Rev.* **128,** 546 (1962).
22. L. H. Nosanow, *Phys. Rev.* **146,** 120 (1966).
23. T. R. Koehler, *Phys. Rev. Lett.* **17,** 89 (1966); **18,** 654 (1967).
24. H. Horner, *Z. Phys.* **205,** 72 (1967).
25. P. F. Choquard, *The Anharmonic Crystal*. Benjamin, New York, 1967.
26. N. Boccara and G. Sarma, *Physics* **1,** 219 (1965).
27. H. Horner, *J. Low Temp. Phys.* **8,** 511 (1972); and in *Dynamical Properties of Solids*, Vol. I, G. K. Horton and A. A. Maradudin (eds.). North Holland, Amsterdam, 1974.
28. B. H. Brandow, *Ann. Phys.* **74,** 112 (1977).
29. C. M. Varma and N. R. Werthamer in *The Physics of Liquid and Solid Helium*, Vol. I, Chap. 6, K. H. Benneman and J. B. Ketterson (eds.). Wiley, New York, 1976.
30. W. Thomlinson, J. Eckert, and G. Shirane, *Phys. Rev.* **B18,** 1120, (1978).
31. C. Stassis, G. Kline, W. A. Kamitahara, and S. K. Sinha, *Phys. Rev.* **B17,** 1130, (1978).
32. C. C. Ackerman and R. A. Guyer, *Ann. Phys.* **50,** 128 (1968).
33. H. Beck, in *Dynamical Properties of Solids*, Vol. II, G. K. Horton and A. A. Maradudin (eds.). North Holland, Amsterdam, 1975.
34. V. J. Minkiewicz, T. A. Kitchens, G. Shirane, and E. B. Osgood, *Phys. Rev.* **A8,** 1513 (1973).
35. H. R. Glyde and V. V. Goldman, *J. Low Temp. Phys.* **25,** 601 (1976).
36. R. C. Pandorf and D. O. Edwards, *Phys. Rev.* **169,** 222 (1968).
37. D. J. Thouless, *Proc. Phys. Soc.* **86,** 893 (1965).
38. A. K. McMahan, *J. Low Temp. Phys.* **8,** 115 (1972); **8,** 159 (1972).
39. M. G. Richards, in *Advances in Magnetic Resonance*, Vol. 5, p. 305, J. S. Waugh (ed.). Academic Press, New York, 1971.
40. A. K. McMahan and J. W. Wilkins, *Phys. Rev. Lett.* **35,** 673 (1975).
41. W. P. Halperin, C. N. Archie, F. B. Rasmussen, R. A. Buhrman, and R. C. Richardson, *Phys. Rev. Lett.* **32,** 927 (1974).
42. T. C. Prewitt and J. M. Goodkind, *Phys. Rev. Lett.* **39,** 1283 (1977).

Hysteresis

Paul L. Donoho

The phenomenon of hysteresis, in which changes in a material property, such as the magnetization M, do not follow reversibly changes in a driving field, such as the magnetic intensity H, is observed in numerous situations in many materials. Among the best-known examples of hysteresis are the dependence of M on H in ferromagnetic and ferrimagnetic materials, the dependence of polarization on electric field in ferroelectric materials, and the dependence of strain on stress, for large stress excursions, in most materials. In the case of ferromagnetic materials, the irreversible behavior of M as a function of H arises principally from irreversible effects in the motion of the walls of the magnetic domains which make up a given specimen. A typical M-H plot, including a so-called hysteresis loop, is illustrated in Fig. 1. An initially unmagnetized specimen is subjected to an applied field H, which starts from a value of zero and undergoes

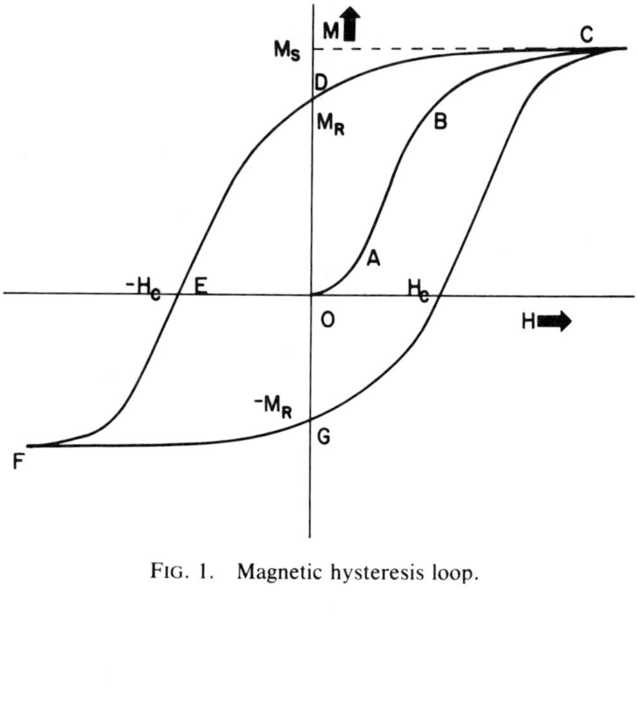

FIG. 1. Magnetic hysteresis loop.

cyclic behavior. The magnetization then traces out the curve *OABCDEFGCDE*.... The behavior of M is essentially reversible during the portion *OA*, for which the domain walls undergo small, reversible motion. For the portion of the path *AB*, however, the *M-H* curve is not reversible as a result of the larger motion of the domain walls, which is strongly influenced by the effects of lattice imperfections such as inclusions, dislocations, grain boundaries, etc. The portion of the curve *BC* represents a region in which the magnetization within each domain is reversibly rotated into the direction of the applied field as M approaches its saturation value M_s. As H is reduced along the portion *CD*, the magnetization decreases reversibly, reaching its remanent value M_r at $H=0$. The applied field is then reversed in direction along the portion *DE*, and M reaches zero at $H=H_c$, the coercive field. Along most of the path *DEF*, irreversible domain-wall motion is again dominant. The path *FGC* is essentially identical in nature to *DEF*. The complete loop *CDEFGC* is called a hysteresis loop, and further cycling of H will lead to repeated tracing of this loop.

For permanent-magnet applications, it is desirable that M_r and H_c be as large as possible, in order to permit the fabrication of magnets of high flux density which cannot be easily demagnetized. Materials exhibiting large values of H_c are generally called "hard" magnetic materials. Values of H_c as large as 2×10^6 A/m have been observed in the recently developed rare-earth-cobalt materials. For applications involving oscillating fields (transformers, motors, etc.), it is desirable that H_c be as small as possible, since the area enclosed by the hysteresis loop is equal to the work done on the specimen per cycle by the applied field. A small value of H_c will, therefore, lead to low power dissipation in the material in ac applications. Materials with low values of H_c are generally called "soft" magnetic materials. Values of H_c as low as 0.16 A/m are found in materials such as Supermalloy. Intermediate values of H_c are often of interest in applications such as the design of magnetic memory devices, and it is interesting to note that almost any desired value of H_c can be obtained from the wide range of available magnetic materials.

See also FERROELECTRICITY; FERROMAGNETISM; MAGNETIC DOMAINS AND BUBBLES; MAGNETIC MATERIALS; MAGNETS (PERMANENT) AND MAGNETOSTATICS.

Ice

Peter V. Hobbs

Water forms more solid phases than any other known substance. If liquid water is frozen under normal atmospheric conditions, or water vapor is deposited at temperatures below 0°C and above about −80°C, the water molecules are arranged in orderly repetitive positions to form a crystalline solid with hexagonal symmetry that is called *hexagonal ice, ice Ih*, or simply *ice* (Fig. 1).

The deposition of water vapor onto a substrate between about −80 and −130°C also produces a crystalline solid, but in this case the structure is cubic and it is known as *cubic ice* or *ice Ic*. In addition, at least eight high-pressure crystalline forms of the water substance have been identified; they are referred to as ice II, ice III, etc. up to ice IX. The phase diagram for the solid phases of water is shown in Fig. 2 and structural data on all the crystalline polymorphs of ice (except ice IV, which is very unstable) are listed in Table I. If water vapor is deposited onto a substrate below about −140°C, a noncrystalline solid known as *amorphous ice* is produced.

The water molecule remains intact in all the polymorphs of ice, although the OHO bond angle and OH length vary somewhat. Ice is held together as a solid by the hydrogen atoms, which lie between pairs of electronegative oxygen atoms, that is, it is a *hydrogen-bonded* material. The hydrogen bond plays an important role in many aspects of chemistry, physics, and molecular biology. For example, the hydrogen bond is responsible for the action of glues and adhesives, for the adhesion of dirt to skin, and for the structure of proteins. Indeed, man himself is composed, for the most part, of hydrogen-bonded materials. Because the crystalline phases of water represent the most complete hydrogen-bonded materials, their study has thrown light on more complex hydrogen-bonded systems.

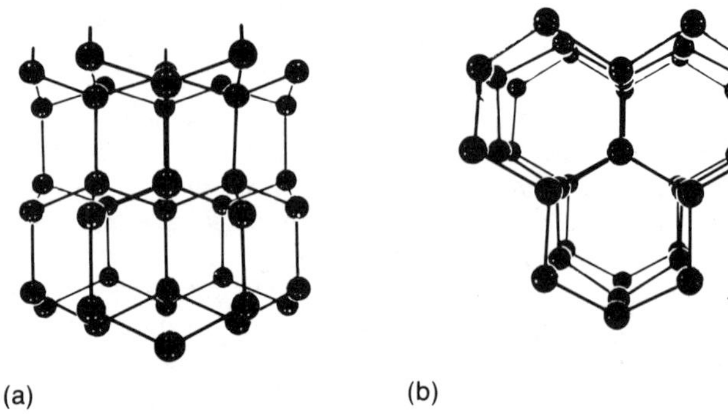

FIG. 1. The arrangement of the oxygen atoms in ice Ih. (The rods joining the oxygen atoms represent bonds. The hydrogen atoms are not shown.) (a) View perpendicular to the *c* axis. (b) View along the *c* axis.

Many of the physical properties of ice are quite unique and have been investigated for hundreds of years. It is only in recent years, however, that some of these properties have begun to be understood in terms of the molecular structure of ice. For example, ice has an unusually high static relative permittivity (97 at 0°C compared to 88 for liquid water at 0°C). It also provides a good example of electrical conduc-

Table I Structural Data on the Polymorphs of Ice

Polymorph	Crystal system	Space group	Cell dimensions[a] (Å)	No. of molecules in a unit cell	No. of nearest neighbors	Distance of nearest neighbors[a] (Å)	O–H–O angles[a] (deg)	Hydrogen positions	Density[a] ($mg\ m^{-3}$)
Ice Ih	Hexagonal	$P6_3/mmc$	$a_0=4.50;c_0=7.32$	4	4	2.74	109±0.2	Disordered	0.931
Ice Ic	Cubic	$Fd3m$	$a_0=6.35$	8	4	2.75 at −130°C	109.5	Disordered	0.93
Ice II	Rhombohedral	$R3$	$a_0=7.79;\alpha=111.3^0$	12	4	2.75–2.84	80–128	Ordered	1.18
Ice III	Tetragonal	$P4_12_12$	$a_0=6.73;c_0=6.83$	12	4	2.76–2.80	87–141	Disordered	(1.16)
Ice V	Monoclinic	$A2/a$	$a_0=9.22;b_0=7.54;$ $c_0=10.35;\beta=109.2$	28	4	2.76–2.87	84–128	Disordered[b]	
Ice VI	Tetragonal[c]	$P4_2/nmc$[c]	$a_0=6.27;c_0=5.79$	10	4	2.80–2.82	76–128	Disordered[d]	1.31
Ice VII	Cubic	$Pn3m$	$a_0=3.43$	2	8[e]	2.95	109.5	Disordered[f]	(1.49)
Ice VIII	Tetragonal	$I4_1/amd$	$a_0=4.80;c_0=6.99$	8	8[e]	2.96 (hydrogen bonded) 2.80 (nonbonded, close neighbors) 3.15 (nonbonded, distant neighbors)	109.5	Ordered	1.49
Ice IX	Tetragonal	$P4_12_12_1$	$a_0=6.73;c_0=6.83$	12	4	2.76–2.80	87–140	Ordered[g]	1.16

[a] Reduced to values at atmospheric pressure and −163°C unless noted otherwise.
[b] Becomes partially ordered when quenched at low temperatures.
[c] Recent measurements show that at −163°C ice VI is probably orthorhombic with space group *Pmmn*.
[d] May be partly ordered at all temperatures and become ordered when cooled slowly below −150°C.
[e] Four of the nearest neighbors are hydrogen-bonded to central molecule.
[f] Partly orientationally ordered.
[g] Proton ordering is not quite complete.

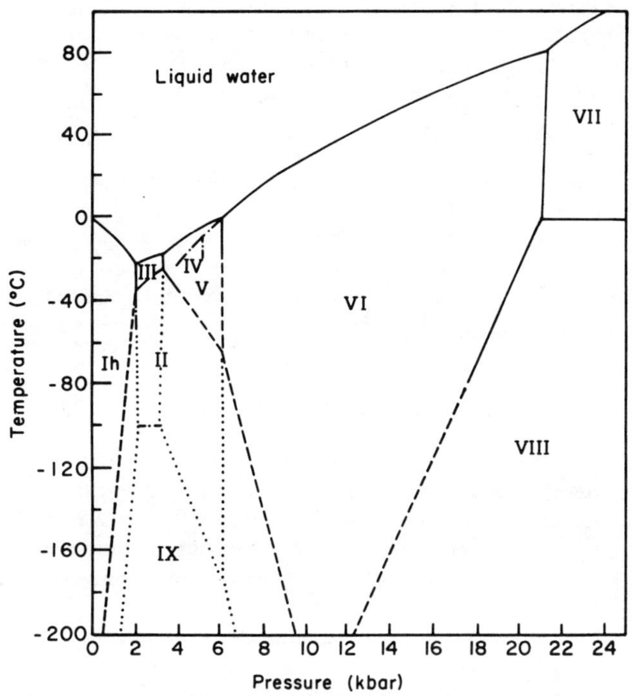

FIG. 2. Phase diagram of the solid phases of water: ———, measured stable lines; ·–·–·–·, measured metastable lines; – – –, extrapolated or estimated stable lines; ····, extrapolated or estimated metastable lines. Ice IV is metastable in the region of stability of ice V; the indicated field for ice IV is inferred from the D_2O system. Ice Ic and vitreous ice are not indicated.

tion by transfer of protons, which have mobilities in ice that are ten times greater than those characteristic of normal ionic conductors. In fact, interesting parallels can be drawn between protonic conduction in ice and electronic conduction in metals and intrinsic semiconductors. This analogy is supported by the fact that the thermoelectric power of ice is comparable to that of some metals.

The optical properties of ice are also unusual. Apart from being transparent, it has the lowest index of refraction (1.31 for the sodium D lines) of any known mineral. It is a doubly refracting, uniaxial, and optically positive crystal. Its birefringence (i.e., the difference between the refractive indices for the extraordinary and ordinary waves) is very small (0.0014 for the sodium D lines) and, unlike most materials, decreases with decreasing temperature.

For frequencies from 30 to 300 MHz, the loss of energy by absorption in ice is sufficiently small that radio waves can be propagated through large masses of dry ice or snow. This fact has been utilized to determine the depths and bottom topography of polar ice sheets from radio echo soundings.

When a stress is applied to a piece of polycrystalline ice, it first deforms instantaneously, then deforms more slowly (*transient creep*) and, given time, settles down to a steady creep rate (*secondary creep*); for stresses in excess of 4×10^5 N m^{-2} the creep curve reaccelerates again (*tertiary creep*). The steady-state creep rate $\dot{\epsilon}_{st}$ during secondary creep is related to the stress σ by

$$\dot{\epsilon}_{st}=A\sigma^n \exp(-E/kt)$$

where A is a constant and E an activation energy. At low stresses $n\simeq3$ between -2 and -45°C; $E\simeq1.3$ eV between -2 and -8°C and $E\simeq0.75$ eV below -8°C.

A single crystal of ice may be likened to a pack of cards in which the faces of the cards represent the parallel sheets of molecules (called the *basal planes*) that can be seen in Fig. 1a. Thus, a single crystal of ice undergoes plastic deformation very readily if there is a component of shear acting parallel to the basal planes that causes them to glide over one another. If, on the other hand, there is no component of shear in these planes, deformation is very difficult.

The surface properties of ice are particularly fascinating because ice is one of the few solids that are commonly encountered close to their melting point. The "stickiness" of ice may be explained in terms of sintering (although it may also be a manifestation of thin "liquidlike" layers on ice). The low coefficient of friction between ice and many solid materials makes skiing and skating possible. The heat produced by the frictional rubbing of a ski or skate on ice forms a thin (~1 μm) layer of water, which acts as a lubricant and lowers the frictional force.

Ice is important in several branches of geophysics including glaciology, oceanography (sea ice), meteorology (cloud physics), and astronomy (planetary atmosphere).

See also ANELASTICITY.

BIBLIOGRAPHY

Neville H. Fletcher, *The Chemical Physics of Ice*. Cambridge, London and New York, 1970. (An intermediate-to-advanced text that illustrates some aspects of solid-state physics using ice as an example.)

Peter V. Hobbs, *Ice Physics*. Oxford, London and New York, 1974. (A comprehensive treatment of the subject at the intermediate and advanced levels.)

Elton R. Pounder, *Physics of Ice*. Pergamon, New York, 1965. (An elementary treatment with emphasis on sea ice.)

Insulators

R. A. Anderson

INTRODUCTION

An insulator, or dielectric, could be conveniently defined as a material that does not conduct electricity, if not for the fact that all known materials conduct a measurable amount of electricity. Nevertheless, insulators are easily distinguished from semiconductors and conductors. At room temperature, the conductivity of a good insulator typically is more than 20 orders of magnitude smaller than that of a good conductor, and the difference becomes larger at lower temperatures.

The physics that underlies the vast difference between the electrical conductivity of insulators and conductors is the first topic considered. Following that, discussions are devoted to properties of insulators, including the importance of the insulator–electrode contacts in determining conduction processes. This article concludes with a brief description of some of the ways an insulator loses its ability to insulate when electrically stressed to the point of breakdown.

BAND STRUCTURE AND CONDUCTIVITY

A medium consisting of isolated atoms or molecules cannot conduct electricity because no charged particles are free to flow in response to an electric field. However, if atoms are brought close together, as in a crystalline solid, electrons can move from one atom to another. It would appear, then, that all crystalline solids should conduct electricity.

The distinction between insulators and conductors originates in the quantum-mechanical behavior of electrons. As a consequence of the periodic structure of a crystalline solid, electron energies are restricted to well-defined energy bands, normally separated by bands of forbidden energies. Each band of allowed energies consists of a very large number of closely spaced allowed momentum values, each of which can be occupied by, at most, two electrons having opposite spins. For this reason, there is a finite number of electrons that can be included within an energy band, and the lower-lying energy bands are completely or partially filled with electrons.

Electrical conductivity is crucially dependent on the degree to which the energy bands are filled with electrons. A completely empty band obviously cannot contribute to electrical conductivity. On the other hand, no charge can be transported by a completely filled band: Because all available momentum values are occupied, the net momentum must remain zero in the presence of an electric field. In a partially filled band, however, the boundary between occupied and unoccupied allowed momenta can become displaced in response to an electric field. Charge is transported because the net momentum is no longer zero. The charge carrier in a nearly filled band is the absence of an electron, called a hole. A hole behaves very much as though it were an electron of reversed charge in a nearly empty band.

In most cases for a material to be an insulator all energy bands must be either full or empty. If one or more energy bands are partially filled, the material usually is a metallic conductor. A further requirement for an insulator is that the gap between filled (valence) and empty (conduction) bands must be large enough that few electrons are thermally excited across this band gap. If, instead of a band gap, there is a small overlap between these bands, electrons at the top of the valence band spill over into the bottom of the conduction band. Such materials, of which antimony is an example, are known as semimetals. In the case of a pure semiconductor, such as germanium, the band gap is so small that a portion of the valence-band electrons are thermally excited into the conduction band, thereby creating electrons and holes which can conduct electricity. Such a material is an insulator at very low temperatures. Other semiconductors owe their conductivity to impurities that either remove electrons from the valence band or add them to the conduction band.

At this point the reader may be wondering how the unlikely situation of exactly enough electrons to fill the valence bands can occur at all. The reason it occurs frequently is as follows. Each repeating unit of the crystal structure contributes one allowed momentum value, which can accommodate two electrons, to each energy band. Therefore, bands can be filled without leftover electrons if each repeating unit contains an even number of valence electrons, a situation that is not at all unlikely.

In the preceding discussions, we have considered only crystalline solids. However, the short-range atomic order in amorphous materials is sufficient to establish energy bands similar to those found in solids having highly periodic structure. Examples of insulators, semiconductors, and conductors are found in both crystalline and amorphous solids, as well as in liquids.

INSULATOR–ELECTRODE CONTACTS

The conductivity of an insulator is strongly influenced by interactions between the insulator and the electrodes attached to it. When an insulator and a conductor are brought into contact, electrons diffuse either into or out of the insulator, depending on the particular materials involved, to establish thermodynamic equilibrium.

If electrons diffuse into the insulator, the insulator–electrode contact is called Ohmic, or injecting. Because an accumulation of electrons injected into the conduction band at the insulator surface is established, which can supply current to the bulk, conduction is limited by the bulk conductivity of the insulator. The bulk conductivity itself depends on whether or not the injected charge extends throughout the thickness of the insulator. If this charge is everywhere larger than the intrinsic charge in the conduction band, then the conductivity reflects the properties of the insulator–electrode contacts rather than the intrinsic conductivity, and the conduction process is referred to as space-charge limited.

A blocking contact occurs when electrons diffuse out of the insulator to establish thermodynamic equilibrium. In this case conduction is limited by the rate at which electrons can cross the interface, and the conduction process is referred to as electrode limited. Current–voltage characteristics depend on whether electrons are thermally excited over, or tunnel through, the potential barrier at the insulator surface.

Contacts also may be either blocking or injecting for holes, which introduces the complexity of recombination of electrons and holes if both charge carriers are flowing simultaneously.

TRAPS

All known insulators contain localized energy levels in the band gap called traps, which tend to capture and immobilize the few charge carriers present in the valence and conduction bands. Traps originate from impurity atoms as well as naturally occurring defects. Because trapping affects recombination as well as the flow of carriers, essentially no conduction process escapes being influenced.

PHOTOCONDUCTIVITY

The absorption of light by an insulator may be accompanied by an increase in the electrical conductivity during illumination. Some insulators also display an increased conductivity in the dark as the result of previous exposure to light. These phenomena are called primary and secondary photoconductivity, respectively. Primary photoconductivity occurs when absorbed photons have sufficient energy to excite electrons into the conduction band from traps or from the valence band. Secondary photoconductivity is ascribed to light-induced changes in the occupation of traps near the electrodes, giving rise to local electric fields which assist the injection of electrons into the insulator. A well-known application of photoconductivity is the xerographic duplicating process.

Photoconductivity provides a means of investigating the mobility and lifetime of charge carriers in insulators. For example, a thin layer of carriers can be generated near an electrode by strongly absorbed light, so that the conductivity due to either electrons or holes can be measured.

ALTERNATING-CURRENT CONDUCTIVITY

The amount of alternating current flowing through a capacitor depends on the electrical polarizability of the insulator between the capacitor electrodes. Polarization, however, does not respond instantaneously to changes in applied voltage. Rather, several separate processes are involved, each of which proceeds at a different rate and contributes a frequency dependence to the polarizability of the insulator. Alternating-current measurements are a useful probe of slow processes such as chain motion in polymers and drift of mobile ions, as well as faster electronic processes, and can provide a wealth of information when performed at a series of different temperatures.

ELECTRICAL BREAKDOWN

The most important application of insulators is electrical insulation, without which electricity would be of little use. Failure of insulators under electrical stress, therefore, remains a topic of research interest.

Ideally, electrical breakdown can be classified as either thermal or electronic. Thermal breakdown occurs when the conductivity increases rapidly enough with increasing temperature to permit thermal runaway. Electronic breakdown involves an instability of the conduction electrons, such as a runaway drift velocity which leads to an ionization avalanche, or enhanced electron injection due to trapped charge near an electrode, as in secondary photoconductivity. The ultimate stage of a destructive electronic breakdown event is probably thermal.

In practice, breakdown often occurs at stresses well below the intrinsic dielectric strength of an insulator, and the mechanisms leading to breakdown are complex and poorly understood. For example, insulation on ac power transmission cables is susceptible to breakdown phenomena collectively called treeing, in which branching networks of tubular voids resembling trees slowly develop. Eventually the insulation is penetrated and a damaging discharge occurs. Trees tend to start growing wherever the electric field is enhanced: at cracks or voids in which gas discharges occur, or at metallic inclusions or electrode irregularities. Cable insulation in a moist environment often develops trees that are filled with water.

We consider, finally, another class of breakdown phenomena called surface flashover, in which discharges follow the insulator surface rather than penetrating the bulk. Insulator surfaces exposed to vacuum are particularly susceptible. Although vacuum surface flashover often appears to be the consequence of an avalanche of electrons which multiply on the insulator surface by secondary emission, mechanisms responsible for flashover in other cases are unresolved.

See also CONDUCTION; DIELECTRIC PHENOMENA; ELECTRON ENERGY STATES IN SOLIDS AND LIQUIDS; PHOTOCONDUCTIVITY.

BIBLIOGRAPHY

H. Fröhlich, *Theory of Dielectrics*. Oxford, London and New York, 1958. (A)

T. J. Gallagher, *Simple Dielectric Liquids*. Oxford, London and New York, 1975. (A)

N. E. Hill, W. E. Vaughan, A. H. Price, and M. Davies, *Dielectric Properties and Molecular Behavior*. Van Nostrand, Princeton, NJ, 1969. (A)

C. Kittel, *Introduction to Solid State Physics*, 5th ed. Wiley, New York, 1976. (E)

M. A. Lampert and P. Mark, *Current Injection in Solids*. Academic Press, New York, 1970. (A)

N. F. Mott and R. W. Gurney, *Electronic Processes in Ionic Crystals*, 2nd ed. Oxford, London and New York, 1948. (I)

J. J. O'Dwyer, *The Theory of Electrical Conduction and Breakdown in Solid Dielectrics*. Oxford, London and New York, 1973. (A)

J. G. Simmons, "Conduction in Thin Dielectric Films," *J. Phys. D: Appl. Phys.* **4**, 613 (1971). (A)

J. M. Ziman, *Principles of the Theory of Solids*, 2nd ed. Cambridge, England, 1972. (I)

Internal Friction in Crystals

Andrew V. Granato

A description of the mechanical properties of crystals is formally similar to that for electrical or magnetic properties, with the elastic constant playing a role analogous to that of the dielectric constant or the permeability.

The mechanical behavior (response to applied forces, or strain produced by a given stress) of a crystal is determined by the interatomic forces between the atoms as well as by the motion of defects in the crystal. The interatomic forces act quickly and may be taken to be instantaneous for mechanical tests, whereas the displacements or strain from defect motion requires a finite time. A perfect or ideal crystal can be defined as one for which there are no defects providing energy-loss mechanisms, or internal friction, for mechanical excitations. The elastic strain response ϵ_{el} of an ideal elastic crystal to an applied stress σ is given by elasticity theory as $\epsilon_{el} = \sigma/G$, where G is the elastic modulus of the crystal. When this stress–strain relation is combined with Newton's laws for a volume element of the crystal, plane-wave solutions without attenuation, or energy loss, of the equations of motion are found.

Real crystals contain defects. When the defect contribution to the strain is linear with a unique equilibrium value for a given stress, but is achieved only after the passage of a characteristic time, the crystal is called anelastic. For a sinusoidal applied external stress, the in-phase (real) part of the delayed response (ϵ_R) leads to an effective modulus change ΔG, while the out-of-phase (imaginary) part (ϵ_I) leads to an energy loss, or internal friction Q^{-1}. These are given by

$$\frac{\Delta G}{G}=\frac{G\epsilon_R}{\sigma_0} \quad \text{and} \quad Q^{-1}=\frac{G\epsilon_I}{\sigma_0}, \tag{1}$$

respectively. The response is often of the Debye type for which

$$\frac{\Delta G}{G}=\frac{\Delta_R}{1+(\omega\tau)^2} \quad \text{and} \quad Q^{-1}=\frac{\Delta_R\omega\tau}{1+(\omega\tau)^2} \tag{2}$$

where Δ_R is the relaxation strength, ω is the angular frequency of the applied stress, and τ is the relaxation time for the delayed anelastic component of the strain.

The classic example of an internal friction peak, which occurs when $\omega\tau=1$, is the Snoek relaxation of carbon in bcc iron. In bcc crystals interstitial impurities normally occupy octahedral sites and have tetragonal symmetry. Depending on the particular site chosen, the tetragonal axis may lie along any one of the cube axes x, y, or z. When a stress is applied along one of the Cartesian x, y, or z directions, the equivalence of the sites is lost, and the carbon atoms will jump by diffusion, preferentially occupying the lowest energy sites. This changes the length of the crystal in the direction of the stress, leading to an anelastic strain with a characteristic delay determined by the time of diffusion of the carbon atom between the sites. When a stress is applied along the ⟨111⟩ direction, no effect is observed, since this stress deforms the x, y, and z directions by the same amount. The relaxation strength is determined by the number of interstitials and the direction of the applied stress.

The overall behavior can be classified conveniently with the help of simple models, as illustrated in Fig. 1. For a crystal containing point defects and dislocations, there is a point-defect strain ϵ_p and a dislocation strain ϵ_d in addition to the elastic strain ϵ_{el}. The point-defect strain may be paraelastic (stress-induced rotation between equivalent positions of permanent dipoles with the help of thermal fluctuations, as in the Snoek effect), or diaelastic (induced dipole, which produces an internal strain, as with interstitials in metals). Also, the effects of phonons and electrons can be included within the formalism as point defects.

The dislocation strain is a diaelastic effect requiring a mass term in the general case and is given by

$$\epsilon_d=\Lambda by \tag{3}$$

where Λ is the dislocation density, b the Burgers vector, and y the average dislocation displacement. The displacement y is given by the dislocation equation of motion

$$M\ddot{y}+B\dot{y}+Ky=b\sigma \tag{4}$$

where $M \simeq \rho b^2$ is the dislocation mass per unit length, B is the dislocation viscosity, and K is a restoring-force constant for the dislocation. Comparing Eqs. (3) and (4), we obtain the model values M_d, B_d, and K_d from M, B, and K by

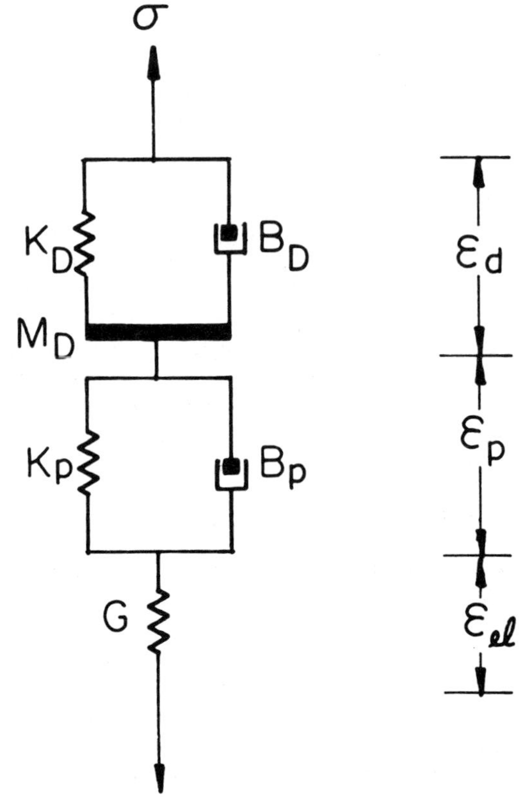

FIG. 1. Schematic representation of strain in a specimen containing point defects and dislocations.

dividing the latter values by Λb^2. For small enough B, the response ϵ_d has a resonance character, but for large B, inertial effects can be ignored and we have a relaxation with

$$\Delta_R=\frac{\Lambda Gb^2}{K_d} \quad \text{and} \quad \tau=\frac{B_d}{K_d}. \tag{5}$$

The coefficients B and K are phenomenological, can arise from many sources representing the dislocation interactions with other defects, and are related to the corresponding quantities appearing for the point-defect strain ϵ_p. In general, any mechanism that damps ultrasonic shear waves in a crystal containing no dislocations will also provide a viscous drag on dislocations, since a moving dislocation generates a shear strain rate in the crystal. Sources of drag B are phonons, electrons, and point defects, while sources of K are the dislocation tension and immobile aligned dipolar point defects. Dislocation internal-friction effects have been detected over a frequency range from 10^{-2} to 10^{12} Hz, and over a point-defect concentration range from 10^{-12} to 10^{-2}.

See also ANELASTICITY; CRYSTAL DEFECTS; ELASTICITY.

BIBLIOGRAPHY

R. DeBatiste, *Defects in Crystalline Solids*. North-Holland, Amsterdam, 1972.

A. S. Nowick and B. S. Berry, *Anelastic Relaxation in Crystalline Solids*. Academic, New York, 1972.

R. Truell, C. Elbaum, and B. B. Chick, *Ultrasonic Methods in Solid State Physics*. Academic, New York, 1969.

Ising Model

Fa Yueh Wu

It is an empirical fact that the magnetization of a permanent magnet diminishes in strength as the magnet is heated, disappearing completely above a certain temperature, called the Curie point. The Ising model is a simple physical model for explaining this phenomenon from a microscopic point of view. Historically, the model was first proposed by Lenz in 1920 and later studied by his student Ernst Ising. For this reason, it is also referred to in the literature as the Lenz–Ising model.

In the microscopic picture, the atoms of a magnetic substance are themselves tiny magnets. The existence of a spontaneous magnetization in the bulk is then explained as a result of an alignment of these tiny magnets, or "spins," due to their mutual interactions. To construct a simple model which simulates this situation, Lenz assumed that (1) the spins are arranged on a regular lattice, (2) each spin can point in only one of two directions, "up" or "down," and (3) there exists an interaction energy $-J$ between two neighboring spins which point in the same direction, and an energy J if they point in opposite directions. This describes the Ising model. Since a physical system attains its lowest energy at absolute zero temperature, for positive J the Ising model will be in a configuration with all spins aligned at absolute zero. It is then hoped that the spin alignment would persist, at least partially, up to some nonzero but finite temperature identified as the Curie point.

The mathematical formulation of the model is as follows: associate a two-valued variable σ_i to the ith spin such that $\sigma_i = +1$ (or -1) denotes an up (or down) spin. The interaction energy between two spins i and j can then be written as $-J\sigma_i\sigma_j$ and the total energy of the system is

$$E = -J \sum_{\langle i,j \rangle} \sigma_i \sigma_j ,$$

where the summation is taken over all neighboring pairs $\langle i,j \rangle$. Then, according to the principles of statistical mechanics, the thermodynamics of the system can be derived from the partition function

$$Z = \sum_{\text{configurations}} e^{-E/kT}$$

where k is the Boltzmann constant and T is the absolute temperature. The summation is taken over all possible spin configurations. The mathematical problem is to find a closed-form expression for Z. Once this is done, the Curie point, if any, will manifest itself in this expression as a point of nonanalyticity in the variable T.

In his dissertation summarized in a 1925 paper, Ising solved this problem for a one-dimensional system (spins arranged on a chain) and found no Curie point. The existence of a spontaneous magnetization in two and three dimensions was established by Peierls in 1936, and in 1944 Onsager published the exact solution of a two-dimensional model. The Onsager solution shows a unique Curie point T_C, near which the specific heat diverges as $-\ln|1 - T/T_C|$. The explicit expression for the spontaneous magnetization, which exhibits a $(1 - T/T_C)^{1/8}$ behavior just below T_C, was derived by Yang in 1952. Although the problems of the three-dimensional Ising model and the two-dimensional model in a magnetic field still remain unsolved, the subject of the Ising model has since been extended to chemistry and metallurgy as well as biology, and developed into a major field of research.

See also FERROMAGNETISM.

BIBLIOGRAPHY

S. G. Brush, "History of the Lenz-Ising Model," *Rev. Mod. Phys.* **39**, 883 (1967). (E)

G. F. Newall and E. W. Montroll, "On the Theory of the Ising Model of Ferromagnetism," *Rev. Mod. Phys.* **25**, 353 (1953). (I)

C. Domb, "On the Theory of Cooperative Phenomena in Crystals," *Adv. Phys.* **9**, 149 (1960). (I)

H. S. Green and C. A. Hurst, *Order-Disorder Phenomena*. Interscience, New York, 1964. (A)

B. M. McCoy and T. T. Wu, *The Two-Dimensional Ising Model*. Harvard Univ. Press, Cambridge, Mass., 1973. (A)

Josephson Effects

D. J. Scalapino

Brian Josephson won the Nobel Prize in 1973 for theoretical work he carried out on the properties of two superconducting metals separated by a thin insulating oxide layer. This structure, a sandwich in which the two superconducting films are the bread with the oxide analogous to a very thin slice of cheese separating the metal films, is now often called a Josephson tunnel junction, and the phenomena associated with the superconducting current flow between the metals are called the Josephson effects. Figure 1 shows a typical junction configuration. Here two cross strips of Pb are separated by a thin oxide layer 10–20 Å in thickness.

In order to understand the Josephson phenomena it is essential to realize that the superconducting state exhibits quantum-mechanical properties on a macroscopic scale. In a superconducting metal, pairs of electrons are bound together by an attractive interaction mediated by the motion of the ionic lattice. According to the basic laws of quantum mechanics, the motion of the center of mass of the pairs is described by a wave. Now, ordinarily the quantum-mechanical waves of the many electrons which make up a metal differ from one another, so that on a macroscopic scale we are not directly aware of the phase properties of the electron wave functions. However, in the superconducting state, the state of lowest free energy corresponds to a situation in which all of the pairs have the same center-of-mass wave function. With this vast number of electron pairs having the same wave structure, it becomes possible to observe the relative change in phase $\Delta\varphi$ of the center-of-mass wave function.

The first Josephson equation states that if there is a phase difference $\Delta\varphi$ between the pair waves in the superconductors on the two sides of a Josephson junction, pairs will tunnel through the oxide giving rise to a supercurrent

$$I = I_1 \sin \Delta\varphi. \tag{1}$$

This is the dc Josephson effect. It predicts, for example, that if a phase difference $\Delta\varphi = \pi/2$ exists between the superconductors on either side of a Josephson junction, then a current I_1 will flow across the junction in the absence of a voltage difference. Naturally, one needs leads on the junction to supply and remove this current, or excess charge will build up giving rise to a voltage which produces further effects which we will discuss below.

Suppose leads from a current source are attached across the junction shown in Fig. 1 and another set of high-impedance leads are used to monitor the voltage V across the junction. The low-temperature $I(V)$ characteristic of the junction will look like that shown in Fig. 2. At low temperatures, the current I passing through the junction can be increased at zero voltage to a critical value beyond which the junction will switch (dashed line of Fig. 2) to a state with a finite voltage. Then as the current is reduced, the voltage will decrease toward zero as shown by the arrow on the solid line in Fig. 2. The zero-voltage current corresponds to the dc Josephson effect and was first reported by P. W. Anderson and J. M. Rowell. The magnitude of the critical current $I_1(T)$ is temperature dependent, vanishing above the superconducting transition temperature T_c and approaching a constant $I_1(0)$ at low temperatures. For a junction composed of the same type of superconducting metal on both sides, $I_1(0)$ is equal to the current which is carried by the junction in the normal (nonsuperconducting) state at a voltage $\pi\Delta/2e$, where Δ is the superconducting energy gap and $I_1(T)/I_1(0) = \tanh(\Delta/2kT)$. A typical Josephson junction will have a current density of order amperes per square centimeter while a very strongly coupled junction can have a current density several hundred times larger.

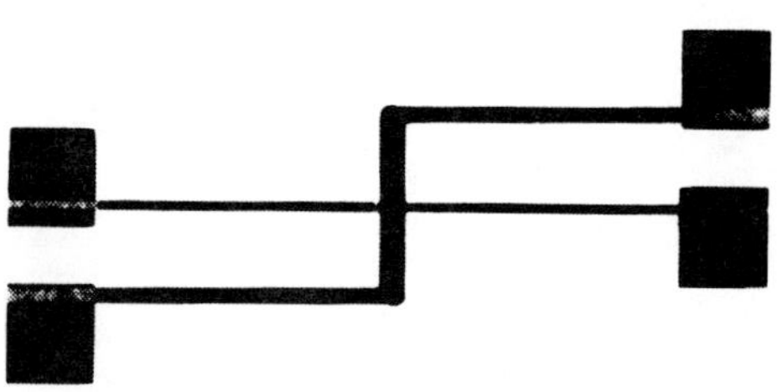

FIG. 1. A typical Josephson tunnel junction formed by two overlapping Pb strips separated by an oxide layer. The glass slide on which this junction is mounted is about an inch square.

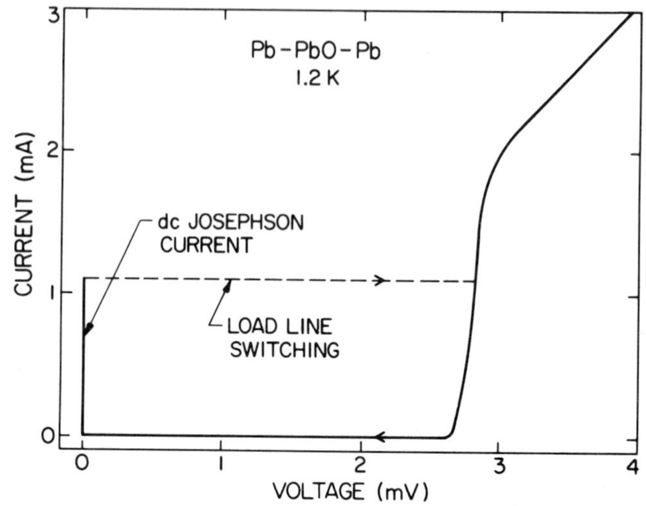

FIG. 2. The $I(V)$ characteristic of a Pb-PbO-Pb Josephson junction showing the dc Josephson current and the quasiparticle current onsetting at a voltage $2\Delta/e$.

If a voltage difference V exists across the junction, quantum mechanics implies that the relative difference $\Delta\varphi$ in the pair phase between the two superconductors changes at a rate set by $2eV/\hbar$:

$$\frac{\partial}{\partial t}\Delta\varphi = \frac{2eV}{\hbar}. \tag{2}$$

Here e is the electron charge, $\hbar$ is Planck's constant divided by 2π, and the 2 arises from the fact that we are dealing with electron pairs. If the voltage is constant at a value V_0, then the phase difference will increase with time, $\Delta\varphi(t) = 2eV_0t/\hbar + \Delta\varphi(0)$, and a current $I = I_1 \sin(2eV_0t/\hbar + \Delta\varphi)$ will oscillate back and forth across the oxide at a frequency $f = 2eV_0/h$. Using the values of e and h, one finds that this frequency is approximately 500 MHz/μV. This is the ac Josephson effect.

In order to observe the ac Josephson effect, Josephson suggested applying an rf field to the junction and looking for current steps in the dc $I(V)$ characteristic. With both a dc bias voltage V_0 and an rf bias voltage $V_1 \cos \omega t$ across the junction,

$$\frac{\partial \Delta\varphi}{\partial t} = \frac{2e}{\hbar}(V_0 + V_1 \cos \omega t)$$

and

$$I = I_1 \sin\left(\frac{2e}{\hbar} V_0 t + \frac{2eV_1}{\hbar\omega} \sin \omega t + \Delta\varphi(0)\right).$$

This current has a dc step whenever $2eV_0 = n\hbar\omega$. Figure 3 shows this step structure for a Josephson junction irradiated by a microwave field with frequency $f \sim 12$ GHz. With the approximate value of 500 MHz/μV, this implies a step spacing of order 25 μV. The ac Josephson effect was first observed in this way by S. Shapiro.

The frequency of the ac Josephson current can be measured with high precision. If, in addition, a careful measurement of the applied voltage V_0 is made, the ratio of the fundamental constants $2e/h$ can be obtained. Measurements of this type were carried out by Taylor, Parker, and Langenberg, who found that $2e/h = 4.835976(12) \times 10^{14}$ Hz/V_{NBS}. Here V_{NBS} is the NBS volt. Given a value of $2e/h$ and the ability to measure frequency, the Josephson relation $f = 2eV_0/h$ can be used to determine V_0. This type of measurement is, in fact, used to define the present NBS voltage standard.

The final Josephson relation concerns the behavior of a junction in a magnetic field. If a magnetic field H is applied parallel to the oxide layer, it again follows from quantum mechanics that the relative pair phase difference between

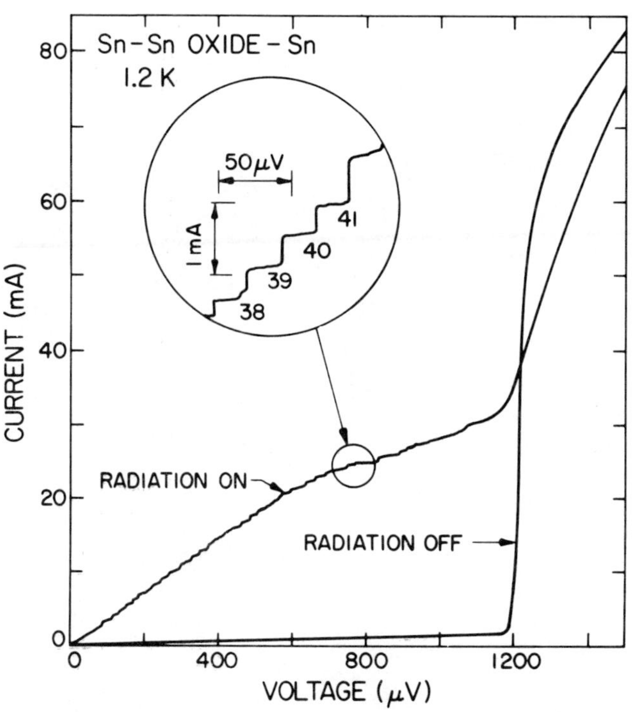

FIG. 3. The $I(V)$ characteristic of a Josephson junction in the presence of a microwave field of frequency f shows vertical steps separated by $hf/2e$.

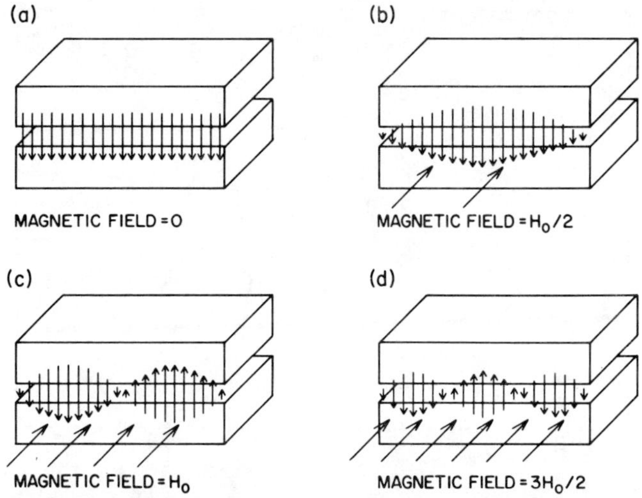

FIG. 4. Effect of a magnetic field on the current density in a Josephson junction.

the superconducting film on opposite sides of the junction varies with the position in the plane of the junction. Specifically, the change of $\Delta\varphi$ with distance is greatest as one moves in the plane of the junction perpendicular to the magnetic field. For identical superconducting films which are thick compared to the penetration depth λ of the magnetic field into the superconducting film, the change of $\Delta\varphi$ over a distance Δx perpendicular to the field H is $(2e/\hbar c)H(2\lambda + l)\Delta x$. Here l is the thickness of the oxide and $\hbar$ is Planck's constant divided by 2π. This means that the current density varies with position. For a junction lying in the x-y plane with a uniform field H in the oxide parallel to the y axis, the current density is

$$j(x) = j_1 \sin\left(\frac{2eH(2\lambda + l)x}{\hbar c} + \Delta\varphi(0)\right),$$

with $j_1 = I_1/A$, where A is the area of the junction.

This effect of a magnetic field H on the current density is illustrated in Fig. 4. In part a the magnetic field is zero, and the current density is uniform. As the field is increased, parts b–d, the current density oscillates in a direction perpendicular to H. The value H_0 is just such that $2eH_0(2\lambda + l)\cdot L/\hbar c = 2\pi$, where L is the length of the junction perpendicular to the field direction. This corresponds to one flux unit $hc/2e$ passing through the junction. When H is equal to H_0, Fig. 4c, the current density makes one complete oscillation and no net current flows across the junction. In Fig. 4d the field has been increased to $3H_0/2$ so that 3/2 wavelengths fit into the junction. In this case a net current smaller than that shown for Figures a or b can flow across the junction.

A plot showing how the critical junction dc current depends on H is shown in Fig. 5. This single-slit-like interference pattern, first observed by J. Rowell, shows in a vivid manner the de Broglie wave properties of electron pairs in superconductors. Thus, by monitoring the Josephson current, it is possible to tell when $(2\lambda + l)LH/(hc/2e) = 2\pi n$. The flux quantum $hc/2e$ corresponds to approximately 2×10^{-7} G/cm^2. Sensitive instruments can measure changes of parts in a thousand in the maximum Josephson current. Thus, Jo-

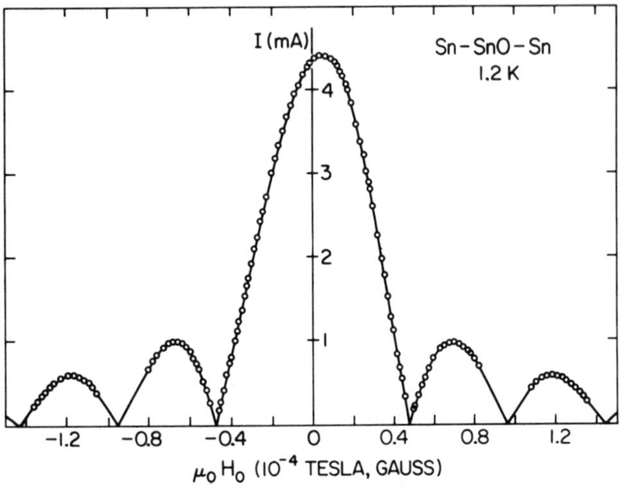

FIG. 5. The critical Josephson dc current versus magnetic field.

sephson junctions provide the means for making flux detectors with sensitivities of 10^{-10} G/cm^2.

Besides their use as magnetometers and voltage standards, Josephson junctions are now used in a variety of devices such as picovoltmeters and high-frequency electromagnetic detectors and mixers. Perhaps their most important applications will be as switching elements in which the junction switches from a zero-voltage current-carrying state to a finite-voltage current-carrying state (see the dashed line of Fig. 2) as the critical current I_1 is exceeded. This switching can occur in times of order tens of picoseconds and with power dissipation less than microwatts. While equivalently fast semiconductor switching elements can achieve similar switching times, their power dissipation is many orders of magnitude larger. Thus the density with which one can pack Josephson junctions can be much higher, reducing the time for electromagnetic signals to pass between them. At present, work on high-speed microcircuit technology using Josephson junctions is an active area of research.

ACKNOWLEDGMENT

The author thanks the University of Pennsylvania tunneling group for the experimental figures used. Support from the Office of Naval Research and the J. S. Guggenheim Foundation is gratefully acknowledged.

See also SUPERCONDUCTIVE DEVICES; SUPERCONDUCTIVITY THEORY.

BIBLIOGRAPHY

P. W. Anderson, in *Lectures on the Many-Body Problems* (E. R. Caianello, ed.). Academic Press, New York, 1964.

P. W. Anderson, *Physics Today*, (November, 1970).

B. D. Josephson, *Phys. Lett.* **1,** 251 (1962).

D. N. Langenberg, D. J. Scalapino, and B. N. Taylor, *Scientific American*, **214**(5), 30 (May, 1966).

Kerr Effect, Electro-optical

Robert E. Hebner, Jr.

The electro-optic Kerr effect is the name given to a modification of the index of refraction of a material by an electric field E. In general, the field induces a difference between two components of the index of refraction, Δn, which can be expressed as

$$\Delta n = \alpha + \beta E + \gamma E^2 + \cdots. \quad (1)$$

For the purpose of this description, the fact that the elements of this equation are tensors has been neglected and only scalar quantities are considered. If only the quadratic term is nonzero, Eq. (1) describes the Kerr effect. Similarly, if the first-order term is the only one which is nonzero, Eq. (1) describes the Pockels effect.

For the Kerr effect, Eq. (1) can be written as

$$n_{\parallel} - n_{\perp} = \lambda B E^2, \quad (2)$$

where $n_{\parallel}$ and $n_{\perp}$ are the indices of refraction of the electro-optic material for light polarized parallel and perpendicular to the electric field, respectively; λ is the wavelength of the transmitted light; B is the Kerr coefficient of the material; and E is the applied electric field.

Measurements of the difference between the indices of refraction are usually made using a laser—a convenient monochromatic light source.

The Kerr effect can be modeled as dealing with two different electric fields: one called the orienting field and the other the sensing field. To appreciate the distinction between the two, it is useful to consider, qualitatively, the behavior of a system on a molecular scale. Assume that the molecules in question have some electrical anisotropy, e.g., a dipole moment. If a low-intensity, linearly polarized light beam is passed through the material, no net effect of the anisotropy is observed because the dipoles are randomly oriented. The electric field associated with this light beam is the sensing field. If a second field, with sufficiently high intensity, is applied, a detectable alignment of the dipoles results. This second field is the orienting field. It should be emphasized that the distinction between the orienting and sensing field is based solely on intensity.

It should be recognized that the Kerr effect is primarily due to molecular reorientation for orienting fields with temporal variations slow compared to molecular relaxation time and due to molecular deformation for high-frequency orienting fields, e.g., high-power laser beams.

Values for Kerr coefficients range from 10^{-21} m/V^2 for helium gas, to 10^{-12} m/V^2 for nitrobenzene, to 10^{-7} m/V^2 for bentonite in water. These depend on temperature, wavelength of the light, and frequency of the orienting field.

The Kerr effect has been used in a variety of applications. It is the basis for a very accurate measurement technique for high-voltage pulses. The Kerr effect has been used to measure the electric field and space-charge distributions in insulating materials. Measurements of the Kerr effect have provided insight into molecular structure and into molecule–molecule interactions in the liquid and gaseous states. Optical shutters and digital light beam deflectors based on the Kerr effect have been constructed using a variety of electro-optic materials.

See also DIELECTRIC EFFECTS; KERR EFFECT, MAGNETO-OPTICAL.

BIBLIOGRAPHY

Elementary

F. A. Jenkins and H. E. White, *Fundamentals of Optics*, 3rd ed. McGraw-Hill, New York, 1976.

R. W. Wood, *Physical Optics*. Dover Publications, New York, 1967.

Intermediate

E. Fredericq and C. Houssier, *Electric Dichroism and Electric Birefringence*. Clarendon Press, Oxford, 1973.

R. E. Hebner *et al.*, Proc. IEEE **65**, 1524 (1977).

C. G. LeFevre and R. J. W. LeFevre, *Rev. Pure Appl. Chem.* **5**, 261 (1955).

C. T. O'Konski, *Molecular Electro-Optics: Part I—Theory and Methods*. Marcel Dekker, New York, 1976.

Kerr Effect, Magneto-optical

W. J. Tabor

The Kerr magneto-optical effect is a name given to a phenomenon that takes place when polarized light is reflected from a magnetized surface. It is named after John Kerr (1824–1907), a Scottish physicist. For example, when a plane-polarized light wave is incident in a perpendicular direction on a polished magnetized body, the polarization direction of the reflected wave is rotated slightly with respect to the polarization direction of the incident wave. The rotation of the plane of polarization is called the Kerr rotation. The phenomenon of the Kerr effect is closely related to the Faraday effect in that both effects involve the interaction of an optical wave with a magnetized medium.

The Kerr effect is quantitatively defined in terms of the following relationships:

$$E_p{}^r = (r_{pp} + \delta)E_p{}^i + r_{ps}E_s{}^i,$$

$$E_s{}^r = r_{sp}E_p{}^i + r_{ss}E_s{}^i,$$

where $E_p{}^i$ and $E_s{}^i$ are the electric vectors of the incident light beam with the subscript p referring to that component in the plane of incidence and s referring to that component perpendicular to the plane of incidence. In Fig. 1, the points *IOR* define the plane of incidence. In a similar way $E_p{}^r$, $E_s{}^r$ define the electric vector of the reflected wave. The terms r_{ps}, r_{sp}, and δ are the Kerr coefficients and the terms r_{pp}, r_{ss} are the Fresnel coefficients which describe the reflection from ordinary dielectrics and metals. Thus, reflection from a magnetized surface is a combination of Kerr and Fresnel effects. In general, the terms r_{ps}, r_{sp}, and δ are functions of the material, magnetization direction, temperature, incident angle, and the wavelength of the light used in the observation.

Three limiting cases of the Kerr effect are generally defined. The first is the polar Kerr effect which takes place

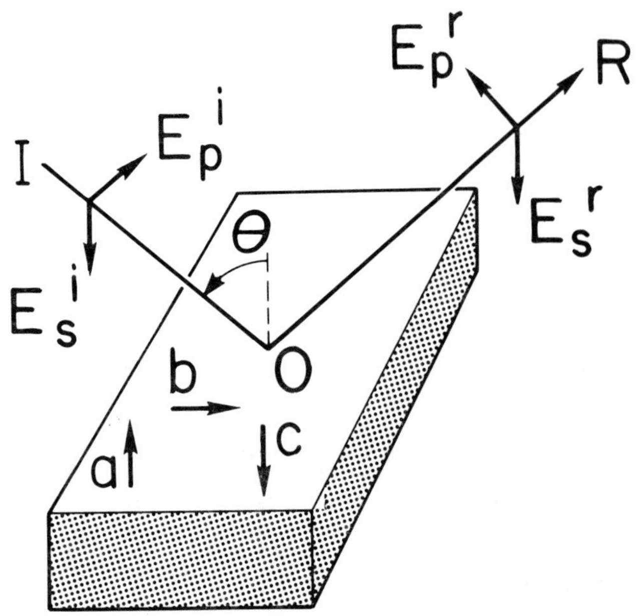

FIG. 1. Possible Kerr effects: (a) polar, where the magnetization vector is perpendicular to the surface; (b) longitudinal, where the magnetization vector is parallel to both the surface of the material and to the plane of incidence; (c) transverse, where the magnetization vector is parallel to the surface of the material and perpendicular to the plane of incidence. *I* is the incident beam of light, *R* is the reflected beam. The plane containing the points *IOR* is the plane of incidence.

when the magnetization direction is perpendicular to the surface of the magnetized body (case *a* of Fig. 1). For this case $r_{ps}, r_{sp} \neq 0$, $\delta = 0$, and the polar Kerr angle Φ_p, which is the angle that the polarization direction of the reflected wave is rotated with respect to the polarization direction of the incident wave, is given by $\tan\Phi_p = r_{ps}/r_{ss}$ for an incident E_s wave.

The second case is called the longitudinal Kerr effect and takes place when the magnetization direction is parallel to the surface of the magnetized body and parallel to the plane of incidence (case *b* of Fig. 1). In this case $r_{ps}, r_{sp} \neq 0$, $\delta = 0$, and the longitudinal Kerr angle Φ_l is given by $\tan\Phi_l = r_{ps}/r_{ss}$ for an incident E_s wave. Although this formula has the same form as the polar case above, it should be remembered that the coefficient r_{ps} will in general have different values.

The third case is called the transverse Kerr effect and takes place when the magnetization direction is parallel to the surface of the magnetized body and perpendicular to the plane of incidence (case *c* of Fig. 1). In this case $r_{sp} = r_{ps} = 0$, $\delta \neq 0$, and instead of a rotation of the plane of polarization, this Kerr effect is a change in the reflection coefficient of the E_p wave depending on whether the magnetization vector points to the right or left of the plane of incidence, δ being positive for one of those directions and negative for the other. The magnitude of the transverse Kerr effect is given by δ/r_{pp}.

In general the Kerr effects are all small; for example, the polar Kerr angles for perpendicular incidence for iron, cobalt, and nickel are typically less than a degree in the visible range of the spectrum. For the same metals and wavelengths the transverse Kerr effect, δ/r_{pp}, is 1% or less.

The Kerr effect is used mostly for the observation of magnetic domains at the surface of a magnetized body. If such a body is illuminated with polarized light and observed through a polarizing filter, the different domains will have varying degrees of brightness.

The Kerr effect is also used to study the properties of magnetic materials that are very opaque to light thereby making them impractical for study by means of the Faraday effect.

See also FERROMAGNETISM; MAGNETIC DOMAINS AND BUBBLES.

BIBLIOGRAPHY

R. Carey and E. D. Isaac, *Magnetic Domains and Techniques for Their Observation*. Academic Press, New York, 1966. (E)

R. Hunt, *IEEE Trans. Magnetics* **5,** 700 (1969). (A)

R. P. Hunt, *J. Appl. Phys.* **38,** 1652 (1967). (A)

W. J. Tabor, *Laser Handbook*, Vol. 1. North Holland, Amsterdam, 1972. Kerr data on a variety of materials can be found in this reference.

Kondo Effect

Alan S. Edelstein

In many cases the temperature dependence of the resistivity ρ of a metal containing a small concentration of magnetic impurities has a minimum. At temperatures below the minimum the resistance increases logarithmically and then becomes temperature independent. This effect is named after J. Kondo who in 1964 first performed a perturbation-theory calculation that correctly predicted $\rho \propto \ln T$. Examples of systems exhibiting this behavior are *Cu*Fe, *Cu*Cr, and *Au*Fe. The magnitude of this resistance increase is usually small, at most several percent, and is certainly not the cause for the great interest in the Kondo effect.

The major reasons for the interest in the Kondo effect are: (1) Many systems which are nearly nonmagnetic exhibit the effect. These systems behave as though they have a stable magnetic moment at high temperatures, but this moment is reduced at low temperatures. Hence, it is crucial to understand the Kondo effect if one wants to understand magnetic moment formation in metals. (2) The complete understanding of this effect requires solving a difficult many-body problem. This many-body problem has required the refinement and development of new theoretical techniques which may be applicable to other problems. (3) Interesting phenomena occur in Kondo systems. For example, it is possible for them to be superconducting and yet not possess a superconducting energy gap. As will be discussed later, some Kondo systems undergo a transition into a superconducting state at one temperature and, at a lower temperature, undergo another transition out of the superconducting state.

Resistivity minima had been observed prior to Kondo's calculation, but how they occurred was not understood. A conduction electron was believed to interact with the magnetic moment via a term in the energy of the system of the form

$$E_{\text{int}} = -2J\mathbf{S}\cdot\mathbf{s}, \tag{1}$$

where J is the exchange integral and $\mathbf{S}$ and $\mathbf{s}$ are the spins of the impurity and the conduction electron, respectively. Equation (1) is called the exchange Hamiltonian. A first Born-approximation perturbation calculation using Eq. (1) of the change in the resistivity $\Delta\rho$ due to the impurity yields $\Delta\rho$ independent of T. Kondo calculated the resistivity using the second Born approximation in a perturbation theory calculation to show that

$$\Delta\rho = c\rho_m[1 + N(E_F)J\ \ln(kT/D)], \tag{2}$$

where c is the impurity concentration, ρ_m is the result of using just the first Born approximation, and $N(E_F)$ is the conduction-electron density of states at the Fermi level E_F. For simplicity Kondo assumed that the conduction electrons occupied a flat band, i.e., N is independent of E, of width $2D$. One sees from Eq. (2) that $\rho\to\infty$ for $T<<T_K$ and $J<0$, where

$$kT_K = D\ \exp[a/JN(E_F)] \tag{3}$$

and a is a constant of order 1 which is different in different theories. Since $\Delta\rho$ cannot diverge, this result shows that perturbation theory fails for $T<<T_K$. The value of T_K can vary drastically depending on the impurity and host. For dilute Mn in Cu, $T_K<0.01$ K, while for dilute Fe in Cu, $T_K\approx 30$ K. The value of T_K is important in that it determines the energy and temperature scale; i.e., $T/T_K \gtrless 1$ and $E/kT_K \gtrless 1$ correspond to ($^{\text{high}}_{\text{low}}$) temperatures and energies, respectively. Much work has been done to develop a theory that is valid at all temperatures. The most promising work is based on renormalization-group techniques. Even with this approach, calculations have been limited to a few properties such as the susceptibility and have not been generalized for $S\neq\frac{1}{2}$.

The exchange form of the interaction energy, Eq. (1), can, in a certain limit, be derived from the more fundamental formulation of P. W. Anderson. Anderson was interested in local magnetic moment formation of dilute 3d transition-element impurities in a nonmagnetic host. In the dilute limit one need only calculate the contribution of a single, isolated impurity. He considered the energy of the total system to consist of a conduction-electron contribution, an impurity contribution, and an interaction term between the impurities and the conduction electrons. The impurity contribution was taken to be a sum over σ of $\epsilon_d n_{d\sigma} + U n_{d\sigma} n_{d-\sigma}$, where ϵ_d is the energy of the d state, $n_{d\sigma}$ is the number of d electrons of spin σ occupying the impurity site, and U, which is positive, is the Coulomb repulsion between d electrons of opposite spin. For a nondegenerate d orbital, double occupancy by similar spins is not allowed because of the Pauli exclusion principle. The interaction term allows an electron to hop on or off the impurity site without changing its spin. This interaction causes the original sharp impurity level to have a width Δ proportional to $|V_{dk}|^2$, where $|V_{dk}|$ is the magnitude of the matrix element involved in the hopping. For $\Delta<<U$ one can transform the terms in Anderson's Hamiltonian to the form of Eq. (1). The exchange integral that then appears in Eq. (1) is the sum of the positive, direct exchange integral J_d and the one that results from transforming the Anderson Hamiltonian, i.e., $J = J_d + J_{\text{eff}}$, where

$$J_{\text{eff}} = \frac{|V_{dk}|^2 U}{E_{df}(E_{df}+U)}, \quad E_{df} = E_d - E_F < 0 \tag{4}$$

The conditions which most favor moment formation are $\Delta<<U$ and $E_{df}/U = -\frac{1}{2}$. Because of intra-atomic exchange, if one adds degeneracy or multiple occupancy to the model, this addition assists moment formation.

For free moments the susceptibility χ per impurity of a substance having a magnetic moment μ per impurity is given by Curie's law, $\chi = \mu^2/3kT$, where k is Boltzmann's constant and T is the absolute temperature. Interactions with the crystal field and the conduction electrons can reduce the value of χ. In Kondo systems for $T<<T_K$, χ is less than the value predicted by Curie's law and approaches a constant χ_0 as $T\to 0$. One can define an effective moment $\mu_{\text{eff}}^2 = 3kT\chi$ as a measure of how these interactions can influence μ. For Kondo systems it has been proposed that the value of μ_{eff}^2 might be reduced by the following three mechanisms:

(1) The crystal field of the neighboring ions.

(2) The formation of a singlet spin-compensated state. In the Kondo problem, the basic energy scale is determined by kT_K. For $T/T_K<<1$ the amount by which the conduction electrons can change their energy is very small compared with kT_K. Numerical calculations using the renormalization group show that the distribution of allowed energy states in this case is similar to that which would exist if $|J|$ in Eq. (1) were to become infinite. Since $T\to 0$ is like $|J|\to\infty$ in Eq. (1), the coupling between a spin-$\frac{1}{2}$ impurity and a conduction electron becomes very strong. This has been interpreted to mean that there is always a conduction electron paired with its spin antiparallel to that of the impurity. It must be paired antiparallel to minimize E_{int} in Eq. (1). Thus, the ground state is a nonmagnetic singlet. It has been suggested that if J is large but finite, $0<T<<T_K$, real transitions out of the ground state are impossible since they require a finite amount of energy. Virtual transitions are possible and the impurity is polarizable. This leads to a repulsive interaction between the conduction electrons and to a finite, temperature-independent value for the susceptibility for $T<<T_K$. Hence $\mu_{\text{eff}}\to 0$ as $T\to 0$. Since a conduction electron is paired with the impurity, it has been suggested that only the net moment is zero in zero external magnetic field, but that its distribution might be as sketched in Fig. 1. The large positive contribution at small r due to the local moment is compensated by the negative, i.e., antiparallel, conduction-electron contribution at large r. Because the impurity is polarizable, the net moment is nonzero in a magnetic field. As we shall discuss, the existence of a large negative conduction-electron polarization is not supported by most experiments. It has been suggested that by exciting the system it is possible to recover the original moment unaffected by the conduction electrons. This effect in principle can be observed in neutron-scattering experiments, but these are difficult experiments since the small concentrations required to avoid interaction effects will result in weak signals.

(3) A finite value for the intrinsic spin correlation time τ, where $\tau\approx h/kT_K$ and h is Planck's constant. The susceptibility

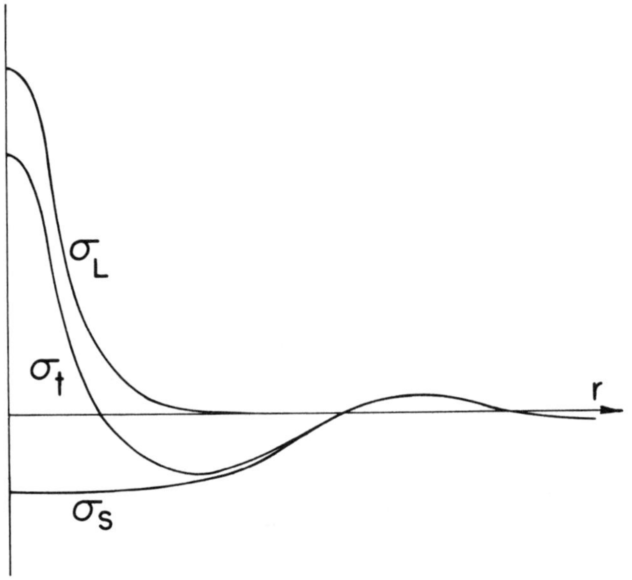

FIG. 1. Sketch of the radial distribution for a singlet spin-compensated state of the local-moment spin density $\sigma_L(r)$, conduction-electron spin density $\sigma_s(r)$, and total spin density $\sigma_t(r)=\sigma_L(r)+\sigma_s(r)$ as a function of r, the distance from the impurity. The net spin in zero magnetic field, the integral of $\sigma_t(r)$ over a unit cell, is zero. The net spin in an applied field is nonzero.

of a dilute impurity is the time average of $\langle S(0)S(t)\rangle$ from 0 to a time $t \approx h/kT$. If a spin only stays correlated with itself a time τ, then for $T>>T_K$ the susceptibility is approximately given by Curie's law, while for $T<<T_K$ the susceptibility is given by $\chi \approx \mu^2\tau/3h$. Hence, $\mu_{eff} \propto (T\tau)^{1/2}$ for $T<<T_K$. Some theoretical work on the Kondo problem has taken this spin-fluctuation view. It is given this name because it is supposed that the interactions with the conduction electrons cause spin fluctuations which limit τ.

Since the Kondo temperature T_K sets the energy scale, it also determines the temperature scale for experimental properties. Thus, one expects these properties to be approximately functions of just T/T_K. These properties also depend on the potential scattering and the crystal-field splitting if the orbital moment is not quenched. No sharp phase change occurs at T_K, since an isolated impurity does not have enough degrees of freedom to permit a sharp phase transition.

A great variety of experiments can be performed. One can vary the impurity, the host, the concentration of impurities, and such parameters as the temperature and the magnetic field. The experiments can be divided into microscopic and macroscopic measurements. For most properties the single-impurity contribution must be proportional to the impurity concentration. For many systems one must employ concentrations less than 100 ppm to observe the single-impurity contribution. Early experimental results were often complicated by interaction effects.

Table I shows the qualitative behavior observed for dilute impurities in various macroscopic measurements. The experimental properties gradually change from their high- to their low-temperature behavior. There is a broad specific-heat anomaly which occurs over several decades in T/T_K and is centered at approximately $T_K/3$. The entropy change derived from the area under a plot of C/T vs. T, where C is the impurity specific heat, is within the experimental uncertainty equal to $R \ln(2S+1)$. This is consistent with a singlet ground state. Theoretical calculations have predicted most of the behavior listed in Table I. For example, renormalization-group calculations of the susceptibility using either the exchange Hamiltonian, Eq. (1), or a special case of the Anderson model ($E_{df}/U=\frac{1}{2}$) discussed earlier yield the results listed in Table I. These calculations differ from earlier ones in that they are valid both above and below T_K. They predict that χ is a universal function of T/T_K. It is interesting that the low-temperature susceptibility of several Kondo systems can be fitted by this function of T/T_K. This agreement is somewhat surprising since these systems do not satisfy the conditions of the calculations, e.g., they are not $S=\frac{1}{2}$ systems. Unfortunately, the properties in Table I are not too useful in distinguishing whether mechanism 2 or mechanism 3, discussed above, is primarily responsible for reducing the effective magnetic moment μ_{eff} at low temperatures since both of these different physical descriptions lead to the properties listed in Table I.

Magnetic impurities drastically affect the properties of superconductors. The electrons in superconductors condense into states in which pairs of electrons are bound and have their spins oppositely directed. Magnetic impurities break this symmetry property and cause a large depression in the superconducting transition temperature T_c. Kondo impurities cause an especially large depression in T_c and gapless superconductivity at very low impurity concentrations. The impurities' tendency to inhibit superconductivity is greatest at $T \approx T_K$. Thus if T_K is somewhat smaller than the superconducting transition temperature T_{c0} of the pure system, it is possible for the system to undergo a transition into the superconducting state at a temperature T_{c1} above which the impurity has its full effect and then undergo a second transition out of the superconducting state at a temperature $T_{c2} \approx T_K$. These effects have been observed in (La,Ce)Al_2 and (La,Th):Ce. There is some question as to whether these systems might undergo a third transition back into the superconducting state for $T<T_K$.

Table I Temperature Dependence of Macroscopic Properties[a]

	Resistivity (ρ)	Thermal conductivity (W)	Susceptibility (χ)	Specific heat (C)
$T>>T_K$	See Eq. (2)	$\propto(1/T)(1+d\ln T)$	$\mu^2/3k(T+T_K)$	—
$T<<T_K$	$\propto(1-bT^2/T_K^2)$	$\propto 1/T$	$\chi_0(1-fT^2/T_K^2)$	$\propto T$

[a] b, d, and f are constants.

Unlike macroscopic measurements, which have not yet distinguished between the singlet spin-compensated state and the finite spin-correlation-time mechanisms, microscopic measurements can in principle make this distinction. Some of the microscopic measurements that have been employed are the Mössbauer effect, γ-ray angular correlation studies, de Haas–van Alphen effect, nuclear magnetic resonance (NMR), and neutron scattering. We will discuss primarily the latter two as applied to the *Cu*:Fe system. Nuclear-magnetic-resonance measurements have been performed on the Cu nuclei, and it is possible to detect separate resonance signals from Cu nuclei which are different distances r_i from the Fe impurities. Signals from the first through fourth nearest neighbors of the Fe impurities have been separated from the main resonance of most of the Cu nuclei. From the temperature dependence of these signals it is possible to conclude that the magnetization or spin density has the form $\sigma(r,T)=f(r)\chi(T)$; i.e., it does not change its shape at the position of these neighbors. This provides evidence against the existence of a compensating antiparallel polarization of the conduction electrons which forms for $T<T_K$. If such a polarization formed, σ would change its shape. The NMR measurements have also determined the value of τ of Fe in Cu:Fe.

Neutron-diffraction studies have been made on Cu:Fe samples containing 400–1000 ppm impurities. These concentrations are too large to permit one to observe only the effects of single impurities of Fe but were necessary in order to have sufficiently large signals. A reasonable extrapolation of the neutron-diffraction data to lower concentrations leads one to conclude that σ has changed its shape for $T<T_K$ and hence provides evidence for the conduction electrons compensating the moment as in Fig. 1. Neutron measurements determine $\sigma(r)$ at smaller distances than the NMR measurements, so these two measurements are not necessarily in conflict. Mössbauer measurements, however, do determine a local susceptibility. Mössbauer measurements on dilute Cu:Fe samples also do not show evidence of a large negative conduction-electron polarization. Neutron-diffraction measurements on Y:Ce also fail to show a large negative conduction electron polarization.

NMR and other neutron measurements have been made to study the dynamics. These measurements are probably consistent with the finite spin-correlation-time mechanism. It is not yet clear whether they are consistent with the singlet-spin-compensated mechanism. Renormalization-group calculations of the dynamics have not yet been performed.

See also CONDUCTION; DIAMAGNETISM.

BIBLIOGRAPHY

C. Rizzuto, *Rep. Prog. Phys.* **37,** 147 (1974); experimental review, I.

A. J. Heeger, *Solid State Physics* Vol. 23, p. 284. (F. Seitz, D. Turnball, and H. Ehrenreich, eds.), Academic Press, New York, 1969; experimental review, I.

J. Kondo, *Solid State Physics* Vol. 23, p. 184. (F. Seitz, D. Turnbull, and H. Ehrenreich, eds.), Academic Press, New York, 1969; theoretical review, I.

G. Grunier and A. Zawadowski, *Rep. Prog. Phys.* **37,** 1497 (1974); theoretical review, A.

H. Suhl (ed.), *Magnetism,* Vol. V. Academic Press, New York, 1973; both theoretical and experimental reviews which include the effect on superconductivity, I and A.

K. G. Wilson, *Rev. Mod. Phys.* **47,** 773 (1975); review of renormalization theory, A.

P. Nozieres, *J. Low Temp. Phys.* **17,** 31 (1974); Fermi liquid description of the problem, I.

C. Kittel, *Introduction to Solid State Physics,* 4th ed., p. 660. Wiley, New York, 1971; E.

Lattice Dynamics

Richard F. Wallis

1. INTRODUCTION

Lattice dynamics concerns the vibrations of the atoms of a solid about their equilibrium positions and the effect of these vibrations on the properties of the solid. In the adiabatic approximation [1] the total wave function of the solid, $\Psi(\mathbf{r},\mathbf{R})$, is written as the product of the electronic wave function, $\psi(\mathbf{r},\mathbf{R})$, and the nuclear wave function, $\chi(\mathbf{R})$, where $\mathbf{r}$ represents all the electronic coordinates and $\mathbf{R}$ all the nuclear coordinates. The term adiabatic signifies that the nuclear motion does not induce electronic transitions. The vibrational motion of the atoms is determined by the nuclear potential energy, which is the electronic energy eigenvalue. The latter is a function of the nuclear coordinates $\mathbf{R}$.

2. LATTICE-DYNAMICAL COUPLING CONSTANTS

We assume that the crystal is in a nondegenerate electronic ground state and that the atoms execute small vibrations about their equilibrium positions. The nuclear potential energy, $\Phi(\mathbf{R})$, may then be expanded in powers of the displacement components of the nuclei from their equilibrium positions:

$$\begin{aligned}\Phi(\mathbf{R}) &= \Phi(\mathbf{R}^{(0)}) + \sum_{l\kappa\alpha} \Phi_\alpha(l\kappa)u_\alpha(l\kappa) \\ &+ \tfrac{1}{2}\sum_{l\kappa\alpha}\sum_{l'\kappa'\beta} \Phi_{\alpha\beta}(ll',\kappa\kappa')u_\alpha(l\kappa)u_\beta(l'\kappa') \\ &+ \tfrac{1}{6}\sum_{l\kappa\alpha}\sum_{l'\kappa'\beta}\sum_{l''\kappa''\gamma} \Phi_{\alpha\beta\gamma}(ll'l'',\kappa\kappa'\kappa'') \\ &\times u_\alpha(l\kappa)u_\beta(l'\kappa')u_\gamma(l''\kappa'') + \cdots .\end{aligned} \tag{2.1}$$

In Eq. (2.1), $u_\alpha(l\kappa) = R_\alpha(l\kappa) - R_\alpha^{(0)}(l\kappa)$ is the αth Cartesian component of the displacement from equilibrium of the κth nucleus in the lth primitive unit cell and the superscript zero denotes the equilibrium configuration. The equilibrium condition requires that the quantities $\Phi_\alpha(l\kappa)=0$. The $\Phi_{\alpha\beta}(ll',\kappa\kappa')$ are the *harmonic* coupling constants and the $\Phi_{\alpha\beta\gamma}(ll'l'',\kappa\kappa'\kappa'')$, . . ., are the *anharmonic* coupling constants. The coupling constants are partial derivatives of the potential energy and therefore satisfy symmetry conditions such as $\Phi_{\alpha\beta}(ll',\kappa\kappa') = \Phi_{\beta\alpha}(l'l,\kappa'\kappa)$. Other conditions [1,2] satisfied by the coupling constants are those imposed by infinitesimal translational invariance and infinitesimal rotational invariance. For the harmonic coupling constants, these conditions take the respective forms

$$\sum_{l'\kappa'} \Phi_{\alpha\beta}(ll',\kappa\kappa') = 0 \tag{2.2}$$

and

$$\sum_{l'\kappa'} \{\Phi_{\alpha\beta}(ll',\kappa\kappa')X_\mu(l'l,\kappa'\kappa) - \Phi_{\alpha\mu}(ll',\kappa\kappa')X_\beta(l'l,\kappa'\kappa)\} = 0 \tag{2.3}$$

where $X_\mu(l'l,\kappa'\kappa) = R_\mu^{(0)}(l'\kappa') - R_\mu^{(0)}(l\kappa)$. Analogous equations are satisfied by the anharmonic coupling constants. Additional conditions on the coupling constants are provided by utilizing the point symmetry and translational symmetry of the crystal [2].

Typically, the nuclear potential energy of a crystal has several physical origins. These include the Coulomb interaction between electrically charged ions, van der Waals interactions, interactions associated with chemical bonds, and the short-range quantum-mechanical repulsion between atoms due to the Pauli principle.

3. EQUATIONS OF MOTION AND NORMAL COORDINATE TRANSFORMATION

The classical equations of motion for the nuclei can be written as

$$M_\kappa \ddot{u}_\alpha(l\kappa) = -\frac{\partial\Phi(\mathbf{R})}{\partial u_\alpha(l\kappa)} \tag{3.1}$$

where the double dot above the u represents a second time derivative. Making use of Eq. (2.1) and retaining only the harmonic terms, we obtain

$$M_\kappa \ddot{u}_\alpha(l\kappa) = -\sum_{l'\kappa'\beta} \Phi_{\alpha\beta}(ll',\kappa\kappa')u_\beta(l'\kappa'). \tag{3.2}$$

These equations corresponding to the various values of l, κ, and α describe a system of coupled harmonic oscillators. Their solution is facilitated by the substitution $u_\alpha(l\kappa) = M_\kappa^{-1/2} \times w_\alpha(l\kappa)\exp(i\omega t)$, which leads to

$$\omega^2 w_\alpha(l\kappa) = \sum_{l'\kappa'\beta} D_{\alpha\beta}(ll',\kappa\kappa')w_\beta(l'\kappa') \tag{3.3}$$

where the quantities $D_{\alpha\beta}(ll',\kappa\kappa') = (M_\kappa M_{\kappa'})^{-1/2}\Phi_{\alpha\beta}(ll',\kappa\kappa')$ are the elements of the *dynamical matrix*. For a nontrivial solution, the determinant of coefficients of the $w_\alpha(l\kappa)$ in Eq. (3.3) must be zero,

$$|D_{\alpha\beta}(ll',\kappa\kappa') - \omega^2\delta_{ll'}\delta_{\kappa\kappa'}\delta_{\alpha\beta}| = 0, \tag{3.4}$$

where the δ_{ij}'s are Kronecker deltas. Equation (3.4) is applicable to any crystal, perfect or imperfect, and specifies the normal-mode frequencies of the crystal. If the crystal contains N atoms, there are $3N$ normal modes and hence $3N$ solutions for ω to Eq. (3.4). We shall distinguish the various normal modes by the index s, $1 \le s \le 3N$. Typically, the normal-mode frequencies range from zero up to the infrared region.

For a macroscopic crystal, Eq. (3.4) is a polynomial equation in ω^2 of very high order. Calculations are simplified for a perfect crystal if we introduce *periodic boundary conditions*, justified by Ledermann's theorem [2], and write

$$\omega_\alpha(l\kappa) = W_{\alpha\kappa}(\mathbf{k})\exp[i\mathbf{k}\cdot\mathbf{R}^{(0)}(l)] \tag{3.5}$$

where $\mathbf{k}$ is the wave vector. Periodic boundary conditions require that the atoms at corresponding points on opposite faces of the crystal have the same displacements. There are $3N$ values of $\mathbf{k}$ that occupy the first Brillouin zone and that specify the independent normal modes [2]. The analog of Eq. (3.4) becomes

$$|D_{\alpha\beta}(\mathbf{k},\kappa\kappa') - \omega^2\delta_{\alpha\beta}\delta_{\kappa\kappa'}| = 0 \tag{3.6}$$

where $D_{\alpha\beta}(\mathbf{k},\kappa\kappa') = \Sigma_{l'}D_{\alpha\beta}(ll',\kappa\kappa')\exp[i\mathbf{k}\cdot(\mathbf{R}^{(0)}(l') - \mathbf{R}^{(0)}(l))]$

is independent of l. If there are r atoms per unit cell, the determinant in Eq. (3.6) is $3r \times 3r$ in size, far smaller than that in Eq. (3.4) and therefore more suitable for computation.

Let us denote the eigenvectors and eigenvalues of $D(\mathbf{k}, \kappa\kappa')$ by $\mathbf{e}_\kappa(\mathbf{k}j)$ and $\omega^2(\mathbf{k}j)$, respectively, where j is the branch index. The normal coordinates, $Q(\mathbf{k}j)$, can be introduced through the transformation

$$w_\alpha(l\kappa) = N^{-1/2} \sum_{\mathbf{k}j} e_{\alpha\kappa}(\mathbf{k}j) Q(\mathbf{k}j) \exp[i\mathbf{k}\cdot\mathbf{R}^{(0)}(l)]. \quad (3.7)$$

The Hamiltonian for the harmonic crystal can be written as

$$H = \tfrac{1}{2} \sum_{\mathbf{k}j} \{|\dot{Q}(\mathbf{k}j)|^2 + \omega^2(\mathbf{k}j)|Q(\mathbf{k}j)|^2\}, \quad (3.8)$$

which consists of the sum of contributions from independent harmonic oscillators. In the quantum-mechanical theory in the absence of anharmonicity, the nuclear wave function is a product of harmonic oscillator wave functions. The energy eigenvalue can be written as

$$E = \sum_{\mathbf{k}j} \hbar\omega(\mathbf{k}j)[n(\mathbf{k}j) + \tfrac{1}{2}] \quad (3.9)$$

where $n(\mathbf{k}j)$ is the harmonic oscillator quantum number for the mode $\mathbf{k}j$. The quantum of excitation energy for a normal vibrational mode is called a *phonon*.

4. *PHONON DISPERSION CURVES*

The solutions of Eq. (3.6) determine the phonon dispersion curves, i.e., the normal-mode frequencies $\omega(\mathbf{k}j)$ as functions of the wave vector $\mathbf{k}$ for the various branches j. As a simple example, we give the normal-mode frequencies for a monatomic linear chain with atomic mass M and nearest-neighbor interactions characterized by the coupling constants $\Phi_{xx}(l,l+1) = \Phi_{xx}(l,l-1) = -\tfrac{1}{2}\Phi_{xx}(l,l) = -\sigma$:

$$\omega(k) = (4\sigma/M)^{1/2} \sin(ak/2) \quad (4.1)$$

where a is the lattice constant. In this case we have only one atom per primitive unit cell and only one branch. For a crystal with r atoms per unit cell, there are $3r$ branches of which three are acoustical branches with $\omega(\mathbf{k}j) \to 0$ as $|\mathbf{k}| \to 0$ and $3r-3$ are optical branches. In crystals of high symmetry with $\mathbf{k}$ in a high-symmetry direction, the branches can be classified as longitudinal (displacements $\mathbf{u}$ parallel to $\mathbf{k}$) or transverse ($\mathbf{u}$ perpendicular to $\mathbf{k}$). The phonon frequencies associated with a given branch are restricted to a finite range or "band." There may be gaps between the various phonon bands.

Phonon dispersion curves can be determined experimentally by means of the coherent inelastic scattering of cold neutrons or by the thermal diffuse scattering of x rays. Inelastic neutron scattering [3] is the more widely used method. If the incident and scattered neutron energies are E_0 and E_s, respectively, the energy, $\hbar\omega(\mathbf{q}j)$, of the phonon created or destroyed in the scattering process is given by conservation of energy as

$$E_0 - E_s = \pm\hbar\omega(\mathbf{q}j) \quad (4.2)$$

where the plus sign refers to the creation and the minus sign to the destruction of the phonon. In addition, conservation of momentum requires that

$$\mathbf{k}_0 - \mathbf{k}_s = \mathbf{q} + \mathbf{G} \quad (4.3)$$

where the wave vectors of the incident and scattered neutrons are $\mathbf{k}_0$ and $\mathbf{k}_s$, respectively, and $\mathbf{G}$ is a reciprocal lattice vector. From experimental values of E_0, E_s, $\mathbf{k}_0$, and $\mathbf{k}_s$, we can determine $\omega(\mathbf{q}j)$ and $\mathbf{q}$ and, hence, the phonon dispersion curves. Results for NaBr [4] are shown in Fig. 1. The gap between the acoustical and optical branches is evident.

5. *SPECIFIC HEAT*

The specific heat associated with the lattice vibrations of a crystal can be calculated in the harmonic approximation by taking the thermal average, $\langle E \rangle$, of the energy given by Eq. (3.9). Replacing the sums over $\mathbf{k}$ and j by an integral over frequency ω, we obtain for the specific heat

$$C_v(T) = \left(\frac{\partial\langle E\rangle}{\partial T}\right)_v = k_B\beta^2 \int_0^{\omega_m} d\omega\, g(\omega)\hbar^2\omega^2 \frac{e^{\beta\hbar\omega}}{(e^{\beta\hbar\omega}-1)^2} \quad (5.1)$$

where $\beta = 1/k_BT$, k_B is Boltzmann's constant, ω_m is the maximum phonon frequency, and $g(\omega)$ is the phonon frequency distribution. In general, $g(\omega)$ must be evaluated numerically using a computer. However, for the case of an isotropic elastic continuum, $g(\omega)$ takes the simple form

$$g(\omega) = \frac{\Omega}{2\pi^2}\left(\frac{2}{c_t^3} + \frac{1}{c_\ell^3}\right)\omega^2, \quad 0 \le \omega \le \omega_m,$$
$$= 0, \qquad \omega > \omega_m, \quad (5.2)$$

where c_t and c_ℓ are the speeds of transverse and longitudinal sound waves and Ω is the volume of the crystal. Substitution of Eq. (5.2) into Eq. (5.1) gives

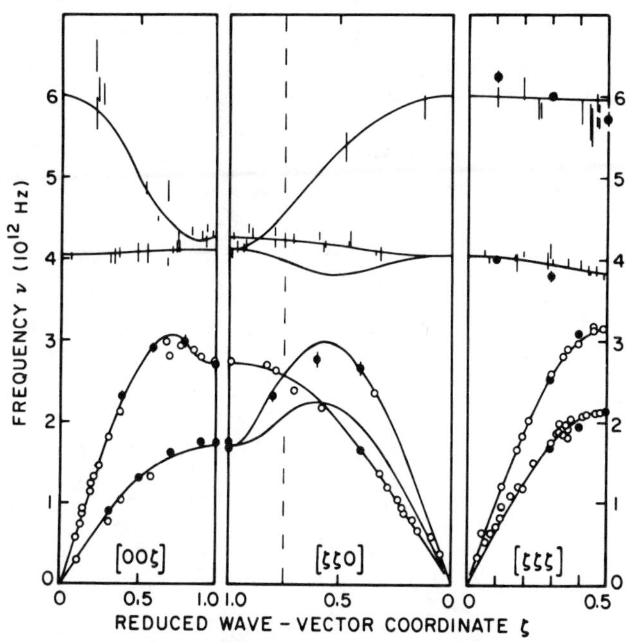

FIG. 1. Phonon dispersion curves for NaBr in the [001], [110], and [111] directions (from Ref. 4). The experimental data are indicated by bars and circles. The solid curves are calculated from a lattice-dynamical model and $\nu = \omega/2\pi$.

$$C_v(T=9Nk_B\left(\frac{T}{\Theta}\right)^3\int_0^{\Theta/T}dx\,\frac{x^4e^x}{(e^x-1)^2} \tag{5.3}$$

where $\Theta=\hbar\omega_m/k_B$ is the Debye temperature [5]. At low temperatures, $T\ll\Theta$, we obtain

$$C_v(T)\cong(12Nk_B\pi^4/5)(T/\Theta)^3, \tag{5.4}$$

which is the "Debye T^3 law." A tabulation of some Debye temperatures is given in Table I. The T^3 law is observed experimentally at sufficiently low temperatures; at higher temperatures, however, deviations from Eq. (5.3) appear because a real crystal is not an elastic continuum. The correct frequency distribution is in general a complicated function of ω and exhibits Van Hove singularities [6], which arise from maxima, minima, or saddle points on the constant-frequency surfaces in **k** space. The frequency distribution for copper [7] is shown in Fig. 2.

Table I. Debye Temperatures of Some Cubic Crystals[a]

Solid	Debye temperature (K)
Sodium	157
Copper	342
Lead	102
Nickel	427
Silicon	647
NaCl	321
LiF	732

[a] From J. S. Blakemore, *Solid State Physics*, 2nd ed., p. 130. Saunders, Philadelphia, PA, 1974.

6. *OPTICAL PROPERTIES*

An ionic crystal can interact with the electromagnetic field through the vibrations of its ions, since the latter are electrically charged and undergo acceleration during a lattice vibration. Optical phenomena are simplest in cubic crystals, which are optically isotropic and can be described by the complex dielectric constant $\epsilon(\omega)=\epsilon_1(\omega)+i\epsilon_2(\omega)$. For a crystal such as NaCl, with two atoms per primitive unit cell, we can show that [1,8]

$$\epsilon(\omega)=\epsilon_\infty+\frac{4\pi e^{*2}}{\mu\Omega_0}\left[\frac{\omega^2-\omega^2+i\omega\gamma}{(\omega_T{}^2-\omega^2)^2+\omega^2\gamma^2}\right] \tag{6.1}$$

where ω_T is the transverse optical phonon frequency of long wavelength, γ is the damping constant, e^* is the transverse effective charge, μ is the reduced mass of the two ions in the unit cell, Ω_0 is the volume of the unit cell, and ϵ_∞ is the limiting value of $\epsilon(\omega)$ for $\omega\gg\omega_T$.

The real part of the dielectric constant, $\epsilon_r(\omega)$, can be written in the absence of damping in the simple form

$$\epsilon_r(\omega)=\frac{\epsilon_\infty(\omega_L{}^2-\omega^2)}{(\omega_T{}^2-\omega^2)} \tag{6.2}$$

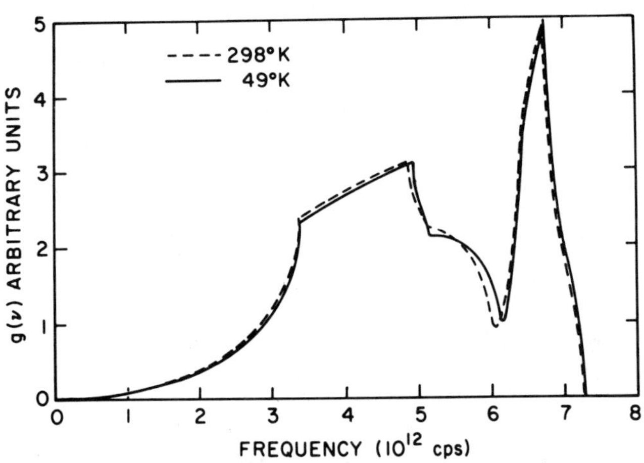

FIG. 2. Frequency distribution function for copper (from Ref. 7). Note that $\nu=\omega/2\pi$.

where $\omega_L=[\omega_T{}^2+(4\pi e^{*2}/\mu\Omega_0\epsilon_\infty)]^{1/2}$ is the longitudinal optical phonon frequency of long wavelength. Without damping the dielectric constant has a pole of ω_T and a zero at ω_L. If we set $\omega=0$ in Eq. (6.1), we obtain

$$\frac{\epsilon(0)}{\epsilon_\infty}=\frac{\omega_L{}^2}{\omega_T{}^2}, \tag{6.3}$$

which is the Lyddane–Sachs–Teller relation. Values of the lattice-dynamical parameters for several crystals are given in Table II.

The complex index of refraction, $n-i\kappa$, is defined by $(n-i\kappa)^2=\epsilon(\omega)$. The reflectivity at normal incidence, R, is given by

$$R=\frac{(n-1)^2+\kappa^2}{(n+1)^2+\kappa^2} \tag{6.4}$$

and is large between ω_T and ω_L, the region of *reststrahlen*. The absorption coefficient, $\eta=2\omega\kappa/c$, is strongly peaked close to ω_T if γ/ω_T is small. The coupling between the electromagnetic field and the transverse optical phonons of long wavelength leads to coupled modes called *polaritons* whose dispersion relation is given by $c^2k^2=\omega^2\epsilon(\omega)$. Experimental information about the polariton dispersion relation can be obtained by Raman scattering [10]. The dispersion curves for polaritons in GaAs are shown in Fig. 3. Note that there are two branches, one above and one below the reststrahlen region.

Table II. Lattice-Dynamical Parameters of Some Cubic Crystals

Solid	ϵ_∞	$\epsilon(0)$	ω_T (cm^{-1})	ω_L (cm^{-1})
LiF[a]	1.9	8.9	307	662
NaCl[a]	2.25	5.9	164	264
KI[a]	2.7	5.1	101	139
ZnS[b]	5.14	8.67	271	352
GaAs[b]	10.9	12.9	269	292
InSb[b]	15.6	17.7	185	197
PbS[b]	17.2	202	65	223

[a] Ref. 8.
[b] Ref. 9.

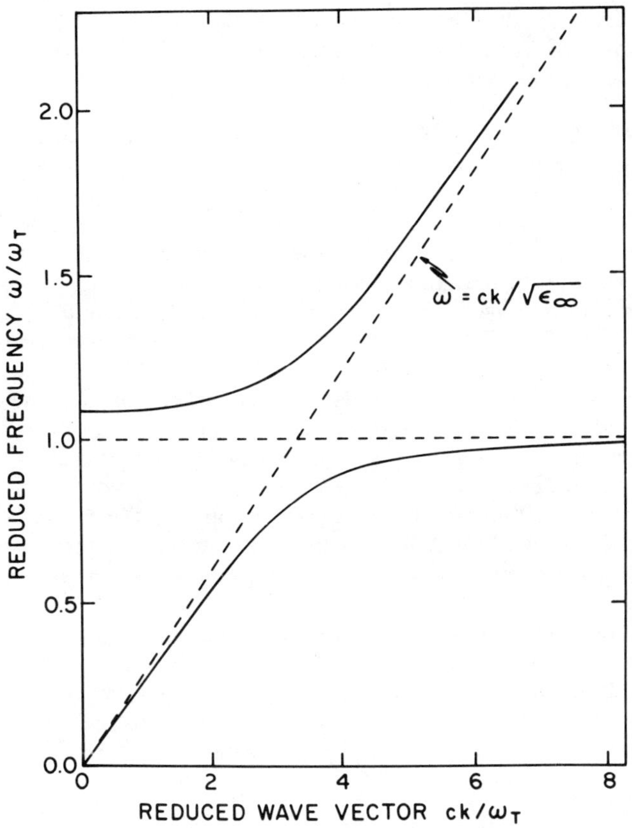

FIG. 3. Polariton dispersion curves for GaAs.

7. IMPURITY MODES AND SURFACE MODES

Associated with an impurity atom in a crystal are changes in atomic mass and coupling constants compared to the perfect crystal. These changes lead to modifications of the normal-mode eigenvectors and frequencies. In accordance with Rayleigh's theorem [2], a decrease in an atomic mass or increase in a coupling constant produces either no change or an increase in the normal-mode frequencies and conversely. Of particular importance are modes whose frequencies lie in regions that are forbidden for the frequencies of the perfect crystal. Such modes are termed *localized modes* because the atomic displacement amplitudes decrease exponentially from the impurity site. For an isotopic impurity of mass M' replacing a host crystal atom of mass M, the localized-mode frequency, ω_l, is specified by the equation [11]

$$\epsilon\omega_l^2 \int_0^{\omega_m} \frac{g(\omega)\, d\omega}{\omega_l^2 - \omega^2} = 1, \quad \omega_l > \omega_m, \tag{7.1}$$

where $\epsilon = (M - M')/M$ and $g(\omega)$ is the frequency distribution of the unperturbed crystal normalized to unity. Typically, a localized mode appears if M' is sufficiently light compared to M.

A crystal surface may be regarded as a defect in which certain coupling constants are set equal to zero. Modes localized at a surface may be derived from both acoustical and optical branches. Long-wavelength surface acoustic modes are known as *Rayleigh waves* [12], whose speed, c_R, in an isotropic medium is given by

$$16\left(1 - \frac{c_R^2}{c_\ell^2}\right)\left(1 - \frac{c_R^2}{c_t^2}\right) = \left(2 - \frac{c_R^2}{c_t^2}\right)^4. \tag{7.2}$$

The speed of Rayleigh waves is always less than c_t and c_ℓ.

8. ANHARMONICITY

The cubic, quartic, and higher-order terms in the displacements in Eq. (2.1) give rise to anharmonic effects such as thermal conductivity, thermal expansion, broadening of optical absorption lines, and ultrasonic attenuation. The coefficient of thermal conductivity, K, for an isotropic crystal can be shown [13] to have the form

$$K = \tfrac{1}{3}Csl \tag{8.1}$$

where C is the specific heat per unit volume, s is the average speed of sound, and l is the phonon mean free path. The coefficient K increases as T^3 at very low temperatures, reaches a maximum, and then decreases as $1/T$ at higher temperatures.

Anharmonicity also leads to a volume dependence of the Debye temperature described by the Grüneisen constant, γ, defined by

$$\gamma = -\frac{V}{\Theta}\frac{d\Theta}{dV}. \tag{8.2}$$

The thermal expansion coefficient, α, can be expressed in terms of γ,

$$\alpha \equiv \frac{1}{V}\left(\frac{\partial V}{\partial T}\right)_P = \frac{\kappa\gamma C_v}{V} \tag{8.3}$$

where κ is the compressibility and C_v is the specific heat at constant volume.

In certain circumstances it is possible to have vibrational waves propagating in an anharmonic lattice without damping. Such waves are known as solitons [14].

See also CRYSTAL DEFECTS; CRYSTAL SYMMETRY; PHONONS; QUASIPARTICLES.

REFERENCES

1. M. Born and K. Huang, *Dynamical Theory of Crystal Lattices*. Oxford, London and New York, 1954.
2. A. A. Maradudin, E. W. Montroll, G. H. Weiss, and I. P. Ipatova, *Theory of Lattice Dynamics in the Harmonic Approximation*. Academic Press, New York, 1971.
3. B. N. Brockhouse, in *Phonons and Phonon Interactions* (T. A. Bak, ed.), p. 221. Benjamin, New York, 1964.
4. J. S. Reid, T. Smith, and W. J. L. Buyers, *Phys. Rev.* **B1,** 1833 (1970).
5. P. Debye, *Ann. Phys. (Leipzig)* **39,** 789 (1912).
6. L. Van Hove, *Phys. Rev.* **89,** 1189 (1953).
7. R. M. Nicklow, G. Gilat, H. G. Smith, L. J. Raubenheimer, and M. K. Wilkinson, *Phys. Rev.* **164,** 922 (1967).
8. E. Burstein, in *Phonons and Phonon Interactions*, (T. A. Bak, ed.), p. 276. Benjamin, New York, 1964.

9. E. Burstein, A. Pinczuk, and R. F. Wallis, *The Physics of Semimetals and Narrow Gap Semiconductors* (D. Carter and R. Bate, eds.), p. 251. Pergamon, Oxford, 1971.
10. C. H. Henry and J. J. Hopfield, *Phys. Rev. Lett.* **15,** 964 (1965).
11. P. G. Dawber and R. J. Elliott, *Proc. Roy. Soc. London* **A273,** 222 (1963).
12. Lord Rayleigh, *Proc. London Math. Soc.* **17,** 4 (1887).
13. C. Kittel, *Introduction to Solid State Physics* 4th ed., p. 224. Wiley, New York, 1971.
14. M. Toda and M. Wadati, *J. Phys. Soc. Japan* **34,** 18 (1973).

Light-Sensitive Materials

H. E. Spencer

Many materials are sensitive to visible and ultraviolet radiation. They include gases, liquids, and solids, and exhibit a wide variety of properties. They possess one common feature, however, in that absorption of visible and ultraviolet photons produces excited electronic states. The fate of the excited states determines the fate of the exposed materials, which may undergo chemical or physical change.

These states can best be discussed in terms of two somewhat different types of materials. In molecular materials, the ground-state properties are characteristic of individual molecules, and the excited states also are molecular in nature. In many solids, however, individual molecules or ions interact strongly and both the ground and excited electronic states are characteristic of the collection of ions or molecules in the solid.

In molecular organic materials, both the ground and excited states, in spectroscopic notation, are usually singlets. Deactivation of the excited singlet back to the ground state, mostly by a radiationless path, but sometimes accompanied by fluorescence, usually dominates unless internal conversion to a triplet state is energetically and kinetically possible. The long lifetime of the triplet state coupled with its high energy relative to the ground state results in a high probability of its reaction with another species.

A specific example of a molecular organic material that has been studied extensively is benzophenone, $C_6H_5COC_6H_5$, in solution at room temperature. It absorbs weakly in the violet and near-ultraviolet region. The excited singlet converts with near unit efficiency to a triplet state that can react with numerous compounds. For instance, if isopropyl alcohol is present in solution, the benzophenone triplet abstracts hydrogen atoms from the alcohol. This is but an initial step; the final products are acetone and benzopinacol. If isobutylene is present instead of isopropyl alcohol, a triplet benzophenone molecule and an isobutylene molecule combine to form a new molecule, an oxetane. Such photochemical reactions often provide the easiest ways of synthesizing certain complex molecules.

Many molecular inorganic materials are photosensitive, also. The gas NO_2 is one, and its photochemistry is important in the chain of events whereby the atmospheric pollutant O_3 is formed. NO_2 is generated as a by-product of high-temperature combustion such as is found in automobile engines. Absorption in NO_2 of ultraviolet radiation of 300- to 400-nm wavelengths from the sun produces an excited state that rapidly decomposes into NO and O. The latter reacts with O_2 to form O_3.

Other examples of photosensitive inorganic materials are found among coordination compounds of transition metals. For example, $K_3Fe(C_2O_4)_3 \cdot 3H_2O$, both as a solid or in solution, is decomposed by visible and ultraviolet radiation. In solution, products consisting of Fe(II) complexes and CO_2 form with high quantum efficiency. One of the most commonly employed actinometers is an acidic solution of $K_3Fe(C_2O_4)_3 \cdot 3H_2O$.

In many solids, absorbed ultraviolet and visible radiation energetically raises electrons from the valence band to the conduction band, leaving behind mobile holes. Often these electronic carriers are detected in photoconductive or photovoltaic experiments. The electrons and holes commonly recombine, often in a radiationless manner, but sometimes accompanied by luminescence, particularly at low temperatures. Proper doping with impurities enhances this luminescence.

Silver halides exemplify photosensitive solids of scientific and commercial importance. The chloride, bromide, iodide, and mixtures of these salts form the basis of many photographic systems. Values of extinction coefficients are high in the ultraviolet region and become lower as they extend into the visible portion. After absorption, the hole–electron pairs that do not recombine reduce silver ions to metallic silver and oxidize halide ions to halogen. For photographic applications of a silver salt, microcrystals of micron or submicron linear dimensions are usually dispersed in a gelatin coating on an inert polymeric substrate. To decrease recombination and thereby increase sensitivity, small amounts of impurities, such as silver, gold, silver sulfide, and gold sulfide, are added to the microcrystals. Sensitizing dyes often are also adsorbed to the microcrystals. Absorption of light by the dyes produces conduction-band electrons in the silver halides, in the usual case.

These photoelectrons form silver centers. By a repeated cycle of trapping of the photoelectrons, and neutralization by interstitial silver ions, silver centers nucleate and grow. Those of sufficient size to act as catalyst sites when the exposed film is immersed in a developer solution are called latent-image centers. In the developer, microcrystals that have been exposed and contain latent-image centers are reduced to metallic centers at a much greater rate, in general, than those that have not been exposed. The final image is either the reduced silver or a dye formed by the reaction of the oxidized developer and a suitable reagent. Many variations of this basic process are utilized.

BIBLIOGRAPHY

D. R. Arnold, N. C. Baird, J. R. Bolton, J. C. D. Brand, P. W. M. Jacobs, P. deMayo, and W. R. Ware, *Photochemistry, an Introduction*. Academic, New York, 1974. (I)

J. G. Calvert and J. N. Pitts, Jr., *Photochemistry*. Wiley, New York, 1966. (A)

T. H. James, ed., *Theory of the Photographic Process*, 4th ed. Macmillan, New York, 1976. (A)

Liquid Crystals

P. G. de Gennes

CLASSIFICATION AND EXAMPLES

In a crystal the building blocks (atoms or molecules) are regularly arranged in a periodic lattice. In a liquid they are completely disordered. With suitably chosen molecules, we can find other states of organization, with less symmetry than in a crystal but more symmetry than in a liquid: they are called *mesomorphic phases* or, more loosely, liquid crystals. The main types known at present are the following [1].

1. *Nematics* [2] are fluids of rodlike molecules with their centers disordered but with a common axis of alignment for the rods (Fig. 1a). A typical molecule giving rise to a nematic phase is methoxybenzilidenebutylaniline (MBBA):

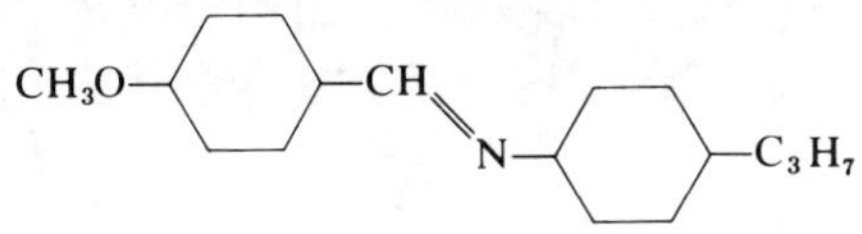

The sequence of phases for MBBA is

$$\text{solid} \xrightarrow{10^\circ\text{C}} \text{nematic} \xrightarrow{43^\circ\text{C}} \text{isotropic liquid.}$$

2. *Smectics A* and *smectics C* [3] are layered systems, each layer behaving like a two-dimensional *liquid*. The A type is optically uniaxial (Fig. 1b), whereas the C type is biaxial (Fig. 1c). Both types can be obtained with certain long organic molecules, such as terephthal-*bis*-*p*-butylaniline (TBBA):

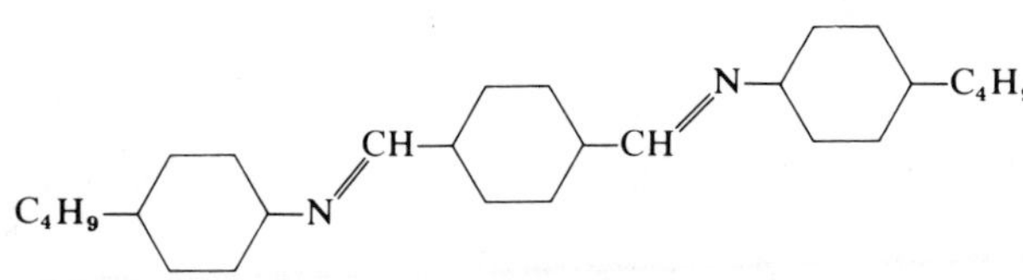

The sequence of phases of TBBA is complex: the main phases are

$$\text{crystal} \xrightarrow{113^\circ} \text{B} \xrightarrow{144^\circ} \text{C} \xrightarrow{172^\circ} \text{A} \xrightarrow{200^\circ} \text{N} \xrightarrow{236^\circ} \text{I}$$

(where I stands for isotropic and N for nematic, and B is another smectic type, to be discussed later). This shows that one compound may have many states in between the crystal and the liquid! This situation is not exceptional; the number of known molecules giving rise to mesomorphic phases is of the order of 2000.

In pure materials like TBBA, the main physical variable that allows us to explore the sequence of phases is the temperature; for this reason the resulting liquid crystals are called *thermotropic*. Another group is obtained with mixtures, where the phases can be obtained by changes in concentration (lyotropic liquid crystals). A major example of lyotropic smectics is found with mixtures of lipid and water [4]. A lipid molecule contains a polar head (hydrophilic) and a hydrocarbon tail (hydrophobic). Many lipid–water systems show a lamellar phase (Fig. 2a) that has the symmetry of a smectic A.

What is the origin of the smectic layers? For lyotropic systems they appear as a natural way of having the heads close to the water and the tails far from it. For thermotropic systems the situation is less clear, but something similar may happen: the molecules usually have an aromatic—rigid—part (Ar) plus terminal chains that are aliphatic (Al). It may be that (Ar)–(Ar) attractions dominate the system's behavior, and they favor a layered system.

3. "Canonic" phases [5] have periodicity in two directions and not in the third; they are found with lipid–water systems, and have the structure shown in Fig. 2b,c: one component is associated in rods, each rod being a one-dimensional fluid tube. The main difference from nematic symmetry is that here the rods are stacked regularly (Fig. 3). There should be a finite shear rigidity for shears in the plane of the figure—but the macroscopic properties of these phases have not yet been studied.

4. *Smectics B* are layered systems, but their x-ray diagrams suggest that each layer is a two-dimensional *solid*. The transition from normal solid to smectic B is often associated with the "melting" of aliphatic chains at the end of the molecule while other parts (the aromatic region in thermotropic systems, or the polar heads in lipid water) remain well ordered: this gives a structure in which successive layers are nearly uncoupled. Smectics B should probably be classified as crystals (with a low resistance to shear).

5. There are many other "exotic" phases of smectics, classified as D, E, F, G. . . . The D phase is a cubic gel, and should be classified as a (very plastic) crystal. The E phase seems to be a variant of the B phase, with less rotational freedom for the molecules. But exotic phases are just beginning to be understood.

The physics of all these liquid crystals has expanded rapidly during the last fifteen years, and cannot be summarized easily. We have chosen here to omit all further discussion of properties at the molecular level (for which there exist many reviews [9,10]) and to focus on certain macroscopic features, which are most unusual.

OPTICAL PROPERTIES OF NEMATICS

Nematics are optically uniaxial, but they flow like liquids. Their optical properties are interesting mainly because the optical axis may be *rotated by weak external agents:* the effects of the walls of the container, of magnetic and electric fields, and of flows are most important in practice.

1. *Alignment by walls:* It is possible to fix the orientation of the optical axis in a nematic slab by suitable treatment of the container surfaces: rubbing in one direction, or covering the surface with suitable detergents, or evaporating metallic films at an oblique angle. It is thus possible to prepare single-domain samples with well-defined boundary conditions.

2. *Alignment by magnetic fields H:* Nematic fluids have

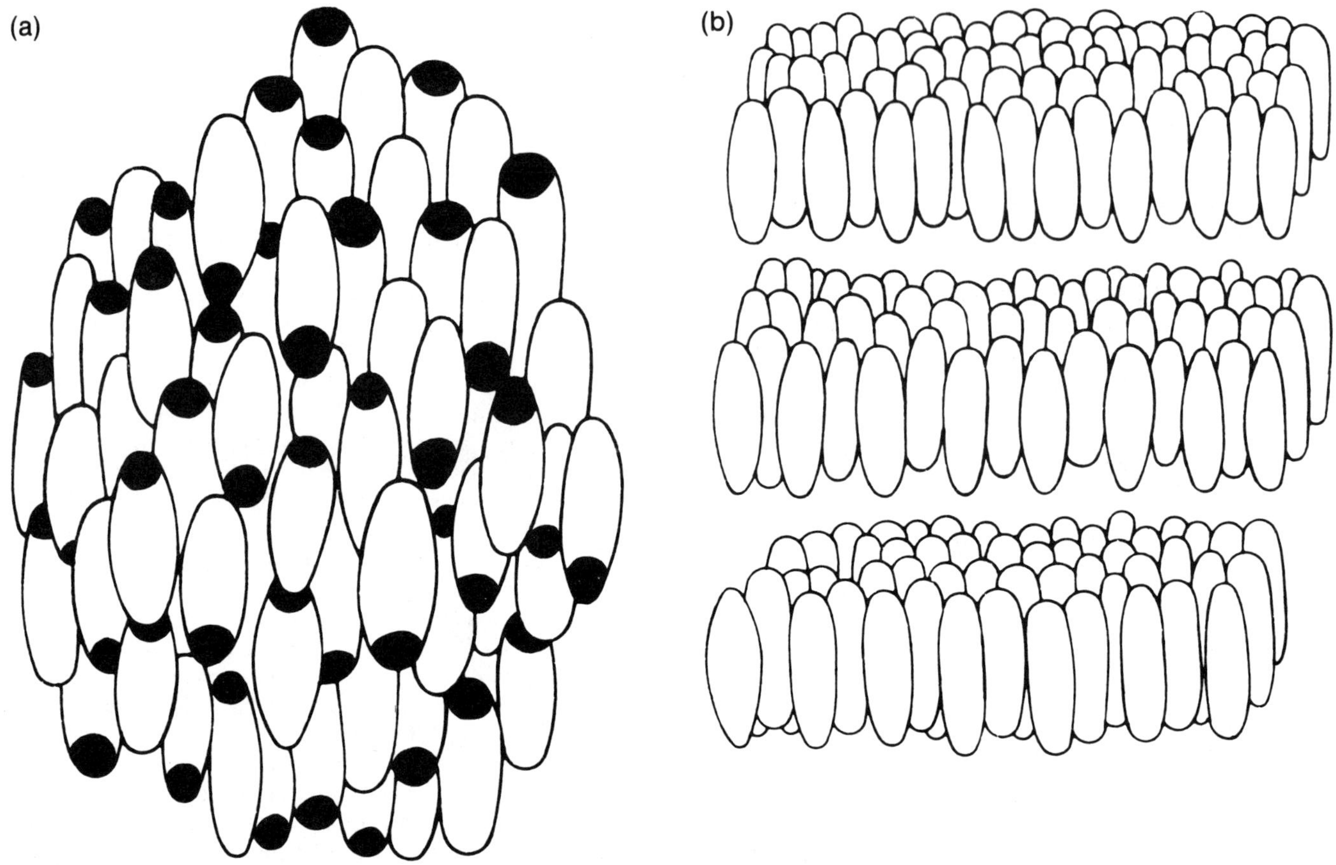

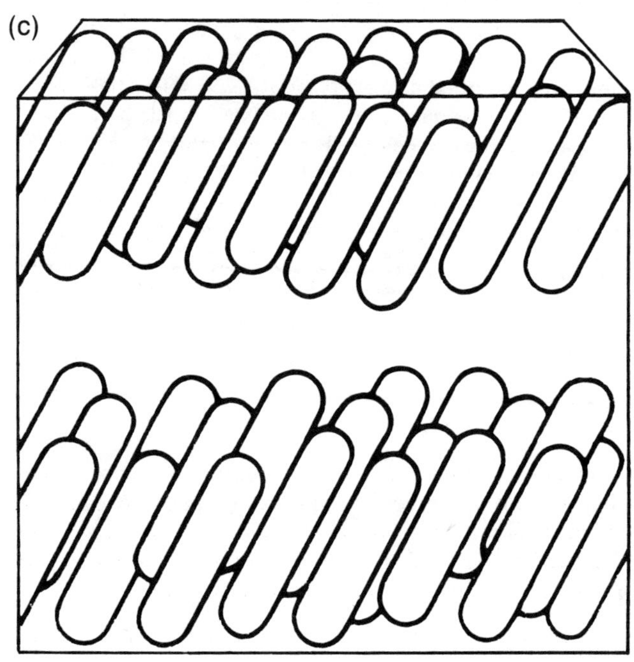

FIG. 1. (a) Nematic phase: the molecules are purposely represented with one black end; note that as many molecules are "up" as are "down"; even if the molecules carry a permanent dipole, the phase is not ferroelectric. (b) Smectic-A phase. (c) Smectic-C phase. Note that in parts (b) and (c) each layer is disordered (liquidlike).

a weak diamagnetic susceptibility χ, and χ is anisotropic. Usually, this anisotropy favors an optical axis parallel to the magnetic field H. With suitable orientation of H, there may be an interesting competition between wall alignment and field alignment, first studied long ago by Freedericksz [6]. Typically, for a slab 50 μm thick, the fields involved are of the order of 10^3–10^4 Oe. The field effects can be monitored very directly by optical observations.

3. *Flows* are strongly coupled to the molecular alignment [7]. In a simple shear flow the long axis of the molecules tends usually to become nearly parallel to the lines of flow. In many cases, a competition can be set up between wall

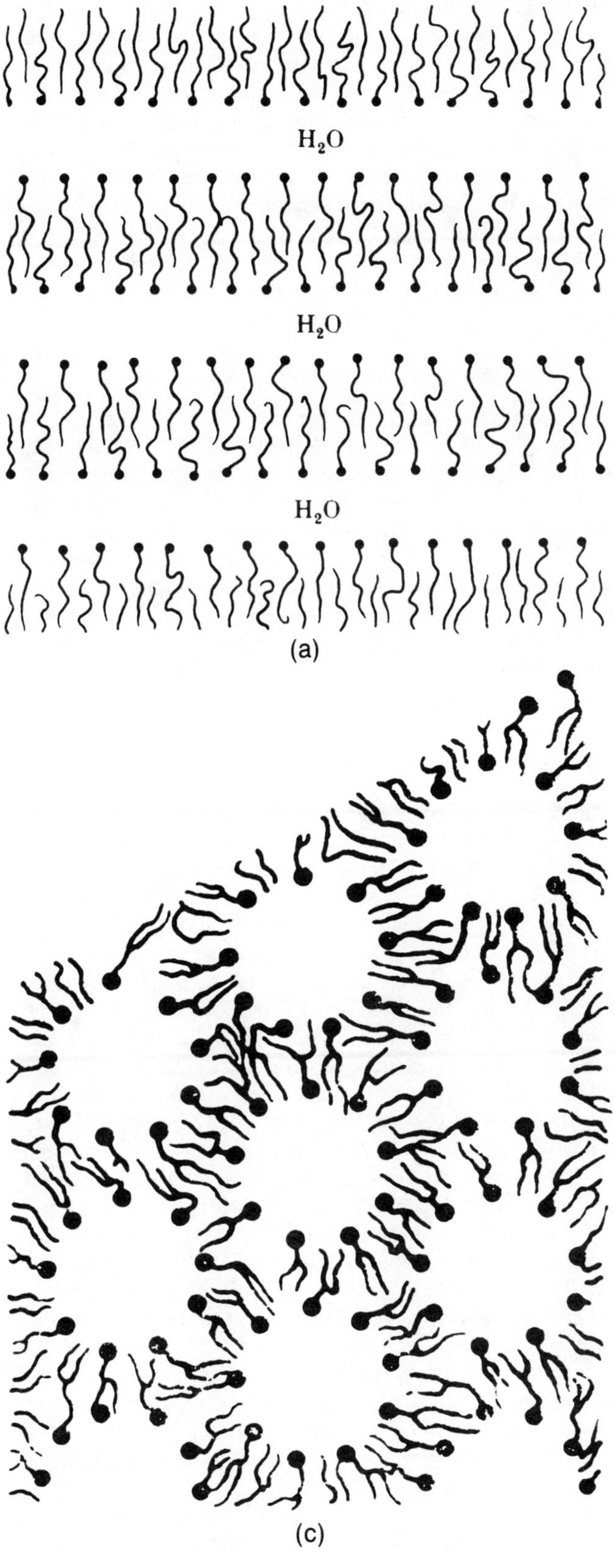

FIG. 2. (a) Lamellar phase of lipid + water systems (smectic A). (b),(c) Hexagonal phases of lipid–water systems (canonic).

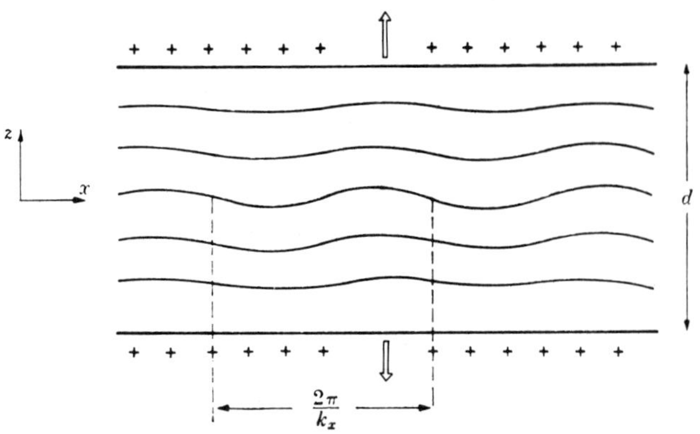

FIG. 3. Instability of smectic-A layers under traction.

alignment and flow alignment, leading to unusual mechano-optic effects and to instabilities even at extremely low Reynolds numbers.

4. *Electric fields E* have remarkable effects on nematics (apart from some rather trivial alignments similar to the H effect discussed above). In particular, when the liquid is slightly conducting ($\sigma \sim 10^{-7}$ mho) the field E can induce an instability: (a) The molecular alignment is slightly deformed by a thermal fluctuation; (b) this reacts on the conduction-current pattern J, and makes J inhomogenous; (c) the inhomogeneous J builds up a space charge Q; (d) the field E acting on Q gives a bulk force in the liquid; (e) the result is a flow pattern, which distorts the alignment as explained in item (3) and *enhances* the original deformation—hence the instability [8]. This process is the basis of various electro-optic devices, where a voltage V (of the order of 10 V) is applied through semitransparent electrodes to a nematic slab (~50 μm thick). When V is high enough, a state of turbulent flow is reached, the alignment is strongly perturbed, and the system scatters light very strongly.

These examples show that the optical axis can be tilted easily by external perturbations. As usual in physics, this also implies that, in the absence of any perturbation, the *spontaneous fluctuations* of the axis are important: they give rise to a scattering of light (in the absence of any field E) that is much higher than for normal liquids: nematics are *turbid* when viewed in thicknesses $\gtrsim$ 1mm.

SMECTICS A

Smectics A are the simplest example of a system with liquid layers, and they have unusual mechanical properties, intermediate between those of crystals and liquids.

Deformability of the Layer Structure

The ideal arrangement is a succession of flat sheets. But the sheets can be deformed very easily. The distance between them is essentially fixed, but they can be moved over large regions since successive layers can slip freely over one another. This shows up in various problems.

Textures: If the boundary conditions do not allow for the ideal flat arrangement, the layers build an interesting deformed texture. The simplest example is a jelly-roll shape (with a singular line on the axis). But there are more complex patterns, where the singular lines become ellipses and hyperbolas ("focal conics") [11].

Undulation modes: An oscillation with a wave vector parallel to the layers does not alter the interlayer distance; thus, the restoring force for these particular modes is extremely weak, and they have a large thermal amplitude, giving rise to a special type of light scattering.

Instability under traction: A smectic slab (typical thickness 100 μm) with the layers parallel to the slab plane is rapidly pulled along the normal to the layers; as soon as the relative displacement of the limiting plates exceeds a few hundred angstroms, the smectic layers *fold* to fill the available space without changing their thickness (Fig. 3).

Flows in a Smectic Structure

At first sight, a smectic appears very viscous—very much like a soap. But the real situation is more subtle. Flow *parallel* to the layers involves only weak friction, very much like in a conventional fluid. Flow *through* the layers requires a pressure gradient, just like the flow of water through a filter. But here the filter and the flowing fluid are one and the same species! The hydrodynamic consequences of these strange features have now been observed in certain weak flows. Strong flows usually imply the creation and motion of many defects (dislocations, focal conics) and are just beginning to be understood.

CHIRALIZATION

Another remarkable example of the deformability of liquid-crystal structures is connected with the effect of solute molecules that are not identical to their mirror image (chiral molecules). A nematic N doped with such molecules becomes twisted into a helical form N* (Fig. 4a), usually known as a *cholesteric* phase [12]. The pitch of the helix is typically in the micron range and thus gives rise to Bragg scattering of optical light. For this reason many cholesterics show beautiful iridescent colors. Also, the pitch (and thus the color) can be very sensitive to various external agents (temperature, fields, chemical contaminants, etc.) and this has been the basis of certain applications [13]. When added to a smectic A, chiral molecules cannot induce a macroscopic twist (because this twist is incompatible with the condition of constant layer thickness). But in a smectic C, chiralization does give a twisted phase C* (Fig. 4b). The C* systems have Bragg reflections and optical rotations very much like N*. They have one further property that was recognized recently [14]: each C* layer is *ferroelectric*, with a permanent moment parallel to the layers and normal to the tilt direction. The ferroelectric moment is usually rather small (of the order of 10^{-2} debyes per mole) because of an easy rotation of the molecules around their long axis. But the existence of a moment in a system that can flow leads to remarkable cross effects.

On the whole, we see that the fluid phases of liquid crystals (nematics and smectics A and C) are beginning to be reasonably well understood. The more rigid B phase and the related exotic phases are in a certain sense less exciting,

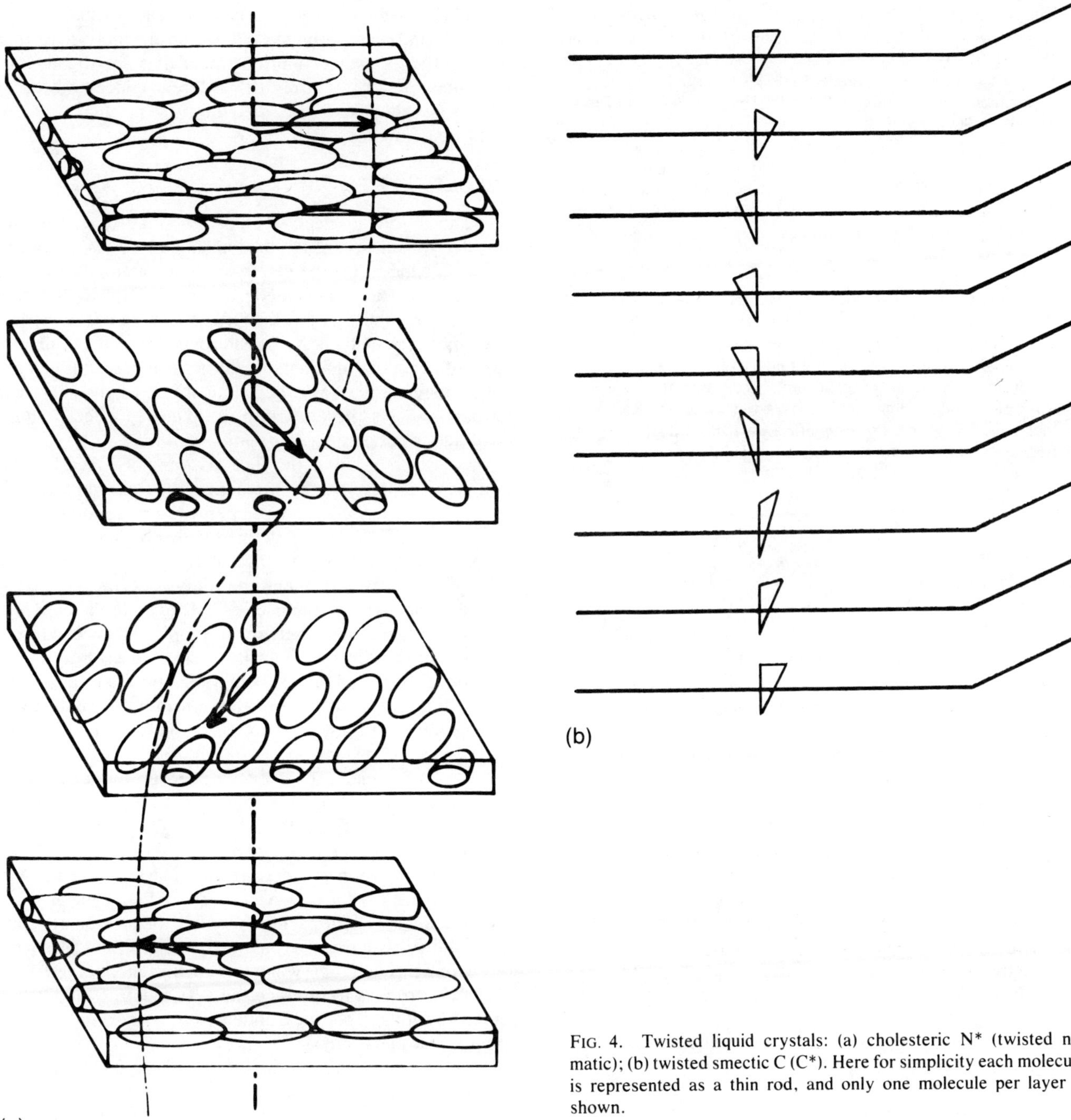

FIG. 4. Twisted liquid crystals: (a) cholesteric N* (twisted nematic); (b) twisted smectic C (C*). Here for simplicity each molecule is represented as a thin rod, and only one molecule per layer is shown.

since they are less deformable; on the other hand, their fundamental symmetries and structures leave many open problems for future work.

NOTES AND REFERENCES

1. The classification of liquid crystals has been constructed mainly by G. Friedel [*Ann. Phys.* **18,** 273 (1922)] using optical observations, and by H. Sackmann and D. Demus [*Mol. Cryst.* **2,** 81 (1966)] using miscibility criteria and textural features.
2. From the greek *nema*, thread; this refers to certain threadlike defects (disclinations) that are found in poorly oriented nematics (see Ref. 6).
3. From the greek *smigma*, soap; certain lamellar phases of the soaps (salts of fatty acids) are smectics.
4. For a review of the polymorphism of lipid–water systems, see V. Luzzati in *Biological Membranes* (H. Chapman, ed.), Academic Press, New York, 1968.
5. From the greek *canon*, rod; this terminology was introduced by F. C. Frank.
6. The continuum elastic theory on which the interpretation of phase effects is based is due to C. Oseen, H. Zocher, and F. C. Frank; see F. C. Frank, *Disc. Faraday Soc.* **25,** 19 (1958).

7. The equations of "nematodynamics" were constructed by J. L. Ericksen and F. M. Leslie. A more general formulation covering the smectics as well can be found in P. Martin, O. Parodi, and P. Pershan, *Phys. Rev.* **A6,** 2401 (1972).
8. The complete explanation is due to W. Helfrich, *J. Chem. Phys.* **51,** 4092 (1969).
9. G. Gray, *Molecular Structure and the Properties of Liquid Crystals*. Academic Press, New York, 1962. A. Saupe, *Angew. Chem. (Int. Ed.)* **7,** 97 (1968). G. H. Brown, W. Doane, and V. Neff, *C.R.C. Crit. Rev. Solid State Sci.* **1,** 303 (1970). P. G. de Gennes, *Thy Physics of Liquid Crystals*, Oxford, London and New York, 1974.
10. For a recent perspective on the field, see the *Proceedings* of the 5th International Conference on Liquid Crystals (Stockholm, 1974) published by *J. Phys. (Paris), Coll. Supp.* **C1,** (1975).
11. The focal conics are always with sets of equidistant surfaces, such as wave fronts in optics. G. Friedel realized this very early, and from observations of focal conics under the microscope he was able to infer—long before any x-ray work (see Ref. 1)—that smectics A were made up of equidistant fluid layers.
12. Because it was found first with cholesterol esters.
13. J. Fergason, *Sci. Am.* **211,** 77 (1964).
14. R. B. Meyer, L. Liébert, L. Strzelecki, and P. Keller. *J. Phys. (Paris), Lett.* **36,** L69 (1975).

Liquid Metals

N. E. Cusack

Liquid metals have much in common with simple nonmetallic liquids. For example, liquid tin at 600°C flows about as easily as water, and it is hard to find significant qualitative differences between, say, liquid sodium and liquid argon in their structures or in their atomic motion as typified by the phenomenon of diffusion. The special character of metallic liquids shows in electrical, magnetic, optical, and all other properties that depend on the dense fluid of free electrons that permeates all metals whether liquid or solid. Indeed, a pure liquid metal can be regarded as a mixture of (i) a fluid of positively charged ions, and (ii) a gas of negative electrons containing one electron per ion in sodium, two in calcium, and so on according to valency. There will therefore be about 10^{28} electrons per cubic meter, and these are loosely called "free electrons" because they can move through the entire system.

The fundamental physical problems of liquid metals can be summarized thus:

1. What is the interaction between the free electrons and the ions?
2. In properties possessed by all liquids (e.g., viscous flow, thermodynamic properties), how do liquid metals behave?
3. In properties possessed by all metals, solid and liquid (e.g., high electrical conductivity), how do liquid metals behave?
4. How does the interaction in (1) determine the properties in (2) and (3)? This is the basic question of liquid-metal physics.

Apart from these questions of physics are all those of liquid-metal chemistry, metallurgy, and technology. The use of liquid metals as coolants in nuclear reactors has focused attention on these problems but we do no more here than mention their existence and comment that one urgent need is to integrate more closely the different concepts used by physicists, chemists, and metallurgists in their thinking about liquid metals.

FORCES BETWEEN IONS AND ELECTRONS

The ions contain electrons closely bound to nuclei. The free electrons flow through the ions under the control of two major influences: (a) the electrostatic attraction of the positive ions; (b) a complicated response to those electrons bound to the nuclei. A powerful development of the modern wave-mechanical theory of solids has made it possible to treat influence (b) mathematically as if it were an electrostatic influence like (a) except that it is a repulsion, which partially cancels the attraction (a). The net effect is quite small and is called a *weak pseudopotential*.

An electron passing through an ion may therefore be thought of as deviated by the appropriate pseudopotential. However, the ion does not suddenly exert an influence as the electron crosses a well-defined ionic boundary. The ions are positively charged and the electron gas is mobile, and therefore ions exert a long-range influence on the gas by attracting electrons to themselves. Meanwhile the electrons are also repelling each other, and a complex balance of forces sets in whose ultimate effect is to envelop each ion in an electron cloud. This cloud, being negative, neutralizes the electrostatic effect of the positive ion rather completely at large distances (say, several atomic diameters) and less completely at small distances.

The cloud of electrons piled up near an ion is called the ion *screening cloud* and the total effect on an approaching electron of the pseudopotential and of the screening cloud is called the *screened pseudopotential*.

In summary we may therefore say that from the point of view of one electron passing through a liquid metal it encounters the screened pseudopotentials of one ion after another and is scattered by them. The total effect of all the ions acting through their weak screened pseudopotentials on all the electrons is to produce the very electron clouds that screen the ions. This idea underlies most current thinking about liquid-metal physics.

Since any two ions interact with the same electron gas, the two ions influence each other through the mediation of the electron gas in addition to any more direct ion–ion interaction there may be. The total ion–ion interaction will therefore depend on the density of the electron gas, i.e., on the volume of the system. This contrasts with nonmetals, where there is no electron gas and the interatomic force is, to a good approximation, independent of the volume.

STRUCTURE AND THERMODYNAMIC PROPERTIES

The structure of crystals is revealed in full by x-ray crystallography. This is not so for liquids. Since the electron–ion interaction is not well enough understood for us to calculate the structure theoretically from first principles, the structure of liquid metals is not known in any detail. What x-ray and

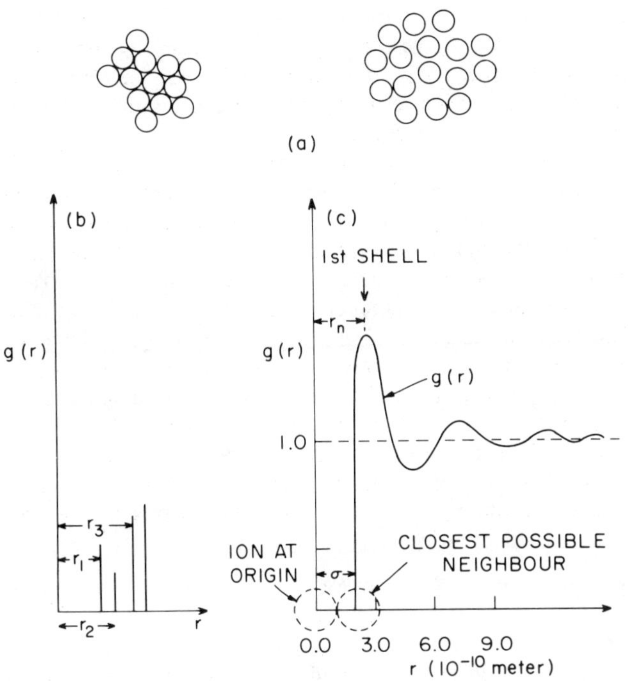

FIG. 1 (a) Schematic regular and irregular arrays of atoms shown in two dimensions. (b) Radial distribution function for a perfect regular or crystalline array. Atoms occur only at definite distances r_1, r_2, etc. (c) Radial distribution function $g(r)$ for a liquid metal. Atoms occur at any distance greater than σ (the effective hard-sphere diameter) with varying probability; r_n is the mean distance of nearest-neighbor shell, which typically accommodates 11–12 neighbors on average.

neutron scattering studies do provide is a quantity known as the radial distribution function, denoted by $g(r)$ and defined as follows: Choose any one ion as origin; then in a spherical shell of radius r, the average number density of the other ions is $g(r)$ times the average number density of the whole liquid. Figure 1 contrasts the structures of liquid and crystalline metals. The function $g(r)$ exceeds unity at the first closely packed shell of neighbors and is zero inside this shell because the neighbors cannot interpenetrate the origin atom. Note that $g(r)$ is an average. The ions are constantly moving about and interchanging neighbors like a milling crowd; $g(r)$ represents the average appearance of numerous instantaneous snapshots of the crowd. To think of the arrangement of ions in a liquid metal as a jumble of billiard balls in a bag is not far from the truth, and a hypothetical system of hard spherical atoms is called the "hard-sphere model" of a liquid. Because hard spheres are comparatively simple objects, it is possible to treat the hard-sphere model mathematically and to derive its thermodynamic properties by statistical mechanics.

It is a somewhat remarkable fact that the structure and some of the thermodynamic properties of liquid metals can be represented quite well by those of the hard-sphere model provided that the diameter of the hypothetical hard spheres is selected by a theoretical process that takes the electron–ion interaction into account. The entropy of metals and the heats of mixing of alloys are examples of thermodynamic properties that in the 1970s were just beginning to be understood in this way.

ELECTRONIC PROPERTIES

Perfectly regular arrays of atoms in space are known theoretically not to have any electrical resistance. Cold crystalline metals are good approximations to this and their resistance is indeed very low. As they are heated, their electrical resistance rises because the ions vibrate about their ordered positions, thus acquiring *thermal disorder*. On melting, the array of ordered positions dissolves and *structural disorder* sets in, represented by Fig. 1c. This causes a further rise of resistance by about a factor of 2 at the melting point. Further heating increases the resistance of liquid metals still more. Thus although the resistance of liquid metals is higher than that of the corresponding crystals, it is not catastrophically so and liquid metals are good conductors. They also have thermoelectric powers and thermal conductivities comparable with those of solid metals.

These properties are now fairly well understood on the basis of wave mechanics. As explained earlier, the electrons are deflected when they collide with the ions because the latter have a weak screened pseudopotential. This is using particle language. In the wave language of wave mechanics, the electron waves are scattered by the ions to a degree depending on the pseudopotential, and the scattered wavelets from the different ions may be in or out of phase, thus reinforcing or canceling each other. This is a typical problem in diffraction theory, well understood in, for example, x-ray crystallography. The combination of pseudopotential and diffraction theory, employing the structural knowledge of Fig. 1c, leads to expressions for electrical resistance, thermoelectric power, and thermal conductivity in quite good agreement with observations in many pure liquid metals and some liquid alloys.

It is in more subtle electronic properties such as the Hall effect, the optical properties, and the magnetic properties that the peculiarities of the liquid state appear. Directly or indirectly these properties depend on the *spectral distribution function*, a quantity expressing the distribution of energy and wavelength among the free electrons. This quantity must derive from (i) the structure and (ii) the electron–ion interaction and, although there are formal theoretical expressions for it, the spectral distribution function is hard to calculate because insufficient detail is known about the structures of liquids. This is one of the great stumbling blocks. Consequently, although Hall effects and optical properties are measurable and do change when a metal melts, there are at present no very good theories of these phenomena in the liquid state. It appears that characteristic features of them, which depend on the crystal structures of the individual metals, simply disappear when the structure dissolves on melting, making liquid metals more uniform among themselves in their behavior than are crystalline metals.

OUTSTANDING PROBLEMS IN LIQUID-METAL PHYSICS

Metallic nature exists because the attraction between the ions and the free electrons, while enough to impede the electrons, is insufficient to bind or localize them. This is a matter of degree; the weaker the interaction, the less the electrical resistance, and the more closely the electrons in the metal resemble an ideally free electron gas. Sodium is probably the most ideal metal in this sense, and the theories referred to earlier work best for the weakest interactions.

As the ion–electron interaction gets stronger, the approximations in the theories become more dubious, the lack of detailed structural information becomes more serious, and the subject enters a region where both theory and experiment are in difficulties. Three cases are liquid transition metals (Fe, Ni, Ce, etc.) where the special character of the ions with unfilled electron shells makes the electron–ion interaction strong and complicated; liquid metals heated under pressure until the density is greatly reduced—this removes their metallic nature altogether and they become liquid semiconductors; certain binary alloys (e.g., of cesium with gold) at certain compositions appear to bind the electrons to the ions so tightly that they are no longer free to conduct electricity.

These three examples typify the present frontiers of the subject and show it merging with the physics of liquid semiconductors and molten salts.

See also CONDUCTION, ELECTRIC; ELECTRON ENERGY STATES IN SOLIDS AND LIQUIDS; METALS; MOLTEN SALTS.

BIBLIOGRAPHY

S. Z. Beer, ed., *Liquid Metals*. Marcel Dekker, New York, 1972. (1–A)

T. E. Faber, *Theory of Liquid Metals*. Cambridge, London and New York, 1972. (1–A)

M. Shimoji, *Liquid Metals*, Academic Press, London, New York, San Francisco 1977.

S. Takeuchi, ed., *The Properties of Liquid Metals*, Taylor and Francis 1973.

R. Evans and D. A. Greenwood, eds, *Liquid Metals, 1976*, Conference Series No. 30, Institute of Physics, London and Bristol.

Low-Energy Electron Diffraction (LEED)

C. B. Duke

Low-energy electron diffraction (LEED) is the coherent reflection of electrons in the energy range 5 eV$\lesssim E \lesssim$500 eV from the uppermost few atomic layers of crystalline solids. A schematic diagram of an apparatus for measuring the configuration and intensities of the diffracted beams is presented in Fig. 1. Diffraction occurs because the wavelength λ (in angstroms) is related to the accelerating voltage V (in volts) via

$$\lambda = 12.2638/V^{1/2}, \quad (1)$$

leading to λ of the order of atomic dimensions for V in the "low-energy" range. The surface sensitivity of the diffraction process arises from the short inelastic-collision mean free paths, 5 Å $\lesssim \lambda_{ee} \lesssim$ 10 Å, for substantial energy losses ($w \gtrsim 5$ eV) by these low-energy electrons. Since most electron detectors exhibit an energy resolution of $\Delta E \lesssim 0.5$ eV, the detection only of elastically scattered electrons (elastic low-energy electron diffraction or ELEED) effectively discriminates against reflected electrons emanating from depths greater than λ_{ee}. The detection of electrons that have experienced only a single energy-loss process (inelastic low-energy electron diffraction or ILEED) also measures the consequences of scattering events occurring within distances of about λ_{ee} of the surface.

The nature of electron diffraction from crystals can be understood most simply by regarding a crystal as composed of geometrically equivalent layers of atoms parallel to a given surface. An incident electron of wave vector $\mathbf{k}$ (i.e., momentum $\mathbf{p} = \hbar\mathbf{k}$) is diffracted from the array of these two-dimensional gratings stacked together to form the crystal. The periodic translational symmetry of these diffraction gratings is manifested in the electron–solid scattering cross sections by virtue of the reflected electrons emerging from the solid in a series of diffracted beams as indicated in Fig. 2. The configuration of the diffracted beams relative to the beam of incident electrons may be predicted from the translational symmetry of the system, which leads to the momentum conservation law

$$\mathbf{k}_{\parallel}' = \mathbf{k}_{\parallel} + \mathbf{g}, \quad (2a)$$

$$k_{\parallel} = (2mE/\hbar^2)^{1/2} \sin\theta \quad (2b)$$

for electrons of energy E incident on a planar surface at an angle θ relative to its exterior normal. The vectors $\mathbf{g}$ are the reciprocal lattice vectors of the two-dimensional atomic diffraction gratings, as indicated in Fig. 2. The only information conveyed by the configuration of the diffracted beams is the translational symmetry parallel to the surface of the two-dimensional atomic diffraction gratings that make up the solid. Determination of the unit-cell structure within each of these gratings and of the packing sequence of the gratings relative to each other requires analysis of the diffracted intensities.

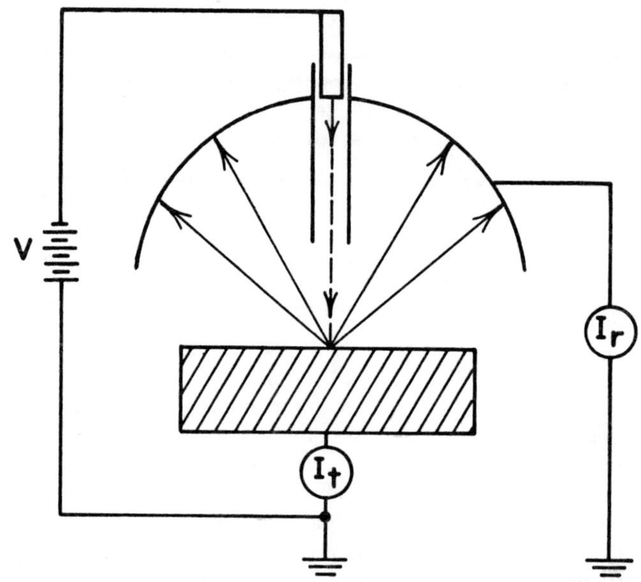

FIG. 1. Schematic diagram of a low-energy electron diffraction apparatus (i.e., $V \sim 100$ V) illustrating that the current drawn from the gun, $I = I_t + I_r$, passes either through the target to ground (I_t) or is reflected back from the target (I_r). [After C. B. Duke, *Adv. Chem. Phys.* **27**, 1 (1974).]

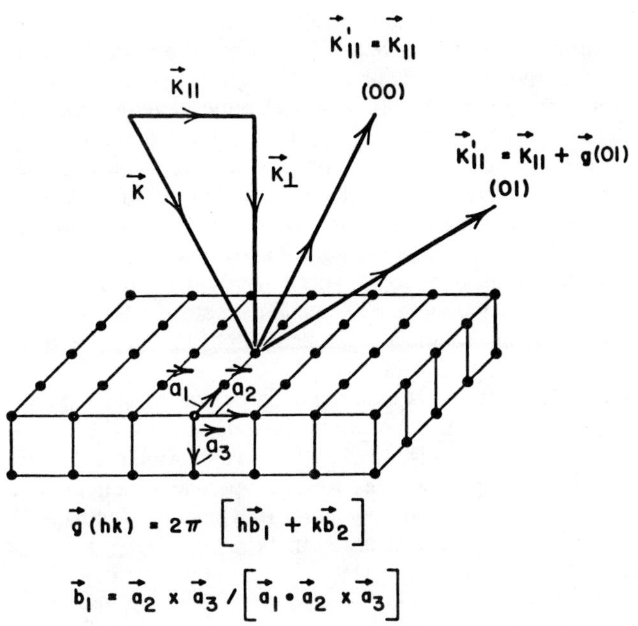

FIG. 2. Schematic illustration of an incident electron beam of wave vector $\mathbf{k}=\mathbf{k}_{\perp}+\mathbf{k}_{\parallel}$, scattered elastically from a single crystal into a state characterized by the wave vector $\mathbf{k}'=\mathbf{k}_{\parallel}'+\mathbf{k}_{\perp}'$. The construction of the reciprocal lattice associated with the single-crystal surface is also shown. The vectors $\mathbf{g}(hk)$ designate the reciprocal lattice vectors associated with the lowest-symmetry Bravais net parallel to the surface. [After C. B. Duke, *Adv. Chem. Phys.* **27,** 1 (1974).]

The observation in 1927 of diffracted beams corresponding to the parallel-momentum conservation law (Fig. 2) was one of the original demonstrations of the wave nature of the electron, a fact that led to the recognition in 1937 of Clinton Davisson as a Nobel laureate. The potential for the use of ELEED as a technique for surface crystallography was recognized in Davisson and Germer's original 1927 paper. This application of ELEED was not achieved, however, until the early 1970s. The reason for this 40-year delay between recognition and achievement lies in the nature of the electron–solid interaction. The fact that it is strong (i.e., $\lambda_{ee}\sim 5$ Å) is necessary for ELEED to be a surface-sensitive spectroscopy. Because it is strong, however, a single-scattering (Born-approximation) calculation, patterned on the analysis of x-ray diffraction from bulk solids, does not suffice to describe the scattered intensities. Since analysis of these intensities is required for surface crystallography, the accomplishment of the latter awaited the construction of an adequate renormalized quantum field theory of the electron-diffraction process that properly describes the occurrence of multiple elastic and inelastic electron scattering events. Such theories were forthcoming during 1969–1971. Surface crystallography via ELEED intensity analysis subsequently has become routine for the low-index faces of clean metals and semiconductors, as well as for simple adsorbed overlayers on metals.

That inelastically scattered electrons as well as elastically scattered ones emerge from the surfaces of crystalline solids in a series of beams was also noted in the original Davisson and Germer paper. As in the case of ELEED, however, the application of ILEED for quantitative surface characterization awaited the construction in 1971 by Duke and Laramore of a proper theory of the inelastic diffraction process. Presently, analyses of ILEED intensities from the low-index faces of clean metals are utilized to determine the energy–momentum relation of electronic surface excitations (e.g., surface plasma oscillations).

Measurements of the configuration and intensity of LEED beams are carried out under ultrahigh-vacuum conditions ($p\lesssim 10^{-8}$ N/m^2) in order to prevent contamination of the surface during the experiment. A typical apparatus consists of the vacuum chamber and pumps, electron gun, electron detector, and goniometer (i.e., a sample holder permitting precise control of the orientation and temperature of the specimen). Most LEED measurements are performed, however, in conjunction with other experiments which determine the elemental composition and electronic structure of the surface under examination. Thus, LEED is best viewed as one of a number of spectroscopic techniques utilized in concert to characterize solid surfaces. Specifically, ELEED is the primary technique for the determination of the atomic geometry of crystalline surfaces, and ILEED is one of many for the assessment of their electronic structure.

See also CRYSTALLOGRAPHY, X-RAY; ELECTRON DIFFRACTION.

BIBLIOGRAPHY

History and General Usage

C. B. Duke, *Crit. Rev. Solid State Sci.* **4,** 541 (1974). (E)
G. Thompson, *Contemp. Phys.* **9,** 1 (1968). (E)

Electron–Solid Interactions

C. B. Duke, *Crit. Rev. Solid State Sci.* **4,** 371 (1974). (E)
C. B. Duke, *Comments Solid State Phys.* **6,** 141 (1975). (I)

ELEED Theory and Applications to Surface Crystallography

C. B. Duke, *Adv. Chem. Phys.* **27,** 1 (1974). (I)
C. B. Duke, in *Proc. 2nd Int. Conf. Solid Surfaces* (H. Kumagai and T. Toya, eds.) (*Jap. J. Appl. Phys. Suppl.* 2, Pt. 2, 1974), pp. 641–651. (A)
C. B. Duke, in *Proc. Int. School of Phys. "Enrico Fermi,"* Course 58 (F. O. Goodman, ed.), pp. 99–213. Editrice Compositori, Bologna, 1975. (I)
J. B. Pendry, *Low Energy Electron Diffraction Theory and Its Application to Determination of Surface Structure.* Academic Press, London, 1974. (A)
T. N. Rhodin and D. S. Y. Tong, *Phys. Today* (October 1975) p. 23. (E)

ILEED Theory and Application to Characterize Surface Electronic Structure

C. B. Duke, *Adv. Chem. Phys.* **27,** 1 (1974). (I)
C. B. Duke, in *Proc. Int. School of Phys. "Enrico Fermi,"* Course 58, pp. 214–249. Editrice Compositori, Bologna, 1975. (I)

Luminescence (Fluorescence and Phosphorescence)

Arthur M. Halpern

Luminescence is the emission of optical radiation (infrared, visible, or ultraviolet light) by matter. This phenomenon is to be distinguished from incandescence, which is the emission of radiation by a substance by virtue of its being at a high temperature (> 500°C). Luminescence can occur in a wide variety of matter and under many different circumstances. Thus, atoms; organic, inorganic, or organometallic molecules; polymers; organic or inorganic crystals; and amorphous substances all luminesce under appropriate conditions. The most straightforward way in which luminescence can be produced is by the application of nonionizing radiation (visible or ultraviolet light) to some form of matter. For example, the organic dye rhodamine-B emits orange light if it is irradiated with green light (or shorter wavelength light). Many papers, fabrics, and laundry detergents contain "optical brighteners" which emit blue or purple light when exposed to ultraviolet radiation. The production of light in this manner is called *photoluminescence*.

Other common types of luminescence and their modes of production are as follows:

Radioluminescence (or scintillation) is produced by ionizing radiation. Some polymers contain organic molecules which emit visible light when exposed to such radiation as x rays, γ rays, or cosmic rays, and thus act as detectors for high-energy radiation. In many biomedical applications scintillation counters are used. The scintillators are organic molecules which emit visible light when they are excited by the decay products of certain radioactive isotopes such as ^3H (tritium), ^{14}C, and ^{18}O.

Electroluminescence is produced by electric fields or plasmas. An example is the gas discharge tube. These are lamps which are filled with a gas such as neon, argon, nitrogen, etc., and contain electrodes which are connected to a high-voltage source. The atoms or molecules are thus ionized, and subsequent ion–electron recombination leads to the production of light. The common fluorescent lamp is a discharge tube in which mercury atoms are electrically excited and then transfer this energy to a material (phosphor) which coats the inside surface of the tube. The phosphor emits most of the observed light. Another example is lightning in which the static electricity generated by a moving air mass ionizes or breaks up the nitrogen and oxygen molecules. The N_2 molecule and some NO which is formed are primarily responsible for the visible light which is then produced. Another example is the now familiar light-emitting diode.

Chemiluminescence is produced as a result of a chemical reaction usually involving an oxidation–reduction process. One common example is the so-called luminol reaction. When luminol, an organic molecule, is mixed with an appropriate reagent containing an oxidizing agent, bright blue light is given off. This light is one of the products of the chemical reaction.

Bioluminescence is the result of certain oxidation processes (usually enzymatic) in biological systems. The firefly and its larva, the glowworm, are well-known examples. These organisms produce a molecule called luciferin and an enzyme known as luciferase. When these molecules are in the presence of oxygen and adenosine triphosphate, yellow-green light is produced. Many other organisms such as bacteria, fungi, and certain marine species are bioluminescent.

Thermoluminescence is activated thermally after initial irradiation by some other means.

Although there is a very wide range of circumstances in which luminescence is observed, there is one common aspect to this phenomenon. When a luminescent atom or molecule emits a photon, there is a change in its electronic structure. When electrons are confined to a region of space which is defined by the positions of the atomic centers of a molecule (or nucleus, in the case of an atom) a set of discrete energy levels is imposed on the system. The way in which the electrons fill up these energy levels (or orbitals) determines the properties of the system. Thus, the electronic structure of a molecule or atom is a representation of how the electrons are arranged in the orbitals.

For example, the *ground state* of a system corresponds to the situation in which the electrons occupy the lowest possible energy levels. The Pauli exclusion principle requires that not more than two electrons occupy one orbital. *Excited states* refer to other arrangements in which the electrons fill the energy levels in such a way that the total energy is higher than that of the ground state. In this context, then, luminescence is the emission of a photon and is a consequence of a transition from an electronically excited state to some other lower energy state, usually the ground state.

Because the electron possesses an intrinsic spin angular momentum, the picture describing electronic states is more complicated. It is found experimentally that when an electron is confined to an atom or molecule (or when it is under the influence of a magnetic field), its spin angular momentum is quantized and can have one of only *two* possible values. If two electrons populate the same orbital, they must have different spin values; in such a case, the electrons are said to be paired. When an excited state is represented, one of the electrons is promoted to occupy a higher (usually vacant) energy level. Thus in the excited state, there are two half-filled orbitals. The electrons in these orbitals can have different spin values, which was the case in the ground state, or they can have the same spin values. The former situation is referred to as a *singlet* state, while the latter denotes a *triplet* state. This is an important distinction because it greatly affects the characteristics of the luminescence which is produced by these excited states.

It is observed experimentally that a triplet state has a lower energy relative to the corresponding singlet state. Moreover, it is known that in the presence of a magnetic field, the triplet state splits into three sublevels whose separation is proportional to the magnetic field strength, hence the term "triplet" state.

Thus there are two different types of luminescence: *fluorescence*, in which emission takes place from an excited singlet state, and *phosphorescence*, in which emission occurs from an excited triplet state. The relative energies of the ground state and various excited singlet and triplet states are schematically portrayed in Fig. 1. In this diagram, each state is also represented by an orbital-energy-level scheme which is filled with four electrons. The upward and downward arrows symbolize the different spin values of the electron.

Because most organic molecules are ground-state singlet systems (all electrons paired), the direct production of an excited singlet state (via photon absorption) is considered

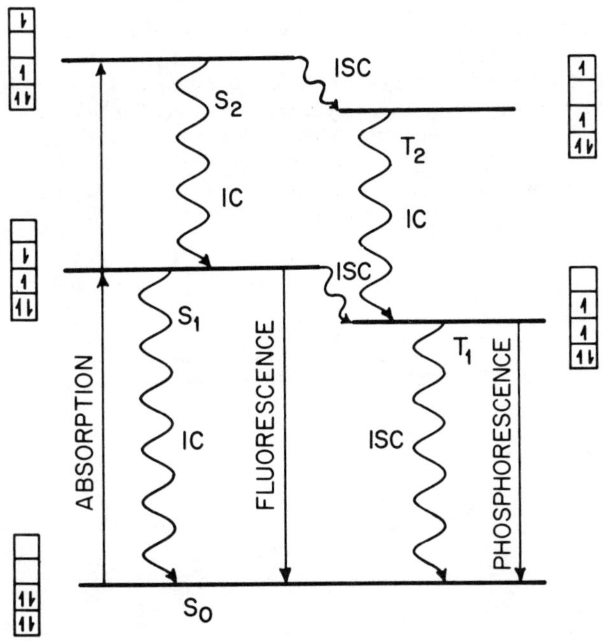

FIG. 1. Schematic electronic-state diagram for a four-electron system. S_0, S_1, and S_2 refer to the ground, first, and second excited singlet states, respectively, while T_1 and T_2 denote the first and second excited triplet states. Alongside each of the electronic state positions is an orbital diagram in which the direction of the arrow symbolizes the electron's spin value. The straight arrows connecting states describe radiative transitions. For example, S_1 or S_2 can be reached from S_0 via photon absorption, and the ground state is produced from S_1 or T_1 by photon emission. The wavy arrows portray radiationless transitions: internal conversion (IC), or intersystem crossing (ISC).

to be a spin-allowed transition. In contrast, the direct production of a triplet state is a spin-forbidden transition because, during this process, there must be a net change in the spin angular momentum of the system. This has the consequence that singlet–singlet transitions occur with much higher probability relative to the singlet–triplet transitions. For organic molecules containing "light atoms" such as H, C, N, O, S, P, and others, the ratio of a singlet–singlet *vis-à-vis* singlet–triplet transition probabilities is about 10^8.

Thus, in photoluminescence, an excited singlet state (e.g., S_1 or S_2 in the diagram) can be directly produced by photon absorption. This excited state can then undergo what are called radiationless transitions. In these processes, other electronic states are reached without the absorption or emission of a photon, and the transitions basically involve the conversion of electronic energy into vibrational (i.e., thermal) energy. The spin-forbidden radiationless transition (i.e., singlet–triplet or triplet–singlet) is usually called *intersystem crossing* (ISC), and the spin-allowed transition (singlet–singlet or triplet–triplet) is referred to as *internal conversion* (IC) (see Fig. 1).It is generally observed that the rate of a radiationless process decreases sharply as the energy gap between the two electronic states increases. Because for most molecules, the energy differences between the excited states are much larger than the gap between the lowest excited state and the ground state, the lowest excited state is usually formed with very high efficiency after optical excitation into any of the higher excited states. Thus, with very few exceptions, fluorescence and phosphorescence take place from the lowest excited singlet and triplet states, respectively.

For most inorganic molecules or organic molecules which contain "heavy atoms" (e.g., bromine and iodine atoms, metal atoms or ions, etc.), the spin-forbidden processes are much less restrictive, and luminescence is frequently called fluorescence even if it involves a change in spin angular momentum. In these cases, the emission is much shorter lived than the typical phosphorescence lifetime of a molecule containing "light atoms." In this sense, then, the distinction between fluorescence and phosphorescence is made on the basis of the luminescence lifetime.

For most organic molecules, however, the phosphorescent lifetimes are generally much longer than fluorescent lifetimes (see above). Because the triplet state is so long-lived, it has a high probability of undergoing a reaction by which the excitation energy is dissipated by some means. If this process occurs as a result of an interaction with another molecule, it is called *quenching*. One common consequence of the quenching of a triplet state by another molecule is energy transfer in which the excitation energy is "picked up" by the quencher. The quencher, now energetically enriched, can undergo some chemical or physical reaction. This is called triplet sensitization: The triplet state of a molecule induces some reaction in another species as a result of energy transfer.

A common mode of quenching of triplet states is the interaction with molecular oxygen. This process takes place with high efficiency not only because the triplet state is intrinsically long-lived, but also because the oxygen molecule has two unpaired electrons in its ground state (i.e., it is a ground-state triplet). The consequence of this is that the quenching reaction is a spin-allowed process. In some cases, this has been shown to result in the production of electronically excited (singlet) oxygen. This energy-enriched, long-lived form of molecular oxygen can act as a powerful oxidizing agent toward some suitable substrate. The application of light in the presence of oxygen is known to result in denaturation of some biological systems. This is called photodynamic action and is known to involve the excited triplet state as a sensitizer.

Triplet states also have a high reactivity toward other molecules which might be present in very small amounts (such as impurities). For this reason, phosphorescence in fluid media is usually very weak because of the high mobility of the molecules and the correspondingly high probability of an encounter between the triplet state and a quencher. The phosphorescence intensity of a system can be significantly enhanced by cooling the system so that the bulk solvent viscosity increases. This results in a decrease in the encounter probability between a triplet state and a potential quencher. Alternatively, phosphorescence can often be readily observed at room temperature if the phosphorescing molecules are dissolved in a rigid polymeric "glass" or some other type of isolating matrix.

See also ELECTROLUMINESCENCE; THERMOLUMINESCENCE.

BIBLIOGRAPHY

R. S. Becker, *Theory and Interpretation of Fluorescence and Phosphorescence*. Wiley-Interscience, New York, 1969. (I)

John B. Birks, *Photophysics of Aromatic Molecules*. Wiley-Interscience, London, 1970. (A)

C. A. Parker, *Photoluminescence in Solutions*. Elsevier, Amsterdam, 1968. (I)

P. G. Seybold, "Luminescence," *Chemistry* **42**, 6 (1973). (E)

G. H. Schenk, *Absorption of Light and Ultraviolet Radiation*. Allyn & Bacon, Boston, 1973. (E)

N. J. Turro, "The Triplet State," *J. Chem. Educ.* **46**, 2 (1969). (E)

F. Williams (ed.), *Luminescence of Crystals, Molecules, and Solutions*. Plenum Press, New York, 1973. (A)

Magnetic Cooling

O. V. Lounasmaa

Adiabatic demagnetization of a paramagnetic salt, often referred to as magnetic cooling, is the oldest method of reaching temperatures significantly below 1 K. This technique has been practiced for over 40 years. The use of nuclear instead of electronic magnetic moments for cooling was pioneered by N. Kurti in 1955. Nuclear refrigeration has recently proven its worth in investigations of the superfluid phases of ^{3}He down to 1 mK and below. Recent reviews about magnetic cooling have been written by Hudson[1] and Lounasmaa.[2]

PARAMAGNETIC SALTS

Consider a solid with paramagnetic ions. In salts suitable for magnetic cooling the interaction energy ϵ of these ions with their crystalline environment and with each other is small in comparison with the thermal energy kT at temperatures near 1 K. Each ion is thus relatively free in a system of randomly oriented dipoles. This results in an entropy $R\,\ell n(2J+1)$ per mole, where J is the total angular momentum quantum number. In practice, only salts of the iron or the rare-earth groups of ions are important for refrigeration. Their paramagnetism is caused by the magnetic moment associated with electrons in the incompletely filled $3d$ and $4f$ shells.

In zero magnetic field the molar entropy of the salt is a slowly varying function of the temperature until ϵ becomes an appreciable fraction of kT. Spontaneous ordering of the dipoles occurs at a temperature $\theta \simeq \epsilon/k$ characteristic of the salt; this reduces the entropy. At very low temperatures the system goes to a ground state of zero entropy as required by the third law of thermodynamics. At higher temperatures, usually above 1 K, the entropy rises over $R\,\ell n(2J+1)$ as the lattice contribution becomes appreciable.

At $T > \theta$ the entropy of the spin system can be lowered by an external magnetic field B which orients the dipoles. For significant polarization, i.e., large reductions in entropy, the magnetic interaction energy between a dipole and the external field should be comparable to or larger than kT.

The qualitative features of the entropy-vs.-temperature diagram of a paramagnetic salt are illustrated in Fig. 1. We assume that, by pumping on a bath of liquid ^{4}He or ^{3}He, point X at the initial temperature T_i and zero magnetic field has been reached. The salt is then magnetized isothermally to an external field B_i (1 T in Fig. 1) along the path $X \rightarrow Y$. The latent heat of magnetization, represented by the rectangular area $XYZ''X''$, is absorbed by the helium bath. Next the salt is thermally isolated by pumping away the exchange gas or by operating a superconducting heat switch and then demagnetized adiabatically to $B \rightarrow 0$ along the path $Y \rightarrow Z$. Considerable cooling results; the temperature T_0 after the demagnetization depends on T_i, B_i, and θ.

The external heat leak then begins to warm up the system along the $B=0$ curve. The amount of heat that the salt absorbs during this part of the experiment is represented by the darkly shaded area in Fig. 1, and it is considerably smaller than the heat of magnetization which was removed

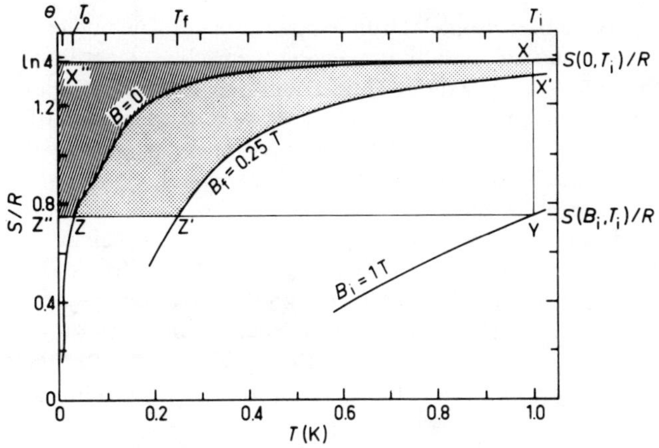

FIG. 1. Entropy diagram of chromic potassium alum illustrating magnetic cooling: $T_i = 1$ K, $T_f = 0.25$ K, $T_0 = 35$ mK, $\theta = 10$ mK. For more details, see text.

at T_i. If demagnetization is stopped at a nonzero external field B_f, the final temperature T_f is higher than T_0 but the amount of heat absorbed (darkly and lightly shaded areas in Fig. 1) during warming along the $B = B_f$ curve from Z' to X' is larger. The warming-up rate is thus slower which may be important in many types of experiment.

These ideas can be put on a more quantitative basis and we refer to the books of Hudson[1] and Lounasmaa.[2] Well above θ the initial and final temperatures, T_i and T_f, and magnetic fields, B_i and B_f, are related by the expression

$$B_i/T_i = (B_f^2 + b^2)^{1/2}/T_f,$$

where b is the effective interaction field between the dipoles. It is found that the entropy reduction is a function of B_i/T_i only.

"BRUTE FORCE" NUCLEAR MAGNETIC COOLING

The basic principle of nuclear magnetic cooling is the same as discussed above, but significant differences occur in practice. As nuclear magnetic moments are about 2000 times smaller than their electronic counterparts, 2000 times larger values of B_i/T_i are required for an equivalent entropy reduction. With $B_i/T_i = 300$ T/K, we obtain an entropy decrease of only 1.25% in copper, which, however, is enough for experiments. Starting temperatures below 15 mK and magnetic fields in excess of 5 T are thus required for nuclear magnetic cooling. These conditions can be reached by dilution refrigeration and by superconducting solenoids.

The advantage of nuclear cooling is that the magnetic ordering temperature θ of nuclear spins is well below 1 μK in a typical metal; for cerium magnesium nitrate, the weakest paramagnetic salt, $\theta = 1.9$ mK. Temperatures in the submicrokelvin region can thus be reached by nuclear demagnetization. Furthermore, it is necessary to separate the magnetic ions of paramagnetic salts with neutral atoms, usually crystalline water, to reduce their mutual interactions. This is not necessary with nuclear spins; a high nuclear entropy per unit volume is thus achieved.

At very low temperatures it is meaningful to speak about two distinct temperatures, the nuclear spin temperature T_n

and the conduction electron temperature T_e. The nuclei approach thermal equilibrium among themselves according to the spin–spin relaxation time τ_2 and the nuclear spins and the conduction electrons according to τ_1, the spin–lattice relaxation time. At low temperatures $\tau_2 \ll \tau_1$ making a separate nuclear spin temperature T_n meaningful.

For most metals τ_1 is of the order of seconds at 10 mK; for insulators τ_1 is days or even weeks. It is thus clear that metals must be used for brute-force nuclear refrigeration. The small τ_1 in metals is due to conduction electrons which act as intermediaries between the nuclear spins and the lattice. Only electrons near the Fermi surface contribute; their number is proportional to T_e making τ_1 proportional to $1/T_e$, i.e.,

$$\tau_1 T_e = \kappa,$$

where κ is the so-called Korringa constant. In practice only copper and indium have been used for brute-force nuclear magnetic cooling.

The unavoidable external heat leak $\dot{Q}$ into the conduction electron system has an important effect on the equilibrium between T_e and T_n. If $\dot{Q}$ is large, the spin–lattice relaxation process is not sufficiently rapid for cooling the conduction electrons adequately, and $T_e - T_n$ will be large. The relationship

$$(T_e/T_n) - 1 = \mu_0 \kappa \dot{Q}/\Lambda(B^2 + b^2)$$

describes the situation quantitatively; here μ_0 is the permeability in free space and Λ is the nuclear Curie constant.

Numerical calculations show that often, in order to stay cold for sufficiently long times and to obtain significant refrigeration of conduction electrons or an external specimen, nuclear demagnetization should not be carried out to $B_f = 0$ but should be stopped at some intermediate field. In particular, the lowest T_e is reached by demagnetizing to

$$B_f(\text{opt}) = (\mu_0 \kappa \dot{Q}/\Lambda)^{1/2};$$

in this case $T_e/T_n = 2$.

Many laboratories already have nuclear refrigerators which are capable of cooling superfluid ^3He below 0.5 mK; the lowest ^3He temperature obtained so far is 0.21 mK. Conduction electrons in copper have been refrigerated to $T_e = 50$ μK. The lowest nuclear spin temperature produced in copper is much less, $T_n = 50$ nK (5×10^{-8} K). Figure 2 is a schematic illustration of a cascade refrigerator for investigating spontaneous nuclear ordering in metals.[3,4]

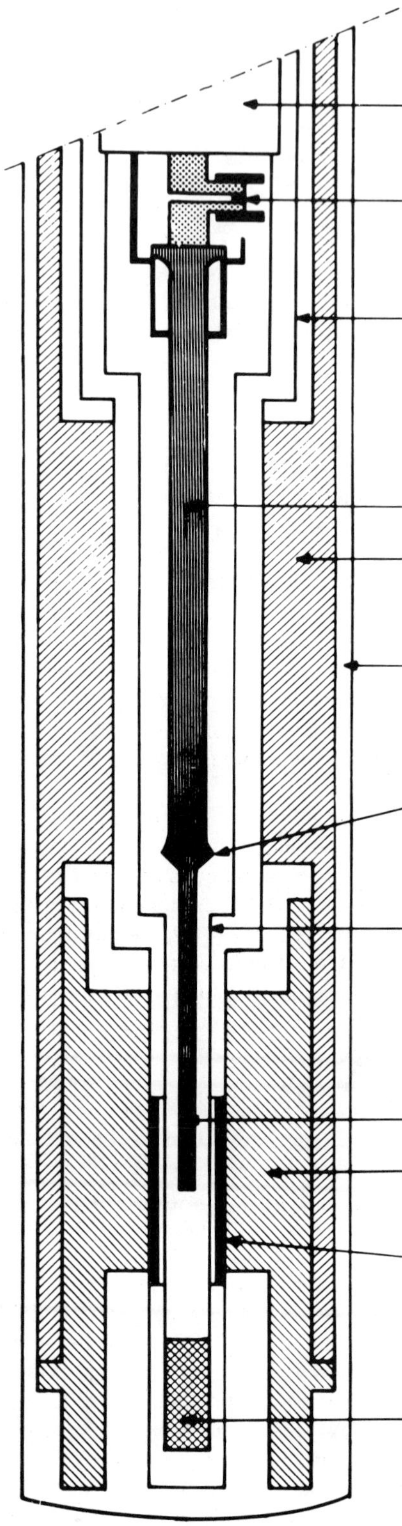

FIG. 2. Schematic illustration of a cryostat[3] in which a dilution refrigerator and two copper nuclear stages operate in series. Demagnetizing from 0.4 mK and 7 T to zero field, a nuclear spin temperature of 50 nK was reached in the second nuclear stage. This is the lowest temperature ever produced.

NUCLEAR COOLING BY INDIRECT METHODS OF POLARIZATION

A somewhat different and promising procedure of refrigeration is the hyperfine-enhanced nuclear cooling method.[5,6] In Van Vleck paramagnetic lanthanide alloys, in which the rare-earth ion occupies a crystal-field-split singlet ground state, an applied magnetic field mixes excited states with the ground state. A large electronic polarizability is thereby induced which produces an enhancement of the external magnetic field at the sites of the rare-earth nuclei. In the best specimens, usually intermetallic alloys of praseodymium, the field is boosted by a factor varying from 10 to 20. For

$PrNi_5$, with modest precooling conditions (B_i=2.0 T and T_i=20 mK), a reduction of 25% in the total nuclear spin entropy can be achieved. With this method liquid ^{3}He has been cooled down to 1.0 mK.

In nuclear demagnetization experiments it is sometimes sufficient to cool just the nuclei (remember $\tau_2 \ll \tau_1$), which makes it much easier to produce ultralow temperatures than in an equilibrium situation. Abragam, Goldman, and their co-workers have cooled the nuclear spins of CaF_2 and LiH specimens, doped with some electronic impurities, below 1 μK. A high initial magnetization, up to 90%, was achieved by dynamic nuclear polarization. Demagnetization was then performed in a rotating frame. The nuclei retained the low temperature, because of their high degree of thermal isolation from the lattice, for a considerable length of time. Very interesting results have been obtained, explained by nuclear antiferromagnetism and nuclear ferromagnetism.[7]

CONCLUSION

Magnetic cooling by means of a paramagnetic salt will soon be a technique of the past with dilution refrigerators taking over the temperature range between 2 mK and 0.3 K. Nuclear refrigeration is the coming method. Techniques developed for studies of ^{3}He will soon be used for other measurements as well.

See also CRYOGENICS; PARAMAGNETISM.

REFERENCES

1. R. P. Hudson, *Principles and Applications of Magnetic Cooling*. North-Holland, Amsterdam, 1972.
2. O. V. Lounasmaa, *Experimental Principles and Methods below 1 K*. Academic Press, London and New York, 1974.
3. G. J. Ehnholm, J. P. Ekström, J.-F. Jacquinot, M. T. Loponen, O. V. Lounasmaa, and J. K. Soini, *J. Low Temp. Phys. 39*, 417 (1980).
4. O. V. Lounasmaa, *Progress in Low Temperature Physics*, **8**, (North-Holland, Amsterdam, in preparation).
5. K. Andres *Cryogenics* **18**, 473 (1978).
6. K. Andres, *Progress in Low Temperature Physics*, **8**, (North-Holland, Amsterdam, in preparation).
7. A. Abragam and M. Goldman, *Rep. Prog. Phys.* **41**, 395 (1978).

Magnetic Domains and Bubbles

G. P. Vella-Coleiro

A magnetic material may be thought of as an aggregation of atomic magnetic moments, or tiny magnets, which are acted upon by a large internal force known as the exchange interaction. This interaction is a quantum-mechanical concept and although it is electrostatic in origin it has no classical analog. Its effect is to couple adjacent atomic magnetic moments parallel to one another in a ferromagnetic material, and antiparallel in an antiferromagnetic one, thus forming regions where the magnetic moments are aligned in the same direction. These regions are known as magnetic domains. The magnitude of the exchange interaction can be appreciated by noting that in iron, a typical ferromagnetic material, it is equivalent to an internal magnetic field with a strength of 4×10^8 A-turns/m (5×10^6 Oe) at room temperature, which is several orders of magnitude larger than commonly encountered magnetic field strengths.

Ordinarily, a sample of ferromagnetic material does not form a single magnetic domain, but rather a collection of domains with the magnetization of adjacent domains pointing in different directions. If the sample were to form a single domain, as shown schematically in Fig. 1a for a rectangular bar, there would be a substantial amount of magnetostatic energy associated with the magnetic field emanating from the surfaces, this energy being proportional to the square of the magnetic field strength integrated over the volume in which it resides. By a break up into two oppositely oriented domains, as in Fig. 1b, the magnetostatic energy is reduced, and further reductions occur with subsequent subdivisions. Each domain is separated from adjacent domains by a transition region, known as a domain wall, which is discussed below. Magnetic energy is associated with a domain wall, and the process of subdivision halts when the reduction in magnetostatic energy is counterbalanced by the energy of the additional domain walls which would be introduced.

In a single-crystal magnetic material the domains are not oriented in arbitrary directions, but they line up parallel to well-defined crystalline axes. This effect is due to the phenomenon of magnetocrystalline anisotropy which causes certain directions in the crystal to be energetically favored over others. For example, in single-crystal iron, whose atoms have a cubic arrangement, the magnetic energy is lowest when the magnetic moments in a domain are oriented parallel or antiparallel to a cube edge direction of the atomic structure, which is therefore known as the easy axis of magnetization. Since in a cubic lattice there are three equivalent cube edges, the domain structure in iron can be rather complicated, the angle between adjacent domains depending on the orientation of the sample. By contrast, a material such as cobalt can exhibit a very simple domain pattern. Single-crystal cobalt has a hexagonal structure and at room tem-

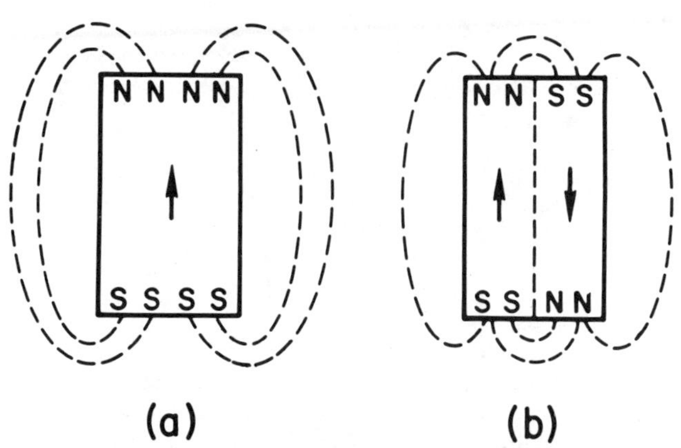

FIG. 1. Magnetostatic energy is associated with the magnetic field (broken lines) which surrounds a magnetic domain due to the magnetic poles (indicated by N and S) on the surfaces. This energy is highest for a single domain, as in (a), and it is approximately halved by subdividing the sample into two domains, as in (b). Further reductions in magnetostatic energy occur as the number of domains increases.

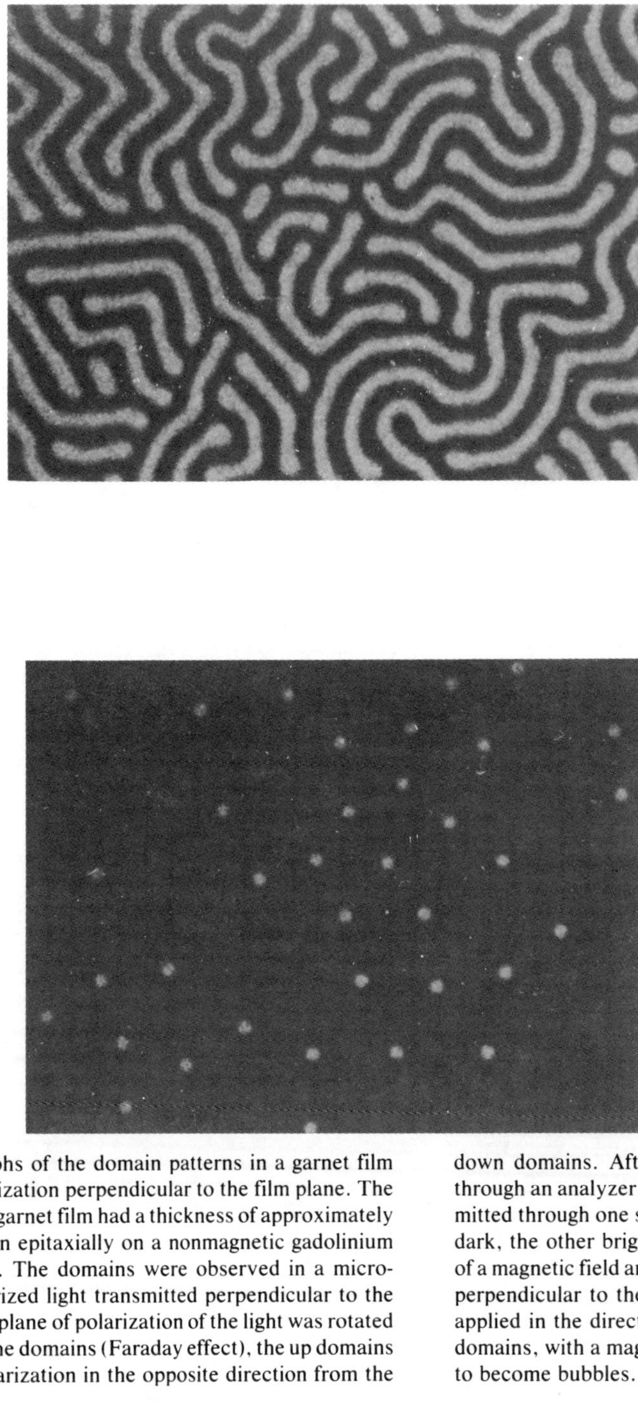

(a)

(b)

FIG. 2. Photomicrographs of the domain patterns in a garnet film with easy axis of magnetization perpendicular to the film plane. The yttrium–samarium–iron garnet film had a thickness of approximately 6.5 μm, and it was grown epitaxially on a nonmagnetic gadolinium gallium garnet substrate. The domains were observed in a microscope by means of polarized light transmitted perpendicular to the plane of the sample. The plane of polarization of the light was rotated by the magnetization in the domains (Faraday effect), the up domains rotating the plane of polarization in the opposite direction from the down domains. After traversing the sample the light was passed through an analyzer which was adjusted to extinguish the light transmitted through one set of domains. Thus one set of domains appears dark, the other bright. Photograph (a) was obtained in the absence of a magnetic field and shows serpentine domains with magnetization perpendicular to the plane of the film. In (b) a magnetic field was applied in the direction opposite to the magnetization in the bright domains, with a magnitude sufficient to cause the "island" domains to become bubbles. The bubble diameter was approximately 6 μm.

perature the easy direction of magnetization is parallel to the hexagonal axis. Thus a plate of cobalt cut with its faces perpendicular to the hexagonal axis forms domains oriented parallel or antiparallel to the plate normal. A similar situation exists in specially prepared ferrite crystals known as garnets, which have achieved great technological importance recently due to their suitability for use in magnetic bubble devices (see below). The domain pattern in such a crystal, made visible by means of the Faraday effect, is shown in Fig. 2a. It consists of serpentine domains with the direction of magnetization alternating between up and down. In the absence of an external magnetic field the areas of the up and down domains are equal, resulting in no net magnetization.

As noted above, the domain walls which separate adjacent domains contain energy, predominantly from two sources, the exchange interaction and the magnetocrystalline anisotropy. The exchange energy density associated with a domain wall is small if the wall is broad, occupying many lattice sites, since this results in adjacent magnetic moments within the wall being nearly parallel. In this case, however, the magnetocrystalline energy density would be large since the wall would contain many magnetic moments pointing in a direction other than an easy axis. Reducing the wall width reduces the number of these moments, thus reducing the magnetocrystalline energy. The equilibrium wall width is therefore determined by a balance between the exchange and the magnetocrystalline energies. The former would like all the moments to be parallel, which would result in a very broad wall, while the latter prefers the moments to lie parallel to an easy axis, which would result in a wall only one lattice constant wide. Somewhere in between these two extremes lies a compromise wall width which minimizes the sum of the two energies. Analysis of the wall structure in an infinite magnetic medium shows that the minimum energy occurs at the wall width which makes the exchange energy density equal to the magnetocrystalline energy density.

Another source of energy which is very important in determining the behavior of domains is an external magnetic field. Just like a magnetic compass needle which orients itself parallel to the earth's magnetic field, magnetic domains try to orient so that the magnetic moments are parallel to an external field. Because of the competing forces arising from the magnetocrystalline anisotropy and the magnetostatic energy, complete orientation only occurs when the magnetic field exceeds a critical value which depends on the magnetic properties of the material and on its shape. For field strengths lower than the critical value, the domains whose magnetic moments are most nearly parallel to the external field are favored over others and thus occupy a larger volume. In the example of Fig. 2a, application of a low-strength magnetic field perpendicular to the plane of the magnetic film, say in the up direction, causes the domains pointing up to expand in area at the expense of those pointing down. The process continues until the critical magnetic field strength is reached, when the sample becomes a single domain pointing up. The situation is modified somewhat by the presence of coercivity which inhibits changes in the net magnetization for external fields lower than some value which depends on the material and on its preparation, such as heat treatment. Values of the coercive field range from less than 10 A-turns/m (0.1 Oe) for materials such as that of Fig. 2, to a few hundred thousand ampere-turns per meter for permanent magnet materials.

Another type of domain can occur if the stripe domains do not extend from one edge of the sample to another, but are isolated, for example by cutting with a magnetized needle at both ends (several of these "island" domains can be seen in Fig. 2a). In the presence of a sufficiently strong magnetic field oriented antiparallel to the magnetic moments in these domains, they shrink into magnetic bubbles, which are actually cylindrical domains with their axes perpendicular to the plane of the magnetic film, see Fig. 2b. Bubble domains are stable over a diameter range of approximately 3 to 1, in a range of applied field strengths of approximately one-tenth the saturation moment of the material. Their stability is determined by an interplay between the magnetostatic energy, the total wall energy, and the external field energy. The magnetostatic energy decreases with increasing radius, while the other two increase, and a stable bubble domain is formed at the radius where the sum of the three energies is a minimum. As the magnetic field is increased, the bubble radius decreases until a point is reached where a further increase in magnetic field energy can no longer be compensated by a decrease in magnetostatic and wall energies; then the bubble becomes unstable and collapses. Conversely, as the magnetic field strength is reduced, the bubble radius increases up to the point where a distortion of the circular shape of the bubble to an elliptical one would result in a decrease of the magnetostatic energy by a larger amount than the increase in wall and magnetic field energies. The bubble thus becomes unstable toward such distortions and the domain runs out into a stripe.

The current interest in magnetic bubbles stems from their potential applicability to memory systems capable of handling large amounts of data at high speed. In a manner analogous to magnetic tape or disc reorders, binary information can be coded in terms of the presence or absence of a domain. Unlike the tape or disc units, however, where the domains are stationary and data are accessed by moving the magnetic material past read/write heads, in a bubble device the bubble material is stationary. Access to the data is achieved by translating the bubbles within the material by means of externally applied magnetic fields. In the presence of a nonuniform magnetic field, a bubble domain moves in the direction in which the magnetic field strength is least since this results in a reduction of the total energy. In a

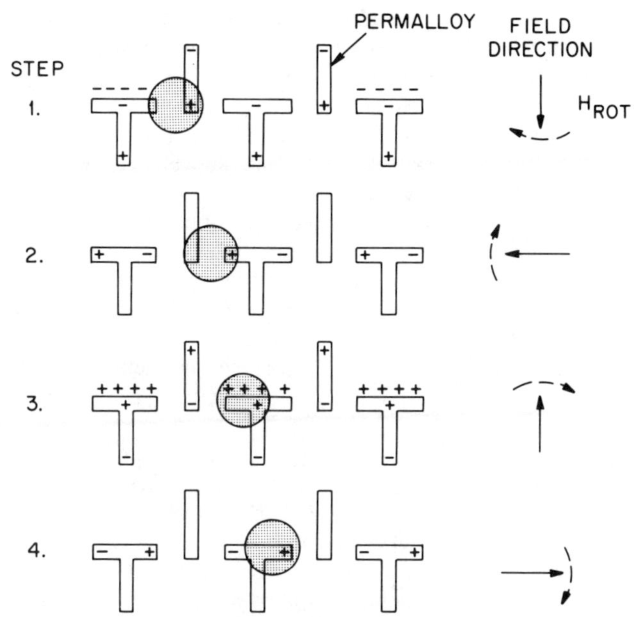

FIG. 3. Schematic representation of bubble propagation in a Permalloy structure consisting of T and I elements. The thin Permalloy elements are in close proximity to the bubble material, to which a perpendicular magnetic field is applied to stabilize the bubble domains, represented by the discs. A rotating in-plane magnetic field H_{rot} is also applied. As the in-plane field rotates from the direction shown in step 1 to step 2, the magnetization of the permalloy elements, indicated by the + and − signs, changes and causes the bubble to move to the right. Futher motion occurs as the field rotates to positions 3 and 4, and when one cycle of rotation is complete the bubble will be in a location equivalent to that in step 1, but one circuit period to the right. A stream of bubbles can thus be propagated in the plane of the film, data being encoded according to the presence or absence of a bubble in the stream.

practical bubble device, a nonuniform magnetic field is obtained by means of an array of high-permeability Permalloy elements in close proximity to the magnetic material, in conjunction with a magnetic field in the plane of the material (see Fig. 3). The in-plane field is separate from, and in addition to, the perpendicular field which is required to stabilize the bubble domains. The Permalloy elements are magnetized by the in-plane field, thereby creating localized nonuniform fields which exert forces on the bubble domains, causing them to seek regions where the local magnetic field has the least value. As the direction of the in-plane field rotates, these regions move along the Permalloy elements, carrying the bubbles with them. A coded stream of bubbles can thus be made to propagate past a detector, which makes use of the magnetoresistive properties of a Permalloy element to produce an electrical signal in response to the passage of the bubble domains. In-plane fields rotating at rates well in excess of 10^5 Hz can be produced readily by means of a pair of orthogonal solenoids, and materials which support bubbles with submicrometer and larger diameters are available. Photolithographic and electron-beam techniques have been developed to produce the Permalloy patterns required for bubble propagation and detection, thus forming the basis for high-speed, high-storage-density magnetic bubble memories.

Another type of bubble memory, currently in the early stages of development, utilizes an array of bubbles, called a bubble lattice, which is stabilized by the magnetic interaction between bubble domains. For domains of similar polarity this interaction results in a repulsive force between domains which, when closely spaced, form a two-dimensional hexagonal array with a spacing approximately equal to twice the bubble diameter. Information coding is achieved by using two types of bubble which differ only in the detailed structure of their domain walls. One bubble type has a wall which contains Bloch lines, i.e., 180° reversals of the magnetization direction, while the other does not. These two types of bubble can be differentiated by the fact that they propagate at different angles with respect to a field gradient, which forms the basis for data retrieval in this type of bubble memory.

Additional information on magnetic domains and bubbles may be found in the references listed below. Reference 1 is a very readable, nonmathematical account of ferromagnetic domains and their observation. Reference 2 contains a more advanced discussion of domain theory as well detailed descriptions of several techniques used for studying domains. References 3 and 4 are textbooks on magnetism which contain several chapters on domains. A review of the theory and device aspects of magnetic bubbles may be found in Ref. 5, while more detailed information is contained in Refs. 6, 7, and 8.

See also FERRIMAGNETISM; FERROMAGNETISM; HYSTERESIS; MAGNETIC MOMENTS.

REFERENCES

1. E. A. Nesbitt, *Ferromagnetic Domains*. Williams and Wilkins, Baltimore, 1962. (E)
2. R. Carey and E. D. Isaac, *Magnetic Domains and Techniques for their observation*. Academic Press, New York, 1966. (A)
3. A. H. Morrish, *The Physical Principles of Magnetism*. Wiley, New York, 1966. (I)
4. S. Chikazumi, *Physics of Magnetism*. Wiley, New York, 1966. (I)
5. A. H. Bobeck and H. E. D. Scovil, "Magnetic Bubbles," *Scientific American*, 78–90 (June 1971). (I)
6. A. H. Bobeck and E. Della Torre, *Magnetic Bubbles*. North-Holland, Amsterdam, 1975. (A)
7. T. H. O'Dell, *Magnetic Bubbles*. Wiley, New York, 1974. (A)
8. A. H. Eschenfelder, *Magnetic Bubble Technology*. Springer-Verlag, New York, 1980.

Magnetic Materials

J. J. Rhyne

INTRODUCTION TO MAGNETIC MATERIALS

Magnetic materials have traditionally been considered as those elements, alloys, and compounds which exhibit an ordered arrangement or correlation among magnetic electron moments or spins. This net magnetic polarization can be either ferromagnetic, in which all spins are aligned parallel; antiferromagnetic, in which neighboring spins are aligned antiparallel; or ferrimagnetic, which is an antiparallel arrangement of spins of two dissimilar atomic species. In some materials, principally the rare earths, or compounds containing rare-earth atoms, more complicated periodic or "spiral" spin orderings are observed which may not be commensurate with the regular lattice periodicity. The details of these magnetic spin arrangements and the exchange interaction mechanisms responsible for stabilizing them are the subject of other works[1,2] and sections in this volume (see, e.g., *Rare Earths, Ferromagnetism*).

Historically, most magnetic materials consisted of, or at least contained, one of the 3*d* transition metals Fe, Ni, or Co if they were ferromagnetic, and Mn or Cr if the materials were antiferromagnetic. The three ferromagnetic elements have magnetic ordering or "Curie" temperatures well above room temperature, and have been studied widely in both the pure elemental state and in a multitude of alloys and compounds such as Permalloy (Fe-Ni), the Alnico (Ni-Co-Al-Cu) magnet alloys, the Fe-Si alloys, and the ferrite spinels (ferrites) MFe_2O_4 (where M is a 3*d* or other transition metal), to name just a few.[3]

Starting in the 1950s intense interest and investigation took place in the rare-earth elements, alloys, and compounds[4–7] which has made a significant impact on the physics of magnetism and magnetic materials as well as providing new areas for applications. For example the RCo_5 (R is a rare earth) and related compounds are used in permanent magnet applications,[8] and the garnets ($R_3Fe_5O_{12}$) and orthoferrites ($RFeO_3$) are being used in bubble-memory applications.[9]

AMORPHOUS ALLOYS

The study of crystalline materials both from the 3*d* transition elements and from the rare-earth elements has provided much of the experimental basis for our understanding of magnetism in the solid state. Recently, however, a new class of materials has emerged, that of amorphous magnetic alloys. These are solids; however, their constituent atoms are not distributed periodically on a lattice as is the case with crystalline solids; rather the atoms are arranged in a random statistical distribution with regard to interatomic spacing, orientation, and type and number of neighboring atoms (coordination number).[10] They can be thought of then as metallic glasses.

There are two primary classes of amorphous-metal magnetic materials: (1) Those that contain dominantly one or more of the $3d$ transition metals Fe, Co, or Ni and (2) those that contain rare earths. The $3d$ transition-metal materials are stable in the amorphous phase only when combined with a metalloid atom (e.g., Si, P, C, or B) in the ratio of 75–85% transition metal atoms to 15–25% of the metalloid. The metalloid atom is often referred to as a "glass former." A wide variety of amorphous alloys from this class have been prepared[10]; for example, various compositions from the systems Fe-Ni-P, Co-P, Fe-P-C, and Fe-Pd-P have been studied. The second class of amorphous alloys is that of the rare-earth–transition-metal alloys, for example Tb_xFe_{1-x} and Gd_xCo_{1-x}, where in this case alloys have been shown to be amorphous for values of x ranging from 2% rare earth up to above 75% rare earth.[11] In general, elemental materials cannot be produced in the amorphous form except possibly at cryogenic temperatures, and they then recrystallize when heated to room temperature.

Amorphous metals have been produced by a variety of techniques[12] including rapid cooling or "quenching" from the molten state, evaporation, and sputtering. In the quenching technique a fine stream of molten metal is directed onto a cold rotating metal wheel or roller which produces thin ribbons or filaments of amorphous metal. Alternatively the so-called "splat cooling" method may be used in which a droplet of liquid metal is quenched on a cold surface using a piston and anvil technique to produce thin sheets of amorphous alloys. In general these methods rely on the very rapid cooling of the metal at rates up to approximately 10^6 deg/sec which inhibits the formation of the crystalline phase and results in the atomically amorphous phase. Amorphous alloys are also produced by electrodeposition or electroless deposition from aqueous solutions and may be produced in bulk form (millimeter dimension) by high-rate sputtering techniques.

MAGNETIC PROPERTIES

There are many properties of magnetic materials (both crystalline and amorphous) which make them both interesting from the point of view of our understanding of the basic underlying physics of magnetism, and also useful in magnetic devices such as computer components, transformers, permanent magnets, etc. The ordering temperature, which is the temperature at which the magnetic material undergoes a phase transition from the magnetically disordered (paramagnetic) state to a magnetically ordered (long-range spin-correlated) state, is a reflection of the strength of the exchange interactions that are responsible for the coupling of the atomic electron spins. The temperature of this phase transition is the Curie temperature for a ferromagnetic or ferrimagnetic material, and the Néel temperature for an antiferromagnetically ordered material. Generally, the higher the transition temperature, the stronger the magnetic interaction.

Another important property of magnetic materials is the magnetic anisotropy,[13] which refers to the energy required to rotate the direction of magnetization from its intrinsic energetically stable direction (called the easy axis) to some other crystallographic direction. This energy density is typically of order 10^3–10^5 J/m^3 in the $3d$ transition metals and alloys, and reaches values more than 100 times higher in rare-earth materials at low temperatures. The variation of the magnetic anisotropy energy with strain provides the mechanism for magnetostriction which is the dilitation in crystalline lattice parameters which accompanies a change in the magnitude or direction of the magnetization.[14] Such magnetostrictions can be as large as 1% in some of the rare-earth materials at low temperature.

These basic properties of magnetic materials and other quantities which are derived from them, such as the coercive force, the remanent magnetization, and hysteresis loops, and which are related to the occurrence of magnetic domains are useful particularly in describing the device properties of magnetic materials.

MAGNETIC PROPERTIES IN THE AMORPHOUS STATE

The occurrence of ordered magnetism at all in the amorphous or glassy state may be first seem surprising. However, it becomes reasonable when one considers that the nature of the magnetic interactions is governed principally by the local environment of the magnetic atom which may not be too different in the amorphous and crystalline phases.

It is instructive to compare the magnetic properties mentioned above in both crystalline and amorphous-phase materials. Detailed experimental comparisons are sometimes difficult to obtain since many amorphous alloys do not have single-phase crystalline analog compounds of the same composition. This is particularly true of the $3d$ transition-metal amorphous alloys (e.g., $Fe_{0.80}P_{0.13}C_{0.07}$); but not of the rare-earth materials which offer many opportunities to compare both amorphous and crystalline materials of the same composition (e.g., $GdCo_3$, $TbFe_2$, etc.).

The effect of the topological disorder in a structure without translational invariance (noncrystalline) is to introduce fluctuations into the exchange interactions which modify the Curie temperature and magnetization. In general the Curie temperature is decreased in the amorphous state compared to the counterpart crystalline phase, often by a large fraction; although there are exceptions (e.g., the rare-earth cobalt series) in which the Curie temperatures actually increase[15] in the amorphous phases as a result of a change in the electronic band structure. In the following we shall discuss in more detail some of the features of both the rare-earth amorphous systems and the transition-metal alloys.

Rare Earth Alloys

The alloys of transition metals Ni, Co, and Fe with the

heavy rare earths Gd through Tm have been studied extensively within the last four years.[11] The atomic structure of these alloys has been examined by x rays and neutron scattering to confirm their amorphous character and these structures have been compared to computer-simulation calculations based on the "dense random packing model for hard spheres."[16]

The rare-earth alloys are found to be strongly magnetic with Curie temperatures in some cases (e.g., the RCo_2 series) higher than the crystallization temperature (approximately 300°C) at which point the alloys spontaneously transform to the crystalline compound phase. The rare-earth alloys are ferrimagnetic with the transition-metal magnetic moments all parallel, but antiparallel to the rare-earth moments. Actually in many cases a canted or conelike structure for the rare-earth spins is observed as a result of a giant local anisotropy interaction of the rare-earth spins which competes with the exchange and attempts to misalign the rare-earth moments. The angular spread in the rare-earth moment can be significantly affected by an external magnetic field which aids the exchange field. This unique combination of a giant local anisotropy field and the exchange field is believed also to account for the giant coercive fields and hysteresis loops observed in the amorphous alloys (see Fig. 1) when they are magnetized by an applied field at low temperatures. The crystalline-phase counterpart compounds exhibit only small coercive fields. Neutron scattering has provided evidence for nonuniformities in the magnetic order, or the occurrence of microdomains of order 100 Å, in the amorphous materials as well as evidence that the systems may not undergo a conventional second-order phase transition at the Curie temperature, but rather one of smeared character over a range of temperature.

There are many other unique features of the rare-earth alloys such as the anomalous variation of Curie temperature with the magnitude of the effective rare-earth spin which will not be discussed here. In addition some of these alloys show considerable promise for computer bubble-memory applications.[17]

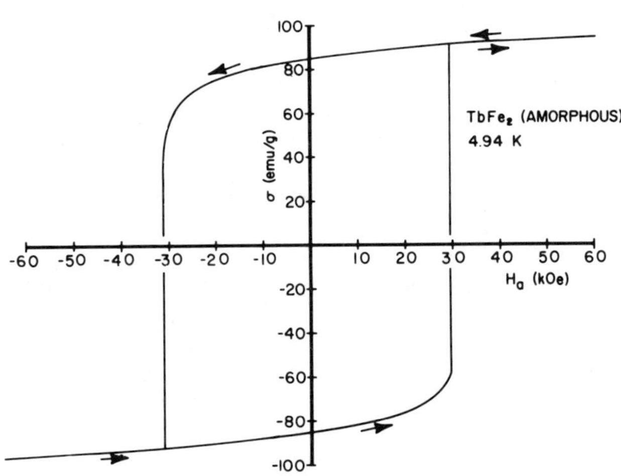

FIG. 1. Hysteresis loops for $TbFe_2$ at 4.9 K showing giant coercive fields of approximately 30 kOe [from J. J. Rhyne, J. H. Schelleng and N. C. Koon, *Phys. Rev.* **B10**, 4672 (1974)].

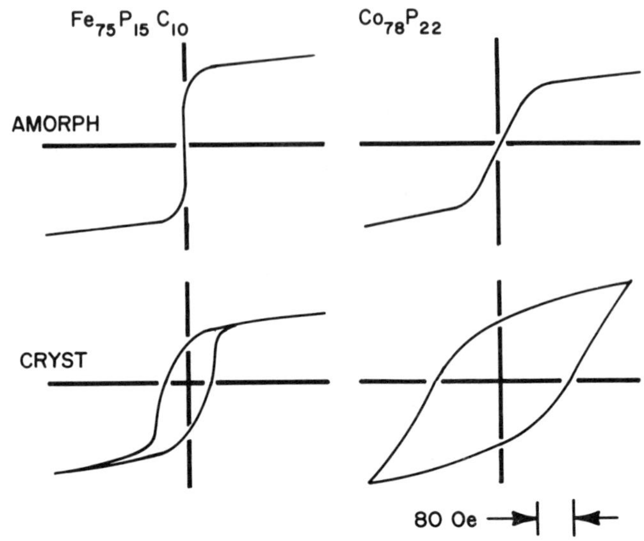

FIG. 2. Hysteresis loops for two crystalline and amorphous transition-metal alloys. Note the very small coercive fields compared to those in the rare-earth alloys shown in Fig. 1 [from G. S. Cargill, AIP *Conf. Proc. Series* **24**, 138 (1975)].

Transition-Metal Amorphous Materials

Amorphous transition-metal-based materials generally have Curie temperatures which are well above room temperature, but which are lower than those of the parent transition metal because of the structural disorder. They also exhibit a lower magnetic moment per atom largely because of the dilution effect of the impurity "glass-former" atoms. Changes in the electronic band structure between crystalline and amorphous states are also important, particularly in alloys containing Ni.

In many respects some of the important properties of transition-metal amorphous materials are on opposite ends of the scale from the rare-earth-based system. These include the magnetic anisotropy and magnetostriction. While the rare earths, except Gd-based alloys, are magnetically hard, the transition-metal alloys are soft, in many cases softer than their crystalline counterparts, which leads, as shown in Fig. 2, to a very low coercive force and remanence.[18,19] Furthermore, the remaining anisotropy can often be further reduced by appropriate heat treatment which at the same time lowers the magnetostriction.[20] Such low-anisotropy and zero-magnetostriction materials have great potential for use in devices which presently use crystalline soft magnetic materials. Amorphous alloys in some sense are more "perfect" than their crystalline counterparts since, although disordered, they do not suffer from such disturbances as crystal grain boundaries, pinning centers, etc., which degrade the performance of the overall material. It is the absence of these imperfections that make it possible to make an amorphous material which saturates easily in low applied magnetic fields and which has very low magnetic "losses."

In the future we may expect to find continued strong interest in the field of magnetic materials in both crystalline and amorphous forms from the viewpoint of providing a window into the basic physics of magnetism and also for useful device applications.

See also FERRIMAGNETISM; FERROMAGNETISM; GLASSY METALS; MAGNETIC DOMAINS AND BUBBLES; MAGNETOSTRICTION; RARE EARTHS; TRANSITION ELEMENTS.

REFERENCES

1. S. Chikazumi, *Physics of Magnetism*. Wiley, New York, 1969.
2. G. T. Rado and H. Suhl (eds.), *Magnetism*. Academic Press, New York, Vol. I (1963), Vol. II A (1965), Vol. II B (1966), Vol. III (1963), Vol. IV (1966).
3. R. M. Bozorth, *Ferromagnetism*. Van Nostrand, Princeton, N.J., 1951.
4. K. Gschneidner and L. Eyring (eds.), *Handbook on the Physics and Chemistry of Rare Earths*. North Holland, Amsterdam, 1979.
5. R. J. Elliott (ed.), *Magnetic Properties of Rare Earth Metals*. Plenum Press, New York, 1972.
6. K. N. R. Taylor and M. I. Darby, *Physics of Rare Earth Solids*. Chapman and Hall, London, 1972.
7. W. E. Wallace, *Rare Earth Intermetallics*. Academic Press, New York, 1973.
8. E. A. Nesbitt and J. H. Wernick, *Rare Earth Permanent Magnets*. Academic Press, New York, 1973.
9. F. B. Hagedorn, "Magnetic Bubbles—for the Non-Specialist," *AIP Conf. Proc. Series* **24**, 1 (1975), and references contained therein.
10. G. S. Cargill III, "Structure of Metallic Alloy Glasses," in *Solid State Physics*, Vol. 30, p. 227, H. Ehrenreich, F. Seitz, and D. Turnbull (eds). Academic Press, New York, 1975.
11. J. J. Rhyne, Ref. 4, Chap. 16.
12. N. J. Grant and B. C. Giessen (eds.), *Rapidly Quenched Metals: Second International Conference*, MIT Press, Cambridge, Mass., 1976.
13. K. Yosida, "The Status of the Theories of Magnetic Anisotropy," *J. Appl. Phys.* **39**, 511 (1968).
14. Earl Callen, "Magnetostriction," *J. Appl. Phys.* **39**, 519 (1968).
15. L. J. Tao, R. J. Gambino, S. Kirkpatrick, J. J. Cuomo, and A. Lilienthal, "Magnetic Properties of Amorphous Gd Co Films," *AIP Conf. Proc. Series* **18**, 641 (1974).
16. G. S. Cargill and S. Kirkpatrick, *AIP Conf. Proc. Series* **31**, 339 (1976).
17. R. J. Gambino, P. Chaudhari, and J. J. Cuomo, "Amorphous Magnetic Materials," *AIP Conf. Proc. Series* **18**, 579 (1974).
18. E. M. Gyorgy, A. J. Leamy, R. C. Sherwood, and H. S. Chen, "Amorphous Alloys as Soft Magnetic Materials II," *AIP Conf. Proc. Series* **29**, 198 (1976).
19. G. S. Cargill III, "Ferromagnetism in Amorphous Solids," *AIP Conf. Proc. Series* **24**, 138 (1975).
20. T. Egami, P. J. Flanders, and C. R. Graham, Jr., "Amorphous Alloys as Soft Magnetic Materials," *AIP Conf. Proc. Series* **24**, 697 (1975).

Magnetic Moments

R. K. Nesbet

The magnetic dipole moment of an infinitesimal current loop carrying electric current J (in units of electrostatic charge times velocity) is defined as

$$d\boldsymbol{\mu} = \frac{1}{c} J\, d\mathbf{S}, \tag{1}$$

where $d\mathbf{S}$ is the oriented surface element enclosed by the current loop and c is the velocity of light. The magnetization $\mathbf{m}$, or magnetic moment per unit volume, in volume element $d\tau$ is defined by

$$d\boldsymbol{\mu} = \mathbf{m}\, d\tau. \tag{2}$$

For distributed currents, the electric current density $\mathbf{j}$ is defined as net current crossing a unit element of surface locally perpendicular to $\mathbf{j}$ within the distribution. When the electric charge and current densities are time independent the vector fields $\mathbf{m}$ and $\mathbf{j}$ can be related by considering surface currents in a plane locally perpendicular to $\mathbf{m}$, which can be taken to have a single component $m_z(x,y)$. Each element $dxdy$ in the surface of constant z can be assumed to define a current J circulating about its periphery. All current between z and $z+dz$ is included in J. The net current through surface element $dzdx$ perpendicular to the current plane is obtained as the difference of adjacent loop currents, using Eqs. (1) and (2):

$$j_y dzdx = J(x,y) - J(x+dx,y)$$
$$= -c\frac{\partial m_z}{\partial x} dzdx. \tag{3}$$

Similarly, the net current through $dzdy$ is

$$j_x dzdy = J(x,y+dy) - J(x,y)$$
$$= c\frac{\partial m_z}{\partial y} dzdy. \tag{4}$$

In arbitrary geometry, these equations are

$$\mathbf{j} = c\,\nabla \times \mathbf{m}. \tag{5}$$

Since Eq. (5) implies

$$\nabla \cdot \mathbf{j} = 0, \tag{6}$$

the equation of continuity for the electric charge density ρ associated with $\mathbf{j}$,

$$\nabla \cdot \mathbf{j} + \frac{\partial \rho}{\partial t} = 0, \tag{7}$$

requires that the charge density be time independent for Eq. (5) to be valid. Stationary charge and current densities will be assumed here.

Each elementary particle or aggregate with intrinsic angular momentum (spin) $\hbar\mathbf{J}$ carries an associated magnetic moment

$$\boldsymbol{\mu} = \gamma_J \hbar \mathbf{J} \tag{8}$$

where γ_J is the gyromagnetic ratio. The magnetic moment of an electron is defined in terms of the Bohr magneton

$$\mu_e = e\hbar/2m_e c, \tag{9}$$

while nuclear magnetic moments are defined in terms of the nuclear magneton

$$\mu_N = e\hbar/2m_p c. \tag{10}$$

Here $-e$ is the charge of the electron, m_e is the electron mass, and m_p is the proton mass.

The magnetic moment of an electron is $\boldsymbol{\mu}_s + \boldsymbol{\mu}_l$, where

$$\boldsymbol{\mu}_s = -g_e \mu_e \mathbf{s},$$
$$\boldsymbol{\mu}_l = -\mu_e \mathbf{l}. \tag{11}$$

Here g_e is the electronic g factor, approximately

$$g_e \cong 2.00232, \tag{12}$$

and $\hbar\mathbf{s}$ and $\hbar\mathbf{l}$ are the spin and orbital angular momentum operators, respectively. The total magnetic moment operator for the electrons in an atom is

$$\boldsymbol{\mu} = -\mu_e(\mathbf{L} + g_e\mathbf{S}), \tag{13}$$

where $\mathbf{S}$, $\mathbf{L}$ are total $\mathbf{s}$, $\mathbf{l}$, respectively. In a state of total angular momentum

$$\hbar\mathbf{J} = \hbar\mathbf{L} + \hbar\mathbf{S}, \tag{14}$$

in the limit of weak spin–orbit coupling, the atomic magnetic dipole moment is

$$\boldsymbol{\mu}_J = -g\mu_e\mathbf{J}. \tag{15}$$

Here g is the Landé factor

$$g = 1 + \frac{J(J+1) + S(S+1) - L(L+1)}{2J(J+1)} \tag{16}$$

if g_e is approximated by 2.0.

The magnetic dipole moment of a nucleus is defined by the operator for each nucleon,

$$\boldsymbol{\mu} = \mu_N(g_l\mathbf{l} + g_s\mathbf{s}) \tag{17}$$

where for a proton,

$$g_l = 1, \quad g_s/2 \cong 2.793, \tag{18}$$

and for a neutron,

$$g_l = 0, \quad g_s/2 \cong -1.913. \tag{19}$$

Tables of nuclear magnetic dipole moments have been published by Ramsey [5,6] and by Lederer *et al.* [3].

Rotational magnetic moments of molecules have also been discussed by Ramsey [6]. These moments involve coupling of electronic excitation to rotation of the nuclear framework of a molecule. They are of the order of magnitude of the nuclear magneton because the molecular rotation constant contains the reciprocal of the nuclear mass of the molecule.

In atoms and molecules, magnetic hyperfine-structure energy level splittings arise from the interaction between nuclear magnetic moments and the electronic magnetic moment distribution, which is stationary but nonuniform in a quantum-mechanical stationary state. The energy of interaction of stationary current density distributions due to physically distinct particle distributions is given by Maxwell's equations in the form

$$W_{12} = -\frac{1}{c^2}\iint \frac{1}{r_{12}}\mathbf{j}_1(1)\cdot\mathbf{j}_2(2)d\tau_1 d\tau_2, \tag{20}$$

$$W_{12} = -\iint \frac{1}{r_{12}}(\nabla_1 \times \mathbf{m}_1)\cdot(\nabla_1 \times \mathbf{m}_2)d\tau_1 d\tau_2, \tag{21}$$

in terms of the equivalent magnetizations. If $\mathbf{m}_1$ and $\mathbf{m}_2$ vanish outside the region of integration, this result can be transformed by partial integration to

$$W_{12} = \iint_{r_{12}>\epsilon\to 0} \frac{1}{r_{12}}(-\nabla_1\cdot\mathbf{m}_1)(-\nabla_2\cdot\mathbf{m}_2)d\tau_1 d\tau_2 - \frac{8\pi}{3}\int \mathbf{m}_1\cdot\mathbf{m}_2 d\tau. \tag{22}$$

The infinitesimal volume integral excluded from the first term here is $+(4\pi/3)\int\mathbf{m}_1\cdot\mathbf{m}_2 d\tau$, in the limit $\epsilon\to 0$. This is included in the second term, which vanishes if the particle distributions are spatially separated. The second term contributes to hyperfine structure as the Fermi contact interaction between a nuclear magnetic moment and the electronic spin density at that nucleus.

The first term in Eq. (22) is of the same form as the Coulomb interaction, except that electric charge densities ρ are replaced by magnetic monopole densities $(-\nabla\cdot\mathbf{m})$. The integrated monopole moment of any isolated system vanishes, since

$$\int(-\nabla\cdot\mathbf{m})d\tau = 0 \tag{23}$$

by the divergence theorem, if $\mathbf{m}$ vanishes outside a finite volume. The magnetic dipole moment is

$$\boldsymbol{\mu} = \int(-\nabla\cdot\mathbf{m})\mathbf{r}d\tau = \int \mathbf{m}d\tau, \tag{24}$$

in agreement with the definition of $\mathbf{m}$.

A magnetic scalar potential can be defined by

$$\psi_2(1) = \int \frac{1}{r_{12}}(-\nabla_2\cdot\mathbf{m}_2)d\tau_2. \tag{25}$$

Then the first term of Eq. (22) becomes

$$W^m = \int(-\nabla\cdot\mathbf{m}_1)\psi_2 d\tau, \tag{26}$$

analogous to the electrostatic interaction energy

$$W^e = \int \rho_1\phi_2 d\tau. \tag{27}$$

This analogy provides definitions of magnetic multipole moments and polarizabilities formally equivalent to their electrostatic counterparts. Details of magnetic multipole interactions relevant to hyperfine structure interactions have been worked out [1,2,7].

Important differences between electric and magnetic interactions arise from the different nature of the source terms ρ and $-\nabla\cdot\mathbf{m}$. For a wave function of definite parity (inversion symmetry), ρ is a scalar quantity (even parity) while $-\nabla\cdot\mathbf{m}$ is pseudoscalar (odd parity). Spherical tensor multipole moments are defined by integrating the source density multiplied by r^l times a spherical harmonic of order l. Since the spherical harmonic has the parity of l, electric multipole moments of a system with definite parity are nonzero only for even values of l, while magnetic multipole moments are nonzero for odd values. Hence elementary particles or systems with definite inversion symmetry can only have intrinsic electric monopole, quadrupole, etc., moments, and can only have intrinsic magnetic dipole, octupole, etc., moments.

REFERENCES

1. H. B. G. Casimir and G. Karreman, *Physica* **9**, 494 (1942).
2. V. Jaccarino, J. G. King, R. A. Satten, and H. H. Stroke, *Phys. Rev.* **94**, 1798 (1954).
3. C. M. Lederer, J. M. Hollander, and I. Perlman, *Table of Isotopes*, 6th ed., Wiley, New York, 1967.
4. W. K. H. Panovsky and M. Phillips, *Classical Electricity and Magnetism*. Addison-Wesley, Reading, Mass., 1955.
5. N. F. Ramsey, *Nuclear Moments*. Wiley, New York, 1953.
6. N. F. Ramsey, *Molecular Beams*. Oxford Univ. Press, New York, 1956.
7. C. Schwartz, *Phys. Rev.* **97**, 380 (1955).

Magnetoacoustic Effect

J. D. Gavenda

In pure metals at liquid helium temperatures the conduction electrons provide the dominant mechanism for ultrasonic attenuation. The variation in this attenuation caused by an applied magnetic field **B** is referred to as the magnetoacoustic effect. In recent years much has been learned about the properties of conduction electrons in metals through use of this experimental technique.

An ultrasonic wave propagating through a metal sets up an oscillating electric field in the neighborhood of the vibrating ions. The conduction electrons move so as to rapidly screen out this field and, in the process, absorb energy from the sound wave which is then transferred to the thermal motion of the lattice via collisions.

Electronic attenuation is important when the mean free path l of the electrons is at least as large as the sound wavelength λ. This condition is easily satisfied in very pure (99.999%) metals at liquid helium temperatures, for ultrasonic frequencies f greater than about 10 MHz.

When $l \gg \lambda$, only those electrons moving perpendicular to the direction of propagation $\hat{q}$ of the wave are effective absorbers. The explanation lies in the much higher velocity of the electrons than that of the sound wave. In metals the electrons which play a role in transport processes are those moving with the Fermi velocity v_F. Since v_F is several hundred times the velocity of sound v_s, the sound wave can be regarded as frozen in the metal as far as the electrons are concerned. If an electron is moving in the direction $\hat{q}$, it experiences an alternating electric field which does no net work on it. However, if it moves perpendicular to $\hat{q}$ along a crest of the wave, it will stay in a region of constant electric field and thus will have a net amount of work done on it. This explains why the attenuation is caused principally by those conduction electrons moving perpendicular to the direction of propagation of the sound wave.

When a magnetic field is applied to the metal, the situation is drastically changed. The electrons are forced to move in orbits with their direction of motion continually changing. On the portions of their orbits where they move essentially parallel to a sound wave front they absorb energy from the sound field; elsewhere there is very little interaction. The net result is a magnetic-field-dependent attenuation which may be monotonic, oscillatory, or resonant in character, depending on the metal and on the geometry of the experiment.

The most common experimental geometry has the magnetic field **B** perpendicular to $\hat{q}$. In a metal such as potassium which has a spherical Fermi surface, i.e., free-electron behavior, the projections of electron orbits on planes perpendicular to **B** will be circles with radii $R = p/eB$, where p is the component of electron momentum perpendicular to **B** and e is the charge on the electron. As **B** is varied in intensity an electron will absorb energy coherently from the sound field when the diameter of its orbit equals $n\lambda$, where n is an integer. Since the diameter is inversely proportional to B, the attenuation will have oscillations which are periodic when plotted against $1/B$. These periodic variations in attenuation are called *geometric oscillations* to distinguish them from other magnetoacoustic effects. Given values of λ, e, and B, one can solve for the value of p for the electrons causing the oscillations. If the sound frequency is 50 MHz, geometric oscillations will be observed for values of B ranging from about 0.01 to 0.1 T.

For a spherical Fermi surface of radius p_0, we have $p = p_0 \sin\theta$, where θ is the angle between **B** and the electron's velocity. Thus, for a given value of B the diameters of the orbits will vary from zero to a value determined by p_0. The effects of all the orbits are summed to give the total attenuation, but those having θ near 90° will dominate, since there are more of them than in any other angular interval.

Geometrical oscillations are also observed in metals which have nonspherical Fermi surfaces, hence noncircular electron orbits in a magnetic field. The periods of the oscillations are determined by the extremal diameters of the orbits in the direction of propagation, so a Fermi surface can be calipered by measuring periods for a variety of directions of **B** and $\hat{q}$.

In a metal such as copper with a Fermi surface which contacts the Brillouin zone boundaries, open orbits are possible when **B** is perpendicular to the directions ($\hat{G}$, for example) in $\hat{k}$ space for which contact occurs. The open orbit in real space will drift in a direction perpendicular to both **B** and $\hat{G}$. If $\hat{q}$ is parallel to this drift motion, sharp resonance-type peaks in the attenuation will occur when the period of the open orbit equals $n\lambda$. *Open-orbit resonances* are narrow peaks, in contrast to the sinusoidal geometric oscillations, because all electrons on open orbits will have precisely the same period, determined by the crystal lattice spacing.

In an extremely pure metal it is possible to have an electron relaxation time τ comparable to, or greater than, the period of the sound wave. Then it is no longer appropriate to neglect the motion of the sound field with respect to the electron velocity. The sound field at a particular point in the metal will have changed appreciably by the time an electron on a large-diameter closed orbit returns to its starting point. If the period of the sound field equals the cyclotron period of the orbit, *acoustic cyclotron resonance* may be observed. The requirement that τ be greater than $1/f$ is so stringent, however, that acoustic cyclotron resonance has not been widely employed in the investigation of metals.

When the magnetic field is applied parallel, rather than perpendicular, to the direction of propagation of a shear wave, *Doppler-shifted cyclotron resonance* may be observed. The sound field experienced by the electrons has its frequency Doppler shifted by their drift velocity v_B along the magnetic field. Energy is absorbed from the sound field when the electron cyclotron frequency $f_c = eB/2\pi m_c$ is an integral multiple of the Doppler-shifted frequency $f - v_B/\lambda$. For sufficiently large B, the cyclotron frequencies of all the electrons will exceed their Doppler-shifted frequencies and none of them will be able to absorb energy from the sound field. The value of B beyond which there is no attenuation is called the Kjeldaas edge field. It can be related to the Gaussian curvature of the Fermi surface in the direction of $\hat{q}$ and **B**. For nonspherical Fermi surfaces, attenuation peaks will result if there are bands of electrons with the same value of $m_c v_B$, which will thus all experience Doppler-shifted cyclotron resonance at the same value of B.

The types of phenomena discussed so far can all be explained by semiclassical transport theory. For magnetic fields above about 5 T, quantum phenomena related to the de Haas–van Alphen effect appear in the magnetoacoustic effect. Since the magnitude of the attenuation is proportional to the number of electrons at the Fermi energy, quantum oscillations in the density of states will show up as oscillations in the attenuation. For **B** parallel to $\hat{q}$ these effects are particularly pronounced and are termed *giant quantum oscillations*. Although related to the conventional de Haas–van Alphen effect, giant quantum oscillations are caused by just those electrons which experience Doppler-shifted cyclotron resonance. This selectivity makes it possible to single out specific cross-sectional areas for measurement.

All of these phenomena manifest themselves in variations in the velocity of the sound wave, as well as in the attenuation. However, since the small variations in velocity are more difficult to measure and because the interpretation of the results is more complicated, little use has been made of velocity measurements.

The variety of experimental arrangements of propagation, polarization, and magnetic field directions is limited only by the ingenuity of the experimentalist, but the interpretation of the data is relatively straightforward only for the basic situations described above.

See also Cyclotron Resonance; de Haas–van Alphen Effect; Fermi Surface.

BIBLIOGRAPHY

J. D. Gavenda, "Study of the Electronic Properties of Metals by Ultrasonics," Prog. Appl. Mat. Res. **6**, 41–68 (1954). (E)

A. R. Mackintosh, "The Interaction of Long-Wavelength Phonons with Electrons," in *Phonons and Phonon Interactions*, pp. 181–220, Thor A. Bak (ed.). Benjamin, New York, 1964. (I)

A very complete review of all aspects of ultrasonic wave propagation in metals has been published by J. Mertsching, "Theory of Electromagnetic Waves in Metals (III)," *Phys. Stat. Solidi*. **37**, 465–522 (1970). (A) Part I of this series is also relevant: *Phys. Stat. Solidi*. **14**, 3–61 (1966). (A)

Magnetoelastic Phenomena

R. L. Melcher

The magnetic and elastic properties of both crystalline and amorphous solids are, in general, coupled through the magnetoelastic interaction. This coupling arises from the elastic strain dependence of the interactions experienced by the individual magnetic moments associated with the ions or nuclei of the solid. Through these strain-dependent interactions a stress applied to the solid can influence its magnetic behavior and, conversely, a change in the magnetic state of the solid can alter its elastic properties. Numerous physical phenomena, both static and dynamic, result from the magnetoelastic coupling and a variety of experimental techniques have been developed to investigate them.

One of the more familiar phenomena associated with magnetoelastic coupling is static magnetostriction which refers to the change in size or shape of a magnetically ordered solid when its state of magnetization is changed by an applied magnetic field. In addition, magnetoelastic coupling can strongly influence the elastic moduli and concomitantly the dispersion and absorption of elastic waves in a solid. Resonant magnetoelastic coupling occurs in those cases when the magnetic and elastic modes of a solid are tuned to the same frequency. The establishment of thermal equilibrium in a magnetic solid is governed by the coupling of the magnetic and elastic degrees of freedom. Magnetoelastic coupling is important not only in magnetically ordered media (e.g. ferro-, ferri-, and antiferromagnetic) but also in paramagnetic materials including nuclear paramagnetism.

Microscopically, magnetoelastic coupling arises from the strain dependence of any of several interactions which influence the magnetic behavior of the material. Often the dominant contribution to the magnetoelastic interaction is the strain dependence of the crystalline electric field. For the transition-metal ions, this crystal field acts on the electronic orbital motion which in turn is coupled to the spin magnetic moments via the spin–orbit interaction. For the rare-earth ions, the spin–orbit interaction is so strong that the spin and orbital moments cannot be considered separately and the crystalline electric field acts on the moments associated with the total angular momentum of the ion. These types of single-ion magnetoelastic coupling are equally valid for paramagnetic impurities in diamagnetic host lattices and for magnetically ordered media. As the concentration of magnetic ions increases, two-ion contributions to the magnetoelastic coupling can become important. These include the strain dependence of the dipolar and exchange interactions which couple two magnetic ions. The single-ion and two-ion magnetoelastic interactions in magnetically ordered media are not always easily distinguished experimentally and theoretical estimates of the relative magnitudes are used to determine which dominates. In magnetically dilute materials only the single-ion magnetoelastic coupling is significant.

The magnetic moments associated with nuclear spin are coupled to the lattice strain via one of several types of interaction. In magnetically ordered materials, the strain can couple first to the electronic moments which are coupled to the nuclear dipole moments through the hyperfine interaction, thus resulting in an indirect nuclear spin magnetoelastic coupling. In metallic materials the dynamic interaction of the strain with the conduction electrons can cause the generation of an internal magnetic field which couples to the nuclear magnetic dipoles. Finally, nuclear spins which possess an electric quadrupole moment are coupled to the elastic strain through strain-induced changes in the electric field gradient at the nucleus due to the surrounding lattice.

Although the elastic properties of a nonmagnetic solid can be fully described by the symmetric strain tensor, the magnetoelastic interaction has been shown to depend not only on the symmetric strain but also on the antisymmetric elastic rotation tensor. These rotational contributions arise from the torques exerted on the magnetic moments in an anisotropic material by local lattice rotations and/or the applied field. The total interaction Hamiltonian of a magnetoelastic solid takes the general form

$$\mathcal{H} = \mathcal{H}_M(\mathbf{M},\mathbf{H}) + \mathcal{H}_{ME}(\mathbf{M},e_{ij},\omega_{ij}) + \mathcal{H}_E(e_{ij}),$$

where $\mathcal{H}_M(\mathbf{M},\mathbf{H})$ describes the magnetic properties of a rigid solid, $\mathcal{H}_E(e_{ij})$ describes the elastic properties of a nonmagnetic solid, and $\mathcal{H}_{ME}(\mathbf{M},e_{ij},\omega_{ij})$ is the magnetoelastic interaction which couples the two. Here, **M** represents the magnetic variables required to describe magnetic properties of the system, **H** is an applied magnetic field, and e_{ij} and ω_{ij} represent the symmetric elastic strain tensor and the antisymmetric elastic rotation tensor, respectively. Phenomenological descriptions of magnetoelastic phenomena rely heavily on the symmetry properties of the elastic strain, of the surrounding of the magnetic ion or nucleus, and of the interaction itself. Through application of these symmetries, the number of coupling constants required to characterize the magnetoelastic system is reduced to a minimum. Once the explicit form of the Hamiltonian $\mathcal{H}$ is known, both the static and dynamic properties of the coupled system can be derived.

The static properties of the system are calculated from the thermodynamic free energy which is deduced from $\mathcal{H}$. These properties include the magnetostrictive distortion of the lattice, the magnetoelastic contributions to the magnetic anisotropy and the magnetoelastic contributions to the elastic stiffness moduli. In most cases the contributions to the elastic moduli are small; however, there are cases in which the coupling completely dominates the elastic behavior and can even result in lattice instabilities and structural phase transitions which reduce the symmetry of the lattice.

The dynamic properties of the system are calculated with the aid of the coupled equations of motion which can be derived from the Hamiltonian $\mathcal{H}$. The dynamic coupling of the magnetic and elastic modes of the system enables one to use elastic waves as a probe of the magnetic system and the magnetoelastic interaction. Sensitive techniques have been developed to measure the resulting absorption and dispersion of elastic waves. These measurements provide valuable information regarding the coupling between the lattice and the magnetic degrees of freedom and how this coupling leads to the establishment of thermal equilibrium in the solid.

Magnetostrictive transducers, which convert high-frequency electromagnetic radiation into elastic waves, and magnetoelastic delay lines, whose delay time can be varied by changing the biasing magnetic field, are two technological applications which are made possible by magnetoelastic interactions.

See also CRYSTAL AND LIGAND FIELDS; DIAMAGNETISM; ELASTICITY; MAGNETIC MOMENTS; MAGNETOACOUSTIC EFFECT; MAGNETOSTRICTION.

BIBLIOGRAPHY

C. Kittel, *Phys. Rev.* **110,** 836 (1958). (I)

R. C. Le Craw and R. L. Comstock, "Magnetoelastic Interactions in Ferromagnetic Insulators," in *Physical Acoustics*, Vol. 3B, Warren P. Mason (ed.). Academic Press, New York, 1965. (E)

Edmond B. Tucker, *Physical Acoustics*, Vol. 4A, Warren P. Mason (ed.). Academic Press, New York, 1966. (I)

D. I. Bolef, *Physical Acoustics*, Vol. 4A, Warren P. Mason (ed.). Academic Press, New York, 1966. (I)

R. L. Melcher, "Rotational Invariance and Magnetoelastic Phenomena in Anisotropic Media," in *Proceedings of the International School of Physics, Enrico Fermi*, Course LII, Varenna, Italy, E. Burstein (ed.). Academic Press, New York, 1973. (A)

R. L. Melcher, "The Anomalous Elastic Properties of Materials Undergoing Cooperative Jahn-Teller Phase Transitions," in *Physical Acoustics*, Vol. 12, Warren P. Mason and R. N. Thurston, (eds.). Academic Press, New York, 1976. (I)

Magnetoresistance

W. A. Reed

When a conductor carrying a current is placed in a magnetic field its resistance changes (usually increases). This increase is called its magnetoresistance and is due to the Lorentz force acting on the electrons combined with a distribution in the electrons' velocities. The magnetoresistance is only one of several galvanomagnetic effects and a general discussion of all these effects is given by Jan.[1]

The relation between the electric field, current density and magnetic field is given by

$$E_i = \rho_{ij}(B)J_j, \tag{1}$$

where ρ_{ij} is the resistivity tensor. The element $\rho_{ij}(B)$ can be expanded in powers of **B** such that

$$\rho_{ij}(B) = \rho_0 + \rho_{ijk}B_k + \rho_{ijkl}B_kB_l. \tag{2}$$

The magnetoresistance terms of the tensor are contained in the diagonal elements of $\rho_{ii}(B)$ which, from symmetry considerations, only contain terms with even powers of **B**. When the magnetic field is parallel to the current, the resistance change (proportional to ρ_{iiii}) is called the *longitudinal magnetoresistance*, whereas when the field is perpendicular to the current, the change in resistance (proportional to ρ_{iijj} or ρ_{iikk}) is called the *transverse magnetoresistance*. The magnetoresistance is usually expressed as $\Delta\rho/\rho_0$, where $\Delta\rho = \rho(B) - \rho_0(0)$ is the change in resistivity in the field, and ρ_0 is the resistivity in zero field at the temperature of measurement. The magnitude of the transverse magnetoresistance is a function of both the magnetic field and the resistivity and generally follows a reduced Kohler's rule which states:

$$\frac{\Delta\rho}{\rho_0} = F\left(B\frac{\rho_\Theta}{\rho_0}\right), \tag{3}$$

where ρ_Θ is the resistivity at the Debye temperature.

Theories of magnetoresistance generally fit into one of five regimes[2,3] depending on the value of $\omega_c\bar{\tau}$, where ω_c, the cyclotron frequency (eB/m^*c), is 2π times the reciprocal of the time it takes an electron to travel once around its orbit on the Fermi surface and $\bar{\tau}$ is the average relaxation time of the electrons. These regimes are:

(1) The low-field region ($\omega_c\bar{\tau}\ll 1$) is where the time between electron collisions is short and the electron traverses only a small fraction of the Fermi surface before scattering. In this regime the anisotropy of the scattering mechanisms dominates the magnetoresistance. Experimentally the condition $\omega_c\bar{\tau}\ll 1$ is satisfied at high temperatures where phonon scattering is significant and/or in materials with nontrivial amounts of defects or impurities. In this regime the size of $\Delta\rho/\rho_0$ in metals is generally about 0.1% to 1%. "Many-valley" models have been used to successfully fit low-field galvanomagnetic data of semiconductors and semimetals[4] with Fermi surfaces constructed from a few ellipsoids, but these models have not been successfully applied to the more complicated Fermi surfaces usually found in metals.

(2) The intermediate-field region ($\omega_c\bar{\tau}\approx 1$) is a region for which few if any theories of magnetoresistivity exist and is therefore of little interest experimentally.

(3) The high-field ($\omega_c\bar{\tau}\gg 1$) is where the electrons are able to complete many cyclotron orbits between collisions. Thus it is the topology of the Fermi surface and not the details of the scattering mechanisms that dominates the character of the magnetoresistance[5,6]. To obtain the high-field region experimentally the measurements are made at low temperatures ($T<4$ K), in high magnetic fields ($B>10^4$G), and on high-purity single crystals.[7] In metals the high-field magnetoresistance can saturate at $\Delta\rho/\rho_0\approx 1$ when the metal is not compensated (the number of electrons does not equal the number of holes) or when the current and field are perpendicular to an open orbit. However, if the metal is compensated or an open orbit is parallel to the current, the transverse high-field magnetoresistance can become greater than 10^5. Because $\Delta\rho/\rho_0$ is so sensitive to the topology of the cyclotron orbits the high-field magnetoresistance has been extensively used to study the Fermi surfaces of metals.[5,6]

(4) The quantum oscillation region ($\hbar\omega_c\gg kT$ and $\omega_c\bar{\tau}\gg 1$) is where the separation of the quantized energy levels is greater than the thermal smearing. Experimentally this region is the same as the high-field region except that requirements on temperature, field, and crystal perfection are more stringent. The oscillations in $\Delta\rho/\rho_0$, which are periodic in $\mathbf{B}^{-1}$, provide information similar to that gained from de Haas–van Alphen measurements, i.e., the extremal cross-sectional areas of the Fermi surface. In many metals the spacing between Landau levels may become greater than the energy gap between bands and the phenomenon of magnetic breakdown can occur. Since magnetic breakdown can change both the topology of the cyclotron orbits and the state of compensation, it profoundly affects both the oscillatory and monotonic magnetoresistance.[8]

(5) The quantum limit ($\hbar\omega_c\gg\epsilon_F$ where ϵ_F is the Fermi energy) is almost impossible to obtain experimentally in metals except for possibly bismuth, so that experimental studies have been restricted to semiconductors.[9]

The phenomenon of negative magnetoresistance is usually associated with measurements on samples which either are magnetic or contain magnetic impurities, or on samples where the mean free path of the electrons is comparable to the sample dimensions. In these cases the decrease in the magnetoresistance is a result of the magnetic field reducing the amount of scattering by some mechanism such as aligning the spin of magnetic impurities or rotating domains in magnetic samples or reducing the number of collisions with the sample's walls in size-effect measurements. In addition to the usual negative magnetoresistance effects numerous anomalous effects have been reported, usually in the longitudinal magnetoresistance. These have most often proved to be artifacts of the experiment.[10]

See also CONDUCTION, ELECTRIC; CYCLOTRON RESONANCE; DE HAAS–VAN ALPHEN EFFECT; FERMI SURFACE; GALVANOMAGNETIC EFFECTS.

REFERENCES

1. J.-P. Jan, in *Solid State Physics*, Vol. 5, F. Seitz and D. Turnbull (eds.). pp. 1–96, Academic Press, New York, 1957.
2. R. G. Chambers, in *The Fermi Surface*, pp. 100–124, W. A. Harrison and M. B. Webb (eds.), Wiley, New York, 1960.
3. Colin M. Hurd, *The Hall Effect in Metals and Alloys*, pp. 1–50. Plenum Press, New York, 1972.
4. Arthur C. Smith, James F. Janak, and Richard B. Adler, *Electronic Conduction in Solids* McGraw-Hill, New York, 1967.
5. E. Fawcett, *Adv. Phys.* **13,** 139 (1964).
6. W. A. Reed and E. Fawcett, *Science* **146,** 603 (1964).
7. W. A. Reed, in *Methods of Experimental Physics*, Vol. 11, pp. 1–32. R. V. Coleman (ed.) Academic Press, New York, 1974.
8. R. W. Stark and L. M. Falicov, in *Progress in Low Temperature Physics*, Vol. 5, pp. 235–286, C. J. Gorter (ed.). North-Holland Publishing Co., Amsterdam, 1967.
9. E. N. Adams and T. D. Holstein, *J. Phys. Chem. Solids* **10,** 154 (1959).
10. W. A. Reed, E. Blount, J. A. Marcus, and A. J. Arko, *J. Appl. Phys.* **42,** 5433 (1971).

Magnetostriction

B. S. Chandrasekhar

If the magnetization of a solid object is changed, for example, by changing the magnitude of a magnetic field applied to it, then the size of the object changes slightly. This is the phenomenon of magnetostriction, discovered in 1842 by J. P. Joule (who is better known for his experiments on the equivalence of mechanical and thermal energies and whose name is borne by our unit of energy). The effect is small, with the change in length between the unmagnetized and magnetized states being typically a few parts in 10^5; but the effect is both an important strand in the fabric of our basic understanding of solids, and of use in several practical applications.

The origins and detailed characteristics of magnetostriction depend on whether the solid under consideration is ferromagnetic or not. Ferromagnetic magnetostriction has been investigated for a longer time and far more extensively than the other kind, and it also has provided all the practical applications. It occupies therefore the major part of this article. The term "magnetostriction" will be used henceforth without precedent qualifier; the context will make clear which kind it is.

The physical origin of magnetostriction in a ferromagnet like iron or cobalt ferrite is as follows. The energy of a magnetized single crystal depends on the direction of the magnetization, essentially because of the interaction between the spin moments of the electrons which cause the magnetization and the orbital motion of the electrons around the nuclei. This anisotropy energy, for any given direction of the magnetization, is itself dependent on the deformation of the crystal, a deformation which produces its own contribution of an elastic energy to the total energy of the crystal. The minimizing of the sum of these magnetoelastic and elastic contributions to the energy leaves the crystal with a resultant strain, the magnetostriction.

An unmagnetized crystal consists of many microscopic magnetized domains pointing in different directions so that the net magnetization is zero. The application of a strong enough external field, along a direction whose direction cosines referred to the cubic crystal axes are α_1, α_2, and α_3, will produce a single domain extending over the whole crystal and magnetized along that direction, and will also change the length l_0 of the crystal in that direction magnetostrictively by δl. Then one finds that

$$\frac{\delta l}{l_0}=\lambda_0+3(\lambda_1-\lambda_0)(\alpha_1{}^2\alpha_2{}^2+\alpha_2{}^2\alpha_3{}^2+\alpha_3{}^2\alpha_1{}^2),$$

where λ_0 and λ_1 are the magnetostriction constants for a given material. The equation takes a different form for noncubic crystals. For an isotropic polycrystalline sample, one would measure a single magnetostriction constant λ_p, which is approximately equal to $(2\lambda_0+3\lambda_1)/5$.

Magnetostriction constants decrease with increasing temperature, and vanish at the Curie point. Some typical values for cubic materials at room temperature are given below:

Substance	$10^6\lambda_0$	$10^6\lambda_1$	$10^6\lambda_p$
Fe	+21	−21	−7
Ni	−46	−24	−34
$FeO{\cdot}Fe_2O_3$	−20	+78	+40
$Co_{0.8}Fe_{0.2}O{\cdot}Fe_2O_3$	−590	+120	—

In nonferromagnetic substances, where the magnetization is induced by and proportional to the applied field, a similar magnetoelastic coupling, i.e., dependence of magnetization on strain, leads to a magnetostriction which is quadratic in the field. Typically a strain $\delta l/l_0$ of 10^{-6} is produced by a field of 10 T (= 10^5 G). In pure conductors at low temperatures the magnetization shows an oscillatory dependence on field because of the quantization of the orbital motion of the conduction electrons in the field (the de Haas–van Alphen effect). In this regime the magnetostriction also shows an oscillatory behavior with magnetic field, with amplitudes as large as 10^{-5}. These effects have been used to analyze the electronic structure in a number of metals and semiconductors.

Magnetostriction and its inverse effect, viz., the dependence of magnetization on strain, have found several applications using ferromagnets. Magnetostrictive acoustic transducers have been used in echo sounders, fish finders, sonar. They have been employed, because of their ability to transmit considerable ultrasonic energy into the surrounding medium, in ultrasonic cleaning and in processes such as homogenization, emulsification, and particle dispersion and agglomeration. In the field of information processing, magnetostrictive delay lines, filters, and memories have been used.

See also DE HAAS–VAN ALPHEN EFFECT; MAGNETIC DOMAINS AND BUBBLES; MAGNETOELASTIC PHENOMENA.

*BIBLIOGRAPHY**

B. S. Chandrasekhar and E. Fawcett, "Magnetostriction in Metals," *Adv. Phys.* **20,** 775–794 (1971). An advanced review of the effects in nonferromagnetic substances.

B. D. Cullity, "Magnetostriction and the Effects of Stress," in *Introduction to Magnetic Materials,* Chap. 8 Addison-Wesley, Reading, Mass. 1972. Reviews ferromagnetic magnetostriction at an intermediate level.

R. M. Hornreich, H. Rubinstein, and R. J. Spain, "Magnetostrictive Phenomena in Metallic Materials and some of their Device Applications," *IEEE Trans. Magn.,* **MAG-7,** 29–48 (1971).

J. Kanamori, "Anisotropy and Magnetostriction of Ferromagnetic and Antiferromagnetic Substances," in *Magnetism,* Vol. I, Chap. 4 (G. T. Rado and H. Suhl, eds.). Academic Press, New York, 1963. Advanced treatment, mainly theoretical.

Y. Kikuchi, "Magnetostrictive Materials and their Applications," *IEEE Trans. Mag.* **MAG-4,** 107–117 (1968). Review at intermediate level.

* Each of these articles gives numerous detailed references to the original literature.

Magnets (Permanent) and Magnetostatics

A. E. Berkowitz

Permanent magnets are useful because they produce a constant magnetic flux density in their vicinity. This occurs because permanent magnet materials (PMM) can sustain a significant intensity of magnetization in the absence of any external applied magnetic field. In this article, some of the more important properties of PMM will be introduced and their magnetostatic behavior will be reviewed.

Figure 1 shows schematically the intensity of magnetization **M** (dashed arrow) and lines representing the flux density in and around a rod of PMM in zero applied field after a saturating field has been applied in the direction of **M**. The flux lines are shown as emanating from the north and south pole distributions at the ends of the magnet. The concept of magnet poles is an entirely fictitious but extremely useful analytical tool for calculating field strengths around magnetized media. Actually, the fields result from variations in **M**.

The flux density B outside a magnetized body is the gradient of a scalar potential Φ, i.e.,

$$\mathbf{B}=-\nabla\Phi$$

where

$$\Phi = \frac{\gamma}{4\pi} \int \frac{\mathbf{M} \cdot \mathbf{r}^0 \, d\tau}{r^2}.$$

The integration is over the magnetized body, r is the distance to the point at which **B** is being calculated, $\mathbf{r}^0$ the corresponding unit vector, and $\gamma = 36\pi \times 10^9$ N(m/A·s)^2. Volume and surface pole densities (ρ_M and σ_M, respectively) are introduced by a transformation of Φ to the form

$$\Phi = \frac{\gamma}{4\pi} \int \frac{\rho_M}{r} \, d\tau + \int \frac{\sigma_M}{r} \, dS$$

where

$$\rho_M = -\nabla \cdot \mathbf{M}, \qquad \sigma_M = \mathbf{n} \cdot \mathbf{M};$$

dS denotes integration over the surface of the magnetized body with **n** the unit vector normal to the surface. **B** may now be written as

$$\mathbf{B} = \frac{\gamma}{4\pi} \int \frac{(dp_M)}{r^2} \mathbf{r}^0$$

where $dp_M = \rho_M \, d\tau$ or $\sigma_M \, dS$.

Thus the pole strength and the resulting external magnetic flux depend on variations in **M**. Obviously the maximum pole strength would occur at the surface of a uniformly magnetized PMM. Uniform magnetization of PMM generally does not exist because, as depicted in Fig. 1, the fields produced by surface poles *within* the PMM are antiparallel to **M**. These internal fields reduce **M** by demagnetizing the PMM. In the demagnetized state, which is the lowest energy state, the PMM is subdivided into regions called domains whose dimensions are of the order of 1 μm. In each domain, **M** has the saturation value corresponding to the composition of the PMM, but the orientation of **M** among the various domains is such that the net **M** of the PMM vanishes. Thus the measures of a useful PMM are twofold. First, **M** must be high enough so that variations in **M** produce large pole strengths. Second, the PMM must be able to withstand the effects of the internal demagnetizing field as well as the effects of external fields directed opposite to **M**.

To examine the latter property, we consider the second quadrant of the hysteresis loop of a PMM as shown in Fig. 2. Figure 2 indicates the response of **M** and the internal flux

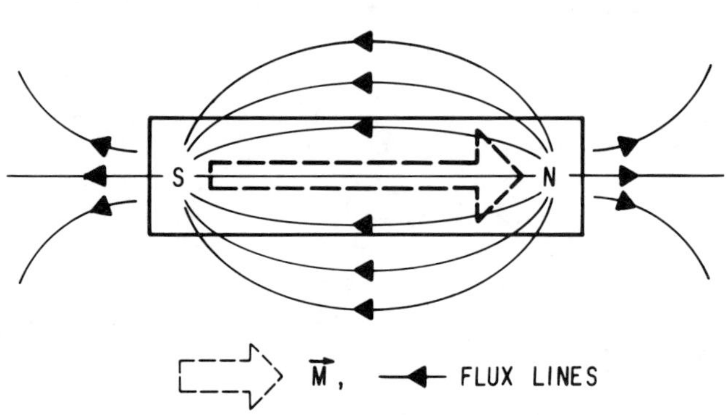

FIG. 1. Magnetization **M** (shown as dashed arrow) and resulting magnetic flux lines in a magnetized permanent magnet in zero applied field.

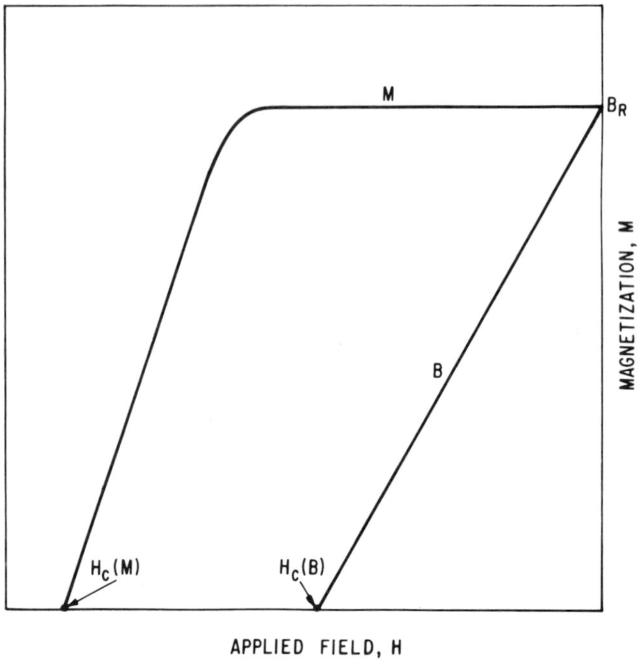

FIG. 2. Second quadrant of the hysteresis loop of a permanent magnet. Field **H** is applied opposite to direction of **M**.

density $\mathbf{B} = \mu_0 \mathbf{H} + \mathbf{M}$ of the PMM to an external applied field **H** where $\mu_0 = 4\pi \times 10^{-7}$ Hy/m. Also indicated are B_R, the remanent magnetization in zero applied field, and H_c, the coercive force. $H_c(M)$ and $H_c(B)$ are the fields required to reduce **M** and **B**, respectively, to zero. It may readily be shown that the maximum external field produced by a PMM is proportional to $(BH)_{max}$. Thus, from Fig. 2 it is evident that a useful PMM has high B_R and H_c, as well as a slow decrease of M in the second quadrant. In terms of the discussion relating to Fig. 1, we note that a high B_R implies a large value of **M** whereas H_c is a measure of the ability of a PMM to withstand demagnetizing fields.

There are two principal mechanisms responsible for high H_c in PMM. The first arises from the manner in which the net magnetization increases in response to an external field. In the demagnetized state, the direction of **M** varies from one domain to another. As an increasing field is applied, those domains whose **M** is parallel, or nearly parallel, to **H** grow at the expense of the other domains. The regions between adjacent domains are called domain walls. The magnetization of a material may, therefore, be viewed as the motion of domain walls. If it requires a large amount of energy (supplied by a magnetic field) to move a domain wall from one region of a PMM to another, then a high H_c results. Impediments to wall motion arise from microstructural heterogeneity in some PMM. Defects, impurities, precipitates, etc. can be energy barriers to wall motion. Both the oldest (carbon steel) and the newest (alloys of Co and rare earths such as Sm) PMM are examples of the operation of this mechanism for high H_c.

The second prominent mechanism for high H_c in PMM is the existence of single magnetic domains. When a particle of magnetic material is small enough (<1 μm) it is always uniformly magnetized; only the direction of magnetization can be changed, not the magnitude. Thus, the magnetization of an assembly of single-domain particles can be changed only by rotation of the magnetization of the particles. If it requires a high field to rotate the magnetization of the particles, then the assembly may have a high coercive force. In all magnetic materials, the magnetization prefers to lie in certain crystallographic directions. It requires a finite energy to rotate the magnetization from an "easy" crystallographic direction to a "hard" one. This energy is called the magnetocrystalline anisotropy energy. If this anisotropy energy is high in a given material, then an aggregate of single-domain particles aligned with their "easy" directions parallel will have a high H_c. This is the operative mechanism in barium oxide ferrite PMM. Another type of anisotropy energy can be provided by the shape of single-domain particles. If the particles are long and thin, the energy will be minimum when the magnetization is along the longitudinal axis. This can be seen from Fig. 1 to arise from the fact that the internal demagnetizing field is minimum in this case because of the r^{-3} dependence of the demagnetizing field. The particles in magnetic recording tape are dominated by this mechanism, as is the family of Alnico magnets. In the latter case, the elongated particles are precipitates in a heterogeneous alloy.

BIBLIOGRAPHY

Magnetostatics

W. F. Brown, Jr., *Magnetostatic Principles in Ferromagnetism.* North-Holland, Amsterdam, 1962.

Permanent Magnets

D. Hadfield, ed., *Permanent Magnets and Magnetism.* Iliffe Books Ltd., London, 1962.

D. L. Martin and M. G. Benz, *AIP Conf. Proc.* **5**, 970 (1972).

Single-Domain Particles

E. Kneller, in *Magnetism and Metallurgy* (A. Berkowitz and E. Kneller, eds.), Chapter 8. Academic Press, New York, 1969.

Many-Body Theory

D. C. Mattis

To most physicists today the many-body problem has connotations of a highly theoretical, quantum-mechanical study of systems of large numbers of interacting particles, such as electrons in a metal, with densities greater than 10^{20} particles/cm^3, or nucleons within the nucleus of an atom or in the nuclear matter of stars, with densities approaching 10^{40} particles/cm^3. But the question may be asked, *how many* bodies are required before we have a problem [1]? In Newtonian mechanics, the three-body problem was insoluble. With the birth of general relativity around 1910 and of quantum electrodynamics in 1930, the two- and one-body problems became insoluble; and within modern quantum field theory, the problem of zero bodies (the vacuum) has become insoluble!

Of course, one goal of many-body physics is to discover to what extent a complex system may be approximated by a collection of individual particles moving more or less independently of one another in the collective field of all of them. A separate, distinct goal is to discover to what extent such "quantum fluids" satisfy appropriate hydrodynamical equations. On closer inquiry, we observe that the interaction of each individual particle with the remaining fluid affects important properties such as its mass, because of the inertia of the "cloud" of interactions that accompany it. Further, properties of pairs of particles, such as their mutual interaction potential, are profoundly modified by the motion of the remaining particles. Thus, when the self-consistent equations of the fluid are derived, these changes must be taken into account *a priori*. It was not until the late 1950s that a proper theoretical framework—the use of second quantization and introduction of the Greens functions of quantum field theory and statistical mechanics—permitted a systematic development of this program.

The mounting popularity of this formerly recondite subject stems from successful applications to the theory of metals, both normal and superconducting; nuclear matter; and quantum fluids such as helium (^{3}He and ^{4}He). But the new techniques have not led to many *exact* results, nor have they been without notable antecedents in classical and in quantum mechanics. It is therefore most instructive to start by an examination of some early versions of the many-body problem.

THE CLASSICAL MANY-BODY PROBLEM

It is well known that Newton's equations of motion for three or more bodies, interacting by inverse-square-law forces, admit no closed-form solution. The difficulties are partly related to a physical question: Is the system in a bound state? For it is possible that one or more of the objects acquires energy from the others, achieves escape velocity, and is ejected from the system. Note also that the potential energy function satisfies Laplace's equation, so that there are no configurations of static equilibrium. Partly, also, the difficulties arise because of the detailed nature of the information that is required. Given a set of initial conditions, perhaps the shape of a galaxy, we wish to integrate the equations of motion of the celestial bodies either backward to the origins of time or forward to the ultimate end—but round-off error and long computation times endemic to multidimensional and numerical integration do not permit many results of this kind to be obtained. Thus the tendency in astrophysics is increasingly to forget about the many-body problem and concentrate on the study of hydrodynamics of "cosmic fluids."

When, however, the forces are not pure inverse-square law [2] it becomes possible for the *total* potential energy function $V(\mathbf{r}_1,\ldots,\mathbf{r}_N)$ of N interacting objects to have configurations of stable equilibrium, at $\mathbf{r}_i=\mathbf{r}_i^0$. Taylor's series expansion of this function up to terms quadratic in the displacements $x_i=x_i-x_i^0$ yields a system of coupled oscillators that can always be solved. The normal modes of regular one-, two-, and three-dimensional lattice structures, first studied by Hamilton [3] in 1839, form the basis for the theory of sound propagation (or "phonons" in the quantized version [4]) and that of the elastic properties of solids. The equations of coupled oscillators can be solved subject to arbitrary initial conditions, even when the masses of the particles are unequal—e.g., when one of the masses is much greater than the rest, as in Brownian motion [5]. Whereas for finite N the correlation functions are all periodic, with the period being Poincaré's recursion time t_P, in the limit $N \to \infty$ when the sum over normal modes must be replaced by an integral, we obtain that $t_P \to \infty$ and that all initial correlations decay inexorably. This explanation of *irreversibility*, then, is one important result of proceeding to the large-N limit.

If coupled oscillators represent the one extreme of a permanently bound system, the hard-sphere gas represents the other of a permanently *un*bound assembly that must be held together by external pressure or by mathematically imposed boundary conditions. In one dimension, where it is known as Tonk's gas of impenetrable rods, the constraints of conservation of energy and momentum together with the limitations of one-dimensional motion permit two colliding rods merely an *exchange* of energies and momenta upon impact, and thus the initial distribution of velocities is permanently maintained. However, the time evolution of N hard disks in two dimensions, or of hard spheres in three, is quite different, since (depending on impact parameter) each collision alters the relative momenta $\mathbf{p}_i - \mathbf{p}_j$. Indeed, after a (finite) characteristic time the entire system "forgets" its initial conditions to approach a well-defined state of thermodynamic equilibrium [6]. Computer simulations of the final state [7] have identified *two* different equilibrium phases: a close-packed solid phase and a low-density gas phase, with the lack of a liquid presumably due to the absence of attractive forces in the model. In any case, the study of such examples has established that the many-body problem is at the root of the science of thermodynamics.

A VARIETY OF COMPLEMENTARY QUANTUM METHODS

Bohr's 1913 theory of the hydrogen atom failed for helium and the heavier elements because it lacked the framework for the many-electron problem. The successor Hartree method improperly distinguished among in-principle indistinguishable electrons. Finally, use of determinantal wave functions by Hartree, Fock, and Slater allowed the construction of a one-particle self-consistent dynamics compatible with the exclusion principle. By the 1930s a substantially correct picture of the electron waves in atoms, molecules, and solids had emerged; the complementary, contemporaneous Thomas–Fermi method sacrificed the individual particles in favor of collective properties of the total *electron fluid* [8]. However, it also became apparent that a significant part of the energy of open-shell atoms, molecules, and metals, of the order of $\frac{1}{2}$ eV per electron, could not be explained by an independent-particle model; this energy came to be known as the "correlation energy," which many-body theory assumed the task of explaining.

The process of selecting relevant operators and establishing an "effective Hamiltonian" governing their interactions is the starting point. Because correlation energy is related to the electrostatic cost of density fluctuations, the construction of an effective Hamiltonian with the use of particle-density operators is indicated. Consider a metal containing noninteracting electrons having all possible momenta $\mathbf{p}$ within a Fermi sphere of radius p_F. It is convenient to take this state as the "vacuum," and for the elementary particles to take electrons only for $p > p_F$, introducing "holes" as the quasiparticles inside the Fermi sphere. The appropriate second-quantized operators a_p, $a_p{}^\dagger$ destroy/create an electron (and are thus defined to vanish inside the Fermi sphere), b_p, $b_p{}^\dagger$ destroy/create a hole (and vanish outside p_F). All fermion operators are subject to the anticommutation relations

$$\{a_p, a_{p'}\} \equiv a_p a_{p'} + a_{p'} a_p = \{b_p, b_{p'}\} = \{a_p, b_{p'}\} = 0,$$
$$\{a_p, a_{p'}^\dagger\} = \{b_p, b_{p'}^\dagger\} = \delta_{p,p'} \quad (1)$$

to ensure that the Pauli principle is exactly obeyed by any state constructed with their help. The particle-density operators

$$\rho_q = \sum_{j=1}^{N} e^{iq\cdot R_j}$$

are, in this formalism,

$$\rho_q = \sum_k (a_{k+q}^\dagger a_k + a_{k+q}^\dagger b^\dagger_{-k} + b_{-k-q} a_k - b^\dagger_{-k} a_{-k-q}), \quad (2a)$$

$$\approx \sum_k (a_{k+q}^\dagger b^\dagger_{-k} + b_{-k-q} a_k). \quad (2b)$$

Terms neglected in approximating (2a) by (2b) are only involved in the scattering of free electrons or holes, of which there are expected to be very few when the potential that creates these free particles is itself weak. We now study the "two-body" interaction potential:

$$\mathcal{H}' = \frac{1}{2V} \sum_q U_q \rho_q \rho_{-q} \quad (V = \text{volume}) \quad (3)$$

with $U_q = 4\pi e^2/q^2$ the Fourier transform of e^2/r or, for neutral weakly interacting fermions, $U_q \approx$ constant for small q, and 0 at large q. The unperturbed kinetic energy operator is

$$\mathcal{H}_0 = \sum \frac{p^2 - p_F{}^2}{2m} (a_p{}^\dagger a_p - b_p{}^\dagger b_p). \quad (4)$$

Examination of the equations of motion of operators such as $a^\dagger b^\dagger$ or ab indicates that no new operators appear if, to approximations (2b), we join

$$[b_{-k'+q'} a_{k'}, a_k{}^\dagger b^\dagger_{-k+q}] \equiv b_{-k'+q'} a_{k'} a_k{}^\dagger b^\dagger_{-k+q}$$
$$- a_k{}^\dagger b^\dagger_{-k+q} b_{-k'+q'} a_{k'} \quad (5)$$
$$\approx \delta_{k,k'} \delta_{q,q'} \quad (|k| > p_F, |k-q| < p_F)$$

in addition to the exact commutations

$$[a_k{}^\dagger a_k, a_{k'}^\dagger b_{k''}^\dagger] = \delta_{k,k'} a_{k'}^\dagger b_{k''}^\dagger \quad \text{and}$$
$$[b_k{}^\dagger b_k, a_{k'}^\dagger b_{k''}^\dagger] = \delta_{k,k''} a_{k'}^\dagger b_{k''}^\dagger. \quad (6)$$

With this algebra, operators carrying momentum q do not mix with operators carrying $q' \neq q$. This constitutes the so-called random-phase approximation (RPA) [9]. The problem is now effectively linearized, i.e., reduced to coupled oscillators, and all that remains is to find the normal modes. One method consists of constructing operators $\Omega_q \equiv \Sigma_k (F_k a_k{}^\dagger b^\dagger_{-k+q} + G_k b_{-k} a_{k-q})$ that satisfy the commutator equations of motion of exact normal modes,

$$[\mathcal{H}_0 + \mathcal{H}', \Omega_q] = \omega \Omega_q. \quad (7)$$

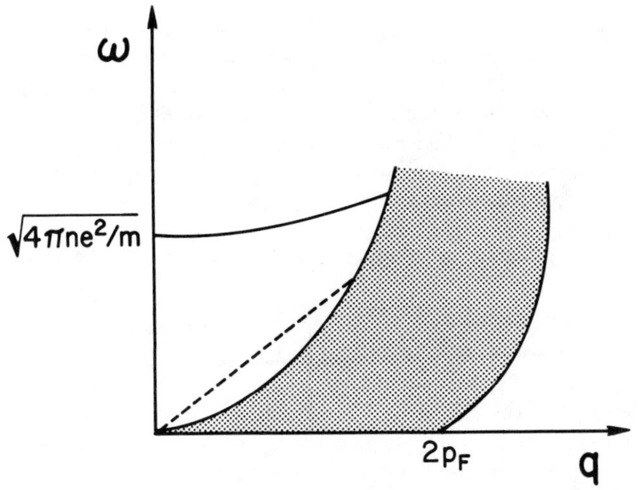

FIG. 1. The excitation spectrum of a Fermi gas (schematic). Shading indicates the continuum and coincides with the excitation spectrum of the unperturbed Hamiltonian, Eq. (4). Solid line shows *plasmon* branch (for $U_q = 4\pi e^2/q^2$), dashed line the *phonon* branch of the weakly imperfect neutral Fermi gas (U_q is constant at small q, vanishes at large q).

Equating coefficients of the ab operators on both sides of this equation leads to a set of linear equations in the F's and G's, the solution of which reduces to a proper-value equation first enunciated obtained by Bohm and Pines [9];

$$1 = \frac{2U_q}{V} \sum_{\substack{p<p_F \\ |p+q|>p_F}} \frac{(p\cdot q/m)+(q^2/2m)}{\omega^2 - (p\cdot q/m + q^2/2m)^2}. \tag{8}$$

This equation possesses two types of solutions: continuum solutions $\omega_{pq} = p\cdot q/m + q^2/2m + O(1/V)$, which interlace [10] the excitation spectrum of $\mathcal{H}_0$, and bound-state "collective" solutions above the highest value in the continuum, $p_F q/m + q^2/2m$. For the latter, now that the summand in (8) is finite, the sum may be replaced by an integration that is evaluated by elementary methods to yield the solutions displayed in Fig. 1. For the Coulomb potential, the quantized unit of this bound-state mode is denoted "plasmon," a stable high-energy charge fluctuation. For the weak finite-ranged potential with finite $U_{q=0}$, the same manipulations yield the "phonon" spectrum $\omega \propto q$, also indicated in the figure.

Landau's 1956 theory of Fermi liquids [11] concerns the energies of quasiparticles on either side of the Fermi level, particularly their dependence on the states of occupation of the other particles. Originally developed for the study of liquid ^{3}He, this theory was ultimately shown to provide the justification, to all orders in the perturbation expansion, of many important features of the independent-particle model. Thus, violent departures from free-particle behavior, e.g., the solidification of ^{3}He (under pressure) or the superconductivity of an electron gas (at low temperature), now have to be ascribed to the failure of perturbation theory to converge, which is a good sign of a genuine phase transition. Subtler examples concern the metal–insulator phase transition that is supposed to occur as atoms in a metal are slowly separated. Beyond a critical interatomic separation of the order of several angstroms, the Coulomb fluctuation energy greatly exceeds the covalent binding responsible for the band structure of the itinerant electrons, and it "pays" for atoms to become neutral and for the dc conductivity to vanish abruptly [12]. A much simplified model due to Hubbard [13] has been studied in connection with such problems; expressed in terms of operators that create or destroy particles at discrete lattice sites R_i, it is

$$\mathcal{H} = -\sum_{\substack{i,j \\ s=\pm 1}} T_{ij} c^{\dagger}_{is} c_{js} + U \sum_i c^{\dagger}_{i+} c_{i+} c^{\dagger}_{i-} c_{i-}. \tag{9}$$

The Fourier transform of the transfer matrix T_{ij} yields the Bloch energies of electrons in the conduction band. The interaction U has been simplified to where it only connects pairs of (necessarily opposite-spin) electrons on the same site. For T_{ij} transferring only between nearest neighbors, in one dimension, the model is *exactly* soluble [14]. In any number of dimensions, rigorous upper and lower bounds to the ground-state energy may be computed [14], as illustrated in Fig. 2. Unfortunately, none of the exact results have yet confirmed the variety of phase transitions (magnetic, insulator, etc.), at various values of the electron density and parameter U, that have been predicted by a number of intuitive or approximate methods.

Bardeen, Cooper, and Schrieffer's 1957 theory of superconductivity [15] shows that the various phenomena of superconductivity come from a most spectacular breakdown of perturbation theory. Formulated in terms of Cooper-pair destruction operators $a_{ps}a_{-p-s}$ or $b_{ps}b_{-p-s}$ and their Hermitian conjugate creation operators, this theory showed that however weak a phonon-induced attractive matrix element $-V_0$ connecting such pairs in the vicinity of the Fermi sur-

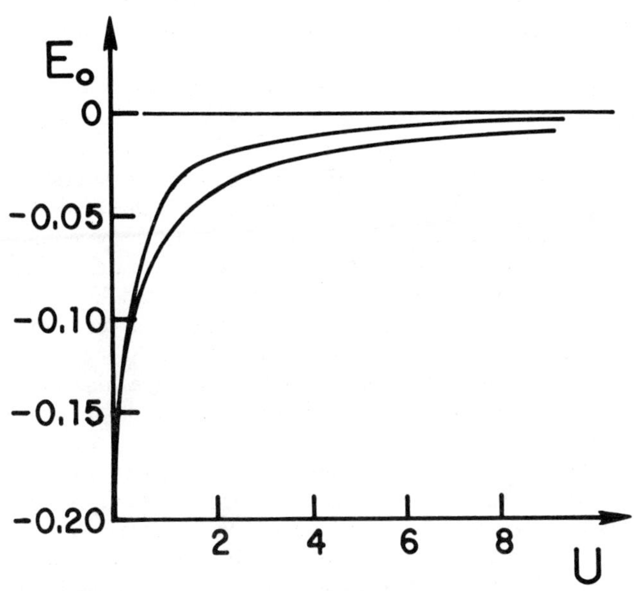

FIG. 2. Rigorous upper and lower bounds to the ground-state energy [14] of the Hubbard model in a half-occupied three-dimensional simple cubic lattice. Aside from indicating much steeper slope for $U<1$ than for $U>1$ (in units of the bandwidth), this sort of calculation appears incapable of determining changes of phase, for which the much more difficult study of Green functions or correlation functions is required.

face might be, there would always be energies $|p^2 - p_F^2|/m < V_0$ for which this interaction was, in fact, *strong* compared to the unperturbed energies. The BCS result consequently finds the ground state and the elementary excitations to be nonanalytic in the coupling constant, with an essential singularity at $V_0 = 0$.

Somewhat related results could already be seen in a 1947 study [16] of the superfluidity of a weakly imperfect Bose gas modeling ^{4}He. Here the field operators a_p, $a_p^\dagger$ satisfy commutation relations $[a_p, a_{p'}^\dagger] \equiv a_p a_{p'}^\dagger - a_{p'}^\dagger a_p = \delta_{p,p'}$ and the Hamiltonian is

$$\mathcal{H} = \sum \frac{p^2}{2m} a_p^\dagger a_p + \frac{1}{V} \sum U_p a_{k_1-p}^\dagger a_{k_2+p}^\dagger a_{k_2} a_{k_1}. \tag{10}$$

In the unperturbed ideal Bose–Einstein condensed state, all particles occupy the $p = 0$ momentum state, so that for weak interactions we may continue to treat $a_0 \sim a_0^\dagger \sim n_0^{1/2}$ as macroscopic parameters and neglect their fluctuations. The largest parts of the remaining interaction Hamiltonian are therefore

$$\frac{1}{V} U_0 n_0^2 + \frac{2n_0}{V} \sum_{p \neq 0} U_p a_p^\dagger a_p + \frac{n_0}{V} \sum_{p \neq 0} U_p (a_p^\dagger a_{-p}^\dagger + a_{-p} a_p) + O(n_0^{1/2}). \tag{11}$$

Both this operator and the kinetic energy are quadratic in the field operators, and thus a difficult problem has once again been reduced to a simpler one in coupled harmonic oscillators. Again, we seek normal modes and obtain, as for the weakly interacting Fermi gas, a linear excitation spectrum $\omega \propto q$ in the long-wavelength limit $q \to 0$. But the two results differ significantly in that there is now no continuum of quasiparticle excitations: The entire excitation spectrum of the weakly interacting Bose gas consists of collective modes.

Many flaws have been noted in the foregoing theory. (a) Neither the van der Waals attractive potential nor the repulsion of the hard cores at short distances admits of a finite Fourier transform, so that the formalism is defective. (b) A study of the dilute limit [17] indicates that it is the scattering length of the interaction, and not the potential, that enters into a correct theory. (c) Penrose and Onsager have noted that, for hard spheres, not more than 8% could be condensed into the $p = 0$ state, making n_0 a very poor approximation to $n = N/V$ in any event. (d) The theory seems unable to predict the existence or nature of *rotons* [18], which have been so important in the phenomenology of superfluidity.

Although normal-mode approximation methods seem to do even worse in two dimensions or one dimension, for reasons associated with the amount of phase space available to long-wavelength fluctuations, by a fortunate compensation many of the interesting many-body problems have been able to be solved in one dimension, through a variety of special methods [19].

GREEN'S FUNCTION METHODS

A vital simplification was brought into quantum electrodynamics by the introduction in 1949 of the graphical methods of Feynman diagrams for the systematic evaluation of various physical quantities in given orders of perturbation theory. It is not certain who first appreciated the relevance of such methods to the many-body problem [20]; however, by the 1960s they had become quite prevalent. One of the essential ingredients in this calculus is the "propagator," also known as the causal Green function [21],

$$G(p, E) = -i \int_{-\infty}^{\infty} dt\, e^{iEt} \langle \Psi_0 | T\{c_p(t) c_p^\dagger(0)\} | \Psi_0 \rangle \tag{12}$$

in which Ψ_0 is the exact ground state of the interacting system, T is the symbol for time ordering, and the time dependence of the operators is given by $c_p(t) = [\exp(i\mathcal{H}t)] c_p(0) [\exp(-i\mathcal{H}t)]$. All observables—occupation numbers, kinetic energy, interaction energy, etc.—can be obtained by suitable manipulations of (12). It is therefore an important quantity to calculate, and a relatively simply one, since only the operators of the pth particle are explicitly involved. We can express G in terms of G_0, the trivially calculable *unperturbed* Green function,

$$G_0 = (E - \epsilon_p + i\delta_p)^{-1}$$

where $\epsilon_p =$ one-particle energy and

$$\delta_p = \pm 0^+ \quad \text{for} \quad |p| \gtrless p_F,$$

by means of Dyson's equation:

$$G(p) = G_0(p) + G_0(p) \sum (p) G(p), \tag{13}$$

which has the formal solution

$$G = \{E - [\epsilon_p + \sum (p)] + i\delta_p\}^{-1}. \tag{14}$$

The proper self-energy part $\Sigma(p)$ must be calculated separately; its real part yields the effect of interactions on the mass of the quasiparticle, while its imaginary part tells about the lifetime. To first order, $\Sigma(p)$ is given by (a) in Fig. 3, for which the rules for the evaluation of such diagrams yield [21]

$$\sum{}_1(p) = \tfrac{1}{2} \left(\frac{1}{2\pi}\right)^3 \int d_3q\, U_q - \frac{1}{(2\pi)^3} \int_0^{|q|<q_F} d_3q\, U_{|q+p|}, \tag{15}$$

which is entirely equivalent to Hartree–Fock theory. Higher-order diagrams yield further corrections, including the finite

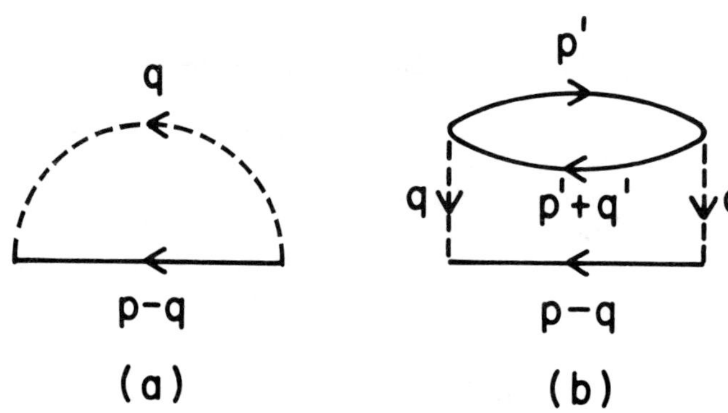

FIG. 3. (a) First-order contribution to proper self-energy of a quasiparticle shown in diagrammatic form. The rules for evaluating it [1] yield Eq. (15). (b) A simple second-order diagram that contributes to the finite lifetime of a quasiparticle [22].

lifetime. By examining these corrections to all orders, Luttinger was able to prove [22] that subject to the convergence of perturbation theory, in three dimensions the imaginary part of $\Sigma(p)$ vanished as $(p^2-p_F{}^2)^2$ as $p\to p_F$, establishing that the quasiparticles near the Fermi surface were sharp and well defined. Use is often made of the so-called Lehmann representation

$$G(p,E)=\int_0^\infty d\epsilon\left[\frac{A^+(p,\epsilon)}{E-(\epsilon_F+\epsilon)+i0^+}\right.$$
$$\left.+\frac{A^-(p,\epsilon)}{E-(\epsilon_F-\epsilon)-i0^+}\right] \qquad (16)$$

in which it can easily be shown that $A^{\pm}>0$. Eq. (16) shows that the causal G has a complicated analytic structure as a function of the complex variable E; this makes the theory awkward to generalize to finite temperature T.

Once perturbation theory ceases to be the overriding consideration, analytically more satisfactory Green's functions can be used, which easily carry over to finite T. There exist at least four species [23]: advanced and retarded, which are analytic in the lower and upper half-planes of complex E, respectively; each of these can be a commutator or anticommutator, the former being most useful for the study of bosons (plasmons, phonons, magnons, etc.) and the latter for Fermi–Dirac type quasiparticles (electrons, holes, etc.). They are

$$G_r(t'-t'')=-i\theta(t'-t'')\langle[A(t'),B(t'')]_\eta\rangle,$$
$$G_a(t'-t'')=i\theta(t''-t')\langle\text{same}\rangle \qquad (17)$$

where $\theta(t)$ is the unit step function, and averages are with respect to the grand canonical ensemble:

$$\langle op\rangle=\frac{1}{Z}\operatorname{Tr}\{ope^{-\beta(\mathcal{H}-\mu N)}\},\quad Z\equiv\operatorname{Tr}\{e^{-\beta(\mathcal{H}-\mu N)}\} \qquad (18)$$

and $\eta=\pm1$ is used to distinguish commutator and anticommutator Green's functions,

$$[A,B]_\eta\equiv AB-\eta BA. \qquad (19)$$

Basically, it is the Fourier transform $G(E)$ that we need:

$$G(E)=\frac{1}{2\pi}\int_{-\infty}^{\infty}dt\,G(t)e^{iEt}, \qquad (20)$$

for we can show [23] that the thermal-averaged correlation functions are given by the discontinuity of the corresponding Green functions on the real axis:

$$\langle B(t')A(t)\rangle=i\int_{-\infty}^{\infty}dE\,\frac{e^{-iE(t-t')}}{e^{\beta E}-\eta}$$
$$[G(E+i0^+)-G(E-i0^+)] \qquad (21)$$

where $G(E\pm i0^+)=G_{r,a}(E)$. By appropriate choices of the operators A and B *all* physical observables can be examined in this manner, once the corresponding Green function is known. The method of choice for obtaining it involves the *equations of motion*:

$$EG(E)=\frac{1}{2\pi}\langle[A(0),B(0)]_\eta\rangle+G^{(1)}(E) \qquad (22)$$

where the equal-time commutator is usually a known constant, and

$$G^{(1)}(E)\equiv-\frac{i}{2\pi}\int_0^\infty dt\,e^{iEt}\langle[C(t),B(0)]_\eta\rangle. \qquad (23)$$

This is itself a Green function, with the operator $C(t)$, given by a commutator,

$$C(t)\equiv[A(t),\mathcal{H}-\mu N], \qquad (24)$$

taking the place of A in $G^{(1)}$.

We may then study the equation of motion of $G^{(1)}$, which in turn yields a new function $G^{(2)}$, and continue this process Q times until either of two conclusions: After the Qth iteration no new functions appear, in which case the set of $Q+1$ linear equations in $Q+1$ Green's functions is closed and may be straightforwardly solved; or the hierarchy is terminated by our approximation of the $G^{(Q)}$. If the approximation consists of replacing $G^{(Q)}$ by its (unperturbed) value in the absence of interactions, then the system of Q equations in the remaining functions is essentially equivalent to Qth-order perturbation theory. But vastly more interesting results have been obtained when $G^{(Q)}$ is treated as a function, generally nonlinear, of the remaining Green functions on the basis of some physically motivated arguments. The system of equations, now nonlinear, has been found to exhibit such behavior as phase transitions at various temperatures and coupling constants, etc.; i.e., to mimic the rich variety of behavior that has come to be associated with the many-body problem.

See also PHONONS; PLASMONS; QUASIPARTICLES; SUPERCONDUCTIVITY THEORY.

REFERENCES

1. Here we quote loosely from Chapter 1 in R. D. Mattuck's delightful *Guide to Feynman Diagrams in the Many-Body Problem* (McGraw-Hill, New York, 1967).
2. The potential function for two rare gas atoms is not r_{ij}^{-1} but, approximately, $V(r_{ij})=Ar_{ij}^{-12}-Br_{ij}^{-6}$; the potential function of open-shell atoms may be more complicated because of covalent bonding and other quantum effects. That an equilibrium configuration always exists (i.e., that collapse of matter is impossible) has been proved by F. J. Dyson and A. Lenard, and most recently and elegantly by E. Lieb and W. Thirring in *Phys. Rev. Lett.* **35**, 687 (1975).
3. Volume 2 of *The Collected Mathematical Papers of Sir William Rowan Hamilton* (Cambridge Univ. Press, London and New York, 1940). Recent advances are discussed in A. Maradudin, E. Montroll, and P. Weiss's "Theory of Lattice Dynamics in the Harmonic Approximation," *Solid State Phys.* Suppl. No. 3 (F. Seitz, *et al.*, eds.; Academic Press, New York, 1963), as well as in the conference proceedings "Molecular Dynamics and the Structure of Solids" (L. Carter and J. Rush, eds.; National Bureau of Standards Special Publication 301, Washington, D.C., June 1969).
4. A familiar topic in all introductory solid-state textbooks. But recent interesting developments in "quantum solids," i.e., solid ^{3}He or ^{4}He isotopes or mixtures, are not so well known; see a recent report by B. Bertman and R. Guyer in *Sci. Am.* August 1967, p. 85.

5. R. J. Rubin, "Statistical Dynamics of Simple Cubic Lattices, Model for the study of Brownian Motion," *J. Math. Phys.* **1**, 309 (1960).
6. A rigorous demonstration that a many-body system subject to arbitrary initial conditions approaches thermal equilibrium (i.e., satisfies the famous "ergodic hypothesis" to a greater or lesser extent) exists indeed *only* for the hard-sphere gas. The immensely difficult proof is the work of Y. Sinai, *Dok. Akad. Nauk SSSR* **153**, 1261 (1963).
7. B. J. Adler and T. E. Wainwright, *J. Chem. Phys.* **31**, 459 (1959) and **33**, 1439 (1960). Amazingly, many of their results have been duplicated by an approximate theory in which the coordinates r_i, p_i of the individual particles are sacrificed in favor of an equal number of "collective coordinates," chosen to be the Fourier components of the particle density. These were treated as approximately independent dynamical variables and used in the construction of a self-consistent equation of state by J. Percus and G. Yevick, *Phys. Rev.* **110**, 1 (1958), following a procedure known as RPA that had been found useful in the quantum many-electron problem. This topic is treated elsewhere in the present text.
8. This method, largely discarded since the 1930s, has lately been reexamined and extended by E. H. Lieb and B. Simon, "The Thomas–Fermi Theory of Atoms, Molecules and Solids," *Adv. Math.* **23**, 22–116 (1976).
9. For a discussion of this topic and a collection of many relevant reprints see David Pines's *The Many-Body Problem* (Benjamin, Advanced Book Program, Reading, Mass., 1962).
10. The slight shift in the excitation spectrum of the continuum may be interpreted by scattering theory to be associated with the decay in time of a prepared initial state. This decay is better studied by examining the individual quasiparticles (i.e., the individual electron or hole) than by examining the couple that make up the elementary excitation of the system. For this, the propagator formalism is required, (see Ref. 1). Note that the finite lifetime of elementary excitations does not of itself limit the electrical conductivity, since the scattering is among excitations *all* of which carry momentum q.
11. The original papers in English translation are reprinted in Ref. 9. A simple textbook approach is provided by D. Pines and P. Nozières, *The Theory of Quantum Liquids*, Vol. 1 (Benjamin, New York, 1966).
12. Experimentally, this effect has been seen in metal vapors. Theoretically, a projection-operator method to reduce charge fluctuations to an energetically optimal amount was introduced by M. Gutzwiller, *Phys. Rev.* **134**, A923 (1964), following ideas that date back to E. Wigner, *Phys. Rev.* **46**, 1002 (1934) and *Trans. Faraday Soc.* **34**, 678 (1938). N. F. Mott has contributed many papers to the theory of collective insulator, and there is a nice characterization of this topic in W. Kohn's "Theory of the Insulating State," *Phys. Rev.* **133**, A171 (1964).
13. J. Hubbard, *Proc. Roy. Soc. (London)* **A276**, 238 (1963), **277**, 237, and **281**, 401 (1964).
14. E. Lieb and F. Wu, *Phys. Rev. Lett.* **20**, 1445 (1968). Upper and lower bounds to the ground-state energy of this Hamiltonian in one and three dimensions were computed by W. Langer and D. Mattis, *Phys. Lett.* **36A**, 139 (1971).
15. An especially clear development is found in J. R. Schrieffer's *Theory of Superconductivity* (Benjamin, New York, 1964).
16. N. N. Bogoliubov, apparently following a suggestion of D. Landau to have the two-fluid hydrodynamic theory of superfluidity placed on a more microscopic basis; the relevant papers are reprinted in Ref. 9.
17. T. D. Lee, K. Huang, and C. N. Yang, *Phys. Rev.* **106**, 1135 (1957); also E. Lieb, *Phys. Rev.* **130**, 2518 (1963).
18. Discussed by R. P. Feynman, *Phys. Rev.* **94**, 262 (1954); R. P. Feynman and M. Cohen, *Phys. Rev.* **102**, 1189 (1956).
19. See the text and reprint volume covering many areas of the many-body problem and statistical mechanics by E. Lieb and D. Mattis, *Mathematical Physics in One Dimension* (Academic Press, New York, 1966).
20. The earliest referenced papers in this general area include K. Husimi, *Proc. Phys. Math. Soc. Japan* **22**, 264 (1940); H. Koppe, *Ann. Phys.* **9**, 423 (1951); H. B. Callen and T. A. Welton, *Phys. Rev.* **83**, 34 (1951), L. van Hove, *Physica (Utrecht)* **21**, 901 (1955); T. Matsurbara, *Prog. Theor. Phys.* **14**, 351 (1955); followed by others too numerous to list.
21. Indicated by a directed solid line in diagrams such as Fig. 3. The rules for evaluating these diagrams and the resulting physical observables may be found in Ref. 1.
22. J. M. Luttinger, "Analytic Properties of Single-Particle Propagators for Many-Fermion Systems," *Phys. Rev.* **121**, 942 (1961).
23. For an exceptionally clear introduction to the use of these retarded and advanced Green functions consult D. N. Zubarev in *Sov. Phys. Usp.* **3**, 320 (1960).

Mechanical Properties of Matter

George A. Alers

INTRODUCTION

Mechanical properties describe the change in shape of matter when external forces are applied. Examples include the simple bending of a beam, the propagation of sound waves, the permanent deformation of metals into useful shapes, and the flow of liquids or gases around obstacles. For matter in the liquid and gaseous states, the usual force is the hydrostatic pressure and the deformation is a change in volume and a flow of the material from one point to another. For matter in the solid state, both tensile and shearing type forces are available to produce the corresponding elongations and shear distortions.

DEFORMATION OF SOLIDS

A solid is said to be *elastic* if the amount of deformation is directly proportional to the applied force. This implies that the process is completely reversible and independent of the way in which the final force or its direction of application has been achieved. A solid is said to be *anelastic* if there are small, additional displacements that depend on the rate with which the forces are applied. For most metals and ceramics, these time-dependent effects are small, but they play the important function of giving rise to the dissipation of vibrational energy that causes the damping of vibrations in machines and oscillating mechanical systems. In materials such as plastics and rubbers, the time-dependent contributions to the deformation are large and these types of solid matter are often called *viscoelastic* materials. The term *plastic solid* or *plastic deformation* describes those materials in which the displacement is a nonlinear, irreversible function of the applied force. Examples are the continuous and permanent deformation of metals by large forces and the response of organic polymers and glasses at elevated temperatures.

Mathematical Description of Elasticity in Solids

The forces that deform solids are described in terms of a *stress tensor* σ_{ij}, which represents a force F_i in direction i applied to an area ΔA_j whose normal is in the j direction; that is,

$$\sigma_{ij} = \lim_{\Delta A_j \to 0} F_i/\Delta A_j$$

and has the units of force per unit area. Obviously, when $i=j$, the force and the area normal are parallel, hence the stresses σ_{11}, σ_{22}, and σ_{33} are called *normal stresses* and work to move the area ΔA parallel to its normal. When $i \neq j$, a *shear stress* results in which the area ΔA is slipped or sheared in either of the two orthogonal directions that are perpendicular to the normal.

The deformations are described in terms of a *strain tensor* ϵ_{ij}, which measures the change in separation between two points in a solid divided by their original separation. Thus, a bar whose original length L becomes stretched to a length $L+\Delta L$ is said to have a strain of magnitude $\Delta L/L$. Thus, the strain has no units. For the general case, the deformation is described by a *displacement vector* which has Cartesian components u, v, and w and whose magnitude and direction describe the displacement of each point in the solid. This means that when a point at position X is displaced in the X direction by an amount $u(X)$ and its neighboring point at $X+\Delta X$ is displaced by $u+(\partial u/\partial X)\Delta X$, the strain in the X direction is $\partial u/\partial X$. The *normal strains* along all three coordinate axes are, therefore,

$$\epsilon_{xx} = \frac{\partial u}{\partial x} = \epsilon_{11}, \quad \epsilon_{yy} = \frac{\partial v}{\partial y} = \epsilon_{22}, \quad \epsilon_{zz} = \frac{\partial w}{\partial z} = \epsilon_{33},$$

and describe the deformation of a cube into a parallelepiped with unequal but orthogonal edges. To describe the shearing of a cube into a parallelepiped without orthogonal edges, the shear strains are

$$\epsilon_{xy} = \epsilon_{yx} = \frac{1}{2}\left(\frac{\partial u}{\partial y} + \frac{\partial v}{\partial x}\right) = \epsilon_{12} = \epsilon_{21},$$

$$\epsilon_{yz} = \epsilon_{zy} = \frac{1}{2}\left(\frac{\partial v}{\partial z} + \frac{\partial w}{\partial y}\right) = \epsilon_{23} = \epsilon_{32},$$

$$\epsilon_{zx} = \epsilon_{xz} = \frac{1}{2}\left(\frac{\partial w}{\partial x} + \frac{\partial v}{\partial z}\right) = \epsilon_{31} = \epsilon_{13}.$$

Geometrically, the shear strain ϵ_{ij} measures the angle by which the edges of a square in the ij plane is tilted away from 90° to form a parallelogram in that plane.

For elastic solids, the stress and strain are linearly related by a constant whose value is called the *elastic constant* of the material. Since the stress and strain are tensors of second rank, the constant of proportionality is a fourth rank tensor. Thus,

$$\sigma_{ij} = \sum C_{ijkl}\epsilon_{kl}$$

and

$$\epsilon_{ij} = \sum S_{ijkl}\sigma_{kl}$$

where the C_{ijkl} are called the *elastic moduli* and the S_{ijkl} are called the *elastic compliances*. It has become customary to contract this notation using the symmetry properties of the elastic tensors by applying the rule 11→1, 22→2, 33→3, 23→4, 13→5, and 12→6 so that stress and strain become six-component vectors and the elastic constants become second rank tensors with 36 elements. Thus,

$$\sigma_i = \sum C_{ij}\epsilon_j \quad \text{or} \quad \epsilon_i = \sum S_{ij}\sigma_j, \; i,j = 1 \text{ to } 6.$$

The inherent symmetry of the stress and strain relations (invariant to interchange of i and j) reduces the number of independent elastic tensor components to 21 for the most general case. Further reductions occur when the symmetry of the atomic structure of the particular material is introduced. Orthorhombic crystals have nine constants, tetragonal crystals have six constants, hexagonal crystals have five constants, and cubic crystals have three constants. *Isotropic solids,* which exhibit no variation of elastic properties with direction, have only two independent elastic moduli. These two constants appear in the literature with different names and describe different common distortions. The Lamé constants, λ and μ, are related to the C_{ij} tensor elements by $C_{11}=C_{22}=C_{33}=\lambda+2\mu$, $C_{12}=C_{23}=C_{13}=\lambda$, $C_{44}=C_{55}=C_{66}=\mu$, all other $C_{ij}=0$. The Young's modulus E and the Poisson's ratio ν describe the distortions of a long thin rod subjected to a tensile stress along its length:

$$E = \frac{\text{tensile stress}}{\text{lengthwise strain}} = \frac{\mu(2\mu+3\lambda)}{\lambda+\mu},$$

$$\nu = \frac{\text{lateral contraction (strain)}}{\text{lengthwise strain}} = \frac{\lambda}{2(\lambda+\mu)}.$$

The bulk modulus B and the shear modulus G describe the change in volume resulting from the application of a hydrostatic pressure and the shear strain resulting from a shear stress, respectively.

Internal Friction and Damping in Elastic Solids

For a perfectly elastic solid in which stress and strain are linearly related and completely reversible, the energy stored in the solid when under stress is exactly returned when the stress is removed. Thus, a vibrating solid will continue to oscillate between deformed and undeformed states indefinitely. To describe the dissipation of elastic energy that causes the vibrations of a solid to die away, anelastic or time-dependent relaxation phenomena must be introduced. The common way of doing this is to generalize the relation between stress and strain with the addition of the first time derivatives of the stress and the strain:

$$\sigma + \tau_1 \frac{d\sigma}{dt} = M_R\left(\epsilon + \tau_2 \frac{d\epsilon}{dt}\right)$$

The coefficients τ_1 and τ_2 have the physical significance of relaxation times which describe the time constant with which the stress or strain relaxes after the sudden application of a fixed strain or stress, respectively. The coefficient M_R is called the relaxed modulus because it corresponds to the ratio of stress to strain after relaxation has occurred. Another modulus, M_U, which defines the ratio of stress to strain at times short compared to τ_1 or τ_2, is called the unrelaxed modulus. If the stress and strain are periodic functions of time with angular frequency ω, then the tangent of the phase angle θ by which the strain lags behind the stress is given by

$$\tan\theta = \frac{M_U - M_R}{M} \frac{\omega\tau}{1+(\omega\tau)^2},$$

where $M=(M_U M_R)^{1/2}$ and $\tau=(\tau_1\tau_2)^{1/2}$. The quantity $\tan\theta$ is a measure of the degree of damping in a vibrating system and is related to the logarithmic decrement δ, the elastic energy dissipated per cycle ΔW from a maximum stored elastic energy of W, the quality factor Q of the vibrating system at resonance, and the $1/e$ decay distance for the attenuation of a propagating wave of wave length λ, by the equations

$$\tan\theta = \frac{\delta}{\pi} = \frac{1}{2\pi}\frac{\Delta W}{W} = Q^{-1} = \frac{1}{\pi}\alpha\lambda.$$

These relationships show that the damping is large when the difference between the unrelaxed and relaxed modulus is large and that the maximum damping occurs when the vibrational frequency ω is equal to the reciprocal of the relaxation time τ.

The physical origin of the relaxation process may be a redistribution of molecular segments on a polymer; the diffusion of atoms between adjacent sites in a crystal; the rearrangement of heat, electric polarization, or magnetization within the sample; and many other stress-induced, microscopic motions. Some phenomena have broad ranges of relaxation times so there is always a relaxation process that matches the vibrational frequency and thus a high level of damping is observed over a broad range of frequencies. Some relaxation processes are thermally activated and the relaxation time depends on the temperature with a form

$$\tau \propto \exp(H/RT),$$

where H is an activation energy and R is the gas constant.

Strength of Solid Materials

For small stresses, the elastic properties of solids discussed above provide a sufficient description. At higher stresses, irreversible phenomena take place which cause either the fracture of the part or its permanent distortion. The maximum stress for which there is no permanent deformation or fracture is called the *elastic limit* or the *proportional limit*. *Brittle* materials fracture into separate pieces abruptly at this stress while *ductile* materials exhibit a time-dependent extension which is not recovered upon unloading. A graph of stress versus strain for a ductile material shows a deviation from linear behavior at the elastic limit and a rapidly increasing degree of strain for each additional increment of applied stress. The stress level at which the strain deviates from the linear relationship between stress and strain by 0.2% is often called the *yield stress* (or more exactly the 0.2% yield stress) and is used in engineering design to designate the maximum stress to which the material can be safely subjected before any permanent set is produced. The slope of the graph of stress versus strain above the yield stress is called the *work-hardening coefficient* and the maximum value of the stress achieved during the deformation is called the *ultimate stress*. For static loads that exceed the elastic limit, the extension or strain will continue to increase as a function of time giving rise to the *creep* of the solid.

When viewed on a microscopic scale, these irreversible changes in shape of a solid can be described in terms of the motion of irregularities in the atomic structure of the material called *dislocations*. A dislocation is a line in the crystal lattice about which the atom rows are distorted from their normal crystallographic positions. By moving from one point in a specimen to another, a dislocation produces a unit of permanent shear distortion in the specimen. The yield stress can be viewed as the stress level at which a detectable number of dislocations become mobile. To strengthen a solid (i.e., to make the yield point occur at a higher stress level) alloying with foreign atoms or precipitating second phases of other chemical species or different phases of the parent material within the solid provides obstacles which retard dislocation motion. When the dislocations are too firmly fixed, the solid will be elastic to very high stresses but then it is likely to break abruptly by the sudden growth of a crack.

DEFORMATION OF GASES AND LIQUIDS

The formal definitions of stress and strain needed to describe the deformation of elastic solids are greatly simplified when gases and liquids are described. Since these forms of matter do not support shear stresses (escept for viscous effects) the stress tensor elements σ_{ik} for which $i \neq k$ are zero and those principal stresses for which $i=k$ are equal to one another since the medium is isotropic. Thus for liquids and gases

$$\sigma_{11} = \sigma_{22} = \sigma_{33} = -P,$$

where P is the pressure and has a negative sign because it conventionally acts inward. Likewise the strain components with $i \neq k$ vanish and those with $i=k$ add to describe the relative volume change produced by the pressure P. Thus, for liquids and gases, the relation connecting stress and strain becomes

$$P = -B(\Delta V/V),$$

where B is the bulk modulus and V is the volume of matter involved. For liquids, the bulk modulus B is only slightly smaller than it is for solids and for many problems of flow, an assumption of incompressibility ($B=\infty$) can adequately describe the flow behavior.

For gases, the compressibility is high (B very small) and the volume changes that accompany pressure changes are large enough that they must be considered. In fact, the bulk modulus of a gas cannot usually be considered as a constant independent of volume, and the question of whether or not the deformation occurs at constant temperature (isothermal) or at constant heat content (adiabatic) has a marked effect.

The usual problem found in the mechanics of those liquids which can be assumed incompressible is to calculate the velocity of each volume element in the material and to construct the path followed by a representative point in a flow pattern. Thus, the velocity vector replaces the displacement vector as the principal variable to be related to the external forces. Since the velocity of the volume element is important, its inertia as measured by its density, ρ, and its acceleration, $d^2\mathbf{R}/dt^2$ (where $\mathbf{R}$ is its position vector), occupy a prominent position in the equations of motion. In general, a volume element of an incompressible fluid acted upon by a body force **G** per unit mass and a local pressure P obeys the equation of motion

$$\rho \frac{d^2\mathbf{R}}{dt^2} = \rho\mathbf{G} - \text{grad } P.$$

If the velocity vector $\mathbf{V} = d\mathbf{R}/dt$ that satisfies this equation is such that curl $\mathbf{V} = 0$, the flow is said to be *irrotational*. Furthermore, if the body force **G** can be derived from a potential function ϕ, then Bernoulli's equation follows which states that

$$\frac{V^2}{2} + \phi + \frac{P}{\rho} = \text{constant},$$

where V is the magnitude of the velocity at the point where ϕ and P satisfy the equation. This equation is useful because it can relate the velocity and pressure at a location in the stream where they are not known to a more distant location where they can be determined easily.

In order to describe a *viscous* liquid, the equations of motion for a volume element must be modified by introducing, in addition to the pressure, a shear stress whose magnitude is proportional to the velocity gradient across the element

$$P_{ij} = -\eta \frac{\partial V_i}{\partial X_j}$$

which introduces the coefficient of viscosity η relating the shear stress P_{ij} to the rate of change of the i component of the velocity with respect to a change in the j position coordinate.

See also ANELASTICITY; CRYSTAL DEFECTS; DEFORMATION OF CRYSTALLINE MATERIALS; ELASTICITY; FATIGUE; INTERNAL FRICTION IN CRYSTALS; RHEOLOGY.

BIBLIOGRAPHY

H. B. Huntington, *The Elastic Constants of Crystals*. Academic Press, New York and London, 1958.

C. Kittel, *Introduction to Solid State Physics*, 3rd ed., Chaps 3–5 and 19. Wiley, New York, 1966.

W. T. Read, Jr., *Dislocations in Crystals*. McGraw-Hill, New York, 1953.

Arnold Sommerfeld, *Mechanics of Deformable Bodies*. Academic Press, New York, 1950.

Metal–Insulator Transitions

N. F. Mott

Solid materials are divided rather sharply into metals, which contain free electrons able to carry a current, and nonmetals such as sodium chloride or glass, which are insulators. It might be said that there are intermediate materials, semiconductors, containing only a few electrons, but in fact, for low temperatures such that bound electrons cannot be freed by heat motion, the division is still sharp. For metals the electrical conductivity tends to a finite value as the temperature is lowered, depending on the purity and degree of perfection if the metal is crystalline; for nonmetals, including semiconductors, the reverse is the case, the conductivity tending to zero. For the purpose of this article a metal is defined as a material that has this the former property; this includes materials such as the "metallic" oxides (for instance, CoO_2) and also very heavily doped semiconductors. All these must contain electrons which are free to move, even at the lowest temperature.

Before the advent of quantum mechanics, it was quite unclear why the electrons in metals were free and in nonmetals not free, though it was known from the Hall effect, soon after the discovery of the electron, that the number of free electrons in a metal such as silver approximated one per atom. However, the formulation of wave mechanics by Schrödinger and its application to electrons in solids by Bloch, Peierls, and Wilson gave a convincing explanation in the early 1930s. Essentially it was assumed that in both metals *and* insulators the electrons are in a sense free, that is to say that they could be described in crystals by the Bloch wave functions

$$\psi = e^{ikx}u_k(x,y,z), \tag{1}$$

which are now a very familiar part of solid-state physics. Here e^{ikx} represents an electron moving with the wave vector k in the x direction, and u_k is a function having the periodicity of the crystal lattice. These wave functions describe an electron with an infinite mean free path; scattering, and with it a finite electrical resistivity, arises only because of deviations from a perfect lattice due to thermal vibrations or impurities. Moreover, the effect of one electron on another was entirely neglected. The insulating property arose because, first of all, the electrons described by the wave function (1) suffer Bragg reflection within the crystal, and this leads to the gaps of forbidden energy, in which no solution of the Schrödinger equation of type (1) exists; and also because the electrons obey Fermi–Dirac statistics, according to which not more than two electrons can occupy one quantum state. The "gaps" lead to the conclusion that the number $N(E)dE$ of states allowed for an electron in the range of energy dE can have the forms shown in Fig. 1. This quantity is known as the density of states. Bands of energy states, containing (for a cubic lattice) two states per atom, are formed, which may or may not overlap. For a monovalent metal, for instance silver, the lowest band is half filled (Fig. 1a). For a nonmetal a filled band is separated from an empty band by the gap ΔE. For a divalent metal (for instance calcium) the first and second bands must overlap as in Fig. 1c.

From this model it was clear that, if two overlapping bands could be separated, for instance by pressure or by change of composition in an alloy, a "metal–insulator transition" would occur; and this has been achieved for the divalent metal ytterbium under pressure. However, there was little if any discussion in the literature on this subject before World War II, apart from a seminal paper by Wigner[1] in 1938 which will be described later. The author's[2] paper of 1949, however, was the first to point out that if two bands overlap, a *small* number of electrons and "holes" in the valence band is not possible, because of the Coulomb attraction between them, which would lead to their forming bound pairs (excitons). It is a result of quantum mechanics that, if two par-

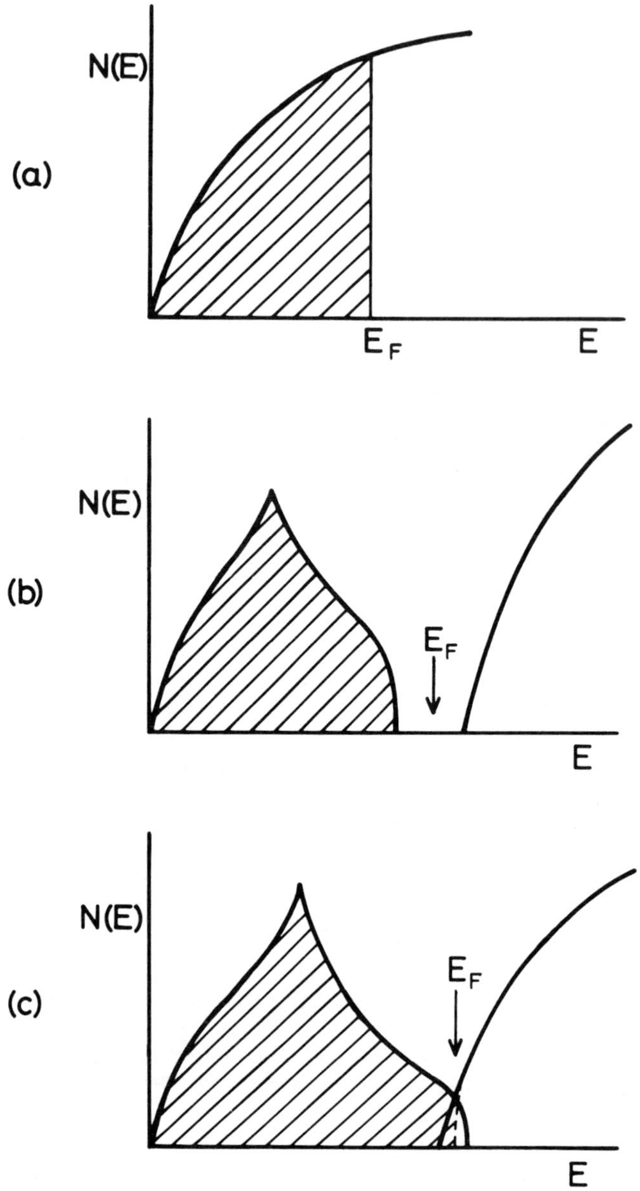

FIG. 1. Density of states in crystalline solids: (a) monovalent metal, (b) nonmetal, (c) semimetal.

ticles attract with a force that behaves as const/r^2 at large distances r, their ground state is always bound. A discontinuous change in the number of electrons and holes could thus be predicted, from zero to a value high enough to screen out the Coulomb attraction sufficiently to prevent pair formation. The prediction of a discontinuous change turned out to be correct, but the nature of the state in which pairs formed was obscure. These were called "excitonic insulators," and after many rather speculative papers about them, which have not been very fruitful in predicting properties of actual materials, a seminal paper by Brinkman and Rice[3] made it possible to look at the problem in the following way. The paper starts with the properties of "electron–hole droplets"; it is known that if electrons and holes (that is vacancies in a valence band) are formed in crystalline germanium, they condense into metallic droplets, and that these droplets have lower energy than bound pairs (excitons). In a droplet in equilibrium the number n of electrons and holes per unit volume is given by

$$n^{1/3}a_H \sim 0.1,$$

where a_H is the hydrogen radius $\hbar^2\kappa/me^2$, with κ the dielectric constant. The corresponding energy gained is

$$E_0 \sim 0.1 me^4/\hbar^2\kappa^2.$$

The numerical constants depend both on the band form and also on the approximation made in any calculation of this kind. The application to metal–insulator transitions is as follows: If the band gap ΔE decreases to the point where $\Delta E = E_0$, then an electron–hole gas will form discontinuously, filling the whole volume. When $\Delta E > E_0$, the material is a nonmetal; as soon as $\Delta E = E_0$, it becomes metallic, even though the bands do not overlap.

It would be very interesting if this discontinuous change from metal to nonmetal could be observed experimentally. This is difficult to do. One trouble is that, if it is attempted to change the gap by, for instance, hydrostatic pressure, there will *necessarily* be a discontinuous change in volume, and the transition will occur in an unstable region. This is because, if one plots free energy F against volume, there will be a kink in the curve at the point P where the transition occurs as in Fig. 2. Therefore, the discontinuous change in volume will be from A to B in the diagram. In the same way, if one changes the gap by varying the composition of an alloy, the region where the transition occurs is unstable. However, it might be possible to observe it for quenched alloys.

A very interesting example of a metal–insulator under pressure, which does show a discontinuous change of volume but none of structure,[4] is that of amorphous germanium, which goes metallic at a pressure of 60 kbar, but which reverts to its original form when the pressure is removed, so that any change of structure is precluded. This is perhaps

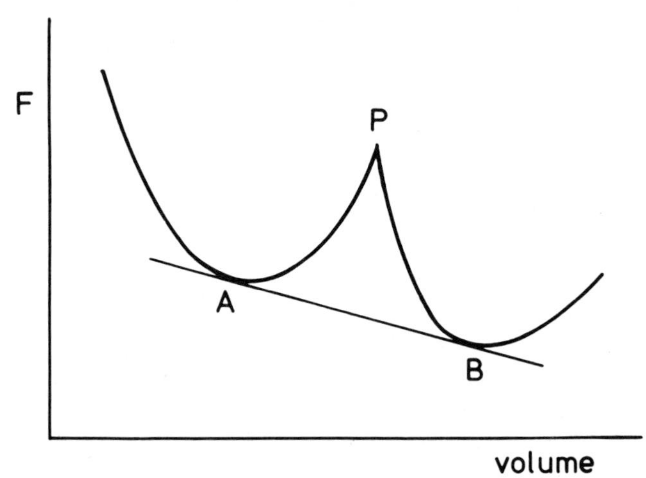

FIG. 2. Free energy F as a function of volume near a metal–insulator transition.

a case of the kind of transition illustrated in Fig. 2. In crystals, however, there is nearly always some change in the structure. In fact, some metal–insulator transitions seem to be entirely due to this change of structure. VO_2 is an example; a pairing of the V ions is thought to allow the formation of a full and empty band (as in Fig. 1), but at 400 K the pairing disappears, so that a half-full metallic band results.[5]

An example where a discontinuous band-crossing transition may be observed is the system $(Ti_{1-x}V_x)_2O_3$. Ti_2O_3 has a very small gap (~0.1 eV) between two d bands, which disappears above about 400 K, leading to a gradual metal–insulator transition, accompanied by an increase in the c/a ratio of the corundum structure. The addition of vanadium[6] also increases c/a, and for $x>0.02$ the material is metallic, showing a conductivity independent of T. But the increase in c/a is not nearly as great as that occurring at 400 K in pure Ti_2O_3; 10% of vanadium is necessary for this. So it looks as if the transition to metallic behavior occurs before the bands overlap, and that this is when $\Delta E = E_0$. Such an interpretation is controversial and others have been given (see the review cited by Van Zandt and Honig).

The so-called Mott transition[2] is similar to the band-crossing transition of Fig. 1, but occurs when each atom contains *one* electron, so that at first sight one would expect a metal as in Fig. 1. In 1938 Wigner[1] proposed that an electron gas of very low density must crystallize, the repulsion between a pair of electrons keeping them apart. Each electron would then carry a magnetic moment, its spin being unpaired. In the same way, for a crystalline array of one-electron atoms at a sufficient distance from each other, each atom will carry a moment and the moments are arranged antiferromagnetically. $TiCl_3$ is an example, the Ti^{3+} ion containing one electron. One can think of the antiferromagnetic lattice as splitting the band and providing an insulating state, as first pointed out by Slater.[7] The two bands are often called "upper and lower" Hubbard bands. The essence of the Mott transition is that, as the atoms are brought closer together, there should be a discontinuous transition to the metallic state, which may or may not be still antiferromagnetic. NiS_2 seems to be an example of the former case,[8] V_2O_3 of the latter,[9] though the interpretation is controversial, as is indeed much else in this field.

Metal–insulator transitions of this kind have been extensively investigated theoretically, following the introduction by Anderson[10] and by Hubbard[11] of a Hamiltonian which allowed analytical treatment (the Hubbard Hamiltonian).

A classic example of the "Mott transition" is the behavior of a heavily doped semiconductor such as silicon containing phosphorus. For low levels of doping, the conductivity tends to zero at low temperatures, the extra electron on the phosphorus atom being firmly bound by the Coulomb attraction of the charged phosphorus ion (P^+). At a concentration n given by

$$n^{1/3} a_H = 0.26,$$

the system goes metallic; a_H is here the Bohr radius $\hbar\kappa / m_{eff} e^2$ for an electron in the field of the donor. This result can be derived in various ways. But there appears to be no discontinuous change in the energy ϵ required to free an electron; it drops continuously to zero as the concentration of phosphorus increases. This is doubtless because the phosphorus atoms are arranged at random and not in a crystalline array. This brings us to a new kind of metal–insulator transition, which occurs *only* in noncrystalline systems. This we call an "Anderson transition."

A series of investigations which began with P. W. Anderson's paper[12] of 1958 entitled "Absence of diffusion in certain random lattices" have led to the conclusion that the conduction band in any noncrystalline system has at its extremities a range of so-called "localized states," which are essentially electron traps. They will have a continuous range of energies, and their wave functions will overlap; but the essence of Anderson's work is to show that, in spite of this overlap, an electron cannot move from one of these states to another without exchanging energy with lattice vibrations; its motion is thermally activated. An energy called the "mobility edge," and denoted by E_c, separates localized states from nonlocalized states. We consider then a material that would be metallic if there were no disorder, so that in Fig. 1a states are occupied up to a limiting energy E_F (the Fermi energy). The question now is whether E_F lies above or below the mobility edge E_c. If it lies above it, the electrical behavior is metallic; that is to say the conductivity varies little with temperature as $T \to 0$, the limiting value being finite. If it lies below E_c, conductivity is activated; charge transport is by "hopping," an electron moving from one state to another by absorbing or emitting phonons. For this kind of process the conductivity σ is expected to obey the relationship

$$\sigma = B \exp(-A/T^{1/4}).$$

The dependence on $T^{1/4}$ arises because the jump distance increases at low temperatures, the electron jumping to another state, even if far away, for which it need absorb as little energy as possible from phonons. This kind of charge transport is known as "variable-range" hopping.

If the number of electrons in the conduction band, or alternatively the degree of disorder, can be changed in such

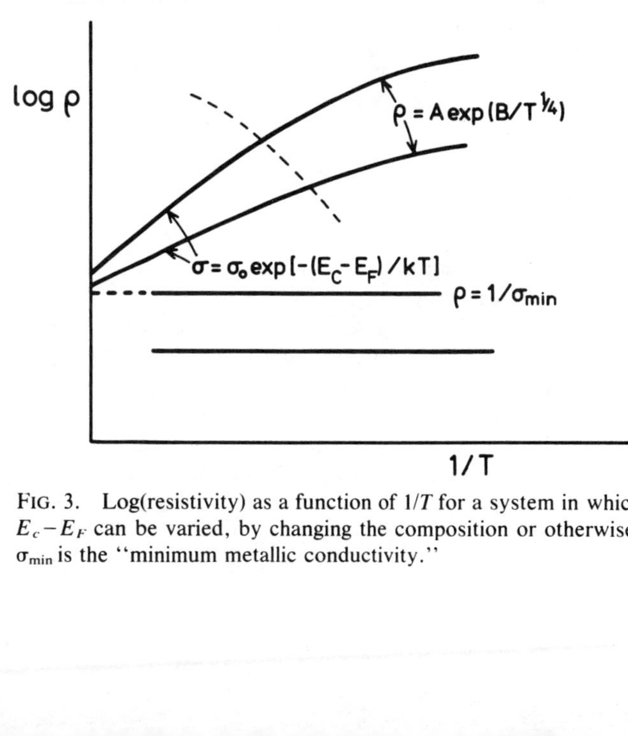

FIG. 3. Log(resistivity) as a function of $1/T$ for a system in which $E_c - E_F$ can be varied, by changing the composition or otherwise. σ_{min} is the "minimum metallic conductivity."

a way that the sign of $E_F - E_c$ changes from positive to negative, then a transition from metallic to nonmetallic behavior will occur. The resistivity as a function of temperature is expected to appear as in Fig. 3. When E_F lies below E_c, conduction at low T will be by variable-range hopping, at high T by excitation to the mobility edge; in the latter range of T, conduction will be of the form

$$\sigma = \sigma_{\min} \exp[-(E_c - E_F)/kT].$$

The value of σ just before an activation energy occurs has been called by the present author the "minimum metallic conductivity" $\sigma_{\min}$. There are many forms of this kind of transition in the literature, recently reviewed by the present author and co-workers.[13]

See also ELECTRON ENERGY STATES IN SOLIDS AND LIQUIDS; ELECTRON–HOLE DROPLETS IN SEMICONDUCTORS; EXCITONS; INSULATORS; METALS.

REFERENCES

1. E. Wigner, *Trans. Faraday Soc.* **34,** 678 (1938).
2. N. F. Mott, Proc. Phys. Soc. A **62,** 416 (1949).
3. W. F. Brinkman and T. M. Rice, *Phys. Rev.* **B7,** 1508 (1973).
4. K. Tamura, J. Fukushima, H. Endo, S. Minomura, O. Shimomura, and K. Asaumi, *J. Phys. Soc. Jpn.* **36,** 558 (1974).
5. A. Zylbersztejn and N. F. Mott, Phys. Rev. **B11,** 4383 (1975).
6. G. V. Chandrashekhar, Q. Won Choi, J. Moyo, and J. M. Honig, *Mat. Res. Bull.* **5,** 999 (1970).
7. J. C. Slater, *Phys. Rev.* **82,** 538 (1951).
8. F. Gautier, G. Kroll, M. F. Lapierre, P. Panissod, C. Robert, G. Czjzek, J. Fink, and H. Schmidt, *Phys. Lett. A* **53,** 31 (1975).
9. D. B. McWhan and T. M. Rice, *Phys. Rev. Lett.* **27,** 941 (1969).
10. P. W. Anderson, *Solid State Phys.* **14,** 99 (1963).
11. J. Hubbard, *Proc. Roy. Soc.* **281,** 401 (1964).
12. P. W. Anderson, *Phys. Rev.* **109,** 1492 (1958).
13. N. F. Mott, M. Pepper, S. Pollitt, R. H. Wallis, and C. J. Adkins, *Proc. R. Soc. A* **345,** 169 (1975).

BIBLIOGRAPHY

The author's book on the subject is *Metal-Insulator Transitions*. Taylor & Francis, London, 1974.

Other aspects of the subject are reviewed by

N. F. Mott and W. D. Twose, *Adv. Phys.* **10,** 107 (1961).

N. F. Mott and Z. Zinamon, *Rep. Prog. Phys.* **33,** 881 (1970).

L. L. Van Zandt and J. M. Honig, "Theories Pertaining to the Semiconductor-Metal Transition in Crystals." *Ann. Rev. Mat. Sci.* **4,** 191 (1974).

Metallurgy

Bruce Chalmers

In earlier days the term metallurgy was used to refer to the technology of the sequence of processes by which a useful object was produced from metal which usually was, at the outset, a component of a chemical compound found in the earth's crust as an ore. More recently, the term has come to describe the newly emerging branch of science that is concerned with exploring and understanding the structure of metals and alloys at the microscopic and submicroscopic levels, the relationship between the structure of a metal (or alloy) and its properties, and the way in which the structure of a particular sample is determined by the chemical composition and its thermal and mechanical history, and the exploitation of this understanding to produce the properties that are called for by the increasingly stringent demands of sophisticated technology.

It is, therefore, convenient to divide the whole field of metallurgy into process metallurgy and physical metallurgy. The former deals with the technology of extracting metals from their ores, by chemical, thermal or electrochemical processes, refining and alloying, and producing a metallic object by casting the molten metal in a mold of the appropriate shape and then, if necessary, deforming it mechanically by forging, rolling, extrusion, or otherwise to bring it to the required final shape.

Physical metallurgy is chiefly concerned with the properties of the final product as distinct from its chemical composition and its shape. The science of physical metallurgy is based on the concept that the properties of an alloy depend only on its structure, and that the structure is determined only by the chemical composition and the thermal and mechanical history of the sample. The structure is studied at two levels: the microscopic and the atomic. At the microscopic level, the structure is defined by the phases that are present and by the shapes and sizes of the crystals of each phase; at the atomic level, the structure relates to the basic crystal structure of each phase, the disposition of the atoms of the various elements within that crystal, and the ways in which the crystals depart from this ideal structure. Thus an alloy may consist of two or more distinct types of crystal, each with its own characteristic structure and composition; one or two of these crystal types may form a continuous phase (the matrix) while the others may be in the form of discrete particles, which could be needlelike, platelike, or equiaxed in shape. These characteristics can be determined by optical microscopy, as the typical size of the crystal is larger than the limit of resolution of the optical microscope, 1 μm. The study of structure at this level is called "metallography." Each type of crystal has its characteristic crystal structure (face centered cubic, body centered cubic, etc), and all the elements present in the alloy are to be found in each crystal, although usually not in the same proportion as in the alloy as a whole; for example, an alloy of iron and carbon (steel) containing 0.2% (by weight) of carbon consists of crystals of iron containing about 0.002% carbon (ferrite), and crystals of iron carbide, Fe_3C (cementite) with about 6.7% of carbon. Thus a phase may have the basic crystal structure of one of the elements that is present, with atoms of other elements present in it in solid solution. It is then described as a "terminal phase." The solute atoms may substitute for atoms of the parent phase at sites which, in the perfect crystal, would be occupied by atoms of the parent element. The substitutional atoms may be dispersed randomly within the crystal, or they may exhibit an ordered arrangement, which can either be of short range, consisting of very small ordered domains, or long range, in which the regularity of site selection extends over many atoms. Substitutional solid solutions do not occur over extended ranges of composition when the diameters of the atoms differ by more than about 15%. The solid solution may also be interstitial, when the solute atoms are accommodated between the atoms that define the structure of the crystal; this usually occurs only when the solute atoms have a diameter not more than about one-half of that of the host. A phase may, on the other hand, have a structure that differs from those of all of its component elements; it is then described as an "intermediate phase." Some intermediate phases have compositions in which atoms of the constituent elements are present in simple proportions (these are referred to as intermetallic compounds); other intermediate phases owe the stability of their structures to geometrical factors that exist when the sizes of the atoms are related in specific ways; others, to a preferred ratio of the number of valence electrons to the number of atoms; and for some, electronic and geometrical factors combine to stabilize the structure. The study of the equilibrium distribution of the elements that are present between the various phases, as a function of temperature, leads to the construction of "phase diagrams" or "equilibrium diagrams." The equilibrium diagram is a thermodynamic statement of the most stable distribution of the elements that are present; the phases that are actually present, and the distribution of the atoms in them as well as the sizes and shapes of their crystals, however, depend on the thermal history of the sample, because the processes that allow the structure to approach equilibrium are either the diffusion of atoms within the solid, which is inherently a slow process, or a cooperative change of structure, which does not take place without a large driving force in the form of a substantial departure from equilibrium.

"Heat treatment" is the term that describes the sequence of processes designed to produce the nonequilibrium structure that provides the required properties. For example, steel is heat treated by first heating it to a temperature at which the equilibrium structure is face-centered-cubic crystals of iron with carbon in interstitial solid solution. If the steel were cooled slowly, so as to maintain equilibrium, it would transform into body-centered-cubic crystals of iron containing very little carbon, and crystals of iron carbide (Fe_3C); if the steel is "quenched," that is, rapidly cooled by plunging into water, the separation into two phases is suppressed because there is insufficient time for the carbon to diffuse to form crystals of iron carbide, and instead the carbon remains where it is and a phase, martensite, which does not appear in the phase diagram, is formed. Martensite is an extremely hard and brittle material, and for most purposes for which hardened steel is needed it is necessary to temper the martensite by heating it to an intermediate temperature. During tempering the structure of the steel is allowed to change toward equilibrium to a carefully controlled extent to achieve the required compromise between hardness and toughness. Another example in which heat treatment provides a controlled approach to equilibrium is the age hardening of some aluminum alloys; the alloy is first "solution treated" at a relatively high temperature at which the alloying elements, such as copper, go into solid solution. Slow cooling would allow the copper largely to separate in the form of crystals of $CuAl_2$; rapid cooling prevents this, but subsequent aging at an intermediate temperature allows the copper to achieve the first stage of its separation, in the form of "zones" that endow the alloy with very high strength.

It has been long known that, while an ideally perfect crystal of a metal should be stronger than any known materials, real crystals are almost always very easily deformed; the reason is the presence in the crystal of imperfections called dislocations, which permit the sliding of one part of the crystal over another to take place progressively, by the motion of the dislocations, which is easy, rather than simultaneously, which would require a very high stress. In order to strengthen a metal, therefore, it is necessary to restrict the movement of the dislocations; there are several ways in which this can be done, of which the two most important are by the introduction of very large numbers of dislocations and by the introduction of obstructions which inhibit their motion. When a metal is deformed, the dislocations that are present not only move, but they also increase in number; typically $1 cm^3$ of a metal crystal would contain 100 km of dislocations, but if it were deformed, for example by rolling, this could increase to 10^7 km/cm^3. Dislocations interact with each other in such a way as to increase the force required to move them, and an effect of deforming a metal or alloy, therefore, is to increase its resistance to further deformation, that is, its hardness. This is the process of work hardening, which can be negated by annealing, which is heating to a temperature at which the dislocations are reduced in extent as a result of the mobility of the atoms. The second method of hardening a metal is by making it heterogeneous on a sufficiently fine scale. Concentrations of solute atoms that strain the crystal locally, or precipitated particles of a second phase, provide obstacles to the motion of dislocations so that the strength is increased. A third method of hardening an alloy is to convert it into a crystal structure which is inherently resistant to the passage of dislocations, as is the case with martensite. The strength depends not only on the internal structure of the individual crystals, but also on the shape, size, and distribution of the crystals of the various phases that are present; it follows that the composition of an alloy and the processes, such as casting, that determine the shape, size, and distribution of the crystals, interact with subsequent processes such as work hardening and heat treatment to determine the properties of the final product.

Some of the most challenging demands on the skill of the physical metallurgist are in the development of alloys that will withstand high stresses for long times at high temperatures in a highly oxidizing environment. This requires an understanding of the processes which allow slow deformation to take place over long periods of time at high temperatures and the ways in which these types of deformation can be stopped or retarded, and it requires the selection of materials that are highly resistant to oxidation.

See also ALLOTROPY AND POLYMORPHISM; ALLOYS; CRYSTAL DEFECTS; CRYSTALLOGRAPHY, X-RAY; METALS.

BIBLIOGRAPHY

R. W. Cahn (ed.) *Physical Metallurgy*. North Holland, Amsterdam, 1965.

J. Wulff, H. F. Taylor, and A. J. Shaler, *Metallurgy for Engineers*. Wiley, New York, 1952.

Metals

Stanley Engelsberg

A large majority of elements in the periodic table crystallize in the metallic state. The properties that essentially characterize a metal are its high electrical and thermal conductivity, its ductility and malleability, and its luster. These properties may in part be deduced from the type of binding that characterizes a metal.

In molecular and ionic crystals the binding can be discussed using classical concepts. That is, once we have the forces between the pairs of particles, which in most cases are derived on the basis of quantum mechanics, we can just add together the potentials of all the pairs of particles to obtain the total energy. In the case of metallic binding the picture is completely different. Here the electrons in the conduction band are no longer tied to a single atom, but effectively overlap all the ions in the crystal. The calculation is quantum mechanical by nature in all of its stages. If we consider the wave functions for the atoms in the metal, we find that at the known interatomic spacings, the last maximum in the radial part of each atomic wave function almost touches that for its nearest neighbors. Because of this overlap in the metal, the crystal's properties are quite distinct from those of, say, a gas of the constituent atoms. On the other hand, in ionic and molecular crystals, many of the physical properties (e.g., optical, dielectric, and magnetic) may be predicted from those of the constituent atoms, ions, or molecules.

One of the important ideas in the Wigner–Seitz theory of the binding of metals was to join the atomic wave functions smoothly with zero slope at the position between nearest neighbors. This new boundary condition on the wave function for the lowest electron state of the conduction band lowers the kinetic energy contribution and is an important part of the cohesive energy of metals.

In nonmagnetic metals and in the absence of an external field, this lowest state can accommodate two electrons, one with spin along the direction of quantization, the other opposed. To obtain the other allowed states for the elctrons in the conduction band we must solve the problem of the motion of electrons in a periodic potential, which is the subject matter of the band theory of solids. The electrons fill the lowest energy states up to the Fermi energy.

That there are available energy states just above the last filled state means that the application of the smallest electric field will accelerate the electrons. For a perfect pure crystal, in the absence of atomic vibrations, which can be made exceedingly small at very low temperatures, the electrical conductivity of a metal would be infinite. At low temperatures the resistivity of a metal is governed by its purity and perfection. That is the case until the metal reaches its superconducting transition temperature, below which temperature there is no measurable resistance. There are still a number of metals that have not been found to be superconducting at the lowest attainable temperatures. Most notably these include the best high-temperature conductors, i.e., the alkali metals and the noble metals. Those metals that undergo transitions to a magnetically ordered state are also nonsuperconducting.

That there are states available to the electrons just above the Fermi surface means that no matter how small the energy of a photon of light that you shine on a metal, the light can be completely absorbed. This is the basic reason for the opacity of metals in the visible and lower-energy part of the spectrum. It is directly connected with the good electrical conductivity of metals.

At all finite temperatures the electrons in fact do not just fill up the allowed energy states, stopping at the Fermi energy. In addition they have a thermal smearing in a small energy interval about the Fermi energy of width $k_B T$ (Boltzmann's constant times the absolute temperature) that is about 1/40 of an electron volt at room temperature. That there is a finite density of states (number of allowed states per unit energy interval) for the electrons at the Fermi surface leads to an electronic contribution to all of the equilibrium thermodynamic properties of metals. For example, there is an electronic contribution to the specific heat of metals, which is linearly proportional to the temperature. There is also a small contribution to the magnetic susceptibility of metals, which is paramagnetic for elements including and to the left of the transition elements in the periodic table. Starting with the noble metals and for those following, a diamagnetic susceptibility is observed. The oscillations of the diamagnetic or orbital part of the susceptibility as a function of the magnetic field give a good deal of information about the metals' Fermi surface. In a large number of metals that have incomplete $3d$ or $4f$ shells, the electronic interactions lead to ordered magnetic phases.

See also CRYSTAL BINDING; ELECTRON ENERGY STATES IN SOLIDS AND LIQUIDS; FERMI SURFACE; CONDUCTION; SUPERCONDUCTIVITY THEORY; SUPERCONDUCTING MATERIALS; PARAMAGNETISM; DIAMAGNETISM; DE HAAS–VAN ALPHEN EFFECT; FERROMAGNETISM; FERRIMAGNETISM; MAGNETIC MATERIALS.

BIBLIOGRAPHY

N. Ashcroft and D. Mermin, *Solid State Physics*. Holt, Rinehart & Winston, New York, 1976.

F. Seitz, *Modern Theory of Solids*. McGraw-Hill, New York, 1940.

Metals, Liquid *see* Liquid Metals

Mössbauer Effect

G. J. Perlow

This effect which bears the name of its discoverer, Rudolph Mössbauer, was first described in a publication in 1958. It is the recoil-free emission and absorption of nuclear gamma rays and is a very useful special case of nuclear resonance fluorescence. The recoil-free process results in emission of γ-rays with natural or nearly natural linewidth which allows the observation in the γ-ray spectrum of the interaction between the nucleus and its atom. The phenomenon occurs only in solids or extremely viscous liquids.

The phenomenon is illustrated in Fig. 1. A γ-ray, shown as resulting from a nuclear excitation following the β decay of a radioactive parent, can be removed from a collimated

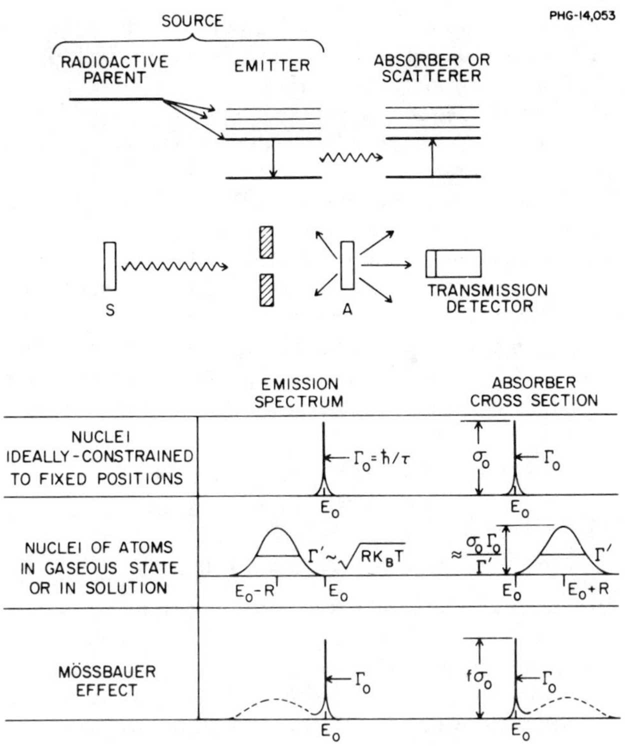

FIG. 1. Description of the Mössbauer effect.

beam by resonantly re-exciting another of the same species. In an *idealization* of the process, the emitter and the target nuclei are constrained from moving. The energy distribution of both the emitted radiation and the absorber cross section has the natural width $\Gamma_0=\hbar/\tau$, where τ is the mean lifetime of the excited state. Resonant cross sections are high, of order $\lambda\!\!\!^{-2}$, where $\lambda\!\!\!^{-}=$ wavelength/ 2π. For energies less than about 100 keV, the nuclear events dominate over excitation processes in the electronic shells of the atom and could be observed readily. Such nuclear states may typically have mean lifetimes of a few nanoseconds or longer, and therefore widths less than $\approx 10^{-6}$ eV. Indeed some are very much narrower.

The situation has been over-idealized. There is no method for rigidly constraining the nuclei. Before Mössbauer's discovery, the following description would be appropriate: A γ-ray with energy E carries momentum E/c, and the emitting nucleus (in fact the whole atom) recoils with energy $R=E^2/2Mc^2$. This energy is lost to the transition. A similar shortfall of energy occurs at the absorber, which cannot absorb the quantum unless it also recoils. In addition there is thermal motion which, because of the Doppler effect, spreads both the emission intensity and the absorption cross section out to a width $\Gamma'\approx(RK_BT)^{1/2}$, where T is the Kelvin temperature and K_B is Boltzmann's constant. The resonant absorption is then due only to the small overlap between the two distributions.

Mössbauer's remarkable discovery was that the spectrum of radiation emitted from nuclei in *solids* is actually a composite of the idealized and the unconstrained case. The same is true for the distribution of the absorption cross section. For relatively low-energy transitions, some fraction f of the events occur without recoil and as if the nucleus were exactly at rest. The emission or absorption widths have the natural value. Except for a small but important chemical effect there is no energy shift. The sharp line is superimposed on a spectrum of recoil events but on the scale in which one uses Mössbauer spectroscopy, the latter forms a featureless and flat background. The recoil-free fraction f depends sensitively on the energy, and on the temperature and chemical form of the sample.

A technique, also due to Mössbauer, that displays the narrow recoil-free line is the method of velocity or Doppler spectroscopy. If the source and absorber are moved relative to one another with relative velocity v (taken as positive when the two are approaching), then the energy as seen by the absorber is shifted by Ev/c. By varying v, typically a few millimeters per second, the resonance can be scanned. The counting rate is a minimum at the velocity of exact resonance, and on either side of resonance the background is due to recoil events and to irrelevant radiation that had been incompletely rejected by the detector.

Figure 2 is a velocity spectrum for ^{182}W taken at $T=4.2$ K. The transition is between the first excited and the ground state. Here $E=100$ keV, $\tau=2.0$ nsec, and $\Gamma_0=3.3\times10^{-7}$ eV or 1.0 mm/sec. Because the width enters into the spectrum twice, the minimum observable value (for lines of the Lorentz shape as here) is $2\Gamma_0=2.0$ mm/sec. The radiative parent, 115-dcay ^{182}Ta, is made by neutron irradiation of a tantalum metal foil. Its β decay produces the excited emitter nucleus ^{182}W. The latter is then present as a dilute impurity in a tantalum host lattice. The absorber is a foil of tungsten metal in which only the mass-182 isotope whose relative abundance is 26.4% is effective. It is important to note that the substance studied is ^{182}W, not the parent ^{182}Ta. Following a radioactive decay, the daughter atom adjusts its electronic shells to its new nuclear charge in a time short compared to τ in almost all of the cases studied so far.

There are at present 104 transitions in 86 isotopes of 44 elements for which the Mössbauer effect has been observed. A condensed list is given in Table I. The most important is

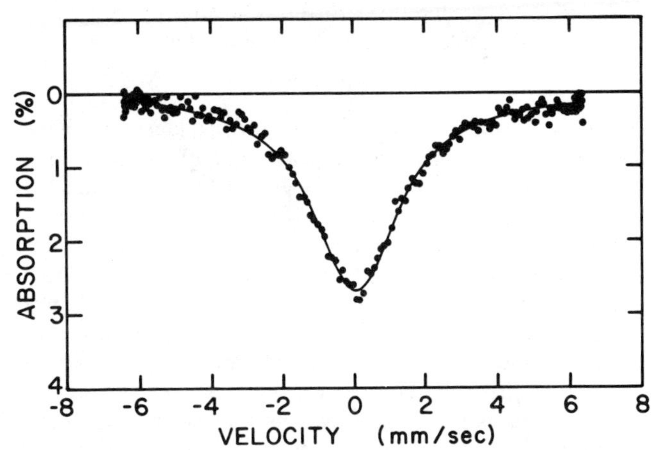

FIG. 2. The Mössbauer effect in ^{182}W. The scatter of the points is due to the statistics of counting. The smooth curve is a Lorentzian, fitted to the data by least squares. The observed width is 3 mm/sec to be compared to the natural width $2\Gamma_0=2.0$ mm/sec. The excess is caused by absorber thickness.

Table I Elements Having at least One Mössbauer Isotope

Potassium	Iodine	Terbium	Osmium
Iron	Xenon	Holmium	Iridium
Nickel	Cesium	Erbium	Platinum
Zinc	Barium	Thulium	Gold
Germanium	Lanthanum	Ytterbium	Mercury
Krypton	Praseodymium	Lutetium	Thorium
Technetium	Neodymium	Hafnium	Protactinium
Ruthenium	Promethium	Tantalum	Uranium
Tin	Samarium	Tungsten	Neptunium
Antimony	Europium	Rhenium	Plutonium
Tellurium	Gadolinium	Dysprosium	Americium

the 14.4-keV transition in ^{57}Fe, because of the key role played by iron in materials science, magnetism, chemistry, and biology. Some of the characteristics of this resonance are $\tau=140$ nsec, $2\Gamma_0=9.3\times10^{-9}$ eV or 0.19 mm/sec. The cross-section at resonance is $\sigma_0=2.4\times10^{-18}$ cm^2. This is about 500 times larger than the cross section for all electronic processes in the atom for radiation of this energy. The internal conversion coefficient of the state is about 8, i.e., the state decays eight times more often by internal atomic processes than by γ-ray emission. Most transitions with energies appropriate for Mössbauer spectroscopy, decay more often by internal conversion than by radiation. The target nucleus is more of a true absorber than a scatterer.

Although there is an interesting classical analog, the recoil-free phenomenon is basically quantum mechanical.

Before the emission (or absorption) of the quantum, the lattice is in some complicated internal state. An impulse is given to a single atom, and if that causes a change of energy state, the result is a shift of the γ-ray energy. In the contrary case the internal lattice state remains unchanged, and except for an exceedingly small energy shift due to the macroscopic recoil of the sample as a whole, the line is neither shifted nor broadened. It is the fraction f of these in which we are interested. The details of the calculation are complex and are omitted here.

Various treatments lead to the following instructive intermediate formula for the recoil-free fraction:

$$f=e^{-k^2\langle x^2\rangle_{av}}. \tag{1}$$

Here $k=1/\lambda\!\!\!^{-}$ and x is the momentary displacement of the nucleus along the direction of observation of the gamma ray. The displacement is due to lattice vibrations—the recoil does not specifically enter. The $\langle\ \rangle_{av}$ brackets denote a time average over a period long compared to τ. The energy enters via k, since $E=c\hbar k$, and as $\langle x^2\rangle_{av}$ increases with temperature, the Mössbauer fraction is increased with lower E and lower T. Even at $T=0$, however, $\langle x^2\rangle_{av}$ is not zero because there is an unremovable "zero-point" motion. The fraction f is very small if, in the course of a few nuclear lifetimes, the atom moves more than a wavelength or so from its initial position. Thus the Mössbauer effect is characteristic of solids, where the atomic motion is bounded, or of liquids of such extremely high viscosity that the atom, although free, does not diffuse rapidly enough. For an emitter dissolved in water at room temperature, the typical diffusion distance is perhaps 1,000 times too great. Liquids in which the diffusion is small enough to display the Mössbauer effect appear quite solid by ordinary inspection.

Equation (1) also shows that the effect is favored for heavier stiffer lattices as these have smaller values of $\langle x^2\rangle_{av}$ at a given temperature.

A slight variant of Eq. (1) due to Waller and Debye describes the coherent part of x-ray and neutron scattering from crystals and predates the Mössbauer effect by many years. The quantity f is sometimes called the Debye–Waller factor. Formulas for it in which $\langle x^2\rangle_{av}$ is evaluated with the Debye model of lattice vibrations, can be found in the literature of x-ray diffraction. In an approximation suitable for $T<0.3\Theta$, where Θ is the Debye temperature of the material, the recoil-free fraction is

$$f=\exp\{-(3R/2K_B\Theta)[1+(2\pi^2/3)(T/\Theta)^2]\}, \tag{2}$$

which reduces to

$$f=e^{-3R/2K_B\Theta}\quad\text{for } T\ll\Theta. \tag{3}$$

In Eq. (2), $(T/\Theta)^2$ comes from the temperature-dependent atomic motion, while in Eq. (3) only the zero-point motion of the atom remains. It is the only term of importance at liquid helim temperature for almost all materials. Equation (3) predicts the value $f=0.93$ for ^{57}Fe in its own lattice at low temperature.

APPARATUS

In order to obtain a spectrum, one needs a radiation source, a motion apparatus ("drive"), an absorber or scatterer (much more usually the former), a radiation detector that furnishes electrical pulses proportional in amplitude to the energy of the quantum absorbed, a pulse amplifier, a single-channel analyzer that can select the desired signal and reject others, and a method of storing the counts. In addition, a cryostat suitable for lowering the temperature of source and absorber separately or together may be required. About 10 Mössbauer nuclides can be used at room temperature. Most of the remainder require liquid helium temperature (4.2 K).

The preparation of sources is one of the more difficult aspects of Mössbauer spectroscopy. The more common ones, however, are commercially produced. Experiments can also be done in which the source is the target of an accelerator beam, or is produced by the beam and implanted into a host by the reaction momentum. The short time required for equilibrium of the atomic shells makes these methods practical.

Most of the interesting Mössbauer spectra show multiplet structure, a composite of splittings in the energy states of source and absorber. To make interpretation easier it is preferable when possible to have one of the two, usually the source, unsplit. This is done by judicious choice of the host lattice.

The most common drive method is an electromechanical system in which the velocity can be made to follow a prescribed wave form, usually a sawtooth or a sinusoid. The address of the electronic storage medium (a time-mode analyzer, or a general purpose computer) is advanced at uniform rate and synchronously repeated with the period of the wave form. The counts are thereby sorted by the velocity at which they occurred and a velocity spectrum is accumulated over the course of many periods. However, all of the spectra displayed here were made with mechanical drive systems in the author's laboratory.

Radiation detectors are usually proportional counters, scintillation counters, or semiconductor detectors depending on the energy and the need for energy resolution.

Let us consider, with the aid of illustrations, some of the kinds of measurements that can be made by Mössbauer spectroscopy.

THE LINE POSITION

On the scale of Mössbauer measurements, the energy of a nuclear transition depends on the structure of the whole atom and thereby on the electronic structure of the lattice. Some of the electrons, those in *s* states, have non-vanishing density, $\rho(0)$, at the nucleus. The electronic charge within the nuclear volume causes the nuclear ground state whose mean-square proton charge radius is $\langle r^2\rangle_{av}$ to be increased in energy above that of a point nucleus by an amount proportional to $\rho(0)\langle r^2\rangle_{av}$. An excited nuclear state with the same charge radius would suffer the same increase and the transition energy would be unaltered, but if a difference $\Delta\langle r^2\rangle_{av}$ exists, there will be an energy charge proportional to $\rho(0)\Delta\langle r^2\rangle_{av}$. This is a chemical shift since for a given nuclear transition the quantity $\rho(0)$ is sensitive to chemical state. It is generally called the isomer shift in analogy to the isotope shift of atomic spectroscopy whose origin is similar. It is a useful indicator of ionic charge state because even if *s* electrons are not themselves transferred in forming a particular chemical bond, the gain or loss of, for example, *p* or *d* electrons changes the shielding of *s* electrons and thereby $\rho(0)$. In Fig. 1 we note that although the emitter and the target atoms are in different hosts, the shift is barely observable. The reason is that both of the nuclear states are members of the same rotation band and have very nearly the same intrinsic structure. In contrast, the shift between high-spin ferrous and ferric compounds in ^{57}Fe spectroscopy is typically about 0.8 mm/sec.

There are two other interesting causes of line shift, generally small compared to the isomer shift. The first is the Pound-Josephson effect. It is a second-order Doppler shift proportional to $\langle v^2\rangle_{av}$, the mean-square velocity of the atom about its equilibrium position. It can be thought of as a relativistic time dilatation which lowers the γ-ray frequency by a factor $(1-\langle v^2\rangle_{av}/c^2)^{1/2}$. It shows up as a temperature effect, but is also present whenever one compares the transition in two hosts of different Debye temperatures. The second is the gravitational red shift. An excited nucleus at a height z above an absorber has an extra potential energy gzE/c^2 because of the mass associated with the excitation by the Einstein relation $E=mc^2$. This appears as an addition to the transition energy. It is a red shift if the absorber is above the source, a blue shift otherwise. For the 93-keV radiation from ^{67}Zn, the shift is Γ_0 for each 4.5 m of height. The effect can also be observed in the centrifugal field produced by a rapidly spinning rotor which holds source and absorber at different radii.

MULTIPLET STRUCTURE OF THE SPECTRUM—MAGNETIC EFFECTS

A nuclear state with angular momentum quantum number $I>1/2$ possesses a magnetic dipole moment $\boldsymbol{\mu}$. In a magnetic field $\mathbf{B}$, the Zeeman interaction $-\boldsymbol{\mu}\cdot\mathbf{B}$ splits the state into its $2I+1$ magnetic substates. If there is no electric quadrupole interaction (see below), these are equally spaced in energy between $-\mu B$ and $+\mu B$, the former lying lowest. The γ-ray transitions between the components of the excited state and those of the ground state are observed as a multiplet spectrum.

An important source of magnetic fields is internal to the atom. An unpaired electron density at the nucleus is a magnetization because the electron has an intrinsic magnetic moment. The density and its radial variation produce a magnetic field at the nucleus, much as the magnetization and its divergence produce a field in macroscopic materials. It is called the Fermi contact field. In the ions of the transition metals, there can be a net spin in the partially filled *d* shell. The *d* electrons have no density at the nucleus but produce a contact field by polarizing *s*-electron shells. The orbital motion also produces a magnetic field. It is of somewhat minor importance in *d*-shell ions, but large in the rare earths.

Internal magnetic fields of whatever origin are generally called "hyperfine fields." In order for them to be observed by Mössbauer spectroscopy, they must persist in direction for a time long compared to the precession time of the nuclear spin. In a ferromagnet or in an antiferromagnet the direction of the internal field is preserved in space and is seen. We illustrate with two cases, ferromagnetic iron (Fig. 3) and antiferromagnetic IrF_6 (Fig. 4). In the latter case the ordering temperature lies between the values of 4.2 and 27 K where the two spectra shown are taken. There are eight possible transitions between the magnetically split $I=3/2$ and $I=1/2$ states that characterize the Mössbauer transitions in

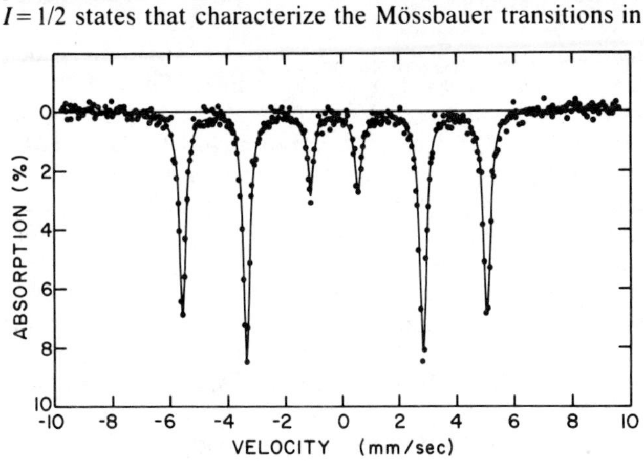

FIG. 3. Velocity spectrum of ^{57}Fe at room temperature. The source is ^{57}Co in a copper host. The absorber is 7.5 μm of pure, annealed, iron foil of normal isotopic abundance (2.2% ^{57}Fe). An average of several runs yields $B=332.3\pm0.2$ kG for the hyperfine field, and $\mu_{ex}/\mu_{gnd}=1.7132$ for the ratio of the excited to ground-state magnetic moments. Other experiments show that B is negative.

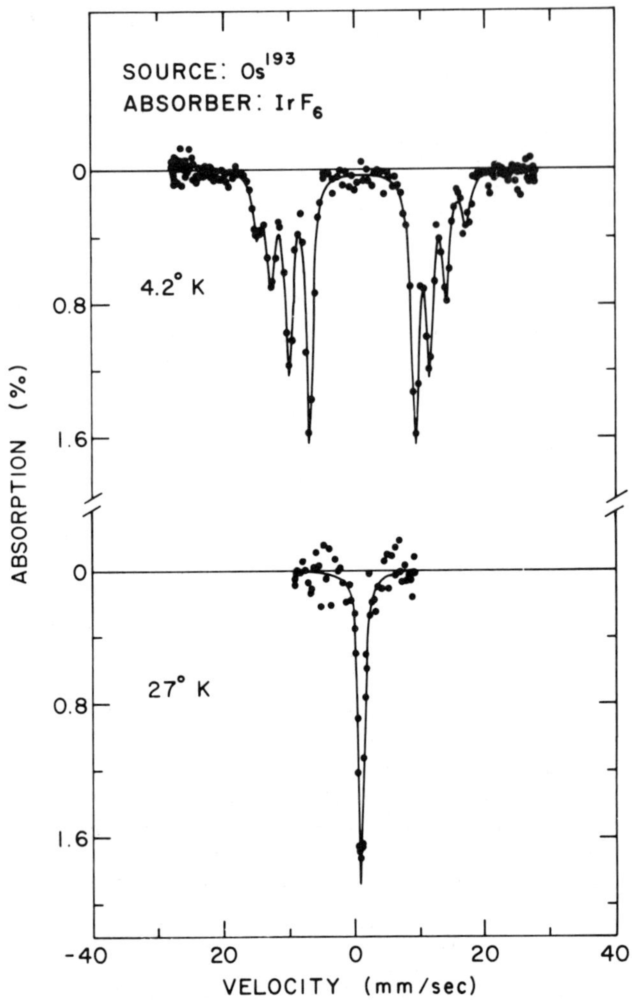

FIG. 4. Spectrum of ^{193}Ir as IrF_6, below and above the antiferromagnetic ordering temperature. Comparison of the split spectrum with one for Ir in iron shows that the observed ratio of magnetic moments depends on the host material because of the existence of a large hyperfine anomaly.

both isotopes. In ^{57}Fe the internal nuclear structure is such that the radiation is almost purely magnetic dipole ($M1$), and two of the possible eight lines (those with $\Delta m = \pm 2$) are forbidden. The ^{193}Ir decay is a mixed multipole ($E2$ and $M1$) and all eight lines are seen. The line intensities depend on the mixing ratio and on the angle between the internal field and the direction of observation. The isomer shifts between source and absorber are visible.

Hyperfine fields may lie either parallel or antiparallel to the bulk magnetization. The behavior of the magnetic splitting with external magnetic field sorts out the two cases.

An ion in the paramagnetic condition has an internal field that fluctuates in direction through interaction with neighboring spins and with thermal vibrations of the lattice. A typical ferrous ion for example shows no magnetic splitting above its ordering temperature as the field averages to nearly zero in the time of a Larmor precession. Ferric iron, however, because it has no orbital angular momentum, interacts less strongly with the lattice. If it is somewhat dilute and the temperature is low, a split spectrum occurs that reflects the electronic state of the ion in its surroundings. Many applications of Mössbauer spectroscopy to biology arise from analysis of such spectra in iron-bearing proteins.

ELECTRIC EFFECTS

Nuclear states with $I \geq 1$ have electric quadrupole moments, that is to say the charge does not have spherical symmetry. The quadrupole moment interacts with the gradient of the electric field to split an otherwise degenerate manifold of substates, or, if there is already magnetic splitting, to alter the intervals. While the electric field is quite small at a nucleus, the same cannot be said of its derivatives. The largest are produced by the electronic structure of the parent atom. Electric charges on other atoms also produce field gradients at the nucleus of the atom in question but their effect is smaller because of the greater effective distance. However, an antishielding effect (Sternheimer shielding) magnifies their influence. If the charge distribution has cubic or spherical symmetry, the field gradients with respect to any three mutually perpendicular axes must be equal, and by Laplace's equation which states that their sum vanishes, each must do so separately. Thus electric quadrupole splitting is a property of less-then-cubic symmetry.

In the case of nuclei with odd mass number, I is an odd half-integer, and the states are always left with a twofold degeneracy by a quadrupole interaction. Thus instead of $2I+1$ different energy levels one sees half this number. However, if the nucleus has even mass number there need be no

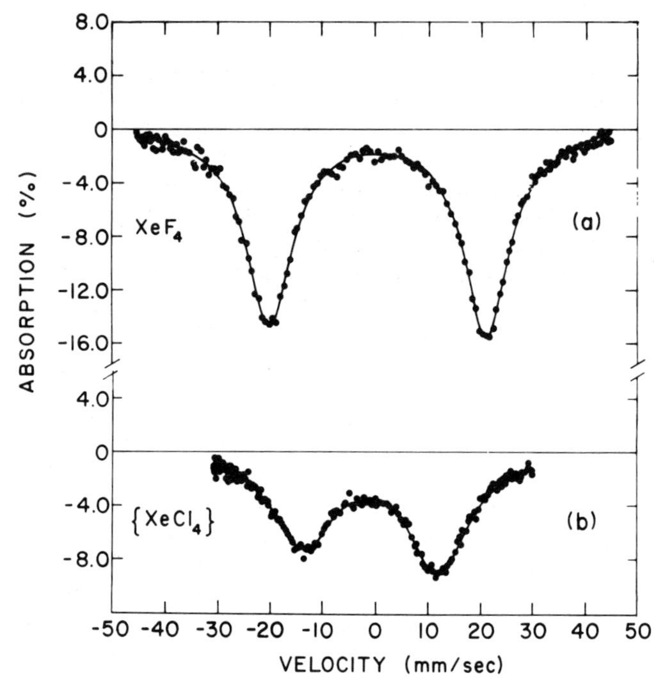

FIG. 5. Quadrupole splitting in XeF_4 and $XeCl_4$. In the former an unsplit source of ^{129}I as sodium paraperiodate ($Na_3H_2IO_6$) is used. In the latter, the unsplit absorber is sodium perxenate (Na_4XeO_6). These both have an octahedron of oxygen ions about the central atom and hence locally cubic symmetry. After the decay of the ^{129}I, both ions are in fact identical, as may be affirmed by the absence of isomer shift when comparing them in the same experiment.

degeneracy if the local symmetry is low enough. We illustrate quadrupole splitting with Fig. 5(a) for ^{129}Xe in the compound XeF_4. The excited $I = 3/2$ state splits into a doublet, while the ground $I = 1/2$ state is unsplit. The resulting spectrum has two lines. The structure of XeF_4 is a square plane of fluorines about a central xenon. The analysis of the spectrum combined with other information shows that the bonds are made by transferring about $0.75e$ from the $5p_{x,y}$ orbitals of Xe to each F, thereby reducing the initial spherical symmetry of the shell. Figure 5(b) is the spectrum of $XeCl_4$ produced in the β decay of ^{129}I in $KICl_4$ and examined at the instant of formation with the aid of an unsplit absorber. The intensity asymmetry is caused by premeditated alignment of the crystals in the source. The mere existence of the lines shows that the species is stable for at least 10^{-9} sec at 4.2 K.

POLARIZATION

If the splitting is magnetic, and the magnetic field direction is kept fixed (in a ferromagnet it suffices to apply a modest external field to orient the domains), then the multiplet lines are polarized. The polarization of a line in a source may be measured by using a polarized line in an absorber. Linear polarization is found in directions perpendicular to the field, circular polarization in the parallel directions, and elliptical in between.

Polarization is also a characteristic of emission or absorption of quadrupole components if the sample is a single crystal.

APPLICATIONS

We conclude with a brief discussion of some of the uses to which Mössbauer spectroscopy has been put.

Nuclear Physics

The main thrust here has been in evaluating electromagnetic moments of excited states. If both excited and ground state are split, the ratio of magnetic or quadrupole moments of the two is obtained directly from the spectrum. If the ground-state moment is available from other data, the hyperfine field or the electric field gradient is obtained. If the line is narrow enough for an external magnetic field to be useful, the magnetic moment of the excited state is determined redundantly. Isomer shifts have been used to determine the change in nuclear radius between ground and excited state. The measurements are mainly somewhat ahead of the theory. An exception is in the case of some of the rotational states, where the shifts are very small, and the theory is concerned with small perturbations of the same intrinsic state that are caused by the rotation.

Magnetism and Phase Transformations

This is a vast area. Extensive studies have been made of a great variety of alloys mainly with ^{57}Fe, but also with a number of other Mössbauer isotopes. In fact a nominally diamagnetic test probe like ^{119}Sn samples the polarized band structure of a magnetic alloy and shows a transferred hyperfine field. Similar fields are also displayed in non metals—for example, CrI_3 studied with the Mössbauer effect in ^{129}I and ^{127}I. Phase transitions between two magnetic phases, or between a magnetic and a paramagnetic phase, are readily determined by observing the hyperfine field with varying temperature. Of considerable interest are magnetic phenomena associated with dilute impurities in a magnetic host; for example, the very large ionic magnetic moment associated with an iron impurity in certain hosts which was discovered by measuring the hyperfine field as a function of applied external magnetic field. Study of the Kondo condensation phenomenon is done similarly. The magnetic structure of the electronic ground state of dilute rare-earth impurities has been an active field. The equivalence or not of different sites for an ionic test probe shows up as simple or multiple spectral groups. A great variety of probes can be incorporated in different hosts by implantation techniques.

Chemistry

In structural chemistry, a Mössbauer spectrum can often decide between two possible structures, or can assign the correct site symmetry for the ion being studied. As an analytic technique, it can detect a suitable element nondestructively, determine a valence state, frequently determine the actual compound by comparison with known parameters, and with some error tell the quantity. The electronic state of an ion and the populations of the orbitals involved in the bonds can be obtained in favorable cases. It is also useful for studying the chemical species formed in radioactive decay. There is a growing collection of systematic data on variation of Mössbauer parameters over a series of related compounds. The greatest concentration of such studies has been on iron and tin compounds. It is possible to investigate a limited area of solution chemistry by Mössbauer spectroscopy, by working with frozen solutions.

Biology

Iron plays a dominant role as the active element in a variety of biologically important proteins where it exists as Fe^{II} or Fe^{III} in either high- or low-spin forms, distinguishable by their spectra. In the haem proteins, for example, the difference between the oxygenated and deoxygenated state is a change between low- and high-spin Fe^{II}, while ligands other than oxygen can result in the other two possibilities. The analysis of the state of the irons and their electronic structure, including magnetic properties, represents the main direction of Mössbauer spectroscopy of biological materials.

Various

Studies of lattice dynamics can be made by examining the variation of the recoil-free fraction and other parameters, such as line–width, with temperature. Radiation damage and the diffusion of hydrogen in materials are two examples.

Rayleigh scattering by the electrons in a solid can be either recoil-free or not. The recoil-free process is described by Eq. (1) with k replaced by $2k \sin\phi/2$, where ϕ is the scattering angle. The scatterer can be any substance as the scattering is not resonant. The Mössbauer equipment selects the recoil-free part. Rayleigh scattering can be used to study many things, for example diffusion in a solid near its melting point. The difficulty is generally in getting enough intensity.

Various experiments of a fundamental nature have been done with Mössbauer spectroscopy. We mentioned the gravitational and centrifugal red shifts. An ether-drift experiment with a source and absorber mounted on a rotor has shown the ether drift to be less than a few centimeters per second. Experiments to detect an absence of invariance under time reversal have been done by polarization measurements in mixed multipole transitions.

A variety of Mössbauer measurements is based on the coherence properties of the narrow-band radiation. The radiation that is transmitted through a resonant absorber has its amplitude distribution altered in the process. If one makes a lifetime measurement of the state using the filtered radiation and a previous γ-ray to determine zero time, the decay is no longer exponential. Other coherence experiments include total reflection, Faraday rotation, frequency modulation, and amplitude modulation of the radiation. Frequency modulation, for example, by rapidly vibrating a source at angular frequency Ω with a piezoelectric crystal, produces an unshifted line of reduced intensity (the "carrier") and a series of lines (the "sidebands") differing in energy from the carrier by integer multiples of $\hbar\,\Omega$. It has been used as an alternative to and a calibrator for velocity spectra. An interesting time dependence ("quantum beats") in the counting rate occurs if the radiation passes through a resonant absorber and is counted. It can be used to detect small energy shifts, and is caused by interference among carrier and sidebands. The classical description of the Mössbauer effect, mentioned earlier, regards the recoil-free line as a frequency modulation carrier, while the sidebands form a continuum due to the aperiodic thermal motion of the atom. The treatment leads directly to Eq. (1) as an approximation for small amplitude vibrations.

See also FERROMAGNETISM; KONDO EFFECT; MAGNETIC MATERIALS; RARE EARTHS; TRANSITION ELEMENTS.

BIBLIOGRAPHY

A. Abragam, *L'Effet Mössbauer et Ses Applications a L'Etude des Champs Internes*. Gordon and Breach, New York, 1964. A very clear theoretical treatment of the electron-nucleus interactions and a discussion of some experiments.

H. Frauenfelder, *The Mössbauer Effect*. W. A. Benjamin, New York, 1962. A review and a collection of reprints of many of the early important papers.

Chemical Applications of Mössbauer Spectroscopy, V. I. Goldanskii and R. H. Herber (eds.), Academic, New York, 1968. A collection of articles by various authors.

N. N. Greenwood and T. C Gibb, *Mössbauer Spectroscopy*. Chapman and Hall, London, 1971. A comprehensive treatment emphasizing chemical aspects.

Mössbauer Effect Methodology, I. J. Gruverman (ed.). Plenum, New York. A yearly volume of talks given at conferences managed by a commercial sponsor.

H. Lipkin, *Quantum Mechanics*. American Elsevier, New York, 1973. The Mössbauer effect is used as a frequent illustration of quantum-mechanical principles.

Mössbauer Isomer Shifts, G. K. Shenoy, and F. E. Wagner (eds.). North Holland, Amsterdam, 1978.

Mössbauer Effect Data Index 1958-1965, A. H. Muir, K. J. Ando, and H. M. Coogan (eds.). Interscience, New York, 1966. *Mössbauer Effect Data Index*. J. G. and V. E. Stevens (eds.). IFI/Plenum, New York, 1971–1975. Yearly volumes for 1971-1975, and a cumulative index for 1966–76. These volumes are an invaluable guide to the literature. Articles are referenced by isotope and by author. Titles and some descriptive material are also given. Since 1978 the editors have published the *Mössbauer Effect Reference and Data Journal* (10 issues annually).

There have been a number of published proceedings of international conferences. A partial list follows:

(a) *Allerton Park Conference Report*, University of Illinois, June 1960 (unpublished).

(b) *The Mössbauer Effect*, Proceedings of the Second International Conference, Saclay, France, 1961, D. M. J. Compton and A. H. Schoen (eds.). Wiley, New York, 1962.

(c) *M^{III}, Third International Conference on the Mössbauer Effect*, Ithaca, New York, 1963, A. J. Bearden (ed.), *Rev. Mod. Phys.* **36**, 333 (1964).

(d) *Hyperfine Structure and Nuclear Radiations*, E. Matthias and D. A. Shirley (eds.), Proceedings of a conference held at Asilomar, August 1967. North-Holland, Amsterdam, 1968.

(e) *Perspectives in Mössbauer Spectroscopy*, S. G. Cohen and M. Pasternak (ed.). The proceedings of the International Conference on Applications, Ayeleth Hashahar, Israel, 1972. Plenum, New York, 1973.

(f) J. Phys. **35**, C6 (1974), **37**, C6 (1976), **40**, C2 (1979), and **41**, C1 (1980). Proceedings volumes of international conferences, Bendor, France 1974, Corfu, Greece, 1976; Kyoto, Japan 1978; and Portorož, Yugoslavia 1979, respectively.

(g) *Workshop on New Directions in Mössbauer Spectroscopy (Argonne 1977)* AIP Conference Proceedings #38, American Institute of Physics, New York 1977.

Molten Salts

George J. Janz

The term molten salts classically has been associated with the liquids formed by the fusion of crystalline salts of relatively high melting temperatures. In the crystalline state such salts are virtually insulators (i.e., nonconductors), whereas in the molten state the specific electrical conductivity is about a million times larger than that of water. A comparison with some metals, aqueous electrolytes, and molecular solvents is in Table I; the electrical conductance in metals is electronic, whereas for aqueous electrolytes, molecular "solvents," and molten salts it is ionic. Most molten salts conduct electricity so well that the term liquid electrolyte has come to be used almost interchangeably with molten salts and fused salts. Eutectic mixtures are frequently used to attain the molten state at lower temperatures, so as to minimize the problems of "container chemistry" or corrosion. The LiCl–KCl eutectic mixture (mp 325°C) is liquid at temperatures more than 200°C below the melting points of the pure components (KCl, mp 772°C; LiCl, mp 610°C) and has seen very wide application in molten-salts studies.

Table I Comparison of molten salts with some metals, aqueous electrolytes, and molecular solvents

	Temp. (°C)	Specific electrical conductivity ($ohm^{-1}\ cm^{-1}$)	Density ($g\ cm^{-3}$)	Viscosity (cP)	Surface tension ($dyne\ cm^{-1}$)
Metals					
Cu	20	5.8×10^5			
Au	20	4.1×10^5			
Aqueous electrolytes					
NaCl (0.1 N)	25	1.1×10^{-3}			
KCl (0.1 N)	25	1.3×10^{-3}			
Molecular solvents[a]					
H_2O	25	0.5×10^{-6}	1.0	1.0	72.0
CCl_4	25	4×10^{-18}	1.59	0.97	28.0
Molten salts[a]					
NaCl (mp 800°C)	807	3.60	1.55	1.38	117
KNO_3 (mp 337°C)	347	0.66	1.86	2.70	110

[a] Both substances are clear colorless liquids.

The range of interest, from the viewpoints of both established technology (e.g., production of metallic aluminum by the electrolysis of molten cryolite) and developing areas (e.g., environmental technology, energy technology) is illustrated in Table II. The concern with energy resources and environmental problems is reflected by the emergence of molten salts in clean-atmosphere processes, utilization in waste disposal and metal recovery, for solar and thermal energy storage, and as the electrolytes of "super-batteries" (i.e., the high-performance batteries required for electric vehicles and for "load-leveling" applications in electric utilities).

Parallel with the preceding, the understanding of the physics and the chemistry of the inorganic molten state has been advanced through the application of a wide range of physicochemical and electrochemical techniques. Thus for the alkali halides the anions and cations have spherical symmetries and the external electron-shell structures ape those of the rare gases. The entropies of melting are in close correspondence, e.g., argon, 3.0 cal deg^{-1} (g at.)$^{-1}$; sodium chloride, 3.1 cal deg^{-1} (g ion)$^{-1}$, although the former occurs at cryogenic temperatures, whereas the latter occurs at high temperatures (argon, mp −189°C; NaCl, mp 800°C, respectively). In the computer simulation of molten-state dynamics, the molten alkali halides are being used as model systems of molten salts. By contrast, nitrates are examples of a class in which the anions exhibit several sites of differing force-field intensity; the nitrate ion has thus found application spectroscopically as a "probe ion" to sense the nature of nearest-neighbor ion–ion interactions in molten electrolytes. A further class of molten salts are those in which a degree of the molecularity from the solid state is retained. Thus salts such as the mercuric halides, which have a well-defined component of covalent bonding in the crystalline state, apparently retain this molecularity in the molten state. These salts have found application as high-temperature "water-analo-

Table II Molten salts applications in processes and products

Chemical industry	Electrometallurgy and materials science	Technology: Energy	Technology: Environmental	Technology: Solid-state electrochemical
As reaction media in: Halogenation Catalysis Oxidation Cracking Condensation Isomerization Dehydration Organometallics	Metal-extractive electrolyses[a] Electroplating[b] Surface preparation[c]	Electrical Fuel cells (e.g., molten carbonates) High-performance batteries (sodium/sulfur, lithium/sulfur, etc.) Thermal batteries Nuclear Fusion and fission reactor technology Storage Phase-change materials for energy storage (solar, thermal)	Clean atmosphere processes: utilization of molten carbonates to remove SO_2, SO_3 SO_2-free metal extraction: new processes (Cu and Pb extraction) Wastes: dissolution of plastic materials; metal recovery	Single-crystal growth Molten salt semiconductors "Solid" electrolytes Thermal switching

[a] Melts of NaCl, cryolite, and an $MgCl_2$ mixture are used to extract Na, Al, and Mg, respectively; various other melts are used to extract Zr, Ti, Nb, Ta, Mo, Pb, etc.
[b] A melt of Sn salts yield Sn plate.
[c] Molten nitrates remove oxides; various other molten salts "metallide" surfaces.

gous'' solvents; the principles of acid–base concepts, the dissolution of metals and metal oxides, and ligand field coordination chemistry have been extended to studies of the inorganic molten state through the use of such high-temperature water-analogous solvents. ''Organic'' molten salts merit brief mention, if only as illustration of the range of compounds encompassed by this term. Examples of this class are the alkyl ammonium compounds and the metal salts of the organic acids series, i.e., salts with a well-defined ionic component in the crystalline state and with the ''organic'' component built into either the cation or the anion. Thus such salts have been proposed for studies of ''weak-field'' molten electrolytes (i.e., the anions and/or cations being highly shielded by the built-in organic components), and in studies of the process of fusion, to which both organic melting effects and ionic melting effects contribute as part of the overall process.

The thermodynamic principles of molten-salts solution chemistry are possibly the most advanced from the viewpoints of theoretical and chemical physics. Applications of equilibrium emf techniques and calorimetry to studies of the enthalpic properties of mixtures of simple ionic salts have contributed significantly to the fundamental understanding of ionic interactions and deviations from thermodynamic ideal behavior.

See also CONDUCTION.

BIBLIOGRAPHY

M. Blander, *Molten Salts Chemistry*. Wiley (Interscience), New York, 1964. (A)

G. J. Janz, ''Physical Properties and Structure of Molten Salts,'' *J. Chem. Educ.* **39,** 59 (1962). (E)

G. J. Janz, ''Molten Carbonate Electrolytes as Acid–Base Solvent Systems,'' *J. Chem. Educ.* **44,** 581 (1967). (I)

L. Lewsing, ''The Salt of the Earth Joins the War on Pollution,'' *Fortune*, pp. 138–148 (July), 1973. (E)

J. Lumsden, *Thermodynamics of Molten Salt Mixtures*. Academic Press, New York, 1966. (A)

B. R. Sundheim, ''Molten Salts,'' *Int. Sci. Technol.*, p. 42 (November), 1962. (E)

B. R. Sundheim, *Fused Salts*. McGraw-Hill, New York, 1964. (A)

Neutron Diffraction and Scattering

G. Shirane

Neutron-scattering techniques play an important role in studies of condensed matter such as solids and liquids. They complement the better-known scattering by x ray, electrons, and light. Favorable cross sections, in particular those of magnetic origin, permit unique structural and magnetic diffraction studies. Another important factor is the fortunate energy–wavelength relation for thermal neutrons. For example, 13-meV neutrons have a wavelength of 2.5 Å, which is of the order of a lattice spacing in solids. This permits direct observation of elementary excitations such as phonons and magnons.

Most of the current sources for neutron-beam experiments are research nuclear reactors. Many research reactors are now providing neutron fluxes higher than 10^{13} neutrons/$cm^2 \cdot s$ at the source and several reach 10^{15}. When these are combined with more efficient monochromator crystals, we can observe low-energy excitations (<20 meV) in relatively small crystals (~0.1 cm^3). Since elastic Bragg scattering is a much stronger scattering process, smaller crystals can be utilized.

In the scattering process of neutrons impinging on a target crystal, the laws of conservation of momentum and energy must be obeyed:

$$\mathbf{k}_i - \mathbf{k}_f = 2\pi\boldsymbol{\tau} + \mathbf{q} = \mathbf{Q},$$

$$\frac{\hbar^2}{2m}(k_i^2 - k_f^2) = \hbar\omega.$$

here $\mathbf{k}_i$ and $\mathbf{k}_f$ are the initial and final wave vectors, respectively, of neutrons and $\mathbf{Q}$ is the total momentum transferred to the sample. The change of neutron energy during the scattering process corresponds to the excitation energy $\hbar\omega$ with a particular momentum $\mathbf{q}$. The object of a neutron-scattering experiment is to measure the excitation energy $\hbar\omega$, and its intensity, as a function of $\mathbf{q}$. This is schematically illustrated in Fig. 1. When $\mathbf{q}=0$ and $\hbar\omega=0$, the scattering process is totally elastic and we have Bragg scattering at $2\pi\tau$, a reciprocal lattice vector.

ELASTIC SCATTERING

Compared with the more widely used x rays for structural studies, the neutron-scattering technique has two important advantages: (1) The coherent neutron-scattering amplitude b varies little with atomic number, as illustrated for several selected elements in Table I. This is in contrast with the x-ray-scattering amplitude, which is proportional to atomic number. Thus in x rays light elements are hard to observe in the presence of heavier atoms. (2) The magnetic moment of the neutron interacts directly with the unpaired electrons in crystals. Thus we can determine the spin density distribution in magnetic crystals.

One technical advantage of the technique is a transparency of neutrons for some common materials used in cryostats and furnaces. For example, at one extreme, 1 mm of Al metal attenuates a 1-Å neutron beam by only 1%. At the same time, however, there are some elements with extremely high neutron-absorption cross section, such as B, Cd, and Gd.

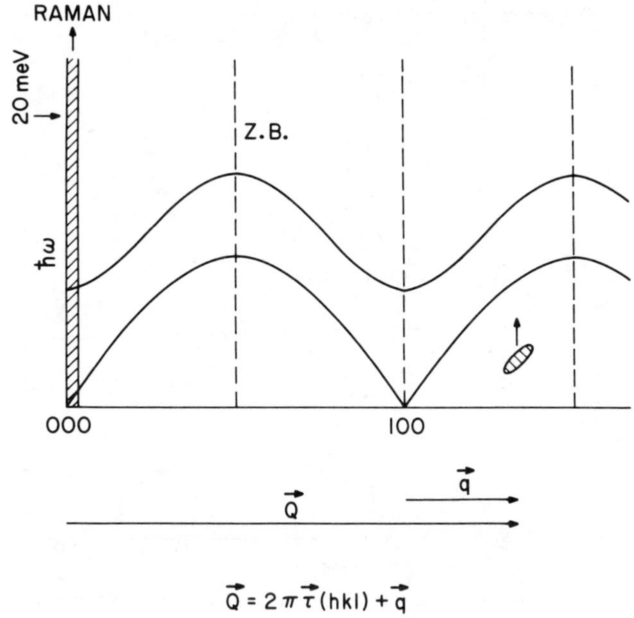

FIG. 1. Schematic picture of range of scattering function $S(Q,\omega)$ covered by thermal neutron scattering. Arrow indicates a "constant-Q" scan, first developed by Brockhouse in 1961. For simple crystals, $Q(100) \sim 1 Å^{-1}$; 1 meV = 0.24 TH$_z$ = 8.1 cm^{-1}.

Nuclear Bragg scattering has been extensively utilized for structural determination involving light atoms, in particular hydrogen-bonded crystals and biological molecules. In this type of study, negative and positive scattering amplitudes for H and D, respectively (see Table I), allow a unique technical approach by partial or selective substitution of hydrogen by deuterium. In a somewhat similar way, the neutron-diffraction technique was selectively utilized in special alloy problems such as MnNi systems.

The magnetic scattering is truly a unique feature of neutron

Table I Scattering Length b for Selected Elements

Nucleus	Atomic number	b (10^{-12} cm)
H	1	−0.37
D		0.67
He	2	0.30
C	6	0.67
N	7	0.94
O	8	0.58
Al	13	0.35
Mn	25	−0.39
Ni	28	1.03
^{60}Ni		0.28
^{62}Ni		−0.87
Cu	29	0.76
Zr	40	0.71
Ba	56	0.52
Pb	82	0.94

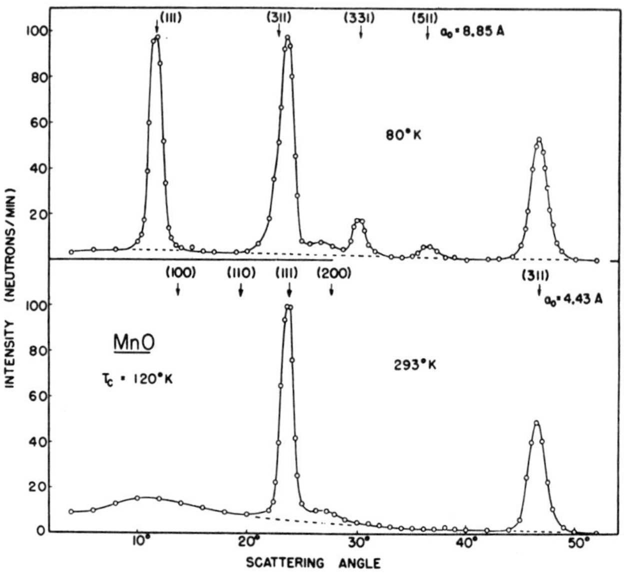

FIG. 2. Neutron powder data above and below the Néel temperature of MnO at 120 K. This is one of the first magnetic cross sections reported by Shull and Wollan in 1951.

studies of crystals. The magnetic scattering amplitude of an atomic spin S is given by

$$p=\left(\frac{e^2\gamma}{mc^2}\right)Sf=0.54Sf\times 10^{-12}\ \text{cm}$$

where f is a form factor for unpaired electrons due to their finite spatial distribution (b is independent of the scattering angle). We can see from Table I that the b's are of the same order of magnitude as p's. An additional factor, which enters into the cross-section formula, is the angle α between **S**, the spin direction, and the scattering vector **Q**. Intensities are proportional to $\sin^2\alpha$. Thus by a comparison of the intensities and positions of several magnetic peaks, we can determine the magnetic spin arrangement. A simple example of magnetic scattering is shown in Fig. 2 for MnO powder. Additional peaks below the Néel temperature at 120 K are due to ordered Mn^{2+} spins.

One important application of the magnetic scattering is the production of fully polarized neutron beams. If the magnetic moment of a monochromator crystal is aligned perpendicular to the scattering vector, the interference effects between the nuclear and magnetic scattering produce the cross sections

$$\sigma=(b\pm p)^2$$

for the two spin states of the incident unpolarized beam. We can select a monochromator, such as $Co_{0.92}Fe_{0.08}$, with $p=b$, which gives a completely polarized beam. Among many advantages of polarized-beam studies, the most obvious is the extremely high sensitivity obtainable, which allows a measurement of a very small magnetic amplitude. On reversing the neutron polarization by a tuned radio-frequency coil, we can measure the ratio

$$R=\left(\frac{b+p}{b-p}\right)^2.$$

Measurable p's are several orders of magnitude smaller compared with the unpolarized-beam technique, where we deal with b^2+p^2. Detailed spin distributions have been mapped out for $3d$ and $4f$ ferromagnetic metals and alloys.

INELASTIC SCATTERING

Thermal neutrons probably offer the most direct way to study the dynamics of condensed matter. A neutron energy range of 80–5 meV corresponds to a wavelength (λ) of 1–4 Å and a wave vector $k=2\pi/\lambda$ of 6.5–1.5 $Å^{-1}$. These are in an ideal range of $\hbar\omega$–Q space for the study of excitations in solids. As schematically shown in Fig. 1, a wide range of Brillouin zones can be covered, since typical $Q(100)=2\pi/a$ is on the order of 1 $Å^{-1}$. Raman scattering covers a much wider energy range with considerably higher resolution but it is limited to the very small region of Q near (000) since $k(\text{light})\sim 0.001$ $Å^{-1}$. The x-ray-scattering technique allows observations over the entire Q range but the energy resolution is too poor for direct measurements of collective excitations.

The most commonly used technique in inelastic studies is triple-axis spectrometry, which allows for scans along any direction in $\hbar\omega$–Q space. This employs a crystal monochromator and analyzer, and the momentum transfer Q is frequently kept constant during the energy scans (see Fig. 1). Because of the relatively weak neutron sources, in comparison with light and x-ray sources, the resolution must quite often be compromised in order to obtain sufficient intensities. It is possible, in favorable cases, to observe excitations ranging from 0.1 to 100 meV.

Phonon dispersion relations are now widely explored for most elements (including solid ^4He) and many simple compounds, such as NaCl and CaF_2. These data in turn have stimulated theoretical interest. The dynamical characteris-

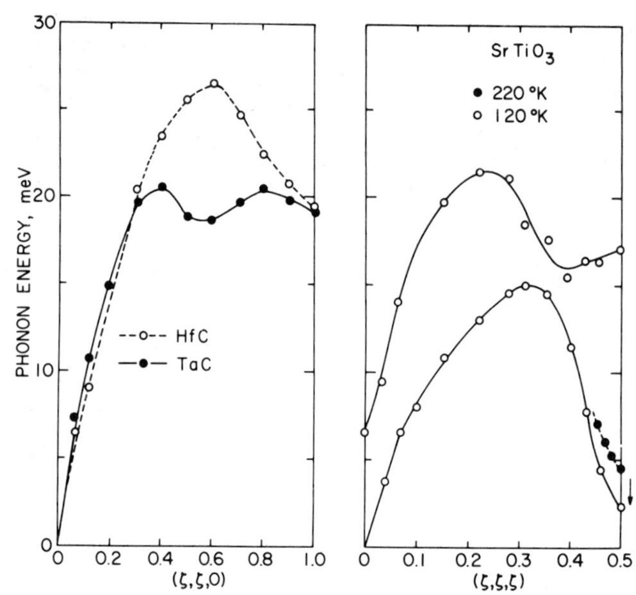

FIG. 3. Two examples of lattice-dynamical studies by triple-axis spectrometry. The electron–phonon interaction is demonstrated for superconductors TaC and HfC and the soft-mode condensation for the 110 K phase transition in $SrTiO_3$.

tics of solids are now reasonably well understood and they constitute an important part of our current knowledge of solid-state materials. This understanding has led to studies of more exotic phenomena in solids. Figure 3 (left side) displays an interesting example in which superconductivity is clearly reflected in the phonon dispersion curve. These two NaCl-type compounds have little mass difference because Ta is situated next to Hf in the periodic table and we would expect the phonon spectra to be nearly identical. On the other hand, TaC is a superconductor with $T_c = 10$ K and HfC has a $T_c < 1$ K. This difference, through electron–phonon interaction, reveals itself in large changes in phonon frequencies.

The neutron-scattering technique has been especially effective in the study of structural phase transitions. One typical example is shown at the right in Fig. 3, where two phonon branches of $SrTiO_3$ are shown. The phase transition at 110 K is caused by a lattice-dynamical instability of a particular type of atomic motion; this is exemplified as a strongly temperature-dependent low-frequency phonon branch, as indicated by the arrow. This mode is called a soft mode because the frequency decreases drastically as the transition temperature is approached. The soft-mode concept represents a basic mechanism of many phase transitions in solids.

The inelastic-scattering technique has also been applied to the study of magnetic excitations, the spin waves of the system. As in phonons, spin waves have been explored for a wide variety of ferromagnets and antiferromagnets. Particularly simple cases are the two-dimensional antiferromagnets, such as K_2NiF_4, where the magnetic interaction is strong within the plane and very weak between the planes. This is directly demonstrated by a large spin-wave dispersion within the plane and very little dispersion for directions out of the plane. Spin-wave dispersions in ionic crystals, such as MnF_2, are now well understood. Itinerant-electron magnets, such as Fe and Cr, are still very much under current investigation. Because the spin waves in these materials have very high energies these studies are quite difficult.

Finally, we should emphasize the role of neutron-scattering studies of magnetic critical phenomena. This topic is probably one of the most extensively explored fields in solid-state physics, both theoretically and experimentally. It offers a simple model system in which basic physical concepts can be tested experimentally without ambiguity. Neutron-scattering studies of magnetic transitions in MnF_2, $RbMnF_2$, and Ni offer typical examples.

Current developments of new techniques in neutron scattering are directed along two paths. One is to improve the energy resolution down to microelectron-volt regions by new types of instrumentation; the other is to extend the upper limit of energy transfer to considerably above 100 meV, by using a pulsed neutron source. At the same time, triple-axis spectrometry will remain one of the basic tools with which to investigate the dynamical aspects of condensed matter.

See also ALLOTROPY AND POLYMORPHISM; FERRIMAGNETISM; FERROMAGNETISM; LATTICE DYNAMICS; NEUTRON SPECTROSCOPY; PARAMAGNETISM; PHASE TRANSITIONS; PHONONS.

BIBLIOGRAPHY

C. Kittel, *Introduction to Solid State Physics*, Chapters 1, 5, 15. Wiley, New York, 1966. (I)

C. G. Shull and E. O. Wollan, "Application of Neutron Diffraction to Solid State Problems," *Solid State Phys.* **2**, 137 (1956). (I)

G. E. Bacon, *Neutron Diffraction*. Oxford (Clarendon Press), London and New York, 1975.

W. Marshall and S. W. Lovesey, *Theory of Neutron Scattering*. Oxford (Clarendon Press), London and New York, 1971.

Neutron Spectroscopy

B. N. Brockhouse

Slow-neutron spectroscopy, the study of slow neutrons inelastically scattered by matter, is now recognized as the most generally useful experimental tool that is available for the study of the atomic motions in condensed matter and of low-lying electronic states in magnetic materials. For the most part the experiments employ thermalized neutrons from fission reactors, though some studies have used neutrons produced from accelerators, particularly from electron linear accelerators.

THEORY AND PRACTICE

The special usefulness of thermal neutrons comes about because they have, at once, wavelengths (λ) of the order of the interatomic spacings in condensed matter, say 1–4 angstroms (Å), and (corresponding) energies of ~80 to ~5 meV, that is, of the normal order of the characteristic quantum energies in condensed matter that are excited at ordinary temperatures.

Neutrons scatter from condensed matter by virtue of two interactions:

1. The neutron–nuclear interaction via nuclear forces leads to (*S*-wave elastic) scattering cross sections (σ) and scattering amplitudes (b), with $\sigma = 4\pi b^2$, that are characteristic of the particular nuclide (in the particular spin state of the combined neutron–nuclide system), and that must be experimentally determined for each nuclide. In most situations the appropriate scattering amplitude is $b_{\text{coh}} \equiv \langle b \rangle$, the average being taken over the distribution of isotopes and spin states of the scattering nuclides of a particular type. Correspondingly, for each element there exists the *coherent scattering cross section* $\sigma_{\text{coh}} = 4\pi\langle b \rangle^2$, which is involved in interference properties of the scattering, and the *incoherent scattering cross section* $\sigma_{\text{inc}} = 4\pi[\langle b^2 \rangle - \langle b \rangle^2]$. These quantities, of course, depend on the isotopic constitution of the specimen, and (in principle, at exceedingly low temperatures) on the temperature. Tables of b_{coh} and other quantities for the elements and for many separate isotopes are available. Hydrogen, with $\langle b \rangle = -0.37 \times 10^{-12}$ cm, vanadium with $\langle b \rangle = -0.048 \times 10^{-12}$ cm, and the strongly capturing isotopes ^{3}He, ^{6}Li, ^{10}B, ^{113}Cd, ^{149}Sm, and ^{157}Gd, for which $\langle b \rangle$ has an important imaginary component, are among the exceptions to the rule that $\langle b \rangle$ is usually real and positive and $\sim 10^{-13}$–10^{-12} cm, with no very clear dependence on the other properties of the isotope or element concerned.

2. The neutron magnetic moment (−1.91 nuclear magnetons) interacts with any local magnetic field in the specimen, in particular with the fields of any atomic magnetic moments such as those connected with the incomplete magnetic shells of iron group transition elements and with rare-earth ions. The resulting magnetic scattering amplitudes (b_{mag}) are of the same general order as the nuclear scattering amplitudes (b), but are calculable from theory. In addition to the direct proportionality of b_{mag} to the magnetic moment of the scattering atoms, b_{mag} is also sensitive to the size and shape of the distribution of the magnetic moment upon the scattering atom, and to the relative orientations of the neutron and atomic magnetic moments and of the vector momentum transfer to the neutron. Since the latter quantities are to some degree controllable, the scattering patterns may contain a wealth of information about the magnetic constitution of the specimen.

Thus slow neutrons are sensitive to almost all nuclear (atomic) excitations, whether in solids, liquids, or gases, as well as to most magnetic (electronic) excitations with energies below, say, 100 meV. Discrimination between different types of excitations depends on the contrast of the scattering amplitudes concerned.

Because of the necessity of maintaining adequately narrow resolution in the incoming (E_0) and outgoing (E') energies and in the angular acceptance of the collimators and detector, the experiments are often intensity limited. Consequently, a bulk of the work has been done at the major reactors of the large national laboratories (e.g., Brookhaven and Oak Ridge in the United States, Chalk River in Canada, Harwell in Great Britain, and now the three-nation Institut Laue–Langevin at Grenoble, France), though important contributions have come also from other institutions. Even so, many experiments are seriously hampered by low intensities and long counting periods (days or weeks). Considered as a source of quanta, even a powerful reactor is a rather dim light bulb. Because of the massive shielding required against fast neutrons leaking from the reactor or accompanying the beam of slow neutrons, the apparatus tends also to be massive. Because of the low flux densities available from reactors, and because also of the comparative transparency of most materials to slow neutrons, use of large specimens (of the order of cubic centimeters) is both required and possible. For studies of crystalline substances single crystals offer enormous advantages over polycrystalline specimens; indeed, use of single crystals is usually essential to the experiment.

Neutron energies are determined from the velocity (v) or from the wavelength (λ) as

$$E=\tfrac{1}{2}m_n v^2=\frac{2\pi^2\hbar^2}{m_n\lambda^2}=\frac{\hbar^2k^2}{2m_n},$$

but not directly. Velocities are determined from the times of flight of the neutrons over known path lengths as measured by mechanical/electronic apparatus. Wavelengths are determined by Bragg scattering from large single-crystal monochromators; beryllium or copper single crystals or pyrolitic graphite, all of special metallurgical character, are most often employed. Special polycrystalline filters (e.g., beryllium, which passes neutrons with $\lambda>4.05$ Å) and pyrolitic graphite filters are also employed.

These elements and others are used in various combinations in the apparatus to select and define E_0 and E' for neutrons (and therefore, also, $\mathbf{k}_0$ and $\mathbf{k}'$).

The neutrons of final energy E' are detected in gas or scintillation counters following capture of the neutrons by means of various nuclear reactions, particularly ^{3}He(n,p)^{3}H, ^{6}Li(n,^{3}H)^{4}He, ^{10}B(n,α)^{7}Li, and ^{235}U(n,fission). The incoming monoenergetic beam of energy E_0 is sampled (monitored) by a "thin" detector of some sort to establish the incoming flux and thus to enable normalization of the scattering patterns and interpretation of them as partial differential cross sections.

The natural variables of a neutron-scattering experiment are two: the vector momentum transfer from the neutron to the specimen,

$$\hbar\mathbf{Q}\equiv\hbar(\mathbf{k}_0-\mathbf{k}')=2\pi\hbar\left(\frac{1}{\lambda_0}-\frac{1}{\lambda'}\right),$$

and the energy transfer,

$$\hbar\omega\equiv E_0-E'=\frac{\hbar^2}{2m_n}(k_0^2-k'^2).$$

In the first Born approximation the partial differential scattering cross section of the specimen per unit solid angle (Ω) is given by an expression of the form

$$\frac{d^2\sigma}{d\Omega\,d\omega}=\frac{k'}{k_0}S(\mathbf{Q},\omega).$$

For most situations the first Born approximation is very good because of the relative smallness (in certain senses) of the interactions between the neutron and the (nuclear and magnetic) scattering centers of the specimen. The only important exceptions are study of the Laue–Bragg elastic coherent scattering by large and relatively perfect crystals (the phenomenon of primary extinction and related effects) and study of the *neutron optics* of materials generally.

Because of the Van Hove relationship, data from neutron-scattering experiments are now usually presented in terms of the reduced variables ($\mathbf{Q}$,ω) rather than in terms of the primary laboratory variables, namely, E_0, E', angle of scattering (ϕ), crystal orientation angles with respect to $\mathbf{Q}$, etc. Often the apparatus is computer controlled, either directly or via cards or tape, and is programmed so that the original measurements are made in terms of $\mathbf{Q}$ and ω (the "constant-$\mathbf{Q}$" method used with triple-axis crystal spectrometers). Figure 1 shows a wave vector diagram (and optical analog) for the constant-$\mathbf{Q}$ method in which the energy distribution of initially "orange" scattered neutrons is studied at constant wave vector transfer $\mathbf{Q}$. The initial and final energy ("color") is determined by Bragg reflection from a single-crystal monochromator.

The functions $S(\mathbf{Q},\omega)$ are Fourier transforms of a class of time-dependent (Van Hove) correlation functions $G(\mathbf{r},t)$, which in some cases have direct intuitive interest and in other cases are useful as checks on statistical mechanical calculations. These correlation functions are therefore sometimes the object of the experiments.

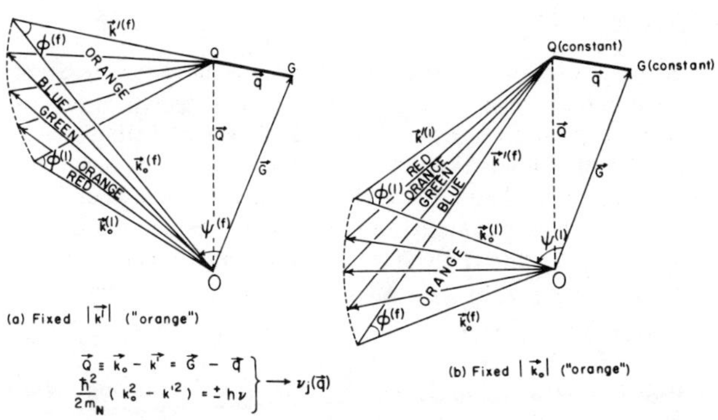

FIG. 1. Reciprocal space diagrams for a triple-axis crystal spectrometer in two modes of operation: In (a), the incident wave vector $\mathbf{k}_0$ is fixed (in the laboratory system) and the analyzing spectrometer is then set to detect a series of different scattered wave vectors $|\mathbf{k}'|$, while the angle of scattering ϕ and the orientation ψ of the single-crystal specimen in the plane are varied in such a way as to keep the wave vector transfer $\mathbf{Q} \equiv \mathbf{k}_0 - \mathbf{k}'$ fixed (in the crystal system of coordinates). Peaks in the detected neutron intensity indicate the phonon (or magnon) frequencies ν_j for the preselected wave vector $\mathbf{q} = \mathbf{G} - \mathbf{Q}$, where $\mathbf{G}$ is a reciprocal lattice vector of the crystal. In (b), the outgoing wave vector $\mathbf{k}'$ is fixed and $\mathbf{k}_0$ varied with similar results. A simple optical analog is also indicated in both diagrams. Both experiments would detect transverse phonons in preference to longitudinal phonons, because $\mathbf{q}$ is nearly perpendicular to $\mathbf{Q}$.

The other major type of experiment involves direct extraction of the dispersion relations for the excitations in the specimen (e.g., frequency/wave vector $\nu(\mathbf{q})$ dispersion relations for phonons or magnons in crystals), by consideration only of energy (ω) and momentum ($\mathbf{Q}$) conservation. This is possible where the scattering function $S(\mathbf{Q},\omega)$ includes components—"spectral lines"—substantially of the form $\delta(\mathbf{Q}-\mathbf{q})\delta(\omega - 2\pi\nu)$. These "lines" of course have intrinsic widths and are broadened by instrument resolution. Often the intrinsic widths of the excitations can be obtained, at least roughly, and thus the "lifetimes" of the excitations induced by anharmonicity, impurities, etc., can be measured.

SUBJECT MATTER AND HISTORY

The field is too extensive to be reviewed adequately here. A recent bibliography (1974) listed about 8000 references to experimental work and to directly related theory in the twin fields of neutron diffraction and neutron spectroscopy. The scientists who have at one time or another worked on these subjects now number in the thousands, with important contributions being made by hundreds of individuals.

The early work on neutron scattering (1935–1950), which established the fields of *neutron diffraction* and *neutron optics* and led to that of *neutron spectroscopy*, is associated preeminently with the name of Enrico Fermi. Other principal figures on the experimental side were E. O. Wollan and C. G. Shull (Oak Ridge National Laboratory), J. R. Dunning, W. W. Havens, Jr., and J. J. Rainwater (Columbia University), D. J. Hughes (Argonne National Laboratory), D. G. Hurst (Chalk River Nuclear Laboratories), and G. E. Bacon (A. E. R. E., Harwell). On the theoretical side principal figures were F. Bloch, O. Halpern, M. H. Johnson, R. Weinstock, G. Placzek, and (later) L. Van Hove and R. P. Feynman. Obviously, these lists are very incomplete.

The first experiments on energy transfers in inelastic scattering date from 1951 (P. A. Egelstaff and R. D. Lowde at Harwell, and Hurst and B. N. Brockhouse at Chalk River). The field of neutron spectroscopy in a practical sense dates from about 1955 (Brockhouse and A. T. Stewart at Chalk River; B. Jacrot at Saclay; Hughes, H. Palevsky, and R. S. Carter at Brookhaven National Laboratory). Other names that should be mentioned include D. G. Henshaw at Chalk River, and K.-E. Larsson and R. Stedman at A-B Atomenergi, Stockholm. An international conference on the subject was held at Stockholm in September 1957, and a series of six larger conferences were sponsored by the International Atomic Energy Agency between 1960 and 1977. The proceedings of these conferences form an important part of the literature of the subject.

Experiments include a considerable variety of types, mainly in the physics and chemistry of solids, liquids, and gases. Recently, some studies of biological interest have also been made. Among the subjects dealt with are the following:

Experimental dispersion curves for the lattice vibrations in most solid phases of most elements (including crystalline ^{4}He) and in a wide variety of the simpler compounds and alloys having up to several atoms per unit cell. The anharmonic properties of lattice vibrations, and the effect of impurities in various concentrations on lattice vibrations.

Experimental dispersion curves for the excitations in liquid ^{4}He ("phonons" and "rotons") at different temperatures and pressures.

The correlations between the atomic motions in a variety of liquid elements and compounds, some dense gases, and in some "amorphous" substances and layered materials.

Evidence for the propagation of high-frequency longitudinal and shear waves and "rotons" in a number of "classical" liquids and in liquid ^{3}He, with partial success. The latter experiment, especially, is very difficult because of the capture of neutrons by ^{3}He.

"Librons" or rotational excitations in molecular crystals.

"Soft modes," and critical effects near the transition temperature, in ferroelectric and related crystals.

The electron–phonon interaction via the lattice vibrations in superconducting and near-superconducting metals.

All the foregoing measurements rely on the scattering of the neutrons by the nuclei of the atoms in the specimen. In magnetic materials analogous situations obtain. The scattering of neutrons by electromagnetic interaction with the electrons allows a nearly equal variety of experiments:

The dispersion curves for the spin waves, "magnons," and the related crystal field spectra have been determined for a wide variety of ordered or partially ordered magnetic systems containing transition-element or rare-earth atoms with a wide variety of magnetic structures—ferromagnetic, antiferromagnetic, ferrimagnetic, helical magnetic structures, etc.—sometimes as functions of temperature up to the Curie or Néel point, and sometimes with added magnetic or nonmagnetic impurities, etc.

The critical magnetic scattering, which sets in near the Curie or Néel point, has been studied, as have shorter-wavelength excitations in paramagnetic materials.
The domain structure existing in certain type-II superconductors has been verified and studied.

Polarized neutrons have not been used for inelastic scattering studies to the same extent as for neutron diffraction, partly because of intensity considerations, partly because the comparative paucity of suitable experiments has had the effect that there exist few spectrometers capable of interesting inelastic experiments with polarized neutrons. Nevertheless, some experiments have been done, including one at Oak Ridge which verified that the magnon carries one unit of angular momentum.

A large number of studies of elastic scattering have given information on the time-independent atomic and magnetic structures of crystals, and some studies of quasi-elastic and of total differential scattering on the asymptotic and instantaneous structures of liquids have been conducted. These works are reviewed elsewhere in this book (NEUTRON DIFFRACTION AND SCATTERING, by G. Shirane).

See also FERRIMAGNETISM; LATTICE DYNAMICS; NEUTRON DIFFRACTION AND SCATTERING; PHONONS.

BIBLIOGRAPHY

Included are an elementary review (Bacon); a readable text at senior undergraduate level (Bacon); a collection of technical review articles (Egelstaff); a very extensive bibliography listing books, articles, and authors (Larose and Vanderwal); an advanced theoretical treatise (Marshall and Lovesey); a monograph on magnetic scattering (Izyumov and Ozerov); a monograph on the wide field of neutron physics (Gurevich and Tarasov).

G. E. Bacon, *Neutron Physics*. Wykeham London, 1969. (E)

G. E. Bacon, *Neutron Diffraction*, 3d ed. Oxford, London and New York, 1975. (I)

P. A. Egelstaff, ed., *Thermal Neutron Scattering*. Academic Press, New York, 1965. (A)

I. I. Gurevich and L. V. Tarasov, *Low Energy Neutron Physics* (translation). Wiley, New York, 1968. (A)

Yu. A. Izyumov and R. P. Ozerov, *Magnetic Neutron Scattering* (translation). Plenum, New York, 1970. (A)

A. Larose and J. Vanderwal, *Scattering of Thermal Neutrons: A Bibliography (1932–1974)*. IFI/Plenum, New York, 1974.

W. Marshall and S. W. Lovesey, *Theory of Thermal Neutron Scattering*. Oxford (Clarendon Press), London and New York, 1971. (A)

Order–Disorder Phenomena

Sandra C. Greer

The phrase "order–disorder phenomena" is used in the literature (1) to refer solely to the superlattice transition in binary alloys; (2) to refer to second-order phase transitions between states of different degrees of order; and (3) to refer to all phase transitions between states of different degrees of order.

In its original and most restrictive usage, order–disorder phenomenon refers specifically to the *superlattice* transition in a binary alloy. Such a transition in a binary solid solution is between (a) a less ordered structure for which each of the components of the alloy (e.g., iron and aluminum) is equally likely to be on a given lattice site and for which there is only short-range order, and (b) an ordered "superlattice" structure in which the components alternate on the lattice and there is long-range order. The less ordered state prevails at high temperatures and the ordered state appears upon cooling.

The extent of long-range order can be designated by the *long-range order parameter*, which is the fraction of the atoms of a component "in place" (i.e., on the alternate lattice sites) in excess of one half (which would the number "in place" for a random arrangement in an equimolar mixture). Let P be the order parameter, a the number of atoms of component A "in place," and N the total number of atoms of component A. Then

$$a = \tfrac{1}{2}(1+P)N$$

and

$$P = \frac{2a}{N} - 1.$$

Thus for the highest degree of order $P=1$, and for complete disorder $P=0$. Even in the disordered state there is still some short-range or local order in that, on the average, an atom of one component is surrounded by atoms of the other component. A *short-range order parameter* σ can be defined as the fraction of nearest-neighbor pairs that consists of unlike atoms, in excess of one half (the fraction for a random arrangement). If g is the number of unlike pairs and r the total number of pairs, then

$$\sigma = \frac{2g}{r} - 1.$$

Thus $\sigma = 1$ for greatest local order and $\sigma = 0$ for local disorder.

The superlattice transition is studied most easily by x-ray and neutron diffraction, which provide the structure as a function of temperature, the long-range order (from the intensities of the superlattice diffraction peaks), and the short-range order (from the diffuse scattering between diffraction peaks).

If the superlattice transition is considered as a function of the composition of the alloy, the compositions and transition temperatures at which the transition is second order are examples of *critical points*. *Second-order transitions* are those at which the second derivatives of the Gibbs free energy G are discontinuous or infinite (e.g., the heat capacity), while the first derivatives of G are continuous (e.g., the volume and the entropy). At a first-order transition the first derivatives of G are discontinuous. At a critical point, the two phases are indistinguishable and can approach one another continuously and there exist anomalies in various thermodynamic properties. Similar critical points occur in many other kinds of physical systems and the phase transitions in them are quite analogous to the superlattice phase transition. Thus, it has become common to apply the designation order–disorder phenomena to all second-order phase transitions that involve a change in the degree of ordering rather than a displacement of molecules (as can occur in some solid–solid crystal transformations). The superlattice transition is a member of this more general class. A number of kinds of phase transitions can be so classified. For example, in a magnet, the electron spins can be aligned (ordered) or randomly oriented (disordered); the magnetization indicates the degree of alignment and is an order parameter, and the Curie point is a critical point. For the liquid–vapor transition, there also exists a critical point at a particular temperature, pressure, and density; the difference between the density of either the liquid phase or the coexistent vapor phase and the density at the critical point is a measure of the amount of (short-range) order and is also an order parameter.

The thermodynamic properties have the same behavior in all these second-order transitions. Experiments show, for example, that at the critical point the specific heat at constant order parameter diverges, and that below the critical temperature the order parameter approaches zero approximately as the cube root of the difference in temperature from the critical temperature. This isomorphism of critical phenomena is most attractive: it means that the same theory can be used to understand many different kinds of phase transitions. The theories of critical phenomena are highly developed and are broadly applicable in the field of order–disorder phenomena.

In 1934 W. L. Bragg and E. J. Williams proposed a theory for the superlattice transition in which the solid is represented by a lattice, at each site of which there is an atom of one component or the other. It is assumed that the attraction between unlike molecules is greater than that between like molecules and that the ordered molecules are subject to a field that is an average of the interactions with all the other molecules. J. D. van der Waals had proposed an analogous theory for the liquid–gas transition in 1873; van der Waals accounted for the molecular interactions by assuming that each molecule is subject to a field that is an average of the fields due to all other molecules and is proportional to the density. Another such theory had also been proposed in 1907 by Pierre Weiss for the paramagnetic–ferromagnetic phase transition; this theory assumes that there is a lattice at each site of which there is an electron spin that has one of two possible orientations and that the energy is lower if adjacent spins are aligned. Again, a "mean field" due to the other molecules is assumed. Thus these theories all make the approximation of a "mean field" at each individual molecule due to all other molecules. At low temperature this mean field is sufficient to maintain order, but as the temperature is increased, the increased kinetic energy of the molecules causes more and more of them to get out of order. As the system becomes disordered, the average ordering field is diminished, so that it is then even easier for the system to become more disordered. (This "avalanche" effect has caused these phase transitions to be called *cooperative* phenomena.)

The methods of statistical mechanics can be applied to determine the thermodynamic behavior of these models. These *mean-field theories* (now called the "classical"—as opposed to the "modern"—theories of phase transitions) give rather good qualitative predictions of the phase transitions, but they fail in some essential ways. They all predict a simple jump discontinuity in the specific heat at constant order parameter at the critical point, whereas experiments on most such phase transitions show a divergence in the specific heat at the critical point. The classical theories predict that the order parameter should approach zero as $|T_c - T|^{\beta}$ where T_c is the critical temperature and β is 0.5, whereas experiments yield β near 0.3. The β is an example of a *critical exponent*. In other ways, too, we find that the classical theories, although useful, are not entirely adequate.

All classical theories have one assumption in common: they assume that the intermolecular forces in the system are sufficiently long ranged that the field acting on a given molecule is, indeed, an average of the forces of all the other molecules. Long-range forces are more amenable to "averaging." In our discussion of the "cooperative" effect near a critical point, we said that the "disordering" of one molecule makes it easier for the next one to be "disordered." This would be especially true for nearby molecules. Thus short-range interactions are especially important near the critical point. Since the two phases become more and more alike as the critical point is approached, it takes less and less energy to change a molecule or group of molecules from one phase to another, and the transition of any given molecule from one phase to another makes it easier for the next one to follow. Thus, fluctuations in the average order parameter occur more and more easily and themselves cause more fluctuations (i.e., the fluctuations are correlated). We can say that the classical mean-field theories ignore the correlations in fluctuations of the order parameter near the critical point. The distance over which the fluctuations correlate is called the *correlation length*. We expect the mean-field theories to work when the intermolecular forces act over a distance much greater than the correlation length, but to fail when the correlation length is comparable to the range of the intermolecular forces. At superconductor and ferroelectric phase transitions, the forces are sufficiently long ranged that mean-field theories do work. For binary alloys, ferromagnets, and the liquid–gas transition, the mean-field theories are inadequate.

A treatment that is more successful than the mean-field models is the *Ising model*. In the Ising model, as in the Bragg–Williams and Weiss models, the phase under consideration is represented by a lattice, at each site of which there is a molecule. The same model can be applied to different phase transitions by modifying the way the two site states are designated. For a binary alloy, the sites would be occupied by one or the other of the two components. For a magnet, the sites would be occupied by states "spin up" or "spin down." The Ising model has even been applied to the liquid–gas transition by designating the two site states as either "occupied" or "vacant"; this is the *lattice–gas* model. For a superlattice model, the interaction energy is taken to be lower between *unlike* molecules than between like molecules. If the energy is lower between *like* molecules, a model of separation into two phases of different compositions results; such "unmixing" occurs both in solid alloys and in liquid mixtures. In the Ising model, as opposed to mean-field models, the nearest-neighbor (short-range) interactions are specifically included. The methods of statistical mechanics can be applied (exactly for two dimensions and approximately for three dimensions) to obtain the thermodynamic properties of the Ising model. The resulting properties agree remarkably well with experimental results in fluids, some magnets, and alloys very near the critical point. For example, the specific heat of the three-dimensional Ising model diverages and the order parameter goes to zero as $|T_c - T|^{\beta}$ with $\beta = 0.31$.

The expression order–disorder phenomenon is often used in a still broader sense to refer to any transition between a state of matter that has some degree of order and a state that has a somewhat lesser degree of order. Such a general definition would certainly include all the second-order transitions mentioned earlier, and even more; it would also include many first-order transitions. For example, the transition between a crystalline solid and its melt is a transition between long-range order and short-range order. Still other transitions that can be called order–disorder transitions according to this broader definition are the conformational changes in biopolymers, the hydrocarbon disordering in biomembranes, the molecular reorientation in plastic crystals, and the molecular alignment in liquid crystals. Lattice models similar to the Ising model have been applied with profit to many of these cases, even for first-order transitions. On the other hand, first-order transitions lack the symmetry of second-order transitions and thus the development of theoretical descriptions of first-order transitions has lagged behind that of descriptions of second-order transitions.

See also CRITICAL POINTS; ISING MODEL; PHASE TRANSITIONS.

BIBLIOGRAPHY

C. Domb and M. S. Green, *Phase Transitions and Critical Phenomena*. Academic Press, New York, 1972. Three volumes of detailed articles by experts. (A)

E. W. Elcock, *Order–Disorder Phenomena*. Wiley, New York, 1956. General discussion, followed by extensive treatment of binary alloys. (E–I)

H. S. Green and C. A. Hurst, *Order–Disorder Phenomena*. Wiley, New York, 1964. Extensive treatment of lattice models. (A)

A. I. Kitaigorodskiy, *Order and Disorder in the World of Atoms*. Springer, New York, 1967. Elementary discussions of a great variety of order–disorder phenomena. (E)

K. Mendelssohn, "States of Aggregation," *Phys. Today*, April, p. 46 (1969). Discussion of order in both position space and momentum space. (E)

H. E. Stanley, *Introduction to Phase Transitions and Critical Phenomena*. Oxford, London and New York, 1971. General discussion of gas–liquid and magnetic critical points with presentation of modern scaling theory. (I–A)

Paramagnetism

J. E. Enderby

Substances with a positive magnetic susceptibility χ_m are said to be paramagnetic; that is, the magnetic moment induced by an applied magnetic field is in the direction of the field. Atomic paramagnetism (as opposed to nuclear effects, to which we return briefly at the end of this article) is electronic in origin and, where it exists, usually overcomes the more fundamental diamagnetism. Electronic paramagnetism is found in free atoms and molecules possessing an odd number of electrons or an unfilled inner shell. (There are in addition a few "even-electron" substances which are paramagnetic, for example, molecular oxygen, but these are somewhat exceptional.) The essential ingredient is that the atoms and molecules of the substance must possess a permanent magnetic dipole moment. A further important contribution to the magnetism of metallic conductors is the fact that the gas of free electrons, which is responsible for the characteristic properties of metals, is itself weakly paramagnetic. This paramagnetism is essentially independent of temperature T, which contrasts with atomic and ionic paramagnetism where $\chi_m \propto T^{-1}$ (Curie's law).

The classical microscopic theory of paramagnetism is due to Langevin. He noted that the tendency for the magnetic field B to align the elementary dipoles is countered by the random orientation induced by thermal agitation. This leads to a result for χ_m which can be written in the form

$$\chi_m = \frac{\mu_0 N m^2}{3kT}, \tag{1}$$

where N is the number of dipoles per unit volume, each of which has moment m, k is Boltzmann's constant, and μ_0 is the permeability of free space. At very low temperatures and high magnetic fields the relationship between magnetization and applied field implied in Eq. (1) becomes more complicated and in fact may cease to be linear.

Paramagnetism can be investigated experimentally by measuring the weak force experienced by a paramagnetic material when placed in an inhomogeneous field. In the Curie method a small volume of material is placed in a region where the magnetic field is varying strongly with distance. The other method frequently used, the Gouy method, employs a long specimen, one end of which is in a field-free region, while the other is placed between the poles of a powerful electromagnet. In each case the force on the sample can be related to χ_m. Experimentally it is found that $\chi_m \propto T^{-1}$ for many substances, although for ferromagnetic materials above their Curie temperature θ, a modified law, originally due to Curie and Weiss and of the form $\chi_m \propto (T-\theta)^{-1}$, is often obeyed.

Quantum-mechanical corrections modify the Langevin formula in several significant ways. Not all directions are accessible to the dipole and the "quantization of space" introduces the notion of an effective dipole moment which depends on the total quantum number. At very low temperatures paramagnetic rare-earth salts show considerable departures from Curie's law. This happens because the effect of the crystalline electric field is to split certain energy levels even when there is no applied magnetic field. Furthermore, the values of χ_m vary with direction for much the same reasons.

The nuclei of many types of atoms possess angular momentum and this in turn gives rise to magnetic moments. Thus an assembly of such nuclear magnets obey laws similar to those for their atomic counterparts; in particular, nuclear paramagnetism obeys Curie's law. However, the contribution from the nuclei to the total susceptibility is only about 10^{-6} of the electronic part and therefore can, for many practical purposes, be neglected.

See also DIAMAGNETISM; FERRIMAGNETISM; FERROMAGNETISM; MAGNETIC MOMENTS; MAGNETS (PERMANENT) AND MAGNETOSTATICS.

BIBLIOGRAPHY

A. Abragam and B. Bleaney, *Electron Paramagnetic Resonance of Transition Ions*. Oxford Univ. Press, London, 1970.

J. A. McMillan, *Electron Paramagnetism*. Reinhold, New York, 1968.

R. S. Tebble and D. J. Craik, *Magnetic Materials*. Wiley, New York, 1969.

R. M. White, *Quantum Theory of Magnetism*. McGraw-Hill, New York, 1970.

D. E. G. Williams, *The Magnetic Properties of Matter*. Longmans, London, 1966.

Phase Transitions

Melville S. Green

Perhaps the most far-reaching principles of physics are its conservation laws, which tell us that the sum of certain quantities—mass, energy, momentum, numbers of atoms of various species—remains constant in the transformations and vicissitudes of matter. Within the limitations imposed by the conservation laws, however, matter may undergo profound changes. Among the most striking of these changes are those called phase transitions. We need not go far from ordinary experience to encounter phase transitions. The precipitation of the water vapor of the atmosphere as rain and the subsequent evaporation of the precipitated water are examples of phase transitions. The characteristic feature is the existence of a substance in two very different forms—in this instance, a palpable liquid form and an impalpable vapor form—and the transformation of one into the other under suitable circumstances. Two other meteorological examples, the precipitation of water vapor in cold weather as snow, and its subsequent evaporation or sublimation, are instances of a solid-to-vapor transformation, while the freezing and melting of natural bodies of water is an example of a liquid-to-solid phase transition. An example from mineralogy is carbon, which may exist in the familiar black amorphous form or in the beautiful crystalline form of diamond. From our human point of view amorphous carbon and diamond are materials of very different economic and esthetic values, but from the dispassionate point of view of physics, they are forms of the same substance. The transformation of amorphous carbon into diamond occurs at temperatures and pressures that until recently could only be produced by geological forces, but the necessary conditions are now produced in the laboratory and in industrial processes.

In all of the examples we have just given the transformation takes place portion by portion. The water vapor in a cloud condenses to raindrops, which fall to the ground and coalesce into lakes and rivers. Diamonds are found in a matrix of amorphous carbon, suggesting that only a portion of the carbon was transformed. Such phase transitions are described as first order. In another type of phase transition the two phases are only infinitesimally different at the transition and the transformation takes place all at once. The magnetization of a permanent magnet will diminish as the temperature increases and become zero at a definite temperature, the Curie temperature. Above the Curie temperature the magnetization stays exactly zero. A more striking example is superfluid helium. Above a well-defined temperature, the lambda temperature, the flow properties of liquid helium are qualitatively no different from those of any other liquid. Immediately below the lambda temperature liquid helium acquires frictionless superflow properties while remaining only infinitesimally different from normal helium. Such phase transitions have been called, for erroneous reasons, second-order phase transitions, but are more properly called continuous phase transitions. Although these two types of phase transitions are seemingly very different, they are in fact aspects of the same phenomenon.

THERMODYNAMICS

The thermodynamics of phase equilibria, in which one or more phases coexist in an equilibrium system, forms a very useful theoretical framework for phase transitions that helps us to understand even nonequilibrium phase-transition phenomena. We will confine ourselves almost exclusively to this theoretical framework in this article.

Two phases in equilibrium, denoted by I and II, may freely exchange the quantities energy, volume, and molecules of each species subject only to conservation of the sum of each quantity. The thermodynamic forces corresponding to each extensive or conserved quantity must be balanced. For the exchange of energy and volume this gives the condition that the temperature and pressure of each phase must be the same:

$$T_{\rm I} = T_{\rm II}, \qquad p_{\rm I} = p_{\rm II}. \tag{1}$$

Moreover, corresponding to the exchange of each species i we have the condition that the corresponding chemical potential must be the same in both phases:

$$\mu_{\rm I}{}^{i} = \mu_{\rm II}{}^{i}. \tag{2}$$

If there are more than two phases, conditions (1) and (2) apply to each pair of phases. Each of the quantities μ^i in each phase may be considered to be a function of the common temperature and pressure and $r-1$ independent variables $x_1 \cdots x_{r-1}$ expressing composition. By a simple counting of variables and equations J. W. Gibbs arrived at his famous phase rule, which determines the maximum number of phases that may coexist in equilibrium and the dimension of the manifold of coexisting states. Gibbs's phase rule states that

$$r + 2 - p = C \geqslant 0 \tag{3}$$

where p is the number of coexisting phases. C is the dimension of the manifold of coexisting phases. Since C may not be negative, p may not be greater than $r+2$.

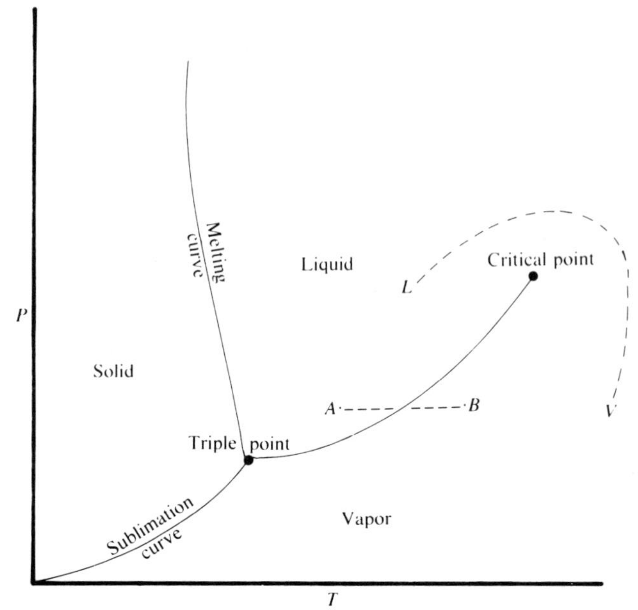

FIG. 1. Manifolds of coexistence for a single-component system with one liquid, one vapor, and one solid phase. The lines of two-phase coexistence are labeled with their conventional names. At a point of coexistence of two phases any quantity of one phase may be in equilibrium with any quantity of the other.

THE CASE OF A SINGLE COMPONENT

The most familiar application of the phase rule is to the case of a single component. In this case the maximum number of coexisting phases is three. This occurs on a manifold of dimension zero, the triple point. Two-phase coexistence occurs along a line. In the most common case, with one solid, one liquid, and one vapor phase, the manifolds of coexistence in the T–P plane are sketched in Fig. 1.

At a triple point any quantity of the three phases may coexist in equilibrium. The triple point of water may be realized in a cell in which water vapor, liquid water, and ice are present in equilibrium at the same time. Such a triple-point cell realizes unique temperature and pressure, which is the fundamental fixed point for both the International Absolute and International Practical Scales of Temperature.

VOLUME CHANGE AND LATENT HEAT

Figure 1 may be considered to be the projection on the T–P plane of the surface $\mu(T,P)$ where μ is the chemical potential of the single species. To each temperature and pressure not on a manifold of coexistence corresponds a unique phase, as indicated in Fig. 1. Although $\mu(T,P)$ must be continuous according to the conditions of equilibrium [Eqs. (1) and (2)], its first derivatives need not be. The differential of $\mu(T,P)$ is

$$d\mu = -S\,dT + v\,dp \tag{4a}$$

or

$$\left(\frac{\partial\mu}{\partial T}\right)_P = -S, \qquad \left(\frac{\partial\mu}{\partial P}\right)_T = v, \tag{4b}$$

where S and v are the entropy and volume per mole, respectively. Thus the discontinuity of the first partial derivatives of μ, which is the origin of the adjective "first-order," is an expression of the fact that the density and other properties of two coexisting phases are different. The change of entropy in a phase change has a simple calorimetric meaning. Since a reversible phase transformation occurs at a constant temperature, the change in entropy per mole is simply the heat absorbed in the process divided by the temperature:

$$\Delta S = S_{\rm II} - S_{\rm I} = \frac{L}{T} \tag{5}$$

where $S_{\rm I}$, $S_{\rm II}$ are the molar entropies of phases I and II and L is the heat absorbed at constant temperature or *latent heat* of the process (latent because not concomitant with a change in temperature).

CLAPEYRON–CLAUSIUS EQUATION

The slope of a coexistence line is related to the entropy change and volume change by an equation that may be derived by differentiating the equilibrium condition, Eq. (2):

$$\frac{dP}{dT} = \frac{S_{\rm II} - S_{\rm I}}{v_{\rm II} - v_{\rm I}} = \frac{L}{T\,\Delta v} \tag{6}$$

where dP/dT is the slope of the coexistence curve. Since sublimation, vaporization, and melting all represent transitions from a less to a more disordered state, the statistical mechanical relation—entropy is equivalent to disorder—explains the fact that the latent heats of these phase changes are positive. Sublimation and vaporization both involve positive volume changes, so the corresponding coexistence curves have positive slopes. The molar volume of a liquid and its corresponding solid, however, are very nearly the same and the volume change may be slightly positive or slightly negative. Water is one of the relatively few cases where the liquid is denser than the solid. This circumstance explains the natural phenomenon that ice forms on the surface of a lake as well as that water melts under pressure.

CONTINUITY OF STATE AND THE CRITICAL POINT

As indicated in Fig. 1, the vapor–pressure curve has a definite end point, the critical point. Thus even though the liquid and vapor phases are physically quite distinct, the corresponding domains in Fig. 1 are connected. The solid region, on the other hand, seems to be completely disjoint from the fluid region. It is thus physically possible to transform a liquid into a vapor without a phase change by following a reversible process (the line L–V in Fig. 1). This is the continuity of the liquid and vapor states.

Along the coexistence curve the difference in density, entropy density, and other properties between the liquid and vapor phases diminishes and disappears entirely at the critical point. An important practical meaning of the critical point may be seen from Fig. 1. For temperatures below the temperature of the critical point (critical temperature) a vapor may be liquefied by compressing without cooling; above the critical temperature this is not possible. Thus fluids with low critical temperatures are difficult to liquefy. Helium, with a critical temperature of 5.1 K, was liquefied by Kamerlingh Onnes only in 1908. The evanescence of the distinction between liquid and vapor in the neighborhood of the critical point is the cause of the very striking phenomenon of critical opalescence. The natural thermal fluctuations produce temporary liquidlike or vaporlike regions throughout the fluid that are large enough to scatter light and cause the otherwise transparent substance to become cloudy or *opalescent*.

CRITICAL SINGULARITIES

The surface $\mu(T,P)$ becomes singular in the mathematical sense at the critical point. In Fig. 2 are sketched a sequence of curves representing the intersection of this surface with a series of planes parallel to its tangent plane at the critical point. The surface is concave downward. $\mu(T,P)$ and its first partial derivatives are continuous at the critical point but its second partial derivatives are infinite there. This means that the specific heat at constant volume and the isothermal compressibility, as well as other important physical quantities, become infinite at the critical point.

Near the critical point $\mu(T,P)$ may be represented in the so-called scaling form

$$\mu(T,P) = \mu_c - S_c\,\Delta T + v_c\,\Delta p + |g_1|^{2-\alpha} F_{\pm}(g_2|g_1|^{-\Delta}) \tag{7}$$

where S_c, v_c, and μ_c are, respectively, the critical entropy,

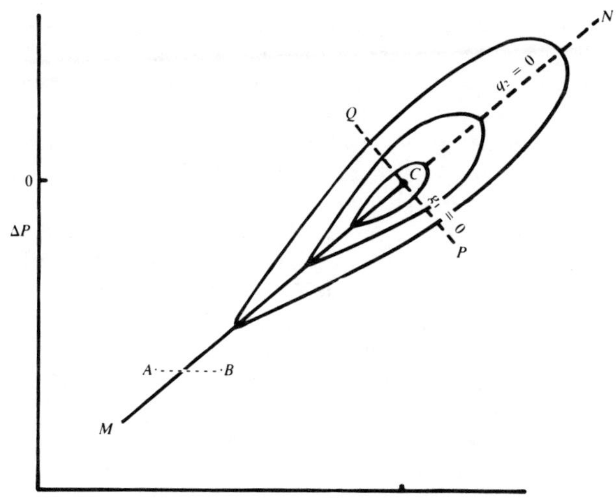

FIG. 2. Critical singularity of the surface $\mu(T,P)$. The closed curves represent the intersection of this surface with a sequence of planes parallel to the tangent plane at the critical point. MCN is the line $g_2 = 0$; QCP, $g_1 = 0$.

volume per mole, and chemical potential, and g_1 and g_2, called scaling fields, may be taken to be linear combinations of $\Delta T = T - T_c$ and $\Delta P = P - P_c$. The indices $\pm$ refer to the cases in which $g_1 \geqslant 0$ and $g_1 < 0$, respectively. The line $g_2 = 0$ is tangent to the vapor pressure curve and to the critical isochore (curve of constant volume equal to critical volume); $2 - \alpha \simeq 1.9$, $\Delta \simeq 1.5$. The line $g_1 = 0$ has a less obvious meaning. One of the important recently discovered features of the critical region is the principle of universality. For fluids this means that the function $F^{\pm}$ is the same for all fluids. The only features of the singularity that change from fluid to fluid are coefficients of ΔT and ΔP in g_1 and g_2 and the parameters of the tangent plane at the critical point, i.e., S_c, v_c, μ_c.

CONTINUOUS PHASE TRANSITIONS

The transformation from one phase to another usually takes place along a path (*A–B* in Fig. 1) that crosses the coexistence curve. When the system reaches the coexistence curve along such a path, it hesitates, as it were, while small portions of the system are transformed successively from one phase to the other. Since the singularity in the coexistence curve is experimentally a simple jump singularity in the derivatives of $\mu(T,P)$, there is no observable precursor or aftereffect of the transition. Along the curve $g_2 = 0$ (*MCN* in Fig. 2), the phenomenon looks very different. For a fluid this path may be easily achieved by confining the system to a vessel of fixed volume. As g_1 is diminished through positive values to zero, the system consists of one fluid phase. At the critical point and below, $g_1 \leqslant 0$, the system divides up without hesitation into equal amounts of liquid phase and vapor phase, which in this region are only slightly different from each other. Since the path $g_2 = 0$ passes right through the singularity, the specific heat and compressibility tend to infinity both above and below the transition.

For a fluid the path $g_2 = 0$ is only one among many achievable paths, and the system clearly consists of two phases separated by gravity for $g_1 < 0$. For certain systems the most natural path or even the only physically possible path is along $g_2 = 0$. This may come about for reasons of symmetry or other reasons. Moreover, for the same reasons the state of the system for $g_1 < 0$ may not be easily identifiable as a mixture of two phases but appears to be a single, albeit somewhat anomalous, phase. The properties of this anomalous mixed phase for $g_1 < 0$ approach those of the single phase, $g_1 > 0$, continuously. For this reason such transitions are called continuous phase transitions. An example is the case of an antiferromagnet such as rubidium manganese fluoride, $RbMnF_3$, where above the temperature 83.02 K, called the Néel temperature, the magnetic moments of the manganese ions are oriented randomly but below the Néel temperature the moments are oriented alternately parallel or antiparallel to some fixed direction. The situation is illustrated in Fig. 3, which represents the lattice of manganese ions in, say, the 001 plane. This lattice may be divided into two interpenetrating cubic lattices of A sites and B sites. The exchange interaction between manganese ions favors nearest neighbors whose magnetic moments are oppositely oriented. Above the Néel temperature this bias of the interaction is manifested only locally, so that the nearer neighbors of any manganese ion are likely to have oppositely oriented moments but the farther neighbors are random. This is called local order. Below the Néel temperature the effect of this bias extends cooperatively throughout the crystal and the moments on the whole A and on the whole B lattice are each oriented in the same way but opposite to each other. This situation may be realized in two distinct but equivalent ways, A up, B down, and B up, A down. These two realizations may be taken to be two phases in equilibrium. Since these two phases differ physically only by a translation of the lattice by one primitive lattice vector, there is no macroscopic agency or field that favors one over the other. Thus the system is restricted to a completely symmetric state that above the Néel temperature is the state of zero average moment on each site but below the Néel temperature is an equal mixture of both phases. We are confined to the line $g_2 = 0$. Because this line passes through the Néel point (critical point), the specific heat becomes infinite, giving rise to the characteristic signature of a continuous phase transition, a specific heat curve that is reminiscent of the Greek letter λ (Fig. 4). On the other hand, properties that depend only on the first derivatives of the thermodynamic potential, such as the magnetization M_s of either one of the sublattices, are continuous.

Other physical systems exhibiting continuous phase transitions are ferromagnets and superfluid helium. In the former the phenomena are somewhat complicated by the effect of magnetic-dipole forces, which make the phenomena somewhat dependent on the shape of the sample, but these may be minimized in appropriately shaped samples. In the case of the normal-to-superfluid transition in liquid helium, the external field that would take us off the line $g_2 = 0$ is zero as a consequence of a very fundamental symmetry principle, gauge symmetry, whose breaking would involve the creation or destruction of helium atoms. Although the examples of continuous phase transitions discussed here have the same basic thermodynamic description, they are physically very different phenomena.

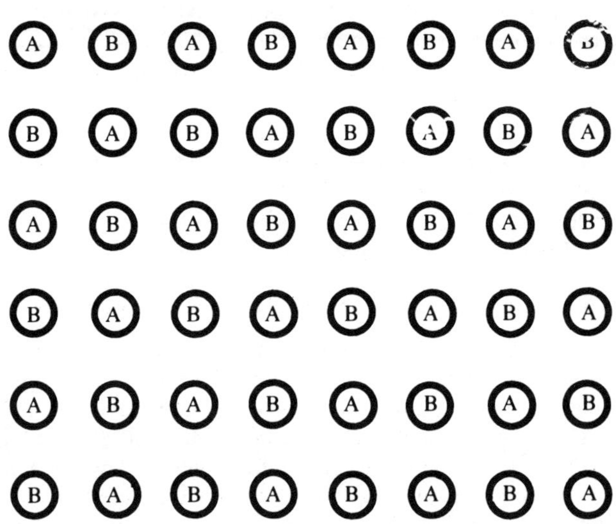

FIG. 3. Manganese ions in the (001) plane of $RbMnF_3$, showing two kinds of lattice sites.

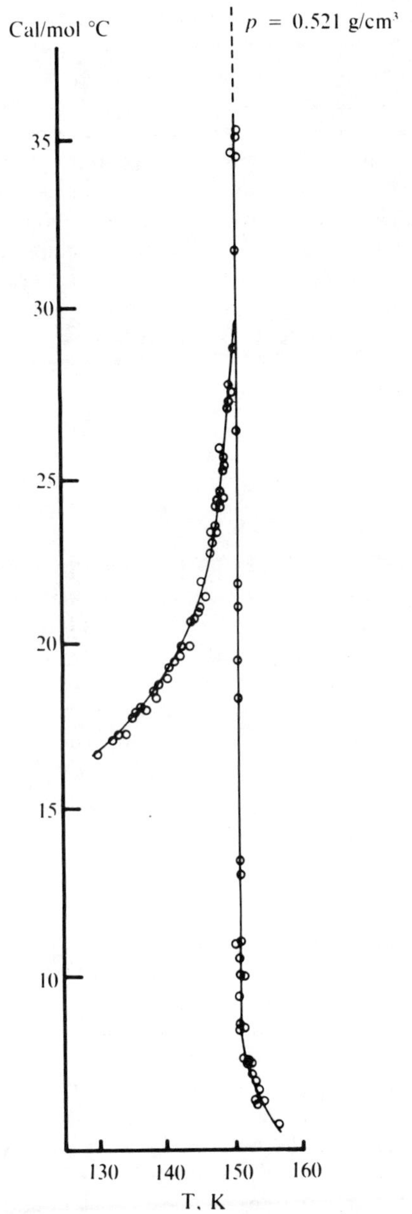

FIG. 4. Variation of C_v of argon with T along critical isochore. [After M. Bagatskii, A. Voronel, and V. Gusak, *Sov. Phys. JETP* **25,** 72 (1963).]

ORDER PARAMETER

Associated with a continuous phase transition is a scalar or vector parameter that determines the physical nature of the transition. It has the value zero in the random state (usually above the transition temperature) and acquires nonzero values in the ordered state (usually below the transition). In the case of a fluid the order parameter is a scalar, the difference in density between the liquid and vapor phases. In the case of an isotropic ferromagnet or antiferromagnet, the order parameter is a three-dimensional vector quantity, the magnetization or the sublattice magnetization. In the case of superfluidity (also superconductivity), the order parameter is a complex quantity with a magnitude and phase angle, which may be interpreted as the wave function of the finite fraction of the particles that have zero momentum. The superflow properties may be considered to be a manifestation of the fact that a wave function with zero momentum cannot be localized. The order parameter also determines the *universality class* of the transition. Continuous phase transitions that have order parameters with the same tensorial character and the same space dimension belong to the same universality class, i.e., have the same singularity in the thermodynamic potential. When the order parameter is multidimensional we note that the ordered state must be considered to consist of a mixture of not merely two but of a continuous infinity of phases in equilibrium.

See also CRITICAL POINTS; FERRIMAGNETISM; FERROMAGNETISM; ORDER–DISORDER PHENOMENA; SUPERCONDUCTIVITY THEORY.

BIBLIOGRAPHY

H. Callen, *Thermodynamics*. Wiley, New York, 1960.

C. Domb and M. S. Green, *Phase Transitions and Critical Phenomena*, Vols. I, III, Va, Vb, VI. Academic Press, London, 1972–1977.

J. W. Gibbs, "Equilibrium of Heterogeneous Substances," in *Collected Works*. Longmans, Green, New York, 1931.

T. Hall, *J. Chem. Educ.* **38,** 484 (1961); *The Chemist,* July, pp. 276–279 (1970).

K. Mendelssohn, *The Quest for Absolute Zero*. McGraw-Hill, New York, 1966.

H. E. Stanley, *Introduction to Phase Transitions and Critical Phenomena*. Oxford, London and New York, 1971.

M. Zemansky, *Heat and Thermodynamics*, 5th ed. McGraw-Hill, New York, 1968.

Phonons

P. G. Klemens

Phonons are quanta of energy of the normal modes of vibration of a crystal lattice or of an elastic continuum. The concept of the phonon is closely analogous to that of the photon in electrodynamics. In each case we consider the quantization of the energy of a single wave. According to classical theory, the possible values of the energy content are continuously variable. According to quantum theory, the energy content of a wave, or of any other harmonic oscillator, of frequency ν can have only a discrete set of values $(N+\frac{1}{2})h\nu$, where N is an integer and h the Planck constant. The zero-point energy $h\nu/2$ is always present and can be regarded as part of the formation energy of the oscillator. The additional energy consists of an integral number N of quanta or phonons. Energy can be added to or removed from the oscillator only in integral numbers of phonons.

The concept of phonons refers to the quantization of the energy content of each oscillator, not to the nature of the oscillator itself. The oscillator can be a progressive displacement wave in a discrete lattice of atoms (lattice wave), or a displacement wave in an elastic continuum (sound wave), or even a localized vibration. It is not necessary to consider a discrete lattice to have the energy of each oscillator quantized. Phonons can exist in an elastic continuum just as photons exist in a continuous medium or in free space.

The quantization of the energy of each vibrational mode or oscillator has important thermodynamic consequences. While the number of phonons in each oscillator must be, at any instant, an integer N, the average number of phonons in an ensemble of identical oscillators, each of frequency ν, in thermal equilibrium at temperature T is given by the Planck distribution

$$N^0 = (e^{h\nu/kT} - 1)^{-1} \tag{1}$$

where k is the Boltzmann constant. The average energy content of each oscillator (without zero-point energy) is $N^0 h\nu$, and each oscillator contributes to the specific heat of the system an amount $h\nu(dN^0/dT)$. Unless $kT \gg h\nu$, the specific heat of the oscillator falls below the classical value of k. Hence, the specific heat of a solid containing G atoms and having the vibrational energy of $3G$ harmonic oscillators is well below the classical value of $3Gk$ at low temperatures. It is given by

$$C(T) = \int d\nu\, S(\nu) h\nu (dN^0/dT) \tag{2}$$

where $S(\nu)\, d\nu$ is the number of normal modes in the frequency interval $(\nu, \nu + d\nu)$. At low frequencies these modes or lattice waves are similar in character to plane elastic waves in a continuum; at higher frequencies the character and spectral distribution of the lattice wave are increasingly influenced by the crystal structure. At lowest frequencies, $S(\nu) \propto \nu^2$ and $C(T) \propto T^3$ (Debye specific heat). It is one of the objectives of lattice dynamics to relate $S(\nu)$, and hence the thermodynamic properties, to the crystal structure and the nature of the interatomic forces.

The concept of phonons is also important because the phonon can be looked upon, for many purposes, as a particle of fixed energy and momentum moving freely within the solid. Let the displacement field due to a lattice wave be of the form

$$\mathbf{u}(\mathbf{r}) = B(\mathbf{q})\mathbf{E} \exp[i(\mathbf{q}\cdot\mathbf{r} - 2\pi\nu t)] \tag{3}$$

where $\mathbf{u}$ is the displacement of a particle at position $\mathbf{r}$ and time t; $\mathbf{q}$ and ν are the wave vector and frequency, respectively; $\mathbf{E}$ is a unit vector denoting polarization direction (or a set of vectors if there is more than one atom per unit cell); and $B(\mathbf{q})$ is the amplitude. Each phonon contained in that wave acts like a particle of energy $h\nu$ and of "momentum" $h\mathbf{q}/2\pi$, moving with a velocity equal to the group velocity of the wave, i.e., $\mathbf{v} = 2\pi\, \partial\nu/\partial\mathbf{q}$, since in wave motion the group velocity governs the energy flow, and since the phonon is a quantum of energy. We can thus describe the transport of vibrational energy within a solid in terms of a flow of phonons and use the concepts of the kinetic theory of gases. The phonon can also be partly localized by quantizing not a single wave (3), but a wave packet.

The thermal conductivity of an insulator, or the lattice component of the thermal conductivity of solids that contain free electrons, is given by

$$\kappa = \tfrac{1}{3} C v l \tag{4}$$

where C is the lattice specific heat, v the average phonon velocity, and l the average phonon mean free path, or wave attenuation length. In a structurally perfect solid with truly harmonic interatomic forces, there would be no energy exchange between the lattice waves, and the phonon mean free path would be infinite. The interchange of energy among lattice waves can be pictured as the creation of some phonons and the destruction of others. The mean free path is limited by lattice defects, which scatter phonons from one wave into another, and by the anharmonic component of the lattice forces, which, to lowest order, leads to three-phonon interactions. In such an interaction process, a phonon of mode $\mathbf{q}$ and one of mode $\mathbf{q}'$ are destroyed and a phonon $\mathbf{q}''$ is created, or vice versa. Energy must be conserved in the process, so that

$$\nu + \nu' = \nu'', \tag{5}$$

while the uniformity of the medium requires that

$$\mathbf{q} + \mathbf{q}' = \mathbf{q}''. \tag{6a}$$

Equation (6a) expresses the conservation of momentum of the interacting phonons, just as momentum is conserved when free particles collide. These momentum-conserving processes (also called normal processes) are the only ones allowed in an elastic continuum. In a discrete lattice, however, additional processes are possible. Let $\mathbf{b}$ be an inverse lattice vector of the crystal, i.e., all properties of the static lattice can be expressed as Fourier series of the form $\Sigma_{\mathbf{b}} F(\mathbf{b}) \exp(i\mathbf{b}\cdot\mathbf{r})$. The additional three-phonon interaction processes satisfy a wave-vector selection rule

$$\mathbf{q} + \mathbf{q}' = \mathbf{q}'' \pm \mathbf{b} \tag{6b}$$

and are called Umklapp processes (because the net momentum of the phonon triplet is "flipped over"). Each such process is a three-phonon interaction and a Bragg reflection combined. Thus the phonon momentum is not always conserved and is not a true momentum. Although momentum can be ascribed to a sound wave in a continuum, it is not possible in a crystal to apportion its overall momentum uniquely between the lattice and the lattice waves.

A theory of lattice thermal conductivity can be set up in terms of the phonon mean free path, limited by phonon–phonon interactions and by phonon–defect scattering, and using Eq. (4) or its appropriate spectral generalization.

Phonons can also interact with other excitations in a crystal, particularly electrons. If we treat the electron as a wave of wave vector $\mathbf{k}$, where $h\mathbf{k}/2\pi$ is its momentum, it can be scattered from state $\mathbf{k}$ into $\mathbf{k}'$ with the emission of a phonon $\mathbf{q}$, or vice versa. Again there are normal and Umklapp processes, satisfying

$$\mathbf{k} = \mathbf{k}' + \mathbf{q}(\pm\mathbf{b}). \tag{7}$$

These electron–phonon processes limit the electron mean free path; the latter enters into the description of electronic conduction processes. When the interaction involves mainly normal processes and the sum of the momentum of the electron and phonon gas is conserved, electrons and phonons exert a mutual dragging force on each other that becomes evident in the phonon-drag component of the thermoelectric power.

In principle, all mobile excitations can interact with phonons. Neutrons passing through the solid can be scattered by phonons analogously to (7). Their change in energy and momentum can be measured and attributed to $h\nu$ and $h\mathbf{q}/2\pi$ of the phonon. Inelastic or phonon-assisted neutron scattering is thus a powerful tool to determine the phonon dispersion curves, i.e., the dependence of ν on $\mathbf{q}$; at the same time these processes strikingly confirm the phonon concept.

Spin–lattice relaxation processes involve the emission or absorption of individual phonons at special centers in the solid. These can be studied either by monitoring the magnetic state of impurities or defects, or by measuring their effect on phonon flow (ultrasonic attenuation, thermal resistance).

See also ACOUSTOELECTRIC EFFECT; LATTICE DYNAMICS; QUASIPARTICLES.

BIBLIOGRAPHY

R. Berman, *Thermal Conduction in Solids*. Oxford, London and New York, 1976. (I)

J. S. Blakemore, *Solid State Physics*. Saunders, Philadelphia, PA, 1969. (I)

C. Kittel, *Introduction to Solid State Physics*, 4th ed. Wiley, New York, 1971. (E)

P. G. Klemens, "Thermal Conductivity and Lattice Vibrational Modes," in *Solid State Physics*, Vol. 7, pp. 1–98. Academic Press, New York, 1958. (A)

A. A. Maradudin, "Theoretical and Experimental Aspects of the Effects of Point Defects and Disorder on the Vibration of Crystals," in *Solid State Physics*, Vol. 18, pp. 274–420; Vol. 19, pp. 1–134. Academic, New York, 1966. (A)

J. A. Reissland, *The Physics of Phonons*. Wiley, New York, 1973. (I)

J. M. Ziman, *Electrons and Phonons*. Oxford, London and New York, 1960. (A)

Photoconductivity

Clarence R. Crowell

Photoconductivity is bulk electrical conductivity induced by visible light or infrared radiation. Photoconductivity has frequently been confused with photo(electric) emission, which is light-induced injection of carriers across a boundary that may be a solid–solid or solid–vacuum interface, and the photovoltaic effect, which involves the conversion of electromagnetic energy to electrical energy (e.g., a solar cell) and where the motion of light-generated minority carriers toward a solid–solid rectifying junction develops an electrochemical potential difference. The adjective photoelectric applies to all three phenomena [1]. A pure photoconductor should have nonrectifying (Ohmic) contacts that are not illuminated.

Photoconductivity was first reported by Smith [2] for selenium and has since been observed in most solids. In semiconductors and insulators, photoconductivity is described as intrinsic or extrinsic if it is excited, respectively, by photons of energy greater than or less than the energy band gap. Appreciable photoconductivity is, however, always associated with defect- or dopant-induced energy levels within the band gap.

Photoconductive gain G is the number of carriers that transit a structure per absorbed photon; G equals the ratio of the system response time τ_0 to the carrier transit time T_r. T_r varies inversely as the applied voltage until, in the absence of shallow traps or sensitizing centers, it equals the dielectric relaxation time τ_d. T_r determines the bandwidth ΔB of the detector response. When shallow traps are added, G is increased, but the maximum of $G\Delta B$ is invariant and a significant figure of merit. If sensitizing centers, a fraction f of which are charged with the carrier species, are added in numbers greater than the shallow traps, then [3]

$$G\Delta B = (2\pi\tau_d f)^{-1}.$$

Sensitization for visible light may be quenched by infrared radiation. Another figure of merit is the detectivity,

$$D^* \equiv \frac{S/N}{P_D}\left(\frac{\Delta B}{A}\right)^{1/2}$$

where S/N is the signal to available noise power ratio for a radiant power density P_D on a detector area A.[4]

See also ELECTRON ENERGY STATES IN SOLIDS AND LIQUIDS; PHOTOELECTRON SPECTROSCOPY; PHOTOVOLTAIC EFFECT.

REFERENCES

1. *IEEE Standard Dictionary of Electrical and Electronic Terms*. Wiley, New York, 1972.
2. W. Smith, *Nature* **7**, 303 (1873).
3. A. Rose, *Concepts in Photoconductivity and Allied Problems*. Wiley-Interscience, New York, 1963.
4. H. Levinstein, *Semicond. Semimetals* **5**, 3 (1970).

Photoelectron Spectroscopy

Dean E. Eastman

The external photoeffect—in which electrons are ejected from matter which is irradiated—was discovered by Hertz in 1887 and led Einstein, in 1905, to postulate his quantum theory of light. However, it was not until after the mid-1940s that the utility of photoelectron spectroscopy—which is based on the external photoeffect—was recognized as an important technique for studying the electronic structure of all phases of matter [1–3]. Since then, tremendous progress both in understanding fundamental photoemission processes and in technological development has been occurring. Traditionally, several subfields of photoelectron spectroscopy have developed using different radiation sources, spectrometers, etc. These include (i) x-ray photoelectron spectroscopy (XPS or ESCA) using x-ray sources [1], (ii) molecular photoelectron spectroscopy (MPS) using resonance lamps [2], and (iii) ultraviolet photoelectron spectroscopy (UPS), which has been applied to solids and surfaces [3–5]. Currently, with the advent of new sources such as synchrotron radiation, etc., the distinctions between these subfields are vanishing.

For all types of photoelectron spectroscopy, photons of a given energy are used to excite electrons above the vacuum energy level ϕ of a sample, and some of these excited electrons are emitted into vacuum, where their energy distribution is measured with an electron energy analyzer (Fig. 1). For a solid sample, there are several parameters which can be varied, namely, the photon energy $h\nu$, polarization $\hat{\epsilon}_p$, and direction $\hat{\Omega}_i$ of the incident radiation, and the energy E and direction $\hat{\Omega}_f$ of the emitted electron. Thus the photoelectron emission intensity per absorbed photon, N, depends on all these parameters and we can write $N = N(E, h\nu, \hat{\Omega}_f, \hat{\epsilon}_p, \hat{\Omega}_i)$. In practice, for most forms of photoelectron spectroscopy what is measured is the electron energy distribution $N(E, h\nu)$ averaged over a range of emission angles for one or more fixed photon energies. We show the energetics of the photoemission process in the energy level diagram in Fig. 1, where several valence and core electron levels are schematically shown for a nickel surface with an adsorbed layer of oxygen. A photon of energy $h\nu$ is absorbed, exciting one of the electrons from a filled valence or core level, and thereby giving rise to a distribution of excited electrons (dashed lines). A small fraction of these electrons are excited sufficiently close to the surface (typically 5–20 Å)

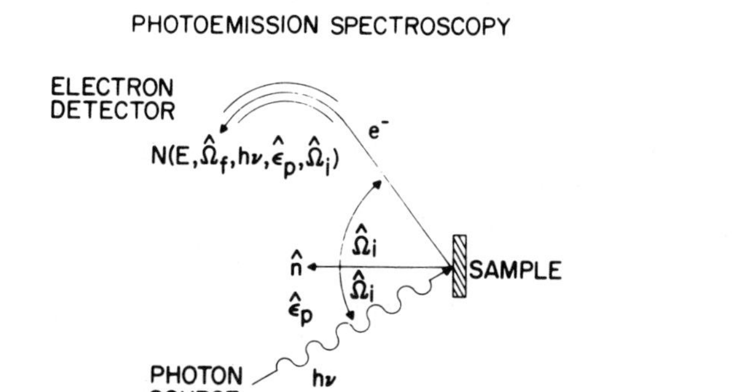

FIG. 1. Schematic experimental photoemission spectroscopy arrangement showing parameters that can be varied (upper), and schematic energy level picture of the photoemission process (lower).

that they can escape the solid without loss of energy. The kinetic energy distribution of these "primary" electrons is measured (cross-hatched in Fig. 1) together with a smooth background of scattered secondary electrons.

Such an energy distribution directly reflects the energy level spectrum of the sample; namely one has the simple relation $E_i = h\nu - IP_i$, where IP_i is the ionization potential energy of the ith ion state of the system and E_i is the corresponding measured kinetic energy. For free atoms and molecules, vibrational and rotational excitations accompany each ion state, and one has the relation $IP_i = I_i + \Delta E_{\text{vib}} + \Delta E_{\text{rot}}$, where I_i is the adiabatic electronic ionization potential and ΔE_{vib} and ΔE_{rot} are vibrational and rotational excitation energies. For XPS and UPS studies of solids, the Fermi energy E_F, rather than the vacuum level, is the customary energy reference level, and one writes $IP_i = E_B^{(i)} + \phi$ where ϕ is the work function (vacuum level minus E_F) and $E_B^{(i)}$ is the electronic binding energy of the ith level. In UPS studies of valence electrons, binding energies are often called initial energies, i.e., $E_i = -E_B^{(i)}$, where E_i is usually used in a one-electron energy-band sense ($E_i < 0$ for occupied valence states.). We next describe several examples of XPS, MPS, and UPS applied to solids and molecules so as to illustrate a select few of the many types of applications of photoelectron spectroscopy.

An XPS spectrum of the CO molecule including both the valence and core electron levels is shown in Fig. 2 [6]. The molecular-orbital configuration of CO is $[(\text{O}1s)^2(\text{C}1s)^2(1\sigma_g 2s)^2(2\sigma_u 2s)^2(1\pi_u 2p)^4(3\sigma_g 2p)^2]$, and six corresponding orbital energy levels are seen in Fig. 2, with binding energies of 542.6, 296, 38, 19.7, 17, and 14 eV, respectively (i.e., vertical ionization potentials measured relative to the vacuum level). These corresponding to removing one electron from each of the six orbital levels. Several additional shake-up lines are seen between 20 and 35 eV that correspond to "shake-up" excitations involving the simultaneous excitation of two or more electrons from the outer three molecular orbitals. The energy resolution (~0.4 eV) for the spectrum in Fig. 2 is representative of the best presently achievable with XPS.

Much higher resolution is easily attained at lower energies using He and Ne resonance lamps, as is seen in Fig. 3 for

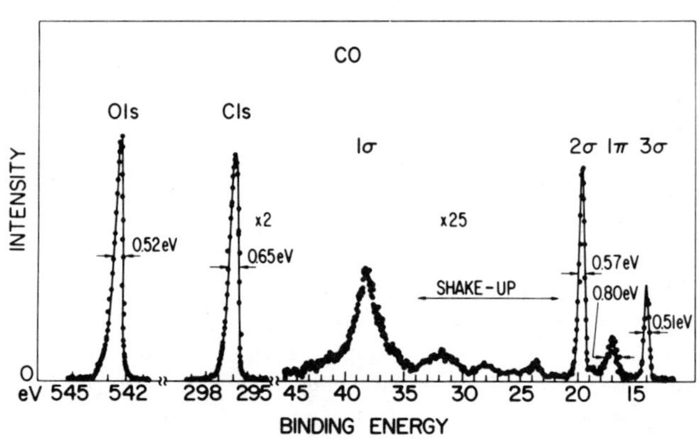

FIG. 2. X-ray photoelectron spectrum for molecular CO. A monochromatized Al $K\alpha$ x-ray source was used. (Reprinted with permission from Ref. 6.)

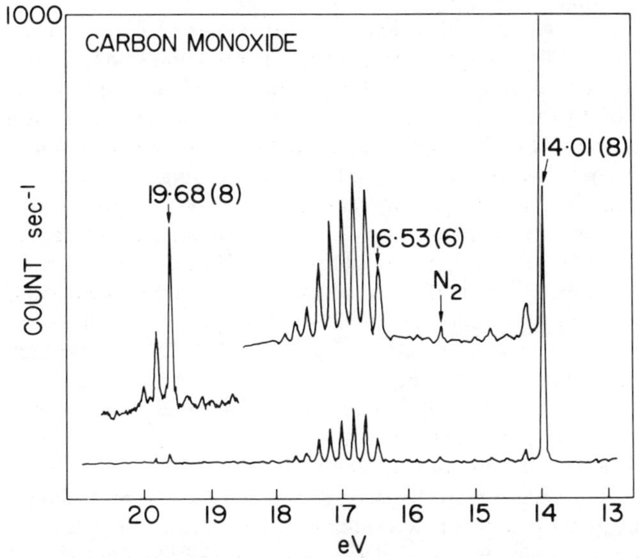

FIG. 3. Photoelectron spectrum for the outer three molecular orbitals of CO using a He I (21.2-eV) resonance lamp. (Reprinted with permission from Ref. 2.)

the MPS spectrum for the outer three valence orbitals $[(2\sigma_u 2s)^2(1\pi_u 2p)^4(3\sigma_g 2p)^2]$ of CO [2]. Here the adiabatic ionization potentials are indicated, and numerous vibrational bands are seen on the high-binding-energy side of each ionization potential. The $(3\sigma_g 2p)^2$ level ($IP = 14.01$ eV) exhibits weak vibrational structure while the $(1\pi_u 2p)^4$ level exhibits strong vibrational structure; this occurs because the 3σ orbital is a nonbonding orbital while the 1π orbitals are bonding orbitals and thus couple electronic and vibrational excitations. In general, the valence molecular-orbital ionization potentials are sensitive to the electronic wave functions, and thus yield direct information concerning the electronic structure which determines the chemical and physical nature of the molecule.

An important general application of x-ray photoelectron spectroscopy is the study of chemical shifts by measuring shifts in binding energies for a core level of an atom in different chemical environments. Indeed, K. Siegbahn and his collaborators, who first developed x-ray photoelectron spectroscopy, coined the widely used acronym ESCA, which is an abbreviation for electron spectroscopy for chemical analysis [1]. An ESCA example showing chemical shifts in the carbon 1*s* levels in ethyl trifluoroacetate ($CF_3COOCH_2CH_3$) is shown in Fig. 4 [6]. All four carbon atoms have quite distinct 1*s* binding energies, with the C1*s* level of the CF_3 fluoro group being shifted by about 8 eV compared with the CH_3 methyl group. This occurs because of the much larger charge transfer onto fluorine relative to hydrogen, and reflects the much larger electronegativity of fluorine. The measurement of such chemical shifts in molecules, solids, and surfaces gives direct information concerning the oxidation states, charge transfer for various chemical bonds, etc. [1].

For solids, valence electrons interact and form energy bands. An important area of application for XPS and UPS is the study of the electronic structure of such valence bands, which determine the physical and chemical properties of

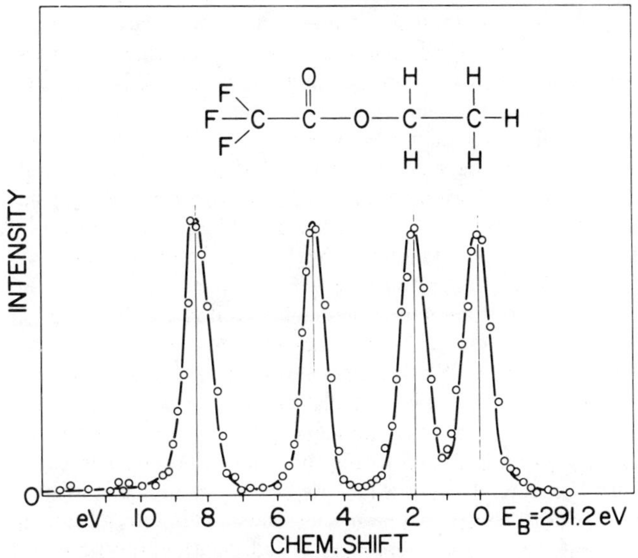

FIG. 4. X-ray photoelectron spectrum (ESCA) for ethyl trifluoroacetate showing the chemical shifts of the C1*s* electron levels for carbon atoms having different chemical environments. (Reprinted with permission from Ref. 6.)

solids. XPS and UPS spectra for Au are shown in Fig. 5, where they are compared with a theoretical density of states [5]. The main features of the *d* bands that extend from −2 to −8 eV below the Fermi energy E_F are resolved in the two photoemission spectra, and are replicated in the calculated density of states. The low emission intensity extending from E_F to −2 eV corresponds to the *s* band of Au. By comparing with photoemission data such as in Fig. 5, accurate theoretical energy-band models can be determined, which can then be used to calculate many physical and chemical properties of interest.

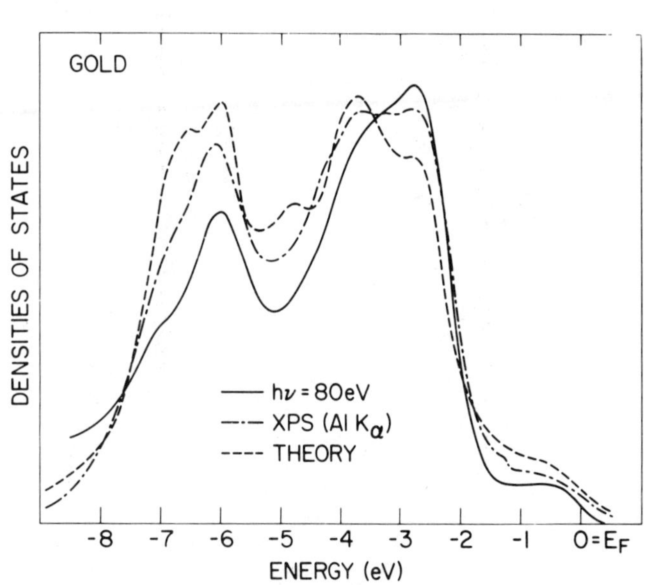

FIG. 5. X-ray and ultraviolet ($h\nu = 80$ eV) photoelectron spectra for the valence bands of Au together with a calculated density of states [5].

In ultraviolet photoelectron spectroscopy, if one uses photon energies less than about 40–60 eV, spectra become more complicated than shown in Fig. 5. At such energies the measured spectrum resembles a quantity usually called the energy distribution of the joint density of states, rather than the total density of states such as shown in Fig. 5. This behavior occurs because the electron states of the solid have a well-defined crystal momentum, and both energy and crystal momentum are conserved in the optical excitation process. In this case, since the photon has negligible momentum (i.e., a wavelength long compared with atomic dimensions), the only electron–hole pairs that can be excited are those which have the same crystal momentum and conserve energy. Consequently, the UPS photoemission spectra change with photon energy and contain both electron energy and momentum information. A set of $h\nu$-dependent spectra for polycrystalline copper are shown in Fig. 6, where they are compared with theoretical spectra [5]. As shown in Fig. 6, the photoemission spectra exhibit d-band emission features between -2 and -5 eV which change markedly with photon energy. These spectral features, which are due to direct interband transitions that conserve energy and crystal momentum, are seen to be described quite well by the set of theoretical spectra, which were calculated using an optimized Korringa–Kohn–Rostoker band model.

Even more detailed features than shown in Fig. 6 can be obtained by studying angle-resolved photoemission from single-crystal samples. In this case, the crystal momentum parallel to the surface ($\mathbf{k}_{\parallel}$) is conserved in the photoemission process [the final-state wave function is identical to a low-energy electron-diffraction (LEED) wave function] and thus both the energy and $\mathbf{k}_{\parallel}$ momentum of various electron states can be directly determined. The polarization $\hat{\epsilon}_p$ and angle of incidence $\hat{\Omega}_i$ of the radiation are also important parameters, especially in the case of angle-resolved photoemission, and can yield additional information concerning the symmetry of various electron states. Thus for angle-resolved photoemission, the emission intensity $N = N(E, h\nu, \hat{\Omega}_f, \hat{\epsilon}_p, \hat{\Omega}_i)$ depends on several independent parameters, all of which can be varied to yield information concerning the electronic structure of bulk and surface states.

Photoelectron spectroscopy, especially UPS, has become a powerful tool for determining the electronic structure of surfaces, e.g., intrinsic surface states on semiconductors and metals and sorbed atoms and molecules on various surfaces. The sensitivity to observe surface states and monolayer or submonolayer quantities of adsorbates results from the very short (typically ~ 5–15 Å) escape depths of the photoejected electrons that are measured. An application to chemisorption is depicted in Fig. 7, where photoelectron spectra for

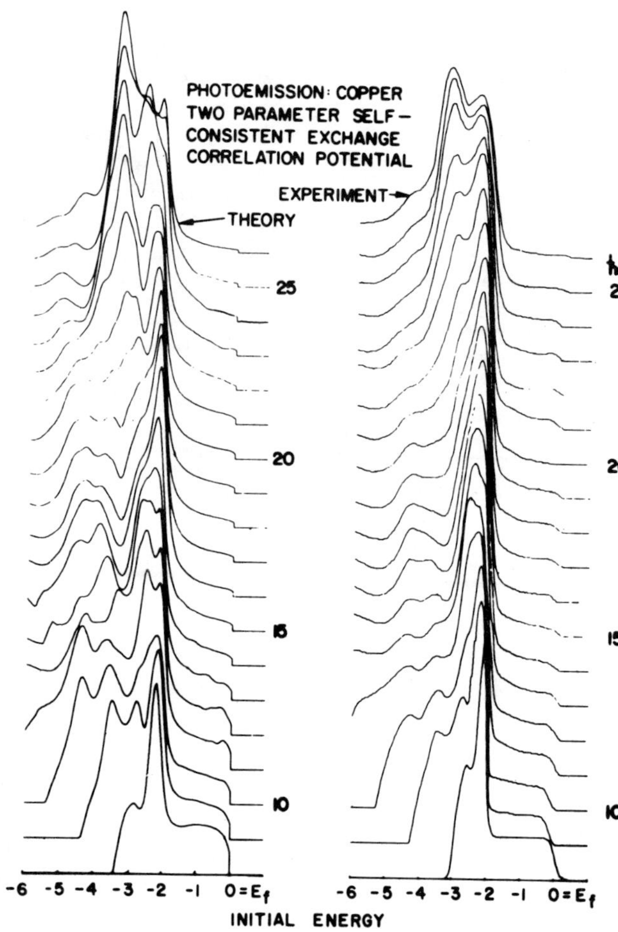

FIG. 6. A set of ultraviolet photoelectron spectra for copper with $8 \leq h\nu \leq 26$ eV together with calculated spectra [5]. Strong $h\nu$-dependent trends due to direct interband transitions are seen (see text).

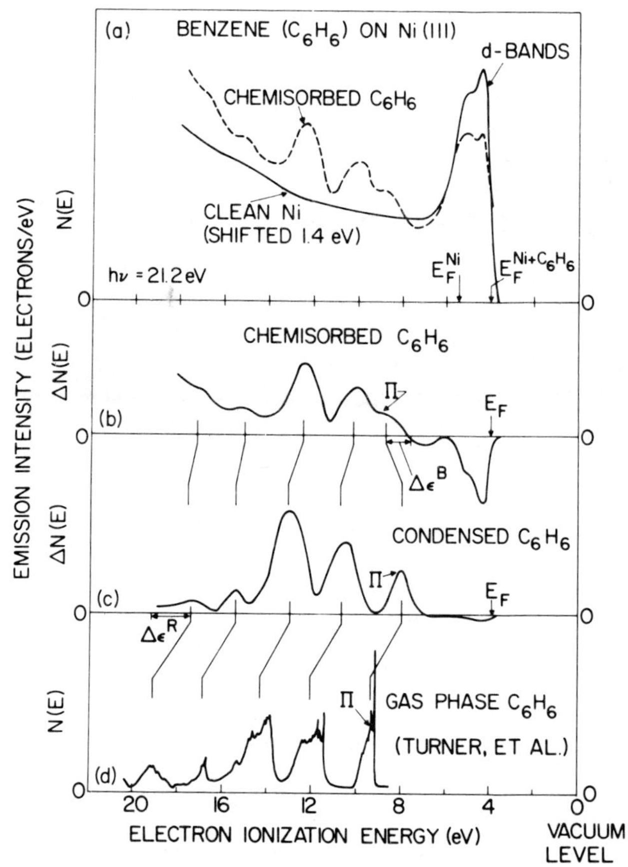

FIG. 7. Photoelectron spectra ($h\nu = 21.2$ eV) for clean Ni(111), chemisorbed benzene on Ni(111), condensed benzene, and gas-phase benzene [7].

clean single-crystal Ni(111) and for Ni(111) with a layer of chemisorbed benzene (C_6H_6) are shown, together with spectra for several layers of condensed benzene and for gas-phase benzene [7].

Figures 7(b) and 7(c) show the difference spectra for chemisorbed C_6H_6 + Ni(111) minus clean Ni(111) and condensed C_6H_6 + Ni(111) minus clean Ni(111), respectively. For both spectra, the five lower-lying peaks are associated with benzene molecular-orbital energy levels. Several messages concerning the application of photoelectron spectroscopy to chemisorption are illustrated in Fig. 7, namely, the use of photoemission as a "fingerprint" of the adsorbed molecular species, the large changes in ionization energies of a nonbonding nature (relaxation shifts) that occur for adsorbed species, and the importance of π-d bonding in hydrocarbon chemisorption. Concerning the "fingerprinting" aspect, the similarity of the spectrum for chemisorbed benzene to those for condensed and gas-phase benzene—in particular the lowest four energy levels which involved nonbonding σ orbitals—implies that the adsorbed benzene molecule remains structurally intact. At elevated temperatures (above ~470 K) benzene is observed to dehydrogenate and leave a carbonaceous surface residue. Concerning relaxation shifts, all the nonbonding lower-lying σ orbitals for chemisorbed benzene have ionization energies relative to the vacuum level which are reduced from those of molecular benzene by about 2 eV. This relaxation shift is due to increased screening of the excited hole state of the molecule by electrons from the metal substrate. Concerning the observation of π-d bonding, if one compares the spectrum in Fig. 7 for chemisorbed benzene with that of condensed benzene, one observes that while the lowest-lying four σ orbitals have decreased ionization energies (i.e., increased screening), the ionization energy of the uppermost doubly degenerate π orbital is increased. This shift signifies π-d bonding, which appears to be generally involved in the strong chemisorption bonds for unsaturated hydrocarbons on transition metals.

In summary, photoelectron spectroscopy has developed over the past couple of decades into a powerful tool for probing the electronic structure of gases, liquids, solids, and their interfaces. In this brief article, only a few of the many applications and aspects of photoelectron spectroscopy [1,2, 4,5,8] have been touched upon. The rapid development and increase in the range of applications of photoelectron spectroscopy is continuing at present, especially with the recent advent of new and improved intense continuum synchrotron radiation sources.

See also Adsorption; Electron Energy States in Solids and Liquids; Surfaces and Interfaces.

REFERENCES

1. K. Siegbahn *et al.*, *ESCA Atomic, Molecular, and Solid State Structure Studied by Means of Electron Spectroscopy*. Almqvist and Wiksells Boktryckeri Ab, Stockholm, 1967.
2. D. W. Turner *et al.*, *Molecular Photoelectron Spectroscopy*. Wiley-Interscience, London, 1970.
3. L. Apker *et al.*, *Phys. Rev.* **74,** 1462 (1948).
4. D. E. Eastman, in *Techniques of Metals Research VI*, pp. 413–479 (E. Passaglia, ed.). Interscience, New York, 1972.
5. D. E. Eastman, in *Proceedings of the IV International Conference on Vacuum Ultraviolet Radiation Physics*, Hamburg, 1974, pp. 417–449 (E. E. Koch *et al.*, eds.). Pergamon-Vieweg, 1975.
6. U. Gelius *et al.*, *A High Resolution ESCA Instrument with X-ray Monochromator for Gases and Solids*. Uppsala University Report UUIP-817, 1973.
7. J. E. Demuth and D. E. Eastman, *Phys. Rev. Let.* **32,** 1123 (1974).
8. Several additional general references on photoelectron spectroscopy are (a) E. W. Plummer, "Photoemission and Field Emission Spectroscopy," in *Interactions on Metal Surfaces* (R. Gomer, ed.). Springer-Verlag New York, 1975; (b) Thomas A. Carlson, *Photoelectron and Auger Spectroscopy*. Plenum Press, New York, 1975.

Photoionization

David L. Ederer

Photoionization is a process where a photon of frequency ν and energy $h\nu$ interacts with an atom or molecule to produce an ion and one or more electrons. The quantity h is Planck's constant and has a magnitude of 6.626×10^{-34} J s. The interaction can be represented by the equation $h\nu+M\rightarrow M^{m+}+me$, where $m\geq1$. Since ionization energies (i.e., the minimum energy necessary to remove an outer electron from the system) lie in the range 8×10^{-19}–4×10^{-18} J, the corresponding minimum frequency of the photon to produce ionization is 1.3×10^{15}–6×10^{15} s^{-1}. This frequency corresponds to wavelengths* between 2000 and 500 Å (1Å = 10^{-10} m). Not all photon absorption processes produce ionization. In molecules, for example, *dissociation* occurs with high probability at low photon energies. In this process neutral or excited fragments are produced. As the photon energy increases above threshold it eventually exceeds the binding energy of the electrons in inner shells of the atom or molecule. (These shells are labeled K, L, M, etc., counting from the most tightly bound electrons.) Thus a very energetic photon, e.g., an x ray, can remove any electron whose binding energy is less than the photon energy.

The probability that an energetic photon will produce ionization is measured by the photoionization cross section σ, which has the dimension of an area. The number of ions N_I produced by N photons in a thin layer of atoms or molecules x meters thick is given by

$$N_I = Nn\sigma x, \tag{1}$$

where n is the number density of atoms or molecules. Thus the number of ions produced in a gas column of length L is obtained by integrating Eq. (1) to obtain

$$N_I = N[1-\exp(-\sigma nL)]. \tag{2}$$

From Eq. (2) the maximum number of ions that can be obtained by the primary photoionization process is just equal to the photon number N. However, in the case of molecules some of the photons may produce *dissociation*, reducing proportionally the number of ions produced. The magnitude of the photoionization cross section σ is a function of the photon energy and at the ionization threshold typically is of the order of the cross-sectional area of the atom or about

* The photon wavelength is related to the frequency by the formula $\lambda=c/\nu$, where c is the velocity of light.

10^{-21} m^2. To illustrate the functional dependence of the photoionization cross section on the photon energy $h\nu$, the photoionization cross sections [1] of two atoms, helium and xenon, are shown in Fig. 1. Note that in the case of helium the cross section decreases monotonically, while in xenon there are abrupt changes in the magnitude of the cross section at photon energies corresponding to the binding energies of the L- and M-shell electrons. These discontinuities are called *x-ray absorption edges*. At photon energies corresponding to the binding energy of some of the N-shell electrons there is no abrupt change in the cross section; rather it rises to a prominent maximum as the photon energy increases. At other photon energies there are minima and secondary maxima as well as narrow features (width<0.1eV) due to a photoionization process called *autionization*. The detailed study of these features in xenon and other atoms and molecules yields insight about the fundamental properties [2] of the electron cloud surrounding the nuclei.

The total photoionization cross section described in the preceding paragraphs includes the direct production of a multiply charged ion and several electrons. However, this cross section is typically one or two orders of magnitude smaller than the cross section for single ionization. In the x-ray region *Compton scattering* has an appreciable cross section. This process produces ionization but the photon is inelastically scattered rather than absorbed. Other related topics that may be of interest to the reader are: *predissociation, preionization,* the *Auger effect, shake-up,* and *shake-off.*

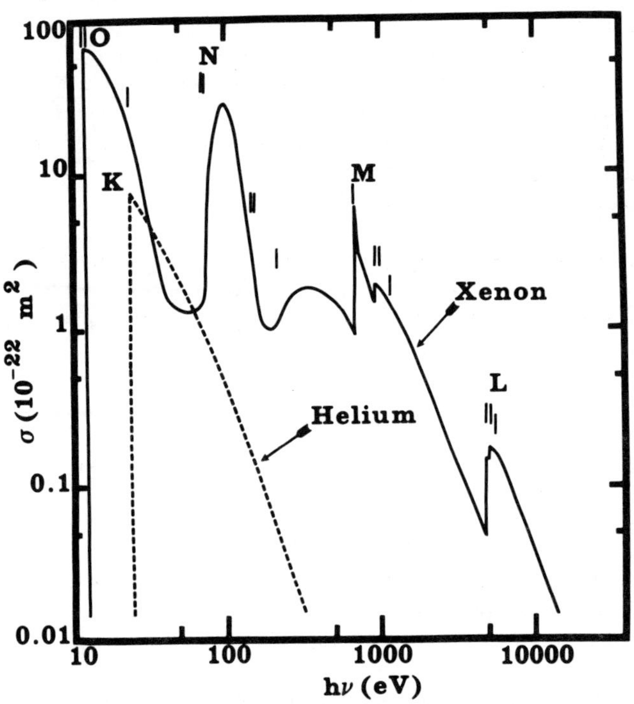

FIG. 1. The photoionization cross section of helium and xenon. The photon energy is expressed in units of electron volts (1 eV = 1.602×10^{-19} J). The binding energy of the subshell electrons (K, L, M, etc.) is denoted by the tic marks. The K binding energy of xenon (not shown) is about 34 keV and the maximum cross section at the K absorption edge is 7×10^{-25} m^2.

See also AUGER EFFECT.

REFERENCES

1. Photoabsorption cross section compilations can be found in the following articles: R. D. Hudson and L. J. Kieffer, *Atomic Data* **2,** 205 (1971); J. H. Hubbell, *Atomic Data* **3,** 241 (1971).
2. Geoffrey V. Marr, *Photoionization Processes in Gases.* Academic Press, New York, 1967.

Photovoltaic Effect

John J. Hall

The *photovoltaic effect* is the voltage produced between two dissimilar materials when their common junction is illuminated with radiation. It was first reported in 1839 by Becquerel, who observed that a potential difference was produced between two identical electrodes immersed in an electrolyte when but one of them was illuminated. Today, interest centers on the photovoltaic effect produced in totally solid systems, for which several important applications are found; among these are the photocell, which utilizes the photovoltage produced across a metal–semiconductor junction as a measure of the intensity of the incident light, and the solar cell, which uses the photovoltage produced by an irradiated *p-n* semiconductor junction as a source of power.

When light or other radiation is absorbed in matter, the energy of the absorbed photons or particles may be transferred to the electron system of the material, resulting in the creation of charge carriers. These may be electron–ion pairs in an electrolyte, for example, or electron–hole pairs in a semiconductor. If the carriers are created in a region of varying electrical potential, they will be accelerated by the resultant electric field and a current, the photocurrent, will flow. In photovoltaic systems, the electric field occurs at the interface of two dissimilar materials. The origin of this field is the difference in the chemical potential, or Fermi level, of the electrons in the two isolated materials. When they are joined, the system approaches a new thermodynamic equilibrium state which may only be achieved when the Fermi level is equal in the two materials. This occurs by the flow of electrons from one material to the other until a voltage difference is established between the two whose potential is just equal to the initial difference of the Fermi levels. It is this "contact potential" at the material interface which is the source of the electric field driving the photocurrent.

The photovoltaic effects produced at metal–semiconductor or semiconductor *p-n* junctions provide important and instructive examples. If a voltage V is applied across such a junction, in the absence of illumination, the resultant current may be expressed by the diode equation:

$$I = I_s[\exp(eV/kT) - 1],$$

where I_s is the reverse saturation current, and kT/e is the absolute temperature expressed in volts (300 K = 0.026 V). If the junction region is now illuminated to produce a constant photocurrent I_l, the total current through the junction becomes

$$I = I_s[\exp(eV/kT) - 1] - I_l.$$

The maximum photovoltage due to I_l is produced when the circuit is open and the total junction current is zero, under which condition we may easily solve for the open circuit photovoltage

$$V_{oc}=(kT/e)\ln(1+I_l/I_s).$$

In real devices, the photocurrent is often several orders of magnitude greater than the reverse saturation current, so that photovoltages many times kT/e may be obtained. However, under conditions of constant illumination, I_l/I_s is a sufficiently strong function of the temperature that photocells and solar cells ordinarily show negative temperature coefficients of the photovoltage.

See also ELECTRON ENERGY STATES IN SOLIDS AND LIQUIDS; PHOTOCONDUCTIVITY; PHOTOELECTRON SPECTROSCOPY; SEMICONDUCTORS.

BIBLIOGRAPHY

M. Altman, *Elements of Solid-State Energy Conversion*, Chap. 4. Van Nostrand-Reinhold, New York, 1969. (E)

H. J. Hovel, "Solar Cells," in *Semiconductors and Semimetals*, Vol. 11, R. K. Willardson and A. C. Beer (eds.). Academic Press, New York, 1975. (A)

S. M. Sze, *Physics of Semiconductor Devices*, Chap. 12. Wiley-Interscience, New York, 1969. (I)

Piezoelectric Effect

Donald F. Nelson

The *piezoelectric effect* is a linear, reversible electromechanical interaction occurring in materials possessing certain symmetry properties. The *direct piezoelectric effect* is the production of an electric polarization by a strain; the *converse piezoelectric effect* is the production of a stress by an electric field. The direct effect is linear in the strain; the converse effect is linear in the electric field. Twenty of the thirty-two crystal classes are piezoelectric. All 20 piezoelectric classes lack a center of symmetry. The one remaining noncentrosymmetric crystal class, denoted O (Schönflies) or 432 (Hermann–Mauguin), possesses other symmetries which combine to exclude piezoelectricity. Important piezoelectric crystals include quartz, Rochelle salt, $(NH_4)H_2PO_4$ (ADP), $LiTaO_3$, and $LiNbO_3$. Other materials, e.g., electrically poled ferroelectric ceramics composed of $PbZrO_3$ and $PbTiO_3$, that possess less structural regularity than crystals may exhibit piezoelectricity.

HISTORY

The direct effect was discovered by Pierre and Jacques Curie in 1880. On the basis of thermodynamic reasoning G. Lippmann predicted the existence of the converse effect and that its interaction strength would be identical to that of the direct effect. This prediction was later verified by the Curie brothers.

The first practical use of piezoelectricity was made during World War I when P. Langevin produced ultrasonic waves in water surrounding a quartz plate (now called an *electromechanical transducer*) forced to vibrate by the application of an alternating voltage. Shortly after the war W. G. Cady invented the *piezoelectric resonator* which exploits the interaction of the piezoelectrically driven mechanical resonance of a crystal upon the electrical driving circuit. One application of this is in the frequency control of an electrical oscillator, another is in making efficient band-pass electrical filters.

MATHEMATICAL CHARACTERIZATION

The direct piezoelectric effect produces a contribution $e_{ijk}s_{jk}$ to the linear polarization P_i,

$$P_i=\epsilon_0\chi_{ij}E_j+e_{ijk}s_{jk}, \qquad (1)$$

where e_{ijk} is the *piezoelectric stress tensor*, s_{jk} is the strain tensor, χ_{ij} is the linear electric susceptibility tensor, E_j is the electric field, and ϵ_0 is the permittivity of free space. Summation over repeated tensor subscripts is implied. The converse effect produces a contribution $-e_{kij}E_k$ to the linear stress t_{ij},

$$t_{ij}=c_{ijkl}s_{kl}-e_{kij}E_k, \qquad (2)$$

where c_{ijkl} is the stiffness tensor. The latter equation may be solved for the strain s_{mn},

$$s_{mn}=s_{mnij}t_{ij}+d_{kmn}E_k, \qquad (3)$$

where $d_{kmn}\equiv s_{mnij}e_{kij}$ is the *piezoelectric strain tensor* and s_{mnij} is the compliance tensor.

Both piezoelectric tensors possess interchange symmetry on their last pair of indices,

$$e_{kij}=e_{kji}, \quad d_{kij}=d_{kji}. \qquad (4)$$

This allows them to possess at most 18 independent components and to be recorded as 3×6 matrices. To do the latter, the last two tensor indices are contracted to one index having values 1, 2, 3, 4, 5, 6 corresponding to 11, 22, 33, 23 or 32, 13 or 31, 12 or 21, respectively. To make matrix multiplication equivalent to the scalar product of tensors, the value of d_{kl} is one-half of d_{kij} when $l=4, 5, 6$; no such factors are needed for e_{kl}.

SECONDARY PIEZOELECTRIC EFFECTS

A succession of the direct and converse piezoelectric effects produces a piezoelectric stiffening of the stiffness tensor. For an acoustic wave propagating in the direction **a** the effective stiffness tensor becomes

$$c_{ijkl}^{\text{eff}}=c_{ijkl}+\frac{a_m e_{mij}a_n e_{nkl}}{\epsilon_0 a_p \kappa_{pq} a_q}, \qquad (5)$$

where κ_{pq} is the dielectric tensor. Piezoelectricity also alters the dielectric tensor to be

$$\kappa_{pq}^{\text{low}}=\kappa_{pq}+(1/\epsilon_0)e_{pmn}d_{qmn} \qquad (6)$$

for frequencies well below the mechanical resonances of the crystal. Piezoelectricity also leads to indirect contributions to the electro-optic effect, the elasto-optic effect, etc.

See also CRYSTAL SYMMETRY; ELASTICITY.

BIBLIOGRAPHY

W. G. Cady, *Piezoelectricity*. Dover, New York, 1964 (two volumes). (I) This has been the standard reference source since its first publication in 1946.

D. F. Nelson, *Electric, Optic, and Acoustic Interactions in Dielectrics*. Wiley, New York, 1979. (A) Chapters 10, 11, and 17 show that Eqs. (1) and (2) hold for nonpyroelectric piezoelectrics when components are referred to either the spatial (laboratory) or material coordinate systems; for pyroelectrics they are true only in the material coordinate system and only when the spontaneous electric field has been cancelled by extrinsic surface charge.

J. F. Nye, *Physical Properties of Crystals*, Chap. VII. Clarendon, Oxford, 1957. (E)

Plasmons

David B. Chang and *Betsy Ancker-Johnson*

A plasmon is a quantum of the charge-density oscillations that occur in a plasma, i.e., in an electrically neutral collection of charges in which some of the charges are free to move in response to their Coulomb interactions. Plasmons involving charges in solids occur in metals, semimetals, and semiconductors, and they have been studied in large atoms and molecules. The plasmons can be classified according to the types of charges involved (electrons or holes, valence or conduction); the number of mobile charged components; the degree of departure from thermal equilibrium; the degree of degeneracy; and the dimensional constraints on the collective oscillations.

Some plasma oscillations in solids involve all the valence-band electrons. The corresponding plasmon energy is of the order of 10 eV for most solids and is given approximately by $E=\hbar(4\pi n_v e^2/m)^{1/2}$, where $\hbar$ is Planck's constant divided by 2π, n_v is the density of valence-band electrons, e is the electronic charge, and m is the free-electron mass. The expression is valid when E is larger than valence-band–unoccupied-band energy differences, e.g., for Al, Mg, Be, Si, and the alkali metals. For semiconductors and doped semimetals, in addition to these high-energy plasmons, there are low-energy plasmons associated with conduction-band electrons. These plasmons have energies ranging from hundredths to several tenths of an electron volt, depending on the temperature and the amount of doping. In metals, the plasma oscillation frequencies are much higher than collision frequencies, while in semiconductors and doped semimetals, the collision and conduction-band plasma oscillation frequencies can be comparable.

The high-energy plasmons have been observed in metals, semimetals, and semiconductors through the energy-loss spectra of high-energy (typically 1–10's of keV) electrons scattered by the plasmons and through the inelastic scattering of high-frequency electromagnetic radiation (for instance, Cu $K\alpha$ x rays). In heavy atoms, photoabsorption cross sections in the UV to soft-x-ray region can be reasonably well explained in terms of the excitation of high-energy atomic plasmons. The low-energy plasmons in semiconductors and doped semimetals have been extensively studied by Raman scattering [typically of photons from neodymium-doped yttrium aluminum garnet (1.2 eV) and CO_2 (0.12 eV) lasers]. These studies give information on the plasma densities and plasmon dispersion properties and lifetimes resulting, e.g., from Landau damping, collisions, and interactions of the plasmons with each other and with acoustical and optical phonons. In addition, radiation from plasmons has been detected from surfaces, and plasmons have been indirectly observed through their effects on electron tunneling and on photoemission spectra.

A relatively incompressible plasma occurs when only one charge component is free to move, with the other charges tied to lattice positions: This one-component plasma occurs in simple metals, in extrinsic semiconductors, and in ionic crystals at temperatures where only one ion species is mobile. A relatively compressible plasma occurs when charge carriers of both signs are free to move: These multicomponent plasmas occur in pure semimetals, in intrinsic semiconductors, and in semiconductors and insulators into which equal densities of oppositely charged carriers have been injected or created by impact ionization. When a magnetic field B is applied to the plasma, the types of electromagnetic waves which the plasma can support at frequencies lower than the cyclotron frequency $\omega_c=eB/mc$—with c denoting the speed of light—depends markedly on whether the plasma is compressible or incompressible. In the former case, Alfvén waves are replaced by helicon waves with frequency inversely proportional to the square of the wavelength.

In semiconductors and insulators, the energy required to produce one-component plasmas is of the order of 10^{-3}–10^{-1} eV, and that required for the production of multicomponent plasmas is of the order of 0.1–5 eV. In metals and semimetals, free carriers exist at $T=0$ K. Consequently, solid-state plasmas in thermal equilibrium are attainable at room temperatures (0.025 eV) and below. Depending on the material and temperature, these equilibrium plasmas can be either degenerate (e.g., in metals, in semimetals, and in some heavily doped semiconductors) or Maxwellian (e.g., in some semiconductors). Nonequilibrium solid-state plasmas can be produced by a variety of means: by electron–hole injection, by electric-field-induced impact ionization (e.g., in InSb at 2×10^4 V/m), and by single- or multiphoton processes (e.g., plasma in PbTe produced by CO_2 laser irradiation).

In a Maxwellian plasma, the electric potential from a charge is effectively screened out at distances greater than the Debye length $\lambda_D=(\epsilon kT/4\pi ne^2)^{1/2}$, where ϵ is the dielectric constant, k is Boltzmann's constant, T is the absolute temperature, n is the electron density, and e is the electronic charge. In a degenerate plasma, the electric potential from the charge is effectively screened out at distances greater than a Fermi–Thomas length $\lambda_{FT}=(\epsilon E_F/6\pi ne^2)^{1/2}$, where E_F is the Fermi energy, although a small oscillatory tail dropping off as $1/r^3$ does exist at large distances $r\gg\lambda_{FT}$. These screening lengths in solid-state plasma range from the order of an angstrom to a micron. Consequently, solid-state plasma properties can often be calculated by assuming the plasma to be infinite in extent. However, some interesting effects occur when the boundedness of the plasma is taken into account. For instance, at the surface between a plasma and material of dielectric constant ϵ_0 a surface plasmon of energy $E_s=h\omega_p/(1+\epsilon_0)^{1/2}$ exists, which is confined to within a screening length of the surface. Surface plasmon detection in the energy-loss spectra of fast electrons passing through thin solid films, in optical reflection, in LEED and Auger electron spectroscopy, and in tunneling of electrons through thin insulating layers, has proved useful for surface diagnostics.

Plasmons are responsible for a variety of physical effects. In metals, for instance, the electron plasma determines not only the electrical transport properties of the crystal but also the cohesive energy, crystal structure, and phonon frequencies. The Van der Waals force between two blocks of metal at small distances of separation L varies as L^{-3}, and is due to surface plasmons. The Van der Waals force between two conducting chains varies as $L^{-3}(lnL)^{-3/2}$, having a much longer range than the the L^{-6} force between nonconducting chains. This long range is a consequence of the spatial dispersion of the conduction-electron plasmon modes in a chain. Pi electrons in cyclic polyene molecules have similar sound-wave-like plasmon dispersion relations, so these carcinogen-like molecules may also exhibit very long-range Van der Waals interactions.

See also ELECTRON ENERGY STATES IN SOLIDS AND LIQUIDS; QUASIPARTICLES.

BIBLIOGRAPHY

J. A. Ball, J. A. Wheeler, and E. L. Fireman, "Photoabsorption and Charge Oscillation of the Thomas-Fermi Atom," *Rev. Mod. Phys.* **45,** 333–352 (1973).

David B. Chang, Robert L. Cooper, James E. Drummond, and Allan C. Young, "Calculation of Van der Waals Forces in Systems with Nonlocal Conductivity," *J. Chem. Phys.* **59,** 1232–1242 (1973).

Maurice Glicksman, "Plasmas in Solids," in *Solid State Physics* (H. Ehrenreich, F. Seitz, and D. Turnbull, eds.), Vol. 26. Academic Press, New York, 1971, pp. 275–427.

P. M. Platzman and P. A. Wolff, "Waves and Interactions in Solid State Plasmas," in *Solid State Physcis*, (H. Ehrenreich, F. Seitz, and D. Turnbull, eds.) Suppl. 13. Academic Press, New York, 1973.

Polarizability

Charles W. Drake

The polarizability is a microscopic electric property of dielectric media which relates the macroscopically observable electric susceptibility to the properties of the molecules or atoms comprising the media. In macroscopic terms, we may consider an electrically neutral dielectric to have electric dipole density **P** induced in the medium when an external electric field is applied (because of the negative and positive charge separation). As a function of the electric field **E** a component of **P** can be expanded in general as

$$P_i = \sum_j a_{ij}E_j + \sum_{j,k} b_{ijk}E_jE_k + \cdots.$$

It is found in practice that at normal temperatures and electric field strengths only the leading term is necessary. For isotropic substances, **P** is parallel to **E** or $a_{ij} = \chi_e \delta_{ij}$, where $\delta_{ij} = 1$ or 0 as $i=j$ or $i \neq j$. χ_e is the electric susceptibility of the dielectric and is a constant with respect to geometric properties. Therefore

$$\mathbf{P} = \chi_e \mathbf{E}. \tag{1}$$

For anisotropic substances, such as many crystalline solids, there are at most six independent a_{ij}'s. We will confine ourselves to the isotropic case. If the medium is considered to be made up of N molecules per unit volume of a single kind, we may write $\mathbf{P} = N\langle \mathbf{p} \rangle_m$, where $\langle \mathbf{p} \rangle_m$ is the average effective dipole moment of a molecule. The polarizability α, or more specifically the molecular polarizability, can be defined as the dependence of the electric field on the molecular dipole moment or

$$\langle \mathbf{p} \rangle_m = \alpha \mathbf{E}_l. \tag{2}$$

$\mathbf{E}_l$ is the electric field at the position of the molecule. This local field will not in general be identical to the macroscopic field **E** which appears in Eq. (2). The local field $\mathbf{E}_l$ may be set equal to **E** for gases at low pressures, with the criterion that the distance between molecules be large compared to molecular dimensions. This criterion holds for gases near normal pressure and temperature. In more dense media the method due to Lorentz (Ref. 1, pp. 94–96; Ref. 3, sec. 4.5; Ref. 4, pp. 47–49) is usually employed; this approximation replaces $\mathbf{E}_l$ by $\mathbf{E} + 4\pi/3\mathbf{P}$ so that we may relate χ_e and α by Eqs. (1) and (2):

$$\mathbf{P} = N\alpha E_l = N\alpha(\mathbf{E} + \tfrac{4}{3}\pi\mathbf{P}) = \chi_e \mathbf{E},$$

giving

$$\chi_e = \frac{N\alpha}{1 - \frac{4}{3}\pi N\alpha}. \tag{3}$$

The susceptibility χ_e may be related in turn to the dielectric permittivity ϵ by relations among the electric displacement **D**, the polarization **P**, and the electric field **E**, i.e., $\mathbf{D} = \epsilon\mathbf{E} = \mathbf{E} + 4\pi\mathbf{P} = \mathbf{E} + 4\pi\chi_e\mathbf{E}$ so that $\epsilon = 1 + 4\pi\chi_e$. This may be combined with Eq. (3) and solved for α to give

$$\alpha = \frac{3}{4\pi N}\left(\frac{\epsilon - 1}{\epsilon + 2}\right), \tag{4}$$

which is known as the Clausius–Mosotti equation. For substances where $\mathbf{E}_l \approx \mathbf{E}$ (or $\epsilon \approx 1$) the equation simplifies to $\alpha = (\epsilon - 1)/4\pi N$. The relationship provides a method of measuring α, valid for gases, from the macroscopic measurable quantity ϵ.

We have neglected any time dependence in these electrical quantities. We should however write for $\mathbf{D}(\mathbf{r},t)$:

$$\mathbf{D}(\mathbf{r},t) = \mathbf{E}(\mathbf{r},t) + \int_0^\infty G(\tau)\mathbf{E}(\mathbf{r},t-\tau)d\tau. \tag{5}$$

This shows that **D** depends not only on the present value of **E** but also on all values of **E** at prior times. We may use a Fourier transformation on Eq. (5) and get

$$\mathbf{D}(\mathbf{r},\omega) = \mathbf{E}(\mathbf{r},\omega) + 4\pi\chi_e(\omega)\mathbf{E}(\mathbf{r},\omega),$$

where

$$4\pi\chi_e = \frac{1}{\sqrt{2\pi}}\int_{-\infty}^{\infty} G(\tau)e^{i\omega\tau}d\tau,$$

the Fourier transform of $G(\tau)$. We therefore are dealing with Fourier amplitudes of **E**, **D**, χ_e, ϵ, and α. The discussion here is limited to frequencies much less than any characteristic frequencies in the molecules.

In order to discuss an appropriate model for a molecule, we must consider two polarizability mechanisms. If the molecule has an electric dipole moment in the absence of an electric field in the reference frame in which the molecule is at rest, the molecule is classified as a polar molecule. There will not be a macroscopic polarization, since the dipole axes of the molecules will be randomly oriented, i.e., $\langle \mathbf{p} \rangle_m$ is zero. This is consistent with a general quantum-mechanical theorem that any isolated molecule (or any other quantum-mechanical system which is bound by the electromagnetic interaction) having a definite value of total angular momentum cannot have odd electric multipole moments if there is no degeneracy of states of different parity. An electric field will, however, tend to align the molecules, producing a net polarization and an effective $\langle \mathbf{p} \rangle_m$ parallel to the field direction. This corresponds quantum-mechanically to mixing molecular states of opposite parity by the electric field.

The effective molecular dipole moment arising from the orientation of the permanent dipole moment of polar molecules should become completely aligned or saturated in sufficiently high electric fields. At ordinary fields and temperatures, saturation is prevented by the thermal motion of the molecules, which tends to produce random dipole orientation. The polarization may be calculated by finding the average angle between the applied electric field and the molecular dipole axis using the Maxwell–Boltzmann temperature-dependent distribution of directions. The result is

$$\langle \mathbf{p} \rangle_m = p_0 \left[\coth\left(\frac{p_0 E}{kT}\right) - \frac{kT}{p_0 E} \right], \tag{6}$$

where p_0 is the permanent dipole moment, k is the Boltzmann constant, and T is the absolute temperature. Equation (6) is known as the Langevin formula. For $p_0E/kT \ll 1$ the formula reduces on the right to $p_0^2E/3kT$.

In nonpolar molecules, which do not have permanent electric dipole moments, induced moments can be produced in an electric field by distortion of the electronic charge distribution. This mechanism also applies to monatomic molecules or atoms and of course will also be present in a polar molecule. This induced polarizability is an additional mechanism producing a molecular polarizability, which can be calculated from various models, from a simple Lorentz atom to elegant quantum-mechanical schemes. The induced polarizability is not temperature dependent.

The polarizability from both effects, electronic deformation α_e and molecular orientation, is

$$\alpha = \alpha_e + p_0^2/3kT \qquad (p_0E/kT \ll 1). \tag{7}$$

From (7) it can be seen that if the polarizability of a gas is plotted as a function of $1/T$, a straight line should result. If the slope is not zero, the molecules must be polar, and from the slope, the permanent dipole moment, p_0, can be found. The intercept with the $1/T$ axis gives α_e.

The polarizability of nonpolar molecules and the permanent dipole moment of polar molecules can be measured on isolated, individual molecules using molecular-beam techniques. Experiments are done using electric and magnetic deflection of the molecules in inhomogeneous fields, and also resonance experiments are performed where the quantum-mechanical energy levels are measured. The tensor nature of α must be reconsidered when studying an isolated molecule. For example, with a diatomic molecule an $\alpha_\perp$ must be defined when the electric field is perpendicular to the intermolecular axis and an $a_\parallel$ when **E** is parallel to the intermolecular axis. Such experiments and appropriate techniques are described in Refs. 5–7.

The permanent-dipole-moment determinations are of prime importance in molecular-structure studies. The electronic polarizability measurements serve as excellent tests for molecular electronic wave function calculations.

See also DIELECTRIC PROPERTIES.

REFERENCES

1. J. R. Reitz and F. J. Milford, *Foundation of Electromagnetic Theory*. Addison-Wesley, Reading, Mass., 1960. (I)
2. G. E. Owen, *Electromagnetic Theory*. Allyn and Bacon, Boston, 1963. (I)
3. J. D. Jackson, *Classical Electrodynamics*, 2nd ed. Wiley, New York, 1975. (A)
4. W. F. Brown, "Dielectrics," in *Handbuch der Physik*, Vol. XVII. Springer-Verlag, Berlin, 1956. (A)
5. N. J. Ramsey, *Molecular Beams*. Oxford University Press, Oxford, 1955. (A)
6. P. Kusch and V. W. Hughes, "Atomic and Molecular Beam Spectroscopy," in *Handbuch der Physik*, Vol. XXXVII/1. Springer-Verlag, Berlin, 1959. (A)
7. T. C. English and J. C. Zorn, "Molecular Beam Spectroscopy," in *Methods of Experimental Physics*, Vol. 3, Pt. B (D. Williams, ed.). Academic Press, New York, 1974. (A)
8. T. L. Weatherly and Q. Williams, "Electric Properties of Molecules," in *Methods of Experimental Physics*, Vol. 3 (D. Williams, ed.). Academic Press, New York, 1962. (A)
9. J. H. Van Vleck, *The Theory of Electric and Magnetic Susceptibilities*. Oxford University Press, Oxford, 1932. (A)

Polaron

Jozef T. Devreese

A charge placed in a polarizable medium will be screened. Dielectric theory describes the phenomenon by the induction of a polarization around the charge carrier. The induced polarization will follow the charge carrier when it is moving through the medium. The carrier together with the induced polarization is considered as one entity and was called a polaron by L. Landau.

A conduction electron in an ionic crystal or a polar semiconductor is the prototype of a polaron. Fröhlich proposed a simple model Hamiltonian for this entity through which the dynamics of the electron is treated quantum mechanically. The polarization, carried by the longitudinal optical phonons, is represented by a set of quantum oscillators with frequency ω and the interaction between the charge and the

Table I

Name	Type of Carrier	Polarization Coupling Mode
Large polaron (Fröhlich)	Conduction electron (hole) in the effective-mass approximation	Longitudinal optical phonons
Piezopolaron	Conduction electron (hole)	Piezoelectric polarization carried by the acoustic phonons
Electronic polaron	Conduction electron (hole)	Polarization of the core states, described by excitons
Small polaron	Conduction electron (hole) in the tight-binding approximation	Longitudinal optical phonons
Magnetic polaron	Conduction electron (electron spin)	Nuclear spin of ions

polarization field is linear in the field. This model (which up to now has not been solved exactly) has undergone extensive investigations. The strength of the electron–phonon interaction is related to a dimensionless coupling constant α. When the coupling is weak (α small) the energy (ΔE) involved in the formation of a polaron, i.e., the self-energy, is proportional to α ($\Delta E = -\alpha h\omega$). The polaron mass m^*, which can be measured by cyclotron resonance experiments, is larger than the mass m of the charge carrier without self-induced polarization,

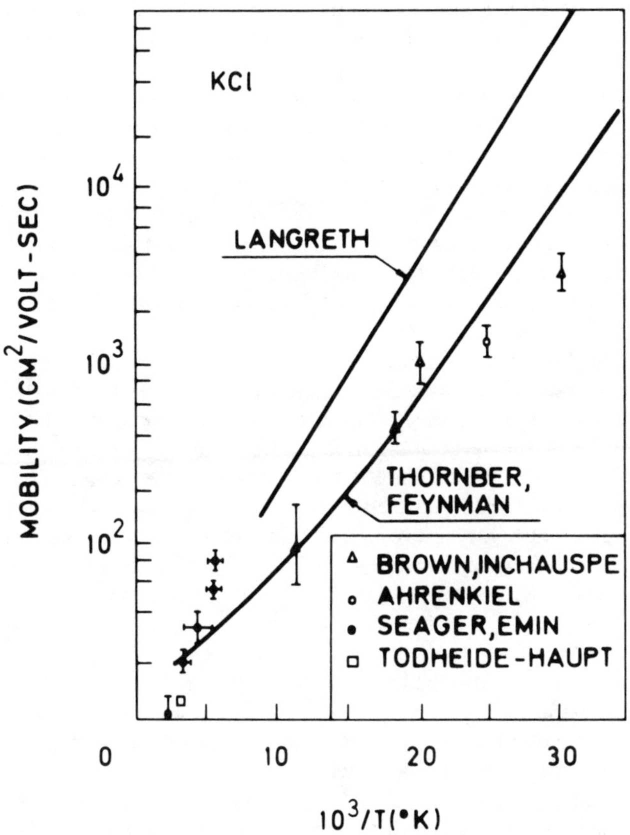

FIG. 1. Comparison of experimental and theoretical mobilities for KCl. The points shown by triangles and by open and closed circles are all photo-Hall data. The squares are drift mobility data. The two theoretical curves were drawn for $m/m_e = 0.412$ and $\Theta = 300$ K [From F. C. Brown, *Point Defects in Solids*, Vol. I (J. Crawford and I. Slipkin, eds.). Plenum, New York, 1972.]

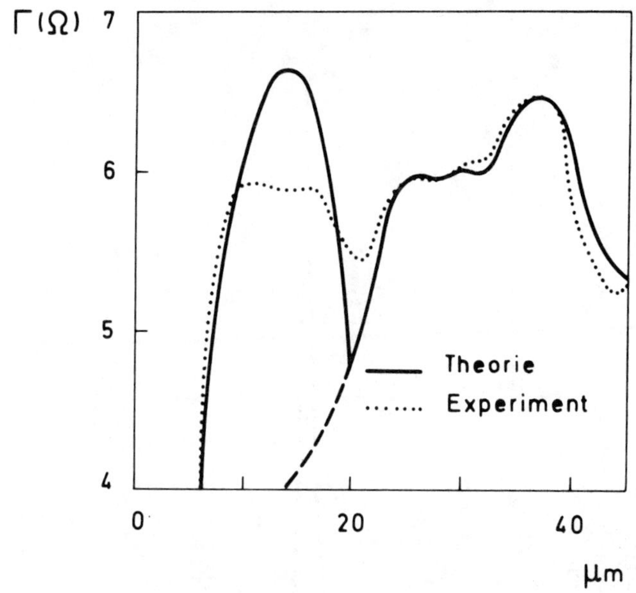

FIG. 2. Infrared optical absorption in CdO as measured by Finkenrath, Uhle, and Waidelich [see *Solid State Commun.* **7**, 11, (1969)]. The absorption for $\lambda < 20$ μm is predominantly due to polaron effects.

$$m^* = m(1 + \alpha/6).$$

When the coupling is strong (α large), a variational approach due to Landau and Pekar indicates that the self-energy is proportional to α^2 and the polaron mass is proportional to α^4.

Feynman used an exactly solvable model for the polaron that made possible variational calculations. He simulated the interaction between the electron and the polarization modes by harmonic interaction between a hypothetical particle and the electron. Using the solutions of this model, he was able to handle the Fröhlich Hamiltonian variationally. The resulting expressions for the self-energy and polaron mass are valid for arbitrary coupling and are quite accurate. Physically more directly accessible properties of the polaron, such as mobility and optical absorption, have been investigated. In the framework of the Feynman approach, this information can be obtained from a frequency-dependent dissipation function $\chi(\Omega)$. For relatively low temperatures T the mobility takes the form

$$\mu = \left(\frac{w}{v}\right)^3 \frac{3}{4\alpha\beta} e^{\beta} \exp\{(v^2 - w^2)/w^2 v\}, \tag{1}$$

where $\beta = 1/kT$, and v and w are functions of α deriving from the Feynman model. Experimental work of F. C. Brown and co-workers on alkali halides and silver halides indicates that

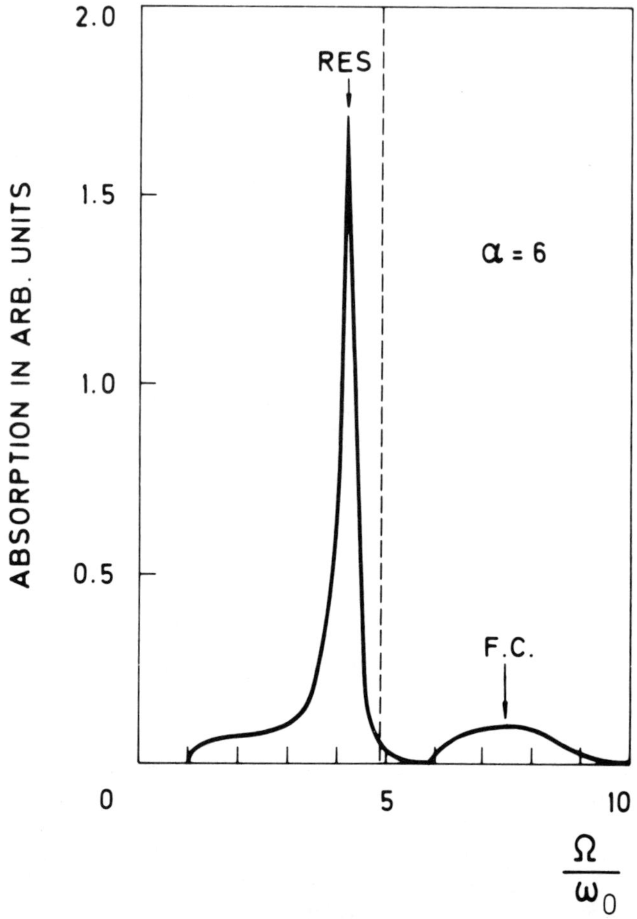

FIG. 3. Optical absorption Γ of polarons at $\alpha = 6$. The RES (relaxed excited state) peak is very intense compared with the Franck–Condon peak. The frequency $\Omega = \nu$ is indicated. (See Ref. 2.)

the mobility obtained from Eq. (1) describes the experimental results quite accurately (Fig. 1).

Calculating the optical absorption from the dissipation function, one remarks that the polaron can absorb energy which is reemitted as phonons. This effect was measured in CdO (Fig. 2). When the coupling increases, the polaron can make transitions toward a relatively stable excited state (Fig. 3).

The optical absorption $\Gamma(\Omega)$ at frequency Ω takes the form

$$\Gamma(\Omega) \propto \frac{\operatorname{Im} \chi(\Omega)}{[\Omega^2 - \operatorname{Re} \chi(\Omega)]^2 + \operatorname{Im}^2 \chi(\Omega)}. \quad (2)$$

The physical properties of polarons in a Coulomb potential and in a magnetic field, and their magneto-optical properties, have also been studied in detail. The generalizations of the Fröhlich polaron concept are displayed in Table I.

See also CYCLOTRON RESONANCE; DIELECTRIC PROPERTIES; PHONONS; QUASIPARTICLES.

REFERENCES

1. C. G. Kuper and G. D. Whitfield (eds.), *Polarons and Excitons*. Oliver and Boyd, Edinburgh, 1963.
2. J. T. Devreese (ed.), *Polarons in Ionic Crystals and Polar Semiconductors*. North Holland, Amsterdam, 1972.

Positron Annihilation in Condensed Matter

Stephan Berko

The positron is the antiparticle of the electron. When injected into ordinary matter it interacts electromagnetically with its surroundings prior to annihilating with an electron into gamma (γ) rays. The positron annihilation lifetime depends on the density of electrons at the site of the positron, and the annihilation γ's carry information about the energy, momentum, and relative spin of the annihilating positron–electron pair. During the last three decades the interaction of slow positrons with matter has been the subject of numerous experimental and theoretical investigations that led to the successful development of the positron-annihilation technique for the study of various problems in physics and chemistry. By measuring the positron lifetime, the energy distribution, and the angular correlation of the annihilation γ's, one learns valuable information about the behavior of the positron–electron pair prior to annihilation, and, by inference, about the electronic properties of the substance under investigation.

The positron (e^+) lifetime, of the order of 10^{-10}–10^{-8} s in liquids and solids, can be accurately measured electronically by using e^+ sources such as ^{22}Na that emit an essentially prompt nuclear γ ray accompanying the e^+. The time-delayed coincidence measurement between this nuclear "$t = 0$ signal" and one of the annihilation γ's yields directly the positron lifetime. The angular correlation between the gamma rays in the 2γ annihilation mode reflects the momentum distribution of the positron–electron pair in the laboratory system; it is the technique most used in the application of positron annihilation for studies in solid-state physics. Typical deviations from collinearity between the two γ's are of a few milliradians. The energy distribution of one of the gammas from the 2γ decay mode also depends, through the Doppler effect, on the momentum density of the annihilating pair.

In the usual positron experiments a beam of high-energy positrons from a long-lived radioactive e^+ source—usually ^{22}Na or ^{58}Co—is injected into the system to be studied. Because of the large ionization and excitation cross section compared to the annihilation cross section, the incident positrons lose their kinetic energy rapidly; annihilation becomes significant below a few eV. The fate of such a slow positron depends on its environment. In gases, liquids, and insulating solids some positrons can capture electrons and form positronium (Ps), the hydrogenic two-body positron–electron system (6.8-eV binding energy). Other bound states such as e^+–atom, e^+–molecule can also be formed, but are not predominant. In a perfect metal crystal Ps is not formed and, after reaching thermal equilibrium, the positron annihilates from a delocalized (Bloch wave-function) state.

The development of the above picture has been made possible because well-known selection rules based on symmetry considerations lead to strong spin-dependent annihilation properties: Annihilation takes place by two γ's when the e^+–e^- system is in a spin-singlet state and by three γ's when in a spin-triplet state. Single-γ annihilation can only take place with nuclear recoil, and has a small probability compared to 2γ and 3γ events. Since the cross section for 3γ annihilation is $\approx 10^3$ times smaller than for 2γ annihilation, the $3\gamma/2\gamma$ yield depends sensitively on the electrons surrounding the positron. If Ps is formed without subsequent collisions, the $3\gamma/2\gamma$ yield is 3:1; if Ps is not formed and the positron annihilates from a "scattering" state in a spin-averaged medium—as in a metal—the $3\gamma/2\gamma$ yield becomes $\approx 10^3$ times smaller. The annihilation lifetime also reflects this spin-dependent cross section. In vacuum, spin-singlet Ps in its ground state annihilates in 1.25×10^{-10} s (2γ's), while the spin-triplet state has a long (1.4×10^{-7} s) lifetime. In a condensed medium the long-lived Ps interacts with neighboring electrons and its "natural" 3γ lifetime can be shortened considerably by 2γ annihilations of its positron with an external electron. Thus a complex decay lifetime spectrum is observed. The 2γ angular correlation also reflects a complex distribution: A broad distribution is characteristic of e^+ annihilation with the bound atomic electrons of the medium, while Ps, if formed, contributes a narrow peak corresponding to its center-of-mass momentum.

The ability to distinguish experimentally between these modes of decay has led to important experiments designed to study the intrinsic properties of Ps, as well as the interaction of e^+ and Ps with atoms and molecules. In gases the study of slow e^+ scattering, of Ps formation, and of Ps scattering is the focus of interest. The study of e^+ and Ps interactions with complex molecules, particularly in liquids, has led to the development of a new branch of chemistry.

In simple liquids, particularly in liquid He, e^+ annihilation exhibits a surprising richness of phenomena. There is strong experimental evidence that in liquid He Ps forms a self-consistent "quantum bubble" prior to annihilation, similar to the bubbles formed by free electrons in He. The angular correlation of the 2γ annihilation can be used to measure the zero-point energy of Ps in such a "self-trapped" state, yielding a value of the bubble radius to be compared to theory.

In organic solids, positron-annihilation experiments reveal substantial Ps formation probabilities through the appearance of complex lifetimes and narrow angular correlation components. The sensitivity of the various lifetime components to temperature, pressure, and crystallinity resulted in the "free-volume" theory of Ps behavior in these solids. The affinity of the positron to low-atomic-density regions was used more recently to study phase transitions in liquid crystals and in polymers. In ionic crystals, besides annihilation from delocalized states, some positrons can be trapped in various defect centers; there is also indication that positronium can be formed, as well as bound states between a positron and negative ions. In oxides the situation is also complex; experiments in oxide powders indicate a large Ps-formation cross section, with Ps diffusing out of the powder grains. In some single crystals, like quartz, a delocalized Ps-like state similar to excitons was observed and studied.

The field where positron-annihilation techniques have led to many important applications is metal physics. In high-quality single crystals, positrons annihilate from delocalized states; because of their charge they are partially excluded from the positive ionic cores and sample preferentially the conduction electrons, annihilating mainly by the 2γ channel. The precision measurement of the small angular deviation from collinearity of the two annihilation γ's reveals the momentum density of the electrons as sampled by the positron. In metals, sharp breaks in these angular correlations correspond to the position of the Fermi surface of the conduction-band electrons. High-resolution angular correlation experiments are being performed to study the shape and size of the Fermi surface in materials where more standard solid-state techniques fail, such as in disordered substitutional alloys. Spin-aligned positrons—as obtained from parity-nonconserving radioactive e^+ decay—can be used to obtain information about the momentum density of ferromagnetic electrons. In simple metals such as the alkali metals or aluminum where the conduction electrons are nearly free, experimental positron lifetime and angular correlation results have been used as a testing ground for various theoretical many-body physics techniques. In several metal systems the behavior of the e^+ is well enough understood, and the e^+-e^- interaction sufficiently well accounted for, to permit the analysis of the 2γ angular correlation data in terms of the electronic band structure, thus testing the nature of the electronic wave functions.

From careful lifetime measurements, it was discovered that positrons in metals can be trapped in lattice defects such as atomic vacancies, dislocations, microvoids, etc. Since in the localized state the positron sees a lower charge density, its lifetime increases, and the energy of the annihilation γ's as well as the 2γ angular correlation reflects a corresponding change in the momentum density. These effects have recently been developed into sensitive tools for metallurgical problems. For example, by observing the change in the positron lifetime spectrum as more and more vacancies are produced when one increases the temperature of a metal, the value of the vacancy formation energy can be obtained. Quenching and annealing experiments have been performed in various metals and alloy systems. Vacancy clustering and microvoid formation due to radiation damage by electrons and neutrons can also be observed by positron annihilation measurements, and several research groups are trying to develop these techniques for routine nondestructive testing of materials.

In closing, the recent developments in producing low-energy (few eV) positron beams should be mentioned. These beams have already yielded interesting results in Ps studies and e^+–atom collision cross section measurements, and will no doubt be used in the future to study the electronic properties of solid-state surfaces.

See also ELECTRON ENERGY STATES IN SOLIDS AND LIQUIDS; FERMI SURFACE; RADIATION DAMAGE IN SOLIDS.

BIBLIOGRAPHY

H. J. Ache (ed.), "Positronium and Muonium Chemistry." Advances in Chemistry Series, No. 175, American Chemical Society, Washington, 1978.

S. Berko, "Positron Annihilation," *Compton Scattering, The Investigation of Electron Momentum Distributions*, Chap. 9, p. 273 (B. Williams, ed.). McGraw-Hill, New York 1977.

M. Doyama and R. R. Hasiguti, "Studies of Lattice Defects by Means of Positron Annihilation," *Crystal Lattice Defects* **4**, 139 (1973).

V. I. Goldanskii and V. G. Firsov, "Chemistry of New Atoms," *Ann. Rev. Phys. Chem.* **22,** 209 (1971).
R. R. Hasiguti and K. Fujiwara, (eds.), *Positron Annihilation, Proceedings of the Fifth International Conference on Positron Annihilation*. Japan Institute of Metals, Sendai, 1979.
P. Hautojarvi, ed., *Positrons in Solids*. Springer Verlag, Berlin, 1979.
A. T. Stewart and L. O. Roellig (eds.), *Positron Annihilation*. Academic Press, New York, 1967.
R. N. West, "Positron Studies in Condensed Matter," *Adv. Phys.* **22,** 263 (1973).

Pyroelectricity

S. C. Abrahams

Pyroelectricity is a property of crystals with a net dipole moment per unit cell (the electric polarization vector **P**) in which a temperature change ΔT produces a proportional polarization change ΔP, and conversely, as given by

$$\Delta P_i = p_i \, \Delta T$$

where p_i is one, two, or three coefficients that together specify the pyroelectric vector **p.** The vector **P** necessarily conforms to the crystal symmetry, and is nonzero only in the ten noncentrosymmetric point groups that possess a unique direction, namely, 1, 2, *m*, 2*mm*, 3, 3*m*, 4, 4*mm*, 6, and 6*mm*. Detection of pyroelectricity is a definitive test for noncentrosymmetry. The total pyroelectric coefficient measured at constant stress is the sum of the coefficient at constant strain (the primary pyroelectric coefficient) and the piezoelectric contribution due to thermal expansion (the secondary pyroelectric coefficient). The ratio of primary to secondary pyroelectric coefficient varies widely with material but is generally in a range on the order of 10^2–10^{-2}. The magnitudes of the total pyroelectric vector **p** and polarization vector **P** are also a function of the material and range from 10^{-2} to 10^{-8} C m^{-2} K^{-1} for **p** and from 1 to 10^{-3} C m^{-2} for **P**. Nonuniform heating causes nonuniform stresses and strains and can give rise to a polarization by means of piezoelectric effects: this is sometimes referred to as tertiary pyroelectricity.

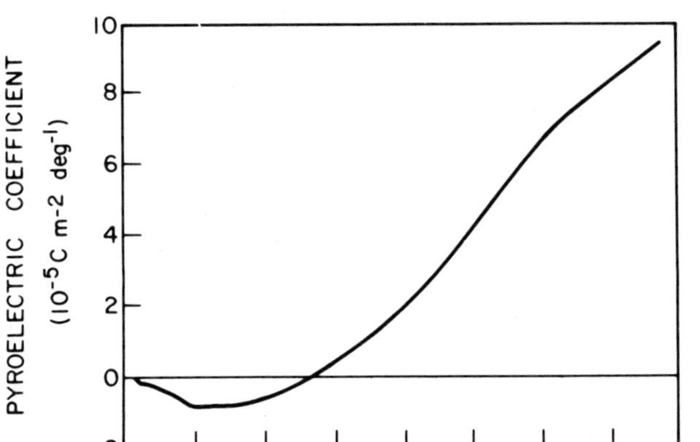

FIG. 1. Temperature dependence of the total pyroelectric coefficient p_2 of lithium sulfate monohydrate measured at constant stress [after S. B. Lang, *Phys. Rev.* **B4**, 3603 (1971)].

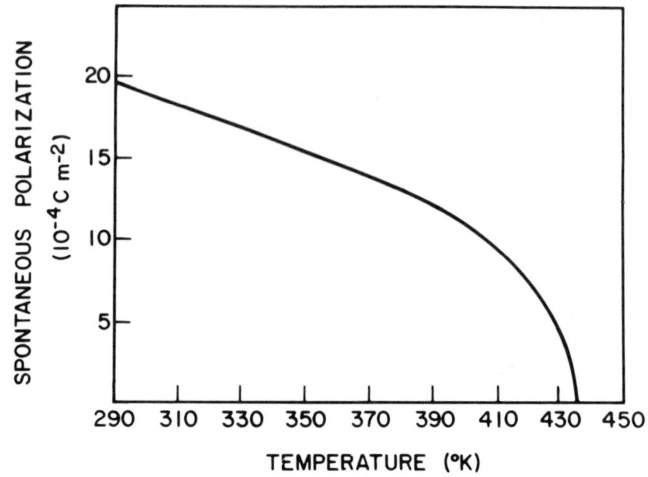

FIG. 2. Temperature dependence of spontaneous electric polarization $\mathbf{P}_s$ in terbium molybdate [after E. T. Keve, S. C. Abrahams, K. Nassau, and A. M. Glass, *Solid State Commun.* **8**, 1517 (1970)].

The temperature dependence of the total pyroelectric coefficient is determined by the combination of primary and secondary coefficients: an example of a resulting dependence is given in Fig. 1. The absolute temperature dependence of **P** can be experimentally determined only if **P** becomes zero at a phase transition or if the sense of **P** can be reversed by application of an electric field. In such cases the crystal is ferroelectric and the spontaneous polarization $\mathbf{P}_s$ generally behaves as in Fig. 2, although it is possible for $\mathbf{P}_s$ to increase with increasing temperature as in Rochelle salt or ammonium sulfate.

The combined use of x ray or neutron anomalous scattering with measurement of the pyroelectric coefficient allows the sense of **p** to be determined in terms of the atomic arrangement, i. e., the absolute sense of the pyroelectric vector. In ferroelectric crystals, the absolute sense of $\mathbf{P}_s$ can be similarly determined. The contribution of each elementary dipole within the unit cell to the total polarization may then be analyzed. Relative changes in atomic position and electronic charge distribution with temperature may consequently be related to the magnitude of **p**. Typical displace-

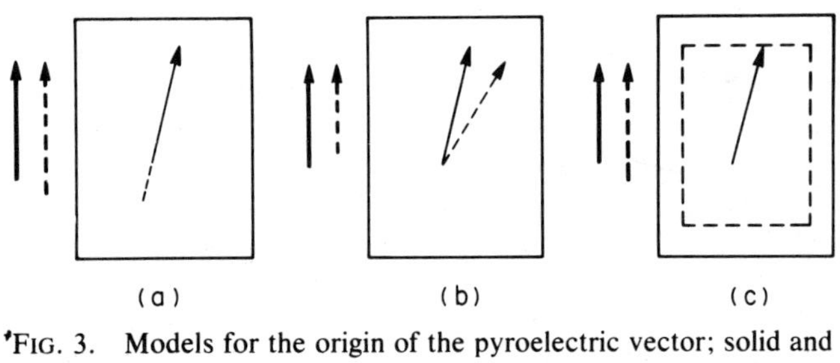

FIG. 3. Models for the origin of the pyroelectric vector; solid and broken lines represent states at different temperatures. (a) Change in *magnitude* of elementary dipole causing a primary pyroelectric effect; (b) change in *orientation* of elementary dipole causing a primary pyroelectric effect; (c) change in *volume* of unit cell causing a secondary pyroelectric effect. The polarization change ΔP is the difference between the polarization vector **P** at the two temperatures, represented by the heavy solid and broken arrows.

ments of ionic sublattices in nonferroelectric but pyroelectric crystals are on the order of 10^{-4}–10^{-5} Å K^{-1}: in ferroelectrics, the displacements may be two or three orders of magnitude greater. Dipolar rotations are on the order of 10–10^{-1} arc min K^{-1}. A representation of the origin of the pyroelectric vector is illustrated in Fig. 3.

Numerous devices using the pyroelectric effect have been designed, including infrared and millimeter wave detectors, temperature sensors, calorimeters, thermal imagers, and television camera tubes. Many pyroelectric materials have been considered for such device applications. Among the more widely used are triglycine sulfate and lithium tantalate single crystals, lead zirconate titanate ceramics, and polyvinylidene fluoride (which can be manufactured in wide sheets with enhanced preferred orientation and polarity).

BIBLIOGRAPHY

Landolt-Börnstein, *Elastic, Piezoelectric, Pyroelectric, Piezooptic, Electrooptic Constants, and Nonlinear Dielectric Susceptibilities of Crystals*, III/11. Springer-Verlag, Berlin, Heidelberg and New York, 1979.

S. B. Lang, *Sourcebook of Pyroelectricity*. Gordon & Breach, New York, 1974.

M. E. Lines and A. M. Glass, *Principles and Applications of Ferroelectrics and Related Materials*. Oxford (Clarendon Press), London and New York, 1977.

J. F. Nye, *Physical Properties of Crystals*. Oxford (Clarendon Press), London and New York, 1957.

S. Ramaseshan and S. C. Abrahams, ed., *Anomalous Scattering*. Munksgaard, Copenhagen, 1975.

K. Toyoda, *Ferroelectrics*. Bibliography in each issue.

Quasiparticles

P. W. Anderson

The concept of quasiparticles in solid-state and many-body physics is closely related to high-energy physicists' idea of a "physical" or "dressed" *renormalized* particle, as opposed to the "bare" or "undressed" particles that appear in the unrenormalized theory. The idea is that very often a many-body system will have excitations that behave like free elementary particles but are actually very complicated in their internal structure, which hides possibly large effects of interparticle interaction. The clearest exposition of the idea is Landau's in his "Fermi liquid theory" [1,2], in which he proves that at sufficiently low temperatures an interacting sea of Fermi particles such as electrons or He_3 atoms behaves in many essential respects as though it were composed of weakly interacting free fermions, with a well-defined, sharp Fermi surface and a constant mass (not equal to the bare one) and spin susceptibility (also not equal to the bare one). He introduces a quasiparticle distribution function $n(p,\sigma)$ for quasiparticles of momentum p and spin σ, and assumes that overall $n(p,\sigma)$ differs very little from the value for a perfectly free Fermi gas at $T \to 0$,

$$\begin{aligned} n_0(p,\sigma) &= 1, \quad p<p_F, \\ &= 0, \quad p>p_F, \end{aligned} \qquad (1)$$

where p_F is the Fermi momentum. We can then write the energy of the total system as a power series in $\delta n = n - n_0$:

$$\begin{aligned} E(\delta n) &= E_0 + \sum_{p,\sigma} E_p \delta n(p,\sigma) \\ &+ \sum_{pp'\sigma\sigma'} F(p,p',\sigma,\sigma')\delta n(p,\sigma)\delta n(p',\sigma') + \dots, \end{aligned} \qquad (2)$$

and Landau shows that only these terms play a role as $T \to 0$. The quantities F depend only on the directions of p and p', not on their distance from the Fermi surface, and can be expressed in terms of a relatively few "Landau parameters," which determine the renormalizations of mass, susceptibility, etc. The important idea is that (2) is valid and useful even though the distribution of actual real, "bare" Fermi particles does not resemble (1) at all, as is true in liquid He_3, for instance.

The concept has been generalized to many other situations, the most important being the BCS theory of superconductivity, where Schrieffer's book [3] gives a beautiful exposition of the quasiparticle concept. In the BCS theory the quasiparticle states are elementary equivalent fermions that are neither pure electrons nor pure holes but a linear combination of both. Nucleons in the shell model of the nucleus can also be understood as occupying quasiparticle states.

In general, the quasiparticle is a fermion; it is a special case of the concept, also due to Landau, of an "elementary excitation" of a many-body system, which may be either a boson (phonon, roton, plasmon, spin wave, etc.) or a fermion. By very general arguments [4] it can be shown that the very lowest-energy excited states of any such system will usually take on the very simple form of an apparently noninteracting gas of free objects, which can be bosons, fermions, or both.

See also PHONONS; PLASMONS; POLARON; EXCITON; SUPERCONDUCTIVITY THEORY.

REFERENCES

1. L. D. Landau, *Sov. Phys. JETP* **3**, 920 (1956); **5**, 101 (1957).
2. For an elegant, simple and complete review, see P. Nozières, *Interacting Fermi Systems*. Benjamin, New York, 1963.
3. J. R. Schrieffer, *Theory of Superconductivity*. Benjamin, New York, 1964.
4. P. W. Anderson, *Concepts in Solids*. Benjamin, New York, 1962.

Radiation Damage in Solids

David O. Welch

The physical properties of solids can be altered considerably by exposure to electromagnetic and particle radiation. The extensive study of this phenomenon since its prediction in 1943 by Eugene Wigner has been motivated principally by its consequences for materials used in the construction of nuclear reactors. More recently, the study of radiation effects in semiconductors and insulators has been stimulated by the deleterious effects of radiation from the Van Allen belts, the solar wind, and the explosion of nuclear weapons upon the properties of solid-state electronic devices in artificial satellites and spacecraft.

Even though the study of radiation damage in solids has largely paralleled the development of nuclear energy and space technology, the first recognition that particle irradiation can disrupt the orderly arrangement of atoms in crystals, and thus strongly affect their physical properties, seems to have been made in 1914 by Hamburg, who suggested that the properties of so-called metamict minerals could be explained by the disruptive action of α-particles emitted by the decay of radioactive isotopes incorporated into the crystal structure of the minerals. At the time of its proposal, Hamburg's hypothesis was not generally accepted. Today a wide variety of experimental and theoretical evidence supports the view that the primary cause of the alteration of the properties of crystalline material by exposure to radiation is the displacement of atoms from their normal locations in the crystal structure to sites that are normally empty in unirradiated crystals, i.e., the creation of lattice defects.

A complete description of the process of radiation damage and its consequences requires an understanding of the mechanisms by which energy is transferred from the radiation to the atoms of a solid, and by which the transferred energy results in the creation of lattice defects. In addition, it is necessary to know the type, distribution, and stability of the defects created and how defects in turn affect the macroscopic properties of materials, such as mechanical strength, electrical conductivity, and optical properties. Ideally, this understanding can be incorporated into a quantitative model that can be used, for example, to predict the evolution of the mechanical properties of a structural member in a nuclear reactor over the lifetime of the reactor. Unhapplily such a comprehensive model does not yet exist, even for such ideal, well-characterized materials as pure, single crystals of copper, much less for more complex materials such as structural steel. However, even though a detailed, comprehensive understanding of radiation damage is not yet complete, some aspects of the problem are quite well understood for specific classes of materials, for example, electron damage to pure, face-centered cubic metals at low temperatures. Furthermore, a significant amount of the difficulty arises because of the lack of a quantitative understanding of the behavior of unirradiated materials in some areas, such as mechanical properties.

THE INTERACTION OF RADIATION WITH SOLIDS

The first step in the radiation damage process is the transfer of energy from the radiation to the solid. An atom of a crystal that receives sufficient energy, typically a few tens of electron volts, will be permanently displaced from its lattice site. If the displaced atom (a "primary knock-on") has sufficient energy, it can in turn displace other atoms ("secondaries"), and so on, creating a cascade of displaced atoms. The mechanism by which atoms receive energy from the radiation depends on the nature of the radiation.

A beam of energetic, charged particles, such as electrons, protons, or fission fragments, loses energy primarily by exciting the electrons of the solid to higher energy levels and by collisions with the nuclei of the atoms of the irradiated solid. At sufficiently high energies, energy loss by electronic excitations dominates, but as the particle slows down, energy transfer to nuclei becomes increasingly important until this mode of energy transfer dominates near the end of the particle's range. In metals, only the energy transferred to the nuclei by the irradiating particles is effective in creating lattice defects. In some insulators, however, such as the alkali halides, lattice defects can be created by the energy released upon the deexcitation of electrons from excited states. Furthermore, in polymers and molecular crystals the state of chemical bonding can be altered by electronic excitations.

Neutrons interact essentially only with the nuclei of the irradiated solid. In addition to elastic collisions in which there is a transfer of kinetic energy, nuclear reactions can also be important in the radiation damage process. Of particular importance are (n,p) and (n,α) processes, since the hydrogen and helium formed by these reactions are effective in stabilizing the atomic displacement damage that results from elastic collisions. As a consequence of the short range of nuclear forces compared with that of the Coulomb interaction, the mean spacing between collisions in which an atom is displaced from its lattice site is substantially larger for a neutron than for a proton of the same energy. This difference is about three orders of magnitude for 1-MeV particles in a typical metal, in which case the mean spacing between displacements for neutron bombardment is about 5 cm.

Energetic photons displace atoms indirectly through the action of electrons excited by Compton or photoelectric processes, while at high photon energy (e.g., greater than about 5 MeV) the creation of electron–positron pairs begins to make a significant contribution to the creation of displacement damage. Direct interaction of the photons and nuclei makes a negligible contribution if the photons are less energetic than about 100 MeV. In contrast to metals, for which photons of γ-ray energy are required to cause atomic displacements, some insulators, notably the alkali halides, suffer displacement damage from photons in the ultraviolet range, in which case the necessary kinetic energy for displacement is supplied to the atoms by radiationless deexcitation of electrons from states in the conduction band. In metals the energy must be supplied by Coulomb scattering of fast electrons excited by the irradiating photons.

THE NATURE OF RADIATION DAMAGE IN SOLIDS

The elementary units of radiation damage in simple crystalline solids are normally occupied crystal lattice sites from which atoms are missing (vacancies) and atoms located in normally unoccupied positions in the lattice (interstitials). In nonmetals, vacancies and interstitials can exist in a variety of charge states. In addition to such displacement damage, chemical impurities resulting from nuclear transmutations may be present in appreciable quantities. The physical properties of irradiated crystals are determined by the density, distribution, state of aggregation, and electronic state of vacancies and interstitials. These quantities depend on the conditions of irradiation, such as temperature, irradiation intensity, and length of exposure, and on the nature of the radiation. For example, a beam of 1-MeV electrons irradiating a metal produces isolated pairs of vacancies and correlated interstitials, whereas a beam of 1-MeV neutrons produces isolated cascades of radiation damage, about 1000 atomic volumes in size, in which the defect distribution is highly correlated. This difference in defect distribution is attributable primarily to the difference in mean energy of atoms recoiling from their lattice sites following collisions with the irradiating particles. In the case of the 1-MeV electrons, the mean recoil energy is of the order of 50 eV; hence primary recoil atoms usually have insufficient energy to produce further displacements, and each displacement event generally consists of the creation of one interstitial–vacancy pair. However, in the case of 1-MeV neutrons, the mean primary recoil energy is of the order of 100 keV; thus primary knock-ons have ample energy to displace highly energetic secondary knock-ons, thereby generating a cascade of displaced atoms.

The temperature dependence of the state of radiation damage can be illustrated by the behavior of a crystal of platinum irradiated with fast electrons at 4 K and heated slowly to a temperature of several hundred degrees following the irradiation. At the low temperature of irradiation there is insufficient thermal energy for atomic motion; therefore until the crystal is warmed, the state of damage remains as it was created by the radiation: isolated interstitial atoms separated by a small distance from their original lattice sites, now vacant. When the temperature is raised to about 10 K, there is sufficient thermal energy for atomic motion such that the most closely separated interstitials and vacancies begin to recombine and thus be annihilated. By 30 K interstitials can migrate freely through the lattice by diffusion until they either find vacant sites and are annihilated or become trapped at impurity atoms or by other interstitial atoms to form clusters.

As the temperature increases, the interstitial clusters grow in size, reaching about ten or so atoms per cluster by 500 K. At this temperature, thermal fluctuations can supply enough energy for the atoms to exchange places with adjacent vacancies. The vacancies thus become mobile and migrate through the crystal, occasionally combining with, and thereby gradually annihilating, interstitial clusters. By 700 K the radiation damage has been removed through the annihilation of vacancies and interstitial atoms, and the platinum crystal is returned to an undamaged state. In other materials, the details of the recovery from the damaged state created by fast electrons at low temperature can differ somewhat from the events just described for platinum, but the broad details of the recovery process are the same for a wide variety of materials. Studies of recovery characteristics following irradiation with fast electrons at low temperatures have provided much of the existing knowledge of the properties of metals and elemental semiconductors.

In general, the state of damage as a function of temperature and mode of irradiation is determined by a balance between the rate of production of elementary defects by the radiation and the competing processes of defect aggregation and annihilation. The latter processes depend critically on the migratory mobility of defects, which in turn is a sensitive function of temperature, and on the concentration and spatial distribution of defects. An excellent illustration of the complex relationship between these factors is provided by the swelling of metals subjected to a high exposure of fast neutrons at temperatures of the order of half the melting point. Since interstitials and vacancies are quite mobile at these temperatures, it would be naively expected that annihilation processes would readily occur and radiation effects would be minor. Indeed, despite decades of research on the effects of neutrons on the properties of metals (albeit under different conditions), the phenomenon of swelling was completely unexpected when it was discovered in 1966. The resulting volume changes can exceed 10% and pose serious technological problems for the development of fast breeder and fusion reactors. The swelling occurs because under the conditions of the irradiation, vacancies aggregate to form empty cavities or "voids." This occurs largely because, compared to vacancies, interstitial atoms are preferentially attracted to sinks, where they may be absorbed from the crystal lattice, leading to an excess of vacancies and their subsequent aggregation into voids. This preferential attraction is thought to occur because interstitial atoms distort the geometry of the crystal lattice more severely than do vacancies, and it is these distortions that cause the defects to be attracted by internal stress fields surrounding the sinks. In addition, it is thought that helium atoms produced by (n,α) reactions play a critical role in the nucleation and stabilization of voids. Interestingly, all of the properties of vacancies and interstitials that give rise to the formation of voids were known prior to 1966, but the behavior of the system as a whole under the appropriate conditions had not been considered.

The lattice distortions described above, together with the disruption to the periodicity of the lattice by missing or improperly located atoms, result in scattering of the conduction electrons of metals, thus causing an increased electrical resistivity. Furthermore, the scattering of electrons by lattice distortions provides the basis for the study of radiation-produced defects by transmission-electron microscopy. Both of these effects have been the principal tools for the study of radiation effects in metals. Recently studies of the diffuse scattering of x rays by the defect-produced lattice distortions have helped to resolve a long-standing controversy over the structure of interstitials and the interpretation of recovery following irradiation of metals.

Lattice defects in ionic crystals and semiconductors can trap electrons and holes in the valence band, consequently giving rise to well-defined optical absorption and emission and to magnetic resonance effects. Because of these characteristics, defects and thus the radiation damage process can be studied in these materials by a wide variety of experimental techniques. As a consequence, probably more is known about the structure and properties of lattice defects and their production by radiation for the alkali halide family than for any other class of materials. As in the case of metals, the elementary units of damage are vacancies and interstitials, although in insulating materials these can exist in a variety of charge states. For example, in NaCl anion vacancies can be empty or can trap one or two electrons.

The state of radiation damage in nonconductors, as in metals, under given conditions reflects a dynamic balance between the creation of vacancies and interstitials and their aggregation or annihilation. In contrast to metals, however, some nonconducting crystals (Al_2O_3, Ge, and SiO_2, among others) can be made amorphous by sufficiently long exposure to radiation. Another phenomenon peculiar to insulating materials is the production of linear trails of atomic disorder or "tracks" in the wake of massive charged particles. Etching solutions preferentially attack the damaged region, producing large holes that are observable by optical microscopy. The track phenomenon has been exploited for a variety of applications, ranging from solid-state detectors and the production of small holes of controlled size in membranes to geochronology.

Finally, a wide variety of chemical effects can be produced by electronic excitation by ionizing radiation in solid organic and polymeric materials and in pseudostable inorganic materials such as ammonium perchlorate. Some of these effects are decomposition to form gaseous products, polymerization, creation of cross links, and breaking of primary links in high polymers.

See also COLOR CENTERS; CRYSTAL DEFECTS; RADIATION INTERACTION WITH MATTER.

BIBLIOGRAPHY

D. S. Billington and J. H. Crawford, Jr., *Radiation Damage in Solids*. Princeton Univ. Press, Princeton, NJ, 1961.
R. O. Bolt and J. G. Carroll, eds., *Radiation Effects on Organic Materials*. Academic, New York, 1963.
J. W. Corbett, *Electron Radiation Damage in Semiconductors and Metals*. Academic, New York, 1966.
G. J. Dienes and G. H. Vineyard, *Radiation Effects in Solids*. Wiley (Interscience), New York, 1957.
"Radiation Effects on Materials" (report to the American Physical Society by the study group on physics problems relating to energy technologies), *Rev. Mod. Phys.* **47**, Suppl. No. 3, 1975.
E. Sonder and W. A. Sibley, "Defect Creation by Radiation in Polar Crystals," in *Point Defects in Solids* (J. H. Crawford, Jr., and L. M. Slifkin, eds.), Vol. 1. Plenum, New York, 1972.
M. W. Thompson, *Defects and Radiation Damage in Metals*. Cambridge, London and New York, 1969.

Radiation Interaction with Matter

J. H. Crawford, Jr.

As used here the term *radiation* includes both electromagnetic and corpuscular forms. Though *matter* is a term generally encompassing elementary particles, nuclei, atoms, molecules, etc., the emphasis of this article is on solids and the consequences of their irradiation. Transition probabilities associated with excitation and ionization of atomic and molecular systems and their associated electromagnetic spectra are considered only incidentally.

I. ELECTROMAGNETIC RADIATION (PHOTONS)

The types of interaction processes characteristic of photons depend on their energy (E_p) or wavelength. For convenience two ranges will be treated: (a) high energy ($E_p > 10^4$ eV) and (b) low energy ($E_p < 10$ eV). The interval between has become an important research area only recently with the availability of intense monochromatic radiation in the far ultraviolet from synchrotron sources; this spectral region will not be included in this discussion.

A. High-Energy Photons (X Rays and γ Rays) [1–3]

There are three dominant interaction processes, each producing energetic electrons: the photoelectric effect (PE), the Compton effect (CE), and pair production (PP). At the low end of this energy range PE dominates. Photons are absorbed by atomic systems to eject deep-shell electrons with kinetic energies $E_e = E_p - W$, where W is the electron binding energy. Secondary processes are the emission of x rays and Auger electrons as the deep-shell vacancies in the ionized atoms are filled. The PE cross section (probability per unit time per unit photon flux) increases approximately with the fifth power of the atomic number Z and decreases from the deep-shell (K, L, or M) threshold energy with increasing E_p as $E_p^{-7/2}$. For large values of Z the PE cross section can be appreciable even for E_p as large as 1 MeV.

In the 0.1- to 10-MeV range of E_p, interactions are usually dominated by CE which results from the particle nature of electromagnetic radiation and is governed by conservation of energy and momentum. The probability of a Compton scattering event scales with Z of the atomic system involved, but in constrast to PE, for which photons deliver a particular kinetic energy to electrons, CE electrons cover a continuous spectrum from zero to a maximum:

$$E_e(\max) = 2E_p^2(mc^2 + 2E_p)^{-1}, \quad (1)$$

where m is the electron mass and c is the velocity of light. Creation of electron–positron pairs (PP) becomes possible when E_p exceeds 1.02 eV $= 2mc^2$, i.e., twice the rest mass of electron. The cross section for PP, which scales with Z^2, rises slowly from this threshold and reaches values competitive with CE near 10 MeV. The kinetic energy of each component of the pair is $\frac{1}{2}(E_p - 2mc^2)$.

Other less prevalent reaction mechanisms include direct *Compton scattering of photons by the nucleus*, and *photodisintegration of the nucleus*. The first process demands very high values of E_p to become significant and the last is sensitive to the details of nuclear structure.

B. Low-Energy Photons (Infrared, Visible, and Near Ultraviolet) [3,4]

In this range interaction occurs principally by ionization and excitation of the outer orbitals or valence-band electrons. In the vacuum-ultraviolet region ($E_p > 10$ eV) photons are efficiently absorbed by all but the lightest atoms and most molecular and condensed systems. As in the case of PE, the details of the absorption spectra are sensitive to the electronic structure of the absorber, and absorption spectral measurements constitute one of the primary tools for electronic structure investigation. Insulator crystals, the so-called wide-band-gap materials such as alkali and alkaline earth halides and many oxides, are transparent in the visible range. The *absorption edge* or threshold corresponds to the minimum E_p required to excite and/or ionize the constituent atoms of the solid. The first absorption processes result in the creation of *excitons*, i.e., excitations which are analogous to excited states of isolated atoms but which in all but certain halides are not localized but belong to the entire crystal. Just beyond the narrow exciton interval, photoproduction of quasifree electrons and holes (the crystal analog of PP) occurs and photoconductivity results. Recombination of electron–hole pairs and decay of excitons may generate either photons (luminescence) or phonons (thermal lattice vibration). Crystals become essentially opaque beyond the absorption edge. Localized absorption processes can occur at imperfections in insulator or semiconductor crystals since impurity atoms or ions and lattice vacancies and interstitials are frequently ionized or excited by photon energies less than the absorption edge. As the band gap decreases, E_p diminishes, passing through the visible into the infrared. Most commonly used semiconductors exhibit an edge which lies in the near infrared (0.8 eV for Ge) and is indicative of the relatively weak binding of electrons in semiconductors. In metals the band gap goes to zero and photons of all energies can be absorbed to increase the kinetic energy of conduction electrons or holes; this kinetic energy is quickly transferred to the lattice as heat (phonons) by scattering processes.

Photons of suitable energy are capable of exciting vibrations directly in both molecules and crystal lattices. The excitation may occur either by photon absorption or photon scattering (*Raman process*). Since phonon energies lie in the range 10–100 meV, photons well into the infrared are required for the absorption process. The propagating wave must couple to the vibration via its oscillating electric field; hence only those vibrations which generate an ac electric field can be excited. In polar or ionic solids these are the so-called LO or TO phonons and are responsible for the long-wavelength cutoff of infrared transmission in such crystals which is called the *restrahl frequency*. In homopolar materials, there are no electric-dipole vibrations and photon-absorption excitation of polar modes does not occur. *Acoustic* or nonpolar vibrations may be excited in both polar and nonpolar materials by the scattering of visible photons, a process known as Raman scattering which is manifested by the shift of the photon energy by plus or minus the amount of the vibrational energy. Other types of scattering such as Mie or Rayleigh processes do not change the photon energy and will not be discussed here.

C. *Radiolysis* [5]

An important consequence of exposing certain ionic solids to ionizing radiation such as x rays and γ rays is radiochemical decomposition or radiolysis. Of particular interest is the production of lattice vacancies and interstitials in alkali and alkaline earth halides which are manifested by creation of the familiar color centers. In essence these defects are created during the recombination of electrons with halogen molecular ions and the decay of excited halogen pairs (excitons).

II. *CORPUSCULAR RADIATION* [*1,6,7*]

A. *Fast Neutrons*

Neutrons created in the fission or fusion process have energies in the 1- to 20-MeV range and interact with atoms by elastic hard-sphere collisions with their nuclei. This is the process by which neutrons are moderated in a thermal fission reactor. The energy transferred in a hard-sphere collision is given by

$$\bar{T} = \frac{1}{2} T(\text{max}) = \frac{E_n M n}{(M+n)^2}, \tag{2}$$

where E_n is the neutron energy and M and n are the masses of the target atom and the neutron, respectively. $T(\text{max})$ is the energy transferred in a head-on collision. For a typical neutron–atom collision ($E_n = 2$ MeV and $M = 100$ atomic mass units) the kinetic energy of the recoiling nucleus is $\sim 10^4$ eV. Since the displacement energy E_d necessary to eject an atom from its lattice site is in the range 10–50 eV for most solids, not only is the *primary recoil* displaced, but it becomes a particle capable of producing secondary and tertiary displacements and initiates a displacement cascade characterized by a relatively high density of lattice defects (interstitial atoms and lattice vacancies) in a roughly spherical region of ~ 100 lattice distances in diameter. This lattice damage results in stored energy and influences those physical properties (electronic, optical, magnetic, and mechanical) which are structure sensitive.

B. *Charged Particles*

When incident upon a solid, energetic charged particles (protons, deuterons, and heavier ions with energies in the MeV range) lose most of their energy by ionization of lattice atoms, which results in trapped charge and possibly radiolysis (see above) in insulators but which has as the major effect the heating of the solid. A significant fraction of this energy, however, is dissipated via elastic collisions. In contrast to neutrons which undergo hard-sphere collisions, energy transfer results from charge interaction between the moving ion and the nucleus of the target atom, i.e., Rutherford scattering. The primary characteristic of this process is that small energy transfers are strongly favored. Hence the average kinetic energy of an atom recoiling from a proton collision is much less than for a collision involving a neutron of the same energy. Therefore, on the averge the lattice damage will consist of small clusters of lattice defects rather than the extensive disordered regions characteristic of fast neutrons.

C. *High-Energy Electrons*

Because of the rapid increase in mass with energy in the relativistic energy range, MeV electrons are also capable of displacing atoms in solids. This process is a convenient means of both producing isolated interstitial–vacancy pairs and determining the magnitude of E_d. The relativistic expression for the energy transferred in a head-on collision between an electron of energy E_e and an atom of mass M is

$$T(\max)=\frac{2(E_e+2mc^2)}{Mc^2}E_e. \qquad (3)$$

The value of E_d for a particular crystalline solid can thus be determined from the lowest-energy electron capable of inducing a change in some physical property sensitive to displaced atoms. Since absorption of high-energy photons (γ rays) by solids produces electrons with energies in the MeV range (see Section I A), the effect is that of internal electron bombardment; hence, interstitial–vacancy pairs can be created by γ irradiation via the displacement process.

D. Channeling [8,9]

When energetic protons or heavy ions are incident upon a single crystal along some relatively open crystallographic direction, the loss of energy to the lattice is minimized and anomalously large penetration depths are observed. This process is called *channeling* and may involve an open low-index lattice channel or the space between two lattice planes. Channels can be blocked by interstitials and oversized impurities, and the number and identity of these may be investigated by measuring the intensity and the energy of backscattered ions which would be "channeled" in the absence of these blocking imperfections.

See also CHANNELING; EXCITONS; LUMINESCENCE; PHONONS; PHOTOCONDUCTIVITY; RADIATION DAMAGE IN SOLIDS.

REFERENCES

1. D. S. Billington and J. H. Crawford, Jr., *Radiation Damage in Solids*, Chap. 2. Princeton University Press, Princeton, N. J., 1961, (I)
2. O. S. Oen and D. K. Holmes, *J. Appl. Phys.* **30,** 1289 (1959). (A)
3. C. Kittel, *Introduction to Solid State Physics*, 4th ed., Chap. 18. Wiley, New York, 1971. (E)
4. B. DiBartolo, *Optical Interactions in Solids*. Wiley, New York, 1968. (A)
5. E. Sonder and W. A. Sibley, *Point Defects in Solids* (J. H. Crawford and L. M. Slifkin, eds.), Vol. 1, Chap. 4. Plenum Press, New York 1972. (E)
6. B. T. Kelly, *Irradiation Damage to Solids*. Pergamon Press, Oxford, 1966. (E)
7. M. W. Thompson, *Defects and Radiation Damage in Metals*. Cambridge University Press, London, 1969. (I)
8. D. V. Morgan, *Channeling*. Wiley, New York, 1973. (I)
9. S. T. Picraux, E. P. Er Nisse, and F. L. Vook, *Applications of Ion Beams to Metals*. Plenum Press, New York, 1974. (A)

Rare Earths

F. H. Spedding

The rare earths are a group of 17 metallic elements located in the third column and in the extended sixth row of the periodic table. This extension in most periodic charts is shown as a second row six at the bottom of the chart. These elements posed an anomaly in the original Mendelyeev periodic chart, since they all seemed to possess similar properties and to fall in square 57, which was reserved for lanthanum.

The first rare earth to be discovered was yttrium, which was isolated as a rather impure oxide in 1797. During the next 150 years, the discovery of a great many rare earths was reported in the literature, and there was considerable confusion as to just how many rare earths there were, and also as to which elements should be classified as a rare earth. After the x-ray work of Moseley, and with the development of the modern theories of atomic structure, it is now known that 14 elements, starting with element 58 and extending through 71, are somewhat unique. In this part of the periodic table, as the charge on the nucleus is increased with increasing atomic number, the balancing electrons entering the external part of the atom go into an inner ($4f$) subshell. This subshell is well shielded from the electrical and magnetic fields of the neighboring atoms in the condensed state by the completed $5s^2$ and $5p^6$ subshells. The $4f$ subshell can hold 14 electrons, and there are, therefore, 14 elements in this group. They are now generally called the "lanthanides," although for many years the physicists tended to consider these as the true rare earths.

All of the rare earths contain three electrons in their valency shells, and they are, therefore, trivalent. In aqueous solutions they become strong electrolytes. The trivalent ions strongly attract the water molecules, which are dipoles, and thus form a tight sheath of water around the ions. As a result, the properties of the rare earths are much alike in aqueous solutions, and it is this property which gives rise to the general belief that the rare earths are all alike in their properties and resemble the properties of lanthanum. In the case of the metals, the three valency electrons usually become conduction electrons, although, for europium and ytterbium, one of the valence electrons enters the $4f$ subshell, and these metals have only two electrons per atom in their conduction bands. Since the chemical properties of the rare-earth salts are very similar, the chemists prefer to include yttrium and lanthanum with the lanthanide elements when they refer to them as "rare earths." This matter was greatly clarified in 1968 when the International Union of Pure and Applied Chemistry recommended that the elements scandium, yttrium, lanthanum, and the lanthanides be referred to as the "rare earths" while the subgroup of elements from atomic number 58 through 71 be referred to as the "lanthanides."

The rare earths are neither rare nor earths. At the time yttrium was discovered, the Greek influence in chemistry was very strong, and the Greeks classified all materials as a mixture of air, earth, fire, and water. Earths were substances which could not be further decomposed with the heats then available in the laboratory, and the common earths, such as magnesium oxide, calcium oxide, and aluminum oxide had not been decomposed and were thought to be elements. Yttrium oxide and the other rare-earth oxides which were isolated from rare minerals resembled the common earths in properties and were therefore called the rare earths.

Table I Rare earths

Atomic number	Element	Symbol	Atomic weight	Density (g/cm³)	Melting point (°C)	Boiling point (°C)
21	Scandium	Sc	44.956	2.9890	1,541	2,831
39	Yttrium	Y	88.905	4.4689	1,522	3,338
57	Lanthanum	La	138.91	6.1453	921	3,457
58	Cerium	Ce	140.12	6.672	799	3,426
59	Praseodymium	Pr	140.907	6.773	931	3,512
60	Neodymium	Nd	144.24	7.007	1,021	3,068
61	Promethium	Pm	(145)[a]	—	1,168	2,700
62	Samarium	Sm	150.35	7.520	1,077	1,791
63	Europium	Eu	151.96	5.2434	822	1,597
64	Gadolinium	Gd	157.25	7.9004	1,313	3,266
65	Terbium	Tb	158.9254	8.2294	1,356	3,123
66	Dysprosium	Dy	162.50	8.5500	1,412	2,562
67	Holmium	Ho	164.930	8.7947	1,474	2,695
68	Erbium	Er	167.26	9.066	1,529	2,863
69	Thulium	Tm	168.934	9.3208	1,545	1,947
70	Ytterbium	Yb	173.04	6.9654	819	1,194
71	Lutetium	Lu	174.97	9.8404	1,663	3,395

[a] Radioactive isotope separated from fission residues.

They are not rare in nature, since cerium is reported to be more abundant in the earth's crust than tin and lead, and even the rarer rare earths, such as europium and lutetium, are much more abundant than the platinum-group elements. They are always found together, widely distributed in nature, being present to a few parts per million in many rocks, such as the basalt, schists, and silicate rocks. They also occur in concentrated form as mixtures in many minerals such as bastnaesite, monazite, and xenotime.

The rare earths are of great interest to many fields of science, since they serve as powerful tools in checking theories and reaction mechanisms. One can choose a series of compounds whose properties vary across the rare-earth series in a known manner; many of the variables such as crystal structure, except for lattice parameters, can be kept constant, and one can therefore determine whether the mathematical relations between properties are real or rather parameters in arbitrary curve fitting. They are accordingly of interest to geochemists, astrophysicists, chemists, physicists and biologists, for this purpose. While the properties vary very slightly across the series when the rare-earth ions are surrounded by water molecules, as these water molecules are removed, for example, by organic chelates or in anhydrous salts, difference in properties are more marked; finally, in the metals where the difference in properties of adjacent rare earths are small, the difference in properties across the series can be as large as for any other group of elements in the periodic table (See Table I).

The rare earths represent one-sixth of the elements in the periodic table and roughly one-fourth of the metals, and it is therefore possible to find alloys or compounds which have almost any property one desires. In spite of this, for most applications they are rarely used industrially, since other, cheaper materials will serve equally well. However, where they have unique properties, they do find wide application in industry. Thus, for example, they are important in making catalysts for the cracking of petroleum, for abrasive materials in polishing of glass, in "getter" materials for removing impurities from other metals, for phosphorescing materials in TV screens, and in street lights, and for pyrophoric materials in cigarette lighter flints and in weapons. They play an important role in micro devices for electronic circuits and in the preparation of lasers.

See also TRANSITION ELEMENTS.

BIBLIOGRAPHY

R. J. Elliott (ed.), *Magnetic Properties of the Rare Earth Elements*. Plenum Publishing Company, Ltd., London and New York, 1972.

C. M. Lederer, J. M. Hollander, and I. Perlman, *Table of Isotopes*, 6th ed. Wiley, New York, 1967. Includes a review of radioactivity of the rare earth elements.

F. H. Spedding, *Rare Earths*, extensive article for Encyclopedia Britannica, 15th ed., 1974.

F. H. Spedding and A. H. Daane (eds.), *Rare Earths*. Wiley, New York, 1961. A broad review of rare earth chemistry and metallurgy to 1960.

K. Gschneidner and L. Eyring, *Handbook on the Physics and Chemistry of Rare Earths Volume 1: Metals, Volume 2: Alloys and Intermetallics*. North-Holland Publishing Company, Amsterdam, New York, Oxford, 1978.

Reflection High-Energy Electron Diffraction (RHEED)

Richard W. Vook

RHEED (reflection high-energy electron diffraction) is a special case of HEED in which the incident electron beam strikes the surface of a specimen at grazing incidence. Diffraction patterns from monocrystalline, polycrystalline, and amorphous structures can be obtained. The RHEED pattern consists of approximately one half of a transmission electron-diffraction (TED) pattern bounded by a "shadow edge" parallel to the surface. There are, however, important differences, which depend on the angle of incidence, surface topography, and refractive index μ of the diffracting material, that strongly influence the character of the RHEED pattern.

The closest resemblance to half of a TED pattern is obtained in the case of a rough surface, where the pattern is produced by electrons that penetrate only the extremities of the surface protrusions. As the surface becomes flatter and smoother, refraction and surface effects become important. The angle of incidence then determines the depth in the crystal from which coherent diffraction takes place, and the Bragg law becomes

$$(\lambda/2d)^2 = (\mu^2 - 1) + \sin^2\phi \tag{1}$$

where $\mu(=\lambda/\lambda') \simeq 1 + U/2V > 1$; λ and λ' are the electron wavelengths in vacuum and in the crystal, respectively; d is the interplanar spacing; ϕ is the grazing angle of incidence and reflection; U is the mean inner potential; and V is the accelerating voltage. Refraction shifts the reflections toward the shadow edge by an amount

$$Z - Z' = L^2(\mu^2 - 1)/(Z + Z') \tag{2}$$

where L is the distance from the specimen to the photographic plate, and Z' and Z are the positions of the diffracted beam with and without refraction. The method may be used to determine inner potentials.

With very flat surfaces, the penetration of the electron beam in a direction perpendicular to the surface is very small. The surface layer of the crystal that contributes to producing coherently diffracted electrons with detectable intensity may be as little as a few monatomic layers, resulting in symmetric line broadening along a line perpendicular to the surface of the crystal. For monocrystalline materials, diffraction spots are converted to parallel streaks. The flat surfaces needed to see such effects have been obtained by vacuum-deposited thin-film epitaxial growth techniques using as substrates mica or films grown on mica. Mechanically polished faces that have been cleaned and annealed have also been used. A similar effect occurs for faceted monocrystalline surfaces. The streaks are located perpendicular to the facet planes. Their orientation can be used to determine the Miller indices of the crystal plane of the facet.

The sensitivity of RHEED to adsorbed monocrystalline surface layers is comparable to that of LEED, but flat smooth surfaces are required in the former case. Then a few tenths of an average monolayer (around 10^{-8} g/cm^2) can be detected using photographic recording techniques.

RHEED instruments are usually constructed so that the specimen surface can be rotated about its normal by up to 360°. The angle of grazing incidence can be varied by several degrees, and the specimen itself can be translated by at least 0.5–1 cm perpendicular to its surface. The diffraction pattern is usually recorded photographically, either by direct electron impingement on the plate or by photographing the pattern formed on a fluorescent screen. Faraday cages or other electron detectors have also been used for more quantitative work. The camera constant ($L\lambda$) is usually determined from standard specimens. Because of the small angle of incidence, the area of the sample surface from which the diffracted beams arise is much larger than in TED. Consequently, interplanar spacing measurements tend to be less accurate. However, all of reciprocal space can be sampled by rotating the crystal about an axis perpendicular to its surface. Also, the beam intensity at a point on the surface of a sample is much less than in TED, leading to lower heating and radiation damage effects.

See also ELECTRON DIFFRACTION; LOW-ENERGY ELECTRON DIFFRACTION (LEED); SURFACES AND INTERFACES.

BIBLIOGRAPHY

Richard Beeching, *Electron Diffraction*. Methuen, London, 1950. (E)
John M. Cowley, *Diffraction Physics*. North-Holland, Amsterdam, 1975. (A)
Z. G. Pinsker, *Electron Diffraction* (transl. by J. A. Spink and E. Feigl). Butterworth, London, 1953. (I)
H. Raether, "Elektroneninterferenzen," *Hand. Phys.* **32,** 433 (1957). (I–A)
T. B. Rymer, *Electron Diffraction*. Methuen, London, 1970. (E–I)
R. W. Vook, "Microstructural Characterization of Thin Films," in *Treatise on Materials Science and Technology* (H. Herman, ed.), Vol. 4, Academic Press, New York 1 (1974). (E–I)

Relaxation Phenomena

R. Orbach

A physical system disturbed from an equilibrium configuration may relax back to equilibrium by means of contact with a thermal reservoir. The rate of approach to equilibrium, most often characterized by a relaxation time τ, can yield information concerning the coupling between the system and the reservoir, and concerning the microscopic properties of the system itself. Intermediate steps may be involved, and the equilibrium distribution may be only metastable, a position of only relative minimum free energy.

It is convenient in many cases to cast the description of the system and reservoir into the form of a density matrix ρ (see the book by Tolman[1] for a detailed description of the density matrix and its uses). The equilibrium configuration is defined as that configuration for which the off-diagonal density matrix elements, ρ_{nm}, $n \neq m$, are zero, while the diagonal elements are given by the Boltzmann relation,

$$\rho_{mm} = \frac{\exp(-E_m/k_BT)}{\sum_n \exp(-E_n/k_BT)}. \tag{1}$$

Here, E_m is the energy of the mth state, and T is the thermal equilibrium temperature. It proves useful in some cases to define a system temperature T_s (e.g., the spin temperature for nuclear spins in a solid or liquid; see the book by Abragam[2] for details), where T_s may differ from the reservoir temperature, but the system has come into internal equilibrium. In such a case, ρ^s_{nm}, $n \neq m$, vanishes, and

$$\rho^s_{mm} = \frac{\exp(-E_m/k_BT_s)}{\sum_n \exp(-E_n/k_BT_s)}. \tag{2}$$

For systems with finite numbers of energy levels, T_s can even be negative, referring to an inversion of level occupancy. Relaxation to equilibrium with the reservoir takes the route

$$T_s<0 \text{ to } T_s \sim \infty \text{ to } T_s = T.$$

The actual contact between the system and reservoir may not be direct. Conduction-electron spins in a metal typically relax much more rapidly to the kinetic energy of the electrons than to the external environment (the heat bath with which the metal is in contact). This leads to a picture of connected reservoirs of finite heat capacity, ultimately coupled to a very large reservoir of unlimited heat capacity as the relaxation route. Each of the intervening reservoirs may be described by its own local temperature. Thus, nuclear spins interacting with a large external magnetic field may be described in terms of their Zeeman level occupancies (the Zeeman reservoir). In turn, the nuclei may interact with one another via dipolar spin forces. This interaction can relax the Zeeman level occupancies to equilibrium with the "dipolar" reservoir. Finally, the latter may be in contact with a large thermal bath. The specific heats of the first two reservoirs may be comparable, leading to an exchange of energy or relaxation from the Zeeman reservoir to the dipolar reservoir to the thermal bath. The Zeeman relaxation path is thereby indirect, and can even lead to important phase-memory effects when the dipolar reservoir is only weakly coupled to the thermal bath (In such a case, the use of a dipolar temperature is even wrong. The off-diagonal density matrix elements are most certainly not zero. See Abragam[2] for a fuller discussion of this important case.)

Many times, the intermediate reservoir cannot accommodate the energy being dumped into it if its coupling to the thermal bath is weak. This leads to a "bottleneck," and a heating of the intermediate reservoir. An example is the normal electrons (the quasiparticles) in a superconducting metal at finite temperatures. These electrons are in intimate contact with the lattice vibrations of energy $\gtrsim 2\Delta$, where Δ is the energy gap. Excitation of the quasiparticles can occur by any number of methods, one such being the breaking of "Cooper pairs" of the superconducting condensate (an excellent physical description of these processes can be found in the book by de Gennes[3]). The excited quasiparticles relax by generation of lattice vibrations of energy $\gtrsim 2\Delta$, which in turn can break other Cooper pairs. This results in a trapping of these participating lattice vibrations, not unlike the trapping of resonance radiation (see the articles by Holstein[4] and the book by Mitchell and Zemansky[5]). The resultant heating of the intermediate reservoir (in this case the lattice vibrations of energy $\gtrsim 2\Delta$) is relieved only by the relatively slow process of scattering into other vibrations of differing energy (anharmonic scattering), or by direct flow into the thermal bath.

The approach to equilibrium in general depends on the microscopic nature of the system and the character of the equilibrating interactions. For example, the collision term in the Boltzmann equation for the six-dimensional distribution function f (see the article on transport theory in this volume[6]) is often written in the relaxation-time approximation,

$$\left.\frac{\partial f}{\partial t}\right|_{\text{collisions}} = -\frac{f-\bar{f}}{\tau(E)}, \tag{3}$$

where $\bar{f}$ is the equilibrium distribution function and $\tau(E)$ is a relaxation time dependent on the energy of the relaxing particle (gas molecule, electron, etc.). The time $\tau(E)$ can only be defined under special circumstances. For most systems, a relaxation time exists only for quasielastic scattering. Thus, for electrons in metals, $\partial f/\partial t|_{\text{collisions}}$ can be written in the form (3) only for impurity scattering, or upon scattering off of lattice vibrations at temperatures $T > \Theta_D$, the Debye temperature.

Formal expressions for the collision term can be separated into processes which scatter out of the state defined by the position vector $\mathbf{r}_i$ and velocity vector $\mathbf{c}_i$; and for scattering into that state. The subscript i designates the ith particle. The latter processes interfere with the ability to define a $\tau(E)$, and in many cases cause the collision term to generate an integral equation for the distribution function f.

The simple form of (2) must be abandoned when considering relaxation of higher-order multipoles. In magnetic systems, one must deal with the relaxation of the magnetic moment. The characteristics of both magnitude and direction of the magnetization vector require a physical condition to be placed upon the relaxation process. For example, in ferromagnetic resonance[7] the length of the magnetization vector $\mathbf{M}$ may be preserved in the relaxation process, and relaxation may occur to the instantaneous direction of the field experienced by $\mathbf{M}$. The relaxation term in the torque equation for $\mathbf{M}$ is then of the Landau–Lifshitz type:

$$\left.\frac{d\mathbf{M}}{dt}\right|_{\text{relaxation}} = -\frac{\lambda}{M^2}[\mathbf{M}\times(\mathbf{M}\times\mathbf{H})], \tag{4}$$

where λ is a constant and $\mathbf{H}$ is the field experienced by $\mathbf{M}$. The double cross product ensures constancy of $\mathbf{M}$ and relaxation toward $\mathbf{H}$. In the absence of driving terms, the magnetization vector experiences damped precession as exhibited in Fig. 1. An alternate form of the damping, proportional to $d\mathbf{M}/dt$, has been proposed by Gilbert:

$$\left.\frac{d\mathbf{M}}{dt}\right|_{\text{relaxation}} = \frac{\alpha_0}{M}\left(\mathbf{M}\times\frac{d\mathbf{M}}{dt}\right). \tag{5}$$

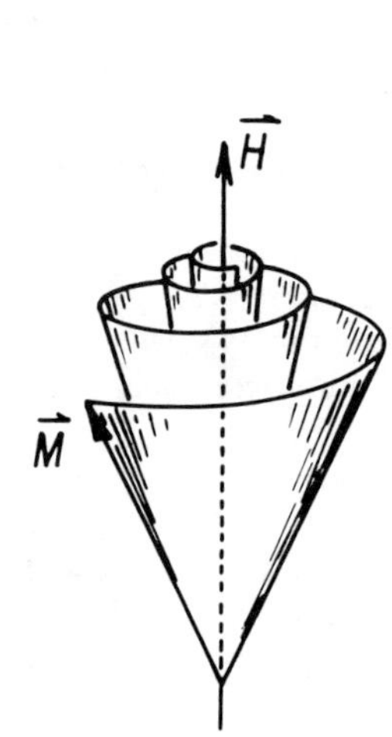

FIG. 1. Damped precession of the magnetization vector $\mathbf{M}$ according to the Landau–Lifshitz relaxation term in the presence of a steady magnetic field, in the absence of alternating fields.

It is a simple matter to express (4) and (5) in the same form if one is willing to alter the free precession constant. For large damping (the constant α_0 large) in, for example, pulsed remagnetization of thin films, Eq. (5) seems to give better agreement with experiment. The time for total magnetization reorientation is proportional to the damping parameter α_0.

Relaxation for a variety of both paramagnetic and ferromagnetic systems can be described in terms of the Bloch conditions:

$$\left.\frac{dM_z}{dt}\right|_{\text{relaxation}} = -\frac{M_z - \bar{M}_z}{T_1}, \tag{6a}$$

$$\left.\frac{dM_{x,y}}{dt}\right|_{\text{relaxation}} = -\frac{M_{x,y}}{T_2}, \tag{6b}$$

where $T_2 \leqq 2T_1$ for sensible response. These famous terms rely for their applicability on the special character of the z direction, as would be the case for a strong steady magnetic field pointing along the z direction, with weak perturbing alternating fields along the perpendicular x and y directions. The quantity $\bar{M}_z$ refers to the equilibrium value of the z component of magnetization. Clearly, Eq. (6) does not conserve the length of the magnetization vector M. Furthermore, it assumes weak alternating driving fields compared to intermoment coupling terms (e.g., dipolar forces). If the converse is true, relaxation may occur to the instantaneous direction of the field experienced by the magnetic system in a plane rotating with the angular frequency of the alternating fields. (This limit is discussed in detail by Redfield.[8]) In more concentrated nuclear-magnetic-moment systems, the longitudinal relaxation time T_1 is caused by fluctuating transverse dipolar interactions, the frequency of fluctuation being equal to the Zeeman precessional rate (for a full discussion of dipolar relaxation, see Abragam[2] and Slichter[9]). The transverse time T_2 is the consequence both of transverse dipolar fluctuations at the Zeeman precessional rate and of longitudinal dipolar fluctuations at zero frequency. The latter are termed the frequency-modulation effect; the former are termed the spin-flip effect.

In more dilute nuclear magnetic and in electronic magnetic systems, longitudinal relaxation is primarily through interaction with the lattice vibrational modes. These give rise to very rapid temperature dependences for the relaxation times. (Experimental examples can be found in the collection of Manenkov and Orbach.[10])

The relaxation times caused by interaction with the lattice vibrations turn out to be very important for determination of the operating characteristics of a variety of physical devices, in particular masers and lasers. The same can be said for atomic or molecular collisions for gaseous devices. The study of relaxation processes is, therefore, important not only for understanding fundamental interactions in gases, liquids, and solids, but also of great practical importance.

See also ELECTRON SPIN RESONANCE; LATTICE DYNAMICS; SUPERCONDUCTIVITY THEORY.

REFERENCES

1. R. C. Tolman, *The Principles of Statistical Mechanics*. Oxford University Press, New York, 1938. (A)
2. A. Abragam, *The Principles of Nuclear Magnetism*. Clarendon Press, Oxford, 1961. (A)
3. P. G. de Gennes, *Superconductivity of Metals and Alloys*. Benjamin, New York, 1966. (I)
4. T. Holstein, *Phys. Rev.* **72,** 1212 (1947). (A)
5. Allan C. G. Mitchell and Mark W. Zemansky, *Resonance Radiation and Excited Atoms*. Cambridge University Press, Cambridge, England, 1961. (I)
6. Sudarshan K. Loyalka, "Transport Theory," in *Encyclopedia of Physics* (R. G. Lerner and G. L. Trigg, eds.), Addison-Wesley, Reading, Mass., 1981. (E)
7. S. V. Vonsovskii (ed.), *Ferromagnetic Resonance*. Pergamon Press, Oxford, 1966; M. Sparks, *Ferromagnetic Relaxation Theory*. McGraw-Hill, New York, 1964. (I)
8. A. G. Redfield, *IBM J. Res. Devel.* **1,** 19 (1957). (A)
9. C. P. Slichter, *Principles of Magnetic Resonance*. Harper and Row, New York, 1963. (I)
10. A. A. Manenkov and R. Orbach (eds.), *Spin-Lattice Relaxation in Ionic Solids*. Harper and Row, New York, 1966. (I,A)

Rheology

F. R. Eirich

Rheology (*rheos*—flow; *logos*—knowledge), the interdisciplinary science of irreversible processes underlying the phenomena of inelasticity and flow of condensed matter, was so named by E. C. Bingham, who also founded the Society, and the Journal, of Rheology. Overlapping with the areas of physical chemistry, solid state, polymer, colloid, surface, tribophysics, biophysics, fluid and applied mechanics, dislocation and plasticity theory, and fracture mechanics, rheology deals with frictional (viscous or diffusional) transport phenomena and any deformations of shape, dimensions, or positions relative to external and internal coordinates, and with the conversion of mechanical energy input into heat (internal friction and irreversible thermodynamics). Rheology is thus bounded by the fields of the mechanics of highly compressible fluids (gases) and by ideal elasticity (full recovery of original shape and invested energy). Since it is the principal science concerned with the stress-, strain-, and strain-rate-dependent behavior of condensed, real and ideal materials, i.e., with viscosity measurements and all kinds of viscoelasticity, it follows that mechanical hysteresis and relaxation; external flow; practically all mechanical testing up to and including failure and fatigue; the chemistry, physics, and engineering of industrial processes; the movements of and within living organisms (biorheology); much of geophysics; and some metallurgy fall within its province, a major division of materials science.

Irreversible processes, be they by macroscopic flow or internal friction, mechanical aging or permanent deformation of solids strained beyond their elastic limit, are the responses of matter being sheared. The simplest case, "simple" shearing flow, proceeds along equidistant, rectilinear (flow) lines; more complicated forms of flow occur in tension, compression, radial, torsional, conical, and biaxial flow, and various forms of extrusion, often affected by the rate and/or frequency of the enforced motions and their duration. Since the volume, and all atomic–molecular motions, are affected by temperature and pressure, all forms of flow must be studied isothermally; temperature changes due to flow as a frictional, nonconservative process have to be made an integral part of rheological studies. Rheology endeavors further to correlate types of macroscopic flow, as expressed by constitutive equations, with atomic, molecular, intermolecular, and domain motions, relating stress and rate of straining with material properties. Ideally, knowing the relevant molecular properties (weight, size, shape, polarity, polarizability, friction factors, etc.), the parameters of the continuum's mechanical equations of motion should be derivable on a molecular basis.

The first constitutive equation of flow, for simple liquids of constant volume in simple shear, was formulated empirically by Isaac Newton, who stipulated that a force f per unit area A on an imaginary stack of horizontal x-z lamellae maintains a constant velocity difference along y, i.e., a shear gradient in steady flow in the vertical x-y plane:

$$\frac{f}{A} \equiv \sigma_{xy} = \eta \frac{dv_x}{dy},$$

where $f/A = \sigma$ is the stress, i.e., a force per unit area, acting along x on imaginary x-z planes in the fluid; η is here a material constant, the coefficient of viscosity, often just called the "viscosity." Its reciprocal, $\phi \equiv 1/\eta$, is called the fluidity. The gradient of the velocity vector, dv_x/dy, is often given the symbol D_{xy} or simply D. Since $dx/dy \equiv \gamma$ is the shear strain, $\dot{\gamma} \equiv d(dx/dy)/dt$ is the rate of shear strain and is identical with D_{xy}. The unit of the viscosity is the poise (dyn s/cm^2); in the new international units, one pascal-second equals 10 poise.

Such "simple" shearing flow, symbolized by the uniform shifting of a deck of cards, is in reality complicated because it entails an internal rotation of angular velocity $\omega = \dot{\gamma}/2$, easily shown by the rotation and change in angles of the diagonals of a square with sides originally parallel to the x-y coordinates. In "pure" shearing motion, on the other hand, diagonals in the x-y plane change lengthwise but stay perpendicular:

$$\sigma_{xy} = \sigma_{yx} = 2\eta\dot{\gamma}_{xy},$$

where now

$$\dot{\gamma}_{xy} = \left(\frac{dv_x}{dy} + \frac{dv_y}{dx}\right).$$

For flow in more than one plane, the stresses and gradients become sets of three-dimensional vectors, i.e., tensors; if one includes a velocity $\mathbf{v}$ which is nonsteady but limited to small Reynolds number (no turbulence), and stipulates constant density, the equation of continuity

$$\nabla \cdot \mathbf{v} = 0$$

and the (still linear) Navier–Stokes equations have to be applied (ρ = density, $\mathbf{F}$ = body force, P = pressure):

$$\rho \frac{dv}{dt} = \rho\mathbf{F} + \nabla P + \eta\nabla^2\mathbf{v},$$

of which Newton's equation is a special case. To solve for a particular flow situation, one has to apply the pertinent equation of motion and to introduce the boundary conditions. Exact solutions exist for the viscometric flows, where the stresses are defined by shear rates only. For example, for the flow of a Newtonian liquid under pressure P through a tube, i.e., the coaxial frictional flow of thin cylindrical lamellae in telescopic motion with an adhering boundary layer of the fluid to the wall, one obtains Hagen–Poiseuille's law:

$$Q/t = \pi PR^4/8\eta l;$$

Q/t is the volume flow rate, R is the diameter of the tube, l is its length; the flow profile (velocity distribution) is parabolic, and the stress at the wall is $\sigma_R = PR/2l$. Other viscometric flows are, e.g., those between parallel moving plates, or a cone and plate, and the slow flow around a sphere, described by Stokes's law:

$$\eta = f/6\pi vr,$$

where v and r are the relative velocity and radius of the sphere, and f is the force acting on it.

Nonviscometric flows are those in which the fate of a flowing volume element is determined by the flow history of its surroundings. This includes most industrial forming applications, e.g., extension by a pulling force f_x during spinning operations:

$$f_x = \lambda d(\epsilon/dx)/dt = \lambda(dv_x/dx),$$

where ϵ is the strain dl/l in the uniaxial tensile direction, x, and $\lambda = 3\eta$ is Trouton's coefficient of viscosity.

Obeying the Navier–Stokes equations are the rather simple, low-molecular-weight liquids whose viscosity is independent of the shear rate, including water. A large body of theoretical rheology deals with more complicated types, so-called non-Newtonian fluids and flows, employing the fundamental principles of stress (or energy) balances, coordinate invariance, etc. For the *memory fluids*, e.g., the stresses are no longer linear functions of the gradients, the viscosities become functions of the stresses or gradients and their derivatives, stresses normal to the flow lines appear, and the gradient history becomes important. In simple shearing flow, one may write for the viscosity function

$$\eta(\dot{\gamma}) = \sigma(\dot{\gamma})/\dot{\gamma},$$

and for the concurrent normal stress differences

$$t_{11} - t_{22} = t_{11} - t_{33} = \sigma_1(\gamma),$$

and

$$t_{22} - t_{33} = \sigma_2(\gamma);$$

which differences represent tensions along the stream lines and are often approximately proportional to $\dot{\gamma}^2$. The general viscosity function for the tube flow is obtained by a log-log plot of σ_R vs $4Q/\pi R^3 \equiv D$, which gives $f(\eta) = \sigma_R/D$ $[\frac{3}{4} + \frac{1}{4}(d \ln D/d \ln \sigma_R)]$. The differential viscosity

$$\eta_{\nabla} = \frac{d\sigma}{d\gamma},$$

and the limit viscosities, η_0 and η_∞ at zero and infinite shear rates, are also used to characterize a rate-dependent viscosity, as is the "power law":

$$\sigma = \eta^0 \left(\frac{D}{D^0}\right)^n D^0,$$

$$\eta = \eta^0 \left(\frac{D}{D^0}\right)^{n-1},$$

with η^0 and D^0 chosen for a standard shear rate, and n is a measure of the deviation from Newtonian flow. Liquids for which η decreases with D are called shear-thinning; when η increases, they are called shear-thickening. Only when Trouton's viscosity increases with stress, or the viscosity increases by cooling or evaporation of solvents, are stable liquid threads formed in extensional flow and fibers can be spun. Shear thickening accompanied by volume expansion is called dilatancy.

True liquids show no lower stress limit for flow. Solids may undergo a variety of extensive, irreversible deformations, collectively called plastic flow (tertiary creep), above a stress yield σ_0. Plastic flow is not homogeneous and not viscometric. In contrast to viscous cases, plastic flow stresses have a low dependence on shear rate and external pressure. However, some two-phase or multiphase solids actually liquefy above σ_0 and then exhibit true, or even viscometric, flow (Bingham plastics):

$$\sigma - \sigma_0 = \eta_T \dot{\gamma}.$$

In tube flow, liquefaction occurs first at the wall when $\sigma_R = \sigma_0$, giving rise to "plug" flow, until σ_0 reaches the axis of the tube. Time-dependent liquefaction under shear and isothermal return from the fluid to the solid state constitutes thixotropy. If the flow of a thin surface layer of any fluid is very different from its flow in bulk, it exhibits a separate surface viscosity.

To model the fluid state on the molecular level, one notes that in the initial or rest state the atoms or molecules move by irregular thermal movements to and from randomly changing positions. In liquids, here defined as dense fluids, the free volume V_f between the molecules is only a fraction (~3–7%) of the total occupied volume. The resulting packing, though random over greater distances or time intervals, shows locally quasiordered domains which by thermal motion fluctuate in dimension and density on a characteristic time scale (see below). External forces cause bulk flow by the superposition of a bias onto the random rest movements, i.e., causing excess directional motion and continuously overbalancing the randomizing molecular collisions. Because of the latter, flow continuously dissipates the energy supplied by input of mechanical energy. (On the other hand, leveling of any excess concentration by thermal motion constitutes diffusion:

$$\eta \propto 1/\mathcal{D},$$

where $\mathcal{D}$ is the coefficient of diffusion.) This model of flow affords a second definition of viscosity, as the ratio of the work dissipated per unit volume and time to the square of the shear gradient:

$$\eta = \dot{W}/D^2.$$

The process of viscous internal (or fluid) friction in bulk flow thus consists of momentum transfer, or drag, between molecular layers of different velocities by cross-diffusion of molecules. For the momentum exchange of the freely moving molecules of gaseous fluids,

$$\eta = m\bar{c}/3\sqrt{2}\pi d^2,$$

where m and d are molecular mass and diameter, and $\bar{c}$ is the average speed. In dense liquids, the molecules are coupled by attraction potentials and move by thermal diffusion steps between positions of the quasilattices. From the perturbation of the distribution function of the rest state by the velocity gradient, and its relation to the pressure tensor, Born and Green found for the viscosity

$$\eta = f(r/V_m)[m\phi(r)^{1/2}]e^{-\phi r/kT},$$

where r is a measure of the intermolecular distances, V_m and m are the molecular volume and mass, and ϕ is the intermolecular potential. Employing the theory of chemical rate processes, on the other hand, Eyring *et al.* based their derivation of η on the existence of typical molecular jump frequencies of activation energies E_{act}:

$$\eta = (V_f^{1/3}/V_m)(2\pi mkT)^{1/2} e^{E_{act}/kT}.$$

The dependence on r, or $V_f^{1/3}$, gives the dependence of η on the external pressure. Both derivations are prototypes of the many molecular theories, some of which are quite successful in predicting the viscosity of simple liquids, and even of mixtures and solutions. Empirically, one often finds the fluidity ϕ_{sol} of mutually miscible liquids to be additive:

$$\phi_{sol} = \Sigma W_i \phi_i,$$

where W_i is the species weight fraction. All derivations must reproduce the observed temperature dependence of the viscosities, approximately represented by Arrhenius' empirical equation

$$\eta = Ae^{E/RT},$$

where E is an activation energy related to the heat of evaporation (mole cohesion) and A is a weakly temperature-dependent pre-exponential factor. A and E are more nearly constant for viscosities measured at constant volume than at constant pressure.

Molecular models permit rationalization of complicated types of flow including the concepts of a time-dependent elasticity and of related relaxation times, τ. In the ideal *elastic* cycle, a forced solid system goes through a continuous series of reversible changes of order of increasingly higher free energy (states of strain) and, on removal of the force, returns to its former state by giving up the imparted work. When the long-range order has broken down by melting, the forced state of higher free energy in flowing liquids, as stated, is one of decreased randomness of motion or positions, maintained against diffusion by the applied force but instantly beginning to return to randomness when the force stops. The difference, though, in the quasiordered states of the domains in the rest and strained state will decline, and the energy dissipate, only as fast as randomizing back diffusion (relaxation) proceeds. Depending on the extent of the imposed order, and on the relative rates of observation and of strain relaxation, one may see ideally fluid or partially elastic, rate- and/or time-dependent, i.e., non-Newtonian, behavior in liquids, the latter the more so, the more complicated the molecular structure. The relaxation time, τ, defined as the time after the cessation of the force required to allow the number of biased molecules n_0 to decay by a first-order process (self-diffusion) to n_0/e, is defined by

$$n = n_0 e^{-t/\tau}.$$

The corresponding time-dependent moduli may be established by observing the stress relaxation at strain γ_0, or tensile strain ϵ_0; $G(t)$ in shear, or $E(t)$, in tension:

$$E(t) = \sigma_t/\epsilon_0 = (\sigma_0/\epsilon_0)e^{-t/\tau}.$$

At constant stress, one observes continuous flow (primary creep), but if a constant weight is applied, the requirement of constant volume (when also $G \approx \frac{1}{3}E$) of an unconstrained sample of constant viscosity reduces the latter's cross section eventually to the point of instability and rupture.

Experimentally, a wide spread of τ's, ranging from milliseconds to thousands of hours in liquids of complicated structure, may be observed. The time-dependent moduli are dimensionally related to τ and η as

$$G = [\eta/\tau],$$

or, experimentally,

$$\eta = \frac{1}{3}\int_0^\infty E(t)dt.$$

Liquids which relax within the time span of the applied force cannot sustain shear or tension. However, when flow is prevented and premature cavitation protected against, tensile forces can be measured even in mobile liquids and are of the order of 10^{11} dyn/cm^2. For viscosities of, e.g., 10^{-1} poises, τ (a measure of domain lifetime)$\approx 10^{-10}$ s, about 1,000 times the period of a thermal oscillation. Similarly, $\tau = \eta_B/K$ has been formulated for the occurrence and relaxation of volume strains (K, bulk modulus; η_B, the viscosity for volume changes). In liquids of high viscosity (small V_f or V_m and/or E_{act}), strained positions may decay so slowly that "frozen strains" persist for long periods, constituting a record of previous thermal or strain history. Where temporary strained states are perturbed by new forces, the new strains will be superimposed. If this is additive, the material obeys Boltzmann's superposition principle; if the duration of the new force was much shorter than the period for the first strain relaxation, the previous strain decay curve will be returned to, a phenomenon often called mechanical memory.

Real materials span the whole range of relaxation behavior as a function of their chemistry, morphology, or of temperature and pressure: "solids" owe their time-independent elasticity and moduli to extensive internal order, which hampers self-diffusion and extends relaxation beyond observation; but sufficient dislocations and other imperfections may move at characteristic stresses, causing "plastic" flow; complete, rate-dependent, energy dissipation is the basic pattern of all "fluids"; combinations of both, exhibiting elements of time-independent and of rate-dependent strained states (including normal stresses) are typical for "viscoelasticity," which is exhibited by two-phase or multiphase materials, or by homogeneous systems of molecules which are capable of time- and stress-dependent states of association. Prominent among them are crystallizable and noncrystallizable macromolecules, especially of the flexible long-chain variety, whose low symmetry and extensive entanglements require cooperative local motions to respond to external forces. Their ordering during cooling and freezing is difficult, so that the decline of V_f with falling temperatures greatly increases the relaxation times. Frequently, this causes, over a relatively narrow volume and temperature range (glass transition temperature T_G, glass volume V_G at $V_f \approx 2.5\%$), even crystallizable molecules to interlock rigidly but in random fashion, and thus to form a "glass." Glass transition is accompanied by large increases in viscosity and elastic moduli, with characteristic values for every material, and its chemically and condition-dependent morphology. Interpretation of the mechanical responses of a material in terms of its infrastructure is called microrheology.

Simultaneously, viscous and linear elastic behavior can be expressed by constitutive equations based on Boltzmann's superposition principle:

$$\sigma_{xy}(t) = \int_{-\infty}^{0} G(t-t')\dot{\gamma}_{xy}(t')dt'$$

with integration from past times t' up to the present time t. Alternatively, this behavior can be expressed by Maxwell's equations and modelled as elastic spring and viscous dashpot, coupled in series or parallel, or correspondingly as electrical capacitance plus resistance. The Maxwell element for elasticoviscous liquids with instant elastic response and subsequent flow (primary creep) is described by

$$\gamma = \frac{d\sigma}{Gdt} + \frac{\sigma}{\eta},$$

while the Voigt element

$$S = G\gamma + \eta\dot{\gamma}$$

fits viscoelastic solids. This equation shows that $\dot{\gamma}$ goes to 0 at $S = G\gamma$, i.e., to strain equilibrium, by a time-dependent retarded approach ("secondary creep"),

$$\gamma_t = (\sigma/G)(1 + e^{-t/\tau});$$

this is followed by a retarded strain recovery after sudden stopping of a tensile stress application at a strain ϵ_0:

$$\epsilon = \epsilon_0 e^{-t/\tau},$$

with τ now termed the retardation time.

Most materials, even though one component, exhibit a spread of τ's attributable to various modes of motion of large molecules and to accidents of morphology. Similarly, inhomogeneous, multiphase, multicomponent materials exhibit very broad relaxation spectra and thus require a combination or set of such equations, with corresponding distributions or sets of moduli and viscosity coefficients or τ's for the description of their mechanical behavior. There are graphical and mathematical methods to analyze creep and relaxation data in terms of creep or relaxation functions, or as spectra of retardation or relaxation times. Bodies with relaxation times shorter than or comparable to the duration of the experiment will dissipate strain energy (input $\frac{1}{2}E\epsilon^2$) during a cycle from zero stress–strain to the final maximal strain and back to zero, so that the return curve of such a cycle shows a lower modulus and lies below the extension curve. The area in between measures W, the hysteresis (loss) per time and volume. When such losses are observed as a function of temperature or strain rate, a maximum is seen at the glass transition, where viscosities, strain rates, and amplitudes, the last decreasing when the former two increase (and vice versa), are intermediate.

Depending on whether solid or fluid domains are preponderant and form the continuous phase, dispersions will have overall properties ranging from solidity to viscoelasticity and Newtonian flow. They include polymers, soaps, latexes, paints, filled elastomers, concrete, muds, biological systems, etc. The viscosity of such dispersions, of colloids, and of solutions of natural and synthetic macromolecules occupies a large chapter of rheology. In sufficient dilution, the contributions of the dispersed particles to the energy dissipation of the flowing dispersion is additive. Einstein based his pioneer calculations on the perturbation of flow of Newtonian continua by the presence of discrete particles, and arrived at the viscosity of dilute suspensions of solid spheres:

$$\eta=\eta_0(1+2.5\phi),$$

where η_0 is the viscosity of the liquid and ϕ is the volume concentration of the disperse phase. This approach has been expanded to cover higher powers of ϕ and also fluid and nonspherical particles; for the latter, the coefficient of ϕ becomes a function of the axis ratio. A characteristic limiting value, the intrinsic viscosity, is defined as

$$[\eta]=(\eta/\eta_0-1)/\phi,$$

for $\phi\rightarrow 0$. However, the coils of long random chains pervade such a large volume ($\propto N^{3/2}$) that the range of dilute solution (viscosity linear with ϕ) is very limited and terms of ϕ^2 and higher power must be considered. Many equations have been developed for concentrated dispersions, most of which can be reduced to the form

$$\eta/\eta_0=1/(1-\phi)^n.$$

For $n=2.5$, the first term of the expansion agrees with Einstein's equation; for critical values of ϕ (e.g., 0.74 for spheres), $\eta\rightarrow\infty$, the dispersion will turn into a pasty, usually thixotropic, solid. Milk and blood are natural multicomponent dispersions as well as macromolecular solutions; clay strata are examples of solid dispersions which may liquefy thixotropically under the mechanical shocks of earthquakes.

Macromolecular chains in solution have been modelled as strings of beads (the monomeric units) of various degrees of stiffness including random coiling. For the flow around individual beads, resistance factors f (which determine also the sedimentation and diffusion rate of dispersed particles) were adopted, e.g., Stokes's law for spheres,

$$f=6\pi\eta r,$$

for a bead of radius r. The flow resistances of the beads of a string or in a swarm are summed up over their number $N=M/m$ (m is the monomer unit and M is the macromolecular weight) and for their hydrodynamic interaction, to obtain a theoretical equation for $[\eta]$. The most frequently used result is that by Flory:

$$[\eta]=\Phi(R_0^2/M)^{3/2}M^{1/2}\alpha^3=KM^{1/2}\alpha^3,$$

where Φ and K are constants, R_0 is the statistical chain end-to-end distance multiplied by factors which account for structural features, and α is a factor depending on the solvent.

More sophisticated theories of macromolecular viscosities introduce spring forces between the beads or chain sections, to account for the extra energy dissipation when in the shear field the coils contract and expand periodically from their equilibrium dimensions with the frequency $\dot{\gamma}/2$. The corresponding coil anisotropy, or the aperiodic rotatory orientation of rodlike molecules in competition with Brownian motion, are further cause for $[\eta]$ to vary with $\dot{\gamma}$ and for the appearance of flow birefringence, offering an opportunity to study macromolecular shapes and internal mobility. Microturbulences induced by long-chain molecules help to reduce fluid friction drag in pipes or around submerged bodies. Frequently, the rheology of macromolecular dispersions and melts also deals with the effects of nonpermanent chain entanglements: for shorter, largely slipping, chains $\eta\propto M$, while for chains of length greater than for a critical molecular weight, $\eta\propto M^{3.4}$. In addition, and especially in the presence of permanent, chemically cross-linked networks embedded in the liquid, the forces cause elastic strain, network slip, and viscous flow encumbered by friction against the net, i.e., a viscoelastic response. Most casting and spinning solutions, plastisols, adhesives, filled elastomers, gels, and biological systems contain networks of some type.

Much insight into the nature of viscoelastic behavior has been derived from research employing short-time, small-strain periodic pulses, e.g., sinusoidal vibrations, and studying phase reflections, wave propagations, and damping. If $\gamma_{xy}(t)=\gamma^0_{xy}\sin\omega t$, with γ^0_{xy} as the maximum amplitude of the strain, ω the frequency, and t the time, the accompanying rate of straining is zero at the peaks and valleys of the curve, and largest at $\gamma=0$, i.e., is 90% out of phase with its strain:

$$\dot{\gamma}_{xy}=\omega\gamma^0_{xy}\cos\omega t.$$

Strain and strain rate evoke simultaneous elastic and viscous stresses which add up to

$$\sigma_{xy}=\gamma^0_{xy}(G'\sin\omega t+G''\cos\omega t)$$

with G' as the elastic (energy storing) and G'' as the inelastic (energy dissipating) or loss modulus. Such composite stresses are then shifted from the in-phase oscillation σ^0_{xy}, and γ^0_{xy}, by some phase angle δ:

$$\sigma_{xy}=\sigma^0_{xy}\sin(\omega t+\delta).$$

It follows that $G'=G^0\cos\delta$; $G''=G^0\sin\delta$; and $G''/G'=\tan\delta$, the "loss tangent"; $G^0=\sigma^0_{xy}/\gamma^0_{xy}$ is the absolute modulus. G' and G'' are frequency dependent and interdependent and, since they act 90° out of phase, can be related by complex notation

$$G^*=G'+iG'',$$

with G^* as the "complex" modulus. The ratio of imparted energy converted into heat to stored energy is $W/W'=(\pi/2)\tan\delta$, and changes with $\tan\delta$ from 0 to 100%, corresponding to the mechanical hysteresis of a transient stress–strain cycle. The greater tan δ, the greater the self-heating of a material during cycling. If the time for sample relaxation is shorter than the stress cycle, the material will remain at some strain level with the cycling superimposed, a condition which aggravates fatigue in highly crystalline materials but reduces it in crystallizable elastomers.

The course of G'' and δ with temperature or frequency reveals much about the molecular response mechanisms of viscoelastic, and thus to some extent of all, materials: at *low* temperatures because of high density and modulus, and at *high* frequencies because of the shortness of time for a cycle, δ and G'' will be small and the response largely elastic. Conversely, for *high* temperatures or *very low* frequencies, the viscoelasticity is reduced to largely viscous behavior, but the energy losses G'' are small because of the low shear rates and stresses; δ will approach 90° and G' will approach 0. Thus, tan δ must peak in between. For a material with a single relaxation time, this will occur at the midpoint of the rise in G' with frequency or falling temperature, which is also the point where the imposed frequency is equal to $1/\tau$; if the inertial effects are included, this is also the point of resonance. The peak of tan δ marks further the most rapid transition from the viscous to the elastic response and is one of the measures of T_G; at higher temperatures, an even steeper rise of tan δ marks the melting range. Molecularly, this means the most rapid changes in the character of molecular, or molecular group, motion, or of morphology, occur during the change from the glassy to the rubbery stage, during changes in domain crystal structure, or fusion.

The course of the storage modulus G' with T and ω indicates the changes in the solid domains of viscoelastic materials. The low-T, high-ω modulus plateau, which for crystalline materials declines only slightly with temperature up to the breakdown at the melting point T_m, shows a marked drop at T_G (up to a factor of 10^5, depending on the chemistry and the amorphous–crystalline ratio) when glassy domains soften, followed by a second plateau or shoulder which depends on the density of the network of chain molecules which survived T_G; if the relaxation times of these network links are sufficiently short, there will be a rapid further drop in G' to the levels of mobile elastic liquids. For long relaxation times, e.g., chemical cross links, the second modulus plateau will be horizontal or even rising, constituting the "rubber range," up to and beyond T_m, followed by a final drop when chemical breakdown occurs at higher temperatures. In some plastics, the cross links break under stress and reform in relaxed positions, while the main molecular chains remain intact, thus showing truly plastic flow of a type which has been called chemorheological.

The reciprocity of the effects of frequency and temperature on the mechanical behavior of viscoelastic materials above the glass transition T_G (or a related reference temperature) means that a desired mechanical change can be made by equivalent *rises* in temperature, or *lowering* of frequency (rate of straining). Furthermore, since major internal mobility begins at T_G, it should be possible to correlate various viscoelastic states by their position on the temperature or frequency scale relative to T_G. Thus, using a system of reduced variables, the behavior of all linear viscoelastic materials becomes comparable at corresponding states. According to this time–temperature superposition principle, sections of log–modulus curves of a material at constant temperature versus log frequency can be shifted by a factor derived by Williams, Landel, and Ferry to merge into a single modulus–frequency curve over a wide range of the frequencies, thus creating "master" curves of moduli which permit prediction of small amplitude rheology under most practical conditions and have become the premise of much viscoelastic engineering.

Whether during cycling or transient force application, the viscoelastic responses become nonlinear for large strain amplitudes, normal stresses appear, and the mathematical and physical interpretations of observed behavior become much more difficult. The typical rubbers (elastomers), molecular networks of long coiled chains, are capable of very high reversible extensions ("high elasticity"), $\propto N^{1/2}$, the number of statistical units in the network link. Energy dissipation around T_G is considerable, but subsides as the usage temperature rises to 50° C or more above T_G. Large strains cause appreciable volume changes, even in shear, requiring constitutive equations which contain not only nonlinear terms, but include also the effects of the bulk modulus. Moreover, large strains tend to render internal and domain organization more anisotropic, so that any description requires the use of strain- or stress-rate dependent parameters. The most frequent consequences of large strains are, first, a modulus or viscosity drop, if an internal structure is broken up or slip is caused by orientation; often followed by a work hardening, if mobile dislocations and entanglements become locked, or a denser packing or crystallization is induced; finally the introduction of crazes and stress concentrations facilitate rupture. It is the microrheology of the domains in front of growing cracks, their work hardening, flow, and energy dissipation with their influence on brittle versus ductile behavior, which determines the strength of (particularly high-molecular-weight) materials. For elastomers, the ultimate properties (critical stress and elongation) are also subject to time–temperature superposition, and hence show a characteristic failure locus in the plot of stress at break versus strain at break, called the failure envelope. A striking response of elastic liquids is found when their dwell time in a force field, say the time of passage of a batch through a piece of processing equipment, is short compared to the relaxation time (small Deborah number). The strain recovery then continues beyond the end of the operation and leads to distortions or even fragmentation of extruded shapes (melt "swell" or "fracture"). Unrecovered strains during cold forming will recover on heating or annealing. The rheology of pastes, doughs, highly and multiply filled binders, and block polymers and composites is further complicated by their inhomogeneous nature. Beside localized energy storage and dissipation, highly nonlinear flow and elasticity of the components and interfaces must be considered and composite, and finite element analysis must be employed.

See also ANELASTICITY; ELASTICITY; INTERNAL FRICTION IN CRYSTALS; RELAXATION PHENOMENA; TRIBOLOGY.

BIBLIOGRAPHY

R. B. Bird, R. C. Armstrong, O. Hassager, and C.F. Curtiss, *Dynamics of Polymeric Liquids*. Wiley, New York, 1977.

R. B. Bird, W. E. Stewart, and E. N. Lightfoot, *Transport Phenomena*. Wiley, New York, 1960.

F. R. Eirich (ed.), *Rheology, Theory and Applications*, Vol I–V. Academic, New York, 1956–1969.

J. D. Ferry, *Viscoelastic Properties of Polymers*, 2nd ed. Wiley, New York, 1970.

C. D. Han, *Rheology in Polymer Processing*. Academic, New York, 1976.

Secondary Electron Emission

P. E. Best

Electrons which emerge from a solid as the result of bombardment by energetic particles are known as secondary electrons. Here the term is used for situations in which the incident or primary particles are electrons. The secondary yield δ is the number of electrons emerging from the solid for each incident electron that enters it. One characterization of the secondary-emission properties of a sample is the plot of δ vs K, the kinetic energy of the incident electrons at the surface. If δ_m is the maximum value of δ, occurring at energy K_m, then plots of δ/δ_m vs K/K_m have a similar form for all solids (Fig. 1). Values of δ_m range from 0.4 for specially prepared porous surfaces, through 0.5 to 1.8 for metals and about 1 to 20 for insulators prevented from charging, up to 1000 for samples with negative electron-affinity surfaces. Corresponding values of K_m range from 0.2 to greater than 20 keV.

For solids with $\delta_m > 1$ the lower and upper values of K for which $\delta = 1$ are called K_{I} and K_{II}, respectively. These values have particular significance for solids which are electrically isolated, including insulator surfaces. At equilibrium in an electron beam, the net current entering such a sample is zero. To achieve this condition in a chamber at ground potential, the sample becomes charged to a potential which causes K to be zero or K_{II}, depending on whether the initial value of K is less than K_{I} or greater than K_{II}, respectively. For an initial K between K_{I} and K_{II} the value of δ is reduced to unity by a relatively slight positive change of sample potential which prevents the slowest secondary electrons from escaping.

The secondary emission from a solid depends on the bulk and surface electronic structures, the energy and angle of incidence of the primary beam, and the surface topography. A quantitative understanding of secondary electron emission requires solution of the transport equations which describe the cascade processes involved. Qualitatively, the main features in secondary-emission measurements are understood by means of a three-step model in which electrons are excited into high-energy conduction states directly or indirectly by the incident beam; they then move to the surface, finally emerging across the surface barrier into the vacuum. The role that one step plays in the emission is probed by experiments in which variation due to the other steps is a minimum. For the data in Fig. 1 the transport and emergence steps can be represented by an average escape depth which varies little with change of K. The escape depth is a measure of the maximum distance from the surface from which an excited electron can reach the surface and escape into the vacuum. The general form of Fig. 1 can then be understood by use of the concept of penetration depth, which is the maximum distance from the surface at which appreciable excitation of valence electrons can be caused by the incident beam. The penetration depth increases with K. For small K the escape depth is greater than the penetration depth, and a constant fraction of electrons excited by the incident beam can emerge into the vacuum. The total number of valence electrons excited by the incident beam is proportional to K, and so the yield rises linearly with K for $K < K_m$ (Fig. 1). For K greater than K_m the penetration depth is greater than the escape depth. Furthermore, the average distance traveled by incident electrons between inelastic scattering events, the mean free path. increases with increase in K for $K > 200$ eV. Therefore, δ decreases as K increases for $K > K_m$ (Fig. 1).

The wide variation of values of δ_m for different samples is mainly a result of the wide variation of escape depths. A large escape depth implies a large value of δ_m. For a metal, the strong electron–electron interaction causes electrons originally excited to states above the vacuum level to lose energy, and the ability to escape into the vacuum, in a short distance (Fig. 2a). The escape depth is less than 10 nm; δ_m is small. In the insulator (Fig. 2b) electrons in the solid with energy just above the vacuum level have insufficient energy above the bottom of the conduction band to excite a valence electron across the band gap. These electrons lose energy by phonon excitation, a process with a relatively long mean free path and with small energy loss per collision. This insulator will have an escape depth and δ_m larger than those for metals. Figure 2c represents the situation for a *p*-type single-crystal semiconductor covered with cesium to produce a negative electron-affinity surface. For such samples

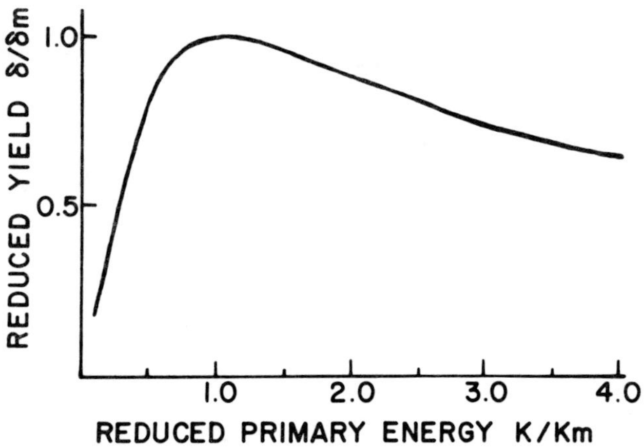

FIG. 1. Plot of reduced secondary yield, δ/δ_m, versus reduced primary energy, K/K_m. This curve has similar form for all solids.

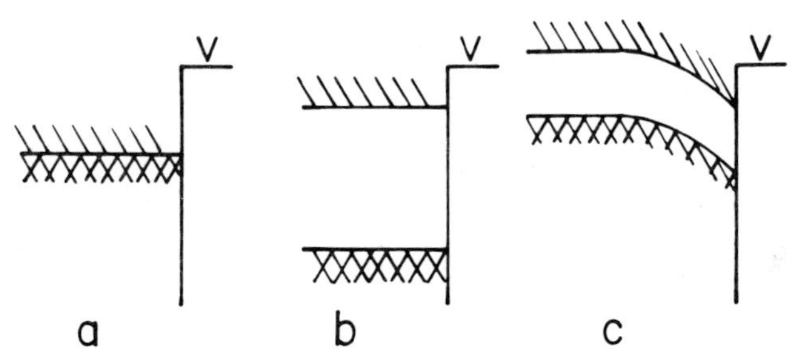

FIG. 2. Simple energy band models for (a) a metal, (b) an insulator, and (c) a semiconductor with a negative electron-affinity surface. The disposition of the energy bands with respect to the vacuum level is the main factor in determining the average escape depth, and hence δ_m. In this figure V is the vacuum level, the hatch marks are at the bottom of the unoccupied band, and the cross-hatching is at the top of the occupied band.

the escape depth can be of the order of microns, the diffusion length for electrons in the bulk semiconductor: δ_m is large. Most studies and applications of secondary emission have been made in the reflection mode. The large escape depth for negative-affinity emitters permits these materials to be used in the transmission mode, in which the secondaries emerge from the face opposite that which the incident beam enters.

A majority of secondary electrons have kinetic energies of less than 50 eV. By far the most prominent peak in the energy spectrum of the secondary electrons is the "slow" peak, which occurs at very low energies, from 1 to 6 eV. Very low-intensity peaks in the energy spectrum can be due to the Auger process, in which electrons are emitted in the radiationless decay of inner atomic vacancies. Auger peaks are used to identify atomic species present at surfaces and to estimate their relative abundance.

Measurements with angular resolution have been made of secondary emission from single crystals. The variation of δ with angle of incidence, at fixed K, is interpreted in terms of a penetration depth that varies with the direction of the incident beam in the crystal. Measurements of angle-resolved energy spectra at fixed input-beam conditions support a theory of excitation in which structure in the density of states is prominent in the energy distribution of the excited electrons. For lower incident-beam energies such measurements have also provided evidence for an emission process in which electrons are excited into surface resonances before emerging into the vacuum.

Secondary emission is essential to the operation of electron multipliers and photomultipliers. The emission is used to distinguish qualitatively between different surfaces in scanning electron microscopy, and is important in the operation of a number of vacuum tube devices.

Acknowledgment. The preparation of this manuscript was supported in part by a grant from the National Science Foundation.

See also AUGER EFFECT; ELECTRON AND ION IMPACT PHENOMENA; ELECTRON ENERGY STATES IN SOLIDS AND LIQUIDS.

BIBLIOGRAPHY

H. Bruining, *Physics and Applications of Secondary Electron Emission*. McGraw-Hill, New York, 1954.

A. J. Dekker, "Secondary Electron Emission," in *Solid State Physics* (F. Seitz and D. Turnbull, eds.) Academic, New York, 1958. Vol. 6, p. 251.

O. Hachenberg and W. Brauer, "Secondary Electron Emission from Solids," in *Advances in Electronics and Electron Physics* (L. Marton, ed.). Academic, New York, 1959, pp. 413–499.

R. U. Martinelli and D. G. Fisher, "The application of semiconductors with negative electron affinity surfaces to electron emission devices," *Proc. IEEE* **62**, 1339–1360 (1974).

N. R. Whetten, "Secondary Electron Emission," in *Methods of Experimental Physics* (L. Marton, ed.). Academic, New York, 1967, pp. 69–84.

R. F. Willis and N. E. Christensen, "Secondary electron emission spectroscopy of tungsten: angular dependence and phenomology," *Phys. Rev. B* **18**, 5140 (1978).

Semiconductor Radiation Detectors

H. W. Kraner

INTRODUCTION

Semiconductors have been actively used as a solid medium for radiation detection for about 20 years. Silicon and germanium have been most extensively (but not exclusively) used not only because of their intrinsic solid-state properties but also because of the high state of development of the arts of producing pure materials. Semiconductors provide a complete solid analog to a gas-filled ionization chamber in which radiation can not only be registered as an event but other properties such as its energy deposition and time of arrival can be directly determined from the ionization produced in a specific sensitive volume. As in a gas-filled chamber, an electric field is used to sweep out and collect the ionization (both positive and negative charges: holes and electrons) caused by the impinging nuclear radiation. The great advantage of a solid over a gas is that it provides a radiation stopping power enhanced by at least the density ratio of $\sim 10^3$. The volume in which the electric field exists is the sensitive region of the detector and is usually configured in either "parallel-plate" geometry with planar electrodes opposing each other at a given spacing or in coaxial cylindrical geometry with the electrodes being inner and outer cylindrical shells. An internal field of these shapes can be produced in a semiconductor as the reverse-bias condition in a diode structure. A material doped such that positive holes are the majority charge carrier is called p type and a material having negative electrons as the majority carrier is called n type. A diode structure is formed by the union of p- and n-type materials which permits current (positive charge) to flow only if the potential of the p-type side is raised (biased) relative to the n-type side. If the device is biased with the p-type side negative with respect to the n-type side, conduction current cannot flow and the bias potential appears across the region of the physical p-to-n transition. The collection field is of the order of the applied bias (V_A, volts) divided by the "depletion" depth W or depth of material depleted of free carriers by the field. In an abrupt p-n junction, W extends from the junction into the material according to

$$W = (2\epsilon\rho\mu V_A)^{1/2},$$

where ϵ (F/cm) is the dielectric constant of the material, ρ is the resistivity (Ω·cm), and μ is the mobility of the majority carrier (cm^2/V·s). Usually the impurity concentration of one side is much greater than the other; for example, if $\rho_p \ll \rho_n$, the active region extends primarily into the n-type side of the device and it is referred to as a p^+n structure. To further the example, in a p^+n silicon device where μ is μ_{e^-}, equal to about 1400 cm^2/V·s at 300 K, and taking $\rho = \rho_n \simeq 500\Omega$·cm with $\epsilon = 12 \times \epsilon_0 = 12 \times 8.85 \times 10^{-14}$ F/cm, we find that a 100-V reverse bias will produce a depletion depth or active region of about 120 μm. This device would be appropriate to detect natural alpha radioactivity.

In addition to the ability to support internal fields, the second critical quality of the semiconductor is that it offers charge-carrier lifetimes (τ) and mobilities sufficient to permit traversals of relatively large sensitive regions (~1 cm). The product $\mu\tau E$ (where E is the electric field) is a length which if greater than the detector dimensions assures full collection of all ionization produced regardless of position in the active region. This is generally the case in Ge and Si. As charge is collected from the active region, an induced charge appears on the terminals of the device. The output rise time is of the order of the active width divided by a carrier saturation velocity of $\sim 10^7$ cm/s.

The amount of charge caused by the ionizing event is related to the energy loss of the event (for example, a gamma ray of energy E, if the gamma ray is totally absorbed in the sensitive volume) by the *average* energy required to create an electron–hole pair, ϵ. Thus, $Q = qE/\epsilon$, where $\epsilon_{Si} = 3.64$ and $\epsilon_{Ge} = 2.96$ eV/ion pair and q is the electronic charge. These values are not greatly changed over a wide range of temperature or for various ionizing projectiles, electrons, protons, alpha particles, etc., and are about one-tenth that for a gas. The average energy loss per ion pair, ϵ, is often taken to be approximately three times the semiconductor band gap (or ionization potential) with some theoretical justification.

The energy resolution of a charge-collecting device is described by the observed spread of energy observations about an average energy loss, E_0, in which the average number of charge carriers collected per event is E_0/ϵ. The distribution E_0 is Gaussian with a full width at half-maximum (FWHM) of $2.35(\epsilon E_0)^{1/2}$. To correct for the fact that the individual ionizations in each event are not independent (but must add up to exactly the total energy loss) a factor F, called the Fano factor, is introduced into the variance, so that the expected full width at half-maximum of a distribution is $2.35(F\epsilon E_0)^{1/2}$, where for small volumes of good material $F \approx 0.1$ but may range from 0.07 to 0.12.

The statistical fluctuations just described must be added in quadrature to the "noise" of the electronic system attached to monitor the collected charge per event. Figure 1 shows the expected energy resolutions for silicon and germanium devices of two important cases: (a) a small silicon detector configured to minimize the electronic or system resolution, taken to be 100 eV, and (b) a much larger germanium detector and system with an electronic resolution of 500 eV. A Fano factor of 0.1 is used and these values of energy resolution are realized in practice. In the first case it is important to note that the silicon detector is capable of detecting radiations down to $\lesssim 1$ keV with an energy resolution of ~120 eV, which permits the distinction of the characteristic x rays of elements over the entire periodic table down to the element sodium and is useful for elemental analysis. It is also interesting to note from Fig. 1 that it is the carrier statistics which limit detector resolution above ~300 keV for germanium detectors where they are most used for gamma-ray spectroscopy.

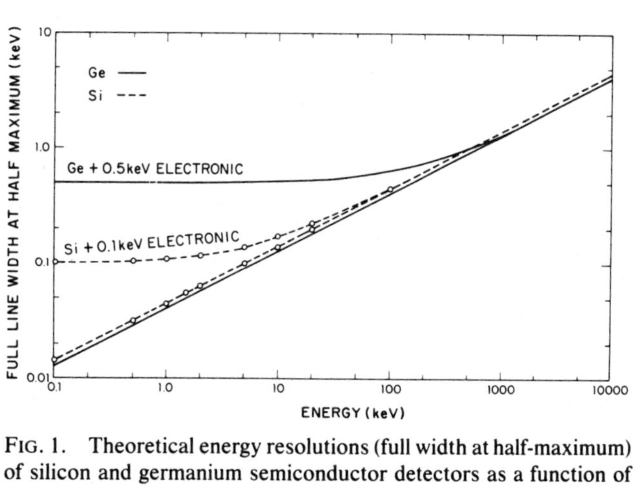

FIG. 1. Theoretical energy resolutions (full width at half-maximum) of silicon and germanium semiconductor detectors as a function of energy loss combined with typical electronic or system contributions.

CONFIGURATIONS

Before describing the several practical configurations of semiconductor devices, two more general concepts must be introduced: the practical requirements of sensitive volume and leakage amount. In a junction detector, the sensitive volume is limited by the depletion depth which in turn is dependent on the applied reverse bias and the resistivity or chemical purity of the available material. Very large sensitive volumes or depletion depths are often desirable, especially for gamma-ray detection, which are not permitted by the practical limits of these parameters. This problem has been surmounted by the technique of lithium-drifting. Lithium is easily diffused into silicon and germanium and is an interstitial n-type dopant. It can be moved through the crystal at moderate temperatures by an electric field. During this process, it pairs with p-type impurities and compensates the crystal to yield material with effectively very large resistivity [noted in Eq. (1)] so that large volumes may be depleted by relatively low potentials. Drifted regions are fabricated up to 1 cm in the planar geometry and up to 2–3 cm in radius in the coaxial geometry.

Large-volume germanium detectors (up to 200 cm^3) are constructed from an entire cylindrical germanium ingot by diffusing lithium from the outer periphery of a p-type ingot to form the outer shell of the coaxial geometry and drifting radially inward. A representative coaxial device is shown in Fig. 2. Thick planar silicon detectors up to 7 mm thick are also constructed by lithium-drifting where the enhanced thickness of the depletion depth is necessary to reduce the device capacitance, thereby improving its noise performance when used as a high-resolution x-ray detector or in order to stop heavy charged particles of several tens of MeV.

Semiconductor devices are operated as reverse-based diodes and draw leakage current with typical diode reverse direction characteristics. The fluctuation of the leakage current is the noise over which a charge pulse must be distinguished and is proportional to the leakage current itself. Near the ambient temperature range the leakage current in silicon and germanium is due to carrier generation by thermal excitation over the semiconductor band gap with a temperature dependence $e^{-E_g/kT}$ where E_g, the band gap, is 1.14 eV in silicon and 0.68 eV in germanium.

Because of the lower band gap of germanium, germanium spectrometers must be cooled to reduce leakage currents to below ~1 nA which gives comparatively negligible current-generated noise. Li-drifted silicon detectors for x-ray spectroscopy with the capability of 100-eV system resolution must also be cooled to yield subpicoampere leakage currents. Convenience has been persuasive in the use of liquid nitrogen as a cryogenic coolant (77 K) which provides an acceptably low temperature for both silicon and germanium devices. Part of the Li-drifting technique for germanium involves a careful final cooling to the liquid nitrogen operating temperature; the device must then remain at this temperature over its operating lifetime. Cooled devices must be contained in an evacuated container called a cryostat to prevent heat input from thermal conduction of a gas ambient. The cryostat vacuum is generally maintained by cryopumping of an internal organic molecular sieve held at the coolant temperature. The clean vacuum condition of the cryostat is also helpful to maintain the very high surface cleanliness of the semiconductor required to keep surface leakage currents low.

In addition to the lithium-drifting of germanium to create nearly intrinsic material and large depletion depths, extremely pure germanium has been produced so that crystals with a net impurity concentration of $\sim 10^{10}/cm^3$ are now available. This value for n^-, a net impurity concentration using $\rho = 1/q\mu n^-$ for Eq. (1), permits a depletion of ~ 1 cm to be achieved with a practicable reverse planar bias of ~ 1000 V. These germanium detectors are called "intrinsic" or "high-purity" germanium and as a convenience may be returned to room temperature after use without harm to the detector. Both coaxial and planar structures have been developed.

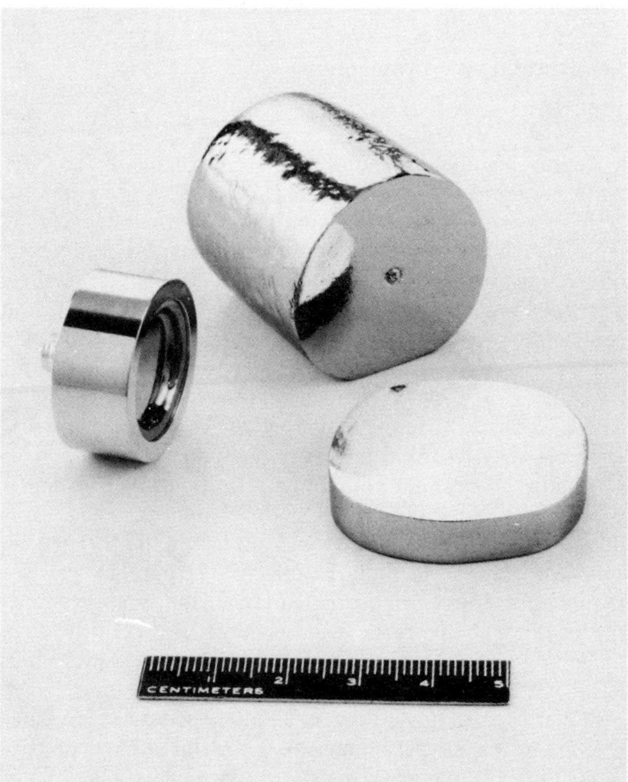

FIG. 2. Representative semiconductor detectors. In the background, a lithium-drifted coaxial germanium detector; to the right foreground, a large-area, planar, high-purity germanium detector; and on the left, a silicon surface-barrier detector.

Silicon detectors formed as *p-n* junctions with bias- and material-dependent depletion depths ranging from 10 to 200 μm are useful for a variety of radiations ranging from fission fragments to high-energy heavy ions and electrons. They are planar structures with the junction formed by either opposite impurity diffusion, ion implantation, or the formation of a Schottky surface barrier. A Schottky surface-barrier junction structure is formed by the combined effects of the work function of a given metallization and the surface states of the semiconductor to yield a given fixed surface charge that is opposed to that of the bulk material forming a junction at the surface. Gold and palladium form *p*-type surfaces on silicon and are used on *n*-type material; aluminum forms an *n*-type surface and is used as an "Ohmic" or nonrectifying contact to *n*-type material or as a weak barrier contact on *p*-type material. Both surface-barrier diodes and ion-implanted junction detectors can yield devices with very thin (<100 Å) entrance windows or inert layers. Evaporated palladium Schottky barriers are also used as rectifying contacts on high-purity germanium detectors.

Figure 2 illustrates the materials used in several typical configurations. In the background is a large cylindrical germanium coaxial detector. The lithium diffusion has been carried out around the periphery; however, after several etching steps in preparation, it is indistinguishable from the rest of the crystal. The central contact for the *p*-type core and the slightly flattened edge which provides good thermal contact to the flat cold surface of the cryostat are noted. In the foreground is a large-area, planar, high-purity germanium crystal with an evaporated palladium p^+ contact, visible on the exposed face by its slightly different reflectivity. The entire volume will be depleted from the lithium n^+ contact on the bottom face. On the left is shown a commercially available silicon surface-barrier detector for charged-particle spectroscopy. The front surface is metallized with gold to form the active junction; the signal is taken from the opposite surface and is available from the small coaxial connector at the rear of the container. Again, the germanium detectors shown here must be placed in a clean, evacuated cryostat and operated at reduced temperatures.

RESULTS

It is impossible in a short article to express fully the impact that high-resolution semiconductor detectors have had on nuclear spectroscopy and related technologies. The energy spectra of several radiations shown in Fig. 3 may give some appreciation in three areas which have been mentioned. These results are typical in that they are not those of the "best possible" system but of systems that are readily attainable in most laboratories.

At the top, the spectrum of alpha particles from the α-decay of ^{241}Am to the several low-lying levels in ^{237}Np is shown as observed with the surface barrier detector of Fig. 2. The system has an electronic resolution of ~ 10 keV and alpha groups separated by only ~ 40 keV are easily observed. The thin ^{241}Am source and the detector were under vacuum to prevent undue α-particle energy loss to an air ambient.

The spectra of several gamma-ray-emitting nuclides are shown in the middle view, measured with a large lithium-drifted germanium [Ge(Li)] coaxial detector. The energy resolution of the 1332.48-keV ^{60}Co peak is 2.2 keV, which is not far from the statistical limit of ~1.5 keV given in Fig. 1. The residual events along the abscissa are due to Compton γ-ray scattering interactions in the detector which do not result in the full γ-ray energy loss; the kinematic Compton "edges" are visible in the ^{137}Cs and ^{60}Co spectra.

The lower spectrum was observed with a cooled, lithium-drifted silicon x-ray spectrometer and associated low-noise electronic system. This spectrometer provides energy resolution (FWHM) as low as 120 eV at low photon energies in the x-ray region. Trace amounts of several elements present in a slurry of ground-up grasses ("orchard leaves") have been detected by their characteristic x-ray emissions when fluoresced by a filtered and "hardened" x-ray source of about 25 keV. The abscissa spans the energy range from 0 to 25 keV, and we clearly see the elements from potassium to lead present in the grass slurry. This target has been standardized and the amounts of Mn, Fe, Cu, and Pb present and observed are known to be 90, 300, 12, and 44 parts per million, respectively. The energy resolution of this system allows the distinction of characteristic x rays down to about sodium ($Z = 12$) in atomic number. This analytical technique is a relatively quick and efficient procedure to measure trace amounts (1–100 ppm) of a broad range of elements nondestructively and without elaborate sample preparation.

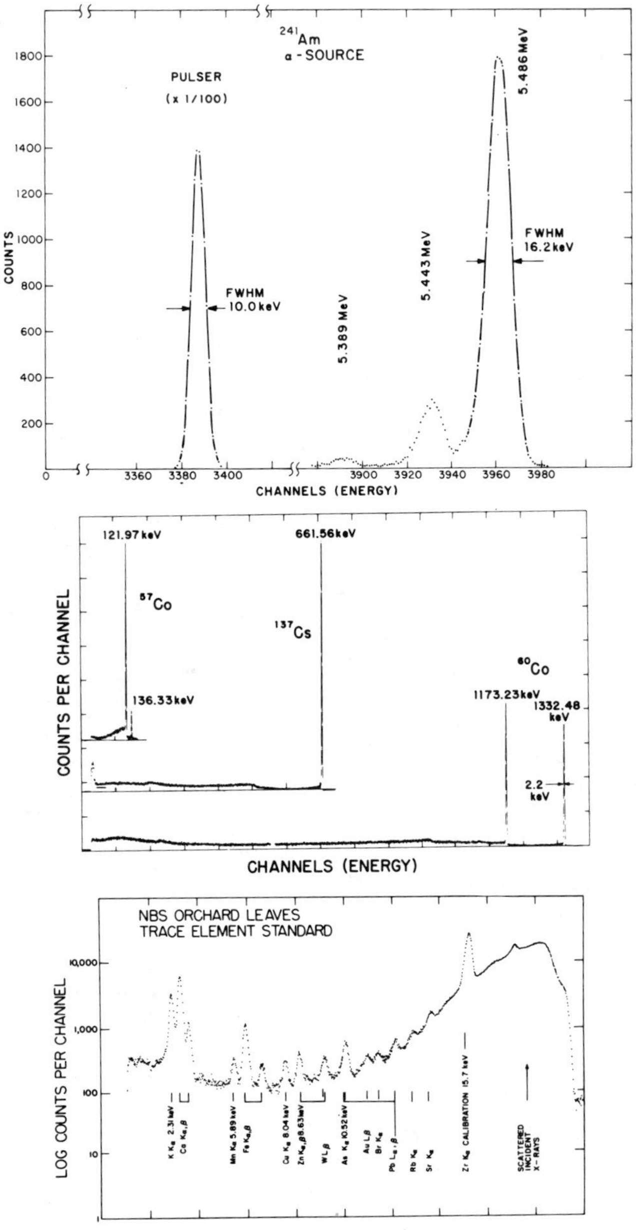

FIG. 3. Typical energy spectra from semiconductor detectors. *Top*: ^{241}Am α particles observed with a silicon surface barrier detector. *Middle*: Gamma-ray spectra of ^{57}Co, ^{137}Cs, and ^{60}Co observed with a large lithium-drifted coaxial germanium detector. *Bottom*: X-ray fluorescence of trace-element standard, NBS "Orchard leaves," taken with a high-resolution lithium-drifted silicon x-ray spectrometer.

In conclusion, it should be mentioned that other semiconductors have received considerable attention but less than wide-spread usage. This group includes the compounds GaAs, CdTe, and HgI_2, which have produced working detectors but spectrometers of poorer energy resolutions than with silicon or germanium. Efforts are generally in the direction of larger band-gap semiconductors of high atomic number to achieve good gamma-ray detection efficiency with a room-temperature device.

See also CRYSTAL DEFECTS; SEMICONDUCTORS; TRANSISTORS.

Semiconductors, Amorphous

Kurt Weiser

Research activity on amorphous semiconductors has been quite extensive for several years [1]. Since 1969 international conferences on Amorphous and Liquid Semiconductors [2] have been held biennially and at international conferences on the Physics of Semiconductors (traditionally crystalline) a respectable fraction of sessions are devoted to amorphous semiconductors [3]. Many other instances which bespeak of the interest in the subject can be cited, and the first question naturally concerns the reason for the interest. Unlike the case of crystalline semiconductors where the intense worldwide research effort was triggered by the discovery of the transistor, one cannot point to any startling new phenomena with device potential in the field of amorphous semiconductors to justify the widespread interest in the subject. For example, the use of such materials in xerography predates the current era of intense research; furthermore, as far as the physics of the process is concerned, in contrast to its technology, it does not depend critically on the amorphousness of the photosensitive films used. While more recent discoveries with device potential have been made, the main interest in the subject remains in the realm of basic physics. Some of the fundamental questions to which only incomplete answers exists concern the band structure of solids lacking long-range order, how current is conducted in such solids, how excess carriers recombine in them, and how structure and electronic properties are related.

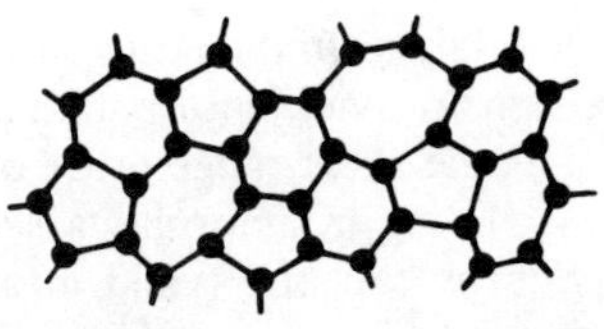

FIG. 1. Schematic diagram of a two-dimensional "ideal" amorphous semiconductor. The structure is "ideal" in the sense that all bonds are completed so that there are no "dangling bonds."

The scope of this article is restricted to covalently bonded materials such as amorphous Si or As_2Se_3. Such materials are called ideal amorphous semiconductors if all atoms are connected by chemical bonds to each other but long-range order does not exist because the bonds are slightly distorted from their crystalline pattern. An ideal two-dimensional amorphous material is shown in Fig. 1. In the case of a single element these bond distortions form the sole source of disorder. In the case of a compound, nonstoichiometry as well as atoms on "wrong" sites will increase the disorder. As an example, in a compound like As_2Se_3 a completely connected network can be constructed even if occasionally an As occupies what should have been a Se site and vice versa. Such a network is still ideal in the sense defined above but more disordered than if the network had merely been a distorted version of the perfect crystal. In practice one may expect that ideal amorphous materials are difficult to obtain and, indeed, amorphous films of germanium and silicon obtained by condensation of the vapor are believed to contain voids and "dangling bonds," i.e., unbonded valence electrons. Amorphous silicon obtained by rf decomposition of silane, on the other hand, appears to be relatively ideal. It should be emphasized that all such statements are based on rather indirect evidence such as, for example, the presence or absence of electron-spin-resonance signals or small-angle scattering of x rays.

A review article of this length obviously cannot include even a superficial description of all the aspects of such a broad field. No mention is therefore made of such interesting but perhaps peripheral topics as lattice vibrations or mechanical properties. The topics selected for discussion are band structure, transport and recombination properties, relation of structure and electronic properties, and device-oriented phenomena. The discussion of these topics will be confined to a summary of the salient experimental facts and theoretical concepts used to deal with them.

The materials generally referred to as amorphous semiconductors are at present drawn from Group-IV elements and from chalcogenides, i.e., Se and Te, and their compounds with Group-V elements such as arsenic. A fundamental difference between the two types of materials exists with respect to their method of preparation. Group-IV elements, e.g., Si or Ge, can be prepared in amorphous form only by condensation of the vapor on a substrate sufficiently cooled to prevent crystallization. (A "cool" substrate may be as hot as 600°C in the case of Si). Amorphous chalcogenides, on the other hand, can be prepared either by condensation of vapor or by quenching the melt. The reason that quenching the melt is not a feasible method for producing amorphous Ge or Si is that these substances have metallic properties in the liquid state. Group-IV elements and chalcogenides also differ with respect to the sensitivity of their electronic properties on the method of preparation and on annealing conditions. In Fig. 2 we show optical properties of Si prepared by different techniques. Differences are also obtained if specimens obtained by a given technique are annealed or are prepared at different substrate temperatures. Clearly, one cannot speak of the absorption constant of amorphous Si without specifying the conditions of its preparation and its thermal history. The electrical resistivities of silicon films prepared by the two main techniques—evaporation and condensation of vapor and radio-frequency (rf) decomposition of silane—are also greatly different, the latter technique yielding films of resistivities several orders of magnitude greater than the former one. The difference has been atributed to one of two causes: (a) When the rf field decomposes the silane (SiH_4), the silicon atoms arrive at the substrate with considerable kinetic energy and are therefore more able to form the ideal amorphous network depicted in Fig. 1. (b) The H atoms produced during the discharge react with "dangling" Si and thus eliminate the unpaired electrons, which presumably determine the electronic properties of the material obtained by condensation of the vapor.

In contrast to Group-IV elements, properties of chalcogenides are relatively insensitive to methods of preparation or to subsequent annealing. As shown in Fig. 3, for example, the optical properties of the typical compound As_2SeTe_2 do not even depend on whether the material is prepared by vapor condensation or by quenching from the melt. The difference between the two classes of amorphous solids in this respect is not well understood. Possibly, the largely two-dimensional nature of the chalcogenides makes it easier for them to attain an ideal amorphous structure during their preparation. In other words, dangling bonds which are expected to be electronically active. as are vacancies in crystals, are readily eliminated by a local rearrangement of atoms. An alternative mechanism for the elimination of dangling bonds is based on the hypothesis that the top of the valence band in chalcogenides is formed from nonbonding

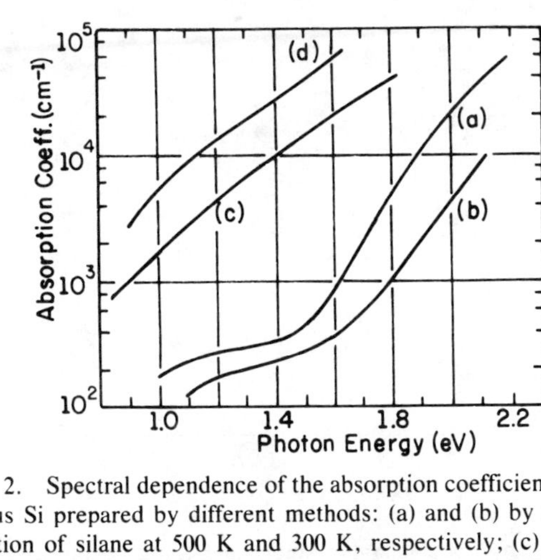

FIG. 2. Spectral dependence of the absorption coefficient of amorphous Si prepared by different methods: (a) and (b) by rf decomposition of silane at 500 K and 300 K, respectively; (c) specimen evaporated by rf sputtering; (d) evaporated specimen. [From R. J. Loveland, W. E. Spear, and A. Al-Sharbaty, *J. Non-Cryst. Solids* **13,** 55 (1973).]

p orbitals of the chalcogen atoms, the bonding *p* orbitals being deeper in the band. It has been argued that it is energetically favorable for an electron in one dangling bond to be given off to another dangling bond thus forming one positively and one negatively charged atom. The plausibility of the argument rests on the idea that the positively charged atom can form a chemical bond with a neighboring chalcogen atom whose nonbonding *p*-orbital electrons are thus transferred into a bonding orbital. The net effect of the process is to combine the electrons of two half-filled nonbonding *p* orbitals into one filled *p* orbital, thereby eliminating unpaired electrons, and to transfer two electrons from a nonbonding orbital into a bonding orbital. The process should give off energy.

A few more words about the structure of Group-IV and chalcogenide amorphous semiconductors are in order. Clearly, the word has a different meaning for such materials from the one associated with crystals. In the latter case x-ray or electron diffraction patterns provide information about the long-range order as well as the short-range order of atoms. Since long-range order is missing in amorphous semiconductors, such diffraction patterns can only yield information about the average environment of an atom. The results of numerous investigations have yielded the following picture: In the case of Group-IV elements the basic tetrahedral building block is preserved and the interatomic distance is the same as in the crystal. Variations in the dihedral angle by which two tetrahedra are connected are responsible for a spread of about 10° in bond angles. The radial distribution function beyond the third nearest neighbors is thus smeared out. This model, in which atoms are joined to one another with variations in bond angles, is known as the random continuous network model and seems to agree well with experimental radial distribution functions deduced after making due allowance for voids. It should be mentioned, however, that several authors have proposed that the experimental data are also consistent with a microcrystalline model. For example, it has been claimed that a model in which building blocks of wurtzite crystals of about 14 Å in diameter are joined to each other in a random manner also reproduces the experimental diffraction data. The preponderance of opinion favors the random amorphous over the microcrystalline model.

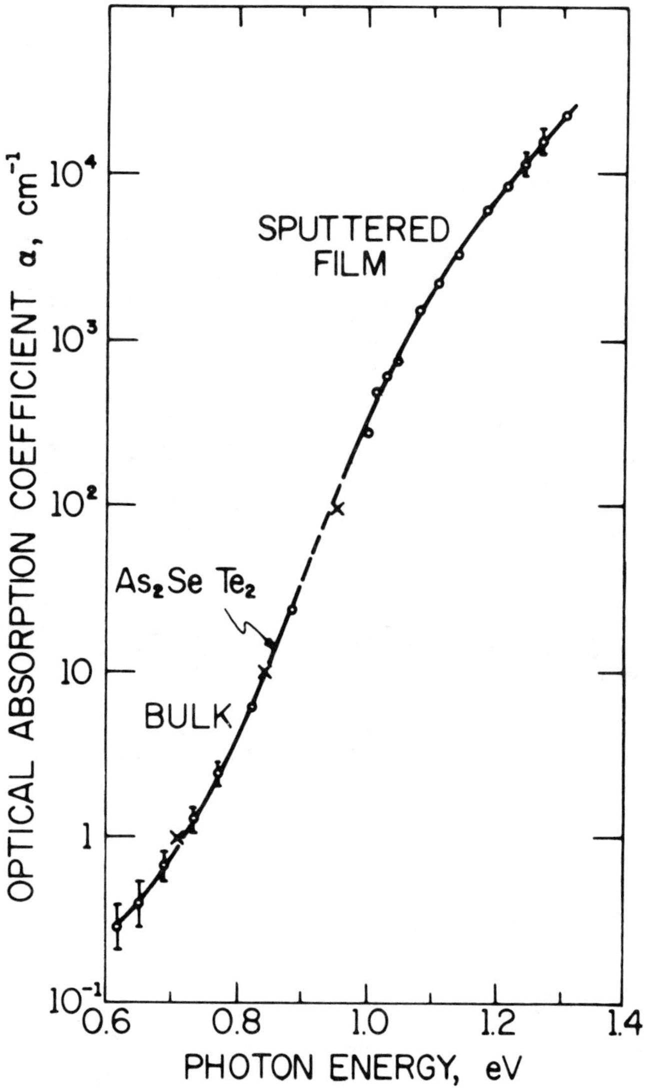

FIG. 3. Absorption constant of amorphous As_2SeTe_2 prepared by vapor condensation (sputtered film) and by quenching of melt (bulk). [From T. C. Arnoldussen, C. A. Menzes, Y. Nakayawa, and R. H. Bube, Phys. Rev. B **9**, 3377 (1974).]

As for the chalcogenides it seems that Se and Te have a random chain structure as expected for their chemical valence states and that in the case of the arsenic chalcogenides each As is surrounded by three chalcogen atoms and each chalcogen by two As atoms. The local order is thus identical to that in the crystals, except for the case of As_2Te_3 whose crystalline structure is different from that of its analogues As_2Se_3 and As_2S_3.

After this brief summary of methods of preparation and their effect on some electronic properties and of the structure of amorphous semiconductors it is appropriate to turn to a brief description of electronic properties. It is an experimental fact that a valence band and a forbidden gap exist in these materials. As in crystalline materials the most readily available evidence for the existence of a forbidden gap comes from a study of the wavelength dependence of the optical absorption. To be sure, the onset of absorption is not abrupt but neither is it for crystals as a result of a variety of causes such as free carrier absorption, impurity levels, strains, etc. The onset of absorption is even less abrupt for amorphous materials, but, as seen in Fig. 4, the dependence of the absorption constant for crystalline and amorphous Se near $\alpha = 1 \times 10^4$ cm^{-1} is fairly similar. (We consider the behavior near this value of α because the photon energy at which $\alpha = 10^4$ can be taken as an approximate value of the bandgap of "direct" crystalline materials; the absorption of light at smaller values of α is attributable to various defects.) Direct evidence for the existence of a valence band in amorphous semiconductors comes from photoemission experiments. In the case of chalcogenides these experiments reveal a remarkable similarity in the density of states as a function of energy for crystalline and amorphous materials. In the case of Ge and Si significant differences are found. Thus, a study of photoemission reveals which features in the den-

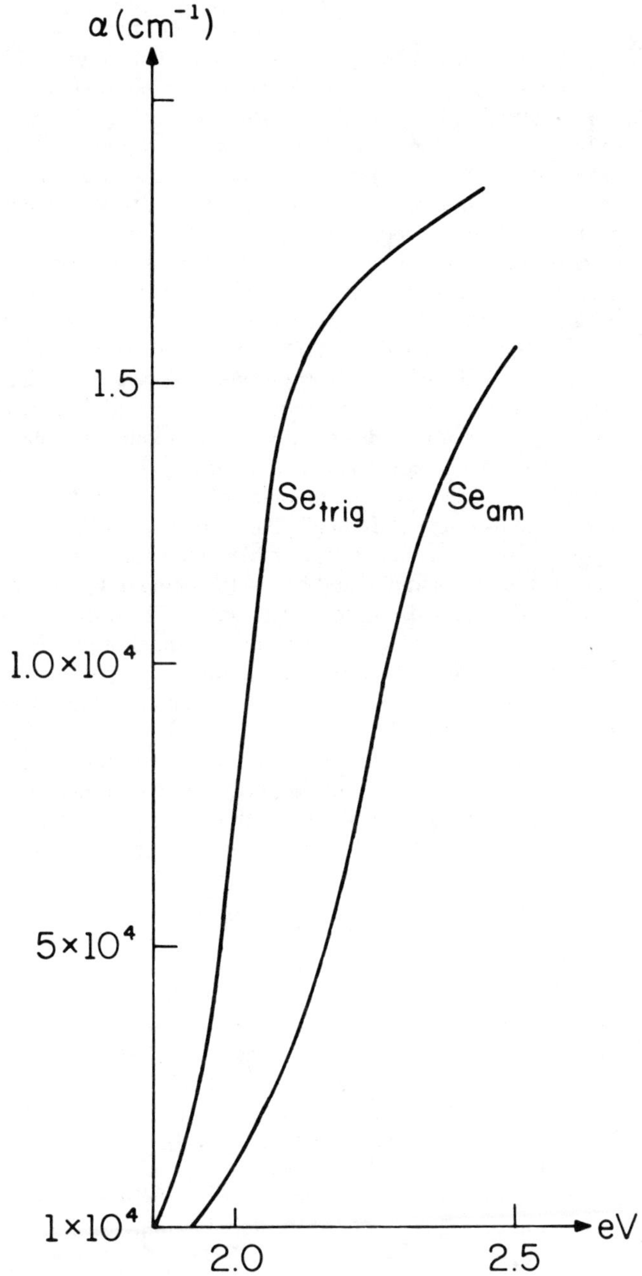

FIG. 4. Spectral absorption constant of crystalline (Se_{trig}) and amorphous selenium; linear plot. [From J. Stuke, *J. Non-Cryst. Solids* **4**, 1 (1970).]

sity of states depend only on short-range order and which on long-range order.

Since one cannot ascribe the existence of a bandgap to the periodicity of the structure it is clear that a chemical approach is needed in order to understand the behavior. In other words, it appears that the chemical bond between two atoms is responsible for the existence of the bandgap. The difference between the isolated chemical bond in a molecule and the collection of such bonds in the solid is that in the latter case the bonding and the antibonding states of the bond are broadened into bands. An energy gap between the two bands persists, however, though its value is different from the energy separation between the bond states. A semiquantitative chemical approach to band formation in amorphous solids has been applied to tetrahedrally coordinated materials such as Ge and Si. The approach starts with the premise that the basis wave functions of the system are the familiar sp^3 orbitals which are known to be responsible for the tetrahedral coordination of these elements. Only two perturbations are assumed to affect these orbitals, namely that caused by the other bonds on a given atom of index i and that caused by the electron in the same bond of index j shared with a nearest-neighbor atom. Thus the Hamiltonian of the system is given by

$$H = \sum_{i,j\neq j'} V_1 |\phi_{ij}\rangle\langle\phi_{ij'}| + \sum_{i\neq i',j} V_2 |\phi_{ij}\rangle\langle\phi_{i'j}|.$$

Here V_1 represents the matrix elements between bonds and V_2 that between neighboring atoms. Even a highly abbreviated account of the procedure for obtaining the wave functions of the system with the above Hamiltonian would be of prohibitive length. Briefly, the following results are obtained: It is found that as long as $V_1/V_2<\frac{1}{2}$ two bands are obtained separated by a bandgap. The perturbation V_2 produces the bandgap. The interbond interaction V_1 determines the bandwidths and produces a partial splitting of the bands into p- and s-like parts, with the top of the valence band being p like and the bottom of the conduction band s like. An important point is that a bandgap will exist in tetrahedrally coordinated materials regardless of the existence of long-range order as long as one uses sp^3 orbitals together with the above Hamiltonian whose physical basis is quite reasonable.

Since a bandgap exists in both the crystalline and the amorphous materials the question arises as to exactly what effect disorder has on the band structure. From the point of view of semiconductor behavior it is the change in properties near the band edges which are of greatest importance. It is generally assumed that for amorphous semiconductors states near the band edges can be separated into extended and localized states as illustrated in Fig. 5. The distinction between the two types of states is rather subtle but for our purposes it suffices to describe them operationally by saying that a carrier placed in an extended state can move through the system without the cooperation of a phonon. A localized state, on the other hand, is characterized by a wave function which falls off exponentially from some origin; a carrier placed in such a state can escape from it only by phonon cooperation. Unlike the case of lightly doped crystalline semiconductors, however, where such states exist as discrete states within the gap, these "bandtail" states are assumed to be continuous. They merge with extended states at an energy which is commonly called the "mobility edge" since it is assumed that the mobility of carriers in localized states is drastically lower than in extended states. Just how deep a bandtail is expected from the theory is not clear. In an ideal amorphous semiconductor with all valence bands satisfied as in Fig. 1 the potential fluctuations must be due to variation in bond angles and interatomic distances. One may expect that such fluctuation can only be of finite extent,

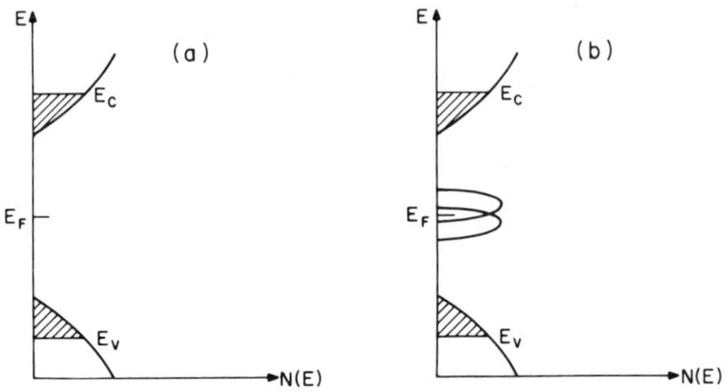

FIG. 5. Density of states versus energy for (a) ideal amorphous semiconductor, and (b) for a nonideal amorphous semiconductor with dangling bonds giving rise to overlapping donor and acceptor states. [E. A. Davis and N. F. Mott, *Philos. Mag.* **22**, 903 (1970).]

i.e., that beyond some critical amount of bending or stretching the bond breaks, so that a "dangling bond" is produced. Thus, the continuous potential fluctuations are likely to be finite, and the extent of bandtailing should therefore also be finite. Unfortunately, there is very little experimental evidence about the depth of these tails since, as discussed below, experiments designed to explore their existence cannot be interpreted unambiguously.

The dangling bonds which are present in a nonideal amorphous semiconductor, however, are expected to produce discrete though broad states within the gap. One may envisage an electron in a dangling bond as a donor electron, ready to be given off to the conduction band; conversely, the dangling band can be thought of as an acceptor producing a hole in the valence band after taking an electron from it. Possibly, the donor and acceptor states overlap giving rise to a broad, partially filled band within the gap as illustrated in Fig. 5b. In addition to such states it is also possible that atoms on wrong sites in a compound amorphous semiconductor may produce states with the gap.

The discussion of electronic properties has so far touched on the existence of a bandgap, on the presumed existence of localized and extended states near the band edges, and on the existence of discrete states within the gap. This scheme, illustrated in Fig. 5, has served as the basis for the interpretation of transport and recombination properties of amorphous semiconductors. Before giving a brief description of these attempts it is well to compare the situation with that in cystalline semiconductors. Clearly, the concept of conduction in both localized and extended states is well established in this case as well since it is known that at sufficiently low temperatures when virtually all carriers in the bands have been frozen out on impurity centers the conduction does not fall abruptly to zero. Conduction in extended states, however, may be expected to be radically different in the two cases. In the case of crystals the current is carried by quasifree carriers, i.e., carriers whose effective mass is different from that of the free electron mass but which move with thermal velocity and a mean free path much longer than an interatomic distance. In the case of amorphous semiconductors, on the other hand, the mean free path of carriers in extended states is expected to be only of the order of a few interatomic distances since for longer distances phase coherence is lost; furthermore, the concept of an effective mass becomes meaningless. Under these circumstances the motion of the carriers can be though of as one of diffusion from atom to atom. The mobility is thought to be inversely proportional to the temperature and to be of order 10 cm^2/V s at room temperature. How is one to investigate the motion of carriers in extended states experimentally? The problem, obviously, is that carriers in extended states are in thermal equilibrium with carriers in localized states so that any experiment thus far designed to measure the mobility of carriers measures some sort of average value of the mobility. For this reason any such experiments (Hall plus resistivity measurements, time-of-flight measurements, photoconductivity and thermopower measurements) do not give direct information about the mobility of carriers in extended states. It is necessary to make assumptions such as that the mobility of carriers in the localized states is completely negligible compared to that in extended states or to choose reasonable values for the effective densities of states above and below the demarcation energy separating extended and localized states. Thus, one of the most interesting problems regarding transport properties of amorphous semiconductors, the mobility of carriers in extended states, has so far eluded direct experimental investigation.

One aspect of the transport theory of amorphous semiconductors which has been borne out by experiment concerns the temperature dependence of the conductivity when the current is transported via carriers in localized states and the Fermi level lies within a band of these localized states. This is the situation depicted in Fig. 5b. Since conduction via localized states can proceed only by phonon cooperation, one might expect that the conductivity is simply thermally activated as is the population of phonons at sufficiently low temperatures. It can be shown, however, that a proper treatment of the percolation problem involved in the motion of a carrier at the Fermi level via the available localized states above it leads to a very different temperature dependence. The key concept in these approaches is that in order to facilitate the motion of the carrier it will not jump necessarily to the nearest empty site. The reason is that such a jump may require more energy than a jump to a more distant site. The average distance will be that which results from a compromise between a tendency to jump to a distant site to minimize the energy and a jump to the nearest site for which the overlap of wave functions is at a maximum. The theoretical temperature dependence for the conductivity is found to be of the form $\sigma \propto \exp[-(T/T_0)^{1/4}]$, where T_0 is a parameter which depends on the density and other properties of the localized states involved. A temperature dependence of this nature has frequently been observed, particularly for Si and Ge films prepared by evaporation and sputtering processes, which are known to contain voids. Presumbly these voids are the source of the localized states.

It should be evident by now that the situation with respect to transport processes is only superficially understood. The same remark can be made about recombination processes in these materials. All one can say from studies of the temperature and intensity dependence of the steady-state and transient photoconductivity around room temperature is that recombination appears to be bimolecular, i.e., that electrons and holes recombine directly rather than through the two-step process characteristic of the Shockley–Read mechanism. Just how the energy of recombination is dissipated is not understood, though it is known that at room temperature the recombination process is not radiative. At low temperatures, on the other hand, the luminescence efficiency is quite high and the recombination appears to occur via recombination centers.

Finally, a few words should be said about device research in this field. Amorphous semiconductors have found their major use in xerography where, as stated earlier, their employment is due mainly to technological factors such as the ability to produce large, uniform thin films, rather than to physical properties unique to amorphous as opposed to crystalline semiconductors. The most unique phenomena of device potential found in amorphous semiconductors so far concern switching effects and reversible optical phenomena induced by light or heat. In many "sandwich" structures consisting of metal electrode–amorphous film–metal electrode, it is observed that upon application of a critical voltage the device switches to a lower voltage state at a current which depends on the resistor in series with the circuit. Such characteristics, though not unique to amorphous materials, make them potentially useful as control devices for display panels. Also of some device potential are certain reversible changes in optical properties brought about by light pulses. It is possible to "amorphize" very small regions, say 2 μm in diameter, of Te-based compositions by impinging an intense, short light pulse on a film. It is believed that such a pulse melts the crystallites and that upon termination of the pulse the molten region cools sufficiently rapidly so as to avoid crystallization. Heating the spots by means of the attenuated laser beam for a longer period, say 1 ms, recrystallizes the region again. This write–erase process can be repeated and these films therefore have potential as an erasable and rewritable optical-storage medium.

Note Added in Proof:

Some important developments in the field of amorphous semiconductors have recently occurred. First and foremost, it was discovered by Spear and Le Comber [4] that amorphous Si can be doped *n* and *p* type by the conventional dopants used for crystalline materials, namely phosphorus and boron. To obtain such an effect the amorphous Si had to be prepared by the rf discharge decomposition of silane that produces material with few states within the gap which mask the introduction of dopants. The next step, not surprisingly, was to produce *p-n* junctions with rather good rectification properties [5]. Shortly thereafter Carlson and Wronsky [6] showed that such *p-n* junctions, or rather *p-i-n* junctions, can be fabricated into solar cells with reasonable efficiency. These authors have also shown [7] that Schottky barrier diodes can be produced with such amorphous Si. Parallel to these efforts a group at Harvard University [8] has demonstrated that similar results can be obtained with amorphous Si prepared by sputtering, as long as the process is carried out in the presence of hydrogen. Clearly, it is the latter element which causes the decrease in the density of gap states by combining with "dangling bonds." These developments, which are being pursued in various laboratories, speak well for the future use of amorphous Si. It appears that at least some of the rich variety of phenomena which exist in crystalline semiconductors may yet be found in their amorphous counterparts and that they will find applications in the laboratory and industry.

Amorphous semiconductors will continue to be an active, interesting area of research partly because most of the basic questions about the field have not been answered adequately and partly because of its device potential.

See also GLASSY METALS; SEMICONDUCTORS, CRYSTALLINE; THIN FILMS.

REFERENCES

1. For an excellent introduction to the subject see N. F. Mott and E. A. Davis, *Electronic Processes in Non-Crystalline Materials*. Clarendon Press, Oxford, 1971.
2. (a) The latest published proceedings of this series of conferences can be found in *Proceedings of the Fifth International Conference on Amorphous and Liquid Semiconductors*. Taylor and Francis, London, 1974. (b) A valuable collection of papers is also found in *Tetrahedrally Bonded Amorphous Semiconductors*. American Institute of Physics, New York, 1974.
3. The latest published proceedings of this series of conferences are those of the *12th International Conference on the Physics of Semiconductors*. Teubner, Stuttgart, 1974.
4. W. E. Spear and P. Le Comber, *Solid State Commun.* **17**, 1193 (1975).
5. W. E. Spear, P. G. Le Comber, S. Kimmond, and H. H. Brodsky, *Appl. Phys. Lett.* **28**, 671 (1976).
6. D. E. Carlson and C. R. Wronski, *Appl. Phys. Lett.* **28**, 671 (1976).
7. C. R. Wronski and D. E. Carlson, *Solid State Commun.* **23**, 421 (1977).
8. W. Paul, A. J. Lewis, G. A. N. Connell, and T. D. Moustakas, *Solid State Commun.* **20**, 969 (1976).

Semiconductors, Crystalline

Heinrich J. Welker and *Walter Kellner*

1. *INTRODUCTION*

Since semiconducting phenomena can be detected in a great variety of matter, including glassy and amorphous solids and even liquids, any definition of semiconductors which includes all matter showing semiconducting phenomena can be only very general and imprecise. In this article, we have preferred to present a rather narrow definition more applicable to the "typical" semiconductors and exclude all other materials from our consideration.

Definition: Semiconductors are crystalline solids with predominantly covalent bonding. They differ from metals in that they have an energy gap $E_g>0$ between the valence and conduction bands. A sufficiently pure semiconductor will show no electrical conduction at $T=0$, while at $T>0$ electronic (electron or hole) conduction occurs due to impurities, thermal activation, or external influence (radiation).

In this definition, the term crystalline excludes any material with an unclearly defined structure. "Predominantly covalent bonding" means that the semiconductor properties disappear gradually as other bonding mechanisms take over. On transition from a group-IV crystal (e.g., Si, Ge) to a III-V (e.g., GaAs), II-VI (e.g., ZnS), and finally I-VII (e.g., NaCl) compound, the bonding changes from covalent to ionic and the crystal structure from a diamond to a NaCl type, the latter being an ionic conductor. Another example that illustrates the importance of crystal structure for the conduction mechanism rather than the position of an element in the periodic table is the radically different behavior of the two modifications of carbon: diamond and graphite. While diamond is a semiconductor with a crystal structure as described below, in the "sheet lattice" of graphite only three electrons per atom form covalent bonds toward neighboring atoms. One electron per atom can move freely within the lattice, resulting in a quasimetallic conductivity.

Although the criterion $E_g>0$ clearly separates metals from semiconductors, it is not a simple matter to distinguish semiconductors from insulators. Indeed, semiconductors may be used as insulating materials (e.g., semi-insulating GaAs or diamond) and many insulators show semiconductor properties (e.g., in photoconductivity).

2. *CRYSTAL STRUCTURE*

Because of the nature of the covalent bond, which is formed by two electrons with opposite spins from two neighboring atoms, the important semiconductor crystals are based on tetrahedral elements: Each atom is surrounded by four equidistant nearest neighbors lying at the corners of a tetrahedron. Figure 1 shows the diamond lattice (IV elements), the zinc-blende lattice (typical for III-V compounds), and the wurtzite lattice. Some of the III-V compounds, e.g., GaN, crystallize in the wurtzite lattice, while many of the II-VI compounds crystallize in both modifications, the zinc-blende and the wurtzite lattices.

In a crystal it is possible to construct a primitive unit cell which has all the symmetry properties of the crystal: the Wigner–Seitz cell. The Wigner–Seitz cell of the reciprocal lattice, lying in the so-called **k** space, is called the first Brillouin zone. For a given set of basis vectors $\mathbf{a}_i$ of the real

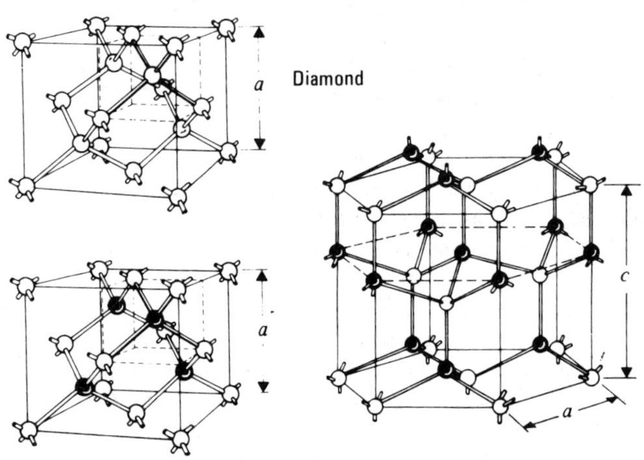

FIG. 1. Lattice structures of important semiconductors.

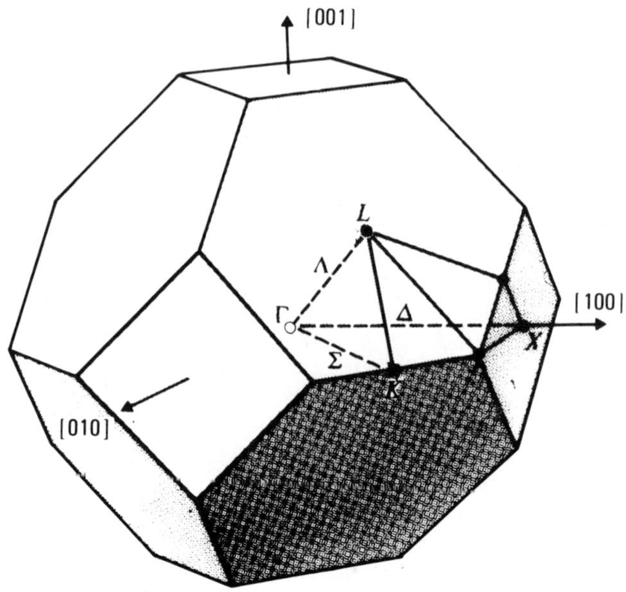

axes: $\Lambda<111>$, $\Delta<100>$, $\Sigma<110>$

points: $\Gamma=\frac{2\pi}{a}(0,0,0)$, $L=\frac{2\pi}{a}(\frac{1}{2},\frac{1}{2},\frac{1}{2})$, $X=\frac{2\pi}{a}(100)$, $K=\frac{2\pi}{a}(\frac{3}{4},\frac{3}{4},0)$

FIG. 2. First Brillouin zone for diamond and zinc-blende lattice showing important symmetry points and lines.

lattice, the basis vectors of the reciprocal lattice $\mathbf{b}_j$ are defined by

$$\mathbf{a}_i \cdot \mathbf{b}_j = 2\pi\delta_{ij}, \quad \delta_{ij} = \begin{cases} 0 & \text{for } i \neq j, \\ 1 & \text{for } i = j. \end{cases} \tag{1}$$

Figure 2 shows the first Brillouin zone for diamond and zinc-blende lattices.

3. *ENERGY BAND STRUCTURE*

To determine the energy band structure of a semiconductor, Schrödinger's equation

$$\Delta\phi_{\mathbf{k}}(\mathbf{r}) + (2m/\hbar^2)[E_{\mathbf{k}} - V(\mathbf{r})]\phi_{\mathbf{k}}(\mathbf{r}) = 0 \tag{2}$$

has to be solved. Bloch's theorem states that for a periodically varying potential $V(\mathbf{r})$, the solutions $\phi_{\mathbf{k}}(\mathbf{r})$ are of the form

$$\phi_{\mathbf{k}}(\mathbf{r}) = e^{j\mathbf{k}\cdot\mathbf{r}} U_{\mathbf{k}}(\mathbf{r}), \tag{3}$$

where $U_{\mathbf{k}}(\mathbf{r})$ is periodic in **r** with the periodicity of the crystal lattice. It can be shown that the energy $E_{\mathbf{k}}$ is periodic in the **k** space. Thus it is sufficient to determine $E_{\mathbf{k}}$ within the first Brillouin zone.

For calculating the band structure of semiconductors the *orthogonalized-plane-wave* method and the *pseudopotential* method are used most frequently.[2] Some results are shown in Fig. 3.[14]

The lower bands, the valence bands, are filled with the valence electrons of the semiconductor, one s and three p electrons. The upper bands, the conduction bands, are empty at $T=0$. The separation between the energy of the lowest conduction band and the highest valence band is

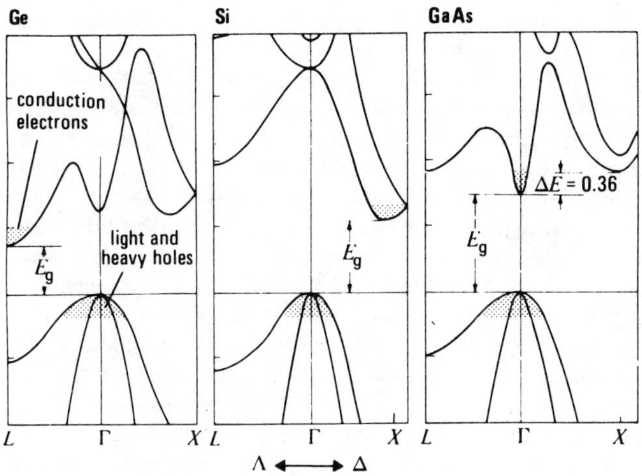

FIG. 3. Band structures of Ge, Si and GaAs. E_g is the band gap. At $T=0$ bands are empty. At $T>0$ electrons accumulate at the conduction-band minimum and holes at the valence-band maximum. (From Refs. 11 and 14.)

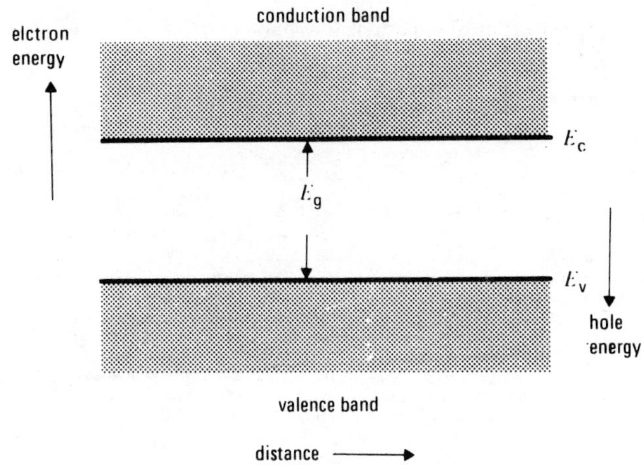

FIG. 4. Simplified band diagram.

called the band gap E_g. No electron states exist in this forbidden energy region.

The two top valence bands can be approximated at Γ ($\mathbf{k}=0$) (for definitions see Fig. 2) by two parabolas of different curvature. This corresponds to two types (*light* and *heavy*) of holes with different effective masses m, given by

$$m=\hbar^2/(\partial^2E/\partial k^2). \quad (4)$$

The most significant difference between the three semiconductors of Fig. 3 is the position of the minimum in the conduction band, lying at L for Ge, on the Δ axis for Si, and at Γ for GaAs. (For definitions see Fig. 2.)

Because of the shift of the conduction-band minimum from $\mathbf{k}=0$, Si and Ge are called indirect semiconductors, while GaAs is called a direct semiconductor. In direct semiconductors band-to-band recombination of an electron and a hole with simultaneous emission of a photon is possible. In indirect semiconductors such a transition is possible only when a phonon is involved, which takes up the difference in the momenta of electron and hole.

Another important difference between direct and indirect semiconductors is the anisotropy of effective mass in indirect semiconductors. While in direct semiconductors the constant-energy surfaces are spheres at the zone center, in indirect semiconductors these surfaces are ellipsoids with their centers lying at $\mathbf{k}\neq 0$. In accordance with (4) this leads to a longitudinal (in the direction of the symmetry axis of the ellipsoid) and a transversal effective mass.

With the aid of an isoelectronic scheme (Table I) a systematic decrease in separation of the valence and conduction bands at corresponding points with the average number of electrons per lattice site can be observed for group-IV elements and for III-V compounds.

Since the subbands of the conduction band meet or overlap at certain points in $\mathbf{k}$ space, no forbidden energy region exists within the band and the band may be considered as continuous in real space. The same is true for the valence band. This forms the basis of a simplified band diagram, where electron energy is plotted versus distance (Fig. 4). This representation is very useful in semiconductor device physics.[10]

Table II lists some properties of important semiconductors. Note that many semiconductors show room-temperature mobilities between 10^3 and 10^4 cm^2/V s, two orders of magnitude above the mobilities of metals. These high mobilities correspond to effective masses which are smaller than the electron rest mass.

Let us illustrate the influence of the band structure on semiconductor devices with a few practical examples: GaAs is highly suitable for high-frequency devices (above 5 GHz) because of its small electron effective mass, equivalent to the strong curvature of the conduction band at Γ (Fig. 3). In Gunn-effect devices electrons are scattered at high elec-

Table I Isoelectronic scheme for IV elements and III-V compounds[a]

N_e	IV Elements			III-V Compounds		
14	Si(3.35)			AlP(~5)		
23			AlAs(3.1)		GaP(2.8)	
32	Ge(0.89)	AlSb(2.3)		GaAs(<u>1.52</u>)		InP(<u>1.37</u>)
41			GaSb(<u>0.8</u>)		InAs(<u>0.42</u>)	
50	Sn(~0)			InSb(<u>0.25</u>)		

[a] Numbers in brackets give the energy gap (in eV) at $T=0$ K for the zone center Γ of the Brillouin zone. Underlined values represent the band gap. N_e is the average number of electrons per lattice site.

Table II Properties of important semiconductors (at 300 K) (from Ref. 19)

	Semiconductor[1]		Crystal-structure[2]	Lattice constant nm	Bandgap[3] eV		Mobilty $cm^2V^{-1}s^{-1}$ Electrons	Holes	Effective mass m/m_0[4] Electrons	Holes	Density $g\,cm^{-3}$	Dielectric constant (static)
Element	Si	14	D	.543	1.12	i	1350	475	0.91/0.19	0.16/0.59	2.33	11.9
	Ge	32	D	.566	0.66	i	3900	1900	1.6 /0.08	0.04/0.3	5.32	16.0
III–V	GaP	23	Z	.545	2.24	i	110	75	1.5 /0.18	0.14/0.86	4.13	11.0
	AlSb	32	Z	.614	1.63	i	200	420	1.5 /0.21	0.11/0.5	4.26	14.4
	GaAs	32	Z	.565	1.43	d	8800	400	0.066	0.4	5.31	13.2
	InP	32	Z	.587	1.34	d	4600	150	0.080	0.8	4.79	12.4
	GaSb	41	Z	.610	0.67	d	4000	1400	0.045	0.06/0.33	5.61	15.7
	InAs	41	Z	.606	0.33	d	33000	460	0.027	0.41	5.67	14.5
	InSb	50	Z	.648	0.16	d	78000	750	0.014	0.4	5.78	17.7
II–VI	ZnS	23	Z	.541	3.6	d	165		0.40		4.09	9
	CdS	32	W	*a* .416 *c* 676	2.42	d	300	50	0.21	0.6	4.82	10
	CdSe	41	Z	.606	1.7	d	800		0.13	0.45	5.74	10
IV–VI	PbTe	87	C	645	0.29	d	6000	4000	0.24		8.16	

1) The figure following the chemical formula gives the average number of electrons per lattice site

2) D: Diamond, Z: Zincblende, W: Wurtzite, C: Cubic NaCl-type

3) i: indirect, d: direct semiconductor

4) Two numbers separated by a slash indicate longitudinal/transversal electron effective mass or light/heavy hole effective mass respectively

tric fields (>3 kV/cm) into the satellite valley. This yields a negative differential conductivity which can be used to generate high-frequency oscillations. In other applications, the band structure is modified for a special purpose (*band-structure engineering*[13]): Red-light-emitting diodes with higher efficiency have been fabricated using $GaAs_xP_{1-x}$ (with $0<x<1$) instead of GaP by changing the band structure from "indirect" toward "direct." Laser diodes use higher band-gap layers ($Ga_{1-x}Al_xAs$) to confine the light and the electrons within the active GaAs layer.

4. *REAL CRYSTALS, DEFECTS, IMPURITIES*

Real crystals differ from the ideal ones discussed so far in many respects. At the surface, the energy band structure is altered by "dangling bonds" which can produce surface states within the energy gap. At $T>0$ thermal vibrations of the lattice atoms, called phonons, occur. When a photon with an energy comparable to the band gap is absorbed, excitons can be created. These are electron–hole pairs which attract each other by Coulomb interaction forming a state within the band gap.

Real crystals also show irregularities in their lattice structure. Some of these *crystal defects* are the following:

Point defects: (a) Intrinsic defects: vacancy, interstitial atom, pair of vacancy and interstitial atoms (Frenkel defect); (b) chemical defects (impurities): shallow impurity, deep impurity, isovalent impurity, pairs of impurities.

Line defects: Edge and screw dislocations (giving rise to "swirls"[15]).

Area defects: Grain boundaries, stacking faults.

The technically most important group among these defects are the substitutional impurities, i.e., impurities on lattice sites. Controlled incorporation of impurities is of fundamental importance in semiconductor device technology. The purest silicon crystals available today contain approximately 10^{12} impurities/cm^3, which is lower by a factor of 5×10^{10} than the density of the Si atoms. The purest semiconductor crystals produced so far have been Ge crystals with less than 10^9 impurities/cm^3. Within the bulk crystal, impurities can be incorporated either from the melt during crystal growth or by nuclear reactions [e.g., $Si(n,\gamma)P^{15}$].

For device applications, monocrystalline layers with the required impurity concentration (*doping*) are grown by epitaxy[16] from liquid phase, vapor phase, or molecular beam. Doped layers with certain doping profiles are also achieved by alloying, diffusion,[17] or ion implantation.[18]

Impurities introduce energy levels within the forbidden band. Donor levels are neutral when filled by an electron and positively charged when empty. Acceptor levels are neutral when empty and negatively charged when filled by an electron. A group-V element (P, As, Sb) on a lattice site in Si tends to "donate" an electron to the conduction band. A group-III element (B, Al, Ga) in Si tends to "accept" an electron to form four covalent bonds, thus leaving a "hole" in the valence band. Donors create *n*-type conduction and acceptors *p*-type conduction.

In III-V compounds group-II elements on III lattice sites act as acceptors and group-VI elements on V lattice sites act as donors. Group-IV elements can act either as donors (on III lattice sites) or as acceptors (on V lattice sites). The latter impurities are called amphoteric.

Shallow impurities are either donors with an energy level close to the conduction band or acceptors with an energy level close to the valence band. The order of magnitude for the ionization energy of a shallow impurity can be calculated from the ionization energy of a hydrogen atom when the free-space permittivity ϵ_0 is replaced by the permittivity ϵ of the semiconductor and the electron mass m_0 by the effective mass m:

$$E_i=\left(\frac{\epsilon_0}{\epsilon}\right)^2\frac{m}{m_0}E_H. \tag{5}$$

E_H is the ionization energy of the hydrogen atom (13.56 eV). For donors these ionization energies are about 0.03 eV for Si and about 0.01 eV for Ge and GaAs. For acceptors the values are 0.015 eV for Ge and 0.05 eV for Si and GaAs.

Impurities with higher ionization energies than those of shallow impurities are called deep impurities. These play an important role as recombination centers or as electron or hole traps; e.g., in bipolar silicon transistors a gold diffusion process is used to reduce the minority-carrier lifetime, thereby yielding shorter switching times. In semi-insulating GaAs chromium is incorporated during crystal growth. Cr acts as an electron trap which eliminates the *n*-type conduction of the original material yielding high resistivity ($\approx10^8$ Ω cm). These GaAs crystals are suitable substrates for high-frequency devices (e.g., Schottky-barrier field-effect transistors, planar Gunn elements). In many other applications, however, the effect of traps is detrimental to device performance (e.g., decrease of current gain in bipolar transistors, low efficiency of light-emitting diodes due to nonradiative recombination).

Isovalent impurities have the same number of valence electrons as the replaced lattice atom. A nitrogen atom on a phosphorus site in GaP forms an isovalent acceptor. Other, more complicated types of impurities are polyvalent impurities, pairs or clusters of impurities, or of impurities plus defects (vacancies, dislocations, etc.).

5. *CARRIER CONCENTRATIONS*

In thermal equilibrium the probability of a state of energy E to be occupied by an electron is given by the "Fermi–Dirac distribution function,"

$$W(E)=\left[1+\exp\left(\frac{E-E_F}{kT}\right)\right]^{-1} \tag{6}$$

where E_F is the *Fermi energy* also called the *chemical potential*. For $E=E_F$, $W(E)$ equals 0.5. For $|E-E_F|\geq 4kT$, $W(E)$ can be approximated by the *Maxwell–Boltzmann distribution function* with an error <2%:

$$W(E)\cong\exp\left(-\frac{E-E_F}{kT}\right) \quad \text{for} \quad E-E_F\geq 4kT \tag{7}$$

and

$$1-W(E)\cong\exp\left(-\frac{E_F-E}{kT}\right) \quad \text{for} \quad -(E-E_F)\geq 4kT; \tag{8}$$

$1-W(E)$ can be regarded as the distribution function for holes.

If the Fermi level is within the energy gap and separated by more than $4kT$ from either band edge, the semiconductor is called *nondegenerate* and the Maxwell–Boltzmann distribution may be used. In this case the electron density n is given by

$$n=\int_{E_c}^{\infty} N(E)W(E)dE \tag{9}$$

where $N(E)\sim(E-E_c)^{1/2}$ is the density of states in the conduction band. n is calculated to be

$$n=N_c\exp(E_F-E_c)/kT, \tag{10}$$

where N_c is the effective density of states in the conduction band:

$$N_c=2(2\pi mkT/h^2)^{3/2}. \tag{11}$$

In a similar way one obtains the hole concentration

$$p=N_v\exp[(E_v-E_F)/kT], \tag{12}$$

where N_v is the effective density of states in the valence band given by a formula similar to (11).

From (10) and (12) the mass-action law is obtained:

$$np=N_cN_v\exp(-E_g/kT). \tag{13}$$

In an intrinsic semiconductor conduction takes place by thermal generation of electron–hole pairs, hence $n_i=p_i$. Since (13) is valid also in this case, we have

$$np=n_i^2. \tag{14}$$

The extrinsic case is characterized by $|N_D-N_A|\gg n_i$ and

$$n=N_D-N_A, \quad p=n_i^2/(N_D-N_A) \tag{15}$$

for n-type conduction and

$$p=N_A-N_D, \quad n=n_i^2/(N_A-N_D) \tag{16}$$

for p-type conduction. A schematic representation of the band diagram, density of states, Fermi–Dirac distribution, and carrier concentrations is given in Fig. 5.

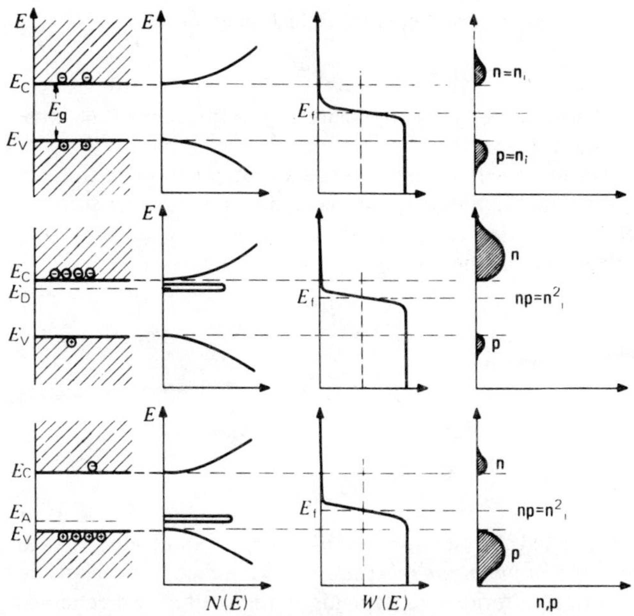

FIG. 5. Schematic band diagram, density of states, Fermi–Dirac distribution, and the carrier concentrations for V intrinsic, n-type and p-type semiconductors at thermal equilibrium. After Ref. 11.

In Fig. 6 the electron concentration for a n-type Si sample is plotted versus $1/T$. The intrinsic case is characterized by the temperature dependence $\exp(-E_g/2kT)$ and the saturation range by Eq. (15). At lower temperatures the Fermi level approaches the impurity level and the probability that an electron occupies the impurity level increases: The impurities are frozen out. The slope in this range is of the order of magnitude of E_i from Eq. (5).

6. *CARRIER TRANSPORT*

For an extrinsic n-type semiconductor the current density

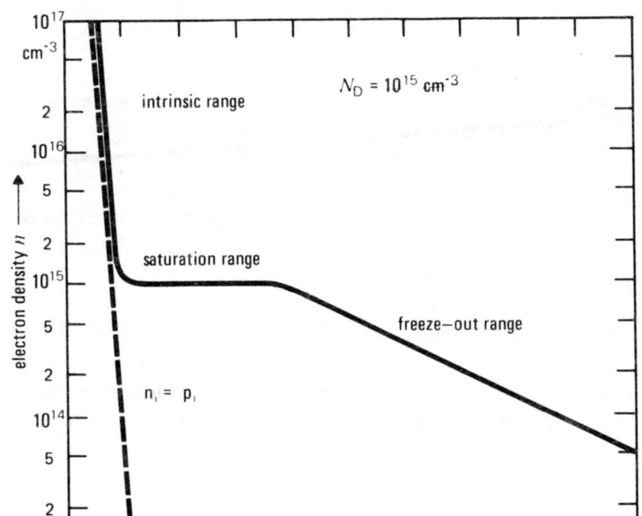

FIG. 6. Electron density as a function of temperature for a Si sample with a donor impurity concentration of 10^{15} cm^{-3}. After Ref. 11.

j due to an external electric field E is given by

$$\mathbf{j} = -nq\mathbf{v}. \tag{17}$$

The drift velocity **v** is opposite to **j** in the case of negative-charge transport.

The electric field causes an accelerated motion of the electrons which is interrupted every time an electron suffers a collision with a lattice atom or any other scattering event. Integration of the equation of motion yields

$$\mathbf{v} = -(q\tau_m/m)\mathbf{E} \tag{18}$$

or

$$\mathbf{v} = -\mu_n\mathbf{E}, \tag{19}$$

with

$$\mu_n = q\tau_m/m. \tag{20}$$

μ_n is called the electron *mobility* and τ_m is introduced as an average *momentum relaxation time*. Relaxation means return to equilibrium. τ_m depends on the scattering mechanism involved. From (17) and (18) it follows that

$$\mathbf{j} = \sigma\mathbf{E}, \tag{21}$$

where the conductivity σ is given by

$$\sigma = nq\mu_n = nq^2\tau_m/m. \tag{22}$$

In a semiconductor there are always both electrons and holes present. Thus, in the more general case of *mixed* conduction (22) has to be replaced by

$$\sigma = q(n\mu_n + p\mu_p), \tag{23}$$

where the *hole mobility* μ_p is given by an equation analogous to (20).

Some of the important scattering mechanisms affecting the mobility are:

Acoustical phonon scattering with $\mu \sim m^{-5/2}T^{-3/2}$;
Ionized impurity scattering with $\mu \sim m^{-1/2}N_I^{-1}T^{3/2}$ (N_I is the ionized impurity density);
Optical phonon scattering with $\mu \sim m^{-5/2}f(T)$, where $f(T)$ decreases monotonically with temperature;
Piezoelectric scattering with $\mu \sim m^{-3/2}T^{-1/2}$, important in compound semiconductors at low temperatures;
Neutral impurity scattering with μ independent of temperature;
Carrier–carrier scattering;
Intervalley scattering;
Intravalley scattering.

When two or more scattering mechanisms act independently from each other the resulting mobility is given by

$$\frac{1}{\mu} = \frac{1}{\mu_1} + \frac{1}{\mu_2} + \cdots; \tag{24}$$

(24) is known as Mathiessen's rule.

Above a certain field strength the charge carriers acquire on average more energy from the electric field than they lose in scattering events, e.g., with optical phonons, which means that the effective electron temperature is higher than the lattice temperature (*warm* and *hot* carriers). In this range Ohm's law is no longer valid and the mobility decreases with increasing electric field. Some experimental curves of velocity versus electric field are shown in Fig. 7. The carrier velocity decreases at low fields with increasing doping concentration. At high fields, however, the drift velocity is essentially independent of doping concentration. Intervalley scattering from the Γ valley into valleys of higher energy is responsible for the peak and the negative differential mobility of the GaAs and InP curve.

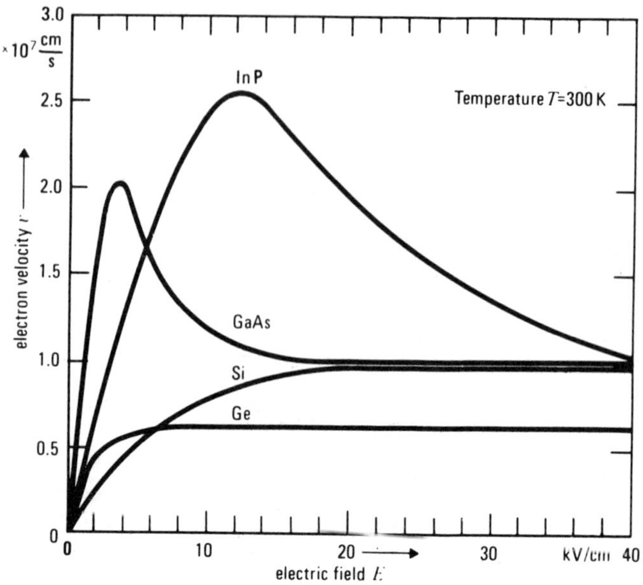

FIG. 7. Electron velocity versus electric field for high purity Ge, Si, GaAs, InP.[20–22]

7. *EXPERIMENTAL METHODS*

A brief list of the methods most frequently used is given in this section.

Conductivity σ or resistivity $\rho = 1/\sigma$ can be obtained from a *four-point probe* measurement (Fig. 8). For a thin sample with probe spacing $s \gg w$ the resistivity is given by

$$\rho = \frac{\pi w}{\ln 2}\frac{V}{I}. \tag{25}$$

In *van der Pauw*'s *method* (Fig. 9) a plane parallel disk with four point contacts at the circumference is employed. For

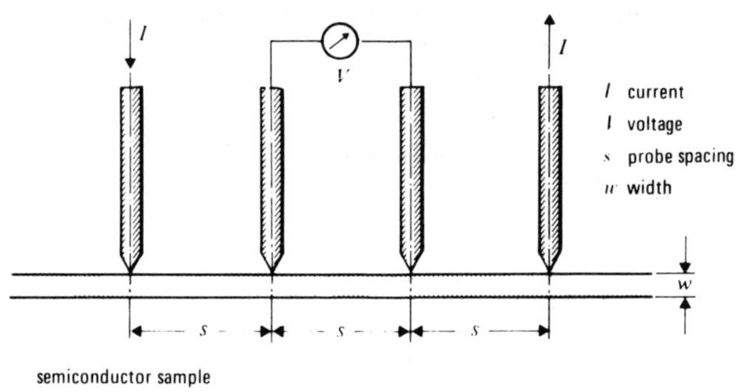

FIG. 8. Setup of four-point probe resistivity measurement.

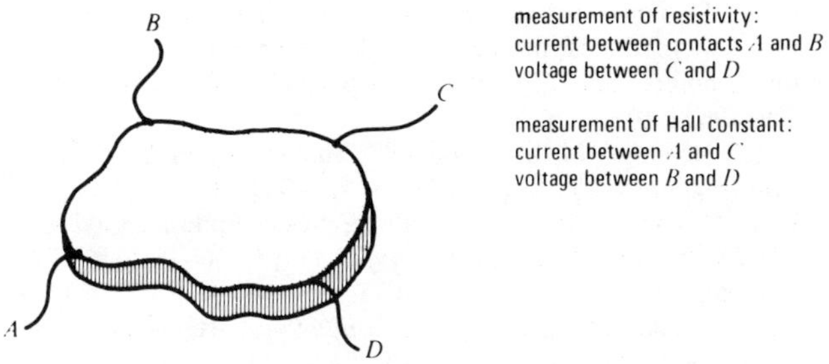

FIG. 9. Plane parallel disk for *van der Pauw* measurements.

measuring resistivity there is a current flow I_{AB} between two neighboring contacts and a voltage drop V_{CD} between the two opposite contacts. Then, by definition,

$$R_{AB,CD} = |V_{CD}/I_{AB}|. \tag{26}$$

In a second experiment

$$R_{BC,DA} = |V_{DA}/I_{BC}| \tag{27}$$

is measured. Van der Pauw proved that

$$\exp(-\pi w R_{AB,CD}/\rho) + \exp(-\pi w R_{BC,DA}/\rho) = 1. \tag{28}$$

From (28) we obtain

$$\rho = \frac{\pi w}{2 \ln 2}(R_{AB,CD} + R_{BC,DA})f, \tag{29}$$

where the *van der Pauw factor f* depends on the ratio $R_{AB,CD}/R_{BC,DA}$. With the same sample the *Hall effect* can be measured when the current flows from A to C and the voltage is measured between B and D. The resistance

$$R_{AC,BD} = V_{BD}/I_{AC} \tag{30}$$

is changed by an amount $\Delta R_{AC,BD}$ when a homogeneous magnetic field is applied perpendicular to the sample surface. The Hall coefficient R_H is then given by

$$R_H = w\Delta R_{AC,BD}/|\mathbf{B}|. \tag{31}$$

The sign of R_H depends on the conductivity type: $R_H<0$ for electron conduction and $R_H>0$ for hole conduction.

Hall-effect and conductivity measurements are used to determine the carrier concentration and mobility. From their temperature dependence the band gap and the impurity concentration may be obtained. In Hall experiments, the magnetic field B is chosen so small that there is no change in conductivity. Under strong magnetic fields, however, a significant increase in resistivity ($\rho \sim |\mathbf{B}|^2$) is observed, the so called *magnetoresistance effect,* which is important in determining the band structure.

Carrier concentration can be obtained from capacitance versus voltage measurements of reverse-biased *Schottky contacts* (rectifying metal–semiconductor contacts):

$$n = \frac{2}{q\epsilon}\frac{dV}{d(1/C^2)}. \tag{32}$$

From the polarity of reverse bias the conductivity type is determined.

The *Seebeck effect* is also used to determine the conductivity type. *Minority carrier lifetime* (e.g., the lifetime of electrons in a p-type sample) is an important parameter in many device applications. It is generally measured using *photoconductivity* experiments, the *photovoltaic effect,* the photoelectromagnetic (PEM) effect, or the Haynes–Shockley experiment. The latter is mainly used to measure *drift velocity*.

The *effective mass* is obtained from *cyclotron resonance,* the *Faraday effect* or the Shubnikov–de Haas effect (oscillations of magnetoresistance at high magnetic fields and low temperatures).

Measurements of *optical absorption and reflection* provide information on band structure and energy levels in semiconductors: *Fundamental absorption* is based on the fact that a semiconductor, while opaque for light with an energy greater than the band gap, becomes transparent for light of lower energy. The absorption spectrum is modified when impurities or excitons have to be taken into account.

In the presence of a strong magnetic field *Landau levels* are introduced in the conduction and valence band. These levels cause a shift of the optical absorption edge. A strong electric field also causes a shift of fundamental absorption (Franz–Keldysh effect).

Table III Materials for bulk effect devices

Gunn element	GaAs InP
Hall generator	Si, InSb, InAs
Magnetoresistor ("field plate")	InSb (InSb with "needles" of NiSb)

8. *SEMICONDUCTOR DEVICES*

Among semiconductor devices, bulk-effect, single-interface, and multiple-interface devices can be distinguished. In

Table IV Materials for diodes and transistors

Name	Interface		
Diodes:	p–n junction	Schottky barrier	MIS
Rectifier	Si	–	–
Varistor	Si	–	–
Varactor	Si	Si, GaAs	Si
IMPATT-Diodes	Si, GaAs	GaAs	–
Tunnel-Diodes	Si, Ge	–	–
Electroluminescent-D.	GaP, GaAs, GaAs . P.	–	–
LASER-Diodes	GaAs/Ga . Al . As	–	–
Solar Cell	Si, GaAs	–	–
Radiation Detectors	Si, Ge, GaAs	GaAs, Ge	–
Transistors:			
unipolar (FET)	Si	Si, GaAs	Si
bipolar	Si, Ge	–	–

this context an interface is created by a *p-n* junction, by a Schottky barrier (metal-semiconductor contact), or by a MIS (metal-insulator semiconductor) structure.

Bulk-effect devices use some of the effects listed above. Thus, a semiconductor can serve as an optical filter, a resistor, a current limiter, a Gunn element (to generate microwave oscillations), a photoconductor, a Hall generator, a magnetoresistor, a thermistor, etc. Table III shows which materials are used today for bulk-effect devices.

Single-interface devices are commonly called diodes. Transistors are the most important group among the multiple-interface devices. When the current flows through a semiconductor with only one type of conductivity (*p*- or *n*-type) the transistor is called unipolar or field-effect transistor (FET). In bipolar transistors, on the contrary, the charge is transported through *npn* or *pnp* structures. Transistors are the basic elements in integrated circuits. Table IV shows which materials are used today for diodes and transistors. Because of its superior materials and device technology in combination with its favorable physical properties, silicon is the dominating semiconductor material. For optoelectronic and microwave applications III-V compounds are becoming increasingly important.

See also CRYSTAL BINDING; CRYSTAL DEFECTS; CRYSTALLOGRAPHY, X-RAY; CYCLOTRON RESONANCE; DE HAAS–VAN ALPHEN EFFECT; ELECTRON ENERGY STATES IN SOLIDS AND LIQUIDS; EXCITONS; GALVANOMAGNETIC EFFECTS; HALL EFFECT; INSULATORS; LIQUID CRYSTALS; MAGNETORESISTANCE; PHONONS; PHOTOCONDUCTIVITY; PHOTOVOLTAIC EFFECT; SEMICONDUCTORS, AMORPHOUS; THERMOELECTRIC EFFECTS; TRANSISTORS.

REFERENCES

1. O. Madelung, "Halbleiter," in *Handbuch der Physik*. (S. Flügge, ed.), Vol. XX, pp. 1–245, Springer, Berlin, 1957. (I)
2. J. C. Slater, "The Electronic Structure of Solids," in *Handbuch der Physik* (S. Flügge, ed.), Vol. XIX, pp. 1–136. Springer, Berlin, 1956. (A)
3. C. Kittel, *Introduction to Solid State Physics*, 4th ed. Wiley, New York, 1971. (I)
4. C. A. Wert and R. M. Thomson, *Physics of Solids*. McGraw-Hill, New York, 1964. (I)
5. O. Madelung, *Physics of III-V Compounds*. Wiley, New York, 1964. (I/A).
6. J. C. Phillips, *Bonds and Band in Semiconductors*. Academic Press, New York, 1973. (A)
7. E. Spenke, *Elektronische Halbleiter*. Springer, Berlin/Heidelberg/New York, 1965. (I)
8. J. Tauc (ed.), *Amorphous and Liquid Semiconductors*. Plenum Press, London/New York, 1974. (A)
9. K. Seeger, *Semiconductor Physics*. Springer, Wien/New York, 1973. (I/A)
10. A. S. Grove, *Physics and Technology of Semiconductor Devices*. Wiley, New York, 1967. (I)
11. S. M. Sze, *Physics of Semiconductor Devices*. Wiley, New York, 1969. (I/A)
12. E. H. Putley, *The Hall Effect and Semiconductor Physics*. Dover Publications, New York, 1968. (A)
13. C. Hilsum, "Band Structure Engineering," *American Institute of Physics Conference Series No. 12*, pp. 77–86. AIP, New York, 1971. (Review)
14. M. L. Cohen and T. K. Bergstresser, "Band Structures and Pseudopotential Form Factors for Fourteen Semiconductors of the Diamond and Zinc-blende Structures," *Phys. Rev.* **141,** 789–796 (1966).
15. W. E. Haas and M. S. Schnöller, "Silicon Doping by Nuclear Transmutation," *J. Electr. Mat.* **5,** 57–68 (1976).
16. H. F. Mataré, "The Electronic Properties of Epitaxial Layers," *Solid State Technol.* 25–34 (Jan. 1976) (Part 1); 36–48 (Feb. 1976) (Part 2).
17. B. I. Boltaks, *Diffusion in Semiconductors. Academic Press, New York, 1963.* (I)
18. J. F. Gibbons, "Ion Implantation in Semiconductors; Part I: Range Distribution Theory and Experiments," *Proc. IEEE* **56,** 295–319 (1968); "Part II: Damage Production and Annealing," *Proc. IEEE* **60,** 1062–1096 (1972). (Review)
19. (a) *Handbook of Electronic Materials, Vol. 2: III-V Semiconducting Compounds* (M. Neuberger, ed.), *Vol. 5: Group IV Semiconducting Materials* (M. Neuberger, ed.). IFI/Plenum, New York, 1971. (b) H. Wolf, *Semiconductors*. Wiley, New York, 1971.
20. C. B. Norris and J. F. Gibbons, "Measurement of High-Field Carrier Drift Velocities in Silicon by a Time-of-Flight Technique," *IEEE Trans. Electr. Dev.* **ED-14,** 38–46 (1967).
21. A. Neukermans and G. S. Kino, "Measurement of the Electron Velocity-Field Characteristics in Germanium Using a New Technique," *Phys. Rev.* **B7,** 2693–2703 (1973).
22. W. Fawcett and D. C. Herbert, "High-Field Transport in GaAs and InP," *J. Phys. C: Solid State Phys.* **7,** 1641–1654 (1974).

Shock Waves and Detonations*

Marvin Ross

A shock wave is a disturbance propagating at supersonic speed in a material, accompanied by an extremely rapid rise in pressure, density, and temperature. Such waves may be generated by the sudden release of a large amount of energy in a limited space, as in the detonation of high explosives (chemical energy), the passage of supersonic aircraft (mechanical energy), or the discharge of lightning bolts in a narrow channel (electrical energy). A shock wave not sustained loses energy through viscous dissipation and reduces to a sound wave (e.g., thunder).

A detonation is a wave in which an exothermal reaction takes place and that moves with supersonic velocity into the undetonated material. In a detonation, a shock wave compresses, heats, and ignites an explosive mixture that gives off sufficient energy and expanding reaction products to sustain the shock. The detonation process requires a shock wave. A flame, or deflagration, differs from a detonation because it propagates at subsonic speeds and is not accompanied by a shock wave.

SHOCK WAVES

Figure 1 illustrates the formation of a one-dimensional, planar shock wave generated by accelerating a piston in small increments from zero velocity to some final constant

* Work performed under the auspices of the U.S. Department of Energy by Lawrence Livermore Laboratory under contract #W-7405-Eng-48.

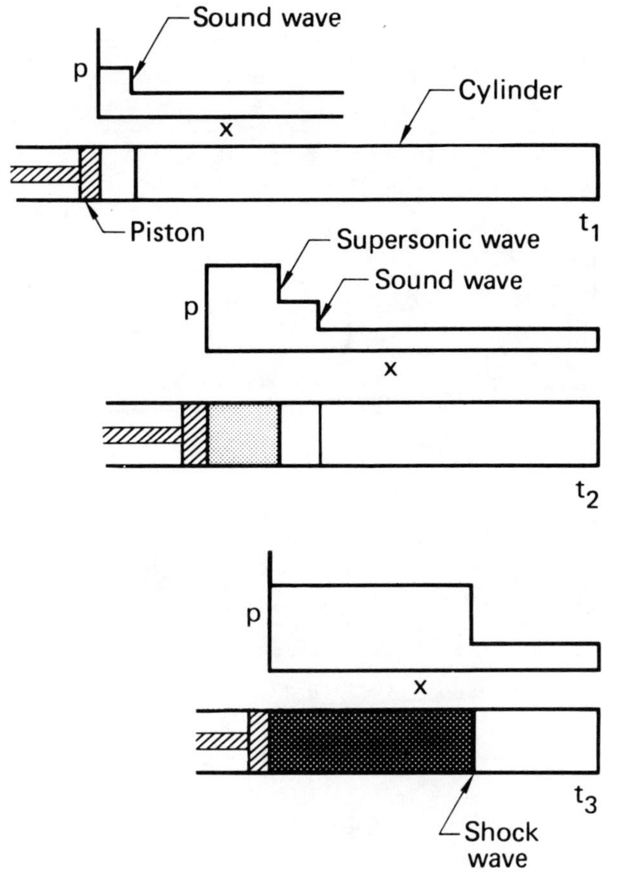

FIG. 1. Formation of a one-dimensional planar shock wave. (Shown by pressure–distance p–x plots at three successive times, t_1 to t_3.)

velocity. The first infinitesimal compression at the piston face results in the propagation of a sound wave. However, subsequent compressions at the piston face take place with the material at higher densities and result in higher local sound speeds. This produces a train of waves in which the first is at the speed of sound in the undisturbed material and the last, closest to the piston face, is supersonic. Because the last wave can catch but not pass the first, all the waves eventually coalesce into a single, steep wave front across which exists a sharp discontinuity in pressure, density, and temperature. The width of the discontinuity is generally a few molecular mean-free-path lengths. Behind the piston, the reverse process of gas expansion creates a rarefaction wave that moves in a direction opposite to the shock wave and piston motion. A rarefaction wave cannot form a shock front.

The change in physical properties across the shock front is described by the Rankine–Hugoniot (R–H) equations: conservation of mass,

$$\frac{V_0 - V}{V_0} = \frac{U - U_0}{S}; \tag{1}$$

conservation of momentum,

$$S^2 = V_0^2 \frac{(P - P_0)}{V_0 - V}; \tag{2}$$

and conservation of energy,

$$E - E_0 = \frac{(P + P_0)(V_0 - V)}{2}, \tag{3}$$

where S is the shock velocity, and E, V, P, and U are, respectively, the energy, specific volume, pressure, and material or piston velocity in the shocked states. The subscript zero indicates the initial states. The R–H equations represent the locus of all final states that can be reached by shock-compressing a material from the same initial state; the resultant curve of pressure against volume is known as a Hugoniot. If the initial state is known, then the final-state properties may be determined from a measurement of any two of these five properties. Of these, the shock velocity is the most commonly measured. Similar equations can be derived to describe conical shocks generated by a sharp-nosed projectile or spherical shocks that travel symmetrically outward in three dimensions. In these cases, the shock wave expands radially and decreases in intensity as the front propagates from its point of origin.

As do other wave phenomena, shock waves reflect, refract, and diffract when colliding with rigid walls, when moving from one medium to another, and when passing geometrical objects. A shock wave traveling in material A will always cross an interface and transmit a shock to an adjacent material B. In addition, a rarefraction wave will be reflected back into A if the impedance of A (initial density × shock velocity) is greater than the impedance of B; or a shock wave will be reflected back into A if the impedance of B is greater. No reflection occurs at equal impedance. For example, a shock wave traveling from a solid to a gas will be reflected from the interface as a rarefaction wave and will be transmitted ahead as a shock wave. In traveling from a gas to a solid, both the reflected and transmitted waves will be shocks.

Shock waves set up in supersonic flows that change direction (Fig. 2) are referred to as oblique. The pressure increases across the shock front because of the extra compression at the compression corner (Fig. 2a) and decreases over a broadened front as the flow expands at the expansion corner (Fig. 2b). Oblique shocks occur on the surfaces of transonic and supersonic aircraft, and because their pressure gradients influence overall structural integrity, they have been studied extensively.

The R–H equations are widely used to study plane shock waves propagating in an ideal gas. For this case, the pressure, volume, and temperature ratios across the shock front are as follows:

$$\frac{P}{P_0} = \frac{2\gamma M^2 - (\gamma - 1)}{(\gamma + 1)}, \tag{4}$$

$$\frac{V}{V_0} = \frac{(\gamma - 1)M^2 + 2}{(\gamma + 1)M^2}, \tag{5}$$

and

$$\frac{T}{T_0} = \frac{PV}{P_0 V_0} \tag{6}$$

where the Mach number M is the shock velocity divided by the speed of sound in the undisturbed medium; γ is the spe-

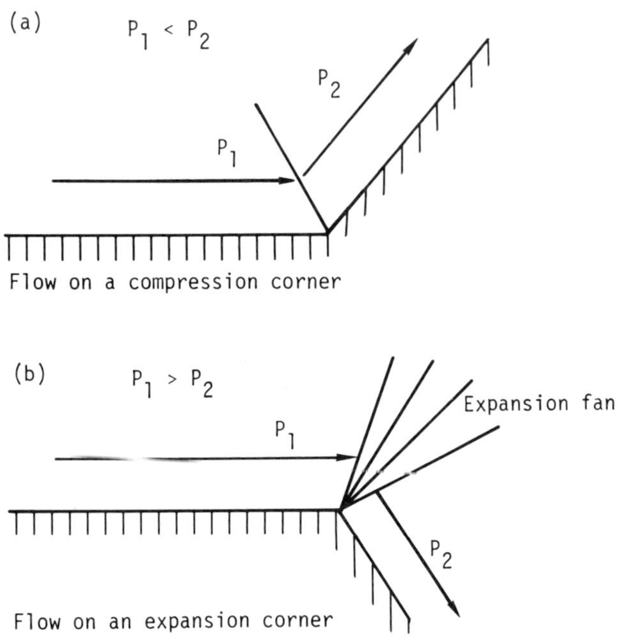

FIG. 2. Formation of oblique shock waves in gases on nonplanar boundaries.

cific heat ratio, C_p/C_v, for the gas. For real gases, the equations must be modified to account for vibration, rotation, electron excitation, and dissociation. A unique feature of shock waves in gases is the high final temperature, up to 15 000 K, that can be generated. This feature is largely responsible for the widespread application of shock tubes to the study of fundamental problems in science and aeronautics. Measurements have been made of chemical reaction rates, vibrational relaxation rates, and the rates and energies of dissociation and ionization.

In aerodynamics, modified shock tubes are used as short-duration wind tunnels to produce the high-Mach-number ($M>16$), high-temperature (6000 K) environments needed to study the gas dynamics encountered by a missile or reentry vehicle—environments that conventional wind tunnels ($M<7$) cannot achieve.

In the shock tube shown in Fig. 3, a one-dimensional shock wave is generated by the sudden bursting of the diaphragm separating a high-pressure driver gas from a low-pressure gas. After bursting, the driver gas acts as a piston and forms a compression wave that rapidly steepens to form a shock front in the low-pressure gas. Simultaneously, a rarefaction forms and travels back into the high-pressure end of the tube. On reaching the end of the tube the shock wave is reflected and moves back into the already shock-heated gas, further raising its temperature, pressure, and density. For the low-density gases used in shock-tube experiments, only the shock velocity need be measured to determine the final P, V, and T.

The shock velocity can be measured by a number of methods that detect the characteristic large, sharply defined changes in material properties. Because the available experimental time is short (10^{-6}s), all instrumentation is designed for fast response. Piezoelectric pressure transducers

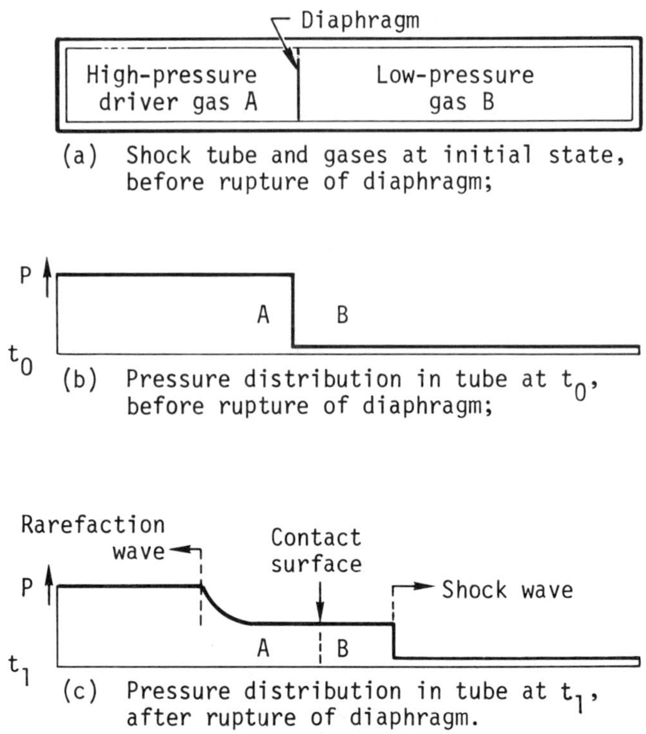

FIG. 3. Generation of a plane shock wave in a simple shock tube.

or heat transfer gauges placed along the tube indicate shock arrival times, and hence velocity. Optical methods are widely used because of their fast response times and noninteractive nature. The large density gradient at the shock front leads to a refractive index gradient that deflects a passing light beam. This deflected light is used by the schlieren method to photograph shock structure and by interferometry to provide quantitative density profiles. If the shock is luminous, a streaking camera can record a continuous measurement of the shock position with time. Chemical composition can be studied with absorption and emission spectroscopy.

The generation of shock waves in solids and liquids, because of their higher density, requires more energy than for generation in gases. This energy can be supplied by explosives or high-velocity guns. The most common method of measuring shock-wave velocities in solids or liquids is the use of electrical probes that short-circuit at the passing of the shock. The high surface luminosity of shock-compressed solids and liquids permits the use of streak cameras.

The relationship between pressure and density is well known for gases but not for liquids and solids. Consequently, most shock-wave experiments on condensed materials have centered on pressure–volume determinations by measuring shock and piston velocity and using the R–H equations. With this method, pressures up to about 1 TPa (10 Mbar) are measured. These data are a valuable extension of static press studies, which are generally limited to pressures below about 50 kPa (500 kbar). Pressures and densities measured for shock-compressed geologic materials are comparable to those found in the earth, and have provided data for geophysical theories.

DETONATIONS

Gases were first detonated in the laboratory in 1881, when an explosive mixture in a long uniform tube was ignited at one end and the initial combustion wave was observed to accelerate rapidly to a high constant speed (detonation velocity) over the remaining length. The detonation velocity is independent of the method of ignition, tube material, or tube diameter (beyond a certain minimum), but is dependent on the chemical composition of the gas mixture. A stable detonation cannot be generated outside certain ranges of composition (called limits of detonability). Typical detonation velocities, temperatures, and pressures in gas mixtures are, respectively, 2000 m/s, 3000 K, and 2 Pa (20 bars), and are relatively insensitive to the initial conditions. For comparison, burning rates are 0.1 m/s. Detonation velocities in gases can be accurately computed with the Chapman–Jouquet (C–J) theory.

For a shock wave without a chemical reaction (Fig. 4), the Hugoniot curve passes smoothly from its initial state (P_0, V_0). This is not so for a detonation, which can be thought of as a two-step process in which a chemical reaction releases energy in a constant-volume explosion and the reaction products are then shock-compressed to some final state. The velocity of this final state is proportional to the slope of the line through the initial and final states (Rayleigh line) and is given by Eq. (2). Because the R–H equations cannot by themselves predict which of the Rayleigh lines (*OA* or *OB*) corresponds to the unique detonation velocity, an additional assumption is necessary. This is provided by C–J theory, which assumes that the detonation wave proceeds at the minimum attainable velocity. This is given by the tangent to the reacted Hugoniot curve drawn from the initial state, the line *OA*. The point of tangency is called the C–J point. Whereas C–J theory is successfully applied to gases, difficulties arise when the detonation properties of solid or liquid explosives are computed. Here, the products form a very dense gas for which the *P–V–E* relationship is not well known and, hence, the computed properties are less accurately predicted. Typical velocities and pressures in solid explosives are 8000 m/s and 30 kPa (300 kilobars).

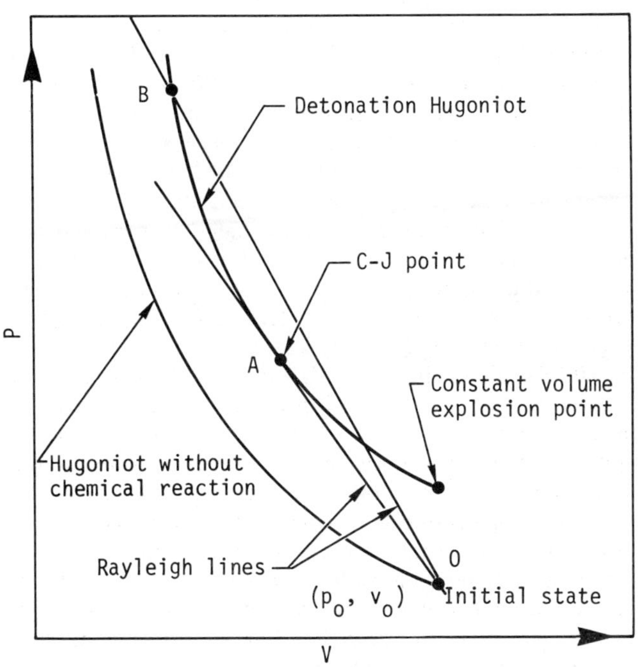

FIG. 4. Hugoniot curve for a detonation.

The detailed structure of a detonation wave is visualized as a shock front followed by a chemical reaction zone. Pressure and density peak at the shock front and fall as the reaction proceeds. In gases at atmospheric pressure, this reaction zone is about 1 mm thick. Beyond this zone, the product gas expands isentropically and acts as a continuously moving piston that propagates the shock at constant velocity. Detonation is initiated by the introduction of a shock wave or a local heat source. The methods used to study detonations are generally the same used for shock waves. Research is directed to determining C–J points, product composition, wave structure, and studying the initiation or onset.

BIBLIOGRAPHY

M. A. Cook, *The Science of High Explosives*. Krieger, Huntington, N.Y., 1971. (I)

I. I. Glass, *Shock Waves and Man*. Univ. of Toronto Institute for Aerospace Studies, Toronto, 1974. (E)

J. K. Wright, *Shock Tubes*. Methuen, London, 1961. (I)

Ya. B. Zeldovich and Yu. P. Raizer, *Physics of Shockwaves and High Temperatures Hydrodynamic Phenomena*, Vols. 1 and 2. Academic Press, New York, 1967. (A)

Solid-State Switching

H. Fritzsche

Switching—that is, the making, breaking, or changing of connections in an electrical circuit or changing the signal level between two distinct values—is the principal operation in electronic computing, information processing, and automatic control. Because of their reliability, speed, cheapness, and small size, solid-state switching devices have replaced mechanical moving-contact switches, relays, and vacuum or gas-filled tubes in most modern technologies.

Many solid-state devices are used as switches although their characteristics are ideally suited for amplification, detection, or modulation of analog signals. This is because of the rapidly increasing use of digital electronics. Digital information is immune to drift and noise problems and can be amplified indefinitely, stored accurately, and handled and processed directly by a computer. The advent of large-scale integration (LSI), that is, the batch fabrication of many electronic circuits and memory arrays on one chip (0.3×0.3 cm^2) containing 10^4–10^5 devices, has removed the last obstacle to a general use of digital electronics, the more complex circuitry which it requires.

The most widely used semiconductor material is silicon, primarily because the quality of its oxide facilitates the device processing steps involving photolithography, etching, diffusion, or ion implantation. GaAs finds important applications for microwave and electron-optical devices. Ge and also chalcogenide glasses are used to achieve special switching-device characteristics. Active research is conducted to develop SiC devices for high-temperature applications, superconducting devices for faster computer logic, and amorphous or crystalline magnetic materials for fast, high-density magnetic bubble memories and shift registers.

Fundamental limitations. A switch must be able to remain in and rapidly change between at least two distinct operating points in response to different input signals. These points may differ in resistance, current, or voltage, and represent OFF and ON or the binary digits 0 and 1 of the Boolean algebra used in computer logic. The least amount of energy needed to change and read a logic state is about $\Delta E = kT$. This yields with Heisenberg's uncertainty relation an ultimate speed $\Delta t \approx 10^{-11}$ s. In practice, circuit and device impedances as well as minority-carrier life or drift times determine the switching speed.

The two switching states in semiconductors correspond to different spatial distributions and local densities of electrons and/or holes. In majority-carrier (unipolar) devices this results in space charges which can be changed rapidly within a dielectric relaxation time t_D (about 0.1 ns for Si having $\rho = 1\ \Omega$ cm). In bipolar devices, injected minority carriers are neutralized within time t_D by an increase in majority carriers. These excess carrier pairs must be swept out by a field which causes a turnoff delay of order 50 ns. Their accumulation is minimized by special circuitry or a Schottky diode clamp in high-speed (5 ns) bipolar logic circuits.

Monostable and bistable switching of a negative-resistance device is illustrated in Fig. 1. The example shows the voltage-controlled, short-circuit-stable, or N-shaped characteristic of a tunnel diode, intersected by a load line at point A. Figure 1(a) describes a monostable switch: A positive input signal shifts the operating point to B and causes switching to another stable point C or D. Removal of the signal causes switching from C to B and restores the initial point A. This is a monostable switch. Figure 1(b) is for a bistable switch: With the same device, a load line is chosen which intersects the characteristics at two stable points A and E. Point F is unstable. After the device has switched from B to C, it cannot return to A upon removal of the signal but remains at point E until a reverse pulse returns it to A. By exchange of coordinates the following remarks apply to current-controlled, open-circuit-stable, or S-shaped thyristor-type characteristics. At high-speed switching, the load line is no longer resistive. Delay and transients are described instead by a dynamic load trajectory governed by a complex and often nonlinear impedance network determined by space charge and by excess carrier drift in the device, its package, leads, and the external circuit.

The following describes the operation of discrete switching devices.

The Schottky diode is a rectifying metal–semiconductor (MS) contact. Interface states and to a small degree the difference in work functions between the metal (M) and the semiconductors (S) give rise to a Schottky barrier of height ϕ, and depletion of majority carriers in the barrier region. The observed current I is the difference between majority carrier flow in opposite directions across the barrier. The

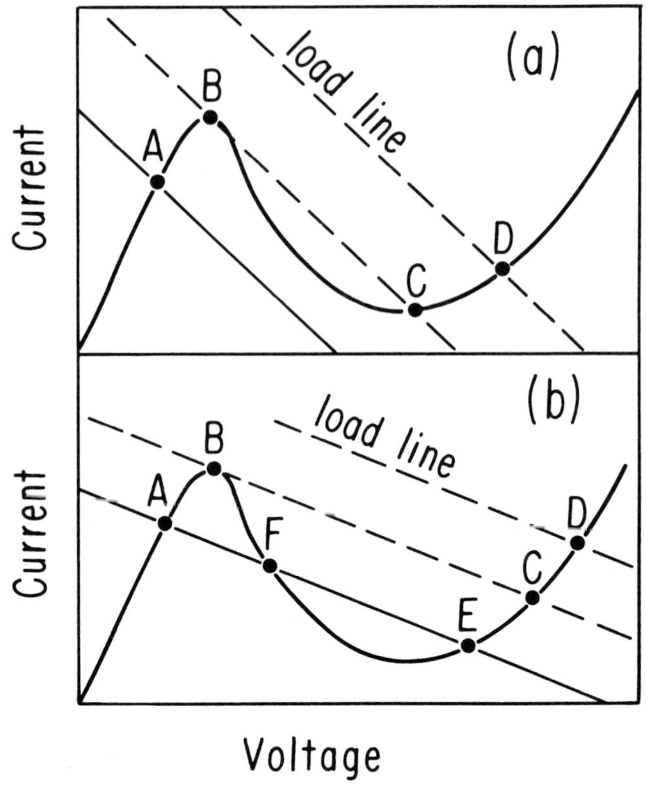

FIG. 1. Tunnel diode (a) as a monostable switch and (b) as a bistable switch, depending on the value of load resistor.

flow from M to S is impeded by $\exp(-\phi/kT)$, the flow from S to M by $\exp[-(\phi - qV)/kT]$ because of the bias potential qV in S. This leads to the rectifier characteristic

$$I = I_r[\exp(qV/kT) - 1], \quad (1)$$

where I_r is the reverse leakage current. Switching times less than 0.1 ns are achieved for small (5 μm diam) devices.

The pn-junction diode is formed at the interface of an n- and p-type semiconductor. Electrons diffuse from the n to the p material and holes from the p to the n material until an internal electric field is built up which causes the resultant drift current to cancel the diffusion current. This prevents further increase of the equal and opposite space charges. The energy barrier in the high-resistance intrinsic junction is about equal to the energy gap and nearly twice as big as ϕ of Schottky diodes. Equation (1) holds also for pn junctions, but with much smaller values of I_r. This leads to a larger forward voltage drop V_F. For a 10-μA current, $V_F \approx 0.25$ V for a Schottky diode and $V_F \approx 0.55$ V for a silicon p-n junction. For $V > 0$, majority carriers flow across the junction. For $V < 0$, they are pulled away from the junction and only the small minority-carrier density contributes to I_r until the breakdown voltage of the diode is reached, where I increases rapidly due to avalanche multiplication of carriers in the high field of the reverse-biased junction. The turnon time of a pn diode is only a few nanoseconds. A turnoff time of microseconds is needed to sweep out the excess minority carriers which are stored in the junction during conduction.

Step-recovery diodes are pn junctions whose graded doping profile minimizes the stored-minority-carrier transit time.

They are used for fast and microwave switching.

The Zener diode is a *pn* junction with a sharp and well-controlled reverse-bias avalanche breakdown voltage (0.2 V to several hundred volts). Its fast switching, voltage sensitivity, and low resistance in the ON state make it useful as an overload and transient suppressor and in control circuits.

The p-i-n diode is a current-controlled microwave switch. It has a thin layer of undoped (intrinsic) region between strongly doped *p* and *n* layers. At microwave frequencies the intrinsic region acts like a capacitor. Forward bias (dc or low frequency) produces a carrier avalanche and makes the diode conductive.

The Esaki tunnel diode is a *pn* junction in which both sides are so heavily doped that the Fermi level lies in the valence band on the *p* side and in the conduction band on the *n* side. At forward bias this overlap of the band edges allows electrons from the *n* side to tunnel quantum mechanically through the junction barrier into empty valence-band states without changing their energy. At higher bias the electron energies are too high for energy-conserving tunneling and the current drops through the negative-resistance range shown in Fig. 1. The current rise at larger biases (>0.4 V) is the conventional forward-bias current of a *pn* junction. Switching time is less than a nanosecond since only majority carriers are involved.

The junction transistor consists of two *pn* junctions in close proximity. Under normal conditions one junction is the forward-biased emitter, the other the reverse-biased collector. The transistor is OFF when both collector and emitter are reverse biased (base current zero). Collector–emitter leakage current is about 10 nA for Si. In its ON state both junctions are forward biased ($V_{CE}<1$ V), injecting carriers into the base region which reaches saturation as a result of space-charge and Ohmic effects. For fast turnoff speeds saturation must be avoided by, for instance, a Schottky diode between collector and base which becomes forward biased before saturation is reached.

The avalanche transistor is a transistor structure with alloyed step junctions and a base-layer resistivity which is larger than that of both the collector and emitter. With increasing V_{CE} avalanche breakdown of the collector occurs ($V_{CE}\approx60$–80 V). The emitter current is first low but with increasing current through the base, the base potential increases, and the emitter becomes forward biased which causes a large emitter–collector current to flow at reduced $V_{CE}\approx30$ V. A negative-resistance region is traversed as the emitter becomes forward biased.

The pnpn transistor has a reverse-biased collector junction, which yields a high OFF-state resistance ($>10^8\ \Omega$) between two forward-biased junctions. After onset of avalanche in the middle junction at the breakdown or switch-ON voltage, the two emitters saturate the two base regions and make the middle junction appear forward biased. A low-voltage ON state results. The device returns to its OFF state when the current falls below a holding current value.

Silicon-controlled rectifier (SCR). Since switching of the *pnpn* transistor is initiated by the condition that the sum of the common base current gains α_{npn} and α_{pnp} becomes unity, rather than by the onset of avalanching, this condition can be achieved and the magnitude of the switch-ON voltage controlled by changing the current of either one of the two forward-biased junctions. This is done in the SCR by an additional terminal to one of the base regions.

The silicon-controlled switch (SCS) has terminals to each of the two base regions of a *pnpn* transistor. One permits control of ON switching, the other controls the holding current and thus OFF switching.

The triac is a SCR device built such that operation with either voltage polarity yields a nearly identical SCR characteristic.

The unijunction transistor (UJT) is an *n*-type semiconductor base with two Ohmic contacts between which is placed a *pn* junction (anode). Forward biasing of the anode injects minority carriers (holes) into the base which decrease the resistance between the anode and the negative base contact. The resultant shift of the base potential below that of the anode increases its forward bias and thus the minority-carrier injection. This leads to an S-shaped, voltage-controlled, negative *I-V* characteristic. The switching speed (ON to OFF) is orders of magnitude lower than that of bipolar transistors because of large charge storage effects in the high-resistivity base region.

Thyristors. The SCR, SCS, and triac are a few of several solid-state switches having thyratron-like S-shaped characteristics. Some can also be light activated. Their low power consumption, trigger sensitivity, and reliability make them useful in many control circuits.

The field effect transistor (FET) is a majority-carrier (unipolar) device, in which the conductance between the source and drain electrodes along a thin semiconductor channel made of *n* or *p* material is controlled by the voltage difference between this channel and an adjacent gate electrode. Insulation between the channel and the gate is provided by a *pn* junction (junction FET) or by a thin oxide layer yielding a metal-oxide-semiconductor structure (MOS) also called insulated gate FET, or IGFET. The gate voltage either enhances or depletes the carrier concentration in the channel and thus alters its conductance between low (OFF) and high (ON) values. The power consumption is very low as it is essentially the change in energy stored on the device capacity. Because of device capacitances the switching speed (particularly OFF to ON) is slower than that of fast bipolar devices. However, the small size of MOS devices, their self-isolation, and ease of manufacture permit large-scale integration of complex digital and memory circuits on one small silicon chip. The work-function difference and hence the switching voltage (about 1.5 V) is decreased to 0.6 V by using *p*-doped polycrystalline Si instead of Al as a gate electrode. Such silicon-gate IGFETs have higher speed and use even less power.

The MNOS memory switch is a MOS device with a silicon nitride dielectric layer (N) between the thin oxide and the gate electrode. Charges trapped in the silicon nitride layer hold the conductance state in the channel and thus provide the memory action; 5–10 V "write" and "erase" pulses of opposite polarity change the charge state of the traps. The memory access time is about 50 ns compared to the 200-ns access time of ferrite core memories. However, the MNOS memory is volatile, which means that it requires circuit voltage to keep its memory.

The Ovonic threshold switch (OTS) contains as its active part a noncrystalline (amorphous) semiconductor. A thin layer of chalcogenide alloy glass, for instance $Te_{50}As_{30}Si_{10}Ge_{10}$, is sputtered or evaporated between two nonreacting electrodes which are less than 1 μm apart. The normal (OFF) resistance is of over 1 MΩ. If an applied voltage of either polarity exceeds the threshold voltage (2–200 V), the OTS switches within 0.1 ns from the threshold point to its ON state. In this state the voltage drop is about 1 V independent of current. As soon as the current falls below a holding current of about 1 mA the OTS returns to its OFF state. Under pulse conditions the ON state is reached after a delay time between 10 and 1000 ns depending on the pulse height above threshold. The OFF state is the normal resistance of the semiconducting glass. In the ON state injection from the contacts, trap filling, and a resultant decrease of the mobility gap produce a conductive region of about 10 μm diameter.

The Ovonic memory switch (OMS) is similar in construction to the OTS except for the use of a bistable chalcogenide glass, for instance $Ge_{14}Te_{83}Sb_3$, which can be reversibly altered by electrical pulses between a resistive noncrystalline state and a conductive microcrystalline state. A 2.5-mA, 1-ms set pulse switches the OMS ON (resistance≈50 Ω); a reset pulse (120 mA, 2 μs) switches it OFF (~50 kΩ). Both memory states remain on removal of circuit voltages. Its nonvolatility, small size, ease of manufacture, and integration with silicon devices offset its relatively slow write–erase cycle time. Read access time is about 50 ns.

Optoelectronic devices use light-emitting and -detecting diodes in close proximity. Light coupling achieves high electrical isolation between input and output.

Integrated logic circuits. The switching elements in digital electronics are no longer discrete devices but instead integrated circuits: gates, clocks, counters, and memory arrays, each containing numerous devices. These integrated elements are used together with drivers, encoders, and other circuitry to prepare the small-, medium-, or large-scale integrated circuits which comprise the random-access memories (RAM), electrically programmable memories (PROM), shift registers, and other peripherals needed to build, for instance, a microprocessor or microcomputer. Even complex computing and data handling systems use only two basic logic operations, counting and gating. The basic logic elements of gating are the operations of coincidence starting, "AND"; of anticoincidence or sorting out, "OR"; and of prohibition, "NOT."

The awesome pace of integrated digital technology aims at reducing cost, achieved by the higher packing density of large-scale integration (LSI), and at increasing speed. Fastest gating (less than 1 ns) is presently possible with bipolar transistors in emitter-coupled logic (ECL). However, lowest cost was achieved by the slower MOS technology with its breakthrough into LSI in 1970. Its gate propagation delay of 100 ns or greater limits its use to medium-speed and non-real-time data processing. At present we witness a new breakthrough in LSI of fast bipolar transistor gates promising delays shorter than 15 ns. This integrated injection logic (I^2L) may possibly lead to low-cost computers.

See also CIRCUITS, INTEGRATED; DIELECTRIC PROPERTIES; MAGNETIC DOMAINS AND BUBBLES; RELAXATION PHENOMENA; SEMICONDUCTORS, AMORPHOUS; SEMICONDUCTORS, CRYSTALLINE; TRANSISTORS.

BIBLIOGRAPHY

Elementary

L. Altman, "The New LSI," *Electronics* **48,** 81 (1975).

A. J. Diefenderfer, *Principles of Electronic Instrumentation*. Saunders, Philadelphia, 1972.

W. D. Haffner, "I^2L, A New Transistor Logic Family," *Popular Electronics* **9,** 56 (1976).

F. G. Heath, "Large-Scale Integration in Electronics," *Scientific American* **222** (2), 22 (1970).

W. C. Hittinger, "Metal-Oxide-Semiconductor Technology," *Scientific American* **229** (2), 48 (1973).

H. V. Malmstadt and C. G. Enke, *Digital Electronics for Scientists*. Benjamin, New York, 1969.

E. W. McWhorter, "The Small Electronic Calculator," *Scientific American* **234** (3), 88 (1976).

Intermediate

W. N. Carr and J. P. Mize, *MOS/LSI Design and Application*, R. E. Sawyer and J. R. Miller (eds.). McGraw-Hill, New York, 1972.

A. S. Grove, *Physics and Technology of Semiconductor Devices*. Wiley, New York, 1967.

D. A. Hodges, *Semiconductor Memories*. IEEE Press, New York, 1971.

S. R. Ovshinsky and H. Fritzsche, "Amorphous Semiconductors for Switching, Memory, and Imaging Applications," IEEE Trans. Elec. Dev. **ED-20,** 91 (1973).

S. M. Sze, *Physics of Semiconductor Devices*. Wiley, New York, 1969.

Advanced

G. Abraham, "Multistable Semiconductor Devices and Integrated Circuits," in *Advances in Electronics and Electron Physics*, Vol. 35, pp. 270–401, L. Marton (ed.). Academic Press, New York, 1974.

H. R. Camenzid, *Designing Integrated Systems*. Von Nostrand, New York, 1972.

R. Cobbold, *Theory and Applications of Field-Effect Transistors*. Wiley, New York, 1970.

A. Goetzberger and S. M. Sze, "Metal-Insulator-Semiconductor Physics," in *Applied Solid State Science*, Vol. *1*, Chap. II, R. Wolfe (ed.). Academic Press, New York, 1969.

"Semiconductors and Semimetals," Vol. 7, R. K. Willardson and A. C. Beer (eds.). Academic Press, New York, 1971.

Superconducting Materials

D. K. Finnemore

Over the past 20 years enormous strides have been made both in our understanding of the phenomena of superconductivity and in our ability to prepare high-quality superconducting materials. Phenomena such as zero electrical resistance, the Meissner effect, flux quantization, vortex motion, and the Josephson effect have been thoroughly studied and are now understood rather well. In addition, with the advent of high-vacuum technology, microfabrication, and sophisticated metallurgical techniques it is possible to prepare materials reproducibly with carefully controlled geometry and predetermined superconducting properties. Gradually enough variables are coming under control to permit a sensible discussion of the design of superconduting materials. The primary purpose of this article is to review several aspects of superconducting-materials research and to discuss briefly some of the ideas and relationships that can be helpful in the design of new materials.

A wide variety of different kinds of solids show superconductivity. After Kamerlingh Onnes discovered the effect in Hg in 1911, there was a long period in which most of the experimental work was directed toward understanding the basic superconducting phenomena and for this work attention focused primarily on the *s-p* band metals such as Pb and Al. These materials could be obtained in very pure form and were well suited for definitive experiments. As the techniques for sample preparation improved, the range of materials which were suitable for well-defined experiments broadened to include transition metals such as Nb and V, semiconductors such as GeTe, laminar metals such as TaS_2, and intercalated laminar metals such as TaS_2-pyridine. Indeed, recently, superconductivity has been discovered in the quasi-one-dimensional polymer $(SN)_x$. At present approximately half the elements of the periodic table and a very wide range of compounds and alloys are known to be superconducting in some range of temperature and pressure. Transition temperatures range from the lowest temperatures attainable, approximately 1 mK, up to the record high of 23 K for Nb_3Ge films. In many cases, a material which is not superconducting at atmospheric pressure, for example the element Sb, will undergo a crystallographic phase change with increasing pressure, and the high-pressure phase frequently can be metallic and superconducting. Superconductivity is a very common phenomenon and occurs in materials with a wide variety of different kinds of chemical bonding.

In spite of the wide diversity of materials which show superconductivity, there is a unified set of ideas, based on the Bardeen–Cooper–Schrieffer (BCS) theory, that forms the basis of our microscopic understanding of various superconducting phenomena. When a substance, say Hg, undergoes a superconducting phase transition, the free engergy of the electron gas changes by only about 10^{-7} eV per atom but there are profound changes in the electronic structure. In the normal state, the electrons behave as independent particles in that one electron can be perturbed without perturbing all the other electrons. In the superconducting state, on the other hand, the electrons respond as a unit because they are locked into a phase-coherent state which can properly be described as a giant molecule of macroscopic dimensions. Rather than acting as independent particles, the electrons respond to a perturbing influence as a coherent, many-body unit. To understand this in more detail it is important to know the origin of the cooperative behavior and to understand how the electrons communicate with one another over such large distances. In the early 1950s it was discovered that T_c was inversely proportional to the square root of the mass of the lattice ions, and this gave the clue that the interaction causing superconductivity was an electron–electron interaction mediated by the exchange of phonons. In this process one electron creates a virtual phonon and another electron, which may be thousands of angstroms away at the time of the creation of the phonon, can move into the region of the phonon and absorb it. The charge fluctuation associated with the phonon is very local in space (a few angstroms) but it is very long range in time (10^{-12} sec) because the ion with its large mass responds relatively slowly. Hence, electrons traveling at the Fermi velocity, v_F, of about 10^8 cm/sec can move from as far away as 10,000 Å in the lifetime of the phonon. With such an exchange of phonons, the electron gas can take advantage of a coherent set of local charge fluctuations and the energy of the electron gas, mostly Coulomb in origin, is lowered. A more detailed analysis would show that the maximum gain in energy occurs if the electron states are occupied in pairs with equal but opposite momentum, the so-called Cooper pair. Below T_c, the electron gas is energetically unstable to the formation of these pairs. With increasing temperature random thermal fluctuations can break up the coherent state and the material reverts to a single-particle behavior of the normal metal above T_c. The electron–phonon interaction is not the only mechanism which could give rise to superconductivity, but careful study using electron tunneling has shown the electron–phonon interaction to be the dominant factor in all of the metals studied so far.

The general picture is this. As a material is cooled through T_c, a fraction of the electrons, n_s, condense into a many-body ground state which has long-range phase coherence or phase rigidity created by the exchange of phonons. Below T_c, the superfluid fraction rises rapidly and reaches 1 at $T=0$. At any finite temperature there are thermally excited electrons (the normal fluid) coexisting with the ground state, and these electrons behave as independent particles similar to the normal-state electrons. The order parameter for the system, Δ, is directly related to the many-body superfluid wave function which in turn determines n_s. Most of the amazing properties of superconductors arise from the rigidity or phase locking of this ground-state wave function.

In the design of a superconducting material, the penetration depth, λ, and the coherence distance, ξ, are extremely important because they control the distribution of supercurrents and magnetic field. Briefly stated, λ is the characteristic distance over which magnetic fields change in a superconductor and ξ is the characteristic distance over which the density of superfluid electrons changes. For example, if a magnetic field is applied to a superconductor, the ground-state electrons respond collectively to provide surface supercurrents to screen out the applied field. In a pure metal the penetration depth, λ_L, varies inversely as the square root of the density of ground-state electrons [Eq. (1) in Table I] and the coherence distance varies as the ratio of the Fermi velocity v_F to the transition temperature T_c [Eq. (2)]. Because n_s is typically 10^{22} cm^{-3} for most metals, λ_L is typically 500 Å for most pure metals. There is a much wider range for ξ_0. For *s-p* band metals where v_F is large and T_c relatively small, ξ_0 may be the order of 10,000 Å. For *d*-band metals, v_F is much smaller and T_c is larger, so ξ_0 may be on the order of 400 Å.

Fortunately both ξ and λ can be varied by the addition of impurities to change the electronic mean free path l. As the material is made dirtier the range over which electrons can communicate via the exchange of phonons is controlled by

Table I Useful relations for design purposes

(1) $\lambda_L=(mc^2/4\pi n_s e^2)^{1/2}$	London penetration depth
(2) $\xi_0=\hbar v_F/1.18kT_c$	BCS coherence distance
(3) $\xi\approx(\xi_0 l)^{1/2}$	Dirty limit ξ_0
(4) $\lambda=\lambda_L(\xi_0/l)^{1/2}$	Dirty limit λ
(5) $\Delta(T=0)=1.76kT_c$	BCS pair potential
(6) $\varphi_0=hc/2e=2.10^{-7}$ G cm^2	Flux quantum
(7) $H_{c1}\approx(\varphi_0/4\pi\lambda^2)\ \ln(\lambda/\xi)$	Lower critical field
(8) $H_{c2}\approx\varphi_0/2\pi\xi^2$	Upper critical field
(9) $\xi(T)\approx\xi_0(1-T/T_c)^{-1/2}$	Temperature dependence of ξ
(10) $\lambda(T)\approx\lambda_L(1-T/T_c)^{-1/2}$	Temperature dependence of λ
(11) $H\approx(\varphi_0/2\pi\lambda^2)K_0(r/\lambda)\quad(r>\xi)$	Magnetic field near vortex (K_0 is modified Bessel function)
(12) $E\approx(\varphi_0/4\pi\lambda)^2\ \ln(\lambda/\xi)$	Line energy of a vortex

the diffusion of electrons into the vicinity of the lattice disturbance and ξ varies as $l^{1/2}$ [Eq. (3)]. The penetration depth, on the other hand, is controlled by the nonlocal character of the current-vector potential relation and varies as $l^{-1/2}$ [Eq. (4)]. Hence, as the material is made dirtier λ increases and ξ decreases.

Broadly speaking there are two classes of superconductors depending on the relative magnitude of λ and ξ. The two classes are distinguished by their magnetic behavior when a long cylindrical specimen is placed in a longitudinal magnetic field. If λ/ξ is less than approximately $2^{-1/2}$ the material is said to be a type-I superconductor in which magnetic flux is excluded for all fields less than the thermodynamic critical field H_c, but penetrates the bulk of the superconductor uniformly for higher fields. For this class of materials, the phase transition at H_c is first order. If λ/ξ is greater than approximately $2^{-1/2}$, the material is said to be a type-II superconductor. Here the flux is excluded up to a lower critical field H_{c1} [Eq. (6)] where flux begins to enter in the form of quantized vortices or fluxoids. As the magnetic field is raised above H_{c1}, the fluxoids increasingly crowd together and the bulk of the material undergoes a second-order transition when the vortex cores begin to overlap at a field H_{c2} [Eq. (7)]. Above H_{c2} superconductivity is quenched in the bulk but can persist within a coherence distance of the surface up to a field $H_{c3}=1.7H_{c2}$ because there are special boundary conditions on the ground-state wave function. In a typical dirty superconductor there are three magnetic field regimes; the Meissner state below H_{c1}, the vortex state $H_{c1}<H<H_{c2}$, and the surface superconductivity state $H_{c2}<H<H_{c3}$.

A very detailed experimental study has been made of the structure of quantized vortices. Essmann and Träuble decorated the surface of a Nb sample in the vortex state with iron spheres that were about 40 Å in diameter and then observed the distribution of iron in an electron microscope by a replication technique. They verified the existence of vortices, which had been predicted much earlier by Abrikosov, and found that they indeed form a triangular array as predicted. Subsequently, Shelten and co-workers studied the interior structure of a vortex in Nb and NbTi alloys using neutron scattering and mapped out the magnetic field distribution. The results show that the vortex has a core of radius approximately equal to ξ. Within the core n_s and Δ are nearly zero and hence the core behaves much like a normal material. Surrounding the core are circulating supercurrents which govern the magnetic field distribution. These fall off in a characteristic distance λ. Although a great deal of work still needs to be done to understand the interaction of vortices with each other and with physical boundaries, the basic features of a free vortex are fairly well understood. A central need is to develop materials in which vortex motion is well controlled.

One of the most pressing problems in superconducting-materials research is to expand the range of materials available for different design applications and, indeed, there is a special need for materials with higher values of T_c. Because the synthesis and characterization of new materials is often both difficult and expensive, it is important to have some reasonable rules to guide the search. For example, it is known on very general grounds that the density of electron states, the density of phonon states, and the strength of the electron–phonon interaction are three interrelated factors which should be large for a high T_c. This is confirmed by extensive electron-tunneling studies in *s-p* band metals which have shown that the Eliashberg equation correctly relates the pair potential to the strength of the electron–phonon interaction. Indeed, the values of T_c calculated from values of Δ measured at low temperatures ($T\ll T_c$) are in good agreement with experiment. In addition, calculations of the electron–phonon matrix elements based on Korringa–Kohn–Rostoker energy bands and wave functions are in fair agreement with experiment for the *s-p* band metals and for the transition metal niobium. Hence for the simpler metals several factors controlling T_c are well understood. Unfortunately, an extension of this work to the high-T_c A-15 compounds is difficult. The experiments are much more difficult because surface contamination plays a more important role and the theory is more difficult because the energy bands are more complicated.

Another guide in the search for new materials is the observation that many high-T_c superconductors show crystal-structure instabilities and soft phonon modes. A detailed understanding of the matrix elements and the dynamical screening for these soft-mode materials is not yet in hand but the high density of phonon states at low phonon energy and the fact that the interaction is very long range in time are favorable factors. Additional work is needed to understand these instabilities well enough to be able to produce and control soft modes by the addition of impurities.

The search for totally new mechanisms for superconductivity, such as the excitonic mechanism, continues to be an active area of research and there are clues that excitons may be important in $NbGe_2$ - Ge mixed-phase materials. Progress has been substantial but much remains to be done.

Over the years, the Matthias empirical rules have been the most productive tools in the search for high-T_c superconductors, and it seems likely that this sort of guide will continue to be an important factor in progress. As research moves to more-complicated materials such as ternary compounds and mixed-phase materials, presumably new guidelines will emerge.

An area of superconducting-materials research which may have enormous industrial application is the development of high-critical-current superconducting composites. Early experience with superconducting magnets showed that the irreversible heating associated with the motion of vortices in a superconductor could lead to unstable flux jumps and a catastrophic return to the normal state. To prevent the motion of the vortices, superconductor–normal-metal composites were developed in which the superconducting material consisted of fine filaments of conductor typically 2–10 μm in diameter in a matrix of normal metal, usually copper. The copper inhibited the motion of vortices and it also provided a low-resistance parallel path for current around a superconducting weak spot. Fabrication techniques for multifilamentary NbTi in a copper matrix are well developed, and this material has been very reliable for magnetic fields in the range of 8 T. At higher fields in the range from 10 to 15 T, Nb_3Sn or Nb_3Ge have the best superconducting properties but they are so brittle that fabrication has been more difficult. A typical process for fabricating the A-15 compounds into multifilamentary wire might involve the drawing and annealing of Nb wires in a bronze matrix and subsequent heat treatment to grow the Nb_3Sn filament by diffusion of the Sn. Substantial success has been gained by these methods but more work in the development of composites is needed.

See also ELECTRON ENERGY STATES IN SOLIDS AND LIQUIDS; EXCITONS; GLASSY METALS; JOSEPHSON EFFECTS; SUPERCONDUCTIVITY THEORY; TUNNELING.

BIBLIOGRAPHY

A. M. Campbell and J. E. Evetts, *Adv. Phys.* **21,** 199 (1972) (or A. M.Campbell and J. E. Evetts, *Critical Currents in Superconductors,* Taylor and Francis Ltd., London 1972). Discussion of the relation of microstructures to the primary of vortices. (A)

P. G. deGennes, *Superconductivity in Metal and Alloys.* Benjamin, New York, 1966. Discussion of type-II superconductivity. (I)

R. P. Huebener and John R. Clem, *Rev. Mod. Phys.* **46,** 409 (1974). Reveiw of type-II research. (A)

R. D. Parks (ed.), *Superconductivity.* Marcel Dekker, New York, (1969). Excellent review articles. (A)

G. Rickayzen, *Theory of Superconductivity,* Interscience, New York, 1965. Introduction to the theory. (I)

B. W. Roberts, Natl. Bur. St. Tech. Note 408. U.S. GPO, Washington, DC, 1966. Table of known superconductors. (E)

J. R. Schrieffer, *Theory of Superconductivity.* Benjamin, New York, 1964. (A)

D. Shoenberg, *Superconductivity,* 2nd ed. Cambridge University Press, London, 1960. Excellent view of the subject before the BCS theory. (E)

M. Suenaga, W. B. Sampson, and C. J. Klamut, *IEEE Trans. Mag.* **MAG-11,** 231 (1975). Discussion of fabrication of A-15 composites. (E)

Superconductive Devices

J. Clarke

The performance, reliability, and importance of superconductive devices have developed rapidly since the middle 1960s. This growth has been stimulated largely by the Josephson effects. Despite the fact that these devices must be operated at liquid-helium temperatures, usually 4.2 K or lower, they are now used for a variety of purposes both in and away from the low-temperature physics laboratory. This article briefly describes the more important devices.

SUPERCONDUCTING QUANTUM INTERFERENCE DEVICES

The most highly developed and widely used Josephson devices are superconducting quantum interference devices (SQUIDs). The two varieties of SQUID, the dc SQUID and the rf SQUID, are the most sensitive devices available to measure magnetic fields, magnetic field gradients, magnetic susceptibilities, and voltages. The dc SQUID consists of two Josephson junctions mounted on a superconducting loop (Fig. 1a). The device shown in Fig. 2a is constructed of thin films; each of the two Nb–NbO_x–Pb junctions is resistively shunted by a gold strip to eliminate hysteresis in the current–voltage (I–V) characteristic. A constant current greater than the critical current (maximum zero-voltage current) of the two junctions biases the SQUID at a nonzero voltage. The critical current and the voltage across the SQUID are periodic in the external magnetic flux, Φ_e, applied to the loop, with a period of one flux quantum, Φ_0. A change in flux, $\Delta\Phi_e$, that is much smaller than Φ_0 can be detected with the flux-locked SQUID shown in Fig. 1a. The feedback circuit generates a current in a coil inserted in the SQUID so that the flux in the SQUID is maintained at a constant value. The voltage V_0 across the resistor R_0 is proportional to $\Delta\Phi_e$.

The rf SQUID consists of a single Josephson junction on a superconducting ring that is inductively coupled to the coil of a cooled tank circuit. The tank circuit is excited at its resonant frequency, typically 30 MHz. The rf voltage across the tank circuit is amplified and rectified to produce a voltage that is periodic in Φ_e. The rf SQUID is then used in a flux-locked loop in a way similar to that of the dc SQUID. Several types of rf SQUID are commercially available; two examples are shown in Fig. 2, parts b and c.

The dynamic range of both rf and dc SQUIDs is typically 10^6 in a 1-Hz bandwidth, the frequency response is typically 0–50 kHz, and the flux resolution is typically $(10^{-5}–10^{-4})\Phi_0$ $Hz^{-1/2}$ for a SQUID inductance of 1 nH. Recent work on the dc SQUID, however, has shown that improvements in sensitivity of two orders of magnitude are possible.

SQUIDs are frequently used in conjunction with a superconducting flux transformer, for example, the gradiometer shown in Fig. 1a. A change in the magnetic field gradient $\partial H_z/\partial z$ produces a supercurrent in the transformer and hence a voltage at the output of the flux-locked SQUID, whereas

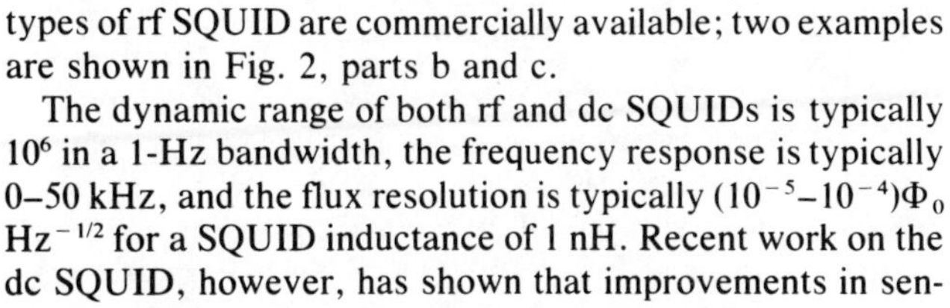

FIG. 1. (a) Flux-locked dc SQUID used as gradiometer; (b) flux-locked rf SQUID used as voltmeter.

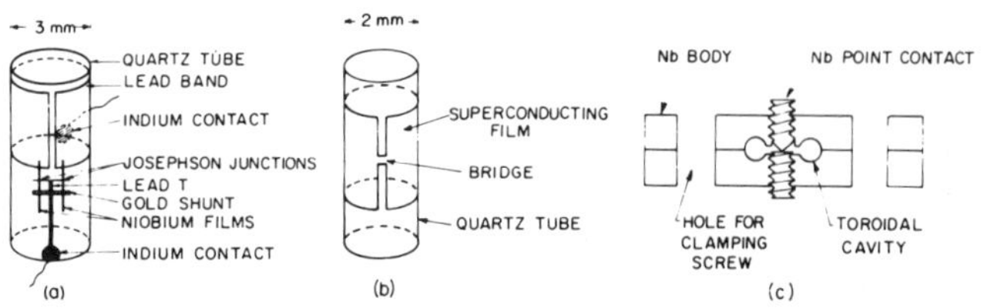

FIG. 2. (a) Tunnel junction dc SQUID; (b) Dayem bridge rf SQUID; (c) point-contact toroidal rf SQUID.

a change in the uniform magnetic field produces no response provided that the two loops are balanced. A typical sensitivity is 10^{-13} T m^{-1} Hz$^{-1/2}$. Other configurations can be used to measure $\partial H_z/\partial x$ or $\partial^2 H_z/\partial z^2$. A flux transformer with a single pickup loop is a vector magnetometer with a sensitivity that can be as high as 10^{-15} T Hz$^{-1/2}$. We can use a configuration similar to that in Fig. 1a to measure magnetic susceptibility by placing the sample in one of the pickup loops. The application of a magnetic field H_z produces an output from the flux-locked SQUID that is proportional to the susceptibility of the sample. A sensitivity of 10^{-9} mks m^{-3} for a 10^3-mm^3 sample in a field of 10^{-2} T is typical. The configuration of Fig. 1b is a voltmeter; in practice, the feedback current is usually applied to the resistor R. The voltage resolution is usually limited by Johnson noise in the resistor: For example, for a resistor of 10^{-8} Ω at 4 K, the resolution is about 10^{-15} V Hz$^{-1/2}$.

SQUIDs have been used for laboratory measurements for over a decade. More recently, SQUIDs have been used as magnetometers and gradiometers at sites far removed from the low-temperature laboratory. These applications have been made possible by the development of portable fiberglass cryostats that use no liquid nitrogen, and that boil off 1 liter or less of liquid helium per day. Applications include magnetocardiography, magnetoencephalography, geophysical surveying, and gravity-wave detection.

DETECTORS OF MICROWAVE AND SUBMILLIMETER RADIATION

Several superconducting devices have been developed to detect electromagnetic radiation at microwave and submillimeter wavelengths. Josephson junction detectors have been operated in several different modes, of which the most promising appear to be the mixer and the parametric amplifier. Point-contact junctions, Dayem bridges, and tunnel junctions have all been used as mixers. The nonlinearity of the I–V characteristic is used to mix the signal frequency f_S with the local oscillator frequency f_{LO} to obtain an intermediate frequency $f_{IF} = |f_S - f_{LO}|$. The voltage oscillations across the current-biased junction at a frequency f_{IF} are amplified with a low-noise IF amplifier. Mixers of this type have been operated at signal frequencies as high as 300 GHz. Higher-order mixing processes in which $f_{IF} = |pf_S - qf_{LO}|$ (p and q are integers) also occur, and are potentially important for frequency comparison. The parametric amplifiers make use of the nonlinear inductance exhibited by a Josephson junction. The most successful devices have been operated with zero static current bias, and with $2f_P = f_S + f_I$, where the external pump frequency f_P, signal frequency f_S, and idler frequency f_I are all approximately equal. Point contacts and arrays of Dayem bridges and tunnel junctions have been operated as amplifiers at 10 GHz, 33 GHz, and 36 GHz. Gains in excess of 10 dB have been observed.

Two other superconducting devices with highly nonlinear I–V characteristics are potentially important as mixers. One is the super-Schottky diode, which consists of a superconductor–semiconductor junction. The other is the superconductor–insulator–superconductor quasiparticle tunneling junction, in which there is a very rapid rise in current at voltages near the sum of the energy gaps in the superconductors. All of these detectors have been tested successfully in the microwave region, but have not yet been operated extensively at millimeter and submillimeter wavelengths, where there is the greatest need for improved sensitivity. Considerable effort will be required to extend the low-noise performance into this frequency range.

The superconducting bolometer represents a quite different approach to the broadband detection of far infrared radiation. A $(4 \times 4 \times 0.05)$-mm single-crystal sapphire substrate is suspended in a vacuum can and maintained at a temperature near the midpoint of the superconducting transition of an aluminum strip deposited on the upper surface. The lower side of the substrate is coated with a conducting film that absorbs radiation incident through a light pipe. The resulting rise in temperature of the bolometer is detected by the change in resistance of the aluminum thermometer. The noise equivalent power at 1.2 K is typically 2×10^{-15} W Hz$^{-1/2}$, a value that is within a factor of 2 of the fundamental limit imposed energy fluctuations. Substantial improvements should be obtained by operating the bolometer in the ^{3}He temperature range.

COMPUTER ELEMENTS

One of the most spectacular potential applications of the Josephson effect is in ultrafast computers. The Josephson devices offer not only very fast switching speeds but also very low dissipation. The low dissipation enables one to achieve a high packing density of junctions and to minimize transit times between junctions. A number of storage and

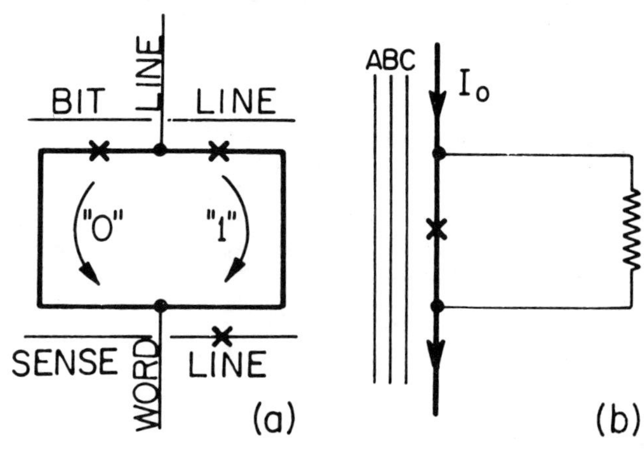

FIG. 3. (a) Memory cell (bit and sense films overlie the nearby films of the cell); (b) logic element (control films A, B, and C overlie the film containing the junction).

logic devices have already been successfully operated by workers at IBM. One type of memory cell (Fig. 3a) consists of a superconducting loop containing two Josephson tunnel junctions with hysteretic I–V characteristics. The experimental configuration of the cell is shown in Fig. 4. A 1 or a 0 is represented by a persistent supercurrent in the clockwise or anticlockwise direction, respectively, and is written in the cell by applying appropriate current pulses to the word and bit lines to drive one of the junctions momentarily into the nonzero voltage state. When the pulses are removed, a persistent supercurrent of the required polarity remains in the ring. Thus, storage requires no external currents and is nondissipative. The contents of the cell can be read nondestructively by applying suitable currents to the word and sense lines. These cells are suitable for operation in an array. The smallest cell yet made has an area of about 900 μm^2, a writing time of less than 80 ps, and a power-delay product of less than 10^{-16} J per writing cycle.

A simple logic circuit is shown schematically in Fig. 3b. A Josephson tunnel junction is connected to two superconducting strip-lines terminated with a matching resistor. The junction is overlaid with three superconducting control lines, A, B, and C, and is biased in the zero-voltage state with a constant current I_0. When a current pulse with the same polarity as the current I_0 is applied to any one of the control lines, the junction switches to a nonzero voltage, and a current is established in the strip line. This process respresents the logical OR operation. The logical AND operation is performed by simultaneously applying current pulses with a polarity opposite to that of I_0 to all three control lines, thereby switching the junction into the nonzero voltage state; pulses with this polarity applied to one or two of the lines do not switch the junction. The logic element can also be used to perform the INVERT and CARRY functions. In practical circuits, a logic delay of less than 200 ps and a power-delay product of about 5×10^{-15} J have been achieved. Adders, multipliers, and shift-registers incorporating these logic elements have been successfully operated.

The switching speed is an order of magnitude faster than, and the dissipation several orders of magnitude smaller than, in semiconducting computer elements. Storage and logic elements that involve the transfer of a single flux quantum are now being investigated, and may be the devices ultimately to be incorporated into a superconducting computer. Within the next few years we can hope to see an ultrafast superconducting computer with dimensions of not much more than 0.1 m.

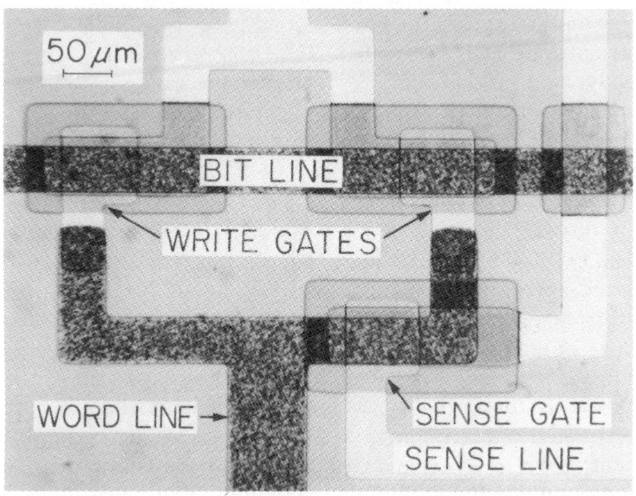

FIG. 4. Photograph of prototype memory cell (courtesy of J. Matisoo, IBM, and H. H. Zappe, *IEE Solid State Circuits*).

STANDARDS

The Josephson effects are being used increasingly in standards laboratories. The most important applications have been the measurement of the fundamental constant ratio e/h, and the maintenance of the standard volt. When a Josephson junction is irradiated with microwaves at frequency f, constant-voltage steps are induced on the I–V characteristic at voltages $V_n = nhf/2e$ ($n = \pm 1, 2, \ldots$). In 1965, Parker and co-workers measured V_n and f to high precision, and determined a value of e/h that differed significantly from the previously accepted value. The new value of e/h was used in a least-squares adjustment of the fundamental constants that resulted in substantial changes in the values of several constants. The standard volt at the National Bureau of Standards and at several national laboratories overseas is now maintained by Josephson junctions. Since frequency can be measured very accurately, it is a simple procedure for different laboratories to compare their standard volts. It should be noted, however, that the standard volt is not *defined* by the Josephson effect.

The Josephson effect has been used in two ways to measure temperatures in the range from 1 mK to 1 K. In the first, a junction shunted with a resistance R is biased at a voltage V_0 by a stable current. The Johnson-noise voltage across the resistor causes the Josephson frequency to fluctuate with an rms bandwidth $\Delta\nu = 4\pi k_B TR/\Phi_0^2$. Thus, by measuring $\Delta\nu$ we can determine the absolute temperature T. In the second method, a superconducting loop containing a resistance R is coupled to a flux-locked SQUID (see Fig. 1a). We determine T by measuring the spectral density of the Johnson-noise currents in the loop, $S(f) = 4k_B TR/(R^2 + 4\pi^2 f^2 L^2)$.

SQUIDs can be used as current comparators to compare two static currents to very high precision and, in a nonfeedback mode, to calibrate rf attenuators over a wide range of frequencies.

PHONON GENERATORS AND DETECTORS

Quasiparticle tunnel junctions have been used to generate and detect phonons. When a quasiparticle is tunnel-injected into a superconductor, it is most likely to make a transition to the gap edge, where it recombines with another quasiparticle to form a Cooper pair, emitting a phonon of energy approximately 2Δ (Δ is the energy gap). The spectrum of phonons emitted thus peaks around 2Δ. The junction can also be used to detect phonons with energies greater than 2Δ. These phonons break pairs to form quasiparticles that increase the tunneling current at a given voltage bias. These techniques are of particular interest in the 10^{11}- to 10^{12}-Hz range. Some degree of tuning can be achieved by varying the energy gap with a magnetic field.

See also JOSEPHSON EFFECTS; QUASIPARTICLES; SUPERCONDUCTIVITY THEORY.

BIBLIOGRAPHY

A number of excellent comprehensive review articles on superconducting devices can be found in *Proc. IEEE* **61**, January (1973) and **68**, October (1980); in *Superconductor Applications: SQUIDs and Machines* (B. B. Schwartz and S. Foner, eds.), Plenum, New York and London, 1977; and in *Future Trends in Superconductive Electronics* (B. S. Deaver Jr., C. M. Falco, J. H. Harris, and S. A. Wolf, eds.), American Institute of Physics, New York, 1978. The *IBM Journal of Research and Development* **24**, March (1980) is devoted to Josephson computer technology. (E, I, A)

J. C. Gallop and B. W. Petley, "SQUIDs and Their Applications," *J. Phys.* **E9**, 417 (1976); a survey of SQUID applications. (E)

H. Kinder, "Phonon Spectroscopy at Ultrahigh Frequencies," *Proc. 14th Int. Conf. on Low Temperature Physics* (Otaniemi, Finland, August 14–20, 1975), Vol. 5, p. 287. North-Holland, Amsterdam, 1975. (I)

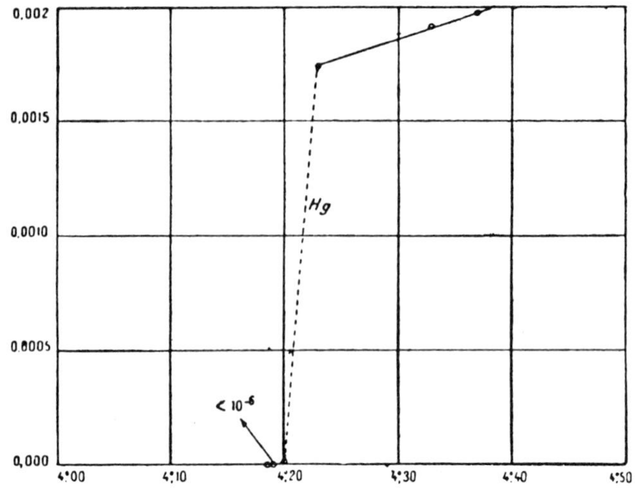

FIG. 1. Resistance of mercury from 4.0 to 4.5 K as measured by H. Kamerlingh Onnes in 1911. This was the first observation of superconductivity. (Reprinted from *Superconductivity*, AIP Selected Reprints.)

Superconductivity Theory

Philip B. Allen

Superconductivity is a low-temperature property of many metals characterized originally (and most dramatically) by the complete absence of electrical resistance (see Fig. 1). Above a characteristic temperature T_c this property abruptly ceases and the metal displays normal metallic resistance. The temperature T_c varies widely from metal to metal, and, unfortunately, is never very high. The highest known value of T_c is 23 K (in Nb_3Ge); most superconductors have T_c much lower (e.g., 3.7 K for Sn).

Another dramatic (and in some ways more fundamental) property of superconductors is their behavior in magnetic fields. It is known from microscopic theory that magnetic fields weaken the ability of a metal to be superconducting. It is energetically possible for superconductivity to exist in a weak homogeneous magnetic field, but there are a variety of ways in which a superconductor can lower the total free energy by creating inhomogeneities in the field, concentrating the field in regions of space where the metal becomes a normal conductor. The specific strategy depends both on sample geometry and on material properties. Most pure superconductors ("type-I" materials) require a relatively large energy to make a boundary between superconducting (*S*) and normal (*N*) states. Thus their strategy tends to be an "all-or-nothing" situation where the field is completely expelled from the sample (Meissner effect) except for a thin penetration region of depth λ (~ 1 μm). This is accomplished by the presence of large diamagnetic supercurrents in the penetration region, which create an internal field $H=B+4\pi M=0$, which completely cancels the external field B. Thus the susceptibility χ is $-1/4\pi$. This is known as "perfect diamagnetism" (see Fig. 2). The experimental observation of the abrupt increase in diamagnetic susceptibility for $T<T_c$ is often a simpler and more reliable signal for the occurrence of superconductivity than is the absence of resistance. The energy expended in expelling the field B is $B^2/8\pi$. At a critical value of B (denoted H_c, often of order 1000 G) there is no longer enough free energy in the superconductor to expel the field, and the sample makes an abrupt (first-order) transition to the normal state. Thus the energy $H_c^2(T)/8\pi$ is a measure of the free-energy difference (or the "condensation" energy) of the superconducting state relative to the normal state. (An abrupt transition at $B=H_c$ occurs only in long, thin samples in axial fields. For other geometries the sample will divide into N and S regions at a lower field—an effect known as the "intermediate state.")

There is another class of materials known as type II (usually alloys) in which the boundary energy between S and N regions is not so high. These materials will show complete flux expulsion only when the applied field B satisfies $B<H_{c1}<H_c$. For fields higher than H_{c1} and less than H_{c2} an intimate coexistence of superconductivity and magnetic field occurs (the "mixed" or "vortex" state). The magnetic field is confined to "flux tubes" each containing one "quantum" of flux, $hc/2e \sim 2\times10^{-7}$ G cm^2. When B approaches H_{c2}, the size of the superconducting regions shrinks continuously to zero and the sample makes a second-order transition to the normal state at H_{c2}. The field H_{c2} is *larger* than the "thermodynamic critical field" H_c, and in some cases can be very large (in excess of 10^5 G. Thus, type-II materials are frequently more useful than type-I for applications in magnets, motors, and generators.

The abrupt disappearance of superconductivity at temperatures T above T_c is not completely mysterious—there is an exact analogy with many other phase transitions, such as the abrupt disappearance of spontaneous magnetism in a ferromagnet at the Curie temperature. However, the absence of electrical resistance is a much more difficult problem. It is *not* explainable either by the classical (Drude–Lorentz) theory, which was available when Onnes discovered superconductivity in 1911, or by the quantum theories which Bloch and Grüneisen developed around 1930 to explain conductivity in normal metals. The mystery was not solved until the BCS (Bardeen–Cooper–Schrieffer) theory in 1957. The BCS theory and its subsequent extensions have provided a very nearly complete description of a remarkable range of physical properties and effects. Prominent among these are the effects predicted by Josephson. On a deeper level the BCS theory has introduced new ideas into the structure of quantum theory which have resulted in a deeper understanding of broken symmetry, and important developments in nuclear theory and field theory. Furthermore, the BCS theory strongly suggests that the instability of the normal electron system (which leads to superconductivity in metals) might well occur in other systems of Fermi particles, such as nuclear matter in neutron stars and liquid ^{3}He. The latter conjecture has been verified by the discovery by Osheroff *et al.* in 1972 of several new phases of ^{3}He at temperatures below 3 mK. Thus the BCS theory of superconductivity can be regarded as one of the most important and far-reaching achievements of modern physics.

Let us approach the microscopic theory by examining a seemingly unrelated problem which will turn out to have much in common with the BCS theory, namely a particularly simple version of the Heisenberg model of a ferromagnet. We have a collection of N particles of spin $\frac{1}{2}$ which may be imagined to be located on the sites of a lattice. There is an external field **B** in the $\hat{z}$ direction. If the spins do not interact with each other, the Hamiltonian is

$$\mathcal{H}_0 = -\sum_i \boldsymbol{\sigma}_i \cdot \mathbf{B}, \tag{1}$$

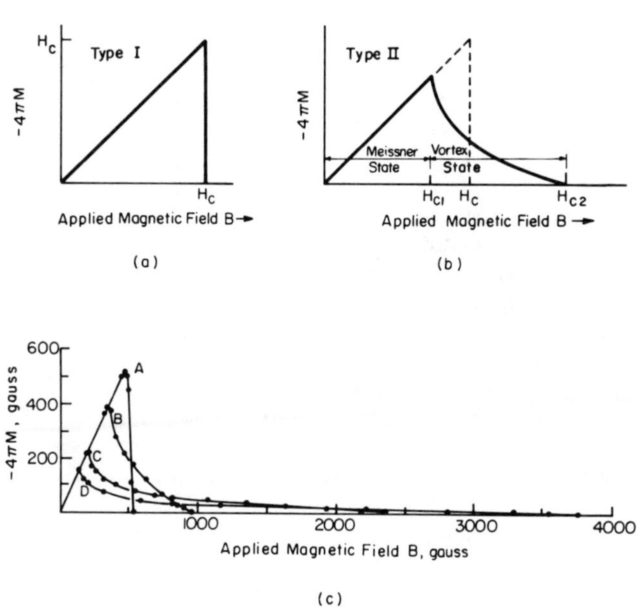

FIG. 2. (a) Magnetization M versus applied field B for a bulk (type-I) superconductor. The slope of 1 corresponds to a susceptibility $\chi = -1/4\pi$. (b) Magnetization curve for a type-II superconductor. The "vortex" state exists for applied fields between H_{c1} and H_{c2}. Below H_{c1} there is complete flux expulsion (Meissner state). (c) Experimental magnetization curves of Pb (A) and Pb-In alloys of increasing In concentration (B,C,D) showing the onset of type-II behavior in the alloys. [These figures are adapted (with permission) from C. Kittel, *Introduction to Solid State Physics*, 5th ed. Chap. 12. Wiley, New York, 1976. The data come originally from J. D. Livingston, *Phys. Rev.* **129,** 1943 (1963).]

where $\boldsymbol{\sigma}_i$ is the spin operator for particle number i. The spin operators can be represented as 2×2 matrices corresponding to the two states, spin "up" $\binom{1}{0}$ and spin "down" $\binom{0}{1}$. Explicit forms for the spin operators are the Pauli matrices,

$$\sigma_x = \begin{pmatrix} 0 & 1 \\ 1 & 0 \end{pmatrix}, \quad \sigma_y = \begin{pmatrix} 0 & -i \\ i & 0 \end{pmatrix}, \quad \sigma_z = \begin{pmatrix} 1 & 0 \\ 0 & -1 \end{pmatrix}. \tag{2}$$

The magnetic moment ($g\mu_B$) has been chosen to be unity for convenience. We can easily solve for the thermodynamic properties of this system. The spins will align with the field insofar as thermal disorder allows. It is an elementary exercise in statistical mechanics to show that the average spin, $\langle\sigma_z\rangle$, is given by

$$\langle\sigma_z\rangle = \tanh(B/T), \tag{3}$$

where the Boltzmann constant k_B is also set equal to 1.

The model becomes more interesting if we let the spins interact with each other, and turn off the external field:

$$\mathcal{H} = -\tfrac{1}{2} J \sum_{ij} \boldsymbol{\sigma}_i \cdot \boldsymbol{\sigma}_j. \tag{4}$$

In real magnets, the exchange interaction J will only couple near-neighbor spins, but we get a simpler problem (and one more closely related to superconductivity) if we assume that all spins are coupled to each other. If we are willing to neglect fluctuations, then at temperature T, the behavior of spin i is equivalent to the system

$$\mathcal{H}_{\mathrm{MF}} = -\sum_i \boldsymbol{\sigma}_i \cdot \mathbf{H}_{\mathrm{MF}}, \tag{5}$$

where $\mathbf{H}_{\mathrm{MF}}$ is the "mean field"

$$\mathbf{H}_{\mathrm{MF}} = JN\langle\boldsymbol{\sigma}\rangle. \tag{6}$$

The solution to (5) and (6) is the same as the solution to (1),

$$\langle\sigma\rangle = \tanh(\langle\sigma\rangle T_c/T), \qquad T_c = JN. \tag{7}$$

This equation describes an elementary phase transition. For temperatures $T>T_c$, the only solution of (7) is $\langle\sigma\rangle = 0$, whereas if $T<T_c$, a new solution $\langle\sigma\rangle \neq 0$ appears. This new solution has lower energy than the old one, and breaks the symmetry of the system. The Hamiltonian (4) is invariant under arbitrary rotations of all the spins. The spontaneous magnetization $\langle\sigma\rangle$ picks out a preferred direction. All directions are equally likely (unless we apply a field) but the system must make a choice. The low-temperature state has lower symmetry than $\mathcal{H}$ does. In arriving at this solution we have made an approximation—namely that each particular spin $\boldsymbol{\sigma}_i$ sees an average environment or "mean field" rather than the particular fluctuations in local environment which actually occur. However, it is also clear that because we have chosen an infinite range for the exchange interaction, the local environment becomes the same as the mean field when N goes to infinity. (We must keep JN finite, of course.) Thus the mean-field solution is exact, and this feature appears in BCS theory also.

Near the critical temperature T_c, the magnetization $\langle\sigma\rangle$ is small. Using the expansion $\tanh x \approx x - x^3/3$, we easily find from (7), for T just below T_c,

$$\langle\sigma\rangle \cong (3/T_c)^{1/2}(T_c - T)^{1/2}. \tag{8}$$

It is a general feature of second-order phase transitions that there exists some "order parameter" (in this case the magnetization, $\langle\sigma\rangle$) which vanishes catastrophically as T increases to T_c. Mean-field theories always give the behavior $(T_c-T)^{1/2}$. The "microscopic" solution such as (7) is model dependent, but near T_c the result (8) always holds, although the coefficient will vary from $(3/T_c)^{1/2}$. Real systems frequently deviate from this, giving some "nonclassical" exponent like $\frac{1}{3}$ (instead of $\frac{1}{2}$) when T is close enough to T_c (in the "critical region"). It is a feature of superconductors that the exponent has the "classical" value $\frac{1}{2}$ unless T is extraordinarily close to T_c. The critical region is only accessible experimentally in lower-dimensional superconducting systems such as thin whiskers or very small particles.

If we are to apply these ideas to superconductors, we must ask what is the order parameter. This is a suprisingly difficult question for which the answer did not fully emerge until BCS theory and after. The simplest and earliest idea was the "two-fluid model" which proposed that the electrons in a superconductor had two choices, "normal" or "superconducting". At $T=0$, all electrons are regarded as superconducting, but as T increases toward T_c, the superfluid fraction x goes to zero. The choice $x(T)=1-(T/T_c)^4$ was made for historical reasons. The square root of x behaves like an order parameter. This theory had considerable success in correlating various physical properties such as critical field, specific heat, and thermal conductivity. Landau and Ginsberg in 1952 unified the phenomenological theories of thermal and electromagnetic properties by assuming that the square root of the superfluid density behaved much like a macroscopic quantum-mechanical wave function, including the property of being complex. This theory, which is completely intuitive and phenomenological, has been used to make many detailed and accurate predictions, most dramatic being the existence of type-II superconductors and the vortex state. This predictive power is all the more astonishing considering that the microscopic nature and origin of the order parameter was largely mysterious. In a very different fashion, Pippard was also able to construct an intuitive phenomenological theory which predicts many of these effects. Since 1957 these early phenomenological theories have been not outmoded but rather put on a firmer footing and given a somewhat deeper interpretation. The BCS theory introduces an "energy gap" which is fundamental to the superconducting state and plays the role of an order parameter.

In a normal metal there is no minimum energy for creating an excitation. An electron can be scattered out of a state $\mathbf{k}$ below the Fermi surface (leaving behind a hole of energy $|\epsilon_{\mathbf{k}}-\epsilon_F|$ where $\epsilon_{\mathbf{k}}$ and ϵ_F are the energy of state $\mathbf{k}$ and the Fermi energy, respectively) and into a state $\mathbf{k}'$ (of energy $|\epsilon_{\mathbf{k}}'-\epsilon_F|$) above the Fermi surface. Since $\epsilon_{\mathbf{k}}$ and $\epsilon_{\mathbf{k}}'$ can be arbitrarily close to ϵ_F, the minimum excitation energy is zero. In a superconductor there is a gap 2Δ in the excitation spectrum. The energy of an electron (or a hole) is $E_{\mathbf{k}}=[(\epsilon_{\mathbf{k}}-\epsilon_F)^2+\Delta^2]^{1/2}$. The minimum energy is Δ for the electron and Δ for the hole. In addition to electron–hole pair excitations, it is possible for the whole Fermi distribution to move coherently by an amount $\delta\mathbf{k}$ in $\mathbf{k}$ space. This new state carries a current and also has a gap for particle–hole creation. Once established the current persists because the mechanisms for decay (i.e., scattering off phonons, impurities, surfaces, or other imperfections) involve particle–hole pair creation which costs energy 2Δ; this energy is unavailable. The idea that superconductivity might be associated with a gap in the excitation spectrum is thus fairly natural. It had been raised as a serious possibility during the 1950s, although the nature and origin of the gap was at that time still obscure.

The key to a development of a microscopic theory was provided by a discovery by Cooper in 1956. He considered a pair of electrons interacting with each other by an attractive interaction potential V. In a vacuum there is a minimum strength that a potential must have for binding to occur. Cooper replaced the vacuum by an inert Fermi sea. The two electrons were assumed not to interact with other electrons, but to be excluded from all states within the Fermi sea. This had the surprising effect of allowing binding at arbitrarily weak interactions. The binding energy of this so called "Cooper pair" is the origin of the superconducting energy gap.

Cooper's work still leaves the nature of the gap mysterious. It is one thing to bind a pair of electrons and quite another to discover how to find a ground state for all the electrons which takes advantage of the binding in a way which is consistent with the Pauli principle. Cooper found that a maximum binding energy occurred when the pair had zero center-of-mass momentum and singlet spin pairing. The spatial part of the pair wave function is thus of the form

$$\psi(\mathbf{r}_1-\mathbf{r}_2)=\sum_{\mathbf{k}}\psi_{\mathbf{k}}e^{i\mathbf{k}\cdot(\mathbf{r}_1-\mathbf{r}_2)}.$$

Because the singlet spin configuration $\chi_{12}=\alpha_1\beta_2-\alpha_2\beta_1$ (where α means spin up and β means spin down) is antisymmetric under particle exchange, the Pauli principle says $\psi(\mathbf{r}_1-\mathbf{r}_2)$ is even under exchange (thus $\psi_{-\mathbf{k}}=\psi_{\mathbf{k}}$). It is clear that if electron 1 is in the state ($\mathbf{k}\uparrow$), then electron 2 must be in the state ($-\mathbf{k}\downarrow$). For shorthand purposes this "pair state" will be denoted by $\mathbf{k}$. In Cooper's problem the role of the attractive interaction V is to scatter the pair from a state $\mathbf{k}$ to another state $\mathbf{k}'$; through this scattering a stable bound pair state ψ is built up out of a coherent superposition of pair states $\mathbf{k}$. Of course any real interaction V would also allow processes where the pair state $\mathbf{k}$ breaks up by having one of the paired electrons scatter to $\mathbf{k}'$ while the other one remains in $\mathbf{k}$. The missing momentum $\mathbf{k}-\mathbf{k}'$ would go to the Fermi sea, for example by creating a hole at $\mathbf{k}''$ and an electron at $\mathbf{k}''+\mathbf{k}-\mathbf{k}'$. However, these processes do not occur in Cooper's problem because he kept the Fermi sea "frozen." A theory of superconductivity must unfreeze the Fermi sea. The BCS theory does this but avoids the problem of pair breaking by omitting all such processes from the Hamiltonian. The BCS Hamiltonian is

$$\mathcal{H}_{\mathrm{BCS}}=\sum_{\mathbf{k}\sigma}(\epsilon_k-\epsilon_F)n_{\mathbf{k}\sigma}-V\sum_{\mathbf{k}\mathbf{k}'}b_{\mathbf{k}'}^{+}b_{\mathbf{k}}, \tag{9}$$

where $n_{\mathbf{k}\sigma}$ is the number operator which counts the number of electrons in the state $\mathbf{k}$ with spin $\boldsymbol{\sigma}$, $b_{\mathbf{k}}$ is the operator which "destroys" the pair $\mathbf{k}$, and $b^{+}_{\mathbf{k}'}$ "creates" the pair $\mathbf{k}'$. Thus the term in (9) proportional to V has the effect of scattering pairs from $\mathbf{k}$ to $\mathbf{k}'$ if the pair state $\mathbf{k}$ is occupied, and vanishing if the pair state $\mathbf{k}$ is unoccupied. A cutoff is also used in the Hamiltonian (9), i.e., only states **k** whose energy is within θ of the Fermi surface are included. The origin of the attractive interaction V and cutoff θ will be discussed later. This Hamiltonian is often called the "reduced" Hamiltonian because only pair-scattering (and no pair-breaking) interactions are kept. This is a vast simplification of the actual state of affairs caused by real interactions. The justification for the simplification is that the reduced Hamiltonian can now be solved exactly. It contains the part of the interaction that caused Cooper's bound state, and this is precisely the part of the actual Hamiltonian which cannot be treated by perturbation theory. The pair-breaking interactions can be added later by perturbative means; or they can be regarded as already approximately accounted for in the construction of the "one-electron" spectrum $\epsilon_{\mathbf{k}}$. In the reduced Hamiltonian the one-electron energies are measured relative to the Fermi energy ϵ_F. It is convenient to define $\xi_{\mathbf{k}}$ as the energy $\epsilon_{\mathbf{k}}-\epsilon_F$.

To solve the Hamiltonian (9) we introduce a basis set and diagonalize $\mathcal{H}_{\text{BCS}}$. The basis set of many-electron wave functions can be taken as products of occupied pair states **k** and singly occupied states $\mathbf{k}'\uparrow$ or $\mathbf{k}'\downarrow$ (where the partner state $-\mathbf{k}'\downarrow$ or $-\mathbf{k}'\uparrow$ is explicitly unoccupied). In the subspace of singly occupied states, the interaction term V vanishes so that $\mathcal{H}_{\text{BCS}}$ is already diagonal. In the subspace of pair states, (9) is not diagonal, but can be represented as a spin Hamiltonian which we know how to diagonalize.

The spin analogy occurs because in the pair-state subspace there are only two possible states for each pair **k**, "empty," which we denote by "spin up," $\binom{1}{0}$, and "occupied," which we denote by "spin down", $\binom{0}{1}$. The operator $b_{\mathbf{k}}$ which destroys a pair has the effect of changing $\binom{0}{1}$ to $\binom{1}{0}$ and changing $\binom{1}{0}$ to 0. Thus $b_{\mathbf{k}}$ has the matrix representation

$$b_{\mathbf{k}}=\begin{pmatrix}0&1\\0&0\end{pmatrix}=\tfrac{1}{2}(\sigma_{\mathbf{k}x}+i\sigma_{\mathbf{k}y}), \tag{10}$$

where we have used the Pauli matrices (2). Similarly, $b_{\mathbf{k}}^{\dagger}$ creates a pair and can be written

$$b_{\mathbf{k}}^{\dagger}=\begin{pmatrix}0&0\\1&0\end{pmatrix}=\tfrac{1}{2}(\sigma_{\mathbf{k}x}-i\sigma_{\mathbf{k}y}). \tag{11}$$

Finally, the number operator $\Sigma_{\sigma}n_{\mathbf{k}\sigma}$ has the eigenvalue 2 when operating on $\binom{0}{1}$ and 0 when operating on $\binom{1}{0}$, or in matrix form

$$\sum_{\sigma}n_{\mathbf{k}\sigma}=\begin{pmatrix}0&0\\0&2\end{pmatrix}=1-\sigma_{\mathbf{k}z}. \tag{12}$$

Now using formulas (10)–(12), the Hamiltonian (9) becomes

$$\mathcal{H}_{\text{BCS}}=-\sum_{\mathbf{k}}\xi_{\mathbf{k}}\sigma_{\mathbf{k}z}-\tfrac{1}{4}V\sum_{\mathbf{kk}'}(\sigma_{k'x}\sigma_{kx}+\sigma_{k'y}\sigma_{ky}). \tag{13}$$

In writing the form (13) we have dropped a constant term $\Sigma_{\mathbf{k}}\xi_{\mathbf{k}}$ (which vanishes anyway since we cut off symmetrically above and below ϵ_F, and $\xi_{\mathbf{k}}$ changes sign at ϵ_F). We have also used the fact that when $b_{\mathbf{k}'}^{\dagger}b_{\mathbf{k}}$ is added to $b_{\mathbf{k}}^{\dagger}b_{\mathbf{k}'}$, the sum is Hermitian so that terms proportional to i have cancelled. Finally, it must be remembered that (13) only describes how (9) operates on pair states. At $T=0$ all electrons occur in pair states, but a higher temperatures there are always thermally excited unpaired electrons and holes. These give an additional term left out of (13) of the form $\Sigma_{\mathbf{k}\sigma}\xi_{\mathbf{k}}n_{\mathbf{k}\sigma}$ where the sum runs only over unpaired states. At any temperature the total contribution to the energy from such states will vanish because there will be equal numbers above and below ϵ_F. However, we must consider the free energy F which differs from the energy by $-TS$ where S is the entropy. The multiplicity of the singly occupied states is the same as of the doubly occupied ones. The result is that including unpaired electrons will double the entropy, leaving everything else unchanged. We can get the same algebraic answer for F by doubling T instead of doubling S. Thus the thermodynamics of (13) is the same as the thermodynamics of (9) at a temperature one-half as large.

We now solve (13) by analogy with our earlier solution of the Heisenberg model. Although **k** is a wave vector, for the purpose of analogy we imagine a one-dimensional spatial variable instead. Then (13) represents an infinite-range "xy model" in a spatially varying transverse magnetic field $\xi_{\mathbf{k}}$. The "xy model" is just a Heisenberg model except that the coupling $\sigma_{iz}\sigma_{jz}$ in the z direction is omitted from the exchange interaction $\boldsymbol{\sigma}_i\cdot\boldsymbol{\sigma}_j$. The range is infinite because all spins **k** are equally strongly coupled. The number of spins involved is the number of **k** states within $\pm\theta$ of ϵ_F, i.e., $N=2N(\epsilon_F)\theta$, where $N(\epsilon_F)$ is the density of states (in energy) at the Fermi level. N is a very large number (going to infinity in an infinite sample) and so we expect a mean-field solution to be exact. The mean-field Hamiltonian is

$$\begin{aligned}\mathcal{H}_{\text{BCS,MF}}&=-\sum_{\mathbf{k}}\boldsymbol{\sigma}_{\mathbf{k}}\cdot\mathbf{H}_{\mathbf{k}},\\ \mathbf{H}_{\mathbf{k}}&=\xi_{\mathbf{k}}\hat{z}+\tfrac{1}{2}V\sum_{\mathbf{k}'}\langle\sigma_{\mathbf{k}'x}\rangle\hat{x}.\end{aligned} \tag{14}$$

The "magnetization" $\langle\sigma_{\mathbf{k}}\rangle$ has a spontaneous component $\langle\sigma_{\mathbf{k}x}\rangle$ in the xy plane caused by the interaction V, and an induced component $\langle\sigma_{\mathbf{k}z}\rangle$ in the z direction caused by the "external field" $\xi_{\mathbf{k}}$. The component of $\langle\boldsymbol{\sigma}\rangle$ in the xy plane can lie in any direction, and we have chosen a coordinate system so that whatever direction the system chooses is labeled the x direction. If the "external field" $\xi_{\mathbf{k}}$ vanishes, we know the solution by direct analogy with (4); the transition temperature would be $VN/2$. The "external field" $\xi_{\mathbf{k}}$ has the effect of severely depressing the transition temperature by pulling the spin orientation out of the xy plane and reducing the energy of spontaneous alignment under the "exchange" interaction V.

The part of the effective field $\mathbf{H}_{\mathbf{k}}$ that lies in the xy plane is the "pair field" which drives the superconducting transition. It is given the name Δ:

$$\begin{aligned}H_{\mathbf{k}x}&=\tfrac{1}{2}V\sum_{\mathbf{k}'}\langle\sigma_{\mathbf{k}'x}\rangle\equiv\Delta,\\ H_{\mathbf{k}z}&=\xi_{\mathbf{k}},\end{aligned} \tag{15}$$

$$\tan\theta_{\mathbf{k}}\equiv H_{\mathbf{k}x}/H_{\mathbf{k}z}=\Delta/\xi_{\mathbf{k}} \tag{16}$$

$$|\mathbf{H}_{\mathbf{k}}|=(\xi_{\mathbf{k}}^2+\Delta_{\mathbf{k}}^2)^{1/2}. \tag{17}$$

The mean field felt by spin **k** lies at an angle $\theta_{\mathbf{k}}$ to the $\hat{z}$ axis given by Eq. (16). According to statistical mechanics, the spin $\langle\boldsymbol{\sigma}_{\mathbf{k}}\rangle$ will line up with the field $\mathbf{H}_{\mathbf{k}}$ exactly as in Eq. (7). Thus we have for the x component

$$\langle\sigma_{\mathbf{k}x}\rangle=\sin\theta_{\mathbf{k}}\tanh(|\mathbf{H}_{\mathbf{k}}|/2T), \tag{18}$$

$$\sin\theta_{\mathbf{k}}=\Delta(\xi_{\mathbf{k}}^2+\Delta^2)^{-1/2}, \tag{19}$$

where we have remembered to multiply T by 2. Finally, we can use Eqs. (17)–(19) to express (15) as a self-consistent equation for Δ,

$$\Delta = \tfrac{1}{2} V \sum_{\mathbf{k}} \frac{\Delta}{(\xi_{\mathbf{k}}^2 + \Delta^2)^{1/2}} \tanh\left(\frac{(\xi_{\mathbf{k}}^2+\Delta^2)^{1/2}}{2T}\right). \qquad (20)$$

This is the solution of our model and is analogous to (7) for the simpler Heisenberg model. It is also the famous BCS equation for the energy gap Δ. That the pair field Δ plays the role of a gap in the electronic spectrum can easily be seen. At $T=0$, all electrons are in pair states. In normal metals ($\langle \sigma_{\mathbf{k}x} \rangle = 0$ and $\Delta = 0$) the axis of quantization is in the $\hat{z}$ direction. The energies of the pair states are $-\boldsymbol{\sigma}_{\mathbf{k}} \cdot \mathbf{H}_{\mathbf{k}} = -\sigma_{\mathbf{k}z} \xi_{\mathbf{k}}$. The negative-energy pairs are all occupied (spin up), and positive-energy pairs are empty. To create an excitation we remove a pair (which costs energy $2|\xi_{\mathbf{k}}|$), and put one electron back in the negative-energy state and one in a positive-energy state. By choosing these states all close enough to ϵ_F, we can make the minimum energy as small as we like. However, in the superconducting state, a "pair field" Δ has mysteriously appeared and rotated the quantization axis away from the $\hat{z}$ direction. To create an excitation, we destroy an occupied pair (i.e., flip a "spin"). This costs energy $2\boldsymbol{\sigma}_{\mathbf{k}} \cdot \mathbf{H}_{\mathbf{k}} = 2(\xi_{\mathbf{k}}^2 + \Delta^2)^{1/2}$. The minimum energy is 2Δ (when $\xi_{\mathbf{k}} = 0$).

The result (20) was derived by BCS using somewhat different techniques. The technique used here was devised by Anderson, and is in some ways the most intuitive method of finding a solution. Alternative methods have also been devised by Bogoliubov, Valatin, Gor'kov, and others. These alternative techniques each provide a somewhat different insight into the meaning of the theory. Descriptions of these methods can be found in the references listed at the end. One special virtue of the procedure used here is that the broken symmetry is clearly manifested. The Hamiltonian (13) is invariant under rotations in the xy plane, but the solution (20) has picked out an arbitrary preferred direction, an azimuthal angle ϕ in the xy plane. All "spins" are oriented in the same direction ϕ. In coordinate space ϕ acquires the significance of the phase $e^{i\phi}$ of the pair wave function $\psi = |\psi(\mathbf{r}_1 - \mathbf{r}_2)| e^{i\phi}$. Of course the absolute value of this phase can have no physical meaning, but it is of crucial importance that each pair have the same phase ϕ. Thus in a macroscopic sample (with no magnetic field inside), if an electron pair at one end of the sample has phase ϕ, pairs at the other end have still the same phase. This is also the phase of the Ginsberg–Landau order parameter. This long-range order of the phase has been called "off-diagonal long-range order" (ODLRO) by Yang. It was the (successful) search for a method to measure this phase which led Josephson to discover the beautiful effects which are named after him.

The solution (20) of the BCS Hamiltonian leads immediately to the explanation and prediction of virtually all measured properties of superconductors (and hopefully still many more to come). Let us briefly examine one property, the magnitude and temperature dependence of Δ. The sum in (20) can be converted to an integral over energy. Because

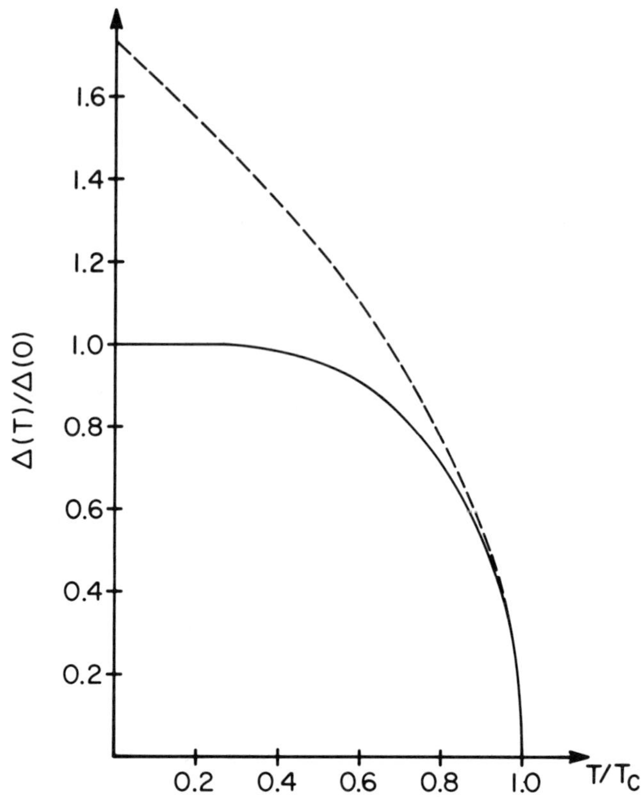

FIG. 3. Reduced value of the BCS energy gap $\Delta(T)/\Delta(0)$ versus reduced temperature T/T_c (solid curve), and approximate form $\Delta(T)/\Delta(0) \approx 1.74\,(1 - T/T_c)^{1/2}$ (dashed curve). To better than 1% accuracy (i.e., the thickness of the line), these curves also represent the magnetization of a spin-$\frac{1}{2}$ ferromagnet in mean-field approximation [Eq. (7)].

the cutoff θ is small, it is reasonable to neglect variation in the electron density of states $N(\epsilon)$; then

$$1 = N(\epsilon_F) V \int_0^{\theta} d\xi \tanh\,[(\xi^2+\Delta^2)^{1/2}/2T](\xi^2+\Delta^2)^{-1/2}. \qquad (21)$$

We have used the fact that the cutoff is symmetric about ϵ_F. At $T=0$, the tanh can be replaced by 1, leaving an elementary integral which can be solved for Δ,

$$\Delta(0) = \theta/\sinh[1/N(\epsilon_F)V] \approx 2\theta e^{-1/N(\epsilon_F)V}. \qquad (22)$$

The approximate form in (22) assumes that the "coupling constant" $N(\epsilon_F)V$ is small. We can also solve (21) to discover the temperature T_c at which Δ goes to zero and superconductivity disappears. This done by setting Δ to 0 in Eq. (21),

$$1/N(\epsilon_F)V = \int_0^{\theta/2T_c} (dx/x) \tanh x$$

$$= \ln(\theta/2T_c)\tanh(\theta/2T_c) - \int_0^{\theta/2T_c} dx \ln x \frac{d}{dx} \tanh x,$$

where the last expression is obtained by integrating by parts. If we assume that $\theta/2T_c$ is significantly bigger than 1, it is reasonable to replace the upper limit of the last integral by ∞; the value of that integral is then $\ln(4\gamma/\pi)$ where $\gamma \cong 1.775$ is Euler's constant. Replacing tanh $\theta/2T_c$ by 1 we find

$$T_c \cong 1.13\theta e^{-1/N(\epsilon_F)V}. \qquad (23)$$

Taking the ratio of (22) to (23) we find a universal relation independent of the coupling constants,

$$2\Delta \cong 3.52 k_B T_c. \tag{24}$$

This relation has been confirmed in many experiments, most notably tunneling as first done by Giaevar. Within the same approximations it turns out that the temperature dependence $\Delta(T)/\Delta(0)$ is a universal function of T/T_c, going to 0 at T_c as $(3.01/T_c)^{1/2}(T_c - T)^{1/2}$. The solution $\Delta(T)/\Delta(0)$ can be obtained numerically, and is graphed in Fig. 3. The resulting curve is remarkably close to the mean-field result (7) for $\langle\sigma(T)\rangle$ in a magnet. The universal behavior of $\Delta(T)$ in BCS theory provides a microscopic explanation for the "law of corresponding states," a phrase denoting the experimental observation of universality in many physical properties of superconductors [$H_c(T)$ for example].

We have not yet considered the origin of the interaction V or the meaning of the cutoff θ. These questions are independent of the mathematical theory described above. An important experimental clue was the discovery in 1950 by Maxwell and Serin and others of the "isotope effect," the dependence of T_c on isotopic mass. This suggested that the electron–phonon interaction might play an important role. Following theoretical work of Fröhlich, Bardeen, and Pines, BCS identified V as the interaction between electrons from lattice polarization, a second-order effect of the electron–phonon interaction. This leads to the identification of the cutoff θ with the Debye temperature Θ_D, and the formula (23) helps explain why the observed values of T_c are so low.

The subsequent development of BCS theory has been a rich field of contemporary physics. One aspect particularly deserving of comment is improvements in the understanding of the nature of the coupling V. The need for refinements of BCS theory was evident almost immediately because of the behavior of the so-called "bad actors" Pb and Hg, which show deviations from the law of corresponding states. Also Giaevar discovered "bumps" corresponding to phonon energies in the differential conductance of superconducting tunnel junctions. These bumps did not have a very natural explanation in the BCS equations. It was soon recognized that the difficulty lay in the BCS assumption of an instantaneous interaction [proportional to $\delta(t-t')$]. Relativistic theory shows that all interactions are retarded at least by $\delta(t-t'-|\mathbf{x}-\mathbf{x}'|/c)$, but the phonon-induced interaction is much more severely retarded than this. Physically what happens is that the first electron at $(\mathbf{x}t)$ interacts with the lattice by a coupling constant M, creating a disturbance which can be Fourier analyzed in terms of phonons. This disturbance develops and propagates in space and time at a rate which is two or three orders of magnitude slower than electron Fermi velocities ($\sim 10^8$ cm/s). In order to interact efficiently with this disturbance, the second electron at $(\mathbf{x}'t')$ must wait for the disturbance to develop. This allows the first electron time to move away. Thus two electrons are able to interact by "leaving messages for each other" in the crystal lattice, never meeting directly (which would be disadvantageous because the direct interaction is dominated by a strong Coulomb repulsion). The form of the actual interaction is

$$V(xt,x't') = -\sum_Q M_Q(x) M_Q^*(x') \times \sin\omega_Q(t'-t) e^{-\gamma_Q(t'-t)} \theta(t'-t).$$

The interaction oscillates in time at frequencies determined by the phonon frequency ω_Q. The interaction responsible for superconductivity occurs primarily during the first half-cycle while V is attractive and before the weakening effects of phonon damping γ_Q and dephasing (caused by dispersion in ω_Q) occur. A generalized version of BCS theory powerful enough to include retarded interactions was first written down by Eliashberg. This theory has a time-dependent pair field. The gap $\Delta(\omega)$, which is measured by an external ac probe at frequency ω, is essentially the Fourier transform of the pair field. The gap is complex, the real and imaginary parts corresponding, respectively, to the parts of the pair field which are in phase and out of phase with the interaction. The gap $\Delta(\omega)$ has bumps when the applied frequency is in resonance with the phonon frequencies. These correspond to Giaevar's bumps in tunneling conductance, because the dc voltage across a tunnel junction behaves like an ac probe at frequency $\hbar\omega = eV$. The careful measurement of these bumps has been developed into a powerful spectroscopic tool by Rowell and McMillan, and has provided detailed confirmation of Eliashberg theory for many superconductors.

See also JOSEPHSON EFFECTS; PHASE TRANSITIONS; SUPERCONDUCTING MATERIALS.

BIBLIOGRAPHY

Some of the most important original papers are reprinted in *Superconductivity, Selected Reprints*, American Institute of Physics, New York, 1964. Other collections of reprints are: D. Pines, *The Many Body Problem*, Benjamin, New York, 1962; N. N. Bogoliubov, *The Theory of Superconductivity*, Gordon and Breach, New York, 1968.

An English translation of the paper by Landau and Ginsberg is available in *Men of Physics: Landau*, Vol. II (D. ter Haar, ed.). Pergamon, New York, 1965.

The Nobel addresses of Bardeen, Cooper, Schrieffer, Giaever, and Josephson are highly recommended, and can be found in *Physics Today* 23–46 (July, 1973); *Science* **183,** 1253 (1974); **184,** 527 (1974).

1. *Elementary* (the algebra of BCS theory is not given).

C. Kittel, *Introduction to Solid State Physics*, 5th ed., Chap. 12. Wiley, New York 1976.

A. C. Rose-Innes and E. H. Rhoderick, *Introduction to Superconductivity*. Pergamon, Oxford, 1969.

H. M. Rosenberg, *Low Temperature Solid State Physics*, Chap. 6. Oxford, London, 1963.

A. W. B. Taylor, *Superconductivity*. Wykeham Publications, London, 1970.

D. R. Tilley and J. Tilley, *Superfluidity and Superconductivity*. Wiley, New York, 1974.

J. E. C. Williams, *Superconductivity and its Application*. Pion Ltd., London, 1970.

2. *Intermediate*

C. Kittel, *Quantum Theory of Solids*, Chap. 8. Wiley, New York, 1963. The pseudospin model is described.

C. G. Kuper, *An Introduction to the Theory of Superconductivity*. Oxford, London, 1968.
E. A. Lynton, *Superconductivity*, 2nd ed.. Methuen, London, 1964.
M. Tinkham, *Introduction to Superconductivity*. McGraw-Hill, New York, 1975.
M. Tinkham, *Superconductivity*, Gordon and Breach, New York, 1965. The original BCS method, the pseudospin method, and the Bogolubov–Valatin method are all described.

3. *Advanced*

P. G. deGennes, Superconductivity of Metals and Alloys. Benjamin, New York, 1966.
R. D. Parks (ed.), *Superconductivity*. Marcel Dekker, New York, 1969.
G. Rickayzen, *Theory of Superconductivity*. Wiley-Interscience, New York, 1965.
J. R. Schrieffer, *Theory of Superconductivity*. Benjamin, New York, 1964.

Surfaces and Interfaces

F. Cyrot-Lackmann and M. C. Desjonquères

Recent years have been a time of increasing interest in the fundamental studies of surfaces and interfaces. This is clearly due to the development of new experimental techniques capable of characterizing the surface at a microscopic level and to the very great importance of many phenomena that occur on solid surfaces. Indeed, it is essential to acquire a deeper knowledge of basic surface phenomena in order to understand in a fundamental way heterogeneous catalysis, epitaxial growth, corrosion, as well as many solid-state electronic devices or biological systems.

The surface of a solid may be defined as the outermost atomic layer including, eventually, foreign atoms. It can be characterized in the same manner as any solid or molecule. One must know the chemical identity of the atoms present and the structural arrangement of these atoms, their vibration states and the energetic and spatial distribution of electrons.

The *structure of the surface* can differ considerably from a perfect termination of a bulk structure (Fig. 1). For a surface free of impurities, the outermost plane may be displaced from its bulk position (relaxation) or it can have a new periodicity (reconstruction). Foreign atoms can be bound to the surface (adsorption and chemisorption) in a finite number or form an overlayer. At the surface of an alloy, the concentration in one of the components can be larger than in the bulk (segregation).

To identify the chemical nature of surface atoms, one principally uses Auger electron spectroscopy. In this method, one uses the fact that electrons ejected from surface atoms as they relax after ionization of a core electron have characteristic energies depending on core levels and, consequently, on chemical identity. This is the usual technique for checking the cleanliness of a surface.

LEED (low-energy-electron diffraction) is used to study surface crystallography. The symmetry of surface structure is derived from that of diffracted electron beams. One has thus found that reconstruction occurs frequently for covalent materials but only for a few metals. For adsorbed layers, a large number of well-ordered periodic arrangements have been found (Fig. 2). The determination of actual surface atom position and distance, relative to inner planes or substrate atoms, is also possible but quite difficult; it has already been investigated by careful analysis of the variations of the intensity of diffracted beams with the energy of incident electrons (intensity profiles) both theoretically and experimentally on several metals (Mo, W, Ni, Fe, etc.). Field ionization micrographs have also been used for a direct observation of the adsorption site of a single adsorbate.

Another problem related to the geometrical arrangement of atoms is the surface stability and the equilibrium shape of a small particle. It obviously depends on the surface tension, i.e., the work, necessary to create a unit area of free surface. The dependence of this quantity on the crystallographic orientation is given by the Gibbs–Wulff form from which the equilibrium shape can be constructed.

The existence of the surface modifies the *vibration spectrum of atoms (phonons)*. Theoretical studies are usually performed neglecting relaxation and reconstruction and with central forces (sometines with angular interactions) between nearest neighbors. One finds

(i) bulk modes, with a modified amplitude and density, which are coupled by the surface;

(ii) surface modes (Rayleigh waves), which are exponentially decreasing from the surface;

(iii) virtual modes (extended Rayleigh waves) with an attentuated propagation along the surface, because of their coupling with bulk modes.

The mean square displacements, specific heat, and entropy have been calculated with a reasonable agreement with experiment (LEED, measurements of the specific heat of small particles, and thermal variation of surface tension, respectively).

Chemical reactions at surfaces, which involve electron sharing or even electron transfer, must certainly depend on

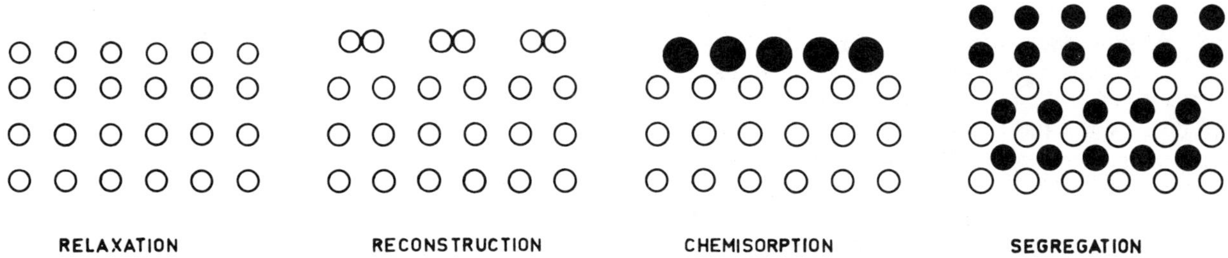

FIG. 1. Some possible structures of a crystal surface.

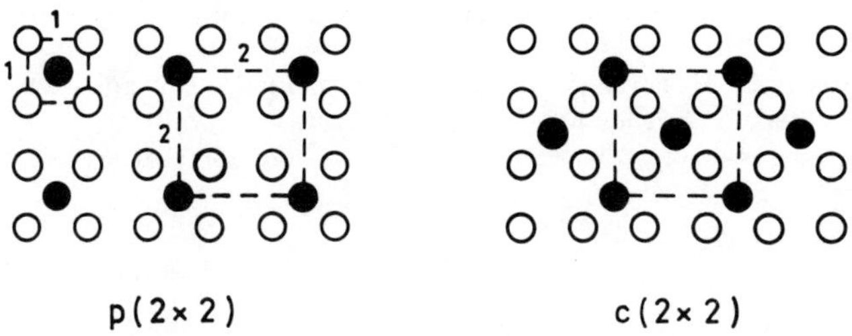

FIG. 2. (100) surface of a cubic crystal showing some frequently observed overlayer structures. The pattern of foreign atoms has a unit mesh twice the size of the (1 × 1) mesh of substrate atoms. They are centered, $c(2 \times 2)$, if the mesh has a foreign atom at its center, or primitive, $p(2 \times 2)$, if it does not.

the distribution in energy of electronic states in the surface monolayer. The knowledge of the *electronic structure* at the surface is then of very great interest, but it is much more difficult to attain than for the bulk because of the loss of symmetry. Qualitatively, one expects that the energy-level spectrum will be composed of the same energy bands because the surface region is only a small fraction of the whole solid. But the modification of boundary conditions can lead, in addition to the perturbed Bloch bulk waves, to the existence of *localized* or surface states at an energy that was forbidden for the bulk, or of *virtual surface states* which are extended states with a strong amplitude near the surface and are then rather localized. These extra states are apparent in the local density of states at the surface, i.e., the density of electronic levels weighted by the magnitude of the wave function at the surface (Fig. 3). Both types of surface states have already been observed on materials such as Al, Si, Ni, W, Mo, although the surface energy-level spectrum is difficult to investigate experimentally. The main techniques are electron spectroscopies (field emission, ultraviolet photoemission, ion neutralization); though they are mainly sensitive to the surface monolayer, they also involve bulk contributions which are not easy to disentangle from surface ones.

In the foregoing, we were concerned with the free surface of solids. Another interesting and related topic is that of

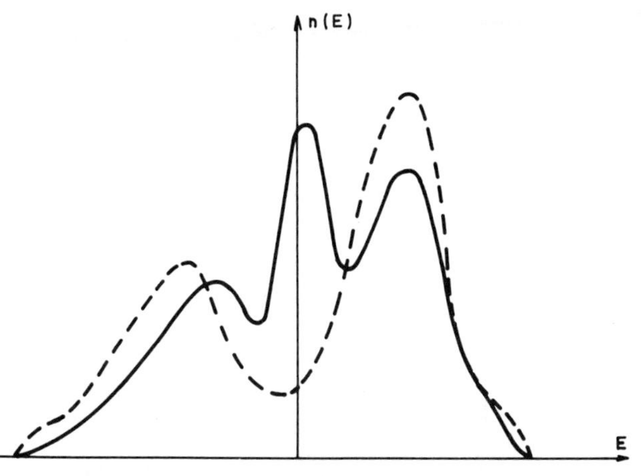

FIG. 3. Schematic local density of states on a (100) surface (solid line) of a bcc transition metal compared to the bulk one (dashed line).

interfaces. This is a very broad field with numerous applications in metallurgy, crystal growth, wetting of surfaces, etc., but where very little is known. We shall restrict ourselves to a few remarks on *interfaces between crystals*. There are, however, other contacts between solids of interest. Thus the *amorphous–crystal interface* covers the problem of the surface treatment of semiconductors, the use of the tunnel effects through the insulating layers in MOS (metal-oxide-semiconductor) devices, and the adhesion of glues of plastics on metals, but it is not well understood. It is the same for the *solid–liquid* contact, which is the key to crystal growth or wetting of surfaces.

When the contact is between two crystals of the same phase, it is called a *grain boundary*. The problem is then defined by the relative position of the two crystals. Grain boundaries can be classified into

(i) *large-angle incoherent grain boundary* where there is no special relationship between the crystals (such a configuration appears clearly in field ion microscopy);

(ii) *low-angle grain boundary* where the two crystals have nearly the same orientation;

(iii) *stacking faults* for structures which can be described as a stacking of close-packed planes;

(iv) *twins* which are contacts of two crystals which can be deduced from each other by a symmetry operation around a plane or an axis.

The existence of these various boundaries plays an important role in plasticity. Large-angle grain boundaries and twins harden the material at low temperature by stopping dislocations. At higher temperature, self-diffusion is faster along the boundary than in the volume and, on the whole, soften the material.

Very little is known about their electronic structure and the possible existence of states localized near the interface. Only some evaluation of the grain boundary energy has been made which can be compared to some measurements deduced from the width of splitting of a dislocation in partials or from the study of stable triple points.

A second type of interface is the *epitaxial interface*. Two crystals in exact epitaxy are in contact along a plane where two of their (small) periods coincide. Cases of perfect epitaxy are known in Guinier–Preston zones as, for example, in Al alloys. The energy of this interface is important in adhesional properties of metals and is probably governed by the sharing of electrons between the two metals.

See also ADSORPTION; AUGER EFFECT; CATALYSIS; CRYSTAL DEFECTS; CRYSTAL GROWTH; ELECTRON ENERGY STATES IN LIQUIDS AND SOLIDS; FIELD-ION MICROSCOPY; LOW ENERGY ELECTRON DIFFRACTION; PHOTOELECTRON SPECTROSCOPY.

BIBLIOGRAPHY

Atomic Structure and Mechanical Properties of Metals, Proceedings, International School of Physics "Enrico Fermi," Course LXI, Varenna, July 8–20, 1974, G. Caglioti, (ed.). North-Holland, Amsterdam, 1976. (I, A)

CRC Crit. Rev. Solid State Mat. Sci., **4**(3), August (1976). (A)

CRC Crit. Rev. Solid State Mat. Sci. **6**(4), October (1978). (A)

CRC Crit. Rev. Solid State Sci., **2**(1), February (1974) (A)

Dynamic Aspects of Surface Physics, Procedings, International School of Physics "Enrico Fermi," Course LVIII, Varenna, 1973, F. O. Goodman (ed.). Editrice Compositori, Bologna, 1975. (I, A)

Homer D. Hagstrum, "Electronic Characterization of Solid Surfaces," *Science* **178**, 275–282 (1972). (E)

Special Issue on Surface Physics, *Physics Today* **28**(4), April (1975).

Thermal Expansion

Richard K. Kirby

When the temperature of a substance is increased, the volume that it occupies is usually increased. A measure of this change is the coefficient of thermal expansion,

$$\beta = \frac{1}{V}\left(\frac{\partial V}{\partial T}\right)_p$$

where V is the volume at temperature T and constant pressure P. In the case of ideal gases the volume at constant pressure is proportional to the average translation kinetic energy per particle, $V \propto \frac{1}{2}mv^2 = \frac{3}{2}kT$, so that the value of β is equal to T^{-1}. At room temperature and atmospheric pressure $\beta \sim 3400 \times 10^{-6}\ \mathrm{K}^{-1}$ for nearly all gases. Because of the short-range structural ordering that takes place in liquids their thermophysical behavior is more complex. As a result the temperature dependence of β usually takes the empirical form $\beta = a + bT + cT^2$. Three examples are given in Table I.

In solids practically all of the thermal energy goes into the motion of atoms around their equilibrium positions in the structure and the thermal expansion of that structure is the result of these vibrations. In addition, the structure can expand differently in different directions depending on the type of crystalline structure. The coefficient of linear thermal expansion in the ith direction is defined as

$$\alpha_i = \frac{1}{L_i}\left(\frac{\partial L_i}{\partial T}\right)_p.$$

The sum of three orthogonal linear coefficients of a crystal is equal to its volume coefficient: $\beta = \alpha_1 + \alpha_2 + \alpha_3$. Some examples are given in Table II.

The temperature dependence of β for solids is very complex but can be approximated by the Grüneisen equation, $\beta = \gamma C_p / VB_s$, where γ is the Grüneisen parameter, C_p is the heat capacity at constant pressure, and B_s is the adiabatic bulk modulus. Since the product γ / VB_s is nearly constant over a wide temperature range, the value of β has about the same temperature dependence as C_p; that is, at low temperatures $\beta \rightarrow 0$ as $T \rightarrow 0$ and at high temperatures β is slowly increasing. It has been found that most metals expand about 7% on heating from 0 K to their melting temperature. Therefore, a metal with a low melting point T_M has a large coefficient of thermal expansion, and vice versa; see Table III.

In the case of metals at $T < 20$ K the heat capacity can be expressed as $C_p = a_{el}T + a_{lat}T^3$, where the two terms indicate the contributions of the conduction electrons and the lattice, respectively. Through Grüneisen's equation the thermal expansion can be expressed in a similar manner:

$$\beta = \frac{1}{VB_s}(\gamma_{el}\alpha_{el}T + \gamma_{lat}\alpha_{lat}T^3).$$

High-sensitivity methods with which β can be determined with a precision of $10^{-10}\ \mathrm{K}^{-1}$ can be used to measure these contributions to thermal expansion as well as those caused by other physical interactions. These include magnetic interactions, the splitting of molecular energy levels, and the resonant tunneling between equilibrium positions in the structure.

The combination of optical and x-ray diffraction techniques to measure thermal expansion has been used to determine the number of thermal vacancies that are generated at temperatures approaching T_M. The number of these vacancies is proportional to the difference between the linear coefficients of the bulk material and the crystal lattice at the same temperature. In the case of copper at the melting point this difference is $1.5 \times 10^{-6}\ \mathrm{K}^{-1}$.

Table I Coefficients of thermal expansion of liquids

Temperature (K)	Water ($\times 10^{-6}\ \mathrm{K}^{-1}$)	Mercury ($\times 10^{-6}\ \mathrm{K}^{-1}$)	Pentane ($\times 10^{-6}\ \mathrm{K}^{-1}$)
273	−64	181.7	1465
283	+85	181.8	1530
293	194	181.8	1610
303	263	181.9	1695

Table II Coefficients of linear thermal expansion of several solids

		200 K	293 K	400 K
NaCl (fcc)	$\alpha_1 = \alpha_2 = \alpha_3$	35.3	39.7	43.0
Fe (bcc)	$\alpha_1 = \alpha_2 = \alpha_3$	10.1	11.8	13.2
SiO_2 (hex)	$\alpha_1 = \alpha_2$	4.9	7.4	8.8
	α_3	10.3	13.6	15.6
In (tetr)	$\alpha_1 = \alpha_2$	39.5	52.9	77.3
	α_3	5.6	−9.6	−34.8
U (orth)	α_1	22	23	24
	α_2	1	0	−1
	α_3	17	19	22
SiO_2 (glass)	$\alpha_{isotropic}$	−0.13	+0.41	0.55
Teflon (polymer)	$\alpha_{isotropic}$	85	>500[a]	150

[a] Phase transition at 293 K causes the coefficient to be very large over about 10 degrees. All values should be multiplied by $10^{-6}\ \mathrm{K}^{-1}$.

Table III Relationship between thermal expansion and melting points

Metal	T_M (K)	β at 293 K ($\times 10^{-6}\ \mathrm{K}^{-1}$)
Lead	601	86.1
Copper	1357	50.1
Platinum	2042	26.7
Tungsten	3653	13.5

See also CERAMICS; METALS.

BIBLIOGRAPHY

Theory

J. F. Nye, Physical Properties of Crystals, p. 322. Oxford, London and New York, 1960.

J. G. Collins and G. K. White, "Thermal Expansion of Solids," in *Progress in Low Temperature Physics,* Vol. IV, p. 450. 1964.

T. H. K. Barron and R. W. Munn, "Analysis of the Thermal Expansion of Anisotropic Solids, Application to Zinc," *Phil. Mag.* **15,** 85 (1967).

F. W. Sheard, "Calculation of the Thermal Expansion of Solids from the Third-Order Elastic Constants," *Phil. Mag.* **3,** 1381 (1958).

G. K. White, "Thermal Expansion at Low Temperatures," *Nature* **187,** 927 (1960).

G. K. White, "Thermal Expansion of Vitreous Silica at Low Temperatures," *Phys. Rev. Lett.* **34,** 204 (1975).

W. E. Schoknecht and R. O. Simmons, "Thermal Vacancies and Thermal Expansion," *Proc. 3d AIP Conf. on Thermal Expansion,* p. 169 (1972).

Measurement Techniques

R. K. Kirby, "Thermal Expansion of Ceramics, Mechanical and Thermal Properties of Ceramics," NBS Spec. Publ. 303, p. 41 (1969).

T. A. Hahn and R. K. Kirby, "Thermal Expansion of a Borosilicate Glass from 80 to 680 K—Standard Reference Material 731," *Proc. AIP Conf. on Thermal Expansion* **17,** 93 (1974).

Data

Thermophysical Properties of Matter (The TPRC Data Series), Vol. 12, *Thermal Expansion of Metallic Elements and Alloys;* Vol. 13, *Thermal Expansion of Nonmetallic Solids.* Plenum, New York, 1976.

Thermionic Emission

I. J. D'Haenens

Thermionic emission is a term used to describe the emission of electrons (and/or ions) from a solid when it is heated in vacuum. Thermionic emission can be viewed as an evaporation process, the temperature dependence of which is accurately described by the Richardson equation:

$$j_R = A(1-r)T^2 \exp(-e\phi/kT), \tag{1}$$

where j_R is the current density of emitted electrons, A is the emission constant, $1-r$ is the transmission coefficient of the surface barrier for electrons, T is the temperature, and $e\phi$ is the work function, which can be interpreted as an electron latent heat of vaporization.

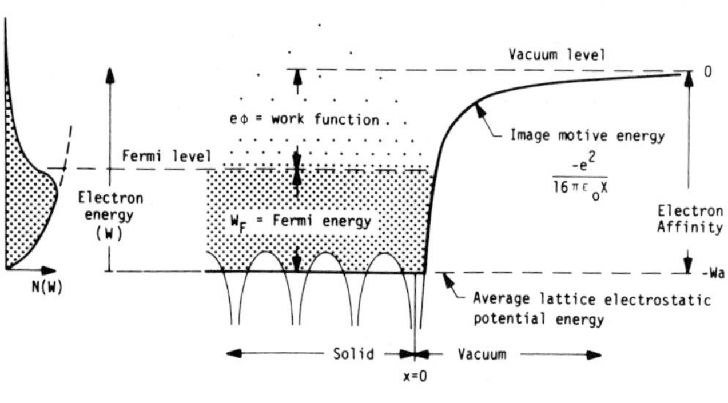

FIG. 1. Sommerfeld model of a metal. (a) Energy distribution of electrons as given by Eq. (2). (b) Boxlike potential well defined by the average lattice electrostatic potential and image motive energies indicated by heavy outline.

The Richardson equation is the fundamental equation of thermionic emission, and can be derived from thermodynamic [1] or quantum statistical [2] considerations of the electrons in the metal. The quantum statistical derivation of the Richardson equation proceeds from a consideration of the thermal population of the electronic states in the metal. The metal is represented in the Sommerfeld approximation, which assumes a boxlike potential well in which the barrier is formed by joining the average lattice potential energy $(-W_a)$ to the image motive* energy $(-e^2/16\pi\epsilon_0 x)$, as represented schematically in Fig. 1. Furthermore, the electrons are assumed to obey Fermi–Dirac statistics, which gives for the energy distribution of electrons in the metal

$$N(W) = 4\pi\left(\frac{2m}{h^2}\right)^{3/2} \frac{W^{1/2}}{[e^{(W-W_F)/kT}+1]}. \tag{2}$$

The current density j will now be given by

$$j = e\int N(W)v_z\,dW, \tag{3}$$

where v_z is the component of electron velocity normal to the surface, and the integration is performed over those electrons which can escape from the metal, i.e., whose normal kinetic energy, $\frac{1}{2}mv_z^2$, is greater than the barrier height, $W_F + e\phi$. To effect the integration, we use the following identities:

$$W = \tfrac{1}{2}mv^2 = \tfrac{1}{2}m(v_x^2+v_y^2+v_z^2); \tag{4a}$$

$$W^{1/2}dW = 2(\tfrac{1}{2}m)^{3/2}v^2dv\,\frac{1}{2\pi} = (2\pi)^{-1}(\tfrac{1}{2}m)^{3/2}dv_x dv_y dv_z; \tag{4b}$$

and the fact that $e^{(W-W_F)/kT} \gg 1$. The current density is thus given by

$$j = 2e\left(\frac{m}{h}\right)^3 \int_{-\infty}^{\infty} e^{-(m/2kT)v_x^2}dv_x \times \int_{-\infty}^{\infty} e^{-(m/2kT)v_y^2}dv_y \times \int_{[2(W_F+e\phi)/m]^{1/2}}^{\infty} e^{W_F-(m/2kT)v_z^2}v_z\,dv_z. \tag{5}$$

* The motion of electrons in the space just outside the surface of metal is greatly influenced by the attraction of each electron to its image charge with a force equal to $-e^2/16\pi\epsilon_0 x^2$. Thus, the electron behaves as if it were moving in an electrostatic field whose potential is the actual potential plus $-e/16\pi\epsilon_0 x$. This fictitious potential is called the *image motive*. The motive is defined as that quantity, the gradient of which at any point gives $1/e$ times the force on an electron. The image motive is not a true electrostatic potential since the definition of true potentials involves a time average of force per unit charge in the limit $q \to 0$, where q is the electric charge at point x. For the image force, $F = -q^2/16\pi\epsilon_0 x^2$, and $\lim_{q\to 0} dF/dq = 0$. In spite of this conceptual difficulty, it is legitimate to use the image motive to calculate the wave function of an electron outside of a metal surface [e.g., J. Bardeen *Phys. Rev.* **58,** 727 (1940)].

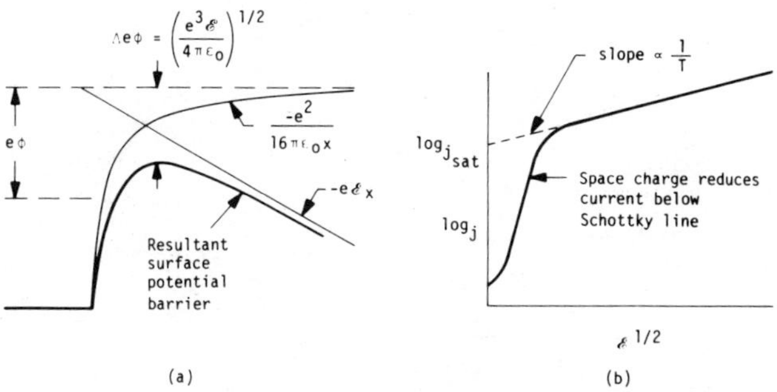

FIG. 2. Schottky effect. (a) Lowering of the potential barrier by image motive and applied fields. (b) Illustrative Schottky plot showing the effect of space charge at low fields, and the definition of saturation current density, j_{sat}.

The first two integrals are of the form

$$\int_{-\infty}^{\infty} e^{-ax^2}dx = \left(\frac{\pi}{a}\right)^{1/2} \rightarrow \left(\frac{2\pi kT}{m}\right)^{1/2}$$

so that

$$j_R = (4\pi mek^2/h^3)(1-r)T^2e^{-e\phi/kT}. \tag{6}$$

Comparison of this result with Eq. (1) indicates that the emission constant A is given by $4\pi mek^2/h^3 = 120.4$ A/(cm K)^2.

The application of an electric field to the surface of a metal results in an increase in thermionic current density. This is referred to as the Schottky effect, and is illustrated in Fig. 2, which shows how the image motive and the applied electrostatic field combine to lower the surface potential barrier. The barrier is lowered an amount

$$\Delta e\phi = \left(\frac{e^3\mathscr{E}}{4\pi\epsilon_0}\right)^{1/2} = 3.8\times 10^{-5}\mathscr{E}^{1/2}\ \text{eV}. \tag{7}$$

This results in a modification of the equation for thermionic current density to

$$j_{RS} = \frac{4\pi mek^2}{h^3}[1-\bar{r}(\mathscr{E})]T^2\exp\left[\frac{-e\phi + (e^3\mathscr{E}/4\pi\epsilon_0)^{1/2}}{kT}\right] \tag{8}$$

$$= j_R\exp\left(\frac{e^3\mathscr{E}}{4\pi\epsilon_0}\right)^{1/2}\left(\frac{1-\bar{r}(\mathscr{E})}{1-\bar{r}(0)}\right), \tag{9}$$

where $1-\bar{r}(\mathscr{E})$ is the transmission coefficient of the modified surface barrier, averaged over the thermionic electron energy. This is known as the Richardson–Schottky equation and predicts the log of thermionic emission current density versus the square root of the applied electrostatic field to be a straight line with a slope that varies inversely with temperature:

$$\log\left(\frac{j_{RS}}{j_R}\right) = \frac{0.1912}{T}\mathscr{E}^{1/2} + \log\left(\frac{1-\bar{r}(\mathscr{E})}{1-\bar{r}(0)}\right). \tag{10}$$

Experimentally, at low values of applied electrostatic field $\mathscr{E}$, the actual current density falls below the Schottky line because of the effect of space charge. The space-charge-limited current in a plane-parallel vacuum diode is given by the Child–Langmuir law:

$$j_{CL} = \frac{4\epsilon_0}{9}\left(\frac{2}{mx}\right)^{1/2}\mathscr{E}^{3/2}, \tag{11}$$

where x is the cathode–anode separation.

Experimental values of the work function ($e\phi$) may be determined from a Richardson plot: a graph of the values of $\log(j_{sat}/T^2)$ vs $1/T$, where the saturation current density j_{sat} is obtained by linear extrapolation of the Schottky line to $\mathscr{E}=0$. In general, for metals a straight Richardson line is obtained. The slope of this line is $-e\phi/2.3k = -5040\phi$, and the intercept extrapolated to the $\log(j_{sat}/T^2)$ axis is defined as $\log A$. Thermionic emission parameters thus obtained are labeled $e\phi_R$ and A_R; A_R contains factors depending on the nonuniformity of real surfaces and the temperature dependence of the work function [2].

See also METALS; SURFACES AND INTERFACES; WORK FUNCTION.

REFERENCES

1. P. W. Bridgman, *The Thermodynamics of Electrical Phenomena in Metals*. MacMillan, New York, 1934. (I)
2. C. Herring and M. H. Nichols, *Rev. Mod. Phys.* **21**, 185–270 (1949). (A)

Thermoelectric Effects

C. L. Foiles

Thermoelectric effects occur when mobile charge carriers in liquids and solids are subjected to the influence of temperature gradients and/or particular electrical potential gradients. In the absence of a magnetic field there are three thermoelectric effects; the Seebeck (discovered in 1822 by T. J. Seebeck), the Peltier (discovered in 1834 by J. C. A. Peltier), and the Thomson (predicted in 1854 and experimentally documented a few years later by Wm. Thomson) effects.

The Seebeck effect is the basis for thermocouples. Consider the circuit shown in Fig. 1. Two different materials, A and B, are joined with their junctions maintained at temperatures T_1 and T_2 while a break in the circuit is maintained at T_0. An open-circuit voltage ΔV is generated if $T_1 \neq T_2$. For homogeneous materials this voltage is independent of the precise details of the temperature gradients and of T_0. Only the difference $T_2 - T_1$ is important. ΔV is linearly proportional to small temperature differences,

$$\Delta V_{AB} = \alpha_{AB}(T_2 - T_1), \tag{1}$$

but it is more generally described as

$$\Delta V_{AB} = \int_{T_1}^{T_2}[S_A(T) - S_B(T)]\,dt \equiv \int_{T_1}^{T_2} S_{AB}(T)\,dT. \tag{2}$$

α_{AB} is the Seebeck coefficient for the AB couple and $S_m(T)$ is the thermoelectric power, or thermopower, of material m. (Note: Thermopower is a commonly accepted misnomer.)

Contrary to an oft stated misconception, $S_m(T)$ is a unique, well-defined property of a single homogeneous material and does not depend on a second material for its existence. This misconception arises from the practical constraint of measuring $S_{AB}(T)$, which necessitates prior knowledge of either $S_A(T)$ or $S_B(T)$ before a separation of terms is possible. Confusion about the sign of thermopower is also common; if $T_2 = T_1 + \Delta T$ causes the polarity shown in Fig. 1, then $S_A > S_B$.

The Peltier and Thomson effects refer to reversible heat processes. Passage of an electrical current density J across an isothermal boundary formed by two different materials causes the generation or absorption of a Peltier heat P_p at the junction. Specifically, if current flow from material A into material B generates heat, then reversing the current direction causes heat to be absorbed. Similarly, passage of an electrical current through a homogeneous material subjected to a temperature difference ΔT produces a Thomson heat P_t. Whether P_t is generated or absorbed depends on the direction of J relative to ΔT. The dependence of both P_p and P_t on current direction permits the unique separation of these heats from irreversible Joule heating, which depends upon J^2 and is always a generated heat. The Peltier coefficient π_{AB} and the Thomson coefficient σ_t are proportionality constants in the expressions

$$P_p = \pi_{AB} J, \tag{3}$$

$$P_t = -\sigma_t J \Delta T. \tag{4}$$

The preceding three coefficients are not independent but are, instead, related through the Kelvin relations

$$\pi_{AB} = T(S_a - S_B), \tag{5a}$$

$$\sigma_{tA} = T\frac{dS_A}{dT}, \tag{5b}$$

which were developed by Thomson (later Lord Kelvin) on the basis of thermodynamic arguments. The original thermodynamic arguments are not rigorous, but the same results emerge from a rigorous analysis using irreversible thermodynamics. Equation (5b) can be integrated to obtain

$$S_A(T) = \int_0^T \left(\frac{\sigma_{tA}(T')}{T'}\right) dT' \tag{6}$$

where $S_A(0)$ is set equal to zero to comply with the third law of thermodynamics. Thus, a careful measurement of the Thomson heat as a function of temperature can be used to determine the absolute thermopower of any material. A measurement of this type provides the basis for absolute values of thermopower and serves as direct refutation of the misconception discussed below Eq. (2).

Thus far the discussion incorrectly implies that thermoelectric effects are scalar quantities. A general analysis relates electrical and heat current densities, **J** and **U**, respectively, to electrical potential and temperature gradients, ∇V and ∇T, respectively, and produces thermoelectric effects which are tensors. For isotropic and cubic materials these tensors reduce to scalars. For anisotropic materials and for materials in the presence of a magnetic field (i.e., thermomagnetic effects) the tensor form is necessary. The study of thermomagnetic effects is becoming increasingly important, but

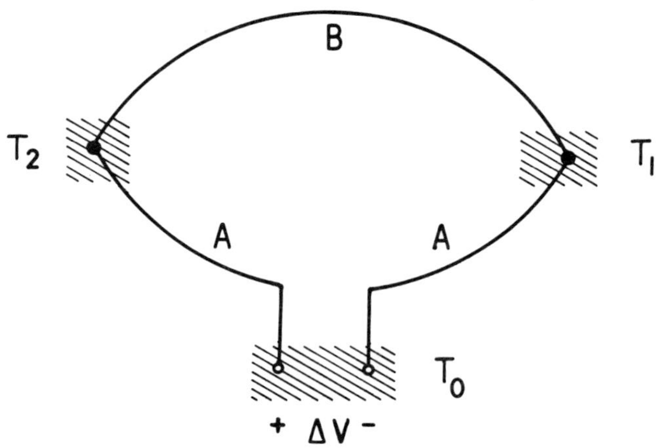

FIG. 1. Thermoelectric circuit from materials A and B. The shaded regions are isothermal regions having the indicated temperatures.

the complexity of the topic precludes any mention or discussion of specific effects.

Turning from the preceding macroscopic viewpoint to a microscopic viewpoint results in thermoelectric effects being associated with deviations from equilibrium distributions of charge carriers. These deviations are known to occur in two ways. Thermal and potential gradients directly affect the distributions and produce "diffusion contributions" to thermoelectric effects. In solids the presence of thermal gradients also causes the phonon distribution to differ from its equilibrium value. This difference means that the phonons contribute to the heat current and that interactions between nonequilibrium phonons and charge carriers alter the distributions of the latter. This alteration produces additional thermoelectric contributions, called "phonon-drag contributions." Both types of contributions have been well documented in semiconductors, semimetals, and metals.

The temperature dependence of thermoelectric effects makes a concise summary of magnitudes impossible. For many materials even the sign of an effect changes with temperature. However, the general pattern of significant, low-temperature phonon-drag contributions which decrease at higher temperatures permits a crude comparison of thermopowers near room temperature. Metals have values of either sign ranging from ~0.1 to nearly 40 μV/K. Semiconductors produce values of either sign from ~1 to 10^3 μV/K. The essential difference in the physics of these materials is the thermal statistics: Semiconductors require Maxwell–Boltzmann statistics while metals require Fermi–Dirac statistics.

One unique situation deserves consideration. The superconducting state, which only occurs in metals at low temperature, is an instance where the thermopower is identically zero.

The utility of thermoelectric effects can be divided into two standard categories, fundamental and applied. A free-electron model predicts that $S(T)$ is related to the sign of the dominant charge carrier, but $S(T)$ is also sensitive to the details of scattering mechanisms. The latter dependence is more important and causes a number of errors in sign prediction. Thus, from a fundamental perspective, thermoelectric and thermomagnetic effects provide complicated but sensitive probes for studying scattering mechanisms, band structure, and Fermi-surface effects. From an applied viewpoint, thermoelectric effects provide a form of thermometry (thermocouples), a mechanism for converting heat energy into electrical energy, and a reversible method for heating or cooling. Metals are generally used for the thermocouple application and semiconductors for the other two applications.

See also ELECTRON ENERGY STATES IN SOLIDS AND LIQUIDS; PHONONS.

BIBLIOGRAPHY

F. J. Blatt, P. A. Schroeder, C. L. Foiles, and D. Greig, *Thermoelectric Power of Metals and Alloys*. Plenum, New York, 1976. (Intermediate level review.)

C. A. Domenicali, *Rev. Mod. Phys.* **26,** 237 (1954). (Excellent treatment from perspective of irreversible thermodynamics, advanced.)

T. C. Harman and J. M. Honig, *Thermoelectric and Thermomagnetic Effects and Applications*. McGraw-Hill, New York, 1967. (Advanced and detailed treatment stressing energy conversion processes.)

R. R. Heikes and R. W. Ure, Jr., *Thermoelectricity: Science and Engineering*. Interscience Publishers, New York, 1961. (Intermediate to advanced treatment emphasizing semiconductors and applications.)

D. K. C. MacDonald, *Thermoelectricity: An Introduction to the Principles*. Wiley, New York, 1962. (Elementary.)

A. H. Wilson, *The Theory of Metals*, 2nd ed. Cambridge University Press, Cambridge, 1953 (Immediate treatment for metals, semimetals, and semiconductors.)

Thermoluminescence

Paul W. Levy

The basic thermoluminescence phenomenology is well illustrated by the following example. It applies to a large fraction of the transparent crystals which can be prepared with one or more different kinds of impurities and to a much smaller fraction of the so-called "pure" crystals. If these crystals are heated soon after they are grown they will not emit light. However, if they are subsequently exposed to a few rad of ionizing radiation and then reheated, light will be emitted. Thus, thermoluminescence is the light which is emitted by a substance when it is heated and which can be attributed to previous exposure to ionizing radiation. Natural quartz, calcite, i.e., $CaCO_3$:Mn, KCl:Tl, Al_2O_3:Cr, and ZnS:Cu are examples of materials in which the thermoluminescence is related to the presence of impurities. Many of the alkali halides exhibit thermoluminescence which is associated with the presence of defects and color centers.

The basic physical processes involved in thermoluminescence are reasonably well understood and are described in several books.[1-3] In the most easily understood case, a thermoluminescent solid must contain two different types of atomic or molecular defects or impurities. One type functions as an electron trap or (electronic) hole trap. The second essential defect or impurity is the center responsible for the emission of light. Consider first the role of the charge-trapping centers. When the crystal is exposed to ionizing radiation, electron–hole ionization pairs are formed. Usually a very large fraction of the ionization-induced charges recombine in milliseconds or less at a variety of defects and impurities termed recombination centers. Often the recombination process is accompanied by light emission, which is classified as fluorescence, phosphorescence, or, collectively, radioluminescence. Important for the thermoluminescence process are the remaining electrons and holes, not undergoing immediate recombination, which are captured (trapped) by defects or impurities, usually referred to as electron, hole, or charge traps. Consider next the centers which emit light or thermoluminescence when the crystal is subsequently heated. The most easily visualized luminescent center, and one of the most commonly occurring, is also a hole trap. When previously irradiated crystals are heated, thermally released electrons "wander about" until they interact with a trapped hole on an emission center. As part of the ensuing electron–hole-recombination, or charge-neutralization, process the electron makes one or more electronic transitions which are accompanied by photon emission. Such centers can be called recombination luminescent centers. A second type of luminescent center is an electron trap which contains a ground state and at least one or more energetic states. When a thermally released electron encounters such a center, it is initially trapped in one of the upper levels. Usually the lifetime of the excited state is very short and the trapped electron quickly makes a transition to the ground state with the emission of light. When the crystal reaches a higher temperature, the trapped electron may be thermally released, perhaps to enter into a second light emission process. However, at higher temperatures it is more likely that the trapped electron will recombine with a thermally released hole, with or without the emission of light. When functioning as an emission center this type of defect may be called a transition luminescent center. Thus, a single type of defect or impurity may be an electron trap, a luminescent center, a recombination center, or some combination of these.

To an appreciable extent there is a duality or reciprocal character associated with the sign of the charge which interacts with the various centers. In the preceding paragraphs the words electron and hole can usually be interchanged without violating physical principles. However, a few of the processes occasioned by this charge interchange procedure would be encountered only rarely.

The defects and impurities responsible for luminescence and thermoluminescence must be able to introduce localized electronic states into the crystal lattice.[4,5] Qualitatively such centers are easily understood. If a normal lattice atom is replaced substitutionally by an atom whose valence is either more or less than the replaced atom, a local charge unbalance is created. Substitutional impurities lacking an electron would tend to trap ionization electrons and impurities containing an excess electron (lacking a hole) would tend to lose an electron (capture a hole). Equally important are lattice defects which create local charge unbalance. They also trap whatever ionization charge restores local charge neutrality. For example, the defect formed by a missing negative ion in an alkali halide, such as NaCl or KBr, tends to trap an electron. The well-known F center is formed when an ionization electron is trapped on a Cl^- ion vacancy in an alkali halide, or on similar negative-ion vacancies in other types of crystals. F centers may be, in the proper circumstances, electron traps (to form dielectron or F' centers), recombination centers, and both transition- and recombination-type luminescent centers.

Another type of center important for thermoluminescence processes is the substitutional impurity having the same valence as the atom, or molecular ion, it replaces. In these cases, the atom functions as an electron trap, as a hole trap, or in both ways. The properties of such centers depend on the detailed electronic interactions between the impurity atoms and the neighboring atoms.[5]

THERMOLUMINESCENCE KINETICS

For simplicity consider the thermoluminescence originating in a solid containing a single type of charge-trapping center and a single type of luminescent center. During irradiation a fraction of the traps will acquire charges. When the sample is heated, the thermally released charges will interact with the luminescent centers and light will be emitted, either by charge capture or by recombination. Let N_0 be the trapping-center concentration and let n_0 be the concentration of trapped charges when the irradiation is terminated. Next, assume that all or a constant fraction of the charges released when the crystal is heated find their way to luminescent centers. Also, assume that each charge interacting with an emission center causes a photon to be emitted. This is equivalent to assuming that nonradiative transitions involving the luminescent center are negligible. Under these conditions, which appear to apply in numerous cases, the light intensity is proportional to the number of charges untrapped per unit time. In turn, the number of charges released per unit time is equal to the trapped-charge population multiplied by the probability that each charge will escape. The escape probability is given by the usual term $s \exp(-E/kT)$, where s is the pre-exponential factor or "attempt-to-escape" frequency, E is the activation energy for thermal untrapping, k is the Boltzmann constant, and T is the temperature in degrees kelvin.

Then the intensity I at any time T is given by

$$I = -\frac{dn}{dt} = (\text{const.})\, nse^{-E/kT}, \tag{1}$$

where n is the trapped charge concentration at time t.

In the usual thermoluminescence measurement the sample is irradiated at a temperature low enough to ensure that the term $\exp(-E/kT)$ is negligible and light emission is undetectable. Subsequently, the sample temperature is increased in a controlled manner as the light emission is recorded. Usually the temperature is increased linearly. There are two reasons for this. First, a precise and reproducible linear temperature increase is the easiest to achieve experimentally. Secondly, a linear temperature program leads to a simple solution of Eq. (1). In this case temperature and time are related by the simple expression $T = T_0 + \beta t$ or $dT = \beta dt$, where β is the heating rate in degrees per unit time. Using this relation one obtains the following expression for the luminescent emission as a function of time, or temperature, for a single type of trapping center:

$$I = n_0 s e^{-E/kT} \exp\left(-\int_0^T \frac{s}{\beta} e^{-E/kT}\, dT\right). \tag{2}$$

The uppermost curve in Fig. 1 is a plot of this expression computed with typical values of the various parameters.

The physical situation leading to Eqs. (1) and (2) occurs when retrapping is negligible—that is, when all but a negligible number of thermally released charges find their way to luminescent centers without undergoing retrapping in a trapping center. Most commonly this occurs when the concentration of luminescent centers greatly exceeds the concentration of trapping centers, or, if the concentrations are approximately equal, the charge-capture cross section of the luminescent center is much larger than the corresponding cross section of the trapping center.

When retrapping is nonnegligible, Eq. (1) must be replaced by one that includes retrapping. This may be done to varying degrees of approximation, some of them leading to equations which have not been solved analytically. The only approximation that has been widely used is based on the following situation.[1] For each charge trapped in a trapping center there is an opposite-sign charge trapped in a recombination or luminescent center. Light is emitted when thermally released electrons find their way, after being retrapped one or more times, to the trapped holes, or thermally released holes interact with trapped electrons. In these cases, the emission is controlled entirely by the competition between thermal untrapping and retrapping and recombination. This approximation leads to the equation

$$I = -\frac{dn}{dt} = (\text{const.})\, n^2 se^{-E/kT}. \tag{3}$$

This equation differs from (1) only in that the n term is replaced by an n^2 term. In this case the dependence of thermoluminescence intensity on temperature obtained with the usual linear temperature rise is given by

$$I = n_0^2 s \exp^{-E/kT} \times N_0 \left[1 + \frac{n_0}{N_0} \int_0^T \frac{s}{\beta} \exp^{-E/kT}\, dT\right]^{-2}. \tag{4}$$

All of the parameters in Eq. (4) have been defined above, including N_0 which is the charge trap concentration. This equation has been used to compute all of the curves in Fig. 1 except the uppermost one, which was computed from Eq. (2).

Figure 1 illustrates many of the principal features of thermoluminescence. Equation (1) and its solution Eq. (2), used to compute the uppermost curve, describe what is usually called *first-order thermoluminescence kinetics*. It is termed first-order because the charge concentration term n appears with the exponent 1 in Eq. (1). Likewise Eq. (3) and its solution Eq. (4), used to compute other curves in Fig. 1, describe *second-order thermoluminescence kinetics*: The concentration term appears with exponent 2 in Eq. (3).

The total area under the first-order kinetic curve is proportional to the initial trapped charge concentration n_0. It

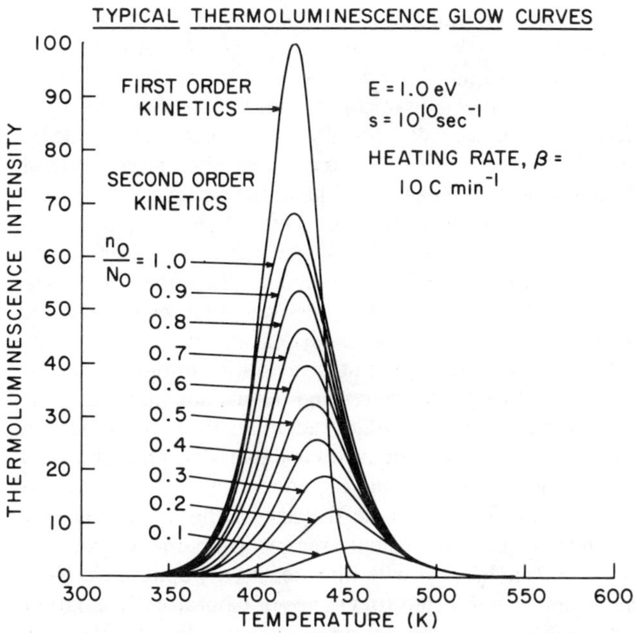

FIG. 1. First- and second-order glow curves containing a single glow peak and computed with "typical" parameters. The first-order curve shape and peak temperature are independent of the initial trap charge concentration. The second-order curve shape and peak temperature depend on the fraction of traps initially populated.

is implicit in the derivation of (1) that the initial charge n_0 is less than the trap concentration N_0. Most importantly, the shape of the first-order curve is independent of n_0, i.e., the shape does not depend on the exposure used to create the initial trapped-charge concentration.

In contrast, the second-order curves depend on the ratio n_0/N_0, which in turn depends on the exposure used to establish the initial trapped-charge population. Also, note that the temperature of the maximum depends on the ratio n_0/N_0. This dependence on n_0/N_0, i.e., the exposure, can be used to differentiate between the two types of kinetics.

The curve describing the emission from a single type of trap is usually referred to as a *glow peak*. A curve containing one or more glow peaks is usually called a *glow curve*. The individual glow peaks are very often identified by giving the temperature of the peak maximum. However, in both first- and second-order kinetics, the peak temperature depends on the heating rate and, as described above, in second-order kinetics the peak temperature also depends on the fraction of traps populated, i.e., the ratio n_0/N_0. Thus, the specification of glow peaks by glow-peak temperatures, while widely used and often useful, must be regarded with continuous attention to the fact that this parameter is a function of heating rate.

KINETIC STUDIES

Much of the research on thermoluminescent materials, in recent decades, has been devoted to determinations of the kinetics and parameters, such as *E*, *s*, etc., which apply to each glow peak. Equations (2) and (4) are not readily converted to forms which are convenient for glow peak analyses. Consequently, a number of analytical procedures have been developed to determine the kinetics and parameters. Most of these are summarized by Chen.[6] However, in recent years computerized numerical procedures have been increasingly more successful. They may employ numerical solutions of the differential equations, e.g., Eqs. (1) and (3), or procedures for fitting the glow-curve equations, e.g., Eqs. (2) and (4), to experimental data. In both cases one starts with an initial estimate of the parameters, often obtained by employing one or more analytical procedures. Next, a computed curve is compared with the data and, by using any of a number of methods, "improved" estimates are obtained. This procedure is repeated until a "best fit" is obtained. Details on one of these procedures and an apparently very successful demonstration of its usefulness have been given by Mattern *et al.*[7]

THERMOLUMINESCENCE MEASUREMENTS

Until recently almost all thermoluminescence measurements were made with apparatus in which a fraction of the light emitted by the sample impinged directly on the photodetector. The spectral-distribution information contained in the emitted light was ignored. Occasionally, emission spectra were obtained from efficient emitters, such as LiF TLD-100 dosimeter material, over the entire range of glow-curve temperatures, or, in a few very favorable cases, the spectra were recorded over a single glow peak. With very few exceptions, the usual thermoluminescence data consisted solely of a curve of photodetector current vs sample temperature or heating time. When the photodetector current is not proportional to the number of light-emitting processes occurring per unit time, such data are not suitable for

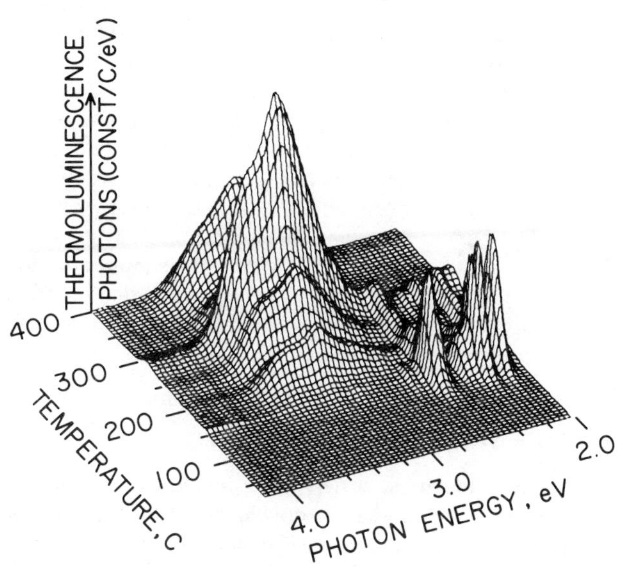

FIG. 2. A "three-dimensional" plot showing thermoluminescence intensity vs. photon energy and sample temperature. The measurements were made on a piece of lavender-colored fluorite, from Korea, after a 10^6 R- ^{60}Co gamma-ray irradiation. This plot shows many of the various combinations of emission bands, line spectra, and glow peaks at various temperatures that can be observed during thermoluminescence measurements.

kinetic determinations. For example, if the emission spectrum changes with temperature, especially in a region where the photodectector sensitivity changes rapidly with wavelength curves of phototube current vs. temperature cannot be expected to provide reliable kinetic analyses. However, for many applications, including thermoluminescence dosimetry, dating of archaeological samples, and the authentication of fine art objects, photodetector current curves are quite adequate.

It is now possible to determine the spectral distribution of the thermoluminescent emission from a large number of materials at closely spaced temperature intervals. There are two reasons for this. In many samples the thermoluminescence intensity can be greatly enhanced by sufficiently large exposure to x rays, gamma rays, and other types of radiation. The second, and most important, is the development of highly sensitive apparatus for making spectral-distribution measurements as the samples are heated. This equipment utilizes large-aperture optics, and the data are recorded digitally with a small computer. In the most favorable cases a 1024-point spectrum can be obtained at approximately 0.2°C temperature intervals. A typical spectrum obtained from such equipment is shown in Fig. 2. In addition, inasmuch as the spectral sensitivity of this equipment can be determined, the recorded data can be converted to curves of absolute emission intensity (or a quantity proportional to the absolute intensity) vs. wavelength or photon energy. From such data glow curves can be obtained for each observed emission band that are eminently suitable for kinetic analysis.

A large fraction of the individual features that are observed in thermoluminescence measurements are illustrated in Fig. 2: (1) The emission spectra contain both broad continuous bands which are characteristic of defects and certain impurities, e.g., in the 3.3-eV region, and sharp-line spectra which are produced by rare-earth elements, e.g., at approximately 2.2, 2.3, and 2.6 eV. Also, there are regions where the sharp-line spectra are superimposed on continuous spectra. (2) Some emission bands occur in certain temperature ranges but not in others. (3) The glow curves corresponding to some of the emission bands differ from those obtained from other bands. (4) When different emission bands produce glow curves in the same temperature range the individual glow peaks often have precisely the same shape. In other words, the glow curves reflect the charge untrapping process and are independent of the type of emission center.

APPLICATIONS

Currently, the most widely used application of thermoluminescence is in radiation dosimetry.[2] Ideally, for dosimetry the thermoluminescence signal, i.e., the glow-peak height or area, should increase linearly with the total exposure. In principle, this can be achieved with a material exhibiting first-order kinetics, as described above, in which the trapped-charged population increases in direct proportion to the total ionization. However, many of the available thermoluminescence dosimeters appear to be subject to a number of difficulties. The response depends on the nature of the radiation, it may be altered by exposure to light or may decay with time, etc. The signal-vs.-dose curve may be nonlinear or display "supralinearity." In the latter case the response curve may increase linearly from "zero" dose to a given dose and, above this value, the slope of the curve increases continuously until a second linear region appears. At the moment, this phenomenon is not understood, although some "explanations" have been suggested.

A second current application is the dating and/or authentication of fine art and archaeological objects. Pottery shards are the best example in this category. When the pottery was originally fired, the natural thermoluminescence was removed from any thermoluminescent materials present; usually these include appreciable quantities of quartz crystallites. After firing, the quartz and other mineral components are continuously subjected to radiations from the potassium, uranium, and thorium impurities and their daughters contained in the object and, in some instances, surrounding material such as soil. To determine the age of the object the thermoluminescence of a small sample from the shard, e.g., 1–50 mg, is measured with high-sensitivity apparatus. Next, the sample is subjected to a series of laboratory irradiations and the intensity-vs.-dose, or sensitivity, characteristics are determined. From the sensitivity curve, it is a simple matter to determine the dose received by the object since it was last heated. Last, the natural radioactive impurity content of the object and, if possible, the immediate surroundings are determined by chemical or activation analysis, counting, or other means. Alternatively, in some instances the exposure rate at the point where the object was found may be determined with dosimeters. Once the radioactive impurity content and the radiation from the surroundings are known the annual dose may be computed. The age is the total exposure acquired by the object since it was fired divided by the annual exposure. At present, accurate age determinations are beset by numerous practical difficulties. However, in many instances the age may be determined accurately enough to determine if an object is authentic. For example, if an object appears to be Mayan, and is purported to have been excavated recently, but exhibits thermoluminescence appropriate for an age of 30 years, it is clearly not authentic.

See also COLOR CENTERS; CRYSTAL DEFECTS; ELECTRON ENERGY STATES IN SOLIDS AND LIQUIDS; RADIATION DAMAGE IN SOLIDS.

REFERENCES

1. G. F. J. Garlick, *Luminescent Materials*. Oxford University Press, Oxford, 1949.
2. J. R. Cameron, N. Suntharalingam, and G. N. Kenney, *Thermoluminescent Dosimetry*. University of Wisconsin Press, Madison, 1968.
3. D. Curie, *Luminescence in Crystals*, translated by G. F. J. Garlick. Methuen and Co., London, 1963.
4. J. J. Markham, *F-Centers in Alkali Halides*. Academic Press, New York, 1966.
5. A. M. Stoneham, *Theory of Defects in Solids: Electronic Structure of Defects in Insulators and Semiconductors*. Oxford University Press, Oxford, 1975.
6. R. Chen. "On the Calculation of Activation Energies and Frequency Factors from Glow Curves," *J. Appl. Phys.* **40**, 570–585 (1969).

7. P. L. Mattern, K. Lengweiler, and P. W. Levy, "Effects of Co^{60} Gamma-ray Irradiation on the Optical Properties of Natural and Synthetic Quartz," *Rad. Eff.* **26,** 237–248 (1975).

Thin Films

Myron Strongin and *D. L. Miller*

INTRODUCTION

In the trivial case, thin films can be used to make very small amounts of a given substance with essentially the same properties as larger samples of the bulk material. The real fascination with the physics and metallurgy of films, however, involves physical phenomena and materials that are unique to films. In this very brief survey we intend to discuss some of the more interesting special properties of films, and mention a few of their important technological applications.

PREPARATION

Films can be prepared in a variety of ways. A traditional way is by electroplating onto metallic substrates. More recent techniques involve evaporation of materials in an ultrahigh-vacuum environment. The evaporant heating source can be either Ohmic or a high-current electron beam. Other common methods involve rf or dc sputtering, which are especially useful techniques when high film–substrate adhesion is required; and chemical vapor deposition, which allows higher deposition rates than the usual sputtering or evaporation techniques. Finally, we mention that anodization provides a way of making oxide films on metal surfaces.

CHARACTERIZATION OF FILMS

Films can be characterized by conventional metallurgical techniques, which involve optical and electron microscopy, and x-ray and electron diffraction. More recently developed surface-sensitive techniques, such as Auger electron spectroscopy [1], secondary-ion mass spectroscopy, and the spectroscopy of photoemitted electrons, can be especially useful for very thin films. Also useful is the technique of Rutherford backscattering [2]. In backscattering experiments, high-energy particles are scattered from the film and an analysis of the energy of the backscattered particles allows a determination of the elements in the sample. This technique, as well as the surface-sensitive spectroscopies, has been developed to give a profile of composition as a function of depth into a film.

STRUCTURAL PROPERTIES OF FILMS

The structural properties of films are intimately connected with the properties of the substrate, deposition rate, gaseous impurities, and substrate temperature. By varying these parameters, unique materials and structures have been stabilized in films.

Probably the most dramatic structural changes in films are those observed in amorphous or microcrystalline metals. These materials are made by depositing metals onto substrates that are cooled to the temperature of liquid helium, which is 4.2 K above absolute zero [3]. At these temperatures the metal atoms lack the thermal energy necessary for extensive diffusion, and extremely disordered structures can be obtained. In the case of Bi deposited at these temperatures it appears that the Bi actually is in an amorphous phase, since at some temperature near 30 K it rapidly crystallizes into its bulk structure [4]. With other materials, grains gradually grow as the temperature is raised, and there is no sharp transformation. These films are probably composed of small grains having the same structure as the bulk material. Finally, it is worth mentioning that many phases can be formed in films that are either difficult or impossible to make by bulk metallurgical techniques; this includes amorphous alloys of semiconductors and metals that at present can only be made in their films.

Another problem involved with film structure is the study of the interaction of metal atoms on various substrates and especially how this interaction affects the structural properties of the film. This constitutes the study of epitaxy [5] and the related problem of film nucleation [6]. In most cases the initial atoms hitting a substrate form small islands of metal about 20–30 Å in diameter. These islands then grow and at some stage they touch and begin to coalesce into larger islands. As the large islands grow together, channels

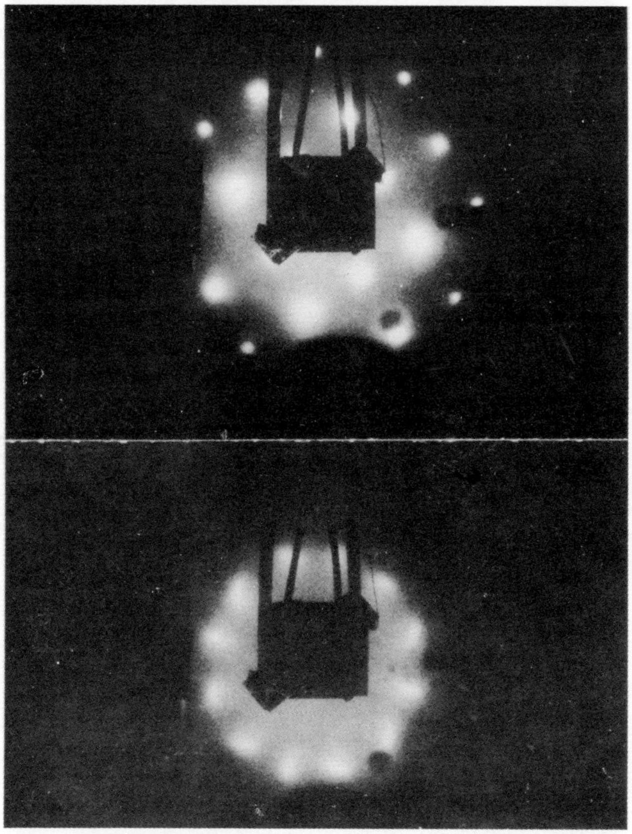

FIG. 1. The LEED pattern from a PbTe single-crystal substrate (top) and an epitaxial Pb film (bottom) grown on this substrate by evaporation of Pb. The Pb film is less than 15 Å thick. The energy of the diffracted electron beam is 94 eV.

are formed that are then covered with newly deposited metal until a uniform film is finally formed. These studies have traditionally been made by using an electron microscope [7]. In recent years low-energy electron diffraction (LEED) has been used to study monolayers of metals on various substrates. This technique has shown that it is possible in certain cases to achieve uniform crystalline monolayers or near monolayers of a metal on a particular substrate. That the diffracted electrons have low energy causes them to interact strongly with the specimen. In some ways this makes a cursory interpretation difficult; however, detailed analysis of the intensity of LEED patterns can give detailed information about atomic positions on surfaces in some cases. A LEED pattern of an epitaxial thin metal layer is shown in Fig. 1.

SOME UNIQUE PHYSICAL PHENOMENA IN THIN FILMS

Besides the convenience of using films in certain applications, there are phenomena that are best studied in, and are even unique to, ultrathin films. Films, of course, provide a way of studying phenomena in approximately two dimensions. We say approximately since the effect of the substrate can be nontrivial. Both magnetic phase transitions and superconducting phase transitions have been studied in this quasi-two-dimensional case. Problems of nucleation and the film–substrate interaction are crucial in the interpretation of these results. For example, results of Ni films and iron–nickel alloys show that the magnetization and Curie temperature stay constant in films down to about 50 Å thick. Below this value various investigators obtain different results; however, the films are granular in this regime and there is no longer an ideal film geometry. In a similar way, experiments where Ni is grown epitaxially on Cu show no magnetic behavior below 7 Å [8], although it is not clear what effect the copper substrate has. A similar situation is found in the investigation of superconducting materials. Below some critical temperature superconducting metals go into an ordered state, where they have zero electrical resistance. As superconducting films approach thicknesses below about 50 Å the transition temperature of some metals show a large decrease [9]. It is likely that the interaction with the substrate plays a role in the degradation of these superconducting properties. It should also be mentioned that phase transitions in two dimensions have been studied in rare gas films and in films of superfluid helium. The recent text by Dash [10] serves as an excellent introduction to this area.

Another class of problems that are of interest involves changes in the transport properties of metal films due to the film boundaries. If the films can be made thin enough so that the distance between internal electron scatterings becomes greater than the film thickness, it is clear that boundary scattering will greatly influence the transport properties [11], and this field has been studied experimentally and theoretically for many years. The problem of how electrons scatter from film boundaries, and the effect of this scattering on transport properties, has also been studied by using magnetic fields, which bend the electron orbits away from the surface, causing a decrease in resistance. This effect has been observed in wires and films.

Another class of problems, somewhat related to the foregoing, involves changes of the transport properties of films due to quantization of electron orbits in the direction perpendicular to the plane of the film. This effect results from the wave nature of the electrons in the film and the fact that the electron waves must fit within the small dimension of the film in discrete units of the electron wavelength perpendicular to the plane of the film. Clearly, if the film is much thicker than the electron wavelength this effect becomes negligible. In the case where the thickness is only a few wavelengths, however, this effect becomes significant. Since the characteristic wavelength for electrons in a metal is only on the order of a few angstroms, or an interatomic spacing, this effect is difficult to observe. In Bi the situation is somewhat different since the films grow on mica in such a way that the electron wavelength perpendicular to the film is on the order of a few hundred angstroms. Films can be made conveniently in this range and the foregoing effects of "size quantization" have been studied in this regime.

In this section we have tried to give some inkling of basic physical phenomena that can be studied with thin films. Of course, a discussion of many unique phenomena in films has been left out, such as the high-field behavior of superconducting films, the optical properties of films, the use of films for studying materials by the technique of electron tunneling, the Josephson effect in superconducting films, the photovoltaic effect in films, adsorption and catalysis on films, amorphous alloys, magnetic films, and many other fascinating phenomena.

UNIQUE MATERIALS USING FILMS

Perhaps the most important modern materials that involve thin film technology are the new multilayer structures involved in integrated circuits [12], where hundreds of components are placed on a silicon chip whose size is of the order of 2.5 mm × 2.5 mm. In some sense this is a unique material that is made by using films, and in fact these devices have revolutionized modern electronic technology. Nevertheless, nearly all the properties of these devices can be understood in terms of the behavior of the bulk materials from which they are fabricated. The very clever thin-film techniques used to make integrated circuits are primarily concerned with reducing device size.

A further important technological use of thin films is found in photovoltaic devices. Again the phenomenon itself is not unique to films; it appears, however, that the economic use of photovoltaic devices could well depend on film technology. The present approaches to photovoltaic device manufacture involve silicon production technology and the technology of producing III–V and II–VI compounds [13]. For GaAs the absorption edge is much sharper than for silicon, and whereas a wafer thickness greater than 10^{-3} cm is necessary for reasonable absorption efficiency with silicon, only about 10^{-4} cm is necessary for GaAs. Hence even though silicon is an intrinsically cheaper material it appears that polycrystalline thin-film GaAs detectors can compete with silicon devices. These considerations are too detailed to be dealt with here, but they essentially involve the conditions on the grain size in these polycrystalline films in relation to their optical absorption lengths and charge-carrier diffusion lengths. Amorphous silicon films have also shown promise for photovoltaic devices when hydrogen is incorporated in the films [14].

In concluding this section we briefly discuss materials that remain superconducting to the highest temperatures, and that promise to be extremely important technologically. These materials have the A-15 structure, and Nb_3Ge, which has the highest known superconducting transition temperature at present, shows an onset of zero-resistance behavior near 23 K and, in some reports, as high as 25 K. Nb_3Sn, which is the easiest niobium-based A-15 superconductor to make, has a transition temperature T_c of about 18 K. An interesting feature of these materials is that higher T_c's appear in less stable materials. While Nb_3Sn can be made by various bulk metallurgical techniques, at present Nb_3Ge has only been prepared by deposition from the vapor phase onto substrates held at a high temperature [15]. Various deposition techniques, such as electron-beam codeposition, sputtering, and chemical vapor deposition, have been used successfully. A problem of further interest in this area is the question whether high-T_c Nb_3Si can be made and stabilized by various deposition techniques. There is speculation that if this material could be formed metastably with the proper A-15 structure, and close to a 3:1 niobium-to-silicon ratio, then a T_c even higher than 23 K could be achieved.

SUMMARY

We have tried to give some idea of unique phenomena that occur in films from the point of view of a solid state physicist. In conclusion we emphasize that films form the basis for many established and developing technologies. In the simplest case, metal layers have always been important for optical coatings, and even as protective coatings for various materials. In the most sophisticated cases, films and their interaction on silicon surfaces form the basis of modern electronic technology. Furthermore, films of silicon, GaAs, and composites of these materials promise to lead to practical photovoltaic devices. It is fair to say that besides being a field of great interest in modern physics, the study of thin films also has contributed vital elements to the most sophisticated commercial technologies.

See also ADSORPTION; AUGER EFFECT; CATALYSIS; ELECTRON AND ION IMPACT PHENOMENA; JOSEPHSON EFFECTS; LOW-ENERGY ELECTRON DIFFRACTION; PHOTOELECTRON SPECTROSCOPY; SURFACES AND INTERFACES; TUNNELING.

REFERENCES

1. R. L. Park, "Chemical Analysis of Surfaces," in *Surface Physics of Materials* (J. M. Blakely, ed.), pp. 377-418. Academic Press, New York, 1975.
2. J. F. Ziegler, "Materials Analysis by Nuclear Backscattering," in *New Uses of Ion Accelerators* (J. F. Ziegler, ed.), pp.75-103. Plenum, New York, 1975.
3. N. V. Zavaritsky, *Dokl. Akad. Nauk SSSR* **86,** 501 (1952); W. Buckel and R. Hilsch, *Z. Phys.* **132,** 420 (1952). Earlier work by Shalnikov on amorphous films dates to the late 1930s.
4. R. Hilsch, *Noncrystalline solids* (V. D. Freschette, ed.). Wiley, New York, 1958.
5. *Epitaxial Growth* (J. W. Matthews, ed.). Academic, New York, 1975.
6. J. P. Hirth and K. L. Moazed, in *Physics of Thin Films* (G. Hass and R. E. Thein, eds.), Vol. 4, pp. 97-136. Academic Press, New York, 1967.
7. A detailed review of this area is provided by the articles by C. A. Neugebauer and S. Mader in *Handbook of Thin Film Technology* (L. I. Maissel and R. Glang, eds.), pp. 8-3 to 8-44. McGraw-Hill, New York, 1970.
8. C. A. Neugebauer, *Z. Angew. Phys.* **14,** 182 (1962).
9. Myron Strongin, R. S. Thompson, O. F. Kammerer, and J. E. Crow, *Phys. Rev.* **B1,** 1078 (1970); D. L. Miller, Myron Strongin, O. F. Kammerer, and B. G. Streetman, *Phys. Rev.* **B13,** 4834 (1976).
10. J. G. Dash, *Films on Solid Surfaces.* Academic Press, New York, 1975.
11. K. Fuchs, *Proc. Cambridge Philos. Soc.* **34,** 100 (1938); E. H. Sondheimer, *Adv. Phys.* **1,** 1 (1952).
12. For a review of the use of thin films in integrated circuits see I. Blech, Harry Sello, and L. V. Gregov, in *Handbook of Thin Film Technology* (L. I. Maissel and R. Glang, eds.), pp. 23-1 to 23-22. McGraw-Hill, New York, 1970.
13. A review of the recent outlook for compound semiconductor photovoltaics is given by J. M. Woodall and H. J. Hovel in *Proc. of the Symposium on Films for Solar Energy, J. Vac. Sci. Technol.* **12,** 1000 (1975).
14. For a review of amorphous semiconductors see David Adler, *Scientific American* **236 (5),** 36, 1977.
15. J. R. Gavaler, *Appl. Phys. Lett.* **23,** 480 (1973); L. R. Testardi, *Proc. IEEE Trans. on Magnetics* **MAG-11,** 197 (1975).

Transistors

R. M. Ryder

A transistor is an electronic amplifying device, using controlled electron streams inside semiconducting materials. The transistor and the electronics derived from it have truly revolutionary importance for human society, even greater than the Industrial Revolution of the 18th–19th century. Through improved communication, electronic control systems, and computers, the "Transistor Revolution" has multiplied the power of man's *mind* many times, just as the Industrial Revolution, based on the steam engine, multiplied the power of man's *muscles* many times. For a review and forecast of the extensive social consequences, the reader is referred to the magazine *Science* for March 18, 1977.

1. PRINCIPLES OF TRANSISTOR OPERATION

Transistors operate by means of novel physical principles, discovered through study of the quantum mechanics of solids. A controlled stream of electronic charges, either negative conduction electrons or positive holes, can travel within a solid semiconductor, usually silicon. "Bipolar" transistors are so called because both kinds of charge carriers take part in the action, but in the "unipolar" transistors only one sign of charge carrier is active. The holes are vacancies in the structure of valence electrons which holds the solid material together. Both holes and electrons respond to an applied electric field by moving with an average velocity characteristic of the material. For small fields, this velocity is proportional to the field, with the constant of proportionality being the mobility of the holes or electrons.

In a typical *npn* bipolar transistor, a thin region of "*p*" type material, conducting by (positive) holes, is constructed between two *n* regions which conduct by (negative) electrons. The electric potential on the "base" (the *p* region) controls the number of charges (electrons) which pass through from the more highly doped *n* region (the emitter) to the less highly doped *n* region (the collector). In a *pnp* transistor, all the polarities are reversed, that is, a stream of holes passes from emitter *p* region to collector *p* region in response to the potential on the thin *n*-type base region. In either case, under suitable conditions the electron stream at the collector can deliver much more power to an external load than is required to activate the base, so that the external load has impressed upon it an amplified replica of whatever signal waveform was applied to the base. Power amplification by a factor of 100 or more is typical for low-frequency transistors.

When the base region is sufficiently thin (a fraction of a micrometer), the speed of the transistor action can be very high, sufficient to amplify or control microwave frequencies up to several times 10^9 Hz. Still higher frequencies, up to more than 10^{11} Hz, can be generated and controlled by semiconductor diodes (two-electrode devices) of various types. Thus the entire presently useful radiofrequency spectrum can be advantageously instrumented with solid-state devices, which have completely superseded vacuum tubes for all low-power applications of receiving type, such as ordinary radio receivers. Power transistors of modest capability, up to a few hundred watts, can also be used for communications frequencies. But for very high-power applications of transmitting type, such as radio broadcasting, vacuum tubes are still necessary to produce the desired tens or hundreds of kilowatts of power.

Field-effect transistors (FETs) are so called because the stream of charge carriers is controlled by a different mechanism, that of induction by the electric field of the control electrode. In an FET, the flow of carriers from a "source" region is attracted or inhibited by the electric field from a nearby "gate" electrode; the third electrode, which terminates the stream of carriers in the triode (three-electrode) device, is called the "drain." For FETs also, amplification results when the amplified replica (output signal) at the drain contains more power than the input signal applied to the gate. FETs are called unipolar because only one of the two types of charge carrier (electron or hole) is involved in the action; the control electrode (gate) is an external metal electrode, not an internal region of opposite conductivity type, like the base of the bipolar unit. Note that the active stream of carriers in the bipolar unit actually passes *through* the interior of the base material, whereas such is *not* the case for the FET, where the gate electrode is an exterior metal.

Depending on just how the gate electrode affects the carrier stream, FETs are further subdivided into two classes. In an insulated gate FET (IGFET), the gate metal is insulated by a thin film between the metal electrode and the semiconductor, so that the flowing charge stream attracted (or repelled) by the gate resides in a thin subsurface layer of the semiconductor, called a "channel." The IGFET is also called a MOSFET (metal-oxide-semiconductor-field-effect-transistor), when the insulating film is (as usual) an oxide. In a "junction FET" (JFET), on the other hand, the gate metal is a rectifying junction contact to the semiconductor surface, which must be reverse-biased to avoid drawing current, and which therefore *repels* carriers from a depletion region channel just below the gate, and thereby also controls the current flowing from source to drain through the channel.

For most communications applications involving amplification, such as radio and telephone amplifiers, bipolar transistors are used because they have greater gain and power capability. JFETs have have some use as low-noise, low-distortion preamplifiers for TV and FM sets; also JFETs in gallium arsenide material (GaAs FETs), which use Schottky-barrier rectifying electrodes as gates, are currently coming into use as microwave amplifiers at frequencies above 2 GHz, up to about 10 GHz (in 1976). IGFETs are not much used in transmission applications because of their high electrical noise, but they are very popular for integrated circuits, especially for logic and computer applications. The reason is that if (as usual) utmost speed is not required and noise is not critical, then the IGFET fabrication process often requires fewer and simpler steps than the bipolars. A typical application is to the handheld calculator, which can tolerate very much slower operation than is needed for large, fast computers, and which needs the simpler processing and high yield that leads to low-cost integrated circuits.

2. *POWER TRANSISTORS*

Most transistors are advantageously very small, economical of both space and power. Often, especially in slow computer applications, they operate at "micropower" levels, of the order of microwatts per transistor. Nevertheless, it is true that when output power is needed, transistors can be built big enough to drive locomotives. Power transistors fall into two classes: (1) linear amplifiers, whose output is a faithful replica of the input; and (2) nonlinear power units, which operate by being triggered on and off, the so-called "switching" mode of operation. It is the latter which can deliver very large output, but only at very low frequencies; they can be thought of as power *controllers* rather than amplifiers, and are very suitable for driving variable-speed electric motors in sizes up to about 1000 kilowatts.

The linear power transistors, intended for use as power output stages of amplification, are simply large-area units, almost always bipolar silicon. They can deliver up to a few hundred watts per unit at audio (audible sound) frequencies, in such typical applications as drivers for loud speakers in public-address systems. When designed for higher frequencies, up to a few times 10^9 Hz, they have to be progressively smaller as the frequency rises, but some amplifiers can still put out 10 W at 2 GHz. For the microwave frequencies especially, there is a complicated fabrication problem in that the emitter and base electrodes have to be very narrow, closely juxtaposed, and still comparatively large in area. Consequently, a large high-frequency power transistor can be thought of as a mosaic of a large number of small transistors, interconnected by metallic leads running over the surface in a complicated form which often looks something like the skeleton of a microscopic fish. Frequently, the emitter and base electrodes are "interdigitated" or interleaved, to keep the emitter–base distance very small; ordinarily ballast resistors are built into the leads or into the emitters to keep the current uniformly distributed, and avoid local high current concentrations which would burn out the transistor.

In other words, a power transistor of this kind is really a special form of integrated circuit, which contains, besides transistors proper, other elements, such as leads, resistors, and heat-distributing elements ("heat sinks"), all built together as a unitary structure. Units of this kind work very well as high-fidelity amplifiers, telephone amplifiers, and small radio transmitters, but are about two orders of magnitude too small for such high-power uses as broadcast radio transmitters.

The nonlinear power units are usually called semiconductor controlled rectifiers (SCRs) or "thyristors," because of the way they operate. They may be thought of as rectifying diodes, capable of producing large direct currents by rectification of ordinary alternating-current power, but which are held "off" for a controllable time by an auxiliary electrode (the "gate"). When the forward-biasing applied alternating voltage becomes sufficiently large, the SCR is triggered "on" and conducts thereafter in the same way as a usual diode power rectifier. In this way, by varying the gate control voltage the SCR can vary the current for a variable-speed electric motor, even in sizes large enough to drive locomotives. Such solid-state control is used today to drive anything needing either manual or automatic control of mechanical motion.

Since the simplest SCRs work under only one polarity of potential, a further development, known as a Triac, has become popular, since it will draw current in either polarity and respond to gating voltage of either polarity. Such units use two SCRs connected together in opposite polarity, and responding to a common gate electrode; the whole structure can be built into a single chip of silicon. Triacs are widely used for driving small motors from electronic control systems.

3. ELECTRONIC DEVICES CLOSELY RELATED TO TRANSISTORS

Amplification is usually obtained via a three-electrode (triode) device, the transistor. But different numbers of electrodes can be used for other purposes, and such solid-state devices find numerous uses in transistor electronics. For example, some two-electrode units (diodes) have extended the usable radiofrequency spectrum to over 100 GHz, while other units can be used to make a variety of electronic sensors for use in electronic control systems. The latter can respond to heat, light, sound, pressure, or other external influences. Some diodes can emit light, and are widely used in displays or signaling; some diodes are lasers, optical generators or amplifiers.

The simple diode *rectifier* permits current flow in only one direction of polarity, thus converting alternating to direct current. Since alternating current is almost universally used for electric power transmission, while direct current is needed at the point of use for many applications (powering electronic systems, motor drives, electroplating, etc.), diode rectifiers find a myriad of uses. Small rectifiers, known as detectors, can detect the presence of radiofrequency waves up to the highest frequencies used in communications. Superheterodyne detection, in which such a detector is driven by a local oscillator at a slightly different frequency, can increase the sensitivity to radio waves by a large factor, of the order of 100,000. Such diodes, called "mixers" because they combine the weak signal with the local oscillator drive, are almost universally used in all radio receivers.

Somewhat similar structures, called "varactors" (variable-reactance diodes), can give ultralow-noise parametric amplification, and for that reason are widely used in radio astronomy and for reception of the very weak signals from communication satellites. Varactors are also useful in power mixers whose object is to produce very accurate control of high frequencies for superheterodyne reception. Still other varactors can produce fine-tuning control for heterodyne oscillators, and are widely used for channel selection in radio receivers.

Diodes for electronic pulse shaping and switching are widely used in logic circuits for computers, as are transistors also. Specified combinations of input pulses are organized to produce specified outputs, resulting in addition, subtraction, multiplication, and division of numbers, as well as other computer logic operations such as switching and routing of information. Similar *p-i-n* "plasma" diodes can generate very high-speed pulses, or switch and control radiofrequency signals.

Microwave diode oscillators of various kinds are preferred local oscillators for superheterodyne reception above about 10 GHz. Among these are "transferred electron" diodes, which utilize the negative resistance consequent to transferring to different energy states of the solid in order to generate microwave oscillations; also "Impatt" diodes, which utilize for the same purpose a combination of transit time of the charge through the device, together with impact ionization of valence electrons by the fast-moving carriers, to produce new carriers and therefore multiply the electrical conductivity. It is the Impatt diodes which have opened up the extended frequency range for radio to above 100 GHz.

Perhaps the best-known diode radiation sensor is the "solar cell," which has been widely used in space vehicles for the generation of electric power directly from sunlight. The light energizes valence electrons so that they become conduction electrons, at the same time leaving mobile holes in the semiconductor material; each solar cell thereby can develop about 1 V, depending on what material is employed. The solar cells are similar in principle to the photocells long used in photography and for such industrial uses as controlling the closing of elevator doors; but the solar cells use high-quality single-crystal material available as a result of the transistor development program. They are therefore substantially larger in size and efficiency, and can produce enough power to operate fairly large electronic systems for use in satellites and space vehicles. Terrestrial use is also possible, and may become large if development to reduce costs is sufficiently successful. The new photodiodes in small sizes are also used as detectors for light beams used in optical communications. Some other types of radiation, such as x rays and nuclear alpha particles, are also detectable by semiconductor diodes.

The light-emitting diodes (LEDs) depend on the fact that in some semiconductors the recombination of holes and electrons can emit light with high efficiency. Numerical displays using this effect have in the mid-1970s become standard in all sorts of electronic equipment such as calculators and measuring instruments, as well as for indicator lights featuring very long useful life and high reliability.

Semiconductor lasers emit light of high spectral purity and high intensity, useful for optical signaling applications; they demand exceptionally high crystal quality, and therefore never could be made or even dreamed of until after adequate methods for making high-quality semiconductor material had been developed for transistor use. The LEDs and lasers are easily modulated by changing the applied voltage, and therefore are likely to find extensive future use in communications and control systems which transmit information via modulated light beams.

In the examples cited, the semiconductor devices are widely useful in the inputs and outputs of electronic systems. One can say that they facilitate communication between electronic systems and the external world. One can also say that the sensors and amplifiers permit the world to transmit information from place to place via electronics. While en route, the information can also be processed by computers which "think" about it. When the information arrives at its destination, electronic actuators can carry out various kinds of appropriate action; or the information can be stored electronically in a memory or library, which is itself also often electronic. By such means, electronics has already greatly increased the power of the human mind.

4. SCIENTIFIC ASPECTS OF TRANSISTORS

New scientific knowledge of solid-state physics was necessary in order to make transistors possible; but this debt to pure science has been repaid many times over by the new scientific knowledge resulting from the transistor program. In addition, the transistor program has led to new scientific instrumentation, to the ubiquitous computers now used for analyzing the information gained, to new, highly pure, and well-controlled materials, and to "high technology" useful for fabricating all sorts of sophisticated syntheses. Only a sampling of a few highlights can be given here.

Viewed as a scientific achievement, the invention of the transistor showed for the first time how to control the interaction of two types of charge (negative electrons and positive holes) in order to achieve amplification via a solid. Much further knowledge of semiconductors has resulted, including the injection, collection, generation, and recombination of holes and electrons; the absorption and emission of light and other radiation; the interaction with vibrational acoustic energy; and the structure of the solid material itself (band theory), including the calculation of its properties from fundamental quantum mechanics.

Along with improved material knowledge has come more powerful use of the material properties in electronic-device applications. As a result of his analysis of the behavior of the first, point-contact transistors, Shockley was able to show the existence of the phenomenon of carrier injection from electrodes into the material, and from that could predict the existence of the "junction" transistors (*npn* and *pnp*, described above). This fine example of successful *prediction* in science was followed by experimental verification, and the very successful bipolar transistors were the result.

One of the most interesting, and for a long time most puzzling, features of the transistors was the behavior of the free surfaces of the semiconductor, and the interfaces with other materials, both metals and insulators. Transistors are very sensitive to surface effects, because the electrical conductivity of the semiconductor typically depends on very small and closely controlled amounts of impurities, called "dopants," often less than one part per million. A correspondingly small amount of charge present on the surface—maybe as little as 1/10,000 of a monolayer of atoms—can completely upset or even invert the conductivity, and produce corresponding drastic changes in the properties of the device. When such a change happens spontaneously at an uncontrolled surface, the device may be regarded as unreliable by the user. Considering that almost any common object not isolated in high vacuum quickly picks up one or several monolayers of adsorbed gas from the ambient atmosphere, one can see that the requirement that the surface be neutral, or at least stable, is often impossibly onerous when the surfaces are exposed to air.

In the early 1950s it was discovered that freshly exposed transistor surfaces were indeed very sensitive, and later it was found that water vapor and sodium ions, both very common contaminants, could produce instability on transistor surfaces. At that point of the development, the only way to keep the surfaces adequately clean was to enclose the transistors in glass–metal hermetically sealed encapsulations. This need meant a great additional expense, and also at the time thoroughly inhibited the development of integrated circuits.

In the late 1950s it was found that a passivating coating of silicon dioxide could give a more stable surface, but even yet, the transistors were sensitive to charged ions on the *outside* of the oxide; also, it was found that sodium and water vapor could actually penetrate through the thin oxide coating, and affect the silicon within. Because of these effects, very small and difficult to control, the manufacture of transistors required dust-free "clean" rooms and carefully filtered air; even so, there was and still is much difficulty in achieving adequate freedom from surface impurities. After further years of effort it was found that surfaces of oxide-coated silicon transistors could be successfully sealed either by metals or by a thin insulating coating of silicon nitride; even then, the outer surfaces had to be protected against electrolytic corrosion. All this took a long time to discover, because the effects are small, occur only slowly, and only under adverse circumstances, such as high ambient humidity or other contamination. But the overall result was eventually to develop, in the best transistors and integrated circuits, a totally unprecedented degree of reliability, very long life and stability in electronic equipment—once the "bugs" are out of it!

During the course of these investigations, it became possible for the first time to make successful field-effect transistors. The FET had been predicted and even patented many years before, but for completely mysterious reasons had never worked. From known electromagnetic theory it was certain that the gate electrode must induce charge in the material, but the induced charge did not show the mobility of the charge ordinarily present. The explanation was provided by Bardeen's theory of surface states, which postulated that the induced charge could not move because it was tied up in immobile surface energy states (more properly called "traps"). This effect could not be circumvented until the invention of the MOSFET by Atalla and Kahng in 1960, in which the traps were held to very small numbers by a surface coating of very clean silicon oxide. Eventually the MOSFETs have proved economical and very successful in integrated circuits, but only after prolonged and expensive experimentation to find ways to keep the surface very clean and fully passivated during and after the many necessary processing steps. Success here was signaled by the spectacular and precipitous fall in prices of electronic calculators in the early 1970s. In all, it took about 40 years and hundreds of millions of dollars to conquer the surface sensitivity of semiconductors, so that MOSFET integrated circuits finally became successful. Even now, it requires superlative care for successful fabrication of such units.

From the scientific point of view, the sensitivity of the surfaces, so troublesome technologically, was a great scientific opportunity to understand surface science. For the first time, a surface became available (the interface between silicon and its oxide) which was stable enough to be measured repeatedly for quantities like numbers and energies of surface charges and surface traps. These studies have helped to elucidate not only transistors but also scientific problems like chemical catalysis.

Another lesson to be drawn here is the complete failure of the Edisonian, or empirical, approach to invention of the FET. Only the scientific approach through detailed theory and experiment was able to discover the surface states and surface charges which led to the successful invention of both bipolar and (later) field-effect transistors.

Another aspect of materials science was the preparation of the semiconductor materials, which had to be single crystals of uniform purity even when doped to less than one part per million with appropriate impurities to control the conductivity. Noteworthy here was the new process of zone refining, in which a liquid zone is caused to move slowly through the material by progressive melting and resolidification, sweeping along as it goes most of the impurities in the material. By this uniquely powerful method, materials as pure as one part in 10^{10} of electrically significant impurities have been obtained. Zone refining is now a standard way of producing ultrapure materials of many other kinds, not restricted to semiconductors.

Growth of single crystals of large size and high crystalline perfection has also been developed for use with transistors. Of particular importance is epitaxial growth, in which a new layer of controlled thickness and impurity content is laid down over a single-crystal substrate of the same material, and acquires the same crystal orientation. Such "epi-layers" are extensively used in silicon transistors and integrated circuits. They may be grown from materials in the vapor phase, from the liquid phase, or from a molecular beam of the semiconductor material.

5. *FABRICATION TECHNIQUES*

Many experimental techniques, originally used in advanced physics research, have been adapted for production of transistors and other modern products. Collectively, these techniques are known as the "high technology" in which the United States led the world in the 1950s and 1960s. Particularly prominent are high-vacuum processes and elaborately instrumented measurements. Some examples are: *vacuum evaporation* (to make thin films of metals or other materials); *sputtering* in gas at low vacuum, to add or remove thin films; *electron beam evaporation*, in which the heating is applied locally by a beam of electrons, thus dispensing with the need to find a crucible material which can hold the liquid metal without contaminating it; and ultraclean techniques, including dust-free rooms, to accomplish assembly operations without injury to minute electrodes or sensitive surfaces. Advanced techniques of analysis include the use of electron beams to produce x rays indicating the chemical composition of test surfaces, neutron activation to generate radioactive counts for analysis, and scanning electron microscopes which use television techniques and electron beams to reveal submicroscopic details of surfaces. Many other computer-aided instruments have achieved great improvements in sensitivity, resolving power, flexibility, capability, accuracy, and sheer quantity of scientific measurements. Such techniques permit fabrication and control of sophisticated structures by using knowledge which once was completely inaccessible to measurement.

To make the complicated microscopic patterns required for transistors and integrated circuits, masking and photolithography are the usual methods. A thin layer of a photosensitive material is laid down over the semiconductor wafer, and then patterned by exposure to light, in a way similar to an ordinary photograph. X rays may be used instead of light, when the pattern is to be especially fine; or computer-controlled electron beams in high vacuum are used, when the pattern is especially complicated, or arbitrarily changeable. After development, the complicated photolithographic pattern becomes differentially soluble, so that after leaching with an appropriate chemical solvent the developed and fixed photosensitive material remains present only in the desired complicated pattern. The underlying material can then be subjected to chemical etching which reproduces the mask pattern in the semiconductor material, or more often in other masking layers over the material. Silicon oxide and silicon nitride masks patterned in this way are often used to produce desired patterns of high-temperature diffused areas or local variations in epi-layer thickness or composition; metal layers are used to produce contacts or interelectrode conductors. Multiple masks are often used, in which the successive patterns must be superposed so that they register with each other to very close tolerances, sometimes to a few micrometers or even a fraction of a micrometer; in integrated circuit use, sometimes as many as 8 or 10 masks must be matched to such tolerances. Considering the close tolerances, the high sensitivity to surface contamination, the elaborate nature of the patterns, and the variety of materials and processes, it is no wonder that such a process is called "high" technology!

6. *CONCLUSIONS*

Transistors and the resulting electronics are beginning a profound, even a revolutionary change in human society, by increasing the power of human communication and human thought. The way in which this has come about is also worthy of intense sociological study of the mechanisms by which science, technology, and society interact; in other words, the societal organization of scientific research and development. Meanwhile, the transistor revolution is beginning to play an indispensable part in improving human knowledge to the point where it may become capable of designing a better society. Knowledge alone is the unique hope for better understanding and better control of the political and economic processes which determine the fate of the world.

See also CIRCUITS, INTEGRATED; ELECTRON BEAM TECHNOLOGY; ELECTRON ENERGY STATES IN SOLIDS AND LIQUIDS; PHOTOVOLTAIC EFFECT; SEMICONDUCTORS, CRYSTALLINE; SURFACES AND INTERFACES.

BIBLIOGRAPHY

A. S. Grove, *Physics and Technology of Semiconductor Devices*. Wiley, New York, 1967.

S. M. Sze, *Physics of Semiconductor Devices*. Wiley, New York, 1969.

Transition Elements

E. Bucher

1. *BASIC CONCEPTS AND DEFINITION*

The Pauli principle states that each electronic quantum state characterized by n, l, m, s (the principal, orbital, magnetic, and spin quantum numbers, respectively) can be occupied by one electron only. Since the orbital state $l=2$ contains $2l+1$ m substates ($m=-2, -1, 0, +1, +2$) each of which can be occupied by $s=+\frac{1}{2}, -\frac{1}{2}$, a transition series is expected to contain $2(2l+1)=10$ elements. Electrons in the $l=2$ state are called d electrons (regardless of all other quantum states); $n=3$ is the lowest principal quantum number for which a d state is occupied. Among all the known chemical elements (naturally occurring and synthetic elements) we can therefore distinguish four series of transition elements: the $3d$ ($n=3$, $l=2$) the $4d$ ($n=4$, $l=2$), the $5d$ ($n=5$, $l=2$), and the $6d$ ($n=6$, $l=2$) elements. The following definition of a transition element is commonly in use: A transition element is a chemical element in its ground state with a partially filled d shell, i.e., with a number of d electrons Z_d restricted by $0 < Z_d < 10$.

A first complication occurs in the $5d$ series where the $4f$ electron shell ($n=4$, $l=3$) is filled up first after the first $5d$ transition element (lanthanum). This leads to a subgroup of 14 elements, called rare earths or lanthanides, among the $5d$ transition elements. Only after the $4f$ shell is filled up are more $5d$ electrons added. The second complication appears in the $6d$ series in which after the initial buildup of the $6d$ shell (Ac, Th) the fourteen $5f$ electrons ($n=5$, $l=3$) appear. These elements form a second subgroup among the transition elements, called actinides. Most of them are synthetically formed transuranium elements.

2. *CHEMICAL PROPERTIES*

We list below the whole complex of transition elements, including $4f$ and $5f$ subgroups:

Group number:	IIIB	IVB	VB	VIB	VIIB	VIIIB		
$3d$ transition series:	Sc	Ti	V	Cr	Mn	Fe	Co	Ni
$4d$ transition series:	Y	Zr	Nb	Mo	Tc	Ru	Rh	Pd
$5d$ transition series:	La–Lu	Hf	Ta	W	Re	Os	Ir	Pt
5d-4f series (lanthanides or rare earths):	La, Ce, Pr, Nd, Pm, Sm, Eu, Gd, Tb, Dy, Ho, Er, Tm, Yb, Lu							
6d transition series (actinides):	Ac	Th						
$6d$-$5f$ series:	Th, Pa, U, Np, Pu, Am, Cm, Bk, Cf, Es, Fm, Md, No, Lr, Ku							

Except for the beginning of the d series, the transition elements are well known for their complex behavior in that they form compounds in many different valence states (e.g., in Ru and Os, all eight valence states have been observed[1]). Among the rare earths, it is sometimes energetically more favorable to convert a $5d$ electron into a $4f$ electron. These divalent rare earths (Sm, Eu, Tm, Yb in general) resemble more closely the preceding Ba in their chemical behavior and do not exhibit transition-element properties, since they are stripped of their only $5d$ electron.

All the elements listed above are metals with relatively high melting points (900–1000°C or higher) and crystallize in relatively simple structures[1] [fcc, bcc, hcp, or sequences thereof (La–Sm)], except Mn and most elements of the actinide series. They show in general a wide range of solid solutions and form numerous intermetallic compounds and complex phases[2,3] (except group III with groups IV, V, VI). In pure bulk form most transition elements are relatively stable in air against corrosion. The lower groups usually form a protective oxide layer (except Eu), whereas the higher group belongs to the noble metals except $3d$ elements Fe, Co, Ni.

3. *PHYSICAL PROPERTIES*

In the atomic state (e.g., in the vapor state) the d energies are relatively sharp. In the condensed state, however, the d levels broaden into an energy band[4] which has a width typically of 2.5–3 eV, 4–4.5 eV, 5–5.5 eV, and 6 eV for $3d$, $4d$, $5d$, and $6d$ elements. Depending on crystal symmetry the d level undergoes a Stark effect, i.e., an energy splitting due to electric fields generated by the charge distribution of the nearest (or higher) neighbors.[5,6] The interaction of this electric potential with the outermost d electrons is so strong that the orbital quantum number is not preserved and the orbital moment is quenched.[4–6] Depending on crystal symmetry, several d subbands may form, which in general overlap.[5,6] Elements in which the highest occupied energy level of the electrons is below the maximum energy of a band represent metals. According to our definition all transition elements exhibit metallic behavior and, therefore, are usually called transition metals (T metals in general or $3d$, $4d$, $5d$, $6d$ metals more specifically).

At low temperature the transition metals show a number of unique features:

(1) The interaction of the d electrons with the lattice vibrations in which the density of states, i.e., the number of allowed states per unit of energy, is involved leads in general to superconductivity, in the transition metals themselves as well as in their alloys and metallic compounds with other transition and nontransition elements.[4]

(2) The occurrence of superconductivity is limited by the strength of the electron–electron interaction. In the $3d$ elements, this leads to ferromagnetism in Fe, Co, Ni; to antiferromagnetism in Mn; and to a rather complex spin-density wave state in Cr below 338°K. All these elements are nonsuperconducting and Ti and V are the only superconductors among the $3d$ elements.

In contrast, all $4d$ elements are paramagnetic; however, the electron–electron interaction in the beginning (Y) and at the end (Pd) of the $4d$ series is still strong enough to prevent superconductivity. These elements show an enhanced paramagnetic susceptibility. In particular, Pd is considered as a model for a paramagnon material, i.e., a material in which electron–electron interactions are important but not strong enough to induce a ferromagnetic transition. All $5d$ elements are also superconducting except the nonmagnetic Yb, Lu, and Pt and the magnetic rare earths in which the magnetic moments arise from the $4f$ electrons. In the latter, the orbital moment is well preserved and the interaction with the crystal-field potential is relatively weak.[5,6] In the analogous $5f$ series the situation is much more complex. Depending on the individual compounds, the properties of $5f$ electrons in the lighter actinides can vary between the limits of complete delocalization into a relatively narrow $5f$ band to strongly localized $5f$ electrons with well-preserved quantum states L, S, J. The actinides are usually very complex because they show several different structures and the physical properties strongly depend on them. Superconductivity has been found in certain phases of Pa, U, and Np, whereas in compounds they may be magnetic. It is striking that in Sect. 2 the transition series contains only eight, rather than ten elements as one would expect from the definition in Sect. 1. This arises from the fact that certain configurations of d and s electrons do not occur in the ground state for energetic reasons which are too difficult to predict from first-

Table I. Chemical Symbols and Outer Electron Configuration Names of all Transition Elements

		Outer atomic configuration	Solid-state configuration in general
$3d$ Series:	Sc = Scandium	$3d^{1}4s^{2}$	$(3d4s)^{3}$
	Ti = Titanium	$3d^{2}4s^{2}$	·
	V = Vanadium	$3d^{3}4s^{2}$	·
	Cr = Chromium	$3d^{5}4s^{1}$	·
	Mn = Manganese	$3d^{5}4s^{2}$	·
	Fe = Iron	$3d^{6}4s^{2}$	·
	Co = Cobalt	$3d^{7}4s^{2}$	·
	Ni = Nickel	$3d^{8}4s^{2}$	$(3d4s)^{10}$

Table I.—(Continued)

		Outer atomic configuration	Solid-state configuration in general
$4d$ Series:	Y = Yttrium	$4d^{1}5s^{2}$	$(4d5s)^{3}$
	Zr = Zirconium	$4d^{2}5s^{2}$	·
	Nb = Niobium	$4d^{4}5s^{1}$	·
	Mo = Molybdenum	$4d^{5}5s^{1}$	·
	Tc = Technetium	$4d^{5}5s^{2}$	·
	Ru = Ruthenium	$4d^{7}5s^{1}$	·
	Rh = Rhodium	$4d^{8}5s^{1}$	·
	Pd = Palladium	$4d^{10}5s^{0}$	$(4d5s)^{10}$
$5d$ and $4f$ Series:	La = Lanthanum	$4f^{0}5d^{1}6s^{2}$	$4f^{0}(5d6s)^{3}$
	Ce = Cerium	$4f^{1}5d^{1}6s^{2}$	$4f^{1}(5d6s)^{3}$
			$4f^{0}(5d6s)^{4}$
	Pr = Praseodymium	$4f^{2}5d^{1}6s^{2}$	$4f^{2}(5d6s)^{3}$
	Nd = Neodymium	$4f^{4}5d^{0}6s^{2}$	$4f^{3}(5d6s)^{3}$
	Pm = Promethium	$4f^{5}5d^{0}6s^{2}$	$4f^{4}(5d6s)^{3}$
	Sm = Samarium	$4f^{6}5d^{0}6s^{2}$	$4f^{5}(5d6s)^{3}$
			$4f^{6}(5d6s)^{2}$
	Eu = Europium	$4f^{7}5d^{0}6s^{2}$	$4f^{7}(5d6s)^{2}$
			$4f^{6}(5d6s)^{3}$
	Gd = Gadolinium	$4f^{7}5d^{1}6s^{2}$	$4f^{7}(5d6s)^{3}$
	Tb = Terbium	$4f^{8}5d^{1}6s^{2}$	$4f^{8}(5d6s)^{3}$
	Dy = Dysprosium	$4f^{9}5d^{1}6s^{2}$	$4f^{9}(5d6s)^{3}$
	Ho = Holmium	$4f^{10}5d^{1}6s^{2}$	$4f^{10}(5d6s)^{3}$
	Er = Erbium	$4f^{11}5d^{1}6s^{2}$	$4f^{11}(5d6s)^{3}$
	Tm = Thulium	$4f^{13}5d^{0}6s^{2}$	$4f^{12}(5d6s)^{3}$
			$4f^{13}(5d6s)^{2}$
	Yb = Ytterbium	$4f^{14}5d^{0}6s^{2}$	$4f^{13}(5d6s)^{3}$
			$4f^{14}(5d6s)^{2}$
	Lu = Lutetium	$4f^{14}5d^{1}6s^{2}$	$4f^{14}(5d6s)^{3}$
	Hf = Hafnium	$4f^{14}5d^{2}6s^{2}$	$(5d6s)^{4}$
	Ta = Tantalum	$5d^{3}6s^{2}$	·
	W = Tungsten	$5d^{4}6s^{2}$	·
	Re = Rhenium	$5d^{5}6s^{2}$	·
	Os = Osmium	$5d^{6}6s^{2}$	·
	Ir = Iridium	$5d^{7}6s^{2}$	·
	Pt = Platinum	$5d^{9}6s^{1}$	$(5d6s)^{10}$
$6d$ and $5f$ Series:	Ac = Actinium	$6d^{1}7s^{2}$	$(6d7s)^{3}$
	Th = Thorium	$6d^{2}7s^{2}$	$(6d7s)^{4}$
	Pa = Protactinium	$5f^{2}6d^{1}7s^{2}$	$5f^{0}(6d7s)^{5}$
	U = Uranium	$5f^{3}6d^{1}7s^{2}$	$5f^{0}(6d7s)^{6}$
			$5f^{1}(6d7s)^{5}$
			$5f^{2}(6d7s)^{4}$
	Np = Neptunium	$5f^{4}6d^{1}7s^{2}$	$5f^{0}(6d7s)^{7}$
			$5f^{4}(6d7s)^{3}$
	Pu = Plutonium	$5f^{5}6d^{1}7s^{2}$	$5f^{1}(6d7s)^{7}$
			$5f^{5}(6d7s)^{3}$
	Am = Americium	$5f^{7}6d^{0}7s^{2}$	
	Cm = Curium	$5f^{7}6d^{1}7s^{2}$	
	Bk = Berkelium	$5f^{8}6d^{1}7s^{2}$	
	Cf = Californium	$5f^{9}6d^{1}7s^{2}$	?
	Es = Einsteinium	$5f^{10}6d^{1}7s^{2}$	
	Fm = Fermium	$5f^{11}6d^{1}7s^{2}$	
	Md = Mendelevium	$5f^{12}6d^{1}7s^{2}$	
	No = Nobelium	$5f^{14}6d^{0}7s^{2}$	?
	Lr = Lawrencium	$5f^{14}6d^{1}7s^{2}$	?
	Ku = Kurchatovium	$5f^{14}6d^{2}7s^{2}$	?

principles consideration (minimization of energy for the whole system of electrons) (see Table I of transition elements with their electron configurations). In the metallic solid state we cannot in general attribute an integral number of *d* electrons and *s* electrons to each element.

In the band picture the character of *s* and *d* electrons is not preserved and their properties are admixed. Therefore, in the metallic state we consider as transition elements those elements with unfilled *d* bands, unlike in the atomic state where an integral number between 1 and 10 *d* electrons is required. This implies that other elements such as Cu and Ag may behave as transition elements in chemical compounds if encountered in a proper valence state, e.g. Cu^{2+} or Ag^{2+} will be in a $3d^9$ or $4d^9$ configuration, respectively, although these elements are not considered as transition elements in the metallic state.

4. APPLICATIONS

The technical applications of transition elements are nearly inexhaustible. Best known are alloys of 3*d* elements for steel production: Cr-Ni alloys with various additions such as Fe, Co, Ti, V, Mn, Nb, Mo, Al to achieve special properties. Ferromagnetic alloys Fe-Co-Ni with various additions are used for transformer cores or electromagnets and permanent magnets. Noble metals Rh, Pd, Ir, Pt are used for electrical switching contacts, catalysis (in particular Pd, Pt); Zr and Hf are used in reactor design for their low neutron absorption. Light-bulb filaments are made from tungsten because of its high melting point. Some of the rare earths find increasing application in improvement of steel quality and for particularly strong magnetic materials ($SmCo_5$). ^{235}U is the well-known reactor fuel, and some transuranium metals like Pu may play a future role in nuclear technology (fast breeder reactors); Nb, Mo, Ta, W also play an important role in high-temperature technology because of their high melting points.

See also ELECTRON ENERGY STATES IN SOLIDS AND LIQUIDS; FERROMAGNETISM; PARAMAGNETISM; RARE EARTHS; SUPERCONDUCTIVITY THEORY.

REFERENCES

1. For description of other properties see e.g. the *CRC Handbook of Chemistry and Physics*, 56th ed. Chemical Rubber Publishing Co., Cleveland, Ohio, 1975. (I)
2. W. B. Pearson, *Handbook of Lattice Spacings and Structure of Metals and Alloys*, Vol. 2. Pergamon, New York, 1967. (I)
3. M. Hansen (ed.), *Constitution of Binary Alloys*, Vol. 1. Mc-Graw-Hill, New York, 1958; R. P. Elliot (ed.), *Constitution of Binary Alloys*, Suppl. 1. McGraw-Hill, New York, 1965; F. A. Shunk (ed.), *Constitution of Binary Alloys*, Suppl. 2. Mc-Graw-Hill, New York, 1970. (I)
4. C. Kittel, *Introduction to Solid State Physics*, 4th ed. Wiley, New York, 1971. (A)
5. M. T. Hutchings, in *Solid State Physics*, F. Seitz and D. Turnbull (eds.), Vol. 16. Academic Press, New York, 1964. (A)
6. A. H. Morrish, *The Physical Principles of Magnetism*. Wiley, New York, 1965. (A)

Transport Properties

David Adler

The transport properties of a solid refer to the response of the constituent electrons and ions in the material to a nonequilibrium force, such as an applied electromagnetic field or a temperature gradient. In addition to direct effects, such as electrical or thermal conductivity, coupled flows, such as thermoelectric or thermomagnetic phenomena, can also be important. The starting point for any discussion of transport properties is the formalism of irreversible thermodynamics, which relates the observed flows (or particles, charge, energy, etc.) to the driving forces (e.g., gradients of temperature, electrostatic potential, chemical potential). Perhaps the most important example of this type of process is electrical conductivity, the flow of charge in an applied electric field; this is the effect with which we shall be primarily concerned in this article.

If we apply a small dc electric field $\mathscr{E}$ to a given material and measure the resulting electrical current flow per unit cross-sectional area J, then the electrical conductivity σ is defined as the ratio of the component of J in the direction of $\mathscr{E}$ to the magnitude of $\mathscr{E}$:

$$\sigma \equiv J_{\mathscr{E}}/\mathscr{E}. \tag{1}$$

In general, the current need not be in the same direction as the field, and we must define a conductivity tensor σ_{ij} by

$$J_i = \sum_{j=1}^{3} \sigma_{ij}\mathscr{E}_j \tag{2}$$

where the subscripts refer to three spatial directions (i.e., x, y, z) chosen in any convenient manner. If the current density is measured in amperes per square meter and the field in volts per meter, the appropriate unit for conductivity is reciprocal ohms per meter ($\Omega^{-1}\ m^{-1}$).

Experimentally, the electric field is ordinarily applied by connecting two ends of the sample to metallic electrodes and imposing a voltage difference $\Delta\phi$ between the electrodes. If the sample has a constant cross-sectional area A, a simple measurement of the current flowing in the circuit gives the conductivity from the relation

$$\sigma = Il/A(\Delta\phi) \tag{3}$$

where l is the length of the sample.

It is the electric current (i.e., a net flow of charge from one end of the sample to another) that is measured in this experiment. Conductivity and current are macroscopic parameters. The importance of electrical conductivity, however, is that it provides information about the microscopic structure of the material. This is because it is the electrons or ions that respond to the applied field by accelerating and acquiring nonzero average velocities, and this response is the source of the current. We describe the effectiveness of any particular charged particle in contributing to the current by evaluating the average component of velocity in the direction of the field per unit applied field, a parameter known as the mobility, μ:

$$\mu \equiv \langle v_{\mathscr{E}} \rangle/\mathscr{E}. \tag{4}$$

The mobility is measured in square meters per volt per second ($m^2/V \cdot s$). As is the case with conductivity, in general, the mobility is a tensor:

$$\langle v_i \rangle = \sum_{j=1}^{3} \mu_{ij} \mathscr{E}_j. \tag{5}$$

Clearly, the contribution of a density n of particles of charge q and mobility μ to the conductivity is

$$\sigma = nq\mu \tag{6}$$

where n is measured in reciprocal cubic meters (m^{-3}) and q in coulombs (ampere seconds, A·s).

It is important to note that experimentally we *measure* the macroscopic parameter σ, but theoretically we *calculate* the microscopic quantity μ. Thus, agreement between a particular model and experiment does not *prove* the applicability of the model, since there are many ways of obtaining the same macroscopic averages from vastly different microscopic motions.

The definition of conductivity in Eq. (1) does not guarantee that σ is independent of field. In general, there cannot be any net current in the absence of an applied field. Thus, a series expansion for the current in terms of the field must begin with the linear term:

$$J_{\mathscr{E}} = \left(\frac{\partial J_{\mathscr{E}}}{\partial \mathscr{E}}\right)_0 \mathscr{E} + \tfrac{1}{2}\left(\frac{\partial^2 J_{\mathscr{E}}}{\partial \mathscr{E}^2}\right) \mathscr{E}^2 + \cdots. \tag{7}$$

If we define the conductivity as the limit of the quantity in Eq. (1) for small fields, then

$$\sigma = \lim_{\mathscr{E} \to 0} \frac{J_{\mathscr{E}}}{\mathscr{E}} = \left(\frac{\partial J_{\mathscr{E}}}{\partial \mathscr{E}}\right)_0. \tag{8}$$

With this definition, σ is independent of field. The small-field regime is the region in which the linear relation

$$J_{\mathscr{E}} = \sigma \mathscr{E} \tag{9}$$

is obeyed. Equation (9) is known as *Ohm's law*, and it is valid for most materials up to fields of the order of 10^5 V/m. At greater fields, higher-order terms in Eq. (7) become significant, and the conduction is called non-Ohmic.

In order to calculate the mobility of any charged particle, we must be able to evaluate its average velocity in the presence of an applied field. Performing any average is possible only if we can determine or estimate the appropriate probability distribution. For the microscopic charged particles in a solid, the best approach is to define the *distribution function* $f(\mathbf{k},\mathbf{r},t)$, which gives the probability that the particle is in the phase-space volume element between $\mathbf{k}$, $\mathbf{r}$, and $\mathbf{k}+d\mathbf{k},\mathbf{r}+d\mathbf{r}$ at time t. Here $\mathbf{r}$ is the position and $\mathbf{k}$ is the wave vector (momentum divided by Planck's constant, $\hbar$). In terms of the distribution function, the mobility follows from Eq. (4):

$$\mu = \frac{\int d\mathbf{k}\; v_{\mathscr{E}} f(\mathbf{k},\mathbf{r},t)}{\mathscr{E} \int d\mathbf{k}\; f(\mathbf{k},\mathbf{r},t)}. \tag{10}$$

In order to obtain the distribution function, we must solve the quantum-mechanical problem for the states available to the charged particle in the solid both at equilibrium and then in the presence of the applied field. The former problem is the core of solid-state physics. If the ions in the material contribute to the conduction, the problem can be solved classically; however, this is appropriate only to nearly perfect ionic crystals, ionic liquids, and plasmas. Much more commonly, electrical conductivity is predominantly due to the motion of electrons, and quantum mechanics is essential.

The states available to the electrons in a solid are calculated by means of the band theory of solids. The results are expressed as the energy E of a state available to an electron with a particular wave vector $\mathbf{k}$. The corresponding velocity can then be determined from the relation

$$\mathbf{v} = \frac{1}{\hbar} \nabla_{\mathbf{k}} E(\mathbf{k}), \tag{11}$$

which is analogous to Hamilton's second equation.

In a gas, the energy levels available to electrons localized on one of the constituent molecules are sharp. There is a large degeneracy, since all the molecules in the gas have the same electronic energy levels. In a solid, the interactions between electrons on closely spaced nearest-neighboring molecules remove this large degeneracy, and the sharp levels spread into quasicontinua called *bands*. According to the Pauli exclusion principle, we obtain the lowest energy state of any material by placing successive electrons into the lowest possible state, one electron per state, until we have accounted for all the available electrons. This is the actual state of the system at a temperature of absolute zero. At finite temperatures, quantum statistics require that the electrons have a probability of occupying any state $\mathbf{k}$ given by the Fermi–Dirac distribution function,

$$f_0(E) = \left\{1 + \exp\left[\frac{E(\mathbf{k}) - E_F}{kT}\right]\right\}^{-1} \tag{12}$$

where E_F is a parameter known as the Fermi energy. E_F is chosen so that the value of n, the total density of electrons in the crystal, is correct; this requirement is

$$n = \int dE\; g(E) f_0(E) \tag{13}$$

where $g(E)$ is the density of electronic states between E and $E+dE$, a function easily derivable from $E(\mathbf{k})$. Equation (13) determines E_F at any temperature.

It is easy to show that a filled band cannot contribute to the electrical conductivity. Independent of the particle velocities, no current can flow in the absence of an applied field; thus, the sum of all the velocities of the electrons in a band must vanish at equilibrium. But in the presence of an applied field, unless sufficient energy is obtained from the field for some electrons to tunnel into higher bands (this is a non-Ohmic effect—the field can always be made small enough to make tunneling very improbable), all that can happen is a rearrangement of the electrons among the same filled states, and the current must remain precisely zero. Thus, if the ground state of a solid is such that its electrons just fill a band, the material is an insulator. Alternatively, if the highest band is partially filled, an applied field induces an electrical current, and the material is metallic. Semiconductors are materials with only a small energy gap between the highest filled and lowest empty states; at finite temperatures, as Eq. (12) shows, some of the electrons must occupy the next highest band, and conduction takes place.

The function $f_0(E)$ defined in Eq. (12) plays the role of the electron distribution function $f(\mathbf{k},\mathbf{r},t)$ in the *absence* of an applied field (i.e., at thermal equilibrium). It is a very difficult problem to find $f(E)$ at a finite value of the field, since electronic energy is no longer conserved (the electrons can gain kinetic energy by being accelerated by the field) and the electronic states now depend on position (since the energy extracted from the field depends on position). However, a simplification is provided by the fact that the conductivity, as defined in Eq. (8), must be evaluated in the limit of *small* applied fields. Thus perturbation theory is applicable, and we can write

$$f(\mathbf{k},\mathbf{r},t)=f_0(E)+f_1(\mathbf{k},\mathbf{r},t) \tag{14}$$

where $f_1(\mathbf{k},\mathbf{r},t)$ is due only to the field, and thus must vanish when $\mathscr{E}=0$.

An electron in a small volume element around the phase-space point $(\mathbf{k},\mathbf{r})$ at time t can leave this volume element for one of three reasons: either because its velocity carries it away from $\mathbf{r}$; because its rate of change of wave vector carries it away from $\mathbf{k}$; or because it is scattered away from $\mathbf{k}$ by crystal imperfections (for a perfect crystal, $\mathbf{k}$ is a good quantum number, and no such scattering occurs). Thus we can write a continuity equation

$$\nabla_{\mathbf{k}}f\cdot\frac{d\mathbf{k}}{dt}+\nabla_{\mathbf{r}}f\cdot\frac{d\mathbf{r}}{dt}=\left(\frac{\partial f}{\partial t}\right)_s \tag{15}$$

where $(\partial f/\partial t)_s$ is the net rate of increase of f due to scattering. Equation (15) is called the *Boltzmann equation*.

In general, the scattering term can be expressed as an integral over f, and Eq. (15) is an integrodifferential equation, extremely difficult to solve. It can be handled by assuming that when the electronic distribution is disturbed from equilibrium by the applied field, the effect of scattering is to restore f to f_0 in such a way that the rate $(\partial f/\partial t)_s$ is proportional to the deviation from equilibrium; i.e.,

$$\left(\frac{\partial f}{\partial t}\right)_s=-\frac{1}{\tau}(f-f_0)=-\frac{f_1}{\tau} \tag{16}$$

where we have used Eq. (14). The parameter τ is called the relaxation time; it is of the order of the time between scattering events. Substitution of Eq. (16) into Eq. (14) yields

$$\mathbf{v}\cdot\nabla_{\mathbf{r}}f+\frac{d\mathbf{k}}{dt}\cdot\nabla_{\mathbf{k}}f=-\frac{f_1}{\tau}. \tag{17}$$

This is just a linear partial differential equation, which can be solved in many cases.

In the presence of a field, the electronic energy becomes

$$E=E(\mathbf{k})-e\phi(\mathbf{r}) \tag{18}$$

where $\phi(\mathbf{r})$ is the electrostatic potential. Since Eq. (12) shows that f_0 depends only on E, then the first term in Eq. (17) can be written (to first order)

$$\mathbf{v}\cdot\nabla_{\mathbf{r}}f\simeq\mathbf{v}\cdot\nabla_{\mathbf{r}}f_0=\mathbf{v}\cdot\left(\frac{\partial f_0}{\partial E}\right)\nabla E. \tag{19}$$

Using Eq. (18) and the elementary electrostatic relation $\mathscr{E}=-\nabla\phi$, we obtain

$$\mathbf{v}\cdot\nabla_{\mathbf{r}}f\simeq e\left(\frac{\partial f_0}{\partial E}\right)\mathbf{v}\cdot\mathscr{E}. \tag{20}$$

In the presence of an applied field, Eq. (18) shows that $dE=-e\mathscr{E}\cdot\mathbf{v}\,dt$; since $dE=\nabla_{\mathbf{k}}E\cdot d\mathbf{k}$,

$$\frac{d\mathbf{k}}{dt}=-\frac{e\mathscr{E}}{\hbar} \tag{21}$$

where we have used Eq. (11). Thus, to first order in $\mathscr{E}$, the second term in Eq. (17) is just

$$\frac{d\mathbf{k}}{dt}\cdot\nabla_{\mathbf{k}}f\simeq\frac{d\mathbf{k}}{dt}\cdot\nabla_{\mathbf{k}}f_0=-e\left(\frac{df_0}{dE}\right)\mathbf{v}\cdot\mathscr{E}. \tag{22}$$

Finally, substitution of Eqs. (20) and (22) into Eq. (17) indicates that

$$f_1=e\mathscr{E}\tau v_{\mathscr{E}}\left(\frac{df_0}{dE}\right). \tag{23}$$

This is the solution of the Boltzmann equation in the relaxation-time approximation. Given the distribution function $f=f_0+f_1$, all transport coefficients can be evaluated. For example, the mobility can be obtained from Eq. (10).

In order to calculate the relaxation time, the predominant scattering mechanisms must be identified. These depend on the particular material being investigated, and often even on the particular sample. The major mechanisms that have been analyzed quantitatively are scattering by thermal vibrations of the ion cores, ionized impurity scattering, neutral impurity scattering, and scattering by vacancies or dislocations.

See also CONDUCTION; ELECTRON ENERGY STATES IN SOLIDS AND LIQUIDS; HEAT TRANSFER; RELAXATION PHENOMENA; THERMOELECTRIC EFFECT.

BIBLIOGRAPHY

N. W. Ashcroft and N. D. Mermin, *Solid State Physics*. Holt, Rinehart & Winston, New York, 1976. (I)

J. S. Blakemore, *Solid State Physics*. Saunders, Philadelphia, 1969. (E)

F. J. Blatt, *Physics of Electronic Conduction in Solids*. McGraw-Hill, New York, 1968. (I)

W. Jones and N. H. March, *Theoretical Solid State Physics*. Wiley (Interscience), New York, 1973. (A)

J. M. Ziman, *Principles of the Theory of Solids*. Cambridge, London and New York, 1972. (I)

Tribology

Warren E. Jamison

Tribology, the science of interacting solid bodies in relative motion, is concerned with both the fundamental nature and the engineering applications of frictional forces and wear. Contact between solid surfaces such as bearings and gears, which are microscopically rough, occurs in discrete regions that deform elastically and plastically until the actual area of contact becomes sufficiently large to support the

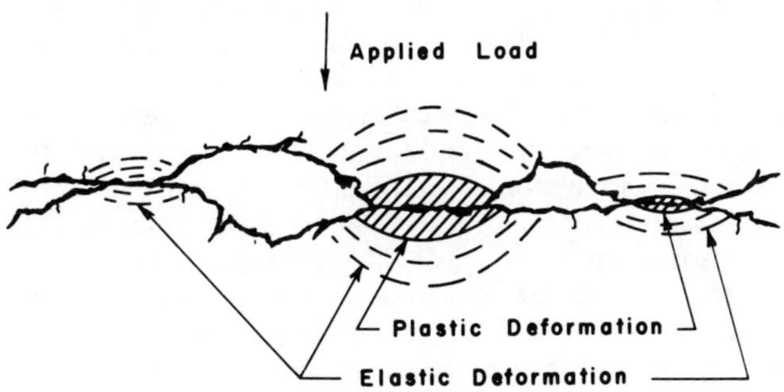

FIG. 1. Schematic of enlarged cross section of engineering surfaces in contact, showing plastic and elastic deformation at contacting asperities.

applied load (Fig. 1). These interactions regulate tangential slip forces in a manner expressed by the three empirical laws of friction:

1. Friction is proportional to the applied load ($F = \mu W$).
2. Friction is independent of the (apparent) contact area.
3. Friction is independent of the sliding velocity.

The first law defines the friction coefficient, μ.

Wear, defined as the loss or transfer of material in a tribological contact, results from various interactive forces, including *adhesion, abrasion* (third-body interactions), *corrosion* (chemically accelerated wear), and *fatigue* (wear in cyclically loaded contacts). In general, the volume of wear (material lost) V is directly proportional to the applied normal load W and the distance slid x, and is inversely proportional to the hardness of the materials p:

$$V = k\frac{Wx}{p}.$$

Lubrication is that part of tribology in which attempts are made to mitigate the effects of friction and wear by imposing films of foreign substances between the contacting bodies. The lubricant may be a solid (graphite, molybdenum disulfide), a chemically active substance (boundary lubricant), or a fluid supplied in such a manner as to support the applied load by externally generated (hydrostatic) or self-generated (hydrodynamic) pressure within the fluid film. Elastohydrodynamic lubrication (EHD or EHL) occurs in highly loaded contacts (e.g., ball bearings, gears) where the change in fluid viscosity due to pressure and the elastic deformation of the surfaces must be considered in analyses.

See also FATIGUE; FRICTION.

BIBLIOGRAPHY

American Society of Lubrication Engineers, *ASLE Proc. Int. Conf. on Solid Lubricants,* 1971. ASLE, Park Ridge, Ill., 1971. Collection of state-of-art papers. (I)

A. Beerbower, *Boundary Lubrication,* AD 747,336. U.S. Army, Washington, D.C., 1972. Very advanced. (A)

E. E. Bisson and W. J. Anderson, *Advanced Bearing Technology.* (NASA SP-38). Washington, D.C., 1964. Descriptive but comprehensive. (I)

F. P. Bowden and D. Tabor, *The Friction and Lubrication of Solids.* Oxford, London and New York, Part I, 1950; Part II, 1964. Comprehensive treatment of experimental investigations. (I)

F. P. Bowden and D. Tabor, *Friction: An Introduction To Tribology.* Anchor Press, New York, 1973. Descriptive paperback. (E)

E. R. Braithwaite, ed., *Lubrication and Lubricants.* Elsevier, Amsterdam, 1967. Description of mechanisms of lubrication and classes of lubricants; little analytical treatment. (I)

P. J. Bryant and M. Lavik, eds., *Mechanisms of Solid Friction.* Elsevier, Amsterdam, 1964. State-of-art papers on friction and solid lubrication. (I)

J. T. Burwell, *Mechanical Wear.* Am. Soc. for Metals, Metals Park, Ohio, 1950. Practical considerations of design. (I)

A. Cameron, *Principles of Lubrication.* Longmans, London, 1966. Theoretical treatment of fluid film lubrication. (A)

A. Cameron, *Basic Lubrication Theory.* Longmans, London, 1971. Shorter version of *Principles of Lubrication.* (I)

M. E. Campbell, J. B. Loser, and E. Sneegas, *Solid Lubricants* (NASA SP-5059). Washington, D.C., 1966. Describes techniques, materials, and applications. (I)

V. N. Contantinescu, *Lubrication in Turbulent Regime.* U.S. Atomic Energy Comm., Washington, D.C., 1968. Theoretical monograph originally published in Rumanian. (A)

D. Dowson and G. R. Higginson, *Elasto-Hydrodynamic Lubrication.* Pergamon, Oxford, 1966. Theoretical treatment with engineering applications. (A)

D. D. Fuller, *Theory and Practice of Lubrication for Engineers.* Wiley, New York, 1956. Comprehensive theoretical and practical treatment at intermediate level. (I)

R. C. Gunderson and A. W. Hart, *Synthetic Lubricants.* Reinhold, New York, 1962. Properties characterization and tabulation. (I)

T. A. Harris, *Rolling Bearing Analysis.* Wiley, New York, 1966. Comprehensive engineering and design of rolling bearings. (I)

E. E. Hondros, *Tribology* (M & B Monograph ME/1). Mills and Boon, London, 1971. Totally descriptive at an elementary level. (E)

F. F. Ling, E. E. Klaus, and R. S. Fein, *Boundary Lubrication: An Appraisal of World Literature.* Am. Soc. Mech. Eng., New York, 1969. Comprehensive survey, annotated bibliography of 2711 publications. (I)

J. J. O'Connor and J. Boyd, eds., *Standard Handbook of Lubrication Engineering.* McGraw-Hill, New York, 1968. Practical aspects of lubrication; available from Am. Soc. Lub. Eng. (E)

A. Palmgren, *Ball and Roller Bearing Engineering.* SKF Industries, Philadelphia, Pa., 1959. Description of bearing types, design, and engineering principles. (I)

O. Pinkus and B. Sternlicht, *Theory of Hydrodynamic Lubrication.* McGraw-Hill, New York, 1961. Comprehensive theory. (A)

T. F. J. Quinn, *The Application of Modern Physical Techniques to Tribology.* (Van Nostrand-Reinhold, Princeton, N.J., 1971. Primarily research techniques. (I)

E. Rabinowicz, *Friction and Wear of Materials.* Wiley, New York, 1965. Development of simple theories relating tribological and mechanical properties. (I)

S. Tolansky, *Surface Microtopography.* Longmans, London, 1960. (I)

U.S. Air Force, *Proc. AFML-MR1 Conf. on Solid Lubricants* (AFML-TR-70-127). U.S. Air Force, Wright-Patterson AFB, Ohio, 1970. State-of-art conference proceedings. (I)

Tunneling

J. M. Rowell

An insulator can be distinguished from a conductor by its inability to carry an electric current, that is, to allow the movement of electrons (see *Electron Energy States in Solids and Liquids*). However, if an insulator is made very thin, this simple distinction begins to fade. For example, if two copper wires are brought close together in a vacuum no current should flow through the vacuum, which is an insulator, until the wires actually touch each other; but when this experiment is carried out, a small current is observed when the wires are still about 50 Å (50×10^{-8} cm) apart, and this current increases rapidly as the wires are moved closer. This phenomenon of current flow through thin insulators, where such a flow would not be expected, is known as electron tunneling.

The concept of tunneling is much more general than this example and applies to all objects meeting a barrier, such as a ball thrown against a wall. In this case, if the wall is thin enough, the ball has a chance of traveling directly through the wall. This alien notion has its roots in quantum mechanics, and was used in the late 1920s and early 1930s to explain a number of phenomena, but was most dramatically illustrated for electrons tunneling through semiconductors by Esaki in 1957 and for electrons penetrating insulators by Giaever in the early 1960s. Giaever also showed that if these electrons tunneled either out of (or into) a superconducting metal (see SUPERCONDUCTIVITY THEORY), then remarkable information about the superconductor could be obtained. In 1962 Josephson predicted that the *pairs* of electrons present in superconductors could similarly tunnel through insulators (see JOSEPHSON EFFECTS; SUPERCONDUCTIVE DEVICES). For their work on electron tunneling, Esaki, Giaever, and Josephson were awarded the Nobel Prize in 1973.

To obtain an idea of how small this effect generally is, consider an electron approaching an insulator which presents a barrier of height h (eV) to its motion and is d Å thick. This is represented in Fig. 1, where the top of the barrier is the energy where electrons can move easily through the insulator, just as a ball can be thrown over a wall if given enough energy. The electron approaching the barrier with energy h (eV) less than the top of the barrier will always bounce back, however many times the approach is made, if it is considered in classical terms. But in quantum-mechanical language one would say that the wave nature of the electron does not stop abruptly at the insulator but decays through the barrier and, if this barrier is thin enough, does not disappear entirely even on the other side. Hence the electron has a chance of appearing on the opposite side. This chance, known as the tunneling probability for the electron through the barrier, can be written approximately as

$$P = \exp(-dh^{1/2}),$$

with d in Å and h in eV. Typically h is 2 eV, hence P is 10^{-62} when d is 100 Å, 10^{-18} when d is 30 Å, and 10^{-6} when d is reduced to 10 Å, that is, still only one in a million when the insulator is a few layers of atoms thick (the spacing between atoms in a crystal is often about 3 Å). With such a small chance for one electron to penetrate the barrier, it is surprising that tunneling can ever be observed, but if the single electron is replaced by a metal with many electrons (about 10^{22} in a cubic centimeter of the metal) continually attempting to tunnel, then a detectable current can be measured. For example, in Fig. 2a the same barrier is shown, but

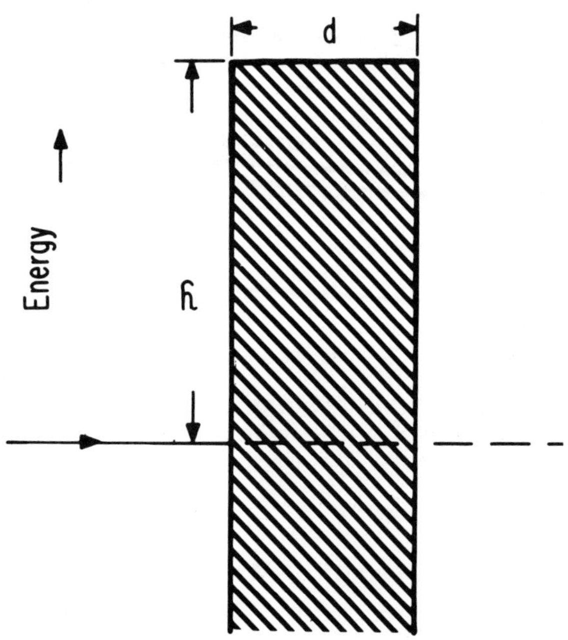

FIG. 1. Representation of an electron meeting a barrier of thickness d and height h above the energy of the electron.

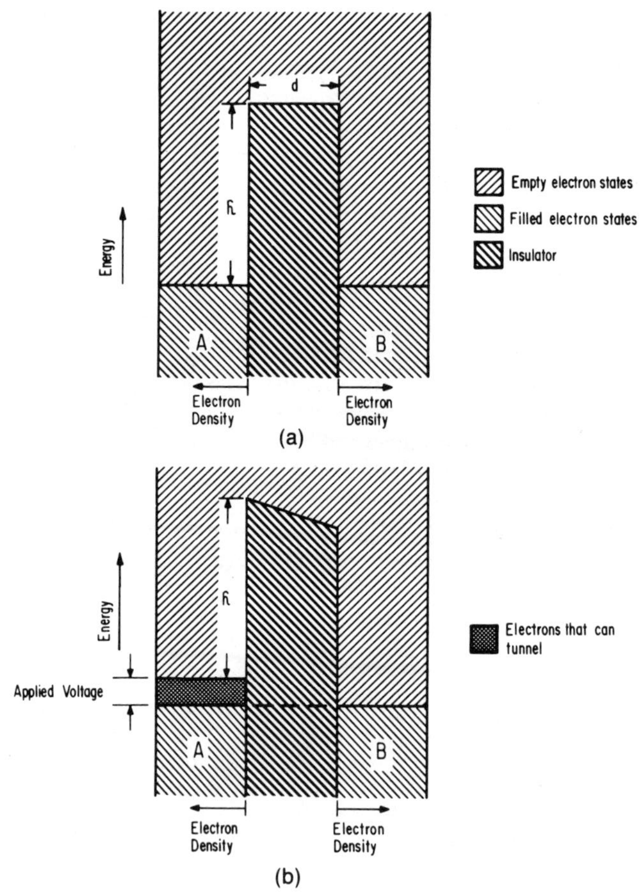

FIG. 2. Schematic of two metals A and B separated by an insulating barrier. The filled and empty electron states are shown. In (a) no potential difference has been applied between A and B, whereas in (b) an applied voltage has shifted the Fermi level of A above that of B. The electrons that can now tunnel are indicated.

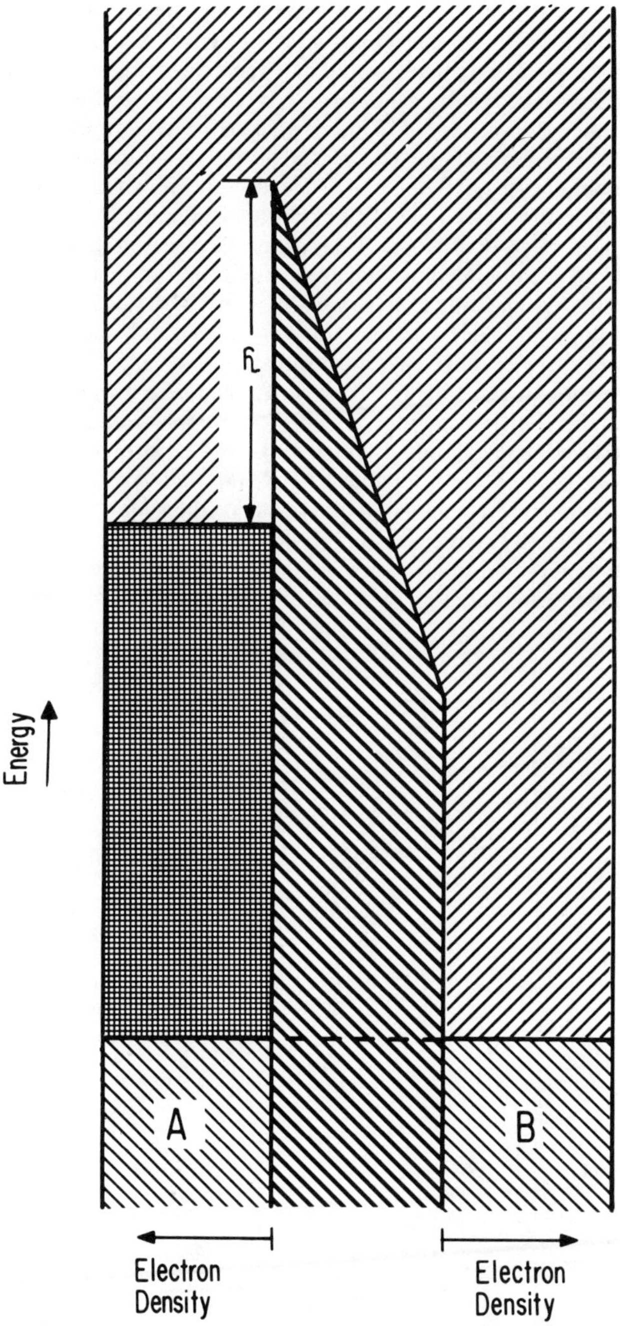

FIG. 3. The effect of applying a very large potential difference, greater than the barrier height, to a tunnel junction of the type shown in Fig. 2. The cross hatchings also have the same meanings as in Fig. 2.

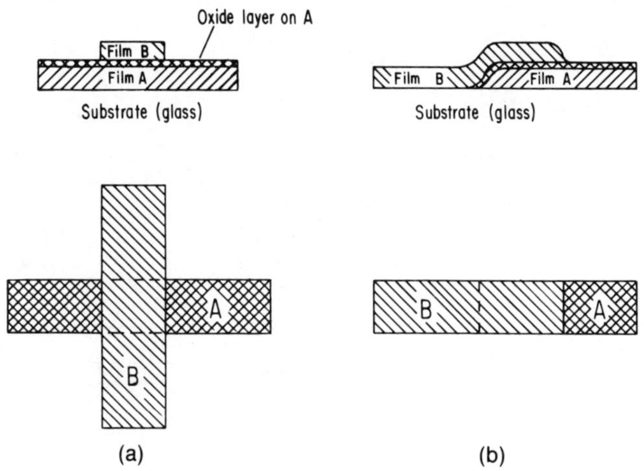

FIG. 4. Two common ways of making thin-film tunnel junctions are indicated, the cross and in-line geometries.

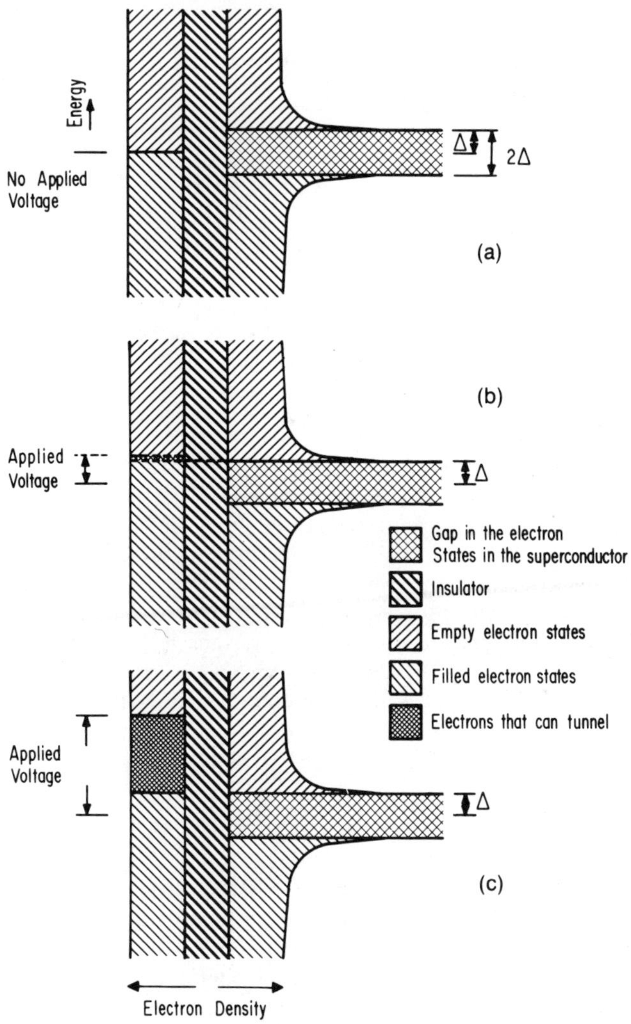

FIG. 5. Representation of a tunnel junction consisting of one normal metal and one superconducting metal, with an energy gap 2Δ. The effect of applied voltage, shown in (b) and (c), is to allow electrons to tunnel into states above the energy gap.

now metals *A* and *B* have a certain density of electron states, which is plotted horizontally against the energy on the vertical axis. The electrons in *A* and *B* occupy the lowest states, and these states are therefore filled up to the "Fermi energy" (see *Electron Energy States in Solids and Liquids*). When no voltage is applied from *A* to *B*, the Fermi levels of *A* and *B* are aligned, as in Fig. 2a. If, by means of an external circuit, a battery is connected across from *A* to *B*, then the Fermi level of *A* is raised relative to that of *B*. The difference in the two levels is equal to the applied potential difference

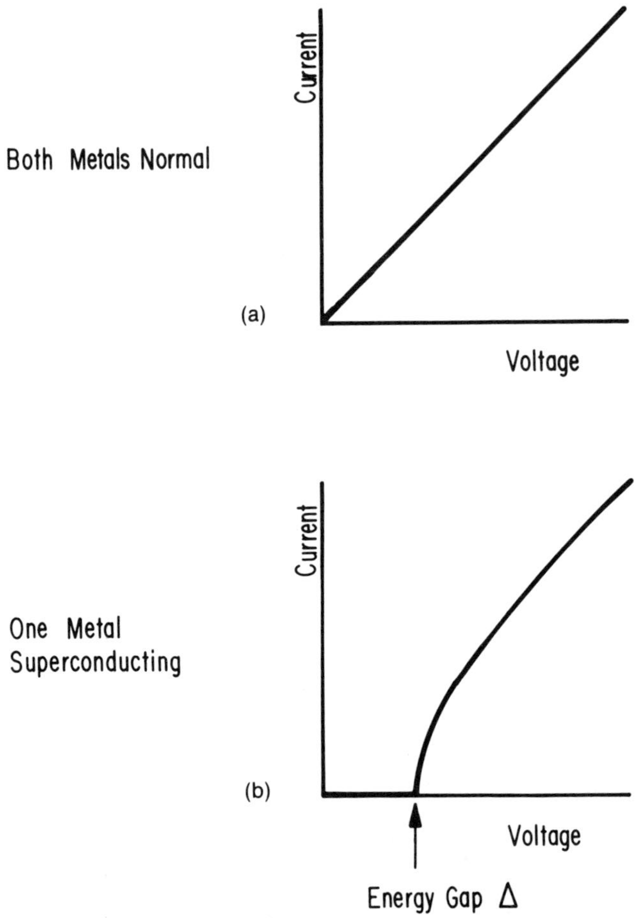

FIG. 6. The tunnel current resulting from a voltage applied to a junction, when both metals are normal (upper trace) and when one metal is superconducting (lower trace).

from the battery, as shown in Fig. 2b. The result is that some electrons in A are now opposite to empty electron states in B. These electrons, indicated in Fig. 2b, each have a chance P of tunneling into B. A tunneling current therefore flows through the insulator, which is the net effect of very many electrons all trying to tunnel, but with any one electron having a minute chance of doing so. As long as the potential difference eV applied between A and B is a small fraction of the barrier height h, then the current density is

$$J=\left[3\times 10^{10}\,\frac{h^{1/2}}{d}\exp(-dh^{1/2})\right]V\ \text{A/cm}^2.$$

If h is again 2 eV and the voltage V is 0.01 V, I is 10^{-55} A/cm^2 when d is 100 Å, and increases to 5×10^{-12} A/cm^2 when $d=30$ Å and to 30 A/cm^2 when d is 10 Å. This illustrates the point that very thin layers of insulator, approximately 10–30 Å thick, have to be made in order to see electron tunneling. If the potential difference eV is larger than the barrier height h, then the barrier is very distorted in shape, as shown in Fig. 3, and some electrons actually tunnel into allowed states in the insulator, rather than into B. As the thickness of the tunnel barrier is then changed by the applied voltage, the current increases exponentially as V is increased.

While it might appear impossible to prepare insulators as thin as 10–30 Å, this is not so as nature is, for once, cooperative. If a metal is heated in a vacuum, it first melts and then "boils," with metal atoms traveling in all directions through the vacuum. The atoms stick to whatever they meet, and if a piece of glass is placed in the vacuum it is soon covered with a film of metal. This process, known as vacuum evaporation, gives metal films with clean surfaces. If such a film, directly after being deposited, is exposed to the air, then it rapidly becomes covered with a layer of oxide which grows by reaction of oxygen with the metal. Oxides are generally good insulators, and Giaever showed that such oxide films were often of the correct thickness to allow electron tunneling. Conveniently, they are formed on a metal, which can then become metal A of Fig. 2 for example. For example, a freshly made aluminum film will grow a layer of aluminum oxide of the correct thickness if it is exposed to air for a few minutes. If a second metal film (B say) is then deposited over this oxide by vacuum evaporation, a sandwich of two metals separated by a thin oxide insulator is produced. This structure is generally called a tunnel junction and two typical

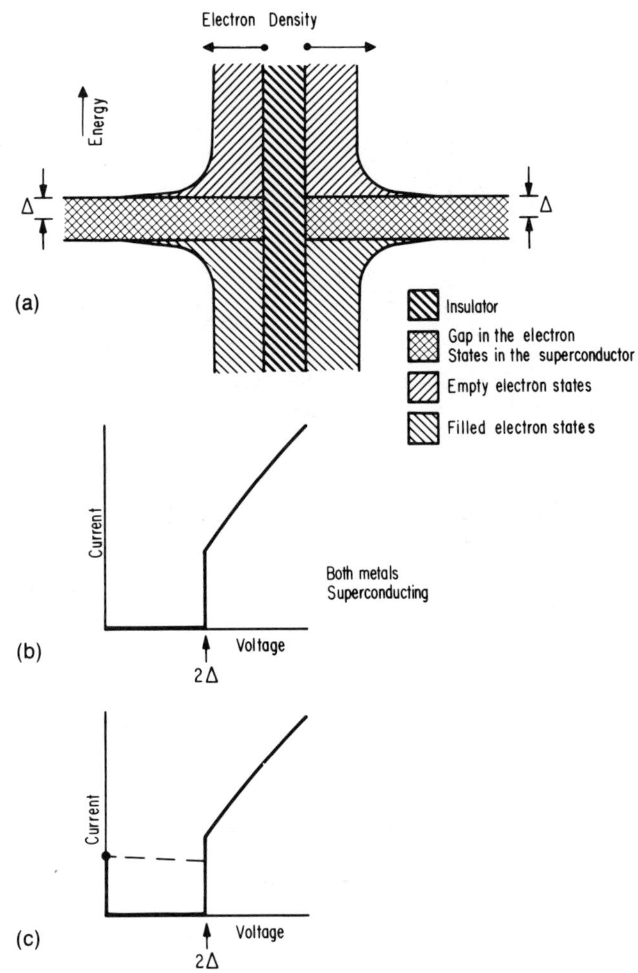

FIG. 7. (a) Representation of a tunnel junction with both metals superconducting. (b) The tunnel current against voltage plot for such a junction. (c) The tunneling of pairs of electrons results in a current flow at zero voltage, as shown.

examples are shown in Fig. 4. In Fig. 4a a cross geometry of two films is indicated, whereas in Fig. 4b the two films have been evaporated in a line. These are the two most common types of tunnel junctions using oxide insulators.

As he increased the voltage across aluminum–aluminum oxide–metal tunnel junctions of the kind shown in Fig. 4a, Giaever saw the two behaviors of current discussed above, the linear dependence at low voltages and the exponential increase at high voltages. However, it was still hard to believe that thin enough oxide layers had been made to allow electron tunneling to be observed. An even more convincing proof that tunneling through the oxide was actually taking place was obtained when one of the metal films was a superconductor. In such metals there is a gap (called the energy gap) in the electron states, as shown in Fig. 5. This gap occurs symmetrically about the Fermi level and has a total width 2Δ. The electrons which would occupy the region of states with Δ below the Fermi level are piled into a very strong narrow peak just below the gap, as shown. Similarly there is a peak of empty electron states just above the gap. In Fig. 5a there is no applied voltage between *A* and *B*. The electrons in the normal metal *A* cannot flow into the gap of superconducting metal *B*, as no electrons are allowed within this gap region. If a battery is now used to raise the potential of *A* relative to *B* by an amount eV which is just greater than Δ, then a few electrons from *A* can tunnel into the large peak in the density of electron states in *B*. As the battery voltage is increased (Fig. 5c), electrons tunnel into states high above the gap and the effect of the gap on the total tunnel current is soon lost. In an experiment, the tunnel current is plotted against the voltage measured from *A* to *B*. If both metals are normal (i.e., not superconducting), the current rises linearly with voltage, as explained above and shown in Fig. 6a. But when metal *B* is cooled enough to be superconducting, then the energy gap opens. At very low temperatures the tunnel current is zero until the applied voltage is close to the energy gap and then the current rises rapidly as the situation of Fig. 5b is reached. Thus with a superconductor to tunnel into, the current depends on voltage as in Fig. 6b. In this way, Giaever observed the effect of the superconductor's energy gap in the tunneling current–voltage plot.

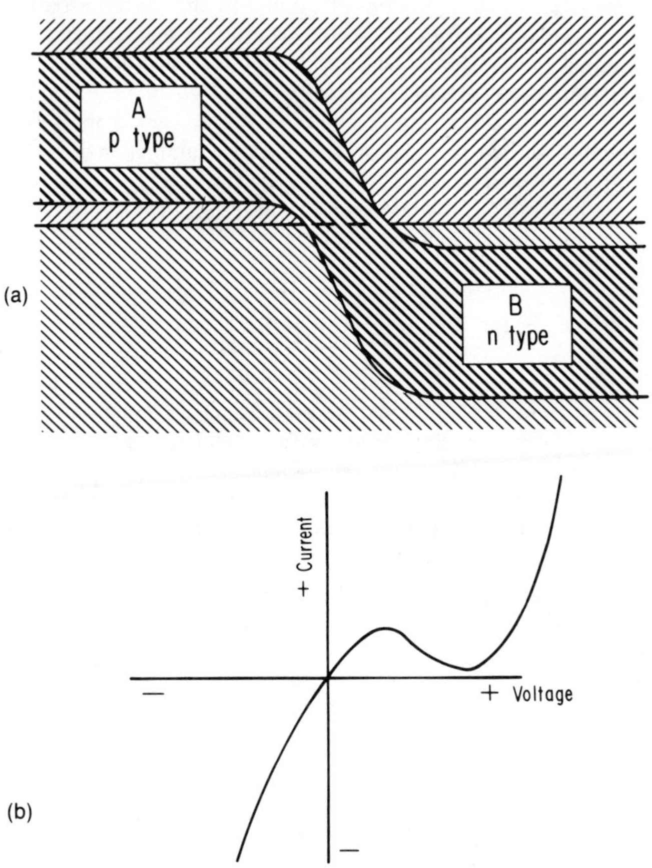

FIG. 8. A schematic of the filled and empty electron states in a junction of *p* and *n* semiconductor layers. The cross hatched regions have the same meanings as in Fig. 2. In the lower plot the dependence of the tunnel current on applied voltage is shown.

From the same arguments used for Figs. 2, 3, and 5, where the voltage from a battery can be regarded as having the effect of sliding the electron density of metal *A* up (or down) relative to *B*, it should now be possible to deduce that if *both* *A* and *B* are superconductors with equal energy gaps (Fig. 7a), then the current will depend on voltage in the way shown in Fig. 7b. However, an even more interesting effect in such a tunnel junction was predicted by B. D. Josephson in 1962. In superconductors there are pairs of electrons; and Josephson showed that, in addition to the tunneling of single electrons discussed above, these pairs of electrons should also be able to tunnel through the oxide without being broken apart. This tunneling of pairs, known as the Josephson effect, results in a number of phenomena, but the simplest is an extra current which flows, up to a certain limiting value, without any voltage appearing across the oxide insulator, as in Fig. 7c. The oxide can be said to have become a weak superconductor, in that it carries a "supercurrent."

The ideas of Figs. 2–5 serve as an introduction to tunneling in semiconductor diodes, which was actually the first system where electron tunneling was observed convincingly. Such a diode is shown in Fig. 8, where the electrons in regions of *p*- and *n*-type semiconductor (see SEMICONDUCTORS) are separated by the region where the material changes from *p* to *n* type. Again, by considering electrons from filled states in *A* tunneling to empty states in *B* when the applied voltage is in one direction, and electrons tunneling from *B* to *A* when the voltage is applied in the opposite direction, it will be seen that the current–voltage plot will have a region of "negative resistance" (where current decreases as voltage increases). This effect is indeed observed in tunnel diodes, also known as Esaki diodes, as illustrated in Fig. 8b.

See also ELECTRON ENERGY STATES IN SOLIDS AND LIQUIDS; JOSEPHSON EFFECTS; SEMICONDUCTORS, CRYSTALLINE; SUPERCONDUCTIVE MATERIALS; SUPERCONDUCTIVITY THEORY.

BIBLIOGRAPHY

E. Burstein and S. Lundquist, (eds.), *Tunneling Phenomena in Solids*. Plenum Press, New York, 1969. (I,A)

C. B. Duke, *Tunneling in Solids*. Academic Press, New York, 1969. (A)

D. R. Lamb, *Electrical Conductive Mechanisms in Thin Insulating Films*. Methuen, London, 1967. (I)

L. Solymar, *Superconductive Tunneling and Applications*. Wiley, New York, 1972; Chapman and Hall, London, 1972. (I,A)

Whiskers

R. V. Coleman

The term *whisker* describes any fibrous growth of a solid, and such forms were studied extensively in relation to the development of microscopic theories of crystal growth, particularly those involving screw dislocations. If crystal growth proceeds by the motion of a dislocation step and if conditions are such that the rate of step generation dominates over the motion of steps away from the dislocation source, then a whisker profile results. This is a simple and elegant explanation for the growth of single-crystal whiskers, but many modifications and alternative theories involving oxidation, solid-state diffusion, stress, recrystallization, or some combination of these have been developed. Impurities may also play a significant role.

Whisker crystals can be produced by vapor deposition, chemical reaction, electrolytic deposition, or oxidation of surfaces, and at vapor–liquid–solid interfaces. A famous form of whisker growth is that of Sn whiskers growing from tin-plated metal where applied stress can enchance the growth rate, and this has led to the term *squeeze whisker*. Low-melting-point metals such as Zn, Cd, Mg, Hg, and K have been grown from the vapor, while higher-melting-point metals and semiconductors such as Fe, Co, Ni, Cu, Au, Ag, Pt, Si, and Ge are often grown by the hydrogen reduction of metallic salts. Whiskers of oxides, carbides, nitrides, metallic salts, graphite, polymers, organic materials, and metallic alloys have been reported.

Stress–strain curves of many whiskers show an e'astic strain region extending to 4 or 5% and a yield stress approaching that expected for a perfect crystal. The high degree of surface and volume perfection characteristic of whiskers has made them useful for studies of plastic flow processes in solids and for studies of a wide range of electric and magnetic properties of solids. Many of these experiments are discussed in the review articles listed in the bibliography, which also contain extensive reference to the original work.

See also CRYSTAL DEFECTS; CRYSTAL GROWTH.

BIBLIOGRAPHY

Growth and Perfection of Crystals (R. H. Doremus, B. W. Roberts, and D. Turnbull eds.). Wiley, New York, 1958, and Chapman and Hall, London, 1958. (A)

The Art and Science of Growing Crystals (J. J. Gilman, ed.). Wiley, New York and London, 1963. (A)

E. M. Nadgornyi, "The Properties of Whiskers," *Sov. Phys.–Usp.* **5,** 462 (1962). (I)

R. V. Coleman, "The Growth and Properties of Whiskers," *Metall. Rev.* **9,** 261 (1964). (I)

C. C. Evans, *Whiskers*. Mills and Boon, London, 1972. (I)

Work Function

T. M. Donovan and A. D. Baer

The work function W is the minimum energy required to remove an electron at the Fermi energy from a solid. Figure 1 shows a simplified model of the surface of metal, with electronic states in the metal occupied up to the Fermi level.

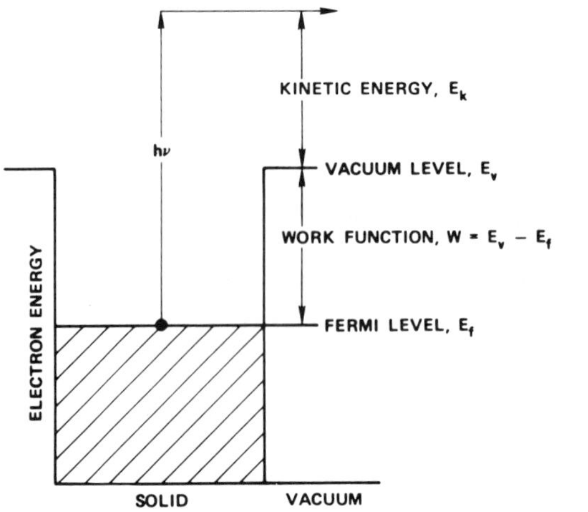

FIG. 1. Simplified model of electrons in a metal showing electron emission.

The work function is

$$W = E_v - E_f,$$

where E_v is the energy of an electron at rest outside the solid, and E_f is the energy of an electron at the Fermi energy inside the solid.

The work function is determined experimentally by measuring the minimum amount of excitation energy required to emit an electron from the solid. There is a high density of filled electronic states near the Fermi level of a metal, and clear-cut theoretical relationships exist which relate photoemission and thermionic emission of electrons from these states to the work function [1–3]. For example, in Fig. 1, if an electron is excited from the Fermi level by a photon of energy $h\nu$, and the electron is emitted from the solid into vacuum without energy loss, then

$$W = h\nu - E_k,$$

where E_k is the kinetic energy of the electron outside the solid.

Since the Fermi level of a semiconductor or insulator generally lies in the band gap, the work function of a nonmetal sample is usually measured by first determining the work function of a metal surface and then measuring the difference between the work functions of the metal and the sample. The difference is equal to the contact potential difference which can be measured directly in a Kelvin probe [2,3] or retarding-potential experiment [4,5].

See also ELECTRON ENERGY STATES IN SOLIDS AND LIQUIDS; PHOTOELECTRON SPECTROSCOPY; THERMIONIC EMISSION.

REFERENCES

1. R. H. Fowler, *Phys. Rev.* **38,** 45 (1931). (I)
2. A. J. Decker, *Solid State Physics*, pp. 211, 230. Prentice-Hall, Englewood Cliffs, N. J., 1957. (E)
3. J. C. Rivière, *J. Solid State Surf. Sci.* **1,** 179 (1969). (A)
4. L. Apker, E. Taft, and J. Dickey, *Phys. Rev.* **74,** 1462–1474 (1948). (I)
5. C. N. Berglund and W. E. Spicer, *Phys. Rev.* **136,** A1030, (1964); **136,** A1044 (1964). (I)

CONTRIBUTORS

SIDNEY C. ABRAHAMS
Bell Laboratories, Murray Hill, New Jersey
Pyroelectricity

DAVID ADLER
Massachusetts Institute of Technology
Transport Properties

GEORGE A. ALERS
Albuquerque Development Laboratory, Albuquerque, New Mexico
Mechanical Properties of Matter

PHILIP B. ALLEN
Department of Physics, State University of New York at Stony Brook
Superconductivity Theory

BETSY ANCKER-JOHNSON
General Motors Technical Center, Warren, Michigan; formerly U.S. Department of Commerce
Plasmons

ANSEL C. ANDERSON
Department of Physics, University of Illinois
Cryogenics

PHILIP W. ANDERSON
Bell Laboratories, Murray Hill, New Jersey, and Physics Department, Princeton University
Quasiparticles

ROBERT A. ANDERSON
Sandia Laboratories, Albuquerque, New Mexico
Insulators

ANTHONY S. ARROTT
Department of Physics, Simon Fraser University
Ferromagnetism

JOHN D. AXE
Physics Department, Brookhaven National Laboratory
Ferroelectricity

A. D. BAER
Michelson Laboratory, U.S. Naval Weapons Center, China Lake, California
Work Function

STEPHAN BERKO
Department of Physics, Brandeis University
Positron Annihilation in Condensed Matter
Positronium

AMI E. BERKOWITZ
Corporate Research and Development General Electric Company, Schenectady, New York
Magnets (Permanent) and Magnetostatics

PHILIP E. BEST
Department of Physics and Institute of Materials Science, University of Connecticut
Secondary Electron Emission

NICHOLAS BOTTKA
Michelson Laboratory, U.S. Naval Weapons Center, China Lake, California
Dielectric Properties

EDWARD A. BOUDREAUX
Department of Chemistry, University of New Orleans
Diamagnetism

BERTRAM N. BROCKHOUSE
Department of Physics, McMaster University
Neutron Spectroscopy

WALTER E. BRON
Department of Physics, Indiana University, Bloomington
Elasticity

E. BUCHER
Department of Physics, University of Constance, Federal Republic of Germany
Transition Elements

MARTIN J. BUERGER
Department of Earth and Planetary Sciences, Massachusetts Institute of Technology
Crystallography, X-Ray
Crystal Symmetry

R. BULLOUGH
Theoretical Physics Division, AERE, Harwell, England
Deformation of Crystalline Materials

JOSEPH CALLAWAY
Department of Physics and Astronomy, Lousiana State University, Baton Rouge
Electron Energy States in Solids and Liquids

BRUCE CHALMERS
Professor Emeritus, Harvard University
Metallurgy

B. S. CHANDRASEKHAR
Department of Physics, Case Western Reserve University
Magnetostriction

DAVID B. CHANG
Advanced Energy Research, Occidental Research Corporation, Irvine, California
Plasmons

HERBERT S. CHENG
Department of Mechanical Engineering, Northwestern University
Friction

JOHN CLARKE
Department of Physics, University of California, Berkeley
Superconductive Devices

ROBERT V. COLEMAN
Department of Physics, University of Virginia
Whiskers

ESTHER M. CONWELL
Xerox Webster Research Center, Xerox Corporation, Rochester, New York
Acoustoelectric Effect

BERNARD R. COOPER
Department of Physics, West Virginia University
Fermi Surface

BERND CRASEMANN
Department of Physics, University of Oregon
Auger Effect

J. H. CRAWFORD, JR.
Department of Physics and Astronomy, University of North Carolina
Radiation Interaction with Matter

ALBERT V. CREWE
Department of Physics, University of Chicago
Electron Microscopy

CLARENCE R. CROWELL
School of Engineering, University of Southern California
Photoconductivity

N. E. CUSACK
School of Mathematics and Physics, University of East Anglia, Norwich, England
Liquid Metals

FRANÇOISE CYROT-LACKMANN
Groupe des Transitions de Phases, Centre National de la Recherche Scientifique, Grenoble, France
Surfaces and Interfaces

J. GREGORY DASH
Department of Physics, University of Washington
Adsorption

P. G. DE GENNES
Collège de France, Paris, France
Liquid Crystals

M. C. DESJONQUÈRES
Groupe des Transitions de Phases, Centre National de la Recherche Scientifique, Grenoble, France
Surfaces and Interfaces

JOZEF T. DEVREESE
University of Antwerp, Antwerp, Belgium
Polaron

I. J. D'HAENENS
Hughes Research Laboratories, Malibu, California
Thermionic Emission

FRANCIS J. DISALVO, JR.
Bell Laboratories, Murray Hill, New Jersey
Allotropy and Polymorphism

PAUL L. DONOHO
Marine Science Institute, University of Texas, Galveston
Hysteresis

TERENCE M. DONOVAN
Michelson Laboratory, U.S. Naval Weapons Center, China Lake, California
Work Function

CHARLES W. DRAKE, JR.
Department of Physics, Oregon State University
Polarizability

G. DRESSELHAUS
Francis Bitter National Magnet Laboratory, Cambridge, Massachusetts
de Haas-van Alphen Effect

MILDRED S. DRESSELHAUS
Department of Electrical Engineering and Computer Science and Center for Materials Science and Engineering, Massachusetts Institute of Technology
de Haas-van Alphen Effect

CHARLES B. DUKE
Xerox Webster Research Center, Xerox Corporation, Rochester, New York
Low-Energy Electron Diffraction (LEED)

POL DUWEZ
Department of Applied Physics, California Institute of Technology
Glassy Metals

DEAN E. EASTMAN
IBM Watson Research Center, Yorktown Heights, New York
Photoelectron Spectroscopy

ALAN S. EDELSTEIN
Department of Physics, University of Illinois at Chicago Circle
Kondo Effect

DAVID L. EDERER
U.S. National Bureau of Standards, Washington, D.C.
Photoionization

F. R. EIRICH
Department of Chemistry, Polytechnic Institute of New York
Rheology

P. R. EMTAGE
Westinghouse Research Laboratory, Pittsburgh, Pennsylvania
Hall Effect

JOHN ENDERBY
H. H. Wills Physics Laboratory, Bristol, England
Paramagnetism

STANLEY ENGELSBERG
Department of Physics and Astronomy, University of Massachusetts, Amherst
Metals

JOHN P. FACKLER, JR.
Department of Chemistry, Case Western Reserve University
Crystal and Ligand Fields

DOUGLAS K. FINNEMORE
Department of Physics, Iowa State University
Superconducting Materials

CARL L. FOILES
Department of Physics, Michigan State University
Thermoelectric Effects

HELLMUT FRITZSCHE
Department of Physics, University of Chicago
Solid-State Switching

J. D. GAVENDA
Department of Physics, University of Texas at Austin
Magnetoacoustic Effect

HENRY R. GLYDE
Physics Department, University of Ottawa
Helium, Solid

BRAGE GOLDING, JR.
Bell Laboratories, Murray Hill, New Jersey
Glass, Nonmetallic

J. GOLOVCHENKO
Physics Department, Brookhaven National Laboratory; Institute of Physics, University of Aarhus, Denmark; Bell Laboratories, Murray Hill, New Jersey
Channeling

ANDREW V. GRANATO
Department of Physics, University of Illinois, Urbana
Internal Friction in Crystals

MELVILLE S. GREEN
Late of Department of Physics, Temple University
Phase Transitions

SANDRA C. GREER
Department of Chemistry, University of Maryland
Order-Disorder Phenomena

THOMAS J. GREYTAK
Department of Physics, Massachusetts Institute of Technology
Fluctuation Phenomena

C. C. GRIMES
Bell Laboratories, Murray Hill, New Jersey
Cyclotron Resonance

JOHN J. HALL
Shearwater Company, New York, New York
Photovoltaic Effect

ARTHUR M. HALPERN
Department of Chemistry, Northeastern University
Luminescence

ROBERT E. HEBNER, JR.
Electrosystems Division, U.S. National Bureau of Standards, Washington, D.C.
Kerr Effect, Electro-optical

ROBERT A. HENLE
International Business Machine Corp., Hopewell Junction, New York
Circuits, Integrated

PETER V. HOBBS
Department of Atmospheric Sciences, University of Washington
Ice

HILLARD B. HUNTINGTON
Department of Physics, Rensselaer Polytechnic Institute
Diffusion

COLIN M. HURD
Division of Chemistry, National Research Council, Ottawa, Canada
Galvanomagnetic Effects

WARREN E. JAMISON
Tribo-Materials Development Corporation, Evergreen, Colorado
Tribology

GEORGE J. JANZ
Molten Salts Data Center, Rensselaer Polytechnic Institute
Molten Salts

CARSON JEFFRIES
Department of Physics, University of California, Berkeley
Excitons

M. N. KABLER
U.S. Naval Laboratory, Washington, D.C.
Electroluminescence

WALTER KELLNER
Siemens AG, Zentrale Forschung und Entwicklung, München, Federal Republic of Germany
Semiconductors, Crystalline

RICHARD K. KIRBY
National Measurement Laboratory, U.S. National Bureau of Standards, Washington, D.C.
Thermal Expansion

S. KIRKPATRICK
IBM Research, Yorktown Heights, New York
Conduction

PAUL G. KLEMENS
Department of Physics and Institute of Materials Sciences, University of Connecticut
Phonons

HOBART W. KRANER
Instrumentation Division, Brookhaven National Laboratory
Semiconductor Detectors

PAUL L. LEATH
Department of Physics, Rutgers University
Alloys

LIONEL M. LEVINSON
General Electric Corporate Research and Development, Schenectady, New York
Ceramics and Refractories

PAUL W. LEVY
Brookhaven National Laboratory
Thermoluminescence
Color Centers

J. D. LITSTER
Physics Department, Massachusetts Institute of Technology
Critical Points

OLLI V. LOUNASMAA
Low Temperature Laboratory, Helsinki University of Technology, Espoo, Finland
Magnetic Cooling

H. LÜTH
Physikalisches Institut der Rheinisch–Westfälischen Technischen Hochschule, Aachen, Federal Republic of Germany
Ellipsometry

L. MARTON
Late of Avances in Electronics and Electron Physics, Washington, D.C.
Electron Diffraction

DANIEL C. MATTIS
Polytechnic Institute of New York
Many-Body Theory

R. L. MELCHER
T. J. Watson Research Center, IBM Corporation
Magnetoelastic Phenomena

RICHARD P. MESSMER
Corporate Research and Development, General Electric Company, Schenectady, New York
Catalysis

D. L. MILLER
Physics Department, Brookhaven National Laboratory
Thin Films

NEVILL F. MOTT
Cavendish Laboratory, Cambridge, England
Metal-Insulator Transitions

ERWIN W. MUELLER
Late of Physics Department, Pennsylvania State University
Field Ion Microscopy

LOUIS NÉEL
Membre de l'Institut, Meudon-Bellevue, France
Ferrimagnetism

DONALD F. NELSON
Bell Laboratories, Murray Hill, New Jersey
Piezoelectric Effect

ROBERT K. NESBET
IBM Research Laboratories, San Jose, California
Magnetic Moments

ARTHUR S. NOWICK
Henry Krumb School of Mines, Columbia University
Anelasticity

RAYMOND L. ORBACH
Department of Physics, University of California, Los Angeles
Relaxation Phenomena

ROBERT L. PARKER
U.S. National Bureau of Standards, Washington, D.C.
Crystal Growth

NORMAN PEARLMAN
Department of Physics, Purdue University
Heat Capacity

GILBERT J. PERLOW
Argonne National Laboratory
Mössbauer Effect

JAMES C. PHILLIPS
Bell Laboratories, Murray Hill, New Jersey
Crystal Binding

ANDRÉ RASSAT
Université Scientifique et Médicale de Grenoble, France
Electron Spin Resonance

WILLIAM A. REED
Bell Laboratories, Murray Hill, New Jersey
Magnetoresistance

JAMES J. RHYNE
National Measurement Laboratory, U.S. National Bureau of Standards, Washington, D.C.
Magnetic Materials

MARVIN ROSS
Lawrence Livermore Laboratory, University of California
Shock Waves and Detonations

LAURA M. ROTH
Department of Physics, State University of New York at Albany
Faraday Effect

JOHN M. ROWELL
Bell Laboratories, Murray Hill, New Jersey
Tunneling

ROBERT M. RYDER
Electron Device Studies Department, Bell Laboratories, Murray Hill, New Jersey
Transistors

DOUGLAS J. SCALAPINO
Department of Physics, University of California, Santa Barbara
Josephson Effects

WALTER SCHEBESTA
Institut für Industrielle Elektronik, Technische Universität Wien, Austria
Electron Beam Technology

JAGDEEP SHAH
Bell Laboratories, Holmdel, New Jersey
Electron-Hole Droplets

GEN SHIRANE
Department of Physics, Brookhaven National Laboratory
Neutron Diffraction and Scattering

F. H. SPEDDING
Ames Laboratory, Iowa State University
Rare Earths

H. E. SPENCER
Research Laboratories, Eastman Kodak Company, Rochester, New York
Light-Sensitive Materials

MYRON STRONGIN
Physics Department, Brookhaven National Laboratory
Thin Films

LYNWOOD SWANSON
Department of Applied Physics, Oregon Graduate Center for Study and Research, Beaverton
Field Emission

W. J. TABOR
Bell Laboratories, Murray Hill, New Jersey
Kerr Effect, Magneto-optical

BARRY N. TAYLOR
U.S. National Bureau of Standards, Washington, D.C.
Constants, Fundamental

EDWARD W. THOMAS
School of Physics, Georgia Institute of Technology
Electron and Ion Impact Phenomena

GEORGE P. VELLA-COLEIRO
Bell Laboratories, Murray Hill, New Jersey
Magnetic Domains and Bubbles

RICHARD W. VOOK
Department of Chemical Engineering and Materials Science, Syracuse University
Reflection High-Energy Electron

RICHARD F. WALLIS
Department of Physics, University of California, Irvine
Lattice Dynamics

GEORGE D. WATKINS
Department of Physics, Lehigh University
Crystal Defects

JOHANNES WEERTMAN
Department of Materials Science and Engineering, Northwestern University
Fatigue

KURT WEISER
Department of Electrical Engineering, Israel Institute of Technology, and Solid State Institute, Haifa, Israel
Semiconductors, Amorphous

DAVID O. WELCH
Department of Energy and Environment, Brookhaven National Laboratory
Radiation Damage

HEINRICH J. WELKER
Siemens AG, Erlangen and Munich, Federal Republic of Germany
Semiconductors, Crystalline

FA YUEH WU
Department of Physics, Northeastern University
Ising Model

INDEX

Main entries are listed in CAPITAL LETTERS with the first page number in **bold face.** References to footnotes give the page number followed by "n." Words with an umlaut over a vowel are alphabetized as if the vowel were followed by an "e."